Methods in Enzymology

Volume 160
BIOMASS
Part A
Cellulose and Hemicellulose

METHODS IN ENZYMOLOGY

EDITORS-IN-CHIEF

John N. Abelson   Melvin I. Simon

Methods in Enzymology

Volume 160

Biomass

Part A

Cellulose and Hemicellulose

EDITED BY

Willis A. Wood

SALK INSTITUTE BIOTECHNOLOGY/INDUSTRIAL ASSOCIATES, INCORPORATED
SAN DIEGO, CALIFORNIA

Scott T. Kellogg

SALK INSTITUTE BIOTECHNOLOGY/INDUSTRIAL ASSOCIATES, INCORPORATED
SAN DIEGO, CALIFORNIA

18439356

ACADEMIC PRESS, INC.
Harcourt Brace Jovanovich, Publishers
San Diego   New York   Berkeley   Boston
London   Sydney   Tokyo   Toronto

ACADEMIC PRESS, INC.
1250 Sixth Avenue
San Diego, California 92101

United Kingdom Edition published by
ACADEMIC PRESS INC. (LONDON) LTD.
24-28 Oval Road, London NW1 7DX

LIBRARY OF CONGRESS CATALOG CARD NUMBER: 54-9110

ISBN   0-12-182061-0   (alk. paper)

PRINTED IN THE UNITED STATES OF AMERICA
88 89 90 91   9 8 7 6 5 4 3 2 1

Table of Contents

Section I. Cellulose

A. Preparation of Cellulosic Substrates

B. Assays for Cellulolytic Enzymes

v

1,4-β-D-Glucan Glucanohydrolases (Endocellulases)

1-4-β-D-Glucan Cellobiohydrolases (Exocellulases)

1,4-β-D-Glucosidases

Cellobiose Dehydrogenases

Assays of Cellulolytic Enzymes and Miscellaneous Enzymes Involved in Cellulolysis

Section II. Hemicellulose

A. Preparation of Substrates for Hemicellulases

B. Analysis of β-Glucan and Enzyme Assays

Contributors to Volume 160

Article numbers are in parentheses following the names of contributors.
Affiliations listed are current.

TERUHIKO AKIBA (82), *The Institute of Physical and Chemical Research, Wako, Saitama 351-01, Japan*

ADOLFO G. AMIT (39), *Département des Biotechnologies, Unité de Physiologie Cellulaire, Institut Pasteur, 74724 Paris Cedex 15, France*

MOTOO ARAI (30), *Department of Agricultural Chemistry, University of Osaka Prefecture, Sakai Osaka 591, Japan*

TOSHIYOSHI ARAKI (72), *Department of Fisheries, Faculty of Agriculture, Kyushu University, Hakozaki, Higashi-ku, Fukuoka 812, Japan*

JEAN-PAUL AUBERT (39), *Unité de Physiologie Cellulaire, Département des Biotechnologies, Institut Pasteur, 74724 Paris Cedex 15, France*

EDWARD A. BAYER (57), *Department of Biophysics, The Weizmann Institute of Science, Rehovot, Israel*

PIERRE BÉGUIN (39), *Département des Biotechnologies, Unité de Physiologie Cellulaire, Institut Pasteur, 74724 Paris Cedex 15, France*

G. BELDMAN (26), *Stichting Nederlands Instituut voor Koolhydraatonderzoek, 9723 cc Groningen, The Netherlands*

ROGER BERNIER JR. (50), *C-I-L Research Centre, Biotechnology Section, Mississauga, Ontario, Canada L5K 2L3*

K. MAHALINGESHWARA BHAT (9), *Rowett Research Institute, Bucksburn, Aberdeen AB2 9SB, Scotland*

PETER BIELY (65, 80, 90), *Institute of Chemistry, Center of Chemical Research, Slovak Academy of Sciences, 843 38 Bratislava, Czechoslovakia*

ROBERT A. BLANCHETTE (20), *Department of Plant Pathology, University of Minnesota, St. Paul, Minnesota 55108*

M. BLUMEL (69), *Department of Biochemistry, Universidade Federal do Parana, Curitiba, Parana, Brazil*

ANNEGRET BORCHMANN (64), *Institute of Wood Chemistry, Federal Research Center of Forestry and Forest Products, 2050 Hamburg 80, Federal Republic of Germany*

GIORGIO CANEVASCINI (10, 52), *Institut de Biologie Végétale, et de Phytochimie, Université de Fribourg, 1700 Fribourg, Switzerland*

MARC CLAEYSSENS (6, 19), *Laboratorium voor Biochemie, Rijksuniversiteit Gent, B-9000 Gent, Belgium*

ANTHONY L. COLE (31), *Department of Plant and Microbial Sciences, University of Canterbury, Christchurch, New Zealand*

MICHAEL P. COUGHLAN (14, 40, 51, 58), *Department of Biochemistry, University College, Galway, Ireland*

CLEMENT K. DE BRUYNE (6), *Laboratorium voor Biochemie, Rijksuniversiteit Gent, B-9000 Gent, Belgium*

R. F. H. DEKKER (54), *CSIRO, Division of Biotechnology, Clayton, 3168, Victoria, Australia*

M. V. DESHPANDE (12), *Department of Chemistry, National Chemical Laboratory, Pune 411 008, India*

V. DESHPANDE (48), *Biochemistry Division, National Chemical Laboratory, Pune 411 008, India*

MICHEL DESROCHERS (50), *Pulp and Paper Research Institute of Canada, Pointe Claire, Québec, Canada, H9R 3J9*

LANDIS W. DONER (17), *United States Department of Agriculture, Agricultural Research Service, Eastern Regional Research Center, Philadelphia, Pennsylvania 19118*

MARY L. DURBIN (37), *Botany and Plant Sciences Department, University of California, Riverside, California 92521*

SHIGENORI EMI (93), *Research Department, Tsuruga Enzyme Plant, Toyobo Company, Limited, Tsuruga, Fukui Prefecture 914, Japan*

T.-M. ENARI (11), *Division of Process Technology, Technical Research Centre of Finland, SF-02150 Espoo, Finland*

JAMES D. ERFLE (7), *Animal Research Centre, Agriculture Canada, Ottawa, Ontario K1A OC6, Canada*

K.-E. ERIKSSON (12, 41, 48, 55, 59), *Swedish Pulp and Paper Research Institute, S-114 86 Stockholm, Sweden*

W. FISCHER (16), *c/o E. Merck, Department Vertrieb Reagenzien, D-6100 Darmstadt, Federal Republic of Germany*

J. D. FONTANA (68, 69), *Department of Biochemistry, Universidade Federal do Parana, Curitiba, Parana, Brazil*

M. GEBARA (69), *Department of Biochemistry, Universidade Federal do Parana, Curitiba, Parana, Brazil*

M. GLENNIE-HOLMES (67), *Agricultural Research Centre, New South Wales, Department of Agriculture, Wagga Wagga, New South Wales, Australia*

JOSTEIN GOKSØYR (36), *Department of Microbiology and Plant Physiology, University of Bergen, N-5007 Bergen, Norway*

DIETER GOTTSCHALK (64), *Institute of Wood Chemistry, University of Hamburg, 2050 Hamburg 80, Federal Republic of Germany*

JOAN HARRINGTON (71), *Biological and Chemical Research Institute, New South Wales Department of Agriculture, Rydalmere, New South Wales 2116, Australia*

SHINSAKU HAYASHIDA (34, 86), *Department of Agricultural Chemistry, Faculty of Agriculture, Kyushu University, Hakozaki, Higashi-Ku, Fukuoka 812, Japan*

JOHN HEPTINSTALL (29), *Department of Biological Sciences, Coventry Polytechnic, Coventry, Warwickshire CV1 5FB, England*

HIROMI HON-NAMI (58), *Technology R & D Division, New Energy Development Organization, Chiba 281, Japan*

KOYU HON-NAMI (58), *Technology R & D Division, New Energy Development Organization, Chiba 281, Japan*

KOKI HORIKOSHI (82), *The Institute of Physical and Chemical Research, Wako, Saitama 351-01, Japan*

MICHAEL A. HULME (13), *Pacific Forestry Centre, Victoria, British Columbia V8Z 1M5, Canada*

B. J. JACOBSON (15), *Pharmacokinetic Drug Analysis Laboratory, Veterans Administration Medical Center, Fargo, North Dakota 58102*

MICHAEL JOHN (84), *Max-Planck-Institut für Züchtungsforschung, Abteilung Schell, D-5000 Köln 30, Federal Republic of Germany*

K. G. JOHNSON (68, 69), *Division of Biological Sciences, National Research Council of Canada, Ottawa, Ontario, Canada, K1A OR6*

GWENNAËL JOLIFF (39), *Département des Biotechnologies, Unité de Physiologie Cellulaire, Institut Pasteur, 74724 Paris Cedex 15, France*

L. JURASEK (83), *Pulp and Paper Research Institute of Canada, Pointe Claire, Quebec, Canada H9R 3J9*

MICHEL JUY (39), *Département des Biotechnologies, Unité de Physiologie Cellulaire, Institut Pasteur, 74724 Paris Cedex 15, France*

AKIRA KAJI (66, 91), *Kagawa University, Hamano-cho, 60-67-209, Takamatsu 760, Japan*

TADASHI KAMIKUBO (27), *Laboratory of Food Chemistry, Faculty of Home Economics, Kobe Women's University, 2-1 Aoyama-Higashi-Suma, Suma-ku, Kobe-shi, Hyogo 654, Japan*

TAKAHISA KANDA (46), *Department of Industrial Chemistry, Faculty of Engineering, Shinshu University, Nagano, Nagano 380, Japan*

MANABU KITAMIKADO (72), *Department of Fisheries, Faculty of Agriculture, Kyushu University, Hakozaki, Higashi-ku, Fukuoka 812, Japan*

KUMPEI KITAMURA (22), *Research and Development Department, Kirin Brewery Company, Limited, 6, Shibuya-ku, Toyko 150, Japan*

R. KLAUS (16), *D-6100 Darmstadt, Federal Republic of Germany*

DIETER KLUEPFEL (18), *Institut Armand-Frappier, Université du Québec, Laval, Québec, Canada H7N 4Z3*

ISAO KUSAKABE (62, 75, 81), *Institute of Applied Biochemistry, University of Tsukuba, Tsukuba, Ibaraki 305, Japan*

ANIL H. LACHKE (87), *Division of Biochemical Sciences, National Chemical Laboratory, Pune 411 008, India*

CHRISTINE M. LADISCH (2), *Textile Science, Purdue University, West Lafayette, Indiana 47907*

M. R. LADISCH (4, 15), *Laboratory of Renewable Resources Engineering, Purdue University, West Lafayette, Indiana 47907*

RAPHAEL LAMED (57), *Center for Biotechnology, George S. Wise Faculty of Life Sciences, Tel Aviv University, Ramat Aviv, Israel*

LOWELL N. LEWIS (37), *Plant Molecular Biology Division, University of California, Berkeley, California 94720*

J. K. LIN (15), *Food and Drug Administration, Cincinnati, Ohio 45226*

WILLIAM A. LINDNER (42), *Biochemistry Department, University of Fort Hare, Alice 5700 Ciskei, South Africa*

LARS G. LJUNGDAHL (58), *Department of Biochemistry, University of Georgia, Athens, Georgia 30602*

AMY C. LO (50), *National Research Council, Division of Biological Sciences, Ottawa, Ontario, Canada K1A OR6*

FRANK G. LOONTIENS (6), *Laboratorium voor Biochemie, Rijksuniversiteit Gent, B-9000 Gent, Belgium*

C. R. MACKENZIE (68, 69, 90), *Division of Biological Sciences, National Research Council of Canada, Ottawa, Ontario, Canada, K1A OR6*

GORDON MACLACHLAN (43), *Biology Department, McGill University, Montréal, Québec, Canada H3A 1B1*

MASAKI MARUI (81), *Seitoku Junior College of Nutrition, Nishishinkoiwa, Katsushika-ku, Tokyo 124, Japan*

RYUICHI MATSUNO (27), *Department of Food Science and Technology, Faculty of Agriculture, Kyoto University, Kitashirakawa, Sakyo-Ku, Kyoto 606, Japan*

MASARU MATSUO (85, 88, 89), *Institute of Applied Biochemistry, University of Tsukuba, 1-1-1 Tennodai, Tsukubashi 305, Japan*

FRANK MAYER (58), *Institute of Microbiology, University of Göttingen, Göttingen, Federal Republic of Germany*

BARRY V. MCCLEARY (8, 60, 61, 63, 67, 70, 71, 73, 74, 76, 78), *Biological and Chemical Research Institute, New South Wales Department of Agriculture, Rydalmere, New South Wales 2116, Australia*

SHEILA MCCRAE (19), *Microbial Biochemistry Department, Rowett Research Institute, Bucksburn, Aberdeen AB2 9SB, Scotland*

ANTHONY MCHALE (51), *Department of Microbiology, Trinity College, Dublin, Ireland*

JACQUELINE MILLET (39), *Département des*

Biotechnologies, Unité de Physiologie Cellulaire, Institut Pasteur, 74724 Paris Cedex 15, France

DANA MISLOVIČOVÁ (65), Institute of Chemistry, Center of Chemical Research, Slovak Academy of Sciences, 843 38 Bratislava, Czechoslovakia

KAIGUO MO (34, 86), Department of Ferment Engineering, Wuxi Institute of Light Industry, Qingshanwan, Wuxi, Jiangsu, China

M. MOBEDSHAHI (4), Lilly Research Laboratories, Lilly Corporate Center, Indianapolis, Indiana 46285

AIDAN P. MOLONEY (40), Department of Biochemistry, University College, Galway, Ireland

YUTAKA MORI (58), National Food Research Institute, Yatabe Tsukuba-gun, Ibaraki 305, Japan

SAWAO MURAO (30), Department of Applied Microbial Technology, Kumamoto Institute of Technology, Kumamoto 860, Japan

KATSUMI NAKAMURA (22), Central Laboratory of Key Technology, Kirin Brewery Company, Limited, 3, Miyahara-cho, Takasaki-shi, Gunma 370-12, Japan

KOTOYOSHI NAKANISHI (81), Institute of Enology and Viticulture, Yamanashi University, Kofu, Yamanashi 400, Japan

THOMAS K. NG (38), Medical Products Department, E. I. duPont de Nemours and Company, Wilmington, Delaware 19898

M.-L. NIKU-PAAVOLA (11), Biotechnical Laboratory, Technical Research Centre of Finland, SF-02150 Espoo, Finland

KAZUTOSI NISIZAWA (46), Department of Fisheries, College of Agriculture and Veterinary Medicine, Nihon University, Setagaya, Tokyo 154, Japan

KUNIO OHMIYA (44, 47), Department of Food Science, and Technology, School of Agriculture, Nagoya University, Chikusa, Nagoya 46401, Japan

KAZUYOSHI OHTA (34, 86), Department of

Agricultural Chemistry, Faculty of Agriculture, Kyushu University, Hakozaki, Higashi-ku, Fukuoka 812, Japan

GENTARO OKADA (28), Department of Biology, Faculty of Education, Shizuoka University, Shizuoka-shi 422, Japan

HIROSUKE OKADA (79), Department of Fermentation Technology, Faculty of Engineering, Osaka University, Suita-shi, Osaka 565, Japan

M. G. PAICE (83), Pulp and Paper Research Institute of Canada, Pointe Claire, Québec, Canada H9R 3J9

RAJKUMAR V. PATIL (32, 49, 53), Department of Biological Chemistry, University of Michigan, Ann Arbor, Michigan 48104

I. N. PAVLOVA (77), Department of Biochemistry and Microorganisms, Institute of Microbiology and Virology, Academy of Sciences of Ukrainian SSR, Kiev, 252143 USSR

A. N. PEREIRA (4, 15), Universidade Tros-os-Montes, Department Microbiologica e Technologica Alimentar, Codex 5001, Villa Real, Portugal

B. PETTERSSON (41, 59), Swedish Pulp and Paper Research Institute, S-114 86 Stockholm, Sweden

L. G. PETTERSSON (12), Institute of Biochemistry, Biochemical Center, S-751 23 Uppsala, Sweden

ROBERTO J. POLJAK (39), Département des Biotechnologies, Unité de Physiologie Cellulaire, Institut Pasteur, 74724 Paris Cedex 15, France

JÜRGEN PULS (64), Institut of Wood Chemistry, Federal Research Center of Forestry and Forest Products, 2050 Hamburg 80, Federal Republic of Germany

F. M. ROMBOUTS (26), Agricultural University, Department of Food Science, 6703 BC Wageningen, The Netherlands

JAI C. SADANA (32, 49, 53), Division of Biochemical Sciences, National Chemical Laboratory, Pune 411 008, India

J. N. SADDLER (1), Biotechnology and Bio-

chemistry Department, Forintek Canada Corporation, Ottawa, Ontario, Canada K1G 325

REIICHIRO SAKAMOTO (30), *Japan Pulp and Paper Research Institute Incorporated, Tokodai, Tsukuba, Ibaraki 300-26, Japan*

TAKASHI SASAKI (56), *Food Resources Division, National Food Research Institute, Tsukuba Science City, Ibaraki 305, Japan*

GEORG SCHMID (5), *Institute of Biotechnology at the Nuclear Research Center, Jülich, D-5170 Jülich, Federal Republic of Germany*

JÜRGEN SCHMIDT (84), *Max-Planck-Institut für Züchtungsforschung, Abteilung Schell, D-5000 Köln 30, Federal Republic of Germany*

H. SCHNEIDER (69, 90), *Division of Biological Sciences, National Research Council of Canada, Ottawa, Ontario, Canada, K1A OR6*

MARTIN SCHÜLEIN (25), *Novo Industri A/S, Novo Allé, DK-2880 Bagsvaerd, Denmark*

I. SHAMEER (67), *Biocon Chemicals Limited, Kilnagleary, Carrigaline, County Cork, Republic of Ireland*

MAXWELL G. SHEPHERD (31), *Experimental Oral Biology Unit, School of Dentistry, University of Otago, Dunedin, New Zealand*

JAIPRAKASH G. SHEWALE (49), *Hindustan Antibiotics Limited, Pimpri, Pune 411 018, India*

SHOICHI SHIMIZU (44, 47), *Department of Food, Science, and Technology, School of Agriculture, Nagoya University, Chikusa, Nagoya 46401, Japan*

ATSUHIKO SHINMYO (79), *Department of Fermentation Technology, Faculty of Engineering, Osaka University, Suita-shi, Osaka 565, Japan*

JOHN C. STEWART (29), *Department of Biological Sciences, Coventry Polytechnic, Coventry, Warwickshire, CV1 5FB, England*

HIROSHI SUZUKI (21), *Institute of Biological*

Sciences, University of Tsukuba, Sakura, Tsukuba City, Ibaraki 305, Japan

KIYOSHI TAGAWA (66, 91), *Department of Bioresource Science, Faculty of Agriculture, Kagawa University, Miki-cho, Kagawa 761-07, Japan*

RIHEI TAKAHASHI (62, 75), *Forest Products Experiment Station, Toyama Forestry and Forest Products Research Center, Kurokawashin, Kosugi-machi, Izumigun, Toyama 939-03, Japan*

V. Y. TAMM (77), *Department of General and Soils Microbiology, Institute of Microbiology and Virology, Academy of Sciences of Ukrainian SSR, Kiev, 252143 USSR*

MICHIO TANAKA[1] (92), *Mitsubishi-Kasei Institute of Life Sciences, Machida-shi, Tokyo 194, Japan*

MITSUO TANAKA (27), *Department of Food Science and Technology, Faculty of Agriculture, Kyoto University, Kitashirakawa, Sakyo-Ku, Kyoto 606, Japan*

MASAYUKI TANIGUCHI (27), *Department of Chemical Engineering, Faculty of Engineering, Niigata University, 8050 Ikarashi-Nino-cho, Niigata-shi, Niigata 950-21, Japan*

RONALD M. TEATHER (7), *Animal Research Centre, Agriculture Canada, Ottawa, Ontario K1A OC6, Canada*

RUDOLF TOMAN (65), *Institute of Chemistry, Center of Chemical Research, Slovak Academy of Sciences, 843 38 Bratislava, Czechoslovakia*

PETER TOMME (19), *Laboratorium voor Biochemie, Rijksuniversiteit Gent, B-9000 Gent, Belgium*

CHOW CHIN TONG (31), *Jabatan Biokimia dan Mikrobiologi, Fakulti Sains dan Pengajian Alam, Sekitar Universiti Pertanian Malaysia, Serdang, Selangor, Malaysia*

TSUNEKO UCHIDA (92), *Mitsubishi-Kasei Institute of Life Sciences, Machida-shi, Tokyo 194, Japan*

[1] Deceased.

HENRYK URBANEK (35), *Laboratory of Enzymology, University of Łódź, Banacha 12/16, 90-237 Łódź, Poland*

HERMAN VAN TILBEURGH (6), *Plant Genetic Systems, B-9000 Gent, Belgium*

A. G. J. VORAGEN (26), *Agricultural University, Department of Food Science, 6703 BC Wageningen, The Netherlands*

MARIA VRŠANSKÁ (80), *Institute of Chemistry, Center of Chemical Research, Slovak Academy of Sciences, 843 38 Bratislava, Czechoslovakia*

U. WESTERMARK (55), *Swedish Pulp and Paper Research Institute, S-114 86 Stockholm, Sweden*

JÜRGEN WIEGEL (64), *Department of Microbiology, and Center for Biological Resource Recovery, University of Georgia, Athens, Georgia 30602*

GORDON WILLICK (50), *National Research Council, Division of Biological Sciences, Ottawa, Ontario, Canada K1A OR6*

DAVID B. WILSON (33), *Section of Biochemistry, Molecular and Cell Biology, Division of Biological Sciences, Cornell University, Ithaca, New York 14853*

PETER J. WOOD (7), *Food Research Centre, Agriculture Canada, Ottawa, Ontario K1A OC6, Canada*

THOMAS M. WOOD (1, 3, 9, 19, 23, 24, 45), *Microbial Biochemistry Department, Rowett Research Institute, Bucksburn, Aberdeen AB2 9SB, Scotland*

TAKEHIKO YAMAMOTO (93), *Faculty of Science, Osaka City University, Sumiyoshi-ku, Osaka 558, Japan*

KUNIO YAMANE (21), *Institute of Biological Sciences, University of Tsukuba, Tsukuba City, Ibaraki 305, Japan*

TUNEO YASUI (81, 85, 88, 89), *Institute of Applied Biochemistry, University of Tsukuba, Tsukuba, Ibaraki 305, Japan*

I. YA. ZACHAROVA (77), *Department of Biochemistry and Microorganisms, Institute of Microbiology and Virology, Academy of Sciences of Ukrainian SSR, Kiev, 252143 USSR*

JADWIGA ZALEWSKA-SOBCZAK (35), *Laboratory of Enzymology, University of Łódź, Banacha 12/16, 90-237 Łódź, Poland*

J. G. ZEIKUS (38), *Michigan Biotechnology Institute, Michigan State University, East Lansing, Michigan 48824*

Preface

Volumes 160 and 161 of *Methods in Enzymology* collate for the first time an array of procedures related to the enzymatic conversion of plant structural biomass polymers into their constituent monomeric units. This collection of methods for the hydrolysis of cellulose and hemicellulose (Volume 160) and of lignin, as well as related methods for pectin and chitin (Volume 161), is timely because of the increasing tempo of investigation in this area. This is in response to an immediate interest in the conversion of biomass monosaccharides into fuel ethanol and the longer term concern for maintaining supplies of liquid fuels and chemicals with eventual petroleum depletion.

Enzymatic treatment of plant biomass involves special methods due to the insolubility of the lignocellulosic complex and other similar polymers. These methods include substrate preparation, measurement of chemical changes, and culturing of organisms that produce the enzymes. Many of the methods are published in applied and special purpose journals not routinely seen by investigators and hence are not highly visible.

The ability to clone genes, transform cells, and express and secrete heterologous proteins in industrially important microorganisms presents opportunities to produce biomass enzymes in large quantity and at low prices. When this capacity is developed, enzymes will not be selected because of better production in a wild-type organism. Instead, the enzymes will be chosen for their superior catalytic capability and compatibility with the conditions of an industrial process. Since genes from various and often obscure organisms may produce enzymes better suited to such purposes, we have attempted to include methods for the preparation of enzymes in each class, for instance endocellulases, from a wide variety of sources so that investigators seeking to develop useful processes may make use of the options available.

We wish to acknowledge the expert secretarial assistance of Ms. Karen Payne in preparation of these volumes.

WILLIS A. WOOD
SCOTT T. KELLOGG

METHODS IN ENZYMOLOGY

EDITED BY

Sidney P. Colowick and Nathan O. Kaplan

VANDERBILT UNIVERSITY
SCHOOL OF MEDICINE
NASHVILLE, TENNESSEE

DEPARTMENT OF CHEMISTRY
UNIVERSITY OF CALIFORNIA
AT SAN DIEGO
LA JOLLA, CALIFORNIA

METHODS IN ENZYMOLOGY

EDITORS-IN-CHIEF

Sidney P. Colowick and Nathan O. Kaplan

Section I

Cellulose

A. Preparation of Cellulosic Substrates
Articles 1 through 8

B. Assays for Cellulolytic Enzymes
Articles 9 through 14

C. Chromatographic Methods for Carbohydrates
Articles 15 through 17

D. Miscellaneous Methods for Cellulolytic Enzymes
Articles 18 through 20

E. Purification of Cellulose-Degrading Enzymes

Cellulases
Articles 21 through 37

1,4-β-D-Glucan 4-Glucanohydrolases (Endocellulases)
Articles 38 through 43

1,4-β-D-Glucan Cellobiohydrolases (Exocellulases)
Articles 44 through 46

1,4-β-D-Glucosidases
Articles 47 through 51

Cellobiose Dehydrogenases
Articles 52 through 55

Assays of Cellulolytic Enzymes and Miscellaneous Enzymes
Involved in Cellulolysis
Articles 56 through 59

[1] Increasing the Availability of Cellulose in Biomass Materials

By T. M. WOOD and J. N. SADDLER

Structural cellulose is a crystalline polymer associated in a matrix with lignin and hemicellulose and as such is highly resistant to enzymatic attack. Pretreatment is therefore necessary, and, not surprisingly, it has been the subject of intensive investigation.

Despite the number of potentially exciting pretreatment methods present in the literature, there is only a limited understanding of how these pretreatments enhance cellulose hydrolysis in lignocellulosic substrates. It is apparent that all pretreatment methods should increase the accessibility of the cellulose component of the biomass substrate to the cellulase enzymes. Since cellulose is a heterogeneous porous substrate its rate of hydrolysis is governed by the number of glucose residues that are accessible to the rather large cellulase enzymes. An effective pretreatment should therefore increase the number of available sites for cellulase action and promote the extensive hydrolysis of the substrate. Many pretreatments result in only very limited improvements in the extent of hydrolysis. It is important, therefore, that the effectiveness of a pretreatment method is not judged by the initial rate of hydrolysis measured, as the cellulose hydrolyzed in the earlier stages of the reaction may differ substantially in accessibility and composition from the majority of the cellulose.

Several excellent reviews[1-3] have already covered methods for increasing the availability of cellulose in biomass materials, and most have separated the different types of pretreatment into physical, chemical, biological, and combinations of these methods (Table I). For all these methods it is assumed that the biomass substrate is already in a form such as wood chips or sawdust which can be readily processed.

Many pretreatment methods involve specialized equipment capable of withstanding high temperatures and pressures. No attempt is made in this chapter to describe such methods: the reader should refer to the relevant literature for full details of these laboratory techniques.

[1] M. A. Millett, A. J. Baker, and L. D. Satter, *Biotechnol. Bioeng. Symp.* **5,** 193 (1975).
[2] M. M. Chang, T. Y. C. Chou, and G. T. Tsao, *Adv. Biochem. Eng.* **28,** 15 (1980).
[3] L. T. Fan, Y.-H. Lee, and D. H. Beardmore, *Biotechnol. Bioeng.* **23,** 419 (1981).

TABLE I
METHODS FOR PRETREATING LIGNOCELLULOSIC MATERIALS TO ENHANCE
ENZYMATIC HYDROLYSIS[a]

Physical	Physicochemical	Chemical	Biological
Irradiation	Steaming	Organosolv	White rot fungi
Pyrolysis	Steam–explosion	Oxidation (ozone, H_2O_2)	Bacteria
Attrition	Autohydrolysis	Gas (SO_2, NO_2, ClO_2)	Enzymes
Ball milling	Ammonia–freeze–explosion	Swelling agents (NaOH, NH_3)	
Hammer milling	Wet oxidation	Cellulose-dissolving solvents	
Roll milling		Acid	
Wetting		Alkali	

[a] All these methods can be used in the laboratory, but some require specialized equipment that can withstand high temperatures and pressure. Only a few of the most promising methods requiring simple laboratory apparatus are described in detail.

Chemical Pretreatments

Most chemical methods are active against the two major deterrents to cellulase hydrolysis, cellulose crystallinity and the lignin barrier. Lignin has been shown to restrict enzymatic and microbial access to the cellulose while crystallinity restricts both the rate and completeness of the enzymatic hydrolysis of the cellulose.[4]

Those reagents that primarily remove the lignin and hemicellulose without significantly affecting the cellulose component include dilute alkali,[5] dioxane,[5] ozone,[6,7] hydrogen peroxide,[8,9] and traditional pulping chemicals. The resulting pretreated material is more accessible and susceptible to enzymatic attack. Other processes, such as organosolv pulping, use organic solvents at high temperatures and pressures to solubilize the lignin.

H_2O_2 and Ozone Pretreatments. There have been several recent reports on the effectiveness of hydrogen peroxide and ozone pretreatments in enhancing the biodegradation of lignocellulosic residues.[6-9] Both these

[4] M. A. Millett, A. J. Baker, and L. D. Satter, *Biotechnol. Bioeng. Symp.* **6,** 125 (1976).
[5] R. W. Detroy, L. A. Lindenfelsar, G. St. Julian, and W. L. Orton, *Biotechnol. Bioeng. Symp.* **10,** 135 (1980).
[6] A. Binder, T. Haltineir, and A. Fiechter, *Proc. Symp. Bioconversion Biochem. Eng., 2nd* p. 315 (1980).
[7] J. Miron and D. Ben-Ghedalia, *Biotechnol. Bioeng.* **23,** 823 (1981).
[8] J. M. Gould, *Biotechnol. Bioeng.* **26,** 46 (1984).
[9] W. C. Neely, *Biotechnol. Bioeng.* **24,** 59 (1984).

reagents attack lignin and the hemicellulose in preference to cellulose. They also disrupt the association between the carbohydrate polymers and lignin, producing a residue with enhanced reactivity to cellulase enzymes. The action of hydrogen peroxide may not require the total removal of the lignin and hemicellulose. It has been shown that the reactivity of aspenwood, which has first been pretreated by steam explosion and subsequent water and alkali extraction, is greatly enhanced by hydrogen peroxide treatment (Brownell and Saddler, unpublished results). Although the exact nature of how the hydrogen peroxide enhances accessibility to the cellulose is not known it is probable that the oxidation mechanism acts in an analogous fashion to that observed after the chlorite treatment of steam-treated, alkali-extracted aspenwood.[10] It was found that the use of a small amount of oxidizing agent, which resulted in the removal of very little lignin, exposed a large area of cellulose to enzymatic attack. It appeared that the lignin removed had been present as a relatively thin film or deposit on a large surface area. Furthermore, removal of lignin with larger amounts of oxidizing agent was apparently from larger, but less spread-out deposits which were not masking as much cellulose surface.

Experimental Method for H_2O_2 Pretreatment.[8] Samples of hammer-milled (2 mm) lignocellulose are pretreated with alkaline peroxide by placing 1 g of the material in 50 ml of distilled water which contains 1% (v/v) H_2O_2. The suspension is adjusted to pH 11.5 with 0.1 N NaOH and stirred gently at 25° for 18–24 hr. Periodic adjustments of the pH are made during the course of the reaction, if necessary. The insoluble residue is collected by filtration, washed with distilled water until the pH of the filtrate is neutral, and then dried at 100° or by solvent exchange (ethanol followed by acetone).

Comments. Delignification of straw by H_2O_2 is not extensive unless the pH of the solution is kept above 10.5: maximum delignification occurs at pH 11.5 or higher. Approximately 17–20% of the hemicellulose is lost at pH 11.5 in the presence of H_2O_2, leaving a residue which can be virtually completely solubilized using commercial cellulase from *Trichoderma reesei*. Complete removal of the hemicellulose can be effected at pH 13; but at this pH it would appear that the presence of H_2O_2 is not absolutely essential. In the absence of H_2O_2 all the hemicellulose remains insoluble after pretreatments up to pH 12. The loss of hemicellulose and lignin from straw in the 1% H_2O_2 at pH 11.5 results in an increase in the cellulose content of the insoluble residue from approximately 36 to 60–70%.

[10] J. N. Saddler, H. H. Brownell, L. P. Clermont, and N. Levitin, *Biotechnol. Bioeng.* **24,** 1389 (1982).

Experimental Method for Ozone Pretreatment.[6] A tank reactor which contains approximately 1 liter of a 5% straw suspension is sparged with a mixture of air and ozone at room temperature. The aeration rate used is of the order of 1 liter/min and the ozone concentration approximately 10 mg/liter. Ozone concentrations at the reactor inlet and outlet can be monitored using a spectrophotometer. Extensive delignification (60%) occurs with an ozone uptake of 20 mg/g. The cellulose component of straw delignified to this extent can be 70% converted to soluble sugars using *T. reesei* cellulase. Further treatment of ozone-treated straw with concentrated (85% w/v) phosphoric acid (20 mg straw/ml) for 5 hr at room temperature followed by the addition of an equal volume of methanol to precipitate the dissolved cellulose fragments, centrifugation, neutralization with 0.05 *N* sodium hydroxide, and washing with water results in a product in which virtually all the cellulose can be hydrolyzed with *T. reesei* cellulase.

Ozonation[7] can also be carried out by passing the gas through ground wheat straw moistened 40% and placed in a glass column. The process is stopped when decolorization of the straw is complete. An average of 1 g ozone/5 g of treated material is reported to be required. The straw is finally freeze dried.

Organosolv Pretreatment. Organosolv pretreatment involves the treatment of lignocellulose with an aqueous solvent in the presence of a catalyst. The catalyst and water hydrolyze many of the carbohydrate bonds in the hemicellulose component as well as the lignin–carbohydrate and lignin–lignin bonds. The solvent results in an organophilic environment in which the lignin dissolves. The primary catalysts are Lewis acids, e.g., $FeCl_3$, $Al_2(SO_4)_3$,[11] while common solvents that have been employed include alcohols, glycerol, dioxane, phenol, and ethylene glycol.[12–16] Organic acids liberated from the substrate during the process accelerate delignification and result in the concurrent solubilization of the hemicelluloses. This process has been shown to be effective on a wide range of substrates, including softwoods; however, difficulties have been encountered in establishing conditions which allow the selective extraction of lignin without dissolution of the hemicelluloses.[17]

[11] K. V. Sarkanen and J. L. McCarthy, *Annu. Rep. Natl. Sci. Found. Grant 77-08979* p. 8 (1977).

[12] S. I. Aronovsky, U.S. Patent 2,037,001 (1934).

[13] S. I. Aronovsky and R. A. Gortner, *Ind. Eng. Chem.* **28,** 1270 (1936).

[14] T. N. Kleinert and K. Tayenthal, *Z. Angew Chem.* **44,** 788 (1931).

[15] T. N. Kleinert, *Tappi* **58,** 170 (1975).

[16] S. M. Hansen and G. C. April, *in* "Fuels from Biomass: A Symposium (D. L. Klass and G. H. Emert, eds.), p. 180. American Chemical Society, Washington, D.C., 1980.

[17] M. T. Holtzapple and A. E. Humphrey, *Biotechnol. Bioeng.* **26,** 670 (1984).

Organosolv pretreatment can be conveniently carried out in the laboratory, but vessels capable of withstanding steam pressures giving temperatures of 160–180° are required. In a typical experiment described by Holtzapple and Humphrey,[17] a reaction tube consisting of a 1 in. diameter by 3 in. long pipe nipple with two end caps is used, and this is placed in a pressure vessel capable of bringing the temperature up to 160° within 5 min. The reaction tube contains wood ground in a Wiley mill to a −20 + 40 mesh size, or 0.015 mm shavings, and is heated for 30 min in an organosolvent mixture comprising 90% butanol and 3.0 N NaOH as a catalyst with a liquid to wood ratio of 10 : 1. Cooling to room temperature is achieved in about 5 min by passing cold water through the pressure vessel. The wood is removed and washed in a sintered glass crucible with acetone and then water.

Under the conditions used, poplar wood was reported to have about 25% of the lignin and about 30% of the hemicellulose removed while cellulose loss was less than 10%. Percentage conversion of the cellulose by cellulase from *Thermomonospora* sp. was approximately 45%; with the cellulase of *Trichoderma* sp. the conversion may be higher.

Optimal pretreatment conditions will vary widely according to the source of the lignocellulose. Extensive preliminary experiments using a variety of conditions are therefore necessary.

Pretreatments Involving Cellulose Solvents

Cellulose solvents act by reducing cellulose crystallinity and disrupting the lignin–carbohydrate association resulting in dissolution of the cellulose and hemicellulose. Cellulose solvents include concentrated mineral acids (H_2SO_4, HCl), ammonia-based solvents (NH_3, hydrazine), aprotic solvents[5,7] (DMSO, sulfur oxides), and metal complexes (cadoxen, cuoxam).[18] Some cellulose solvents transform native cellulose (cellulose I) to the cellulose II form which has been shown to be highly susceptible to hydrolysis by enzymes.[7]

Of the solvents mentioned, those involving the transition metal complexes[18] have been most widely studied. The metal complex solvents form ligands which may be monodentate or bidentate. With a monodentate ligand the principal solvent is ammonia (e.g., cellulose-dissolving ammonia complexes of copper hydroxide and nickel hydroxide (so-called cuoxam, and nioxam, respectively)); with bidentate ligands, solvents include ethylenediamine, biuret, and tartrate.[18] It is reported that both types

---

[18] T. J. Hamilton, B. E. Dale, M. R. Ladisch, and G. T. Tsao, *Biotechnol. Bioeng.* **26,** 781 (1984).

of solvent are most effective when prepared in concentrated (1–2 M) alkali, and it is presumed that this is due to the swelling effect of the alkali on the cellulose. Cadoxen (complex involving Cd^{2+} and ethylenediamine) is a very important cellulose solvent and this is discussed in detail in [2].

Very high conversions (80%) of the cellulose in corn residues have been reported using a ferric sodium tartrate (FeTNa) cellulose solvent and *T. reesei* cellulase.[18]

Experimental Method for Ferric Sodium Tartrate Pretreatment (FeTNa)[18]

Preparation of Solvent. The solvent is prepared by first dissolving 207.1 g of sodium tartrate in 400 ml hot water in a 1.5-liter stainless-steel beaker, cooling the solution to room temperature and adding 81.1 g ferric chloride hexahydrate dissolved in 160 ml water. It is important that these solutions are mixed vigorously using a magnetic stirrer and that they are protected from light. After mixing for 30 min the solution is cooled to 10–15° in an ice bath. At this juncture a solution of 98 g of NaOH dissolved in 200 ml of water is added slowly, making sure that the temperature does not exceed 15°. The yellowish-green solution is transferred to a 1-liter volumetric flask containing 50 g of sodium sulfite and the volume made up to the 1-liter mark with distilled water. Any precipitate of ferric hydroxide appearing at this stage is removed by centrifugation or filtration. Under these conditions the free NaOH concentration in the solution is 1.5 N.

Removal of Hemicellulose. Before using the FeTNa solvent pretreatment, the hemicelluloses in the corn residue are removed by hydrolysis with 5% H_2SO_4 at 90° for 4 hr using a weight ratio of 16:1 (solution: solid). The slurry is agitated throughout the hydrolysis using a magnetic stirrer. Finally, the residue is isolated by filtration and allowed to dry in air after thorough washing with water. The extracted residue at this stage consists of approximately 60% cellulose, 20% lignin, and 18% other constituents. Essentially all the hemicelluloses are hydrolyzed to soluble sugars under these conditions.

Pretreatment. The pretreatment of the lignocellulosic residues left after removal of the hemicellulose is carried out for several hours at room temperature. A volume of 5–14 ml of the solvent is used per gram of lignocellulose. The solvent is finally removed by filtration and water washing.

It will be necessary to determine the optimal conditions for pretreatment for different lignocelluloses by studying the effects of different ratios of solvent/lignocellulose on cellulose conversion using different levels of cellulase activity.

Biological Pretreatments

Biological pretreatment of wood has been primarily concerned with the use of fungi and bacteria that can attack lignin and hemicellulose. In general, soft and brown rot fungi degrade cellulose and hemicellulose in preference to lignin while white rot fungi can remove all three components at the same time or remove the lignin preferentially. Eriksson and Goodell[19] have isolated cellulase-less mutants of the white rot fungi, *Sporotrichum pulverulentum,* which will partially delignify wood without affecting the cellulose component. Unfortunately, there is a significant loss of the xylan and mannan components of the hemicellulose, at the same time as the lignin is oxidized, while the major problem of a lengthy pretreatment time of 2–5 weeks remains.[20,21]

Physicochemical Pretreatment

The two major processes that fall into this category are steam pretreatment and wet oxidation: both require specialized equipment. Wet oxidation involves heating the biomass at around 120° in the presence of water and air or oxygen, under high pressures. During this treatment acids formed by the deacetylation of the hemicelluloses and by oxidation partially hydrolyze the hemicelluloses as well as cellulose and lignin, yielding a residue which is highly accessible.[22]

Steam pretreatment results in the partial decomposition of some of the hemicellulose components to acids, mainly acetic acid, which catalyze the depolymerization of the hemicellulose and lignin. On completion of steaming, the material can either be explosively discharged (steam–explosion) or emptied out after a gradual "bleed-down." It has been shown that the rapid decompression does not greatly influence the accessibility of the cellulose.[23,24] This type of pretreatment also allows the selective fractionation of the three main wood components. After steam treatment the majority of hemicellulose can be removed by water extraction, most of the lignin is removed by subsequent dilute alkali, ethanol, or acetone extraction, leaving a cellulose-rich insoluble residue.[10] The steaming process is usually carried out in a high-pressure vessel at pressures up to 4

[19] K. E. Eriksson and E. W. Goodell, *Can. J. Microbiol.* **20,** 371 (1974).
[20] P. Ander and K. E. Eriksson, *Mater. Org.* **3,** 129 (1976).
[21] A. I. Hatakka, *Eur. J. Appl. Microbiol. Biotechnol.* **18,** 350 (1983).
[22] G. D. McGinnis, W. W. Wilson, and C. E. Mullen, *Ind. Eng. Chem. Prod. Res. Dev.* **22,** 352 (1983).
[23] H. H. Brownell, M. Mes-Hartree, and J. N. Saddler, *Proc. Bioenergy Specialist Meet. Biotechnol., 1st* p. 85 (1984).
[24] H. H. Brownell and J. N. Saddler, *Biotechnol. Bioeng.* **28,** 792 (1986).

MPa and at temperatures of 185–250°. It has been demonstrated to be effective on a wide range of hardwoods and agricultural residues, although softwoods have proved to be more recalcitrant. Steam treatment causes a physical change in lignocellulosic substrates by loosening the plant cell wall structure, increasing the specific surface area and reducing the degree of polymerization of cellulose. Chemical reactions that occur in the steaming process are similar to those that occur in the organosolvent process. It separates the biomass components under relatively mild conditions by cleaving the linkages within the hemicelluloses and between the hemicellulose and lignin (mostly α ether bonds and 4-O-methylglucuronic acid ester bonds to the α carbons of the lignin units) and therefore releasing the lignin from the cell wall matrix. Hydrolysis of α-O-4 and, to a smaller extent, β-O-4 ether bonds in the lignins leads to a lignin of high free phenolic content.

Marchessault and St-Pierre[25] found that the crystallinity of steam-exploded aspenwood was not affected by pretreatment and Dekker and Wallis[26] similarly reported no change in crystallinity of cellulose in sugarcane bagasse following autohydrolysis pretreatment.

Steam treatment under alkaline conditions enhances the accessibility of the cellulose.[26] The alkali contributes to the dissolution of the hemicellulose and lignin, swelling of the cellulose, saponification of the intermolecular ester bonds, and reduction of the degree of polymerization of the cellulose. There have been problems, however, with decomposition of the liberated sugars when this process is carried out at high temperatures. Other reagents which have been added to wood prior to steam treatment include CO_2 and SO_2. Treatment of both softwood and hardwood with SO_2 prior to explosion result in a residue which was more readily hydrolyzed by cellulases.[27,28] Both these reagents increase the acidity during the steaming process consequently enhancing the solubilization and removal of the hemicellulose fraction.

The AFEX (ammonia–freeze–explosion) process involves the pretreatment of lignocellulosic substrates by treating them with liquid ammonia, under pressure, followed by a sudden release which evaporates the liquid ammonia resulting in the subsequent freezing of the exterior of the lignocellulosic fibers.[29] The ammonia in the interior of the fiber begins to

[25] R. H. Marchessault and J. St-Pierre, in "Future Sources of Organic Raw Materials—Chemrawn 1," p. 613. Pergamon, New York, 1980.

[26] R. F. H. Dekker and A. F. W. Wallis, Biotechnol. Bioeng. 25, 3027 (1983).

[27] K. L. Mackie, H. H. Brownell, K. L. West, and J. N. Saddler, J. Wood. Chem. Technol. 5, 405 (1986).

[28] H. Mamers and D. Menz, Appita 37, 644 (1984).

[29] D. E. Dale and M. J. Moreira, Biotechnol. Bioeng. Symp. 12, 31 (1982).

boil causing the fibers to explode which results in reduced particle size. This treatment is reported to reduce the lignin content as well as swelling and decrystallizing the cellulose component. This method is highly effective on agricultural residues and some hardwoods.

Conclusions on Pretreatment Methods

An efficient pretreatment method is one which will allow the easy access of the cellulases to their substrate and enhance the complete solubilization of polymer to monomer sugars. It is apparent, however, that major structural barriers such as the association with lignin and hemicellulose, crystallinity index, degree of polymerization, and specific surface area will all interfere with the mode of interaction between the enzymes and the cellulose molecules. Usually when native cellulose is degraded by enzyme there is always a residual fraction that survives the attack. This material absorbs a significant amount of the original enzyme and restricts the reuse of these enzymes on added, fresh substrate. It is important that a pretreatment method not only enhances the initial rate of cellulose hydrolysis, it should also promote the complete hydrolysis of the cellulose to allow reuse of the enzymes as well as the most efficient use of the lignocellulosic substrate.

[2] Cadoxen Solvolysis of Cellulose

By Christine M. Ladisch

Cellulose is insoluble in water and in most organic solvents as a result of strong inter- and intramolecular hydrogen bonding. A number of cellulose solvents exist although it is difficult to characterize the solvation properties of each because of the broad differences among the cellulosic polymers and fibers themselves, such as size, degree of polymerization (DP), morphology, and impurities.

In general, a desirable cellulose solvent has the following properties: (1) it is clear and colorless, (2) it can be readily prepared or obtained, (3) it is stable over a period of time and can withstand minimal exposure to light and air, (4) it can dissolve cellulose completely (preferably high DP) in a reasonable amount of time, and (5) the dissolved cellulose does not undergo oxidative degradation.

Few, if any, of the known cellulose solvents are considered to have all of the above properties.

Cellulose solvents are best described as belonging to four major categories: (1) strong mineral acids, (2) quaternary ammonium bases, (3) nonaqueous organic solvents such as dimethyl sulfoxide (DSMO) and dimethylformamide (DMF), and (4) metal complexes.

Strong Mineral Acids

Sulfuric (60–72% by weight), hydrochloric (40%), and phosphoric (83%) acids are known to dissolve cellulose.[1] Each of these solvents has a definite concentration range that will dissolve cellulose; out of this range, cellulose merely swells. The ease with which cellulose can be regenerated from solution by a small dilution with water is evidence of the critical concentration range necessary for solvation. Strong acids are rarely used for cellulose dissolution—they are unpleasant to handle, degradation is unavoidable, and some esterification may occur with sulfuric and phosphoric acids.[1]

Quaternary Ammonium Bases

Quaternary ammonium bases have received some attention as cellulose solvents; interest in them diminished after the development of the metal complex solvents. Dimethyldibenzyl-,[2] tetraethyl-,[3] dimethylphenylbenzyl-, triethylbenzyl-, and triethylfuryl-[4] ammonium hydroxide (Triton bases) have been reported as adequate cellulose solvents. A 32–42% base concentration is needed for solubility at 20°. The fastest rate of dissolution using dimethyldibenzylammonium hydroxide was found to be 2–3 hr for a 1% solution of unmodified cotton cellulose.[2] The solution was stable to light and to air, however a change in viscosity over time was noted for concentrated solutions.

Metal Complex Solvents

Iron–Tartaric Acid–Alkali Metal Complex Solvents. Another group of solvents capable of dissolving cellulose is the iron–tartaric acid–alkali metal complex.[5] Also known as EWNN and FeTNa, the complex contains an iron : tartaric acid : alkali mole ratio of 1 : 3 : 6 plus 2–3 N NaOH.

[1] G. Jayme and F. Lang, *Methods Carbohydr. Chem.* **III,** 75 (1963).

[2] E. L. Lovell, *Ind. Eng. Chem.* **16,** 683 (1944).

[3] V. H. Krassig and E. Siefert, *Makromol. Chem.* **14,** 1 (1954).

[4] A. A. Strepikheev, I. L. Knunyants, N. S. Nikolaeva, and E. M. Mogilevsky, *Bull. Acad. Sci. USSR, Div. Chem. Sci. (Engl. Transl.)* **6,** 769 (1957).

[5] G. Jayme, *Tappi* **44,** 299 (1961).

The solvent is green in color, readily dissolves high DP cotton, is stable to atmospheric oxygen, does not cause cellulose degradation, and is used in viscometric and degree of polymerization determinations.

Several variations of EWNN have been reported,[5-7] the most prominent of which is the 1 : 1 : 1 iron–tartaric acid–alkali metal complex.[5] This solvent, a brown solution, is considered to have better cellulose solvating properties than the green 1 : 3 : 6 complex.

Although the solvent power of the EWNN complex is considered to be as good as that of cadoxen (all known cellulose materials can be dissolved without residue formation when a two-stage dissolution procedure is used[5]), there are two major disadvantages associated with this solvent. It is colored and cellulose concentrations above 0.3% are highly viscous.[8,9] The formation of a gel is probable at concentrations above 2% cellulose.

Metal–Alkali Complexes with Biuret. Solutions of copper–alkali–biuret with lithium, sodium, or potassium hydroxide will form a clear, colored solution capable of dissolving cotton linters up to 7–8% (by weight).[9] The dissolved cellulose can be reprecipitated into film or filaments with the addition of a dilute acid/salt mixture. Like the EWNN solvent, the metal–alkali–biuret complex forms a gelatinous solution with cellulose.[9] After storage for several weeks, decomposition of the biuret is indicated by ammonia odors.

Miscellaneous Solvents. Johnson *et al.*[10] have claimed successful cellulose dissolution with a dimethyl sulfoxide/paraformaldehyde (PF) mixture. A clear, colorless, 1–2% cellulose solution (linters, wood pulp, and cotton) is claimed to be obtained by the DMSO/PF mixture; dissolution time is 3 min at 130°.[10,11] The dissolved cellulose may then be reprecipitated by the addition of water or alcohol. Little or no degradation of the cellulose by the solvent has been found. Suggested applications for the DMSO/PF solvent system are for light scattering, sedimentation, and osmometry measurements.[11]

Cellulose has also been dissolved in DMF (*N,N*-dimethylformamide)

[6] G. F. Bayer, J. W. Green, and D. C. Johnson, *Tappi* **48,** 557 (1975).
[7] L. Valtasaari, *Pap. Puu* **4,** 243 (1957).
[8] L. S. Bolotnikova, S. N. Danilov, T. I. Samsonova, and L. D. Turkova, *J. Appl. Chem. USSR (Engl. Transl.)* **35,** 2647 (1962).
[9] G. Jayme, *in* "Cellulose and Cellulose Derivatives" (N. Bikales and L. Segal, eds.), pp. 381–410. Wiley (Interscience), New York, 1971.
[10] D. C. Johnson, M. D. Nicholson, and F. C. Haigh, "Dimethyl Sulfoxide/Paraformaldehyde: A Non-Degrading Solvent For Cellulose," Paper No. 5. Presented at the Institute of Paper Chemistry. Eighth Cellulose Conference, Syracuse, New York, May 1975.
[11] H. A. Swenson, "The Solution Properties of Methylol Cellulose in Dimethyl Sulfoxide," Paper No. 8. Presented at the Institute of Paper Chemistry. Eighth Cellulose Conference, Syracuse, New York, May 1975.

or DMAC (*N,N*-dimethylacetamide), to which N_2O_4 or NOCl has been added.[12] Concentrations of 1–10% cellulose in solution have been obtained; increased concentrations produced highly viscous solutions. Use of N_2O_4 with the DMF or DMAC produces a clear green-blue solution; NOCl produces a clear red-brown solution. The cellulose solution is stable provided moisture is avoided—contact with moist air causes regeneration of the cellulose.[13]

Metal Complexes

The first metal complex solvent, a copper hydroxide/ammonia solution, was introduced by Schweizer in 1857. The solvent had good solvating properties and produced a clear solution but underwent oxidative degradation, was unstable, and the solution was colored (blue).[14] Modification of Schweizer's reagent by substituting ethylenediamine for the ammonia produced a blue solvent with good solvating capacity and slightly better oxidative resistance, but was unstable on storage. This solvent called cupriethylenediamine (Cuen or CED) is sometimes used for cellulose viscosity determinations.[15]

From 1951 to 1971, Jayme[9,16] reported on a series of metal complex solvents for cellulose. Based on the cupriethylenediamine hydroxide solvent, many of the new compounds had different metal ions in place of the copper, such as cobalt, nickel, zinc, and cadmium.

The solvents cobalt–, nickel–, and zinc–ethylenediamine hydroxide did not offer substantial advantages over the copper-based solvents. They were not easily prepared, the nickel and cobalt solutions were colored, oxidative degradation occurred, and the cellulose solubility in zincoxen (zinc-ethylenediamine hydroxide) was low.[9]

Cadoxen. Unlike the zinc, nickel, copper, and cobalt-based solvents, the cadmium–ethylenediamine complex possesses a number of desirable cellulose solvent properties. First introduced by Jayme and Neuschaffer in 1957,[17] tris(ethylenediamine)cadmium hydroxide or cadoxen as it is now widely known, has good solvating properties, is a clear, colorless, nearly odorless liquid, is stable for an almost unlimited time, causes little degradation of cellulose, and is relatively easy to prepare. Like many of the other cellulose solvents, cadoxen is toxic and must be handled with care.

[12] R. G. Schweiger, *Tappi* **57,** 86 (1974).
[13] R. G. Schweiger, *Chem. Ind.* (*London*) **10,** 296 (1969).
[14] D. Henley, *Ark. Kemi* **18,** 327 (1961).
[15] F. L. Straus and R. M. Levy, *Pap. Trade J.* **114,** 31 (1942).
[16] G. Jayme, *Papier* (*Darmstadt*) **5,** 244 (1951).
[17] G. Jayme and K. Neuschaffer, *Makromol. Chem.* **28,** 71 (1957).

Uses of the Solvent Cadoxen. Brown[18] has described cadoxen as "the most useful cellulose solvent both for research purposes and industrial control analyses." Again, the advantages of cadoxen over other available solvents are that it is colorless, relatively easy to handle, dissolves many forms of cellulose quickly, and is stable when exposed to atmospheric oxygen. These properties allow the use of cadoxen in many areas of cellulosic research, such as viscometry,[19,20] refractive index, and diffusion measurements.[14] Cadoxen has also been used to determine the degree of aging of alkali cellulose through viscometric determinations.[21] The molecular weight distribution of cellulose and hemicellulose was determined by GPC using cadoxen as a solvent and elution agent.[22]

Spectrophotometric measurements of cadoxen–cellulose solutions are also possible since cadoxen is clear and colorless. The intrinsic color of wood pulps, dissolved in the solvent, has been determined in order to ascertain the pulps' suitability for the viscose process[23,24] and a quantitative assay for lignin content (0.05 to 3% by weight) in selected pulps[25] has been developed using spectrophotometric techniques.

Within the textile field, cadoxen has been used to study swelling phenomena of fibers,[26-29] the degree of cross-linking in resin-treated fibers,[30] dye content analysis,[31] and cellulose content in fabrics.[32,33] Cadoxen also dissolves protein and has been indicated by Ladisch, Tsao, and Ladisch to have potential in the spectrophotometric (UV) assay of protein in the presence of cellulose.[34] Achwal and Mohite[35] colorimetrically estimated

[18] W. Brown, *Sven. Papperstidn.* **70**, 458 (1967).
[19] L. S. Bolotnikova, S. N. Danilov, and T. I. Samsonova, *J. Appl. Chem. USSR (Engl. Transl.)* **39**, 150 (1966).
[20] L. Segal and J. D. Timpa, *Sven. Papperstidn.* **72**, 656 (1969).
[21] H. Elmgren and D. Henley, *Sven. Papperstidn.* **63**, 139 (1960).
[22] O. Bobleter and W. Schwald, *Papier (Darmstadt)* **39**, 437 (1985).
[23] G. Jayme and K. K. Hasvold, *Papier (Darmstadt)* **20**, 657 (1966).
[24] D. K. Smith, R. F. Bampton, and W. J. Alexander, *Ind. Eng. Chem. Process Des. Dev.* **2**, 57 (1963).
[25] E. Sjostrom and B. Enstrom, *Sven. Papperstidn.* **69**, 469 (1966).
[26] G. M. Evans and R. Jeffries, *J. Appl. Polym. Sci.* **14**, 633 (1970).
[27] G. M. Evans and R. Jeffries, *J. Appl. Polym. Sci.* **14**, 655 (1970).
[28] H. Hampe, B. Phillipp, and J. Baudisch, *Faserforsch. Textiltech.* **23**, 425 (1972).
[29] B. S. Varma and H. C. Bhatia, *Text. Res. J.* **41**, 790 (1971).
[30] W. Schefer, *Text. Res. J.* **41**, 927 (1971).
[31] W. B. Achwal and A. A. Vaidya, *Text. Res. J.* **39**, 816 (1969).
[32] W. B. Achwal and A. A. Vaidya, "Cadoxen: A Cellulose Solvent of Diverse Applications for Control in Textile Processing." Proceedings, Section B, of the ATIRA, BTRA, and SITRA, 24.1–24.19. December 21–23, 1968.
[33] C. M. Ladisch, C. M. Chiasson, and G. T. Tsao, *Text. Res. J.* **52**, 423 (1982).
[34] M. R. Ladisch, G. T. Tsao, and C. M. Ladisch, *Biotechnol. Bioeng.* **20**, 461 (1978).
[35] W. B. Achwal and V. P. Mohite, *Text. Res. J.* **88**, 435 (1972).

direct, acid, metal complex, and reactive dyes on wool dissolved in cadoxen and zincoxen.

Cadoxen has also been used as a pretreatment in order to increase the rate and extent of enzymatic hydrolysis of cellulose. Treatment with cadoxen disrupted the crystalline structure of cellulose, thereby rendering it susceptible to rapid and total hydrolysis to soluble products. The cellulose in pretreated bagasse, cornstalks, and fescue was quantitatively converted to glucose, which showed that the cadoxen pretreatment is also effective for cellulose protected by a lignin seal.

This pretreatment procedure has also been incorporated in a quantitative cellulose analysis.[33,36] Cellulose from various sources was extracted (dissolved in cadoxen and reprecipitated), enzymatically hydrolyzed, and the cellulose content directly calculated from the quantity of sugar produced.

Specific solvent preparation and cellulose dissolution and reprecipitation techniques which give the best dissolution and hydrolysis results follow.

Preparation of the Solvent Cadoxen. Preparation of cadoxen has been described by a number of authors,[37–39] yet all of the procedures described are modifications of Jayme's[17] original procedure. The first cadoxen was prepared by saturating a solution of ethylenediamine in cadmium hydroxide or cadmium oxide, the latter of the two being the simpler procedure. In order to increase the solvating power of cadoxen, Henley[14] added NaOH to the solution. The best solvating properties were obtained when the cadmium concentration ranged from 4.5 to 5.2% by weight, the ethylenediamine from 25 to 30%, and the NaOH from 0.2 to 0.5 M. Unfortunately, the addition of NaOH results in higher oxidative degradation of the cellulose than in pure cadoxen. However, if the cellulose dissolved in cadoxen is stored below 18° the degradation rate is extremely low.[40]

Preparation of the solvent is relatively simple. One of the best descriptions is the preparation of Segal and Timpa.[41] Solvent preparation is described below.

Cadmium oxide should be oven dried at 110° and stored in a desiccator. Sodium hydroxide pellets are added to ethylenediamine and allowed to stand at least 24 hr. The ethylenediamine should then be distilled over sodium hydroxide pellets just prior to use. (The use of freshly distilled

[36] M. R. Ladisch, C. M. Ladisch, and G. T. Tsao, *Science* **201,** 743 (1978).
[37] D. Henley, *Sven. Papperstidn.* **63,** 143 (1960).
[38] H. Reimers, *Papier* (*Darmstadt*) **16,** 566 (1962).
[39] L. Segal and J. D. Timpa, *Sven. Papperstidn.* **72,** 656 (1969).
[40] W. B. Achwal and A. B. Gupta, *Angew. Makromol. Chem.* **2,** 190 (1968).
[41] L. Segal and J. D. Timpa, *Sven. Papperstidn.* **72,** 656 (1969).

ethylenediamine is particularly important for high dissolution power in cadoxen.)

A 28% (by weight) ethylenediamine solution should be prepared using chilled distilled water. After chilling the mixture, 1000 ml is poured into a 2000-ml 3-necked round bottom flask. The flask is placed on ice, the mixture stirred, and 90 g of dried CdO is added through the side neck in 8–9 g portions at 10- to 15-min intervals. The mixture is stirred for 3 hr after all of the CdO has been added (solution turns from brown, to clear, to white) and stored overnight at −5°.

The clear solution (with sediment settled out) is then poured into 250-ml centrifuge bottles and centrifuged (12,000 g) at −5° for 30 min. The clear solution is again poured back into the 3-neck flask and resaturated with CdO. The mixture is stirred until a white precipitate (CdOH) is evident.

Again, the solution is stored overnight at −5° and centrifuged as before. As soon as the mixture reaches room temperature, the clear solution is decanted from the precipitate and centrifuged at 20–22°. It is decanted again and allowed to stand overnight at room temperature to precipitate. The cadoxen is decanted, centrifuged, poured into brown glass bottles, and stored at 5°. If stored cold, the cadoxen will retain adequate dissolution power for up to 1 year. However, if concise solution properties, such as viscometry, are needed the solvent should be as fresh as possible.

Segal and Timpa[41] also recommend the addition of sodium hydroxide to cadoxen. The NaOH is reported to increase dissolution power, but it also increases cellulose degradation.[14,40] The information which follows refers only to cadoxen without NaOH.

Cellulose Dissolution. Cadoxen is reported to dissolve many different forms of cellulose: wood pulps,[14] cotton linters,[37] cotton fibers,[39] rayon,[5] and cellulose derivatives. The cellulosic material is suspended in cadoxen at room temperature and stirred until dissolved. Dissolution times range from 2 min to 2 hr, depending upon the type and molecular weight of the substance, degree of crystallinity, concentration of the solution being prepared, and physical form of the cellulose (powder, loose fibers, etc.). Typical dissolution times for a variety of cellulose sources are given in Table I. Dissolution of higher molecular weight samples is aided by pre-wetting with a small amount of water,[37] dissolving at ambient temperature for 5–15 min, and then storing at 0° for 1 hr.[19] The dissolved cellulose will then remain in solution (at 0°) as long as impurities are avoided. The cadoxen–cellulose solution may be diluted with water and/or cadoxen, however the diluted solution will no longer dissolve additional cellulose.[20] Storage of the solvent at 0° is recommended.[14]

The cellulose should be dried to a constant weight at approximately

TABLE I
DISSOLUTION TIME FOR CELLULOSE IN CADOXEN

Cellulose	Concentration (%)	Time (min)
Avicel	1	2
	5	15
	10	120
Whatman CF-11	1	3
	3	20
	5	120
Cotton		
Unbleached (DP ~3900)	2	13
Bleached (DP ~3700)	2	10

110° and weighed to the nearest 0.1 mg or, preferably, weigh a non-oven-dried specimen and subtract its moisture content from the total weight. The cellulose is then placed in the bottom of a 16 × 150-mm test tube and cadoxen solvent (brought to room temperature) is slowly added. The solution should stand for about 1 min and then be mixed gently using a vortex mixer. Allow the mixture to stand again for 10 min and then place on ice until the cellulose dissolves completely.

Cadoxen solvent dissolves cellulosic materials relatively easily if the procedure is followed exactly as described. Variations in the procedure produce cellulose solutions but generally result in longer dissolution times and limited cellulose concentrations. Examples of alterations in the dissolution procedure, and the consequences thereof, follow.

The cellulose may be added to the solvent (rather than the solvent to the cellulose), however, if any cellulose is retained on the solvent-coated sides of the test tube and is not immediately washed down, a gel will form which then takes several hours to dissolve. Avoid the formation of a gel if at all possible.

It is important to allow the dissolving mixture to stand at room temperature for at least 10 min before being placed on ice. Treatment in cadoxen at room temperature apparently allows the cellulose to swell, which in turn speeds up the dissolution process. Also, the dissolving samples should be put on *ice*, and not into a freezer (10°). Many samples, especially those treated with a cross-linking agent, dissolve easily on ice and partially or not at all when placed in the freezer.

From Table I, it can be seen that some samples take longer to dissolve than others. This is especially true for samples containing noncellulosic material. Dissolution time can be decreased by prewetting the sample with a few drops of a nonionic surfactant, such as 1% Triton X-100 (Rohm

& Haas, Philadelphia, PA) prior to dissolution. Solubilization can be further speeded by milling the cellulose prior to wetting with the surfactant.

Cellulose/cadoxen solutions may be stored in the freezer and are stable for up to 4 months, after which the solutions slowly turn cloudy.

Precipitation of the Cellulose from Solution. Methanol or water will completely precipitate cellulose from cadoxen. Methanol/cadoxen (1 : 1) will yield a more flocculant (and more easily hydrolyzed) precipitate than water/cadoxen. Allow the precipitate to form for at least 10 min before stirring the mixture. One disadvantage of methanol is that it will denature cellulase enzyme and therefore must be removed from the precipitate prior to hydrolysis. Cadoxen does not affect the hydrolysis reaction.

[3] Preparation of Crystalline, Amorphous, and Dyed Cellulase Substrates

By Thomas M. Wood

The number of cellulose substrates used to determine cellulase activity is vast. No attempt has been made in this chapter to be comprehensive. Instead the author has been selective to the extent that only those substrates are discussed which are either widely used or are of special value in some assay procedure. This chapter should be read in conjunction with chapter [1] in this volume.

Many cellulase substrates which are extensively used are commercially available. These include filter paper, Avicel, Sigmacel, Solka Floc, carboxymethylcellulose, and hydroxyethylcellulose. These commercially available substrates are used in cellulase studies without further treatment and are therefore outside the scope of this chapter.

Preparation of Dewaxed Cotton Fibers

Carded cotton (1 g) is extracted in a Soxhlet apparatus for 18 hr with chloroform, with 95% ethanol for a further 18 hr, and then air dried. The solvent-extracted fibers are boiled for 4 hr with a 1% (w/v) solution of NaOH which has been sparged with oxygen-free nitrogen. Nitrogen is also bubbled into the suspension during the whole 4 hr treatment. Oxygen-free nitrogen is obtained by bubbling the gas through a bottle containing Fieser's solution (16 g sodium dithionite, 6.6 g sodium hydroxide, 2 g sodium anthraquinone 2-sulfate, 100 ml water).

Finally, the fibers are washed with 4 liters of hot water, left in cold water (4 liters) overnight, suspended in 500 ml of 1% (v/v) acetic acid, and then washed with water until the pH is neutral. Drying of the fibers is effected by washing with 95% ethanol, followed by acetone and then drying in air or under reduced pressure at 25°.

Comments. Cotton fiber prepared in this way is an excellent substrate for measuring cellulase activity in that it is highly crystalline and requires the complete cellulase system to be present for its rapid dissolution. Cotton fiber is approximately 70% crystalline when determined by the X-ray crystallographic method. The use of cotton fiber as a substrate for cellulases has resulted in many meaningful observations being made on the mechanism of cellulase action, particularly the synergism between components. As currently understood, neither of the major cellulase components (cellobiohydrolase and endo-1,4-β-glucanase) can solubilize cotton fiber to a significant extent when acting in isolation: only the concerted action of these enzymes can achieve rapid hydrolysis.

Cotton is the purest form of naturally occurring cellulose, but it contains several impurities such as cotton wax (0.3–1.0%), pectin, mainly in the outer layers (0.5–1.2%), and protein residues from the protoplasm. Other organic and inorganic residues are present. Any method of purification must aim at removing these impurities without modifying the cellulose by introducing, for example, groups such as carboxyl and carbonyl. Normally, the waxes can be extracted with organic solvents, and pectins and proteins by hot alkali. The alkali treatment is carried out in the absence of oxygen to minimize depolymerization.

About 2% degradation occurs during the alkali treatment. However, as this occurs in a stepwise fashion from the reducing end of the chain, there is really no detectable change in chain length. A few reducing terminal groups are modified during the treatment to D-glucometasaccharinic acid[1] which is stable to further attack by alkali.

Preparation of Hydrocellulose[2]

Any fibrous cellulose (1 g) is suspended in 50 ml 2.5 M hydrochloric acid and heated at 80° for 4 hr, washed with water, neutralized with dilute (1%) ammonia, and washed again with water. Finally, the material is

[1] G. Machell and G. N. Richards, *J. Chem. Soc.* **4**, 4500 (1957).
[2] M. Schuler, L. S. Sandall, and P. Lumer, *J. Appl. Polym. Sci.* **18**, 2075 (1974).

isolated by filtration or centrifugation and dried by solvent exchange (ethanol followed by acetone) or by freeze-drying.

Comments. Under the acid conditions given above, the cellulose very quickly reaches a leveling-off degree of polymerization of between 100 and 250. It is suggested that the amorphous regions of the cellulose are attacked preferentially with the acid treatment described leaving a highly crystalline residue. Hydrocellulose is available commercially as Avicel. Microcrystalline cellulose similar to Avicel can be prepared by blending the undried hydrocellulose at 5% (w/v) solids concentration in a Waring blender. The hydrocellulose which is originally present as millions of aggregates of microcrystals is converted into a preponderance of isolated or free microcrystals. Extensive homogenization (30 min) of hydrocellulose or Avicel results in a colloidal suspension of particles which is reduced in size but shows no decrease in crystallinity. Avicel is reported to be approximately 47% crystalline. Because of its short chain length, Avicel, or the laboratory-prepared hydrocellulose, is useful in assays for endwise acting glucanases.

Preparation of α-Cellulose[3]

Approximately 100 g hammer-milled (2-mm sieve) straw or other lignified plant material is extracted in a Soxhlet with a 2 : 1 mixture of toluene : ethanol and then with 95% ethanol. The air-dried sample is suspended in 1 liter of water in a 3-liter flask in a water bath and heated with stirring to 70–75°. A 11-ml portion of glacial acetic acid is added followed by 32 g of sodium chlorite which is added slowly over a period of several minutes. *Caution!* The whole procedure is carried out in a fume hood. An additional precaution is that a stream of nitrogen gas is passed over the surface on the reaction mixture to prevent the chlorine dioxide which is liberated from forming an explosive mixture. Foaming is controlled by the addition of one to two drops of 1-octanol.

Fresh portions (four in total) of acetic acid (11 ml) and sodium chlorite (32 g) are added at intervals of 20–25 min over a period of 1.5–2 hr. The mixture is then cooled, filtered through a porosity 0 glass sinter, washed with water until free of acid, and dried by solvent exchange (ethanol followed by acetone). The residue (so-called holocellulose) which is partially delignified contains much of the original hemicellulose, and this is removed in the next step.

Suction-dried holocellulose is extracted for 24 hr at room temperature with 18% (w/v) NaOH solution (1 liter) under nitrogen in a Turbula mixer

---

[3] R. L. Whistler and J. N. BeMiller, *Methods Carbohydr. Chem.* **3,** 21 (1963).

or other shaker. Residual solids are recovered by filtration (porosity 0 sintered glass filter funnel), washed with 4 liters of water, 2 liters of dilute acetic acid, a further 4 liters of water, and then dried in air.[1] The extraction should be repeated to yield a product containing 0.5–1% nonglucan carbohydrates.

Comments. A limited amount of degradation of the cellulose occurs during the delignification. However, the chlorite procedure is less drastic in this respect than other laboratory methods used to effect delignification.

Cellulose prepared from straw by the chlorite procedure has a chain length of approximately 2000; wood cellulose prepared in essentially the same way has been reported to have a degree of polymerization of approximately 3000.

A more highly purified cellulose, particularly from wood holocellulose, can be obtained by using hot alkali extraction (40–70°),[3] but the cellulose is more highly degraded. Improved extraction of hemicellulose can also be obtained using a short acid pretreatment (2.5 N H_2SO_4 at 100° for 5 min) before alkali extraction, but this also causes some further depolymerization.

The major portion of α-cellulose is readily attacked by cellulase enzyme components acting alone. However, when this reactive, easily accessible portion of the cellulose is hydrolyzed the rate of hydrolysis then becomes very slow. It appears that the amorphous component of the cellulose which is entangled and overlaid with hemicellulose and lignin in the native material is being preferentially hydrolyzed in the early stages.

Solka Floc is a commercially available wood α-cellulose. In the author's experience Solka Floc may contain up to 17% xylan: in the published specification the values quoted are much lower.

Solka Floc is used extensively as a carbon source for the production of cellulase enzymes.

Preparation of Alkali-Swollen Cellulose

Whatman cellulose powder (CF 11) (1 g) is suspended in 100 ml of 4 N NaOH solution at 10° for 1 hr and then poured into ice-water mixture (2 liters). Washing is effected first by decantation using 1 liter of water and then with a further 6 liters of water using a sintered glass filter (porosity 0). The cellulose is dialysed at 1° for 48 hr against distilled water, treated in a Waring blender for 1 min to disperse the clumps, and washed several times by decantation using 1 liter of water. The cellulose suspension is stored at 1° in water which is 0.05 M with respect to NaN_3. The concentration of the homogeneous suspension can be determined from an aliquot by filtration (porosity 2 sintered glass filter) and weighing the dried (100°)

cellulose. From this figure the concentration can be adjusted to that desired for the cellulase assay.

Comments. Cellulose which is swollen in alkali at a concentration of 4 N is highly susceptible to hydrolysis by cellulase enzymes. At this concentration of alkali complete swelling of the primary wall and the lumen occurs with the formation of a crystal lattice (soda-cellulose I) quite different from that in the original cellulose powder.[4] It seems that after extensive washing, cellulose swollen under the conditions given may exist in the cellulose II lattice, but this point is the subject of discussion. Differences in crystal lattice may be very important factors to be considered in studies on the mechanism of cellulase action.

Preparation of H_3PO_4-Swollen Cellulose (Walseth Cellulose)[5,6]

Avicel (10 g) is suspended in H_3PO_4 (85% w/v) and left with occasional stirring for 1 hr at 1°. The mixture is poured into ice-cold water (4 liters) and left for 30 min. The swollen Avicel is washed several times with cold water by decantation, then with 1% (w/v) $NaHCO_3$ solution. The thick suspension of swollen cellulose is dialyzed at 1° against several changes of water until the dialyzate is neutral, and stored in aqueous suspension (5 mM with respect to NaN_3) at 1°. A short 2 min treatment in a Waring blender provides a homogeneous suspension of cellulose which is readily degraded by many cellulases of different origin. Any lumps can be removed by decantation. The concentration of the cellulose can be determined by weighing a freeze-dried sample.

Cotton fiber can be used in place of Avicel, but the gelatinous acid-swollen cellulose which is formed after 4 hr is difficult to manipulate. Extensive treatment with $NaHCO_3$ solution is required to neutralize the acid. This can best be done by suspending the gelatinous mass in the bicarbonate solution overnight. Treatment with a Waring blender is usually necessary to obtain a usable cellulose suspension which is free from acid. Lumps of partially swollen cellulose can be removed by decantation using large measuring cylinders to provide long sedimentation times. H_3PO_4-swollen cellulose prepared from cotton has a degree of polymerization of approximately 2000: when prepared from Avicel the chain length is approximately 100.

---

[4] J. O. Warwicker, R. Jeffries, R. L. Colbran, and R. N. Robinson, *in* "A Review of the Literature on the Effect of Caustic Soda and Other Swelling Agents on the Fine Structure of Cotton," Pamphlet No. 93, p. 94. Shirley Institute, Didsbury, Manchester, England, 1966.

[5] C. S. Walseth, *Tappi* **35,** 228 (1971).

[6] T. M. Wood, *Biochem. J.* **121,** 353 (1971).

Preparation of Regenerated Cellulose

Cotton fiber (1 g) is suspended in 100 ml 70% (w/v) H_3PO_4 and the mixture heated with stirring at 60° for 4 hr. The clear solution is poured, with stirring, into 1 liter of a mixture of ice and water. The precipitated cellulose is allowed to settle and is then washed by decantation with several liters of water until neutral. NaN_3 is added until the solution is 0.05 M with the bacteriocide: the suspension is kept in this state at 4°. No attempt should be made to dry the cellulose as it has been found to rehydrate only very slowly and to be much less susceptible to cellulase action than never-dried material.

Whatman cellulose powder can be used instead of cotton. The suspension (5% w/v) in H_3PO_4 (85% w/v) is placed in a boiling water bath until the liquid clarifies.

Regenerated cellulose can also be prepared using HCl (40%) or H_2SO_4 (35–50%) at room temperature. The cellulose is stored until it dissolves. It appears that cellulose regenerated after complete dissolution in HCl or H_2SO_4 has a cellulose II lattice: cellulose regenerated from H_3PO_4, however, would appear to exist, mainly, in the native cellulose I crystal lattice. Phosphoric acid appears to degrade cellulose less than other mineral acids. The suspension of regenerated cellulose is useful for turbidimetric assays to determine cellulase activity.

Preparation of Amorphous Cellulose

Whatman cellulose powder (50 g), grade CF11, is milled for 72 hr in 500 ml water with a ball mill. The colloidal suspension is filtered through a sintered glass filter funnel, porosity 1, and made 0.05 M with respect to NaN_3. The concentration of the cellulose suspension is determined on an aliquot by freeze drying.

Ball-milled cellulose is easily degraded by the individual components of the cellulase complex provided the cellulose is not dried. It is a convenient substrate for turbidimetric assays.

Preparation of Dyed Avicel[7]

Avicel (100 g) is suspended in 1 liter of water and heated to 50° in a water bath with vigorous stirring. One liter of a suspension of 1% (w/v) Remazol Brilliant Blue dye is added and the suspension is agitated vigorously for 45 min during which time 200 g of Na_2SO_4 is added in several

[7] M. Leisola, M. Linko, and E. Karvonen, *Proc. Symp. Enzymatic Hydrolysis Cellulose* p. 297 (1965).

portions. At this stage, 100 mg of a 10% (w/v) solution of Na_3PO_4 is added (the pH should be 12) and the agitation is continued for 75 min. The dyed cellulose is recovered by filtration on a sintered glass filter funnel (porosity 1), and washed with distilled water at 60° until the filtrate is colorless. The dyed Avicel is then rinsed with acetone, ether, and dried *in vacuo*.

Dyed Solka Floc can be prepared in exactly the same way.

The procedure for preparing dyed H_3PO_4-swollen cellulose is similar except that the final product is not dried, but kept in aqueous suspension which is 0.02% with respect to NaN_3.

Comments. Dyed celluloses have been used to measure cellulase activity. The method is convenient to use and may replace (with some reservations) assays involving the measurement of reducing sugars released during hydrolysis. These dyed substrates are particularly useful for studies on the kinetics of hydrolysis in the presence of end-product inhibitors such as glucose or cellobiose which give unacceptably high reagent blanks when reducing sugar methods are used.

Preparation of Dyed Carboxymethylceullose[8]

Step 1. Preparation of Diaminoethylcarboxymethylcellulose. Ten grams of fibrous carboxymethylcellulose (CM-23) (Whatman Biochemicals, Ltd., U.K.) is washed successively with 500 ml of 0.5 *M* NaOH, several liters of distilled water, 500 ml of 0.5 *N* HCl, and again with several liters of distilled water. The washed CM-cellulose is suction dried on a Büchner funnel (Whatman No. 4 paper) and 10 g of this is resuspended in 30 ml of water containing 0.7 ml of ethylenediamine. After the pH is adjusted to pH 5.2 with 6 *N* HCl, 3 g of 1-ethyl-3-(3-dimethylaminopropyl)carbodiimide–HCl is added. The suspension is shaken at room temperature for 24 hr. After the first hour of the reaction, the pH is readjusted to 5.2. Finally, the modified cellulose is washed with 4 liters of distilled water on a Büchner funnel using Whatman No. 4 filter paper.

Step 2. Preparation of Trinitrophenylcellulose (TNP-Cellulose). Suction-dried DAE-CM-cellulose (10 g) is resuspended in 30 ml of 0.2 *M* sodium borate buffer, pH 9.0, stirred gently, and 0.6 g of 2,4,6-trinitrobenzenesulfonic acid is added. The pH is adjusted to pH 9.0 with 4 *N* NaOH solution. The container is wrapped in aluminum foil and the reaction mixture mixed gently for 3 hr at room temperature. The fibrous suspension is filtered on a Büchner funnel, washed successively with 500 ml of 0.2 *M* sodium borate buffer, pH 9.0, 6 liters of distilled water, 1 liter of acetone, and dried by suction. The product (3 g approximately) is stable indefinitely when stored in a brown bottle at 4°.

[8] J. S. Huang and T. Tang, *Anal. Biochem.* **73**, 369 (1976).

[4] Preparation of Cellodextrins

By A. N. Pereira, M. Mobedshahi, and M. R. Ladisch

Introduction

Cellodextrins, water-soluble β-1,4 oligomers of glucose with degree of polymerization (DP) between 2 and 7 (Table I), are important substrates for the characterization of cellulases.[1,2] Their contribution to understanding the kinetics of cellulose hydrolysis has been reviewed.[3,4] Cellodextrins can be used to infer features of the active site of carbohydrates and to study induction and repression of enzyme synthesis during fermentation.

Methods

The major methods for obtaining cellodextrins are (1) chemical synthesis and (2) fragmentation of cellulose. Cellobiose, the first member of the series, can be obtained through condensation of 1,6-anhydro-β-D-glucopyranose[5] or 1,2,3,6-tetra-O-acetyl-β-D-glucopyranose[6] with 2,3,4,6-tetra-O-acetyl-α-D-glucopyranosyl bromide or from epicellobiose octaacetate.[7] Trisaccharides, among them cellotriose, and tetrasaccharides, among them cellotetraose, can be prepared from acetylated glycosyl halides of cellobiose or cellotriose and 1,2,3,5-tetra-O-acetyl-α-D-glucopyranose by the Koenigs–Knorr reaction.[8-10] The method is not practical, however, for cellodextrins of higher DP.

Enzymes and acids are the hydrolytic agents which give cellodextrins by fragmentation (hydrolysis) of cellulose. In contrast to the production of maltodextrins from starch, however, methods to produce cellodextrins using enzyme hydrolysis are not as well developed. Enzyme extracts

[1] W. Grassmann, L. Zechmeister, G. Toth, and R. Stadler, *Ann. Chem. (Warsaw)* **503**, 167 (1933).

[2] D. R. Whitaker, *Arch. Biochem. Biophys.* **53**, 439 (1954).

[3] K. Nisizawa, *J. Ferment. Technol.* **51**, 267 (1973).

[4] M. R. Ladisch, J. Hong, M. Voloch, and G. Tsao, "Trends in the Biology of Fermentations for Fuels and Chemicals," p. 55. Plenum, New York, 1981.

[5] V. E. Gilbert, F. Smith, and M. Stacey, *J. Chem. Soc.* 622 (1946).

[6] K. Freudenberg and W. Nagai, *Ber. Dtsch. Chem. Ges. B* **66**, 27 (1933).

[7] W. T. Haskins, R. M. Hann, and C. S. Hudson, *J. Am. Chem. Soc.* **64**, 1289 (1942).

[8] S. H. Nichols, Jr., W. L. Evans, and H. D. McDowell, *J. Am. Chem. Soc.* **62**, 1754 (1940).

[9] K. Takeo, T. Yasato, and T. Kuge, *Carbohydr. Res.* **93**, 148 (1981).

[10] K. Takeo, K. Okushio, K. Fukuyama, and T. Kuge, *Carbohydr. Res.* **121**, 163 (1983).

TABLE I
PROPERTIES OF CELLODEXTRINS

Compound	Molecular weight	Melting point (°C)	Optical rotation $[\alpha]_D$	Solubility in water (deg g liter^{-1})	Diffusion coefficient, $D_0 \times 10^6$ (cm^2 sec^{-1})	Intrinsic viscosity $[\eta]$ (dl g^{-1})	Molecular dimensions[a] (Å) L	d
Cellobiose	342.30	228–230[b]	36.0[b]	147[b]	5.71–5.9[c]	0.027[b]	1.4	0.642
Cellotriose	504.45	201–207[b]	23.5[b]	—	4.84–4.9[b]	0.030[b]	2.0	0.656
Cellotetraose	666.59	252–255[b]	17.3[b]	78[b]	4.2–4.3[b]	0.036[b]	2.6	0.660
Cellopentaose	828.73	264–267[b]	13.8[b]	4.8[b]	3.82[c]	0.038[b]	3.2	0.666
Cellohexaose	990.86	275–278[b]	10.0[d]	—	3.38[c]	0.047[b]	3.8	0.668
Celloheptaose	1152.90	283–286 ± 3[d]	—	—	—	—	—	—
Glucose	180.16	—	—	—	6.75	—	—	—

[a] For extended hydrated cellodextrins (Stuart–Briegleb model).
[b] See Ref. 24.
[c] At 30°.
[d] From Ref. 17.

having predominantly endocellulolytic activity are required and fractionation of large quantities of this cellulolytic complex is difficult.

Production of cellodextrins by HCl hydrolysis of cellulose was reported in 1931,[11] applied,[11,12] and later modified by Miller et al.[13,14] Acetolysis followed by deacetylation is an alternative method. Mild acetolysis conditions have been developed[15] and further explored.[16-18] Another approach based on the known solubilizing and hydrolytic ability of H_2SO_4, has been recently proposed.[19]

Separation of cellodextrins, which are homologous compounds of slightly different molecular dimensions, weight, and melting point (Table I), is a challenge. Microscale separation has been achieved by thin-layer chromatographic processes but larger scale separation preferentially uses liquid chromatography. Suitable chromatographic packings include charcoal and Celite,[2,13,14] charcoal and cellulose,[12] polyacrylamide and dextran,[20-22] silica[23,24] DEAE-Spheron 300,[25] and cation-exchange resin in the Ca^{2+} form.[19,24]

Molecular sieving is the main characteristic contributing to the separation of cellodextrins by gel permeation and chromatography over ion-exchange resin. The presence of unknown contaminants of identical mass but different molecular structure has been recently detected by anion-exchange chromatography.[22] The amount of these compounds changes with the conditions of cellulose depolymerization and, thus, are likely formed by side reactions during hydrolysis or initial steps of neutralization. The preparation of cellodextrins requires well-defined conditions of hydrolysis concentration of acid solution, temperature, time of reaction,

[11] L. Zechmeister and G. Toth, Ber. Dtsch. Chem. Ges. B 64, 854 (1931).

[12] M. A. Jermyn, Austr. J. Chem. 10, 55 (1957).

[13] G. L. Miller, J. Dean, and R. Blum, Arch. Biochem. Biophys. 91, 21 (1960).

[14] G. L. Miller, Methods Carbohydr. Chem. 3, 134 (1963).

[15] K. Hess and K. Dziengel, Ber. Dtsch. Chem. Ges. B 68, 1594 (1935).

[16] E. E. Dickey and M. L. Wolfrom, J. Am. Chem. Soc. 71, 825 (1949).

[17] M. L. Wolfrom and J. C. Dacons, J. Am. Chem. Soc. 74, 5331 (1952).

[18] M. L. Wolfrom and A. Thompson, Methods Carbohydr. Chem. 3, 143 (1963).

[19] M. Voloch, M. R. Ladisch, M. Cantarella, and G. T. Tsao, Biotechnol. Bioeng. 26, 557 (1984).

[20] W. Brown, J. Chromatogr. 52, 273 (1970).

[21] A. Heyraud, M. Rinaudo, M. Vignon, and M. Vincendon, Biopolymers 18, 167 (1979).

[22] K. Hamacher, G. Schmid, H. Sahm, and C. Wandrey, J. Chromatogr. 319, 311 (1985).

[23] M. Streamer, K. Eriksson, and B. Pettersson, Eur. J. Biochem. 59, 607 (1975).

[24] A. Huebner, M. R. Ladisch, and G. T. Tsao, Biotechnol. Bioeng. 20, 1669 (1978).

[25] Z. Hostomska-Chytilova, O. Mikes, P. Vratny, and M. Smrz, J. Chromatogr. 235, 229 (1982).

and purity and nature of acids and solvents to avoid the formation of such compounds.

Dissolution and Hydrolysis of Cellulose in HCl

Cellulose powder is slurried in a saturated solution of HCl (specific gravity >1.19), at room temperature, followed by addition of 0° fuming HCl solution (specific gravity 1.21). Presuspension of cellulose allows wetting of the powder and favors homogeneity. Other authors[12,20,23,24] dissolved the powder directly in ice-cold fuming HCl (specific gravity 1.21). Dissolution of the cellulose powder has been reported in concentrated HCl solution (specific gravity 1.18) at −30°.[26] Hamacher reported only suspension of the cellulose in concentrated HCl (specific gravity 1.19) at 25°.[22] Typically, an HCl concentration of greater than 40% is required for cellulose dissolution. This is usually obtained by bubbling HCl (gas) through reagent grade HCl solution (30–36%, by weight HCl). Because acid/cellulose contacting is strongly exothermic, ice-cold conditions are required to avoid formation of degradation products. This is assured by placing all solutions in the refrigerator prior to carrying out the reactions.

After stirring to obtain a homogeneous, viscous, and yellowish solution [at final concentrations up to 10% cellulose (w/v)], hydrolysis goes on for 1–3 hr at room temperature. Time and temperature will affect the intensity of hydrolysis, relative yield of each cellodextrin, and amount of degradation products. It has been shown[13] that the rate of hydrolysis of cellulose is linear and the yields of cellodextrins maximized at times not exceeding 2 hr. Times and temperatures of hydrolysis have been empirically determined by each author and are strictly related to the concentration of HCl and temperature of the hydrolyzate. Attempts have been made to predict molecular weight distribution of the products of hydrolysis.[12,27] However, in practice, these distributions are difficult to reproduce because of the unpleasant character of HCl solutions, and the limitations of exact control of experimental conditions. In spite of this, use of HCl still represents a preferred method.

Precipitation of High-Molecular-Weight Oligomers

Residual cellulose (soluble in the HCl) and high-molecular-weight oligomers formed during hydrolysis are precipitated by dilution of the

[26] G. Halliwell and R. Vincent, *Biochem. J.* **199,** 409 (1981).
[27] W. Kuhn, *Ber. Dtsch. Chem. Ges. B* **63,** 1503 (1930).

hydrolyzate by addition of ice-cold water.[12-14,20,24,26] Alternatively, soluble cellodextrins can be extracted with distilled water after washing out the HCl from the residual cellulose.[22]

Dissolution and Hydrolysis of Cellulose in H_2SO_4

Avicel (100 g) (Type 101, FMC, DE) is mixed with 50 ml of 80% H_2SO_4 (as 10 ml portions) at 4° to form a very viscous solution.[19] Keeping the flask in ice water, the mixture is constantly mixed with a Teflon stirring rod for 7 min. Next, 100 ml of deionized water is added with mixing, for another minute. The flask is transferred to a 70° water bath and stirred continuously for 14 min, where it becomes less viscous. The reaction is quenched by addition and mixing of 100 ml of absolute ethanol at 4°. The flask is transferred into an ice water bath where about 30–50 g activated carbon (Darco G-60) wet with 50% ethanol is added to the dark brown hydrolyzate mixture and allowed to sit at room temperature for 30–60 min with occasional mixing. Vacuum filtering the slurry and then washing it with 50 ml of 50% ethanol in a large Büchner funnel results in a clear, pale yellow liquid. Approximately 2.4 liters of absolute ethanol is then added to the filtrate resulting in the precipitation of cellodextrins (white in color). The mixture is left in the refrigerator overnight. This is followed with centrifugation for 20 min at 10,000 rpm (4°).[19] We have performed this procedure in our laboratories several times. When care at temperature control was taken by refrigerating all the chemicals (including the cellulose) overnight before starting the procedure, the resulting hydrolyzate is beige in color. This is compared to the previously reported dark-brown mixture obtained when cellulose, at room temperature, was used in place of cellulose cooled to 4°. When the cellodextrins are dried at room temperature, the product is gray crystals which are usually mostly glucose and high-molecular-weight oligomers.

Neutralization

Neutralization increases cellodextrin stability and is required before chromatographic separation. However, the large amounts of salt formed interfere with sugar determinations as well as subsequent chromatographic separation steps. With some exceptions,[12,19] most authors report the neutralization of the hydrolyzate before separation of the cellodextrins.

The most common approach to remove HCl is by direct neutralization using $NaHCO_3$. Two alternatives have been reported. Huebner et al. stripped part of HCl by vacuum (20 mm Hg for 15 min) before salt addition.[24] Hamacher et al. removed HCl by consecutive washes with 1-

propanol and technical grade ethanol.[22] Partial stripping of the acid before neutralization significantly reduces the amount of salts formed, but the HCl fumes must be condensed and trapped. Desalting of the neutralized hydrolysate by Dowex-3 (OH^-) and Dowex W-X8 (H^+), (Dow Chem. Co., Midland, MI) is an alternative and columns of these resins have been previously used.[23]

The original procedure[19] requires the use of significant amounts of ethanol. As a modification, the product may be washed with less ethanol in four repetitions rather than in one step. In each washing step, the gel-like cellodextrins should be well suspended in the fresh ethanol at 4° before each centrifugation (20 min at 10,000 rpm, 4°). After the fourth wash, centrifugation, and decantation of ethanol, the cellodextrin solids are dissolved in about 15 ml of cold, distilled water. The pH of the solution is about 3. Next, 0.1 M Ba(OH)$_2$ solution is added dropwise to bring the pH of the cellodextrin solution up to about 5 (note pH undergoes sudden, large changes). The solution is then centrifuged to remove the sulfate salt formed. The supernatant (a clear, colorless solution) is then decanted and lypholized overnight. The overall product yield is 0.5 to 1.5% based on the starting weight of the cellulose used. A typical chromatogram of the cellodextrin mixture obtained by this method is shown in Fig. 1. The cellodextrin mixture from the HCl method of Miller[13,14] as modified by Huebner et al.,[24] gives infrared (Fig. 2a) and ^{13}C NMR spectra (Fig. 3a) which are similar to spectra of cellodextrins (Figs. 2b and 3b) from the modified H$_2$SO$_4$ procedure of Voloch et al.[19] described here.

Dowex MR-3 may also be used to remove the traces of H$_2$SO$_4$ instead of Ba(OH)$_2$. However, this results in loss of some of the oligomers, especially those of lower DP.[28] We now have unpublished evidence (Dec., 1987) that cellodextrins prepared using H$_2$SO$_4$ may, at times, form slightly sulfated derivatives, which are much more water soluble than underivatized cellodextrins.

Preparative Scale Fractionation of Cellodextrins

The best separation of cellodextrins has been achieved by gel permeation, partition, or ion-exchange chromatography. Partition chromatography of cellodextrins on charcoal and cellulose columns,[12] untreated charcoal and celite columns,[2] and stearic acid treated charcoal-celite columns[13,14,20,26] was the first preparative scale method. Cellulose or celite are used to improve flow characteristics of the columns. Glucose and cellobiose are removed by elution with water or 5–7.5% ethanol. The elution of the cellodextrins, adsorbed to the charcoal surface, is achieved

---

[28] D. G. Bauet and R. E. Lee, *J. Liq. Chromatogr.* **5,** 767 (1982).

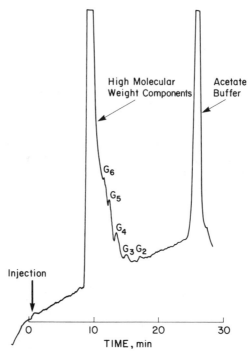

FIG. 1. Chromatogram of cellodextrin mixture obtained by a modified procedure of Voloch *et al.*[19] Sample size of 25 μl; LC column dimensions of 6 mm i.d. by 50 cm long packed with Aminex 50W-X4, H$^+$ form (20–30 μm particle size); eluent flow rate of 24 ml/hr. Eluent is water with H$_2$SO$_4$ added to give 126 μl H$_2$SO$_4$/liter. Column temperature 80°. Sample contains added acetate buffer (to give 16.7 mM sodium acetate).

by increasing the concentration of ethanol in either stepwise or gradient elution. However, this method requires repacking of the column after each run and is characterized by excessively slow flow rates.

Preparative scale separation of cellodextrins by column chromatography on cation-exchange resins (Aminex AG50W-X4 200–400 mesh, BioRad Lab., Richmond, CA) in the Ca^{2+} form, eluted with water, appears to be a good technique.[19,24] It allows relatively fast separation (typical runs of 6 hr) and adequate resolution of the pure components in an all-aqueous solution.

Gel permeation chromatography has also been studied. The thermodynamic parameters and partition coefficients of G$_2$ (cellobiose) to G$_6$ (cellohexaose) on polyacrylamide BioGel P-2 (200–400 mesh) (Bio-Rad Lab., Richmond, CA) are known[20,29] as well as their partition coefficients on

[29] A. Heyraud and M. Rinaudo, *J. Chromatogr.* **166**, 149 (1978).

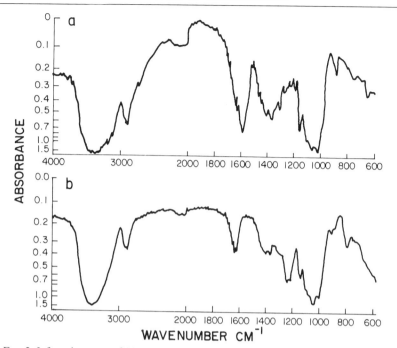

FIG. 2. Infrared spectra of (a) cellodextrin mixture prepared by hydrolysis of cellulose in HCl[24]; (b) cellodextrin mixture prepared by hydrolysis of cellulose in H_2SO_4. Pellets of 80 mg of KBr containing about 4 mg of lyophilized cellodextrin mixture were made and ran on a Beckman IR-33 scanning spectrophotometer.

Sephadex G-15 (Pharmacia Inc., Piscataway, NJ).[20] Low band broadening, good selectivity, and linearity between the logarithm of the distribution coefficients and DP have been found,[20,22,29,30] but the process is time consuming because long columns and low flow rates are required to achieve separation. Elution times of 24 hr are needed for a typical separation. Affinity separation and other techniques are also discussed in chapter [5] by Schmid in this volume.

Checking for Purity

Trace amounts of homologous dextrins, isomers, and derivatives are sometimes present. Hence, the degree of acceptable impurities is defined according to the proposed use. Liquid chromatography may be used to determine the presence of side products in the cellodextrin mixtures (Fig. 4). Preliminary assays may allow the identification of problems and suggest further analytical investigations.

[30] M. R. Ladisch, A. L. Huebner, and G. T. Tsao, *J. Chromatogr.* **147,** 185 (1978).

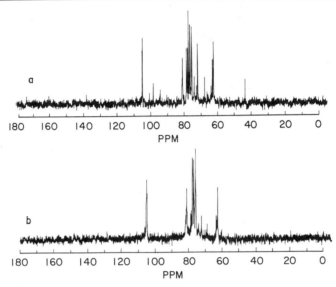

FIG. 3. (a) ^{13}C NMR of 100 mg cellodextrins from modified Miller (HCl) method[24] dissolved in 0.5 ml D$_2$O (also see Table II). Instrument was a Varian XL-200 spectrophotometer with a frequency of 50.304 MHz using broad band homonuclear decoupling. (b) ^{13}C NMR of cellodextrin mixture prepared by modified method of Voloch et al.[19] Conditions same as in (a) (also see Table III).

Preliminary assays are needed to confirm the exclusive presence of glucose and β-1,4-glycosidic linkages in the purified cellodextrins. The assays consist of the analytical determination of the products of hydrolysis of the samples by acid and cellulases. HCl hydrolysis of the purified cellodextrins, under mild conditions [1% (v/v) HCl at 80° for 12 hr] should give rise to glucose. Furthermore, purified cellodextrins should be completely hydrolyzed to glucose and cellobiose by the cellulase enzyme system. Commercially available crude cellulases, however, contain sugar impurities and other carbohydrates that may affect the interpretation of the assay. Consequently, these enzyme preparations are dialyzed or gel filtered prior to use.[19] If these assays give complete hydrolysis with glucose as the only product, an indication of homogeneous cellodextrins is obtained. Incomplete hydrolysis indicates the presence of a derivative of celloologiosaccharide.

Chromatographic Methods for Cellodextrin Identification and Analysis

Analytical chromatography of cellodextirns can be done with or without derivatization. An analytical determination, without derivatization, is

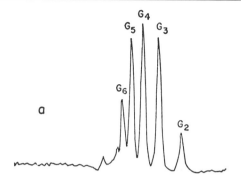

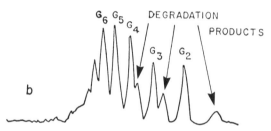

FIG. 4. Liquid chromatogram of cellodextrins having (a) no side products and (b) some side products. Side products formed by localized heating in an anion-exchange bed used in the ion removal step (see Huebner *et al.*[24]). Reprinted with permission. Copyright © 1978 by John Wiley and Sons.

by liquid chromatography over cation-exchange resin in the Ca^{2+} form using water as eluent.[30–32] Concentrations as low as 2 μg of cellodextrins can be measured.

Reversed-phase partition chromatography on alkylated and aminoalkylated silica gels, eluted with aqueous methanol or aqueous acetonitrile, is a common chromatographic technique for the analysis of carbohydrates and has been successfully used in studies with cellodextrins.[33,34] Sensitivity of 20 μg was reported but resolution and column stability is lower than for cation-exchange resins.

Borate-complex anion-exchange chromatography has also been successfully applied to almost all kinds of oligosaccharides. It is also an important analytical tool for the detection and quantification of impurities

[31] M. R. Ladisch and G. T. Tsao, *J. Chromatogr.* **166**, 85 (1978).
[32] G. Bonn, *J. Chromatogr.* **322**, 411 (1985).
[33] E. Gum, Jr., and R. D. Brown, Jr., *Biochem. Biophys. Acta* **492**, 225 (1977).
[34] S. P. Shoemaker and R. D. Brown, Jr., *Biochim. Biophys. Acta* **523**, 133 (1978).

TABLE II
^{13}C NMR Data of Cellodextrins Prepared by
Hydrolysis of Cellulose in $H_2SO_4{}^a$

Index	Frequency	ppm	Intensity
01	5282.66	105.012	59.319
02	5275.82	104.876	75.147
03	4106.29	81.627	20.792
04	4096.17	81.426	30.065
05	4085.22	81.208	22.741
06	4064.28	80.792	64.088
07	3949.84	78.517	32.540
08	3932.06	78.164	22.119
09	3924.21	78.008	36.600
10	3890.62	77.340	102.184
11	3875.94	77.048	23.447
12	3848.92	76.511	100.182
13	3795.40	75.447	113.723
14	3777.44	75.090	40.344
15	3714.11	73.831	20.461
16	3621.29	71.986	29.529
17	3174.45	63.103	29.445
18	3139.46	62.408	64.862

^{a} Spectral lines for TH, 20.00; RFL, 248.3; and RFP, 0.

on cellodextrins.[22] The resins, however, have a tendency to promote sugar conversion reactions.[35]

Once it is shown that the cellodextrins are free of homologous compounds and not contaminated by other substances, the purity of the preparation can be finally checked by the instrumental methods of ^{13}C NMR (21) and infrared spectroscopy.[36] Comparison of the resulting spectra with published data should show similar spectra. Otherwise, a chemically impure cellodextrin should be suspected. A comparison of the infrared spectra of cellodextrins prepared by the HCl (Fig. 2a) and by H_2SO_4 hydrolysis of cellulose (Fig. 2b) demonstrates an unidentified peak at 1210–1260 cm^{-1} in the latter spectrum. Otherwise, the rest of the peaks compare well with published data.[36] ^{13}C NMR spectra of cellodextrins prepared in our laboratory are given in Figs. 3 and 4. The results indicate that the spectrum of the cellodextrins prepared by the modified Voloch method (Table II) (where $Ba(OH)_2$ is used and the product is freeze-dried as described

[35] P. Jandera and J. Churacek, *J. Chromatogr.* **98**, 55 (1974).
[36] H. G. Higgins, C. M. Stewart, and K. J. Harrington, *J. Polym. Sci.* **51**, 59 (1961).

TABLE III
^{13}C NMR Data of Cellodextrins Prepared by
Hydrolysis of Cellulose in HCl[a]

Index	Frequency	ppm	Intensity
01	5286.77	105.093	66.926
02	5275.82	104.876	83.812
03	4944.56	98.291	26.507
04	4081.93	81.143	29.022
05	4070.67	80.919	58.020
06	4064.85	80.803	57.573
07	3949.14	78.503	69.770
08	3938.57	78.293	20.139
09	3924.09	78.005	71.051
10	3891.19	77.351	116.916
11	3862.72	76.785	39.759
12	3850.76	76.548	99.353
13	3806.79	75.673	74.064
14	3796.67	75.472	97.081
15	3721.51	73.978	21.008
16	3714.11	73.831	34.145
17	3711.39	73.777	33.327
18	3654.19	72.640	23.289
19	3632.68	72.212	51.543
20	3621.29	71.986	75.559
21	3405.49	67.696	33.263
22	3207.35	63.757	23.098
23	3182.29	63.259	21.892
24	3174.89	63.112	56.306
25	3140.16	62.422	78.222
26	2196.10	43.655	30.704

[a] Spectral lines for TH, 20.00; RFL, 248.3; and
RFP, 0.

previously) is in better agreement with the literature than those prepared
in our laboratory by the modified Miller method[24] as the latter has an extra
medium size peak at 43.655 ppm (Table III). This peak might indicate the
presence of an impurity since no peak lower than 57 ppm exists in the
literature spectra reported for individual cellodextrins.[21]

Conclusions

Cellodextrins have been proven to be valuable in the study of cellulose
hydrolysis kinetics. Hydrochloric or sulfuric acid can both be used in the
chemical synthesis of cellodextrins. Our work has indicated that the purity of the product obtained via either method varies from trial to trial

using the same procedure. Thus, it is advisable to examine the purity of each batch of product by liquid chromatography, infrared, ^{13}C NMR spectroscopy and ultimately by enzyme hydrolysis. Purified enzyme should yield complete hydrolysis of cellodextrins although this may take 1 to 2 days. Once the mixture proves to be pure, then it may be separated into the individual oligomers via liquid chromatography. Our experience shows that extraordinary attention to experimental detail is required to obtain pure cellodextrins. In the case of cellodextrins prepared by the HCl method, anion exchange resin was found to readily cause formation of by-products. In the H_2SO_4 methods, slightly sulfated cellodextrins are suspected to be formed in some instances. In either case, cellooligosaccharides proved to be more reactive than previously thought. Efforts continue in our laboratory to improve preparation methods for these interesting oligosaccharides.

Acknowledgments

We thank C. Jones and S. Schroeder of the Purdue University Chemistry Department for carrying out ^{13}C NMR analysis (supported in part by NIH Research Grant RR01077). The material in this work was supported by NSF Grant CPE8351916 and USDA special Grants 83-CRSR-2-2250 and 84-CRSR-2-2488.

[5] Preparation of Cellodextrins: Another Perspective

By Georg Schmid

Introduction

Our interest in the preparation of cellodextrins and also xylodextrins originated from the need to obtain defined substrates for the characterization of hydrolytic enzymes. When using sulfuric acid for the initial preparation of oligomer mixtures experiments showed an array of side components using sugar-borate chromatography (unpublished results). Therefore, we started working from the method of Miller et al.[1] Oligomers from the Miller method[1] appear normal by sugar-borate chromatography up to tetraose or pentaose, then the peaks become somewhat distorted.

[1] G. L. Miller, J. Dean, and R. Blum, Arch. Biochem. Biophys. 91, 21 (1960).

Our objective then was to produce the higher oligomers with good purity and on reusable columns.

Experimental Procedures

Preparative Methods

Preparation of Cellooligomer Mixtures.[2,3] Cellulose powder (70 g) was suspended in 700 ml of fuming hydrochloric acid (ρ = 1.19 g/mol) and stirred at 25° for 2–3 hr. The mixture was then slowly added to 5 liters of 1-propanol at room temperature. The precipitate was collected by centrifugation at 5000 g for 5 min and the pellet was resuspended in 1.5 liter technical grade ethanol. This washing step was repeated until the pH of the suspension reached pH 5–6. After a final centrifugation the precipitate was extracted twice with 2.5 liters of distilled water and stirred at 25° overnight. The insoluble material was removed by centrifugation and the soluble cellooligomers of the supernatant were concentrated by rotary evaporation and subsequently freeze-dried to give 3.5 g of product. The depolymerization of cellulose by acetolysis was carried out as described by Miller *et al.*[1]

Size-Exclusion Chromatography (SEC) of Cellooligomers.[2,3] Cellooligomer mixtures were separated on polyacrylamide gel [BioGel P-4 (<400 mesh) (Bio-Rad Labs., Richmond, CA)] columns thermostatted at 65° using double-distilled water as eluent. The dry gel was specially wind sieved to give the desired narrow particle size distribution. Total column length was 210 cm. The original setup was modified to provide closed loop recycle chromatography.[4,5] This resulted in a further improvement in obtainable resolution.[3]

Purification of Oligomers by Affinity Chromatography.[3] Cellooligomers homogeneous according to SEC were further purified by preparative affinity chromatography on a phenyl boronate-agarose (PBA 60, Amicon, Danvers, MA) resin.[6,7] The column (0.9 × 100 cm) was operated at 4° and the bed height was ~90 cm. $(NH_4)_2CO_2$ (100 mM), pH 10.5, served as equilibration and separation buffer. Bound components were washed off the column with distilled water saturated with CO_2 (pH 4.5). The extreme

[2] K. Hamacher, G. Schmid, H. Sahm, and C. Wandrey, *J. Chromatogr.* **319,** 311 (1985).
[3] G. Schmid, doctoral thesis. University of Bonn, Federal Republich of Germany, 1986.
[4] J. Porath and H. Bennich, *Arch. Biochem. Biophys.* **1,** 152 (1962).
[5] K. J. Bombaugh and R. F. Levangie, *J. Chromatogr. Sci.* **8,** 560 (1970).
[6] K. Reske and H. Schott, *Angew. Chem.* **85,** 412 (1973).
[7] Amicon Company, "Boronate Ligands in Biochemical Separations: Applications, Method, Theory of Matrex Gel PBA." Amicon, Danvers, Massachusetts.

pH changes resulted in irreversible shrinkage of the matrix and sometimes even in the formation of cracks, but this did not adversely effect column performance. By using a volatile buffer system, carbohydrates could be directly obtained by freeze-drying of the collected fractions.

Hydrolysis of Glycosylamines.[8] Glycosylamines formed upon affinity chromatography could be completely hydrolyzed on a weakly acidic cation-exchange resin (Amberlite IRC-50, Serva, Heidelberg, FRG) with distilled water as eluent. Columns were operated at room temperature. The residence time on the resin was ~2 hr.

Analytical Methods

Anion-Exchange Chromatography of Cellooligomer Borate Complexes.[2,3] Analyses of oligomers and oligomer mixtures were performed with an automated sugar analyzer, Biotronik ZA 5100 (Biotronik, Munich, FRG) using a 30 × 0.6 cm i.d. glass column thermostatted at 60° and filled with a strong-base anion-exchange resin, Durrum DA-X4-20.[9,10] The carbohydrate–borate complexes were eluted with a two-step borate buffer (buffer A: 0.2 M H_3BO_3, pH 8.15, 0.8 hr; buffer B: 0.4 M H_3BO_3, pH 9.45, 1.8 hr) followed by a regeneration with 10% (w/v) $K_2B_4O_7$ for 0.6 hr and equilibration with 0.1 M H_3BO_3, pH 8.0, for 0.6 hr. Carbohydrates were detected after postcolumn derivatization with orcinol–sulfuric acid reagent by measuring the absorbance at 420 nm. The reducing or nonreducing properties of the cellodextrin components were investigated by using a second sugar analyzer with a different detection system. Contrary to the orcinol–sulfuric acid derivatization, only reducing carbohydrates can be detected by this method which is based on the reduction of a bicinchoninate–Cu^{2+}–aspartic acid reagent.

High-Performance Liquid Chromatography (HPLC) of Cellooligomers.[11] HPLC of oligomers was performed on a cation-exchange resin in the Ag^+ form (HPX-42 A, Bio-Rad Labs., Richmond, CA) using a quasi-continuous analysis method.

Enzymatic and Acid Hydrolysis of Cellodextrins.[3] Samples of cellooligomers obtained at different stages during the preparation were (partially) hydrolyzed with (1) hydrochloric acid (1 mol/liter) at 100° for 1 hr (1 mg sugar/ml acid, neutralization with sodium bicarbonate), (2) α-1,4-(α-1,6)-amyloglucosidase (glucan 1,4-α-glucosidase) from *Aspergillus niger* (Boehringer-Mannheim, FRG) in 0.2 M sodium acetate at pH 4.8 and 37°

[8] M. Biselli, Master's thesis. University of Bonn, Federal Republic of Germany, 1986.
[9] J. X. Khym and L. P. Zill, *J. Am. Chem. Soc.* **74**, 2090 (1952).
[10] R. B. Kesler, *Anal. Chem.* **39**, 1416 (1967).
[11] G. Schmid and C. Wandrey, *Anal. Biochem.* **153**, 144 (1986).

(enzyme concentration, 0.8 U/ml; heat inactivation, 10 min, 100°), and (3) a purified cellodextrin glucohydrolase from *Trichoderma reesei*[12] in 5 mM sodium acetate at pH 4.5 and 50° (enzyme concentration, 0.5–1 mg/liter; heat inactivation, 10 min, 100°).

^{13}C and ^{1}H Nuclear Magnetic Resonance (NMR) and Fast Atom Bombardment Mass Spectroscopy (FAB-MS).[13] NMR spectra were taken at room temperature with a Bruker WM-400 spectrometer (using D$_2$O and DMSO-d_6 as solvents) and negative-ion FAB mass spectroscopy was performed with a Kratos MS 50 machine (glycerol as matrix).

Preparation of Acid-Free Cellooligomer Mixtures[2,3]

For the preparation of acid-free cellooligomers by hydrochloric acid hydrolysis of cellulose it is necessary to remove the acid without producing great amounts of neutralization products such as sodium chloride. A rapid and efficient way to do this is to precipitate the oligomers after partial hydrolysis using alcohol. Whereas the main hydrolysis products, glucose and cellobiose, remain solubilized in the acid–alcohol–water phase, cellooligomers (G$_3$–G$_{12}$) precipitate accompanied by insoluble cellulose. As the oligomers are insoluble in alcohol, the hydrochloric acid adsorbed on the solid phase of cellooligomers and cellulose can be extracted by repeated washing with 1-propanol or ethanol. The powder obtained after water extraction and freeze-drying is a colorless and neutral product, whereas the mixture of oligomers from acetolysis has a yellowish appearance.

High-Resolution Size-Exclusion Chromatography of Cellooligomers[2,3]

A typical elution pattern of cellooligomers separated by high-resolution size-exclusion chromatography (SEC) is shown in Fig. 1a. The excellent separation efficiency achieved on BioGel P-4 columns at 65° (the resolution is in the range of 1.0–1.2 for each component) allows the fractionation of oligomers up to octaose. Recycle chromatography makes the preparation of dextrins up to a degree of polymerization of 12 possible.[3] Rechromatography of individual oligomers at this stage indicates products with uniform molar mass. A plot of the negative logarithm of the

[12] G. Schmid and C. Wandrey, *Biotechnol. Bioeng.* **30**, 571 (1987).
[13] NMR and FAB-MS studies were kindly performed by Drs. V. Wray and L. Grotjahn at the Gesellschaft für Biotechnologische Forschung, Braunschweig, Federal Republic of Germany.

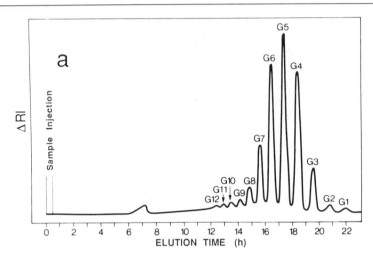

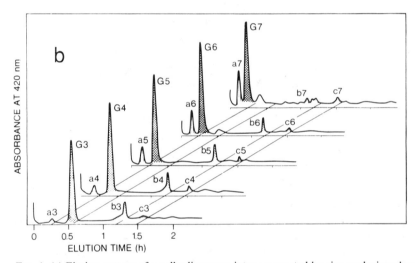

FIG. 1. (a) Elution pattern of a cellooligomer mixture separated by size-exclusion chromatography. Oligomers prepared by hydrochloric acid hydrolysis of cellulose. Resin, BioGel P-4 (<400 mesh); temperature, 65°; column dimensions, 210 × 5 cm; eluent, double-distilled water. (b) Sugar-borate chromatography of cellooligomers homogeneous according to SEC. Oligomers prepared by hydrochloric acid hydrolysis of cellulose. Anion-exchange resin, Durrum DA-X4-20; eluent, two-step borate buffer; column dimensions, 30 × 0.6 cm; temperature, 60°. Reproduced from Hamacher *et al.*[2] with permission of Elsevier Publishing Co.

distribution coefficient versus the degree of polymerization of cellooligomers gives the same linear relationship as reported by various authors.[14,15]

Structural Heterogeneity of Oligomers Homogeneous According to Size-Exclusion Chromatography[2,3]

Analyses of cellooligomers, homogeneous according to SEC, were performed by separating the oligomers as their borate complexes on an anion-exchange resin. Chromatograms obtained for cellodextrins (G_3–G_7) prepared from acid hydrolysis of cellulose are shown in Fig. 1b. It reveals that, in addition to the cellodextrin peaks, there are at least three second-ary components (termed a, b, and c) present. A fourth component starts to build up from cellohexaose onward. Although sugar borate chromatog-raphy indicates that the carbohydrate components are different in their molecular structure, the molar mass of these products has to be the same or nearly the same as for the β-1,4-glucose oligomers. The abbreviations G_3–G_7 are used to denote peaks corresponding to members of the actual cellodextrin (β-1,4-glucose oligomer) series. The products purified by the method of Miller *et al.*[1] have the same elution times as those termed as G_3–G_7. Component (a) was established to be nonreducing by analyses conducted on the second sugar analyzer available. Total acid hydrolysis of cellooligomers (G_3–G_7) yields 97–99% glucose. The composition of the cellodextrin mixture was not influenced by the type of alkanol used for the precipitation step. The composition of secondary components was inde-pendent of the temperature during the precipitation (0–20°). Chromato-grams for individual oligomers produced by acetolysis of cellulose show a number of side components that have elution times very close to those of the actual cellodextrin peaks. Component (b) makes up ~12% of the samples. ^{13}C NMR studies at this stage with cellotriose and cellohexaose gave no information on the constitution of the secondary components. All the main peaks of the NMR spectra could be assigned and were in agree-ment with data available from the literature.[16] Several additional signals of very weak intensity were not identifiable.

Affinity Chromatography of Oligomers on Boric Acid Gels and Identification of Secondary Components[3]

Structurally different oligomers apparently homogeneous according to SEC could be obtained by chromatography on phenyl boronate-agarose.

[14] A. Heyraud and M. Rinaudo, *J. Chromatogr.* **166,** 149 (1978).
[15] T. Sasaki, T. Tanaka, and K. Kainuma, *Nippon Shokuhin Kogyo Gakkaishi* **25,** 538 (1979).
[16] J. C. Gast, R. H. Atalla, and R. D. McKelvey, *Carbohydr. Res.* **84,** 137 (1980).

The secondary components [termed (a) and (b) in Fig. 1b] were identified by a combination of analysis methods: (1) enzyme and acid-catalyzed (partial) hydrolysis of the different oligomers and subsequent analysis of degradation products by sugar-borate chromatography, (2) ^{13}C and ^{1}H NMR spectroscopy, and (3) fast atom bombardment (FAB) mass spectroscopy. In component (b) the glucose unit at the reducing end of the molecule is modified. The only other monomeric sugar after total hydrolysis is fructose. Upon enzymatic hydrolysis of oligomers with a cellodextrin glucohydrolase, that attacks oligomers from the nonreducing end of the chain, fructose is produced once the dimer is degraded. This enolization reaction takes place during the chromatography on BioGel P-4. Component (a) yields only glucose after total hydrolysis. It was found by NMR and FAB-MS studies that these products are a mixture of 1-propyl- and ethylglycosides of oligomers. Interestingly the cellopentaosides have the same elution times during size-exclusion chromatography as cellohexaose, the tetraosides as cellopentaose, etc. These products are obviously formed with the precipitation of the oligomer mixture in alcohol.

Formation and Hydrolysis of Glycosylamines[8]

As could be demonstrated, the β-1,4-glucose oligomers are converted to the respective glycosylamines under the separation conditions on the affinity gel. The molar mass of what was expected to be a cellohexaose, for example, was found to be 989 and not 990 g/mol. The NMR spectra also showed characteristic changes. Glycosylamines could finally be hydrolyzed on a weakly acidic cation-exchange resin to give oligomer products which proved to be homogeneous according to sugar-borate chromatography, NMR spectroscopy, FAB mass spectroscopy, and enzymatic degradation.

Conclusions

To prepare cellodextrins and to conclusively demonstrate their purity one needs to use several analytical techniques and methods to prove homogeneity. The use of sugar-borate chromatography, a relatively slow and therefore today somewhat forgotten technique, proved to be invaluable during our work. We could show that samples that looked perfectly homogeneous by chromatography on cation exchangers in the Ag$^+$ form actually consisted of several different components.[3,8] Enzymatic degradation of oligomers with purified enzymes and spectroscopic methods must also be considered as tools to check the homogeneity of carbohydrate oligomers. At each of the three stages of our preparation scheme we could demonstrate the formation of artifacts. Other authors have reported the use of alcohol for precipitating oligomers and circumventing NaCl production.[17,18] This results in the generation of oligomeric glycosides. The separation of carbohydrate oligomers at elevated temperatures on polyacrylamide columns is a common technique. We demonstrated that Lobry de Bruyn–Alberda van Ekenstein-type rearrangements can occur during elution. The isolation of monomeric sugars in volatile ammonium carbonate buffers has been described by other authors.[6,7] We found that under these harsh conditions glycosylamines are formed. Any separation technique for polyfunctional molecules as carbohydrates must be evaluated with extreme caution toward their potential to result in the generation of secondary components. We also noticed degradation of oligomers upon storage in the desiccator over silica gel during periods of several months.

Acknowledgments

The above results could not have been obtained without the close and fruitful cooperation of many colleagues at the Institutes of Biotechnology in Jülich: M. Biselli, Dr. K. Hamacher, K. Irrgang, and Dr. K.-L. Schimz. I would also like to thank Drs. V. Wray and L. Grotjahn at the Gesellschaft für Biotechnologische Forschung mbH in Braunschweig for performing and discussing the NMR and mass spectra with us.

[17] M. Voloch, M. R. Ladisch, M. Cantarella, and G. T. Tsao, *Biotechnol. Bioeng.* **26,** 557 (1984).
[18] P. Gonde, B. Blondin, M. Leclerc, R. Ratomahenna, A. Arnaud, and P. Galzy, *Appl. Environ. Microbiol.* **8,** 265 (1984).

[6] Fluorogenic and Chromogenic Glycosides as Substrates and Ligands of Carbohydrases

By Herman van Tilbeurgh, Frank G. Loontiens, Clement K. De Bruyne, and Marc Claeyssens

Nitrophenylglycosides are frequently used substrates of carbohydrases. Alternatively the 4-methylumbelliferyl (7-hydroxy-4-methylcoumaryl) derivatives offer a more sensitive (fluorometric) method of detection. Some of these compounds have become commercially available.

We describe the preparation and use of these glycosides in the study of some cellulolytic enzymes. The difference in spectral properties of free 4-

methylumbelliferone and its carbohydrate conjugates allows sensitive and continuous assays of cellulolytic activities in absorption or fluorescence modes.

The preparation and use of 1-thio derivatives with different chromophoric reporter groups are included.

Due to their optical characteristics the chromophoric (fluorochromic) derivatives offer distinct advantages over the use of classical substrates of cellulolytic enzymes since they are sensitive tools for the determination of the number of binding sites, study of association modes, and binding kinetics, breakdown patterns, and inhibition characteristics. Some applications are described below.

Preparation of Substrates and Ligands

Preparation of the 4-Methylumbelliferyl-β-glycosides of the Cellodextrins[1] [MeUmb(Glc)$_n$, n = 1 to 6]

For the isolation of the cellotriose to cellohexaose derivatives powderous Whatman cellulose was acetylated by standard procedures.[2] The cellulose acetate (17 g) was dissolved in 30 ml HBr (50% HBr in glacial acetic acid) and after 3 hr at room temperature the mixture was poured into ice water and extracted with chloroform. The extracts (100 ml) were washed (1% sodium bicarbonate, water) and after evaporation *in vacuo* and drying (sodium sulfate) the resulting syrupy mixture of acetobromo derivatives (10 g) was dissolved in 30 ml acetone and a solution of 6 g 7-hydroxy-4-methylcoumarin, 1.4 g sodium hydroxide, and 14 ml water was added. After 2 hr at room temperature the reaction mixture was evaporated *in vacuo* and dissolved in 50 ml chloroform. After extraction (5% sodium bicarbonate, water) the chloroform was evaporated and the resulting mixture of peracetylated 4-methylumbelliferyl-β-D-glycosides was deacetylated (0.01 M sodium methanolate) overnight. The precipitate was filtered and the filtrate neutralized (Dowex X-50, H$^+$). After evaporation the residue (~5 g) was dissolved in a minimal volume of water (~5 ml) and the mixture was fractionated by gel-permeation chromatography on BioGel P-2 (Bio-Rad) (column dimensions: 2.5 × 150 cm) or silica gel TSK-HW-40 (Merck) (column dimensions: 1.5 × 30 cm). Approximately 1 g of the mixture was applied and the eluates (distilled water, 20 ml hr^{-1}) were monitored continuously at 254 nm.

[1] H. van Tilbeurgh, M. Claeyssens, and C. K. De Bruyne, *FEBS Lett.* **149,** 152 (1982).

[2] M. L. Wolfrom and A. Thompson, *Methods Carbohydr. Chem.* **3,** 143 (1963).

TABLE I

HPLC of the 4-Methylumbelliferylglycosides of
the Cellodextrins [MeUmb(Glc)$_n$]

Parameter	Value
Column material	Rsil (Alltech) or μPorasil (Waters)
Dimensions	1 × 25 cm (preparative) or
	0.39 × 30 cm (analytical)
Isocratic elution	Water/acetonitrile (v/v)
Eluent A	Lower derivatives (n = 1–3): 10/90
Eluent B	Higher derivatives (n = 4–6): 16/84
Flow	1.5 ml min^{-1}
Temperature	Room
Injection volume	100 μl (20 μl analytical)
Detection	313 nm (Waters 440)
Fraction volume	100–200 μl

Further purification of the fractions was achieved by preparative HPLC. Experimental conditions are given in Table I. After lyophilization of the appropriate fractions 10–50 mg quantities of the glycosides (cellotriose to cellohexaose derivatives) were obtained. The hygroscopic powders are chromatographically pure.

Preparation of 4-Methylumbelliferyl-β-D-lactoside (MeUmbLac)[3]

The following experimental conditions were chosen in order to minimize the formation of a second reaction product with unknown structure. Acetobromolactose (3.5 g) dissolved in 100 ml acetone was treated with 1.15 g 7-hydroxy-4-methylcoumarin and 5.7 g tetrabutylammonium hydroxide in 36 ml of water. After standing overnight (room temperature) a procedure similar to that for the cellodextrin derivatives was followed and the 4-methylumbelliferyl-β-D-lactoside was crystallized from methanol: water (3 : 1); the yield was ±1 g. After two recrystallizations a chromatographically (HPLC) pure product was obtained: mp 258°, [α]$_D$ = −38° (0.5, pyridine). Physical proof of its structure was further obtained by ^{1}H NMR analysis.[3]

Concentrations of MeUmb glycosides were determined using the molar absorption coefficient at 316 nm: ε_M = 13,600 M^{-1} cm^{-1}.

[3] H. De Boeck, K. L. Matta, M. Claeyssens, N. Sharon, and F. G. Loontiens, *Eur. J. Biochem.* **131,** 354 (1983).

Preparation of p-Nitrobenzyl- and 2,4-Dinitrophenyl-1-thio-β-glycosides of Cellobiose and Lactose (PNBSC and PNBSL, DNPSC and DNPSL)

The peracetylated 1-mercaptans of cellobiose and lactose (1 g) prepared by a standard procedure[4] were reacted with 3 g p-nitrobenzyl bromide or 3.5 g 2,4-dinitrochlorobenzene in a 20 ml acetone : water mixture (1 : 1) containing 0.5 g potassium carbonate. After agitating 10 min at room temperature the mixtures were neutralized (HAc) and water was added to a total volume of 100 ml. The precipitates were filtered, washed (H_2O), and recrystallized from methanol : acetone (10 : 1); the yields were 1–1.5 g. After deacetylation (0.01 M sodium methanolate) the products were crystallized from methanol. Further purification by HPLC (see Table I) yields chromatographically pure products.

Solutions (pH 5.0) of the 2,4-dinitrophenyl compounds show an absorption maximum at 325 nm ($\varepsilon_M = 10,800 \ M^{-1} \ cm^{-1}$); the p-nitrobenzyl compounds exhibit a maximum at 280 nm ($\varepsilon_M = 8600 \ M^{-1} \ cm^{-1}$).

Assays and Analytical Procedures

Assay of Enzymatic Breakdown of 4-Methylumbelliferyl-β-glycosides of Cellodextrins by Analytical HPLC[1]

Enzymatic assays were at 25° and pH 5.0 (0.05 M sodium acetate buffer). Substrate concentrations ranged between 10 and 2000 μM, and enzyme concentrations between 0.01 and 1 μM. Samples (50 μl) of the reaction mixtures were diluted into 100 μl acetonitrile to stop the reaction and analyzed by analytical HPLC. Experimental conditions were essentially the same as in Table I except that a μPorasil column (Waters) (0.39 × 30 cm) and an injection volume of 20 μl were used. Quantitation at 313 nm of the chromophoric reaction products was performed by measuring peak heights and by appropriate calibrations.

Spectrophotometric and Spectrofluorimetric Assays[1,5]

The release of 4-methylumbelliferone (MeUmb) from the corresponding glycosides can be followed either by difference absorption spectrophotometry at 347 nm[6] or by fluorometric measurements (pH 10.0) at emission wavelengths >435 nm (excitation 335–400 nm).[7] Calibration is

[4] J. Stanek, M. Sindlerova, and M. Cerny, Collect. Czech. Chem. Commun. **30**, 297 (1965).
[5] H. van Tilbeurgh and M. Claeyssens, FEBS Lett. **187**, 283 (1985).
[6] A. C. Rosenthal and A. Saifer, Anal. Biochem. **5**, 85 (1973).
[7] D. H. Leaback, in "An Introduction to the Fluorimetric Estimation of Enzyme Activities," 2nd Ed., p. 24. Koch-Light Labs Ltd., Colinbrook, United Kingdom, 1975.

achieved by using standard solutions prepared from recrystallized MeUmb (from ethanol). The first, spectrophotometric method allows continuous monitoring of the reaction but lacks the sensitivity of the fluorimetric procedure (<1 μM). Continuous fluorometric assay at pH 5.0 is, however, possible but is again less sensitive than the discontinuous assay at pH 10.0.[7]

Any interference of β-glucosidase in the assay of enzymatic activities using the MeUmb(Glc)$_n$ can be minimalized by adding glucono-1,5-lactone (1 mM). Inclusion of cellobiose (5 mM) in the MeUmbLac solution used for the assay of an endocellulase (EG I) prevents the concomitant hydrolysis of this substrate by cellobiohydrolase I (*Trichoderma* sp.).

Kinetic parameters for enzymatic reaction studied in this chapter are obtained either by graphical methods using linearization of the Michaelis–Menten equation or direct fitting of the hyperbola.[8]

Origin and Purification of Enzyme Samples[9,10]

The exo- and endocellulases from *Trichoderma reesei* were purified from extracellular culture fluids either obtained commercially (NOVO, Denmark), from domestic cultures, or gifts (VTT, Finland). Preliminary fractionation of dialyzed (PM-10, Amicon) enzyme samples was obtained by DEAE-trisacryl ion-exchange chromatography (LKB-IBF, France) and further purification was achieved either by affinity chromatography[9] or ion-exchange chromatography on SP-trisacryl ion-exchange chromatography.[10]

Fractions were analyzed for enzymatic activity using the methods described and pooled fractions investigated by gel isoelectric focusing.[5]

Partially purified samples containing cellulolytic enzymes from *Penicillium pinophilum, Trichoderma koningii,* and *Fusarium solani* were kindly donated by Dr. Tom Wood (Rowett Research Institute, Aberdeen).

Ligand Difference Absorption Spectra of 2,4-Dinitrophenyl-1-
thio-β-D-glycosides: Titration of Enzyme-Active Sites[11]

Difference absorption spectra were recorded on a double-beam spectrophotometer (Uvikon 810 with 2 mm slit width), using thermostatted two-compartment mixing (Yankeelov) cuvettes (2 × 0.475 cm) filled with the appropriate concentrations of chromophoric ligand and protein. Con-

[8] M. Sakoda and K. Hiromi, *J. Biochem.* (*Tokyo*) **80**, 547 (1976).
[9] H. van Tilbeurgh, R. Bhikhabhai, L. G. Pettersson, and M. Claeyssens, *FEBS Lett.* **169**, 215 (1984).
[10] R. Bhikhabhai, G. Johansson, and L. G. Pettersson, *J. Appl. Biochem.* **6**, 336 (1984).
[11] H. De Boeck, F. G. Loontiens, and C. K. De Bruyne, *Anal. Biochem.* **124**, 308 (1982).

tinuous titrations were performed using two single-compartment cells (reference and sample) equipped with microstirrers. These were assembled from 1516 E 004 SDC micromotors (4 V, Minimotor S.A., Agno, Switzerland). Openings for two capillary tubings (Teflon) connected to 1-ml Hamilton gas-tight syringes mounted in micrometer burettes are provided (reproducibility of delivered volumes better than 0.1%).

The two cuvettes were filled with an equal volume (e.g., 2 ml) of the two appropriate solutions (buffer and protein) and to both continuously stirred and thermostatted curvettes equal (e.g., 0.1 μl) portions of the common titrant (chromophoric ligand solution) were dispensed manually and in succession. Mathematical treatment of the results was performed by direct hyperbolic fitting of the titration curves.[12]

Fluorescence Spectra of 4-Methylumbelliferyl-β-D-glycosides: Titration of the Quenching by Equilibrium Binding[13,14]

Noncorrected fluorescence spectra were recorded with an Aminco SPF-500 Ratio Spectrofluorimeter equipped with a thermostatted copper cuvette holder and a fluorescence cuvette with stirrer assembly as described above. During titrations a concentrated solution of enzyme was added from a microburette to the continuously stirred solution of the ligand. Readings of the fluorescence (excitation at 316 nm and spectral bandwidth 1 nm) were at 360 nm with a 20-nm bandwidth.

Displacement titrations were performed by first titrating the fluorescence of the indicator ligand and then adding an excess of the competitive ligand to the mixture (from a second burette). The data for both types of titrations were treated as described.[3,14]

Equilibrium Binding Experiments by the Diafiltration Technique[15]

A "forced-flow" dialysis method as described by Blatt et al.[16] has been adapted.[15] A (chromophoric) ligand solution is forced under pressure from a reservoir into a closed, stirred, and thermostatted diafiltration cell, which contains a solution (1–2 ml) of the protein and which is separated from the elution chamber by a semipermeable membrane (PM10, Ami-

[12] Mathematical treatment by linearization as in Ref. 3 or by direct fitting to the hyperbola.[8]
[13] A. van Landschoot, F. G. Loontiens, and C. K. De Bruyne, *Eur. J. Biochem.* **83,** 277 (1978).
[14] H. van Tilbeurgh, L. G. Pettersson, R. Bhikhabhai, and M. Claeyssens, *Eur. J. Biochem.* **148,** 329 (1985).
[15] M. Claeyssens, H. van Tilbeurgh, and C. K. De Bruyne, *Bull. Soc. Chim. Belg.* **94,** 123 (1985).
[16] W. F. Blatt, S. M. Robinson, and H. J. Bixler, *Anal. Biochem.* **26,** 151 (1968).

con). The diafiltrates are analyzed for free ligand (in line spectrophoto-metric analysis) and, for a given system, a binding isotherm can be deter-mined. Calculation of binding parameters was done from blank and binding curves as described.[17]

Applications

Detection and Differentiation of Cellulolytic Enzymes

Example 1: The Cellulolytic System from Trichoderma reesei and Other Fungal Species.[1,5,14,18] The 4-methylumbelliferyl-β-D-glycosides of glucose, cellobiose, cellotriose, and lactose are used to differentiate sev-eral cellobiohydrolase, endocellulase, and β-glucosidase activities in crude *Trichoderma reesei* culture filtrate.[5] Spectrophotometric and fluo-rometric assays (see above) allow simple detection and quantitative mea-surements of eluate activities from chromatographic experiments, as illus-trated in Fig. 1. Rapid screening and visualization after analytical gel electrofocusing become possible also (Fig. 2).

The complexity of the multicomponent cellulolytic system secreted by microorganisms such as *Trichoderma* species is clearly demonstrated. These methods have also allowed rapid screening of activities in *Fusar-ium solani, Trichoderma koningii,* and *Penicillium pinophilum* (Fig. 3).[18] As it appears the latter is particularly rich in β-glucosidase activities, which are sometimes deficient in other species such as *Trichoderma*.

As several components of the *Trichoderma* complex have been puri-fied to homogeneity and their chemical and physical properties deter-mined,[9,10] differences in specificities of these enzymes could clearly be delineated (Fig. 4) using the 4-methylumbelliferylglycosides and the HPLC method described.[1,14] The 4-methylumbelliferyl-β-D-cellobioside and β-D-lactoside are substrates only for CBH I and EG I; cleavage is at the phenolic bond and formation of MeUmbGlc in the case of EG I is due to (self)-transfer reactions.[18] Inhibition by cellobiose is far more evident for the CBH I and can be used to differentiate this activity from the EG I (Fig. 2).[5] MeUmb(Glc)$_3$ is split at the heterosidic bond only by the endo-cellulase III (EG III) and this fact can be used to differentiate the enzyme from the other endo- and exocellulases, e.g., after isoelectric focusing on gels (not shown).[5]

Example 2: Comparison of the Kinetic Parameters of CBH I and EG I for Common Substrates.[18,19] As both enzymes show considerable resem-

[17] W. Yap and R. Schaffer, *Clin. Chem.* **23**, 986 (1977).
[18] M. Claeyssens, unpublished results (1985).
[19] H. van Tilbeurgh, unpublished results (1985).

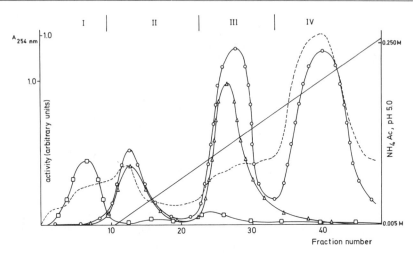

FIG. 1. DEAE-trisacryl ion-exchange chromatography of the cellulase complex from *T. reesei*. Crude culture filtrate (5 ml) was dialyzed against 5 mM ammonium acetate buffer, by gel filtration (BioGel P-6), adsorbed onto a column (1 × 10 cm) of DEAE-trisacryl, and eluted with a 5–250 mM gradient of ammonium acetate, pH 5.0 (room temperature). UV absorbance of the eluates (−−) was followed continuously at 254 nm. Activity measurements (arbitrary units) on 50-μl aliquots of the fractions (5 ml) were performed with solutions of the appropriate 4-methylumbelliferylglycoside (1 ml): (□) β-glucosidase activity; (○) activity against the lactoside; (△) activity against the same substrate but in the presence of 5 mM cellobiose. Fractions I–IV were pooled as indicated. All measurements were at pH 5.0 and 25°.

blances in primary and secondary structure,[20,21] quantitative data for their kinetic parameters for hydrolysis of MeUmb(Glc)$_n$ and MeUmbLac were compared (Table II). Although they share the property of catalyzing the hydrolysis of the lactoside, appreciable differences in K_m and V_{max} values are apparent. Also K_i values for competitive inhibition by cellobiose differ widely (5 × 10^4 M^{-1} for CBH I and 500 M^{-1} for EG I) and this property was exploited to differentiate both enzymes (cf. Fig. 2). Although the EG I cannot hydrolyze the β-1,4 bond in xylobiosides, transfer reactions to xylopyranosides are as common as to glucosides and cellobiosides (self-transfer).[19] In the case of CBH I, several pieces of evidence point to a restricted binding site for small substrates and ligands, whereas for EG I a more extended active site, accommodating several glucosyl residues (donor site), and an acceptor site fitting xylopyranosyl or glucopyranosyl residues are postulated.[19]

[20] R. Bhikhabhai and L. G. Pettersson, *FEBS Lett.* **167**, 301 (1984).
[21] P. Lehtovaara, J. Knowles, L. Andre, M. Pentilla, T. Teeri, I. Salovuori, M.-J. Niku-Paavola, and T.-M. Enari, *Proc. Int. Conf. Biotechnol. Pulp Pap. Ind., 3rd* p. 90 (1986).

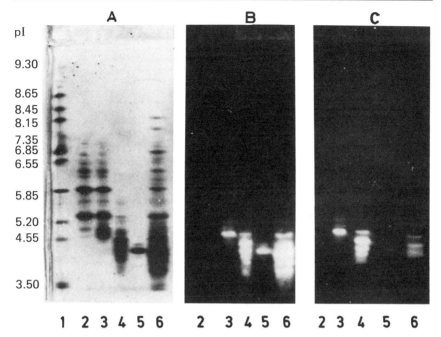

FIG. 2. Analytical isoelectric focusing of the cellulolytic complex from *Trichoderma reesei*.[5] Lane 1: marker proteins (Pharmacia). Lanes 2, 3, 4: fractions I, II, and III from Fig. 1. Lane 5: CBH I (pure). Lane 6: crude mixture. A: Protein staining (Coomassie Blue). B: Activity against MeUmbLac. C: Activity against MeUmbLac in the presence of 10 m*M* cellobiose.

Equilibrium Binding Experiments

Example 1: Binding of Chromophoric 1-Thio-β-D-glycosides of Lactose and Cellobiose to CBH I.[1,16,19] When the glycosidic oxygen is replaced by sulfur, as in PNBSC or DNPSC, these compounds are no longer substrates but are effective competitive inhibitors of the CBH I.

TABLE II
COMPARISON OF KINETIC PARAMETERS FOR SOME SUBSTRATES OF CBH I
AND EG I FROM *Trichoderma reesei*[a]

Substrate	TN^b (EG I)/TN (CBH I)	$K_m{}^b$ (EG I)/K_m (CBH I)
MeUmb(Glc)$_2$	550	90
MeUmbLac	28	5
MeUmb(Glc)$_3$	160	5

[a] See Refs. 1 and 19.
[b] Turnover numbers (TN) (min^{-1}) and K_m values (M) at pH 5.0 (25°).

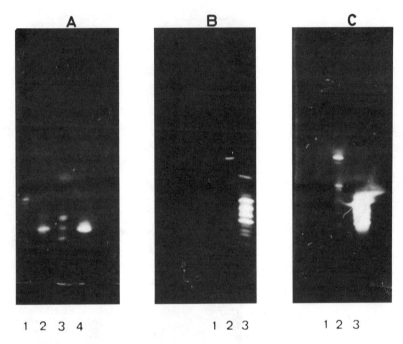

FIG. 3. Analytical isoelectric focusing of the cellulases from *Fusarium solani* (1), *Trichoderma koningii* (2), and *Penicillium pinophilum* (3) A: Activity against MeUmbLac. B: Activity against MeUmb(Glc)$_2$. C: Activity against MeUmbGlc. CBH I from *Trichoderma reesei* is shown in lane 4.

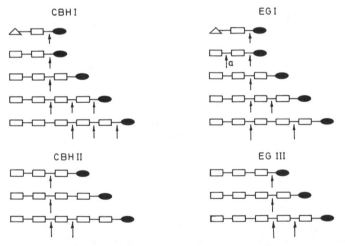

FIG. 4. Specificities of several endo- and exocellulases for the 4-methylumbelliferyl-β-D-glycosides derived from a series of cellooligosaccharides and from lactose. $\triangle$, β-Galactopyranosyl moiety; $\square$, β-glucopyranosyl; $\bullet$, 4-methylumbelliferyl group. Arrows indicate bonds hydrolyzed as deduced from the chromophoric reaction products observed (HPLC analysis). (a) Formation of MeUmbGlc due to self-transfer action of EGI.[19]

An equilibrium binding experiment is illustrated in Fig. 5. One binding site per enzyme molecule and an association constant in agreement with the inhibition constant for the lactoside are found. Experiments with the 2,4-dinitrophenyl-1-thio-β-D-cellobioside yield $n = 1.2 \pm 0.05$ for the number of interacting sites and $K = (1.42 \pm 0.05) \, 10^5 \, (3°)$ for the association constant.

Difference absorption spectra for DNPS(Glc)$_2$ and DNPSLac are shown in Fig. 6. From these results a different microenvironment for chromophore in both ligands could be suggested. Both spectra are however similarly influenced by the additions of lactose or cellobiose. Titrations of the difference absorption spectrum for DNPS(Glc)$_2$ at 330 nm yield the following thermodynamic binding parameters: $K = (7.6 \pm 0.2) \, 10^4 \, M^{-1} \, (22.5°)$, $\Delta H° = -(58 \pm 1) \, \text{kJ mol}^{-1}$, and $\Delta S° = -(103 \pm 4) \, \text{J mol}^{-1} \, K^{-1}$.

Example 2: Binding of Small Ligands to CBH II: Fluorescence Quenching of 4-Methylumbelliferyl Glycosides.[14,16,19] As pointed out (Table I) MeUmb(Glc)$_n$ are substrates for CBH II when $n > 3$. MeUmbGlc and MeUmb(Glc)$_2$ could, however, be used as reporter ligands in a series of binding experiments using fluorescence quenching titrations and

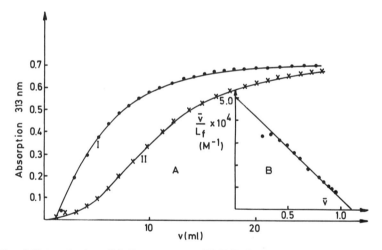

FIG. 5. Determination of binding constants of PNBSL for CBH I by diafiltration.[9] (A) The filtrate is collected in fractions and the concentration of free ligand (L_f) is determined spectrophotometrically (313 nm). The ligand concentration in the stock solution is 200 μM. The amount of bound ligand is computed[17] from a blank (curve I) and a binding experiment (curve II). (B) Scatchard plot: $\bar{v}/L_f = nK - \bar{v}K$; $\bar{v}$ is the degree of saturation as calculated from the amount of bound ligand and protein (170 μM) in the dialysis cell; K is the association constant; n is the number of binding sites. In this case $n = 1.1 \pm 0.1$ and $K = (4.0 \pm 0.4) \times 10^4 \, M^{-1}$.

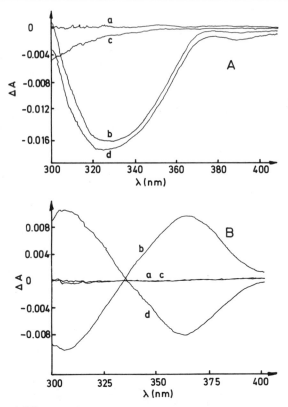

FIG. 6. Ligand difference absorption spectra (ΔA) for DNPSC and DNPSL binding to CBH I. Spectra measured (15.8°) in tandem mixing (Yankeelov) cuvettes. (A) 39 μM DNPSC and 50 μM CBH I: (a) baseline (before mixing); (b) spectrum after mixing of sample cuvette; (c) after mixing of both sample and reference cuvettes; and (d) after adding saturating amounts (>0.1 M) of cellobiose to the reference cuvette. (B) 32 and 50 μM CBH I: (a) baseline; (b) after mixing of sample cuvette; (c) after mixing sample and reference cuvettes; and (d) after the addition of cellobiose (>0.1 M) to the reference.

diafiltration experiments. In both cases evidence for one binding site was obtained. Typically the fluorescence of MeUmb(Glc)$_2$ is quenched in the presence of CBH II and the fluorescence is restored by the addition of a nonchromophoric ligand in excess (Fig. 7). Thus association constants for the indicator ligands could be obtained by direct titration and values for the number of binding sites and association constants of nonchromophoric ligands could be computed from displacement titrations. Typical examples are shown in Fig. 8.

The resulting thermodynamic parameters are gathered in Table III and

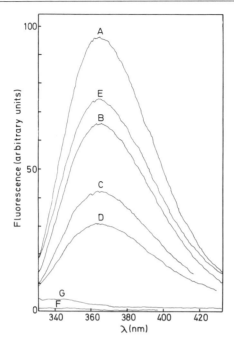

FIG. 7. Quenching of the fluorescence spectrum of MeUmb(Glc)$_2$ by CBH II. Curve A represents the MeUmb(Glc)$_2$ (2 μM) spectrum in the absence of CBH II. Curves B, C, and D show the spectra after the addition of several aliquots of 137.7 μM CBH II. Spectrum D changes to E when solid cellobiose (± 2 mg) is added. When correction is made for dilution, spectrum E is equivalent to spectrum A. Curves F and G represent buffer and protein blanks, respectively. Spectra were measured at pH 5.0 and 6.6°.

a hypothetical binding scheme is proposed as shown below.[14] Interactions of a particular subsite with a glucosyl residue are strongly dependent on the occupation of other subsites. The binding of glucose or glucosides to subsite A induces a conformational transition of the enzyme. This is accompanied by a drastic increase of association constants and kinetic binding parameters of cellobiosides and cellooligosaccharides with the other subsites (van Tilbeurgh *et al.*, unpublished observations).

A		B	C	D
Glc		MeUmb		
		Glc	Glc	Meumb
		Glc	Glc	Glc
Glc	-	(OMe)		

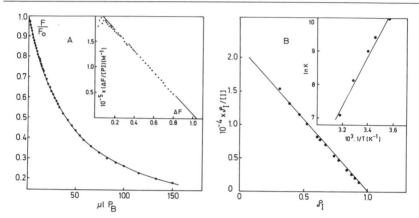

FIG. 8. (A) Continuous titration (pH 5.0, 15.7°) of MeUmb(Glc)$_2$ fluorescence with excess CBH II. Relative fluorescence at 360 nm (F/F_0) of 2.9 μM MeUmb(Glc)$_2$ is plotted versus the volume of CBH II solution (91.4 μM) added (P_B). In the inset ΔF is the dilution corrected decrease in fluorescence and [P] the concentration of free protein obtained by iteration.[3] The slope of the curve yields the association constant: $K = (2.62 \pm 0.03) \times 10^5$ M^{-1}. Extrapolation to the ΔF axis gives the signal change at saturation: $\Delta F_{max} = (1.09 \pm 0.04)$. (B) Determination of the association constant of cellobiose (pH 5.0, 15.7°) by displacement titration.[3] The change in fluorescence (360 nm) of a mixture of 2 μM MeUmb(Glc)$_2$, 30 mM glucose, and 8 μM CBH II is followed when excess cellobiose is added. Data are treated as given.[3] [I] represents the calculated concentration of free cellobiose and ρ_I is the fractional binding ratio.[3] The association constant is calculated from the slope of the curve and the intersection on the abscissa yields the number (n) of binding sites for the competing (displacing) ligand (cellobiose): $K = (1.20 \pm 0.02) \times 10^3 M^{-1}$, $n = (1.00 \pm 0.01)$. In the inset a van't Hoff plot for the binding of cellobiose measured under these conditions is given.

TABLE III

ASSOCIATION CONSTANTS AND THERMODYNAMIC PARAMETERS FOR BINDING OF SMALL LIGANDS TO CBH II FROM *Trichoderma reesei*[a]

Ligand	$K_{25°}$ (M^{-1})	ΔG (kJ mol^{-1})	$\Delta H°$	$\Delta S°$ (J mol^{-1} K^{-1})
MeUmbGlc	$(2.0 \pm 0.1) \times 10^{-4}$	$-(24.5 \pm 0.05)$	$-(33.2 \pm 1.7)$	$-(26.6 \pm 5.8)$
MeUmb(Glc)$_2$	$(2.00 \pm 0.03) \times 10^{-5}$	$-(30.40 \pm 0.05)$	$-(11.6 \pm 1.7)$	63 ± 6
Glucose[b]	$(1.53 \pm 0.02) \times 10^2$	$-(12.00 \pm 0.05)$	n.d.[c]	n.d.[c]
Cellobiose	$(5.40 \pm 0.05) \times 10^4$	$-(15.60 \pm 0.04)$	$-(28.3 \pm 3.3)$	$-(42 \pm 13)$
Cellotriose	$(6.2 \pm 0.1) \times 10^4$	$-(27.40 \pm 0.04)$	$-(34 \pm 2)$	(24 ± 7)

[a] See Refs. 14 and 19.

[b] Values determined at 4.3°.

[c] Not determined.

Additional results of binding kinetics of these small ligands to CBH II were obtained by stopped flow techniques.[19]

Acknowledgments

F.L. and M.C. are indebted to the Belgian NFWO for research grants. M.C. thanks NATO for travel grants.

[7] Use of Complex Formation between Congo Red and Polysaccharides in Detection and Assay of Polysaccharide Hydrolases

By PETER J. WOOD, JAMES D. ERFLE, and RONALD M. TEATHER*

Principle

Evidence for complex formation between cellulose substantive dyes, such as Congo Red (CR), and polysaccharides has been based on changes in viscosity or solubility of polysaccharide,[1,2] and on changes in spectral properties of the dye.[3–5] Since the formation of the complex occurs only with polymer,[3,4] dye binding and the associated changes may be used to monitor hydrolytic enzyme activity. Methods to detect complex formation, and exploitation of this for detection and analysis of β-D-glucanases in solution, by means of insoluble substrate, and by gel diffusion are described.

Materials

Agarose (preferably low electroendosmosis type) and Congo Red (C.I. 22120) may be obtained from any of the main chemical suppliers.[6–8] Congo Red (94% dye content; Aldrich Chemical Co. Inc., Milwaukee,

* For the Department of Agriculture, Government of Canada.
[1] I. Jullander, *Ind. Eng. Chem.* **49,** 364 (1957).
[2] P. J. Wood and R. G. Fulcher, *Cereal Chem.* **55,** 952 (1978).
[3] B. Carroll and J. W. Van Dyk, *Science* **116,** 168 (1952).
[4] P. J. Wood, *Carbohydr. Res.* **85,** 271 (1980).
[5] P. J. Wood, *in* "New Approaches to Research on Cereal Carbohydrates" (R. D. Hill and L. Munck, eds.), p. 267. Elsevier, Amsterdam, 1985.
[6] J. T. Baker Chemical Company, Phillipsburg, New Jersey.
[7] Sigma Chemical Company, St. Louis, Missouri.
[8] Calbiochem-Behring Corporation, La Jolla, California.

WI) was used as a standard to determine dye content in insoluble substrates. Agar was from Oxoid Canada Inc., Nepean, Ontario. O-(Carboxymethyl)cellulose, CMC7H3-SXF (CMC), with a degree of substitution of 0.7, was obtained from Hercules Inc., Wilmington, DE. O-(Carboxymethyl)pachyman (CMP) was prepared from the insoluble $(1\rightarrow3)$-β-D-glucan, pachyman, by the method of Clarke and Stone.[9] Pachyman was a gift from Dr. J. J. Marshall, but is available commercially.[7,8] Oat gum, containing ~80% $(1\rightarrow3)(1\rightarrow4)$-$\beta$-D-glucan, was a gift from the Quaker Oat Co.,[10] but may be prepared from dehulled oats as described by Wood et al.[11] Purified (95–100%) $(1\rightarrow3)(1\rightarrow4)$-$\beta$-D-glucan may be obtained by two successive precipitations with ammonium sulfate followed by precipitation with 2-propanol. $(1\rightarrow3)(1\rightarrow4)$-$\beta$-D-Glucan (lichenan, or from barley), laminaran, and larchwood xylan [an acidic $(1\rightarrow4)$-β-D-xylan] may be obtained commercially.[7]

$(1\rightarrow3)(1\rightarrow4)$-$\beta$-D-Glucan 4-glucanohydrolase (EC 3.2.1.73, lichenase) from *Bacillus subtilis* (enzyme 1) was a gift from B. A. Stone. $(1\rightarrow3)$-β-D-Glucan glucanohydrolase (EC 3.2.1.6, endo-1,3(4)-β-glucanase) from *Rhizopus arrhizus* QM 1032 (enzyme 2) was a gift from E. T. Reese. Cellulase (enzyme 3) from *Trichoderma viride* was from Boehringer-Mannheim (BM) Canada, Dorvall, Qué. (Cat. 238104, Lot 1038402). Activities will be referred to simply as $(1\rightarrow3)(1\rightarrow4)$-$\beta$-D-glucanase, $(1\rightarrow3)$-β-D-glucanase, $(1\rightarrow4)$-β-D-glucanase, and $(1\rightarrow4)$-β-D-xylanase on the basis of the structure of the substrate used.

Spectrophotometric Methods

Detection of Congo Red Binding by Polysaccharides in Solution

Solutions of polysaccharides (1 mg/ml) are routinely tested in the presence of 10 μg/ml CR in dilute buffer.[4] Controls without added dye should also be tested. Under these conditions major (>20 nm) bathochromic (red) shifts in λ_{max} of the absorption spectrum of CR are observed with $(1\rightarrow3)(1\rightarrow4)$-$\beta$-D-glucans, O-(hydroxyethyl)cellulose (HEC), and xyloglucan. Furthermore, the extinction coefficient is increased. Lesser red shifts were observed in the presence of CMC, CMP, galactoglucomannan, glucomannan, and laminaran. The red shift in λ_{max} of the absorption spectrum of CR in the presence of polysaccharides requiring alkali for solution (curdlan, starch, mannan), and polyanionic polysaccharides

[9] A. E. Clarke and B. A. Stone, *Phytochemistry* **1**, 175 (1962).
[10] G. A. Hohner and R. G. Hyldon, U.S. Patent 4,028,468 (1977).
[11] P. J. Wood, D. Paton, and I. R. Siddiqui, *Cereal Chem.* **54**, 524 (1978).

(CMC, CMP, acidic xylans), and some other polysaccharides, may be enhanced in the presence of 1 M NaCl. Enzyme assays utilizing complex formation of CR with (1→3)(1→4)-β-D-glucan (oat β-glucan, lichenan), (1→3)-β-D-glucan (CMP), (1→4)-β-D-glucan (CMC), and (1→4)-β-D-xylans (larchwood xylan) were developed based on these observations. Other substrates, such as HEC and xyloglucan, also may be used. To determine if a particular polysaccharide can bind CR in an exploitable fashion, testing should be done in 0.5–1 M NaCl for maximum dye binding. Since some polysaccharides (e.g., cereal β-glucan) can induce significant wavelength shifts at low concentration (1 μg/ml), caution should be exercised in the use of substrates that are not well defined.

Monitoring Enzyme Action in Solution

The λ_{max} of the absorption spectra of CR (10 μg/ml) increases with increasing concentration of oat β-D-glucan in dilute buffer, and of CMP and CMC in 1 M NaCl, until saturation binding is reached (~10 μg/ml for oat β-D-glucan, 100–200 μg/ml for CMC and CMP). To monitor enzyme reaction in solution, therefore, aliquots of the enzyme reaction mixture are diluted with buffer or 1 M NaCl until the substrate concentration is below the saturation values and CR then added to give a concentration of 10 μg/ml. A wide range of initial substrate concentrations, and hence enzyme concentration, is therefore possible, as illustrated below. The λ_{max} of the difference spectra of CR in the presence of oat β-D-glucan and CMC is ~540 nm, and in the presence of CMP is ~525 nm.

The wavelength shifts and increase in extinction coefficients are readily visible to the naked eye, allowing simple qualitative detection of depolymerization in enzyme digests. Although we have not developed this approach quantitatively, similar CR binding by starch was exploited many years ago in α-amylase assay.[3]

(1→3)(1→4)-β-D-Glucanase. Solutions of enzyme 1 (25 and 50 μl; 0.9 units/ml) were added to oat gum (20 ml; 20 μg/ml) in 0.05 M sodium maleate buffer, pH 6.5. Aliquots (1 ml) were removed in duplicate at intervals (Table I), heated (100°) for 15 min, cooled to room temperature, CR (0.2 ml; 100 μg/ml) added, and the mixture diluted with buffer to 2 ml. Absorbancy (540 nm) was measured. Enzyme alone gave no change in λ_{max} of CR.

(1→4)-β-D-Glucanase. In a similar incubation, enzyme 3 was added to CMC, but this substrate was used at 5 mg/ml. Aliquots (50 μl) of the incubation mixture were added to 4.5 ml of 1 M NaCl, heated (100°) for 10 min, and aliquots (1.8 ml) then mixed with CR (0.2 ml; 100 μg/ml) and absorbancy (540 nm) measured (Table I).

TABLE I
Effect of Incubation of (1→3)(1→4)-β-d-Glucanase with Oat Gum and (1→4)-β-d-Glucanase with CMC on Absorbancy of Sample Minus Absorbancy of CR Alone[a]

Glucanase	Amount of enzyme in incubation mixture	Incubation time (min)					
		0	5	10	20	30	60
(1→3)(1→)-β-d-	1.1 × 10⁻³ units/ml[b]	0.216	0.211	0.198	0.152	0.126	0.051
	2.2 × 10⁻³ units/ml	0.212	0.207	0.162	0.101	0.051	0.011
(1→4)-β-d-	20 μg/ml	0.120	0.110	0.101	0.093	0.091	0.073
	40 μg/ml	—	0.101	0.075	0.040	0.019	0.004

[a] 540 nm.

[b] One unit produces 1 μmol of reducing sugar (as glucose) per min at 37° from oat β-glucan (0.5%) in MES buffer, pH 6.5

Use of Insoluble Substrate in Measurement of (1→3)(1→4)-β-d-Glucan 4-Glucanohydrolase

Cereal β-d-glucans are selectively precipitated by CR from crude extracts of oat or barley flour,[4] and the insoluble CR–glucan complex may be used as an enzyme substrate without prior purification of the cereal β-d-glucan.[5,12]

Oat (or barley) flour (25 g) is stirred for 2 hr at 45° in 0.1 M sodium carbonate buffer, pH 10, the mixture centrifuged (12,000 g, 20 min), and CR (25 ml; 20 mg/ml in H_2O) added to the supernatant with vigorous stirring. The precipitate is collected by centrifugation (12,000 g, 20 min), washed with buffer, water, and 2-propanol, homogenized in 2-propanol, filtered on sintered glass, and dried in air with gentle warming. To prepare substrates which only release significant amounts of dye in the presence of enzyme, the dye content must be reduced to 5–10% for oats and somewhat lower (3–6%) for barley. This is done by further washing with 65–80% 2-propanol before a final wash with 100% 2-propanol. Dye content is determined by dissolving the substrate, with gentle warming, in dimethyl sulfoxide (DMSO)[13] and determining absorbancy at 532 nm (λ_{max}).

To determine enzyme activity, substrate (10–20 mg) is dispersed in buffer, for example 0.05 M 2-(N-morpholino)ethanesulfonic acid sodium salt (MES, pH 6.5) (10–20 ml) and enzyme added. Aliquots of the reaction mixture are removed, centrifuged to remove substrate, and absorbancy

[12] P. J. Wood and K. G. Jorgensen, *J. Cereal Sci.*, in press.

[13] Dyes should be handled with care, especially in the presence of DMSO.

(500 nm) measured. Buffer without enzyme should result in an absorbancy <0.1, preferably ~0.05, after 1 hr.

Normally, absorbancy of the supernatant increases linearly with time, and rate of dye release is proportional to enzyme concentration up to ~2 × 10^{-3} units/ml. This is similar to the enzyme levels used in viscosity assay. The method has been used to assay malt and commercial enzymes. Assay of enzymes with pH optima <5.0 (e.g., malt) may cause difficulties, possibly because of changes in dye solubility and color in this pH range. A number of the substrates which responded well for enzyme 1 were unsuitable for malt enzymes. The substrate used for malt studies was isolated from barley cultivar Nordal and contained 5.3% CR.

Gel Diffusion Assay

Principle

Radial diffusion of enzyme into a substrate-bearing gel slab allows determination of hydrolase activity[14] provided a simple distinction between substrate and product is available. Diffusion zone diameter is proportional to the logarithm of enzyme concentration. This simple system allows assay of many samples with minimum labor and expense and is applicable to small sample volumes and activities.[15] Commercial kits are available which allow rapid and reproducible preparation of gel plates (Miles Scientific, Naperville, IL).

Use of CR binding by β-D-glucans in a gel diffusion assay was originally recommended for screening of seeds and microorganisms.[16] The method has been applied to measurement of β-D-glucanase in commercial enzyme preparations,[17,18] determination of malt (1→3)(1→4)-β-D-glucanase,[19,20] and to screen and monitor purification of cellulolytic microorganisms and their enzymes.[21] Specific examples are described below.

General Aspects of the Use of the Gel Diffusion Assay

Preparation of Gel Plates. Substrate (oat gum, CMP, CMC) is dissolved (12.5 mg/ml) in the buffer in which the enzyme is to be tested, is

[14] J. Dingle, W. W. Reid, and G. I. Solomons, *J. Sci. Food Agric.* **4**, 149 (1953).
[15] W.-B. Schill and G. F. B. Schumacher, *Anal. Biochem.* **46**, 502 (1972).
[16] P. J. Wood, *Carbohydr. Res.* **94**, C19 (1981).
[17] P. J. Wood and J. Weisz, *Cereal Chem.* **64**, 8 (1987).
[18] M. J. Edney, H. L. Klassen, and G. L. Campbell, *Poultry Sci.* **65**, 72 (1986).
[19] H. L. Martin and C. W. Bamforth, *J. Inst. Brew.* **89**, 34 (1983).
[20] G. K. Buckee, *J. Inst. Brew.* **91**, 264 (1985).
[21] R. M. Teather and P. J. Wood, *Appl. Environ. Microbiol.* **43**, 777 (1982).

clarified by centrifugation if necessary, and an aliquot (1 ml) added to 24 ml buffer or, for predyed gel, to 23 ml buffer to which 1.0 ml CR (1 mg/ml in water) has been added. Agarose (125 mg) is then added and the mixture heated to boiling with stirring. The hot mixture is poured into square (9 × 9-cm) plastic disposable petri dishes and allowed to cool and set (2 hr) to a gel thickness of ~4 mm. Wells of 4 mm diameter are cut using a cork borer (no. 1) and the gel removed by suction or forceps. Enzyme solutions (10 μl) are pipetted into the wells, lids placed on the dishes, and diffusion allowed to proceed, conveniently overnight (18 hr) at room temperature. Gels without dye are then stained as described below.

Staining Procedures. Staining procedures differ slightly depending on the substrate being used. The solutions are applied and removed by pouring. Reagent volumes and times are not critical, requiring only sufficient volume to cover the surface to a depth of 2–3 mm. Stained plates can be preserved for later observation by flooding with 5% (v/v) acetic acid or 1 *M* HCl. The specific procedures suggested are tabulated below.

Substrate	Reagent	Time
CMC and CMP	CR, 1 mg/ml	15
	1 *M* NaCl	10
	1 *M* NaCl	10
Lichenan, oat gum	CR, 0.5 mg/ml	10
Xylan	1 *M* NaCl	20
	CR, 1 mg/ml	15
	1 *M* NaCl	10
	1 *M* NaCl	10

Measurement of Diffusion Zones. Diameters are conveniently determined from the average of measurements by vernier calipers at perpendicular directions. Radii may also be used. For controls, gels should be run without substrate and with heat-deactivated enzyme, but note heat stability of some enzymes (see below). For CMP and CMC, diffusion zones, prior to salt application, should not be readily visible. The edges of the diffusion zones are easily distinguished. However, annular regions of different stain intensity may develop; for example, oat gum with CR incorporated in the gel shows an intensely stained ring at the diffusion zone boundary. Measurements should be to the edge of the cleared zone.

Linearity and Reproducibility of the Gel Diffusion Assay and Its Relation to Viscometric Measurements. Six to eight levels of enzyme were applied to six plates, each level applied once to a randomly assigned position on each plate. Results for enzymes 1–3 in 0.05 *M* MES buffer, pH 6.5, and (1→3)-β-D-glucanase from 6-day germinated barley in 0.05 *M*

TABLE II

REGRESSION ANALYSIS OF GEL DIFFUSION ASSAY OF
$(1{\rightarrow}3)(1{\rightarrow}4)$-$\beta$-D-GLUCANASE, $(1{\rightarrow}4)$-β-D-GLUCANASE, AND
$(1{\rightarrow}3)$-β-D-GLUCANASE AND BARLEY USING OAT GUM, CMC,
AND CMP AS SUBSTRATES[a]

Enzyme	Regression m	Coefficients and standard errors[b]			Correlation coefficient	p value
		SE	c	SE		
1	0.625	0.011	0.911	0.020	0.994	<0.005
2	0.489	0.006	1.986	0.006	0.997	<0.005
3	0.565	0.009	2.253	0.010	0.995	<0.005
Barley	0.618	0.008	0.770	0.013	0.997	<0.005

[a] Enzyme 1, $(1{\rightarrow}3)(1{\rightarrow}4)$-$\beta$-D-glucanase; enzyme 2, $(1{\rightarrow}4)$-β-D-glucanase; enzyme 3, $(1{\rightarrow}3)$-β-D-glucanase (cellulose); $n = 36$–48.
[b] Diameter $= m$ log[enzyme] $+ c$.

MES buffer, pH 5.5, are summarized in Table II. The levels of $(1{\rightarrow}3)$-β-D-glucanase in barley were monitored during germination over 6 days by gel diffusion and viscometry. The correlation between diameter and the logarithm of the rate of change of reciprocal specific viscosity was 0.96 ($p <$ 0.005).

General Considerations for Gel Assay. In most cases zone clearing is equally detected with dye incorporated into the gel or with staining after enzyme diffusion, but since there is a possibility of dye inhibition of enzyme, for screening purposes we normally added CR after diffusion. The results of Table II showed no significant differences between plates ($p > 0.25$) but, as recommended by others,[14,15] there is sufficient variability to warrant inclusion of standards, particularly for long-term or between operator comparisons of preparations of similar activity. The linear range varies with different enzymes,[15,17] for example, 400-fold for barley $(1{\rightarrow}3)$-β-D-glucanase and 1000-fold for enzyme 1, but these large differences in activity are represented by a relatively small range in diameters of <2 cm. Diffusion zone diameters increase with time and temperature, decrease with increase in substrate concentration, and are pH sensitive. Different enzymes have different diffusion rates. These factors should be considered and controlled if possible.

Applications of Gel Diffusion Assay

Screening Commercial Enzyme Preparations. Results from a survey of commercial enzyme preparations for the presence of three different β-D-glucanases are shown in Table III. It should be noted that some en-

TABLE III
(1→3)(1→4)-β-D-GLUCANASE, (1→3)-β-D-GLUCANASE, AND (1→4)-β-D-GLUCANASE ACTIVITIES IN
COMMERCIAL ENZYME PREPARATIONS AND HEAT-TREATED CONTROLS AS SHOWN
BY GEL DIFFUSION[a]

| | | | Diameter of diffusion zone (cm) | | | | | |
| | | | Oat Gum | | CMP | | CMC | |
Enzyme	Source	Supplier	U	H	U	H	U	H
α-Amylase[b]	Hog pancreas	Sigma	0	0	0	0	0	0
α-Amylase[c]	Bacillus subtilis	Sigma	2.36	1.00	0	0	0	0
α-Amylase[c]	Aspergillus oryzae	Sigma	1.20	0	1.39	0	1.54	0
Amyloglucosidase[d]	Aspergillus niger	BM[e]	0.80	0.72	0	0	1.95	1.59
Cellulase[d]	Trichoderma viride	BM	1.56	0.54	1.50	0	2.46	1.45
Glucose oxidase[d]	A. niger	BM	0	0.83	0	0	2.09	1.58
Peroxidase[d]	Horseradish	BM	0	1.38	0	1.41	0	1.42

[a] U, Untreated enzyme; H, heat-treated enzyme (1 hr at 100°).
[b] Suspension supplied, diluted 10-fold.
[c] 5 mg/ml.
[d] 10 mg/ml.
[e] Boehringer-Mannheim.

zymes continue to show clearing zones after vigorous heat treatment. This may indicate heat stability and often represents only a small fraction of initial activity. However, artifacts may be responsible, as is probably the case with the peroxidase and glucose oxidase which showed clearing only after heat treatment.

Measurement of β-D-Glucanase and β-D-Xylanase Activity following Column Chromatography. The routine analysis of chromatographic column fractions from extracellular preparations of the rumen cellulolytic bacterium, *Ruminococcus flavefaciens*, is presented as a particular example of the value of the gel diffusion method for assaying a large number of samples. Differences in detail from the above method, arising from the large number of samples and the different system under study, are described.

Substrates (0.1%) in agarose (0.8%) are prepared in batches of multiples of 12.5 ml. Agarose and substrates (CMC, CMP, lichenan, or xylan) are dissolved in 25 mM potassium phosphate buffer (pH 6.8) containing 1 mM dithiothreitol[7] (KPB–DTT buffer) by heating to a boil and the solution then dispensed (12.5 ml) into a 60-ml serum bottle. Sealed bottles are autoclaved for 15 min at 15 psi steam pressure (120°). Once sterilized, bottles of agarose–substrate can be stored until required.

Disposable square petri dishes (9 × 9 cm) are placed on a leveled surface and the contents of an agarose–substrate bottle, liquified by heating to 100°, are poured in to provide a 1.5–2.0 mm layer of agarose which is allowed to cool and solidify before the plate is moved. Petri dishes are finally covered to prevent drying of the medium. A template is prepared for location of wells by tracing the petri dish on paper and drawing a 5 × 5 grid (25 wells) such that wells are equidistant from each other and from the walls of the dish. Wells are cut with a 2-mm blunted hypodermic needle and the agarose plug is aspirated with a smaller needle. Wells of this size hold approximately 5 μl of sample. Once samples have been introduced dishes are covered and incubated at 37° overnight or for shorter times if samples are particularly active.

The enzyme preparation used to establish standard curves for the various substrates was from the culture fluid of a pure strain of a cellulolytic rumen bacterium (as described in the legend to Fig. 1). The hydrolytic activities in the preparation were determined with CMC, lichenan, xylan, and laminaran[7] as substrates using a reducing sugar assay[22] to quantitate end products. Substrates (10 mg/ml) were incubated in KPB–DTT buffer for 0, 30, and 60 min with sufficient enzyme to produce 100–500 μg of reducing sugar (glucose standard). Laminaran was used in the tube assay for (1→3)-β-D-glucanase since it was more readily available than CMP. (Laminaran was not used in the gel diffusion assay because the diffusion zones were not easily seen.) For the diffusion assay the enzyme preparation was diluted with KPB–DTT buffer such that 5 μl incubated overnight (~18 hr) gave diffusion zones ranging in radius from 3 to 12 mm. The assay at each enzyme level was run in triplicate on separate petri dishes. Xylan binds CR relatively weakly and diffusion zones may be difficult to distinguish. The relationships of log activity, as determined by reducing sugar measurements, to radius of diffusion zone had correlation coefficients of 0.99. To simplify relating radius of diffusion zones to enzyme activity, as obtained from column fractions, it is convenient to generate a table of activities relating to a range of radii (e.g., 1–12 mm at 0.2 mm intervals) so activity can be read directly.

The 360 assays to provide the data shown in Fig. 1 used four petri dishes (50 mg of each substrate) and required less than 8 hr of actual working time to complete.

Product Inhibition of CMC Hydrolysis Demonstrated with the Gel Diffusion Assay. The demonstration of possible inhibition of (1→4)-β-D-glucanase activity by end products (i.e., glucose, cellobiose, or other reducing sugars) is difficult when the assay for enzyme activity is based on production of reducing sugars. Using the gel diffusion assay the inhibi-

[22] N. Nelson, *J. Biol. Chem.* **153**, 375 (1944).

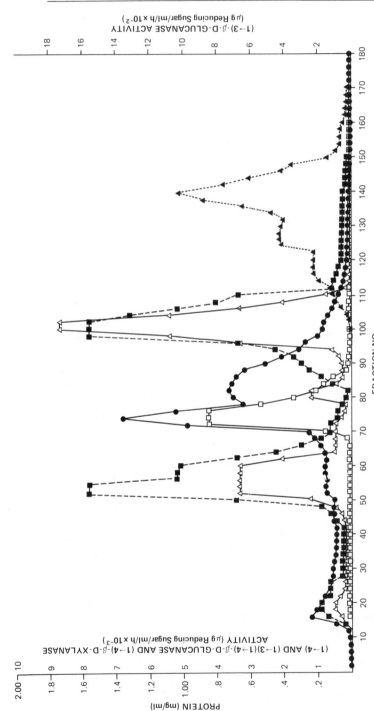

FIG. 1. Chromatography of crude (1→4)-β-D-glucanase from *Ruminococcus flavefaciens* grown on milled filter paper. The culture fluid of the bacteria was concentrated on a 30,000 molecular weight exclusion membrane (Millipore Ltd., Mississauga, Ontario). The supernatant after centrifugation (100,000 *g* for 2 hr) was applied (143 mg protein) to a 2.6 × 50 cm column of Fast Flow DEAE-Sepharose CL-6B (Pharmacia Chem. Co., Dorval, Que.). Elution was with a linear gradient of 25 m*M* potassium phosphate buffer (pH 7.8) containing 1 m*M* dithiothreitol (600 ml) and 1 *M* NaCl in the same buffer (600 ml). Five-microliter aliquots of even numbered fractions (6.15 ml) were placed in wells of petri dishes containing the substrates indicated in the figure. The activities for each substrate were calculated from regression equations. Protein was determined by ultraviolet absorbance [W. E. Graves, F. C. Davis, Jr., and B. H. Sells, *Anal. Biochem.* **22**, 195 (1968)] using the equation: protein (mg/ml) = ($A_{224.0}$ − $A_{233.4}$)0.225 where the constant had been established with bovine serum albumen as a standard. (△) CMC; (□) xylan; (▲) lichenan; (■) CMP; (●)

TABLE IV
Glucose and Cellobiose Inhibition of
$(1\rightarrow4)$-β-D-Glucanase from $R.$ $flavefaciens$
with CMC as Substrate

Concentration (mM)	Inhibition of $(1\rightarrow4)$-β-D-glucanase[a] (%)	
	RF-1	RF-2
Glucose		
0[b]	0	0
100	33	42
200	39	42
300	41	41
Cellobiose		
0[b]	0	0
50	39	53
100	71	55
200	80	54

[a] $(1\rightarrow4)$-β-D-Glucanases (RF-1 and RF-2) correspond to the activities recovered in fractions 48–65 and 90–110, respectively, in Fig. 1.
[b] Activity in the absence of glucose and cellobiose was 0.05 and 0.12 μmol reducing sugars/ml/hr for enzymes RF-1 and RF-2, respectively. Each value is the mean of five replicate assays on different plates.

tion of $(1\rightarrow4)$-β-D-glucanase from $R.$ $flavefaciens$ (RF-1 and RF-2) by glucose and cellobiose could be demonstrated (Table IV). Glucose and cellobiose[7] were added to the agarose–CMC substrate and plates prepared in the manner described. The results demonstrate that glucose is less inhibitory than cellobiose and that enzyme RF-1 is more susceptible to cellobiose inhibition than is RF-2.

Detection of β-D-Glucanase and β-D-Xylanase Activity in
   Polyacrylamide or Agarose Electrophoresis Gels

Procedures for the use of the CR staining technique for the detection in $situ$ of glucanase and xylanase enzyme separated in agarose or polyacrylamide gels have been published.[23,24] The following protocol differs only in detail from these procedures.

[23] P. Beguin, $Anal.$ $Biochem.$ **131,** 333 (1983).
[24] C. R. Mackenzie and R. E. W. Williams, $Can.$ $J.$ $Microbiol.$ **30,** 1522 (1984).

Detection of Enzyme Activity in Polyacrylamide Gels

The enzyme separation gel is laid on top of a thin agarose gel containing the test substrate. The separation gel may be based on any separation system which does not inactivate the enzyme(s) of interest (i.e., continuous and discontinuous buffer systems, sodium dodecyl sulfate and other commonly used detergents, and isoelectric focusing systems are all compatible with the staining system). The agarose gel containing the test substrate is most conveniently cast in the same apparatus as the separation gel, i.e., between two glass plates using 1–1.5 mm spacers. The gel normally contains 0.8% (w/v) agarose, 0.1% (w/v) substrate, and a buffer system appropriate to the enzyme system being studied (e.g., 25 mM potassium phosphate, pH 6.8, for enzymes from *Bacteroides succinogenes*). After the agarose gel has solidified one glass plate is carefully removed by sliding it off, leaving the agarose gel supported by the second glass plate. The polyacrylamide gel is laid onto the agarose gel. The gels are then incubated at 37° for 5 min–16 hr (a shorter time is generally sufficient), the polyacrylamide gel is removed (it may itself be stained to detect proteins or other materials in the sample), and the agarose gel is stained with CR as described above, using the remaining glass plate to support the gel during the manipulations required.

Detection of Enzyme Activity in Agarose Gels

Because agarose electrophoresis gels are difficult to handle, the substrate-containing gel is poured as a thin overlay directly onto the separation gel. The overlay consists of 0.4% (w/v) agarose, 0.1% (w/v) substrate, and an appropriate buffer system. It is important that the separation gel on its support plate be carefully leveled, that it be prewarmed to prevent the overlay from solidifying too quickly (37–42° is appropriate), and that the overlay be relatively hot (50°) to ensure that it will remain liquid long enough to spread evenly over the top of the separation gel by tilting the plate holding the separation gel. The combined gel is incubated as required and then stained with CR as described above.

Use in Detection of β-D-Glucanase and β-D-Xylanase Secreting Organisms

The sensitivity with which β-D-glucanase and β-D-xylanases can be detected using the CR staining technique makes this method particularly useful in the isolation and characterization of microorganisms producing these enzymes.[21] The substrates which are readily used include CMC, CMP, lichenan, and larch wood xylan. Organisms can be grown within or on an agar matrix at relatively high densities (200–400 colonies/plate). The test substrate can be incorporated into the growth medium or applied in a thin agarose overlay either at the time of inoculation or at a later stage in the growth of the culture. Staining may be carried out at any time at which a zone of hydrolysis has been formed which is adequate for detection, but not so large that it becomes unclear which colony is the source of the enzyme. (It is possible with some substrates to incorporate the CR in the substrate-containing medium before inoculation with the microorganisms, but the result is not as satisfactory.) Determination of the appropriate incubation time will be necessary for any given species and growth medium.

Analysis of Microbial Populations

The analysis of populations of microorganisms for the presence of species capable of degrading specific polysaccharides can be carried out using any agar-based growth medium that will support the growth of the organisms. Best results are obtained by growing the colonies within an agar matrix, which reduces the problems caused by the spreading growth morphology and high motility of at least some species in most ecosystems, and by cross-contamination of colonies during staining. As an example of this procedure we will describe the analysis of a rumen bacteria population (obligate anaerobes) using a serum bottle technique. The advantage of this method for the analysis of a microbial population is that the very thin layer of agar medium in which the bacteria are grown restricts the growth of the fast-growing species, allowing slower growing species to form recognizable colonies and to express glucanase activities at detectable levels.

Growth conditions are those described by Teather and Wood.[21] Medium 98-5,[25] containing 1.4% agar (Oxoid Bacteriological Agar No. 1 is used for all growth media described), is modified by the addition of the test substrate (1 mg/ml). Medium (3.8 ml) is dispensed anaerobically into each 30-ml serum bottle and the bottles sealed with butyl rubber stoppers. After autoclaving, the bottles are cooled to 45°, inoculated with appropriate dilutions of rumen fluid, and then placed on a bottle roller while being cooled under running water to produce a thin, uniform agar layer. The bottles are incubated for 16–72 hr at 39°. The average colony size should be small (<0.5 mm). Various incubation times should be tried (e.g., stain some bottles at 16 hr, some at 40 hr, and some at 72 hr) as the level of enzyme production will vary greatly within the population. The medium

[25] M. P. Bryant and L. M. Robinson, *J. Dairy Sci.* **44,** 1446 (1961).

in the bottle can be stained essentially as described above, but if it is desired to isolate the organisms producing clearing zones then the staining solutions must be made up in a reduced sterile dilution fluid such as that described by Bryant and Robinson.[26]

Screening a Gene Bank for a β-D-Glucanase or β-D-Xylanase

Screening for the expression of a cloned gene in an organism such as *Escherichia coli* can be carried out at a much higher colony density than can screening in a mixed population where considerable variation in growth rate, level of enzyme production, and motility can be expected. Best results are obtained when the colonies are grown within the agar medium. The substrate is normally applied as a second agar overlay above the overlay containing the cells. This system has several advantages: (1) because of the second overlay the colonies are all below the surface of the medium, so colony growth rate is uniform, (2) the substrate is in a very thin layer (less than 1 mm) which results in a sharply defined boundary for the zone of digestion and high sensitivity, and (3) the basal medium layer, which may be quite thick, allows extended incubation of the plates if low levels of expression of the glucanase are expected.

A suitable protocol to screen a *B. succinogenes* gene bank, cloned in pUC8 in *E. coli,* for glucanase gene expression is as follows. Petri plates (9 cm diameter) are prepared containing 20–25 ml of the basal medium (LB medium: 10 g Tryptone, 5 g yeast extract, and 10 g NaCl/liter, plus 1% agar and 50 μg/ml ampicillin). The growth overlay medium (same as LB above, except that the agar concentration is reduced to 0.5%) is conveniently autoclaved in batches of 40 ml, sufficient for 10 plates. The substrate overlay contains 25 mM potassium phosphate buffer, pH 6.8, 1 mg/ml of the test substrate, and 0.4% agarose (0.7% agar can be substituted, but agarose gives better results with substrates such as xylan that bind CR weakly). The overlay media are melted and held at 42° (growth medium) and 50° (substrate medium). One milliliter of a dilution of the gene bank stock, containing about 3000 colony-forming units, is added to 40 ml of the growth medium (at 42°). A 10-ml pipet, warmed briefly in a bunsen burner flame, is used to transfer 4 ml of the inoculated growth medium to each petri plate, where it is quickly spread over the agar surface by tilting the plate. The plates are placed on a level surface until all the overlays (which may include a number of 40 ml batches) have been poured and have solidified. The substrate medium (50°) is then similarly applied as a second overlay (4–5 ml/plate). When this second overlay has

---

[26] M. P. Bryant and L. M. Robinson, *Appl. Microbiol.* **9**, 96 (1961).

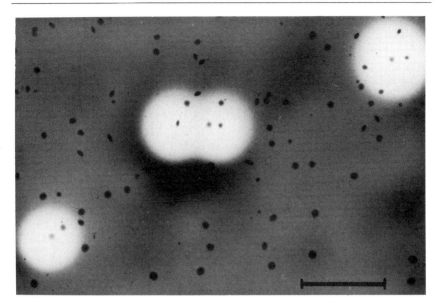

FIG. 2. Detection of bacterial colonies with $(1\rightarrow3)(1\rightarrow4)$-$\beta$-D-glucanase activity. *Escherichia coli* HB101 transformed with plasmid pUC8 carrying insert DNA from *Bacteroides succinogenes* was plated in an LB agar overlay on LB agar plates and overlaid with agarose containing lichenan as described in the text. The plates were incubated for 14 hr at 37° before staining with CR. The colony density was about 400 colonies/plate (bar = 5 mm).

solidified, the plates can be inverted and incubated at 37° for the required period of time. In most cases, overnight incubation (16–24 hr) is sufficient (Fig. 2), but in some cases it has been necessary to extend the incubation period to 72 hr. The plates are stained as described above.

Characterization of Isolated Clones

Replica plating techniques can be used to screen a set of bacterial isolates for glucanase and xylanase specificity quickly and easily. Bacterial colonies are inoculated in a grid pattern on a master plate (25 colonies/plate). After a suitable period for growth (for *E. coli* grown on LB medium at 37°, normally 16 hr) the colonies are replica plated onto a set of test plates, the number of replicas depending on the number of substrates to be tested. The test substrate (1 mg/ml) may either be incorporated in the basal medium or introduced as an overlay after the replicated colonies have been allowed to grow.

If the substrate is incorporated in the basal medium the plates should be kept thin (15 ml of medium in a 9-cm petri dish). The replica plates are normally grown for about 16 hr before staining as described above. The colonies may tend to wash off during staining or 1 M NaCl treatment. If the colony washes off during the NaCl treatment the result will be an apparent clearing zone where the colony prevented free access of the dye to the substrate. In cases where enzyme activities are extremely low, so that this unstained area could be confused with an enzymatically cleared zone, it is necessary to wash off the bacterial colonies[27] prior to staining of the plate.

If the substrate is incorporated in an overlay poured over a well-grown replica plate (4 ml/plate, using the substrate overlay medium described above) the sensitivity of enzyme detection is greater and less substrate is required. However, this approach is practical only when the bacterial colonies are coherent enough to withstand the pouring of the overlay. The incubation time required to detect enzyme activity is normally less than 2 hr. If extended (longer than a few generation times) incubation periods are required to detect enzyme activity, it is also necessary to suppress growth in the overlay by adding an appropriate antibiotic(s).[21] After incubation the plate is stained as described above.

Acknowledgment

Contribution No. 683 of the Food Research Centre and No. 1399 of the Animal Research Centre, Agriculture Canada, Ottawa, Ontario, K1A 0C6.

[27] N. R. Gilkes, D. G. Kilburn, R. C. Miller, and R. A. J. Warren, *Biotechnology* **2,** 259 (1984).

[8] Soluble, Dye-Labeled Polysaccharides for the Assay of Endohydrolases

By BARRY V. McCLEARY

A range of methods has been developed for the assay of polysaccharide endohydrolases and these include viscosimetric[1,2] and nephelometric[3] methods and procedures based on the measurement of increase in reducing sugar equivalents[4,5] and on the rate of release of soluble, dye-labeled fragments on hydrolysis of chromogenic polysaccharide sub-

[1] K. E. Almin, K. E. Eriksson, and B. Pettersson, *Eur. J. Biochem.* **51,** 207 (1975).
[2] D. T. Bourne and J. S. Pierce, *J. Inst. Brew.* **76,** 328 (1970).
[3] M. Nummi, P. C. Fox, M. L. Niku-Paavola, and T. M. Enari, *Anal. Biochem.* **116,** 133 (1981).
[4] M. Somogyi, *J. Biol. Chem.* **195,** 19 (1952).
[5] M. Lever, *Biochem. Med.* **7,** 274 (1973).

strates.[6] Methods which take advantage of the specific interaction of a particular polysaccharide with another component, e.g., starch with iodine[7] or barley β-glucan with Congo Red dye,[8] have also been employed. An enzyme-linked reaction employing nitrophenylmaltotetraose[9] has been developed for the assay of α-amylase. Each of these assays can be used to advantage in certain circumstances. Thus, viscosimetric assays are particularly useful for detecting trace quantities of an endoenzyme,[10] whereas assays based on the measurement of increase in reducing sugar equivalents give a direct measure of the number of glycosidic bonds cleaved. Assays based on the use of chromogenic (dye-labeled) substrates have several advantages over more conventional assays including specificity and simplicity in use. However, since dyeing generally reduces the solubility of the polysaccharide, most commercially available dye-labeled substrates are insoluble and have the inherent disadvantages of heterogeneity in the assay tube and the difficulties associated with dispensing a solid substrate routinely with accuracy.

Soluble, chromogenic substrates can be prepared by selecting a branched, highly soluble polysaccharide as the starting material. Treatment of amylopectin with Reactone Red yields a soluble product[11] as does dyeing of carob galactomannan with Remazol Brilliant Blue R (RBB).[12] But in some cases a polysaccharide with the desired solubility properties is not available as a natural product. This limitation can be overcome by chemically modifying naturally occurring polysaccharides (to increase solubility) before dyeing.[13,14] Our methods for the preparation of soluble dye-labeled substrates for the assay of β-D-mannanase (EC 3.2.1.78, mannan endo-1,4-β-mannanase), endo-1,4-β-D-glucanase (cellulase, EC 3.2.1.4), endo-1,3(4)-β-D-glucanase (lichenase, EC 3.2.1.73, malt β-glucanase), and α-amylase (EC 3.2.1.1) are described below. Substrates for the assay of endo-1,4-β-D-xylanase (EC, 3.2.1.8)[15] and endopectinase[16] have been prepared by other authors.

[6] H. N. Fernley, *Biochem. J.* **87**, 90 (1963).
[7] W. Banks and C. T. Greenwood, "Starch and Its Components," p. 211. University Press, Edinburgh, Scotland, 1975.
[8] P. J. Wood, *Carbohydr. Res.* **94**, C19 (1981).
[9] K. Wallenfels, B. Meltzer, G. Laule, and G. Janatsch, *Fresenius Z. Anal. Chem.* **301**, 169 (1980).
[10] B. V. McCleary, R. Amado, R. Waibel, and H. Neukom, *Carbohydr. Res.* **92**, 269 (1981).
[11] A. L. Babson, S. A. Tenney, and R. E. Megrew, *Clin. Chem.* **16**, 39 (1970).
[12] B. V. McCleary, *Carbohydr. Res.* **67**, 213 (1978).
[13] B. V. McCleary, U.S. Patent 4,321,364 (1982).
[14] B. V. McCleary, *Carbohydr. Res.* **86**, 97 (1980).
[15] P. Biely, D. Mislovičová, and R. Toman, *Anal. Biochem.* **144**, 142 (1985).
[16] D. R. Friend and G. W. Chang, *J. Agric. Food Chem.* **30**, 982 (1982).

RBB–Carob Galactomannan for the Assay of β-D-Mannanase

Preparation of Low-Viscosity Carob Galactomannan

Two hundred grams of gum, locust bean (Sigma Chemical Co. G0753) is suspended in 3 liters of water containing 0.2 g of cellulase preparation (Sigma C7502) at 60°. This is stirred with a spatula to give a thick paste. After incubation for 30 min at 40° the paste is homogenized in a Waring blender and then incubated at 40° for a further 30 min. (β-Mannanase present in the cellulase preparation causes a significant decrease in the paste viscosity during this incubation period.) The slurry is again homogenized in a Waring blender, incubated in a boiling water bath until the temperature reaches 90° (to inactivate β-mannanase) and then centrifuged at 3000 g for 15 min. The clear supernatant solution is poured into two volumes of ethanol (95% v/v) whereupon the galactomannan precipitates as a white fibrous mass. This material is collected on a nylon screen, washed by resuspension in aqueous ethanol (60% v/v), and then dried by solvent exchange with ethanol and acetone and stored *in vacuo;* the yield is 60%.

Preparation of RBB–Carob Galactomannan

One hundred and twenty grams of low-viscosity carob galactomannan is dissolved in 1.6 liters of water at 60°. To this is added 24 g of RBB dye and 160 g of anhydrous sodium sulfate which is dissolved by stirring over 5 min. Thirteen grams of trisodium phosphate is then added and stirring is continued for a further 2 hr at 60°. The solution is then cooled and dialyzed overnight against flowing tap water. On treatment of this solution with 2 volumes of ethanol, the dyed galactomannan which precipitates from solution is recovered on a nylon screen and washed with 66% (v/v) aqueous ethanol until most free dye is removed. The polymer is dissolved again in water at 60° and reprecipitated by the addition of 2 volumes of ethanol. This step is repeated until all free dye is removed. The precipitated galactomannan is then washed with ethanol and acetone and dried *in vacuo;* the yield is 110 g. The RBB to anhydrohexaose ratio is approximately 1 : 50.

For use as substrate, 1 g of the polysaccharide is dissolved in 80 ml of water at 60°, with vigorous stirring over 15 min. Ten milliliters of 3 M sodium acetate buffer (pH 5.0) is added and the volume is accurately adjusted to 100 ml. In the presence of 0.02% sodium azide, the solution is stable at 4° for at least 12 months and shows no significant tendency to settle from solution over this period.

Assay of β-Mannanase

Enzyme preparation (0.5 ml) containing 0–0.5 units of β-mannanase activity per 0.5 ml is incubated with RBB–carob galactomannan substrate solution (1.0 ml, 1% w/v) for 5–20 min at 40°. The reaction is terminated and the high-molecular-weight substrate is precipitated by the addition of ethanol (3 ml). The mixture is stirred, allowed to equilibrate to room temperature for 10 min, and centrifuged at 1000 g for 10 min. The enzyme reaction is monitored by increased absorbance (590 nm) of the supernatant solution.

Advantages and Limitations of the Assay

The described assay for β-mannanase using RBB–carob galactomannan substrate is simple to use and is highly specific for β-mannanase. It is of particular value in measuring this activity in crude enzyme preparations and microbial culture filtrates and for rapidly screening column chromatographic eluates. The assay can be standardized with pure β-mannanase[12] such that absorbance increase at 590 nm can be directly related to international units of enzyme activity.[17] However, each batch of dyed substrate needs to be standardized (degree of dyeing varies from batch to batch) and a standard curve needs to be prepared for each particular β-mannanase being assayed. α-Galactosidase (EC 3.2.1.22) and β-mannosidase (EC 3.2.1.25) do not interfere with the assay. β-Mannosidase from *Helix pomatia* has no detectable action on the polymer. In contrast, α-galactosidase can remove over 70% of the D-galactosyl residues from the substrate, but this is not accompanied by any release of dye-labeled fragments, nor does this treatment render the substrate significantly more susceptible to hydrolysis by β-mannanase.[12]

Carob galactomannan is employed in this assay because the degree of D-galactose substitution (Gal : Man = 23 : 77) is sufficient to impart solubility without interfering with the hydrolysis by β-mannanase. More highly substituted galactomannans (e.g., guar galactomannan, Gal : Man = 38 : 62) are resistant to hydrolysis by β-mannanase and thus make poor

---

[17] The activity in the β-mannanase preparations used to standardize the RBB–carob galactomannan assay procedure is measured by incubating enzyme preparation (0.05 ml) with 0.2% (w/v) carob galactomannan solution (0.50 ml) in 0.1 M sodium acetate buffer (pH 5.0) at 40° for 2–10 min. The reaction is terminated and the increase in mannose reducing sugar equivalents is determined using the Nelson/Somogyi[4] reducing sugar assay. One unit (U) of activity is defined as that amount of enzyme which will release 1 μmol of reducing sugar equivalents per minute at a defined pH and incubation temperature.

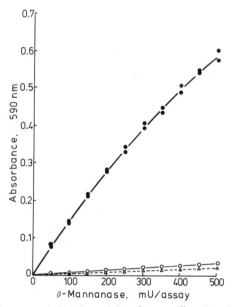

FIG. 1. Standard curves relating the activity of *Aspergillus niger* β-mannanase on carob galactomannan to absorbance increase (at 590 nm) on hydrolysis of RBB–carob galactomannan (●), RBB–guar galactomannan (○), and RBB–mannan (△). Incubations are performed at 40° for 10 min.

substrates. Mannans yield a highly insoluble, highly resistant substrate on dyeing[12] (Fig. 1).

RBB–CM-Cellulose for the Assay of Endo-1,4-β-D-glucanase

Preparation of RBB–CM-Cellulose

Fifty grams of *O*-(carboxymethyl)cellulose 4M6SF (CMC-4M, Hercules Inc.) is added to 1 liter of hot (~50°) water containing 100 mg of Sigma crude cellulase preparation (Sigma C7502) and stirred with a spatula to give a thick paste. The pH is checked and, if necessary, adjusted to 4.5 by addition of 1 *M* HCl. The solution temperature is maintained at 50° for 30 min during which time there is a significant viscosity decrease. The solution is blended vigorously to remove any lumps and then treated with 100 g of anhydrous sodium sulfate, 10 g of Remazol Brilliant Blue R, and 10 g of trisodium phosphate and the temperature raised to 70°. The solution is stirred and maintained at 70° on a hotplate magnetic stirrer for 2 hr.

While still hot, the solution is treated with 1.5 volumes of ethanol to precipitate the dyed polysaccharide which is then recovered on a nylon screen and excess liquid removed by squeezing. This material is redissolved in hot water by homogenizing in a Waring blender and then recovered by precipitation with 1.5 volumes ethanol. This process is repeated until the washings are essentially colorless. The polysaccharide is then dissolved in 2 liters of water and dialyzed against flowing tap water for 16 hr. The polysaccharide is precipitated from solution by the addition of 2 volumes of ethanol and a sufficient volume of 1 M KCl to induce precipitation of the dyed polymer. This material is collected on a nylon screen and dried by solvent exchange with ethanol and acetone and storage *in vacuo;* the yield is 45 g; RBB : anhydrohexaose ~1 : 50.

For use as substrate, 2 g of RBB–CM-cellulose is sprinkled into 80 ml of vigorously stirring hot water. On dissolution, 5 ml of 2 M sodium acetate buffer (pH 4.5) is added, the pH is adjusted to 4.5, and the volume to 100 ml. In the presence of 0.02% sodium azide, the substrate is stable for more than 12 months at 4°.

Assay of Endo-1,4-β-D-glucanase

Enzyme preparation (0.1 ml) is incubated with 0.5 ml of RBB–CM-cellulose substrate solution in 0.1 M sodium acetate buffer (pH 4.5) for up to 10 min at 40°. The reaction is terminated by the addition of 2.5 ml of a precipitant solution that contains 80% ethylene glycol monomethyl ether, 0.3 M sodium acetate buffer (pH 5), and 0.4% zinc acetate. This mixture is vortexed for 10 sec, stood at room temperature for 10 min, and centrifuged at 1000 g for 10 min. The absorbance of the supernatant is measured at 590 nm.

Standardization of the Assay

The assay procedure is standardized using partially or highly purified endo-1,4-β-D-glucanase (cellulase) devoid of β-glucosidase and cellobiohydrolase. *Trichoderma viride* cellulases I and II,[18] *Penicillium emersonii* cellulases I and II,[18] and *Aspergillus niger* cellulase[19] were prepared as described previously. The activity in these preparations was quantitated by incubating enzyme preparation (0.05 ml) with 0.5 ml of 1% CMC-4M in 0.1 M sodium acetate buffer (pH 4.5) at 40° for 2–10 min. The reaction was terminated and the increase in glucose-reducing sugar equiv-

[18] B. V. McCleary and I. Shameer, *J. Inst. Brew.* **93**, 87 (1987).
[19] B. V. McCleary and M. Glennie-Holmes, *J. Inst. Brew.* **91**, 285 (1985).

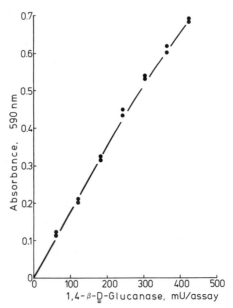

FIG. 2. Standard curve relating the activity of *Aspergillus niger* 1,4-β-glucanase on CMC-4M to absorbance increase (at 590 nm) on hydrolysis of RBB–CMC-4M. Incubations are performed at 40° for 10 min.

alents determined using the Nelson/Somogyi reducing sugar assay. A standard curve relating hydrolysis of RBB–CM-cellulose to units of enzyme activity on CM-cellulose for *A. niger* cellulase is shown in Fig. 2.

Application of the Assay Procedure

The assay procedure employing RBB–CM-cellulose is sensitive and highly specific for endo-1,4-β-D-glucanase (cellulase). Assays based on the use of this substrate are approximately 40 times more sensitive than those employing commercially available Cellulose Azure (Calbiochem Cat. No. 219481).[14] The substrate is resistant to hydrolysis by β-glucosidase; no increase in absorbance of the supernatant solution (after addition of precipitant and centrifugation) was observed on incubation of substrate (0.5 ml) with purified *Aspergillus niger*[20] β-glucosidase (0.1 ml, 5 U) at 40° for 2 hr. The assay can be used to measure enzyme activity in solutions containing high concentrations of reducing sugars and in culture solutions

[20] B. V. McCleary and J. Harrington, this volume [71].

or cell-free extracts. A very similar substrate[21] has been used to stain for enzyme activity in electrophoresis gels.

RBB–CM-Amylose for the Assay of α-Amylase

Preparation of RBB–CM-Amylose

One hundred grams of potato amylose (Sigma A9262) is added to a solution of 9.6 g of chloroacetic acid in 300 ml of ethanol (95% v/v) in a 1-liter Quickfit flask and brought to rapid reflux on a hotplate magnetic stirrer. A solution of 8 M sodium hydroxide (40 ml) in ethanol (300 ml) is added dropwise over 15 min and the slurry maintained under rapid reflux conditions for a further 15 min. The slurry is then poured onto a scintered glass funnel on a Büchner flask and washed under vacuum with 1 liter of 80% (v/v) aqueous ethanol. The product is then dissolved in 1 liter of hot water (60°) by stirring with a spatula. The solution is very viscous but viscosity decreases significantly on addition and dissolution of 100 g of anhydrous sodium sulfate. Lumps are removed by homogenizing the solution in a Waring blender. The solution is then stirred and heated at 60–65° on a hotplate magnetic stirrer and 20 g of Remazol Brilliant Blue R dye and 10 g of trisodium phosphate added, and stirring and heating continued for 2 hr. While still hot, the solution is treated with an equal volume of 95% (v/v) ethanol and the rubbery precipitate which forms on stirring is collected on a fine nylon screen. Excess liquid is removed from the precipitate by squeezing. The precipitate is redissolved in 2 liters of hot water by homogenizing in a Waring blender and precipitated again by adding 1 volume of ethanol. This process is repeated until all free dye is removed. The precipitate is then dissolved in 2 liters of water and dialyzed against flowing tap water overnight. The dialyzed solution is treated with 2 volumes of ethanol and sufficient 1 M KCl to cause precipitation. The precipitate is washed with ethanol (twice) and acetone and dried *in vacuo;* the yield is 85 g; RBB : anhydrohexaose ~1 : 50; CM degree of substitution ~0.12.

For use as substrate, 2 g of RBB–CM-amylose is suspended in 80 ml of hot water (80°) and stirred for 15 min. The suspension is cooled to room temperature and treated with 5 ml of 2 M sodium acetate buffer (pH 5.0) or 10 ml of 1 M sodium phosphate buffer (pH 6.9), the pH is adjusted accordingly, and the volume adjusted to 100 ml. In the presence of 0.02% sodium azide or if overlain with a few drops of toluene, the substrate is

[21] P. Biely, O. Marković, and D. Mislovičová, *Anal. Biochem.* **144,** 147 (1985).

stable for at least 12 months at 4°. Although this substrate is not completely soluble in buffer solutions, it remains completely suspended for up to 1 hr and can be dispensed accurately and reproducibly.

Assay of α-Amylase

Serum preparation is assayed with the substrate buffered at pH 6.9 whereas cereal and microbial enzymes are assayed employing the substrate buffered at pH 5.0. The assay procedure is exactly as described for endo-1,4-β-glucanase.

Standardization of the Assay

The assays are standardized using porcine pancreatic α-amylase (Sigma A2643)[14] or cereal α-amylases purified by substrate affinity chromatography. The activity of these preparations on soluble starch is determined using the Nelson/Somogyi reducing sugar assay.[4] Enzyme preparation (0.05 ml) is incubated with 0.5 ml of 1% (w/v) soluble starch in 0.1 M sodium acetate buffer (pH 5.0) or sodium phosphate buffer (pH 6.9) at 40° for 2–10 min. The reaction is terminated and the increase in glucose-reducing sugar equivalent determined.

Application of the Assay Procedure

The described assay procedure employing RBB–CM-amylose is specific for α-amylase; neither α-glucosidase, glucoamylase, nor β-amylase gives any release of dye-labeled fragments soluble after addition of the precipitant solution. α-Amylase in crude culture filtrates and in extracts containing high concentrations of reducing sugars can be assayed directly. A typical standard curve relating the action of porcine pancreas α-amylase on RBB–CM-amylose to units of enzyme activity on soluble starch is shown in Fig. 3.

RBB–CM-β-Glucan (Barley) for the Assay of Endo-1,3(4)-β-D-glucanase

Malt β-glucanase (endo-1,3(4)-β-D-glucanase; EC 3.2.1.73, lichenase) activity is an indicator of malt quality and has a functional role in the depolymerization of barley β-glucan [(1→3)(1→4)-β-D-glucan] during seed germination[22] and during the mashing process of beer production.[23] Measurement of β-glucanase in malt extracts by conventional reducing sugar assays is not possible due to the high concentration of reducing

[22] A. M. MacLeod, J. H. Duffus, and C. S. Johnston, *J. Inst. Brew.* **70,** 521 (1964).
[23] C. W. Bamforth, *Brew. Dig.* **57,** 22 (1982).

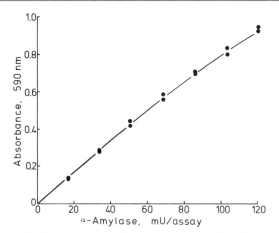

FIG. 3. Relationship between release of chromogenic material of low degree of polymerization from RBB–CM-amylose and α-amylase activity (mU) on soluble starch. Incubations are performed at 40° for 10 min.

sugars present. The assay procedure most commonly employed is based on measurement of the decrease in viscosity of a solution of barley β-glucan.[2] However, this assay is laborious and interlaboratory reproducibility is poor.[24] It was considered that an alternative assay procedure could be developed based on the use of dye-labeled barley β-glucan as substrate.[18] In preliminary experiments it was observed that the dyeing of barley β-glucan with several dyes, including Remazol Brilliant Blue R, produced an insoluble rubbery product which could be dissolved at high temperatures but which precipitated from solution at temperatures below 50°. In the preparation of the presently described substrate this problem is overcome by lightly carboxymethylating the glucan before dyeing, to produce a substrate soluble even at 4°. To achieve the substrate concentrations required to give the desired reaction kinetics it is essential to partially depolymerize the glucan. In the current method, this is achieved by controlled hydrolysis by lichenase.

Preparation of Low-Viscosity Barley β-Glucan

One kilogram of barley (var. Bandulla) flour which had been milled to pass a 1-mm screen is suspended in 2 liters of 90% (v/v) aqueous ethanol and heated in a steam bath for 10 min after the alcohol begins to boil. The slurry is poured onto a fine nylon screen and excess liquid removed by squeezing. The solid is further washed with another 2 liters of 90% (v/v)

[24] G. K. Buckee, *J. Inst. Brew.* **91,** 264 (1985).

aqueous ethanol and then suspended in 4 liters of demineralized water and stirred to give a homogeneous slurry. Forty milliliters of heat-treated (85° for 90 min) Hitempase α-amylase preparation [Biocon (Australia) Pty. Ltd., Boronia, Victoria] is then added and the slurry incubated in a steam bath for 90 min, during which time the temperature of the slurry increased to 85–90° and the starch is completely depolymerized to small oligosaccharide fragments. (After 30 min the slurry is homogenized in a Waring blender to remove lumps.) The temperature is adjusted to 60° and on addition of 3 units of lichenase and thorough mixing, the slurry is incubated at 60° for 40 min and then at 90° for 10 min to inactivate the lichenase. The temperature of the slurry is lowered to about 60° by the addition of 1 liter of water, the solution is homogenized in a Waring blender, and the slurry filtered through a fine nylon screen and squeezed to recover as much liquid as possible. The solid residue is discarded. The recovered solution is centrifuged at 3000 g for 20 min and the relatively clear, yellow-colored supernatant is treated with 200 g of ammonium sulfate per liter of solution and stored at 4° for 16 hr. The precipitate which forms is recovered by centrifugation at 3000 g for 15 min, suspended in 2 liters of 20% (v/v) aqueous ethanol, and stirred for 15 min. Insoluble material is recovered by centrifugation and washed again with 20% (v/v) aqueous ethanol to remove most of the remaining proteinaceous material. The insoluble material, which is mainly barley β-glucan, is washed with ethanol and then added to 4 liters of vigorously stirred water at 80–90°. After 30 min the solution is homogenized in a Waring blender and then centrifuged at 3000 g for 20 min. The clear, colorless supernatant is added to an equal volume of ethanol (95%) and the white precipitate which forms is collected by centrifugation at 3000 g for 10 min, washed with ethanol (twice) and acetone, and dried *in vacuo;* the yield is 50 g. The recovered β-glucan is completely devoid of starch and maltosaccharides, has a protein content of ~0.6%, and an intrinsic viscosity of ~2 dl/g in 0.5 M KCl at 25°. The glucan is completely hydrolyzed to low degree of polymerization oligosaccharides by *Aspergillus niger* cellulase or *Bacillus subtilis* lichenase.[19] Acid hydrolysis of the polysaccharide followed by gas–liquid chromatography of the alditol acetates showed that glucose was the major sugar present (>96%).

Preparation of RBB–CM-β-Glucan (Barley)

Thirty grams of low-viscosity barley β-glucan is added to a 1-liter Quickfit flask containing 3.0 g of chloroacetic acid dissolved in 420 ml of 95% ethanol. The solution is brought to rapid reflux on a hotplate magnetic stirrer and a solution of 4 M sodium hydroxide (25 ml) in ethanol (60 ml) is added dropwise over 15 min. Stirring under rapid reflux conditions is continued for a further 15 min and then the slurry is poured onto a sintered glass funnel and washed exhaustively with 2 liters of 80% (v/v) aqueous ethanol. The solid product is suspended in 1.5 liters of vigorously stirring demineralized water at 70°, stirred for 15 min, and then homogenized in a Waring blender to give complete dissolution. This solution is then treated with 150 g of anhydrous sodium sulfate, 6 g of Remazol Brilliant Blue R dye, and 7.5 g of trisodium phosphate and mixed on a hotplate magnetic stirrer at 70–75° for 2 hr. While still hot, the reaction mixture is centrifuged at 3000 g for 5 min and the supernatant discarded. The pellet is dissolved in 1.5 liters of water at 80° by mixing on a hotplate magnetic stirrer. The dyed β-glucan is precipitated from solution by addition of 1.5 volumes of ethanol with vigorous stirring followed by 2 M NaCl solution until a precipitate formed. After standing for 10 min the dyed glucan is recovered by filtering the solution through a fine nylon screen. Excess liquid is removed by squeezing the rubbery product. This process of dissolution and precipitation is repeated until the washings are essentially colorless. The dyed glucan is then dissolved in 1.5 liters of water and dialyzed against flowing tap water for 16 hr. The concentration of RBB–CM-β-glucan (barley) is determined by lyophilizing about 10 ml of the solution and the concentration of the remaining solution is adjusted to 1% w/v and sodium azide added to a final concentration of 0.02% w/v. Under these conditions the substrate is stable at 4° for at least 12 months. Recovery of RBB–CM-β-glucan (barley) was 80–85%.

Assay of Malt β-Glucanase

Enzyme preparation (0.5 ml) in 40 mM sodium acetate/sodium phosphate buffer (pH 4.6) is incubated with 0.5 ml of 1% (w/v) RBB–CM-β-glucan (barley) substrate solution at 30° for 10 min. The reaction is terminated by adding 3.0 ml of a precipitant solution that contains 80% ethylene glycol monomethyl ether, 0.3 M sodium acetate buffer (pH 5), and 0.5% zinc acetate. This mixture is vortexed for 5 sec, stored at room temperature for 5 min, centrifuged at 1000 g for 10 min, and the absorbance of the supernatant measured at 590 nm.

The activity of endo-1,4-β-D-glucanases (cellulases) on this substrate is measured using the same procedure as for malt β-glucanase. A similar procedure is employed to assay Bacillus subtilis endo-1,3(4)-β-D-glucanase (lichenase) except that the sodium acetate/sodium phosphate buffer employed is adjusted to pH 6.5.

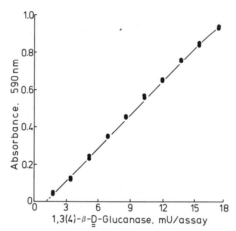

FIG. 4. Standard curve relating the activity of 1,3(4)-β-glucanase I from malt on barley β-glucan to absorbance increase (at 590 nm) on hydrolysis of RBB–CM-β-glucan (barley). Incubations are performed at 30° for 10 min.

Standardization of the Assay Procedures

Activity of the various enzymes on RBB–CM-β-glucan (barley) is standardized against the action of these enzymes on barley β-glucan (5 mg/ml) measured using the Nelson/Somogyi[4] reducing sugar assay. The same pH and temperatures of incubation are employed in both assays. For malt β-glucanase, the assay is standardized (Fig. 4) using malt β-glucanase I recovered from chromatography of malt extract on CM-Sepharose CL-6B at pH 4.5.[25] This enzyme is devoid of β-glucosidase and cellulase activities. Lichenase was purified by affinity chromatography on DEAE-cellulose[19] and cellulases were purified as previously described.[18,19]

Applications of the Assay Procedure

The currently described assay procedure allows the accurate and reliable assay of malt β-glucanase in crude malt extracts and of lichenase and cellulases in crude microbial culture filtrates and commercial enzyme preparations.

[25] J. R. Woodward and G. B. Fincher, *Eur. J. Biochem.* **121**, 663 (1982).

[9] Methods for Measuring Cellulase Activities

By THOMAS M. WOOD and K. MAHALINGESHWARA BHAT

The physical heterogeneity of the substrate and the complexity of the cellulase enzyme system present those wishing to measure cellulase activities with formidable problems. Faced with the use of substrates which are ill defined, and an enzyme system which consists often of a multitude of enzymes acting in synergism in a manner not yet fully understood, the enzymologist has developed a bewildering number of assays in an attempt to throw some light on the complex enzymatic interactions involved in the breakdown of cellulose. This state of affairs, compounded by the existence of a plethora of arbitrary units of activity, has made comparison of quantitative data obtained in various laboratories impossible. This in turn has hindered our understanding of the mechanism of cellulase action and has delayed the commercial development of these enzymes for the generation of glucose feedstock from lignocellulose wastes.

In an attempt to rectify the situation the Commission on Biotechnology (IUPAC) in December 1984 published a number of standard procedures for the measurement of cellulase activities.[1] Some of these recommended procedures have been readily accepted by the biotechnologists to whom they were directed, and it seems likely that valuable comparative data will now accrue. However, to the enzymologist wishing to develop a better understanding of the mechanism of cellulase action the recommendations are too restrictive. It is only by evaluating the various enzymes of the cellulase system in as many ways as possible, using a wide variety of cellulose substrates of differing degrees of polymerization, of crystallization, and of polydispersity, will any further advances be made in our understanding of the complex enzymatic interactions involved in the breakdown of cellulose by microbial cellulases.

This chapter therefore includes, in addition to the IUPAC recommended procedures, a number of alternative assay methods (Table I) which may help toward a better evaluation of cellulase components and their complex interactions that result in the solubilization of cellulose. Some of these alternative methods are included for practical considerations; many, however, are included because of the nature of the unresolved problems.

[1] Commission on Biotechnology, International Union of Pure and Applied Chemistry, *in* "Measurement of Cellulase Activities" (T. K. Ghose, ed.). Biochemical Engineering Research Centre, Indian Institute of Technology, New Delhi, India, 1984.

TABLE I
METHODS OF MEASURING CELLULASE ACTIVITIES

Enzyme	Substrate	Assay
Complete cellulase	Cotton	Solubilization
		Estimation of cellulose in residue[a]
		Reducing sugars released
		Weight loss[a]
		Loss in tensile strength
	Filter paper	Solubilization: release of reducing
	Hydrocellulose	sugars[a]
	Avicel	
	Solka Floc	
	Dyed Avicel	Release of dyed soluble fragments[a]
Cellobiohydrolase	Avicel	Solubilization: release of reducing
(exocellobiohydrolase,	Hydrocellulose	sugars[a]
exocellulase, Avicel-	Dyed Avicel	Release of dyed cellobiose
ase)	Amorphous cellulose	Solubilization: release of reducing
		sugars or decrease in turbidity[a]
	Cellooligosaccharides	Increase in reducing power or
		analysis by HPLC
Endo-1,4-β-glucanase	Carboxymethylcellulose	Release of reducing sugars[a]
(CM-cellulase,	Hydroxyethylcellulose	Decrease in viscosity
endoglucanase,	Cellooligosaccharides	Increase in reducing power or
endocellulase)		analysis by HPLC
	Cotton	Swelling in alkali[a]
	Amorphous cellulose	Solubilization: release of reducing
		sugars[a]
		Decrease in turbidity[a]
β-Glucosidase	o- or p-Nitrophenyl-	Release of o- or p-nitrophenol[a]
	β-D-glucosides	
	Salicin ⎱ glycosides	Release of glucose
	Esculin ⎰	
	Cellobiose	Release of glucose[a]
	Cellooligosaccharides	Increase in reducing power

[a] Assays described in the text.

Studies on the extracellular enzymes from a number of cellulolytic fungi, particularly *Trichoderma* species such as *Trichoderma viride*, *Trichoderma reesei*, and *Trichoderma koningii*, have provided us with more of our basic information on the mechanism of cellulase action.[2] However, other mechanisms may exist in the bacteria, which have been less well studied. In this chapter the conditions of pH and temperature

[2] K. E. Eriksson and T. M. Wood, *in* "Biosynthesis and Biodegradation of Wood Components" (T. Higuchi, ed.), p. 469. Academic Press, New York, 1985.

given for the various assays are those used for *Trichoderma* cellulases. Thus, the assay procedures will normally have to be modified for cellulase enzymes from other sources after consideration of parameters such as pH and temperature optimum, and linearity of activity with respect to time, substrate, and enzyme concentration.

The enzymes listed in Table I are those currently understood to exist in cellulases from fungal sources. There are, however, considerable differences in opinion regarding the specificities of the various enzymes and the roles they play in the dissolution of cellulose in its various forms. Despite these differences some generalization regarding the modes of action and properties of individual components comprising the cellulase system can be made, and it seems pertinent to make some reference to these at this stage to enable the reader to make critical evaluation of the various assay procedures and facilitate interpretation of the results.

There are three main types of enzyme found in cellulase systems that can degrade crystalline cellulose: exocellobiohydrolase (EC 3.2.1.91, 1,4-β-D-glucan cellobiohydrolase), endo-1,4-β-D-glucanase (EC 3.2.1.4, 1,4-β-D-glucan glucanohydrolase and β-glucosidase (EC 3.2.1.21, β-D-glucoside, glucohydrolase), or cellobiase.[3] Exocellobiohydrolases are found as major components in some cellulase systems, but are absent from most. All enzymes appear to exist in multiple forms which differ in their relative activities on a variety of substrates.

Endo-1,4-β-glucanases. Endo-1,4-β-glucanases hydrolyze cellulose chains at random to produce a rapid change in degree of polymerization. Substrates include carboxymethylcellulose, $H_3PO_4^-$ or alkali-swollen (amorphous) cellulose, but crystalline cellulose such as cotton fiber or Avicel is not attacked to a significant extent. Hydrolysis of amorphous cellulose yields a mixture of glucose, cellobiose, and other soluble cellooligosaccharides. The rate of hydrolysis of the longer chain cellooligosaccharides is high, and the rate increases with degree of polymerization: glucose and cellobiose are the principal products of the reaction. Transglycosylation is a feature of some, but not all endoglucanases. Some endo-1,4-β-glucanases act in synergism with the cellobiohydrolase isolated from fungal cellulases to solubilize crystalline cellulose: some however, do not.

Exo-1,4-β-glucanases. Exo-1,4-β-glucanases of the fungi act by removing glucose or cellobiose from the nonreducing end of the chain. Cellobiohydrolase is the most common enzyme. Most cellobiohydrolases appear to release small amounts of glucose from cellulose. Cotton fiber is not attacked to a significant extent, but H_3PO_4-swollen cellulose is hydrolyzed with a characteristic slow fall in degree of polymerization. Carboxy-

[3] T. M. Wood, *Biochem. Soc. Trans.* **13**, 407 (1984).

methylcellulose and cellobiose are not substrates but cellobiose and longer chain cellooligosaccharides are hydrolyzed with the rate increasing with increasing degree of polymerization. Avicel is a substrate that has proved to be very useful for isolating and measuring cellobiohydrolase.

β-Glucosidases. These are not strictly speaking cellulases but they are, nevertheless, very important components of the cellulase system in that they complete the hydrolysis to glucose of short-chain cellooligosaccharides and cellobiose which are released by the other enzymes. Cellulase systems containing low levels of β-glucosidase have poor saccharifying power because of the inhibition of endoglucanase and cellobiohydrolase by cellobiose.

β-Glucosidases hydrolyze cellooligosaccharides at a rate that decreases with increasing degree of polymerization, but cellulose is not attacked. Some β-glucosidases can hydrolyze aryl-β-glucosides but not cellobiose. Other characteristics are that they are not specific for the 1,4-β-linkage, and they possess transferase activity that acts on glucose units to form other sugar molecules such as dimers, trimers, and higher oligosaccharides.

Sugar Analyses

Many of the methods used for assaying cellulase activity involve the measurement of either reducing sugar or total sugar released into solution. There are a number of published assay methods for each type of analysis, but only a few are commonly used in cellulase studies. Of note in this respect are the dinitrosalicylic acid (DNS) method[4] and the Somogyi–Nelson method[5] for the estimation of reducing sugar, and the phenol–H_2SO_4[6] and the anthrone–H_2SO_4[7] methods for estimating total sugar. The authors are prejudiced in favor of the phenol–H_2SO_4 method, and only that is detailed here. A method for estimating reducing sugar involving the use of ferricyanide–ferric alum is extremely sensitive (1–14 µg range); however, this sensitivity makes it difficult to use.

Somogyi–Nelson Method for Determination of Reducing Sugar[5]

Reagents

Na_2SO_4 (anhydrous), 360 g
Rochelle salt, 24 g

[4] G. L. Miller, R. Blum, W. E. Glennon, and A. L. Burton, *Anal. Biochem.* **1,** 127 (1960).
[5] N. Nelson, *J. Biol. Chem.* **153,** 376 (1944).
[6] M. Dubois, K. Gilles, J. K. Hamilton, P. A. Rebers, and F. Smith, *Anal. Biochem.* **28,** 350 (1956).
[7] R. G. Spiro, this series, Vol. 8, p. 4.

Na$_2$CO$_3$, 48 g
NaHCO$_3$, 32 g
CuSO$_4 \cdot$ 5H$_2$O, 8 g
Ammonium molybdate, 100 g
Concentrated H$_2$SO$_4$, 84 ml
Na$_2$H arsenate, 12 g
Glucose standard solution (20 mg/100 ml)

Preparation of Somogyi Reagent I. Anhydrous sodium sulfate (288 g) is dissolved in 1 liter of boiled distilled water followed by 24 g Rochelle salt, 48 g sodium carbonate, and 32 g sodium bicarbonate. The solution is diluted to 1600 ml with boiled distilled water and stored at 27°.

Preparation of Somogyi Reagent II. Sodium sulfate (72 g) is dissolved in 300 ml boiled distilled water followed by 8 g copper sulfate. The solution is diluted to 400 ml with boiled distilled water and stored at 27°.

Preparation of Nelson Reagent. Ammonium molybdate (100 g) is dissolved in 1.8 liters of distilled water. To the solution 84 ml conc. H$_2$SO$_4$ is added followed by a solution of 12 g of sodium arsenate dissolved in 100 ml distilled water. The solution is stored at 37° for 24–48 hr in a brown glass bottle and then at room temperature.

Assay Procedure. A 2 ml sample of a solution consisting of 4 volumes of Somogyi reagent I and 1 volume of Somogyi reagent II (mixed immediately before use) is pipetted into a 15-ml tube along with a solution containing reducing sugar (5–200 μg) and water to give a total volume of 4 ml. The solution is boiled in a water bath for 15 min, cooled, 2 ml of Nelson reagent added, and the solution mixed carefully on a Vortex mixer. Finally, 4 ml of distilled water is added and the solution mixed by inversion. The absorbance is read at 520 nm and then translated (after subtraction of absorbance of reagent blank) into glucose (equivalent) using a standard graph obtained by plotting micrograms glucose added against absorbance.

Comments. (1) The method is sensitive over the range 5–100 μg glucose. (2) Cellobiose gives about 90% of the reducing volume of an equimolar amount of glucose.

Dinitrosalicylic Acid Method for Estimating Reducing Sugar[4]

Reagents

Dinitrosalicyclic acid, 40 g
Phenol, 8 g
Sodium sulfite, 2 g
Rochelle salt, 800 g
All are dissolved in 2 liters of 2% (w/v) NaOH solution and then diluted to 4 liters with distilled water
Glucose standard (2.0 mg/ml)

Procedure. Diluted sugar solution (1.5 ml) to be assayed is mixed with 3 ml DNS reagent, placed in a boiling water bath for 5 min, cooled to room temperature, and the absorbance read at 540 nm. The absorbance values (after subtraction of the reagent blank) are then translated into glucose equivalent using a standard graph obtained by plotting glucose (0.1–3.0 mg) against absorbance.

Comments. (1) The method is not sensitive below 100 μg glucose. For accurate determination of low concentrations of reducing sugars 100 μg glucose should be added to each sample. (2) The color develops only under alkaline conditions: acidic samples should therefore be neutralized. (3) If glucose is used as standard, values for equimolar amounts of cellobiose will be low.

Phenol–Sulfuric Method for Estimating Total Sugar[6]

Reagents

5% (w/v) aqueous solution of phenol (stored at 4°)
98% H_2SO_4

Procedure. Suitably diluted sugar solution (1.0 ml) is pipetted directly into the bottom of a test tube (1.5 × 15 cm). Phenol reagent (1.0 ml) is added in the same way followed by 5 ml of concentrated H_2SO_4 which is pipetted directly onto the sugar solution from a fast-flow pipet or suitable dispensor. The solution is mixed immediately and allowed to cool before the absorbance is read at 490 nm. The absorbance values (after subtraction of reagent blanks) are then translated into glucose equivalent using a standard graph obtained by plotting glucose (5–150 μg) against absorbance.

Measurement of Total Cellulase Activity

The activity of the complete cellulase complex (cellobiohydrolase, endo-1,4-β-glucanase, and β-glucosidase or cellobiase) can be measured using crystalline celluloses such as cotton fiber, filter paper, or Avicel. Cotton fiber or filter paper prepared from cotton fiber are considered to be the best substrates by many; others favor Avicel. All of these substrates contain a high degree of hydrogen bond order, but in the case of native cotton this order has not been disrupted as a result of chemical or mechanical treatment.

Avicel is essentially an aggregate of microcrystals. It is prepared (see chapter [3]) from wood α-cellulose by an acid treatment to remove the amorphous component and a blending treatment which fragments the crystallites into colloidal particles and causes them to coalesce. Despite the fact that Avicel is obviously an artifact it has proved to be extremely useful for assaying cellulase activity.

Measurement of Total Cellulase Using Cotton Fiber

The residual cellulose left after enzymatic hydrolysis is estimated using $K_2Cr_2O_7$–H_2SO_4 oxidizing agent. Alternatively, the reducing or total sugars released into solution can be measured using a variety of colorimetric methods.

Assay Involving the Estimation of Residual Cellulose with K_2CrO_7–SO_4 Reagent[8]

Reagents

$K_2Cr_2O_7$–H_2SO_4 solution [0.5% (w/v)] is prepared by dissolving 5 g of $K_2Cr_2O_7$ in 20 ml distilled water with gentle heating. The solution is cooled and diluted to 1 liter with concentrated (98%) H_2SO_4. The reagent is kept in a glass stoppered bottle at room temperature. Kept under these conditions the reagent is stable for several months

0.2 M sodium acetate buffer, pH 4.8

0.05 M NaN_3 solution

Standard glucose solution [4.0% (w/v) solution]

Procedure. The incubation mixture consists of 2 mg of dewaxed cotton fiber (see chapter [3]), 2.5 ml of 0.2 M acetate buffer, pH 4.8, 0.1 ml of 0.05 M NaN_3, and distilled water and enzyme solution to give a total volume of 5 ml. Incubation is allowed to proceed for 7 days at 37° in a stoppered (silicon rubber) conical centrifuge tube (15 ml capacity) graduated at 1.8 ml. The residual cellulose is sedimented by centrifugation and the supernatant is removed through a glass tube connected to a water pump until a volume of 1.8 ml is left. Exactly 10 ml of distilled water is added and the tube is centrifuged again. The washing procedure is repeated four times to remove all soluble sugars. The residual cellulose is estimated colorimetrically as follows: 4 ml of 0.5% dichromate–H_2SO_4 reagent is added, the tubes are covered with marbles, and then heated on a boiling water bath for 30 min. The tubes are cooled immediately, diluted to 25 ml with distilled water, and the absorbance read at 600 nm. The absorbance is translated into glucose (after subtraction of reagent and enzyme blanks) using a standard graph prepared with 0.1–2.5 mg of glucose. Finally, the glucose figure is converted to anhydroglucose (multiply by 0.9) and the percentage hydrolysis calculated.

[8] T. M. Wood, *Biochem. J.* **115**, 457 (1969).

Comments. The advantages of the method are as follows. (1) As the sample of cotton used in the 5 ml reaction mixture is very small, end-product inhibition by the soluble sugars generated is minimal even when extensive hydrolysis has been effected. (2) Under the assay conditions used none of the individual components of the cellulase system shows significant activity when acting independently: synergistic activity between the components is therefore easily detected and evaluated.

The disadvantages of the method are as follows. (1) The plot of percentage solubilization against enzyme concentration is not linear. (2) Larger samples of cotton fiber cannot be used as the $K_2Cr_2O_7-H_2SO_4$ reagent is suitable only in the range 0.1–2.5 mg glucose.

Cotton fiber can be replaced by Avicel which is easily pipetted (2.5 ml) from a stirred suspension (0.08% w/v) in sodium acetate buffer, pH 4.8.

Assay Involving the Estimation of Reducing Sugar. The reaction mixture consists of 10 mg of dewaxed cotton fiber or Avicel [pipetted from a stirred aqueous suspension (2 ml of 0.5% w/v)], 2.5 ml of sodium acetate buffer, pH 4.8, 0.1 ml of 0.05 M NaN_3 solution, and water and enzyme to give a total volume of 5 ml. The incubation is allowed to proceed at 37° for 18 hr. The mixture is centrifuged and the reducing sugar or total sugar in the supernatant is estimated.

Comment. There are many similar assays in the literature. In many, if not most cases, incubation is at 50° and for 1–2 hr only. In these cases, initial rates of reaction are often being measured and these may not be particularly meaningful in terms of attack on the crystalline cellulose component.

Measurement of Total Cellulase Activity Using Filter Paper[1,9]

This assay is recommended by the Commission on Biotechnology (IUPAC) for the measurement of activity of total cellulase or true cellulase activity. The assay is based on estimating a fixed amount (2 mg) of glucose from a 50 mg sample of filter paper.

Reagents

0.05 M citrate buffer, pH 4.8 [dissolve 210 g citric acid monohydrate in 750 ml distilled water and add NaOH (50–60 g) until the pH is 4.3. Dilute to 1 liter and check pH. If necessary further NaOH can be added until pH is 4.5. This solution is 1 M citrate buffer, pH 4.5: when diluted to 0.05 M, the pH should be 4.8]

[9] M. Mandels, R. Andreotti, and C. Roche, *Biotechnol. Bioeng. Symp.* **6,** 17 (1976).

Dinitrosalicylic acid reagent (DNS) (see under Sugar Analyses for
details of preparation)
Glucose stock solution (10 mg/ml)
Strips of Whatman No. 1 filter paper (1 × 6 cm)

Procedure. One milliliter of citrate buffer 0.05 *M*, pH 4.8 is added to a
test tube of approximately 25-ml capacity followed by 0.5 ml of enzyme
suitably diluted in the same buffer and a 1 × 6-cm strip of Whatman No. 1
filter paper which has been curled round a glass rod. At least two dilutions
of enzyme must be used. One dilution should be capable of releasing
slightly more than 2 mg of glucose (equivalent) using the DNS reagent; the
other should release slightly less. The reaction mixture is incubated at 50°
for 60 min and 3 ml of DNS reagent added to stop the reaction. The
suspension is mixed well and the tubes are transferred to a boiling water
bath for exactly 5 min. Cooling is effected in cold water. After 20 ml of
distilled water is added the contents of the tubes are mixed by inversion of
the tubes several times. Finally, the tubes are allowed to stand for at least
20 min to allow the pulp to settle, and the color formed is read in a
spectrophotometer at 540 nm. Enzyme blanks, reagent blanks, and glucose standard solutions must be treated in exactly the same way.

Calculation of Unit of Activity. The linear glucose standard is obtained
by plotting glucose (1–4 mg) against the absorption at 540 nm. Using this
standard, the absorbance values of the tubes containing enzyme, after
subtraction of enzyme and reagent blanks, are translated into milligrams
of glucose produced. The dilutions are then converted into enzyme concentrations

$$\text{Concentration} = \frac{1}{\text{dilution}}, \text{ i.e., } \frac{\text{volume of enzyme in dilution}}{\text{total volume of dilution}}$$

and the concentration of enzyme that would have released exactly 2 mg of
glucose is calculated by plotting the glucose liberated against enzyme
concentration. Filter paper units of activity (FPU/ml) are equal to

$$\text{FPU} = \frac{0.37}{\text{enzyme concentration to release 2.0 mg glucose}}$$

It should be noted that these units are not international units which are
based on initial rates of reaction. The units of activity should therefore be
expressed simply as FPU, i.e., filter paper units.

The FPU is derived as follows:

$$1 \text{ IU} = 1 \text{ } \mu\text{mol/min of glucose equivalent released}$$
$$= 0.18 \text{ mg/min of glucose}$$

The FPU is based on the release of exactly 2.0 mg of glucose equivalent, i.e., 2/0.18 μmol from 0.5 ml of diluted enzyme in 60 min

$$2 \text{ mg glucose} = \frac{2}{0.18} \times 0.5 \times 60 \ \mu\text{mol/min/ml}$$
$$= 0.37 \ \mu\text{mol/min/ml}$$

The amount of enzyme that releases 2.0 mg glucose contains 0.37 units and therefore

$$\text{FPU (units/ml)} = \frac{0.37}{\text{enzyme concentration required to release 2.0 mg glucose}}$$

Comments. 1. The conditions and criteria of the filter paper assay must be followed rigidly. It is not acceptable to use values of glucose released which are much lower than 2.0 mg (or much higher) and extrapolate to the critical release of glucose. Some consider that these stipulations make the assay difficult to use, but the assay has received wide acceptance and has made possible meaningful comparisons of data obtained from a number of laboratories.

2. Filter paper hydrolyzing activity can be seriously affected by the amount of β-glucosidase/cellobiase enzyme in cellulase preparations. The presence of relatively large amounts of cellobiose in the hydrolysate is known to greatly affect the color yield obtained with the DNS reagent. Indeed, on a molar basis the color yield from cellobiose is 1.5 times that of glucose. However, this problem is common to all methods for determining reducing sugar and therefore should not be taken to suggest that the DNS reagent is particularly suspect in this respect. The research worker should, however, be aware of this problem and be cautious of the interpretation of the results obtained.

3. The critical release of 2 mg glucose in the assay (i.e., 4% hydrolysis of the substrate) ensures that some of the crystalline component of the cellulose is hydrolyzed. There are many cellulases known that can hydrolyze the amorphous easily hydrated cellulose component but are unable to hydrolyze the crystalline component. These cellulases are characterized by a fast initial rate of reaction on filter paper or other highly ordered cellulose and then almost complete cessation of the release of soluble sugars. Extrapolation of such data, which are in essence based on initial rates of reaction, clearly would be very misleading.

4. Several modifications of the method of measuring filter paper activity are used. Thus, Eveleigh and Montenecourt[10] have replaced the 1 × 6

[10] B. S. Montenecourt, D. E. Eveleigh, G. K. Elmund, and J. Parcells, *Biotechnol. Bioeng.* **20**, 297 (1978).

cm (50 mg) Whatman No. 1 filter paper strips used as substrate by 0.5 in. filter paper antibiotic assay disks 740E (Scheicher & Schuell, Keene, NH). Calculation of activity of filter paper units or disk units was carried out as detailed by Mandels *et al.*[9] One disk unit apparently is equal to approximately 2.8 FPU. The modification was introduced to reduce some of the tedium of cutting out innumerable filter paper strips and to reduce the number of small fibers that appear in suspension in the Mandels assay for filter paper hydrolyzing activity.

    5. Toyama *et al.*[11] have developed a method of measuring total cellulase using filter paper activity. The assay is based on the time in minutes (t_m) needed to disintegrate 1×1-cm filter paper square (Toya filter paper No. 2) in a 1% (w/v) enzyme solution in 0.1 M acetate buffer pH 5.0 placed in an L-type glass tube (inner diameter 15 mm, length 155 mm, height of neck 5.5 mm) and shaken at 54 rpm at 40°. The time in minutes needed to disintegrate the filter paper completely into short fine fibers is noted, and the units of activity calculated as follows:

$$\text{Units per mg} = 30,000/t_m$$

It is suggested that approximately 20,000 of these units of activity are equivalent to 1 FP unit.

    This method suffers from the disadvantage that (1) the tubes must be continuously observed, and (2) some of the individual components of the cellulase complex can disintegrate filter paper into fine fibers although not producing significant release of reducing sugars. Cellobiohydrolases I and II of *T. reesei* have been shown to have this property, but cellobiohydrolase I and II of *Penicillium pinophilum* do not.

Measurement of Total Cellulase Activity Using Dyed Avicel[12]

    Reagents

        Suspension of dyed Avicel (see chapter [3] for preparation) containing 100 mg/3 ml
        0.05 M citrate buffer, pH 4.8

    Procedure. A 2 ml sample of diluted enzyme solution is pipetted into a test tube along with 3 ml of dyed Avicel suspension from a vigorously stirred suspension. The reaction mixture is incubated for 1 hr (with or without agitation) at 50°, after which the tube is placed in a boiling water

[11] N. Toyama and K. Ogawa, *in* "Bioconversion of Cellulosic Substances into Energy, Chemicals and Microbial Protein" (T. K. Ghose, ed.), p. 305. IIT, Delhi, India, 1977.
[12] M. Leisola and M. Linko, *Anal. Biochem.* **70,** 592 (1976).

bath for 5 min to stop the reaction. The hot suspension is filtered immediately through Whatman No. 1 filter paper. We have found that centrifugation after heating at 100° is also satisfactory.

After cooling the supernatant the absorbance of the clear solution is measured at 595 nm.

Unit of Activity. The activity is expressed in arbitrary units of absorbance (e.g., absorbance of 0.1–1 unit of activity).

Comments. (1) The method appears to be somewhat variable in reproducibility, the absorbance values obtained being dependent on filtration temperature, type of filter used, and, of course, the extent to which the various glucose residues of the Avicel have been dyed. Reproducibility using different batches of dyed Avicel is therefore poor. (2) Assays involving the use of dyed Avicel are extremely useful when testing the properties of the cellulases in the presence of ions or reducing agents that would interfere with the reducing sugar or total sugar methods of analyses. (3) A major advantage of the assay method is that unlike those assays involving measurements of reducing sugar, the ratio of cellooligosaccharides: cellobiose : glucose in the supernatant does not affect the result. For this reason the assay is useful for measuring the effect of end-product inhibitors on the kinetics of hydrolysis of an insoluble substrate by cellulases.

Measurement of Exocellobiohydrolase Activity

There is no substrate specific for cellobiohydrolase. However, the activity of the purified enzyme can be measured using Avicel. Indeed, Avicelase is now commonly regarded as being synonymous with cellobiohydrolase. Avicel has certainly proved to be a useful substrate for measuring cellobiohydrolase activity, but there is a regrettable tendency to regard even the smallest release of reducing sugar from Avicel to indicate the presence of cellobiohydrolase activity, thus ignoring the fact that some endo-1,4-β-glucanases have been shown to hydrolyze the substrate to a limited extent. Enzymes with little activity on the soluble carboxymethylcellulose (a substrate for endo-1,4-β-glucanase), but showing relatively high activity on Avicel, have been identified and assayed as cellobiohydrolases in column effluents collected during fractionation studies involving chromatography. Enzymes attacking both Avicel and carboxymethylcellulose are complete cellulases.

Amorphous cellulose and cellooligosaccharides are substrates for purified cellobiohydrolases but they are also substrates for endo-1,4-β-glucanases. Attempts have been made to assay exoglucanase activity in crude enzyme preparations by measuring the relative rates of hydrolysis of cellotetraose and cellobiose.

Measurement of Cellobiohydrolase Activity Using Avicel

Reagents

1% (w/v) suspension of Avicel (PH 101) in 0.1 M sodium acetate buffer, pH 4.8

Procedure. The reaction mixture, which consists of a 1% (w/v) suspension of Avicel (PH 101), enzyme, and water to give a total volume of 2.0 ml, is incubated at 50° for 2 hr. The tubes are centrifuged and the supernatant analysed for soluble sugars using Somogyi–Nelson (reducing sugar) or phenol–H_2SO_4 (total sugar). A calibration graph prepared with known amounts of cellobiose is used to determine the reducing sugar present, or, alternatively, a glucose standard graph is used and the reducing sugar expressed as glucose equivalent. A standard graph prepared with known amounts of glucose is used to estimate the total sugar when the phenol–H_2SO_4 reagent is employed.

Unit of Activity. Units of activity can be expressed in international units, micromoles/minute.

Comments. (1) Deshpande *et al.*[13] have reported an assay for the selective determination of exoglucanases (cellobiohydrolases) in enzyme systems also containing endoglucanases and β-glucosidases. The assay is based on the observation that the exoglucanases tested specifically hydrolyze the agluconic bond of *p*-nitrophenyl-β-D-cellobioside to yield cellobiose and *p*-nitrophenol. β-Glucosidase activity, which hydrolyzes both the agluconic and holosidic bonds, is inhibited by gluconolactone. Interference by endoglucanase, which acts to both bonds, must be allowed for by previous standardization of the assay with purified endoglucanase from the same cellulase system. It is unlikely that this method will find general application as it is clearly dependent on (a) the availability of purified endoglucanase and (b) the assumption that all types of endoglucanases normally found in cellulase systems have the same relative activities on agluconic and holosidic bonds. (2) Dyed Avicel (see chapter [3] for preparation) can be used as a substrate using a procedure similar to that recommended for measuring total cellulase activity. This method is particularly useful in studies on the kinetics of hydrolysis of purified cellobiohydrolase in the presence of end-product inhibitors, glucose or cellobiose which make the determination of reducing sugar as total sugar released by enzyme action difficult because of high blank values.

Measurement of Cellobiohydrolase Using H_3PO_4-Swollen Cellulose

The soluble sugars released are measured using colorimetric assays for reducing sugar or total sugar.

---

[13] M. V. Deshpande, K.-E. Eriksson, and L. G. Pettersson, *Anal. Biochem.* **138,** 481 (1984).

Reagents

0.4% w/v suspension of H_3PO_4-swollen cellulose (see chapter [3] for preparation) in 0.1 M
sodium acetate buffer, pH 4.8

Procedure. A 5 ml suspension of 0.4% (w/v) H_3PO_4-swollen cellulose is pipetted into a 15-ml conical centrifuge tube. After centrifugation and careful removal of 3.4 ml of the supernatant with an automatic pipet, the residue is mixed with 0.04 ml of 0.05 M NaN$_3$, and then enzyme and water to give a total volume of 2 ml. The mixture is incubated at 50° for 1 hr, cooled, centrifuged (300 g for 5 min) immediately, and the total sugar (phenol–H_2SO_4 method) or reducing sugar (Somogyi–Nelson method) in the supernatant determined.

Unit of Activity. Units of activity can be expressed in international units when the amount of hydrolysis is small.

Comment. As both endo-1,4-β-glucanase and cellobiohydrolase can solubilize H_3PO_4-swollen cellulose this assay is really of value only for characterizing purified cellobiohydrolase. The specific activity of cellobiohydrolase on amorphous cellulose is much lower than the specific activity of endo-1,4-β-glucanase.

Measurement of Endo-1,4-β-glucanase Using Carboxymethylcellulose

The soluble derivative of cellulose, carboxymethyl (CM)-cellulose, is widely used for the assay of cellulase activity. In reality it is specifically the endo-1,4-β-glucanase that is being measured with this substrate, it having been established (1) that purified exocellulases (cellobiohydrolase and glucohydrolase) have little or no action, and (2) that there is no apparent synergism between exocellulases and the endoglucanases. There is still a regrettable tendency in some publications, however, to regard an enzyme system that can attack CM-cellulose as a true cellulase. It must be remembered that CM-cellulose is not a natural substrate and attack on this fully hydrated soluble cellulose derivative cannot be interpreted to indicate that the same enzyme system will be capable of attacking crystalline cellulose.

Activity on CM-cellulose can be determined by measuring the increase in reducing power of the solution or the fall in viscosity. Strictly speaking, however, these methods cannot always be regarded as alternatives. An enzyme producing only one or two cleavages at sites remote from the end of the CM-cellulose chain would produce a large change in the viscosity (which is a parameter related to chain length), while the reducing power of the solution would not be affected significantly. Conversely, an enzyme attacking from the end of the CM-cellulose chain would produce little change in the viscosity of the solution per unit increase in reducing power. As a multiplicity of endo-1,4-β-glucanase enzymes is found in most culture filtrates, and these vary in their mode of attack on CM-cellulose, it is clear than in certain circumstances the use of only one type of measurement may lead to erroneous conclusions. Both the viscosity and reducing sugar method of determining CM-cellulase activity are therefore listed.

Measurement of Endo-1,4-β-glucanase Using CM-Cellulose and the Reducing Sugar Method of Analysis. There are two commonly used methods of measuring and calculating endo-1,4-β-glucanase activity which involve the estimation of reducing sugar released from CM-cellulose. One is based, like the filter paper assay for total cellulase activity, on equal conversion of the substrate; the other is based on initial rates of reaction. The former is the procedure recommended by the Biotechnology Commission of IUPAC and is designed to facilitate the comparison of data by biotechnologists in various laboratories. This procedure should be carried out exactly as recommended. The method based on initial rates of reaction permits more licence, but the user should be aware of the problems that exist in the interpretation of the data.

1. Commission on Biotechnology Recommended Method[1]

Reagents

2% (w/v) carboxymethylcellulose CMC 7L2 (Hercules Inc., Wilmington, DE 19811) in 0.05 M sodium citrate buffer, pH 4.8 (see under filter paper assay for description of preparation of buffer)

Dinitrosalicylic acid (DNS) reagent (see under methods for measuring reducing sugar for details of preparation)

Procedure. A 0.5 ml sample of enzyme, diluted with citrate buffer, is added to a test tube of volume at least 25 ml. Several dilutions are made of the enzyme such that one dilution will release, from the CM-cellulose substrate, slightly more than 0.5 mg of glucose in the reducing sugar assay using DNS reagent and the other slightly less. The enzyme solution is heated in a water bath for 5 min at 50°, 0.5 ml substrate solution is added, and the solution is mixed well and then heated in the incubator for 30 min. DNS reagent (3.0 ml) is added and the solution is boiled for exactly 5 min in a vigorously boiling water bath. Finally, 20 ml distilled water is added and the solution mixed by completely inverting the tube several times so that the solution separates from the bottom of the tube at each inversion. The absorbance is then measured at 540 nm and, after subtraction of enzyme and reagent blanks, translated, using a linear glucose standard prepared by plotting glucose used against absorption at 540 nm, into micrograms of glucose produced during the enzyme reaction.

Calculation of Unit of Activity. The various enzyme dilutions used to obtain the release of approximately 0.5 g of glucose are converted into enzyme concentrations (1/dilution) and the concentration of enzyme which would have released exactly 0.5 mg of glucose is estimated from a plot of glucose liberated against enzyme concentration.

$$\text{Units of activity/ml} = \frac{0.185}{\substack{\text{enzyme concentration required} \\ \text{to release 0.5 mg glucose}}}$$

It should be noted that these units of activity are not international units although they are derived from the IU as follows:

$$1 \text{ IU} = 1 \text{ } \mu\text{mol of hydrolysis product released per min}$$
$$= 0.18 \text{ mg/min glucose}$$

The critical amount of glucose in the assay is 0.5 mg.

0.5 mg glucose = 0.5/0.18 μmol (which is produced by 0.5 ml of diluted enzyme in 30 min)

$$= \frac{0.5}{0.18 \times 0.5 \times 30} \text{ } \mu\text{mol/min/ml}$$
$$= 0.185 \text{ } \mu\text{mol/min/ml (i.e., IU/ml)}$$

Therefore, the estimated amount of enzyme solution [i.e., the critical enzyme concentration (ml/ml) which releases 0.5 mg glucose in the enzyme–substrate reaction] contains 0.185 IU. Hence,

$$\text{Units of activity/ml} = \frac{0.185}{\text{critical enzyme concentration}}$$

Comment. Although CM-cellulose is widely used for the assay of endo-1,4-β-glucanase activity there are some problems attendant on its use when the enzyme is being measured over a range of pH values. As CM-cellulose is an ionic substrate its properties change with pH. Nonionic substrates such as hydroxyethylcellulose have therefore been recommended in these circumstances. An additional assay procedure for endoglucanase based on hydroxyethylcellulose is given in the publication of the Commission on Biotechnology.

2. Method Based on Initial Rate of Reaction

Reagents

Somogyi reagents I and II
Nelson reagent
0.1 *M* sodium acetate buffer, pH 5.0

1% (w/v) carboxymethylcellulose solution, prepared as follows: carboxymethylcellulose (1.5 g) of degree of substitution 0.5–0.7 is
dissolved as far as possible in 100 ml distilled water with heating
and stirring, cooled, and centrifuged to remove any residue. An
aliquot is taken and the concentration of the substrate is determined after freeze drying and weighing. This step is necessary only
when CM-cellulose of low degree of substitution is used: CM-
cellulose of degree of substitution 0.7 is normally completely soluble. If necessary the concentration of the substrate is adjusted to
1% and the solution made 0.05 M with respect to NaN_3. This
solution can be kept at 4° for approximately 7 days with little
detectable change in the reducing power of the solution
  Glucose standard solution (20 mg/100 ml)
  Procedure. CM-cellulose (1.0 ml) reagent and acetate buffer (0.5 ml)
are mixed and brought to 50° in a water bath. Diluted enzyme (0.5 ml) is
added, the solution mixed, and returned to the incubator for 15 min. The
reaction is stopped by the addition of 2 ml of the mixture of Somogyi I and
II reagent (see under Sugar Analyses). The solution is boiled in a water
bath for 15 min, cooled, and 2 ml Nelson reagent added and the solution
mixed on a Vortex mixer. Finally, 4 ml of water is added, the suspension
mixed, centrifuged at 300 g for 20 min, and the absorbance of the supernatant read at 520 nm. The absorbance is translated, after substration of
enzyme and reagent blanks, into micrograms of glucose equivalent released using a standard graph obtained by plotting the amount of glucose
used against absorbance.
  Unit of Activity. The unit of activity is the amount of enzyme required
to liberate, under the conditions of the assay, 1 μmol/min reducing sugar
expressed as glucose equivalent.
  Comments. 1. Despite the insensitivity of the dinitrosalicylic acid
(DNS) reagent for estimating the reducing sugar released, many favor its
use in the CM-cellulase assay. In our view the use of the DNS method is
not advised when purification of the enzymes of the cellulase complex is
being undertaken. Many cellobiohydrolase and β-glucosidases have been
reported to be prepared free from endo-1,4-β-glucanase (CM-cellulase) on
the basis of assays involving the DNS reagents: some of these results may
be suspect. The insensitivity of the DNS reagent to reducing power less
than 100 μg of glucose equivalent can be circumvented by addition of 100
μg of glucose to each assay, but this is not reported to be done in most
cases. The use of the Somogyi–Nelson reagents which are sensitive over
the range 5–150 μg of glucose equivalent is the best method to use, in our
view. However, the centrifugation step required to remove the precipitate
of unhydrolyzed CM-cellulose is considered to be a disadvantage by

some: the availability of a centrifuge fitted with a trunnion carrier rotor is clearly a necessity.

If extreme sensitivity is wished in the CM-cellulase assay the ferricyanide–ferric alum method[14] (covering the range 1–14 μg of glucose equivalent) can be used, but it is frequently difficult to select suitable enzyme dilutions to obtain reducing sugar values in the narrow range 1–14 μg.

2. The degree of substitution of the CM-cellulose used must be reported in order to permit comparison of the results with that of other investigators. If the degree of substitution is not known it can be determined by the method of Eyler et al.[15]

3. The degree of substitution has been shown to have a profound effect on enzyme attack. Only unsubstituted residues are attacked and it would appear that at least two[16] (or in some cases, three[17]) contiguous unsubstituted residues are required for enzyme action. In CM-cellulose of degree of substitution 0.7, there will be relatively few sites that can be attacked. It appears, however, that a reasonably stoichiometric response can be obtained up to 3–4% degradation,[18] but it is recommended that reducing sugar assays should be within the limits of 2% degradation using a substrate with a degree of substitution of 0.7.

4. The degree of polymerization of the CM-cellulose is not too important for the reducing sugar method of assaying CM-cellulase activity, but it is recommended for uniformity that CM-cellulose of medium viscosity be used.

Measurement of Endo-1,4-β-glucanase Using CM-Cellulose and the Viscosity Method of Analysis[19]

Reagents

0.5% (w/v) solution of CM-cellulose (degree of substitution 0.5–0.7; medium viscosity, in 0.05 M sodium acetate buffer, pH 5.0)

Procedure. The CM-cellulose solution (6.0 ml) is preheated for 30 min in a water bath at 30° and then pipetted into an Ostwald viscometer (water flow time of 15 sec at 30°) in a water bath accurately controlled at 30° ($\pm$0.1°). The enzyme reaction is started by adding 1 ml of diluted enzyme solution (preheated to 30°) followed by immediate mixing. Flow rates are

[14] T. M. Wood and S. I. McCrae, *Biochem. J.* **128**, 1183 (1972).
[15] R. W. Eyler, E. D. Klug, and F. Diephuis, *Anal. Chem.* **19**, 24 (1947).
[16] M. Holden and M. V. Tracey, *Biochem. J.* **47**, 407 (1950).
[17] M. G. Wirsck, *J. Polym. Sci.* **A6**, 1965 (1968).
[18] D. E. Eveleigh, *Can. J. Microbiol.* **13**, 727 (1967).
[19] G. Canavascini, M.-R. Condray, J.-P. Roy, R. J. G. Southgate, and H. Meier, *J. Gen. Microbiol.* **110**, 291 (1979).

determined at intervals of 10 min from the addition of enzyme. By plotting the rate of increase of the reciprocal of the specific viscosity

$$\frac{t - t_0}{t_0} = \eta \hat{s}p = \text{specific viscosity}$$

against enzyme concentration a linear relationship should be obtained.

Unit of Activity. The unit of activity is chosen arbitrarily from the linear relationship, enzyme concentration/rate of increase of reciprocal of the viscosity of the CM-cellulose solution.

Comments. (1) A convenient viscometer which is suitable for the screening of enzyme fractions collected during chromatographic separations can be constructed from capillary tubing shaped like a bulb pipet.[20] The tube is surrounded by a water jacket through which water is circulated at 30°. The flow time of water between marks above and below the bulb should be approximately 30 sec. After a suitable incubation period, a portion of the enzyme–CM-cellulose reaction mixture can be sucked into the viscometer tube at intervals from tubes sitting in a water bath at 30°. (2) Reduction in viscosity is very sensitive since a random break in a cellulose chain at points remote from the end of the chain may cause a large fall in the viscosity with little increase in reducing power. (3) A method for evaluation of the enzymatic activity of a cellulase on CM-cellulose has been developed which permits the determination of the activity in absolute terms, i.e., the number of bonds broken per unit time.[21]

Measurement of Endo-1,4-β-glucanase Using Amorphous Cellulose

Endo-1,4-β-glucanase can hydrolyze amorphous cellulose in which the hydrogen bonding of the native cellulose has been disrupted by ball-milling or by chemical treatment. This activity can be estimated either by measuring the sugars released from H_3PO_4-swollen cellulose (see chapter [3] for preparation) using a procedure similar to that given under estimation of total cellulase activity or by a turbidimetric assay[22] in which the decrease in turbidity of a finely divided suspension of the cellulose obtained by ball-milling is measured. Because most crude enzyme preparations contain a variety of endoglucanases which differ in their relative activities on CM-cellulose and on amorphous cellulose, this assay should not be regarded as an alternative to the assay for endo-1,4-β-glucanase

[20] T. M. Wood and S. I. McCrae, *Carbohydr. Res.* **117**, 133 (1977).
[21] K. E. Almin and K.-E. Eriksson, *Arch. Biochem. Biophys.* **124**, 129 (1968).
[22] M. Nummi, P. C. Fox, M.-L. Niku-Paavola, and T.-M. Enari, *Anal. Biochem.* **116**, 133 (1981).

involving measurement of the attack on the soluble carboxymethylcellulose (CM-cellulose).

CM-cellulose is not a natural cellulose and there are many problems
associated with its use and in the interpretation of the results in terms of
attack on native cellulose. Some consider therefore that amorphous cellulose is a better substrate for measuring endo-1,-4-β-glucanase activity.

Reagents

Amorphous cellulose prepared by milling Whatman CF11 cellulose
powder (see chapter [3] for preparation). The concentration of the
cellulose is determined by the anthrone–H_2SO_4 method[7] for determining total sugar and adjusted to obtain a suitable extinction at
620 nm in a cuvette in a spectrophotometer.

Procedure. Suitably diluted enzyme (200 μl) is incubated with 3 ml of
cellulose suspension in 0.05 M citrate buffer, pH 5.0, at 50°, and the
absorbance at 620 nm measured after exactly 10 min against a citrate
buffer blank.

Unit of Activity. The unit of activity can be defined as the change in
absorbance per minute during the early stages under the reaction conditions used. Linearity of response to enzyme and substrate concentration
and to time must be established for each new cellulose preparation.

Comments. (1) The method is also suitable for measuring cellobiohydrolase and total cellulase activities. It is a more convenient method to
use than those involving the measurement of reducing sugar released or
residual cellulose, but it has not been widely employed. (2) The substrate
can be completely and rapidly dissolved by some cellulases when the
cellulose is sufficiently finely divided to be colloidal. (3) A ball-milling
process involving the use of 95% ethanol appears to provide a very finely
divided suspension of the substrate.[22]

Measurement of Endo-1,4-β-glucanase Using Trinitrophenylcarboxymethylcellulose (TNP-CM-Cellulose)[23]

CM-cellulose is chemically modified to include trinitrophenyl groups
(see chapter [3] for preparation) which absorb at 340 nm. The CM-cellulose used to prepare the TNP derivative is fibrous and insoluble. Enzyme
action effects solubilization of short-chain fragments which carry the
chromophoric TNP groups. Because the substrate is insoluble, the activities measured may not be compared with activities measured using soluble CM-cellulose. This method of assay is useful in conditions where the

[23] R. Huang and T. Tang, *Anal. Biochem.* **73**, 369 (1976).

culture medium contains reducing agents, as used for example in the preparation of anaerobic bacteria.

Reagents

1% (w/v) suspension of trinitrophenyl-CM-cellulose in 0.1 M citrate buffer, pH 4.8

Procedure. A 2 ml sample of the suspension of TNP-CM-cellulose is pipetted into a 15-ml conical centrifuge tube and heated to 50° in a water bath. Diluted enzyme (0.5 ml) is added to the suspension, mixed, and placed in the water bath for 1 hr. The tubes are removed, boiled in a water bath for 5 min to destroy the enzyme, cooled, and centrifuged at 300 g for 10 min. The absorbance of the supernatant is read at 340 nm in a spectrophotometer using a reference sample incubated without enzyme.

Unit of Activity. The unit of activity can be arbitrarily chosen, for example, to correspond to an increase of 1.0 absorbance unit/hr.

Measurement of Endo-1,4-β-glucanase Using the Alkali-Swelling-Centrifuge Test[24]

A sensitive measure of endo-1,4-β-glucanase (or total cellulase) activity is provided by the alkali-swelling-centrifuge test developed by Marsh et al.[24] It was found that if cotton fibers are immersed first in culture filtrates from cellulolytic microorganisms and then in 18% NaOH, there is an increase in uptake of alkali relative to the control. Enzymes causing changes in the cotton fiber that result in an increased capacity for uptake of alkali measured in this way, are said to show S-factor or swelling-factor activity.

Reagent

18% (w/v) NaOH solution

Procedure. The incubation mixture contains 100 mg of native cotton fiber, 2.0 ml of acetate buffer of pH 5.0, and enzyme and water to give a total volume of 10 ml. The cotton is removed after 1 hr at 27°, partially dried on filter paper, and placed in 20 ml of 18% (w/v) NaOH solution. The mixture is shaken for 30 sec by hand and replaced in the incubator. After 1 hr the swollen cellulose is removed, placed in a sintered glass microfilter funnel (BTL type K832H, porosity 2), and centrifuged for 30 min at 340 g in a 15-ml centrifuge tube on an M.S.E. Super-Medium centrifuge (8 × 15 ml Trunnion carrier). The top of the filter funnel fits neatly into the top of the centrifuge tube allowing the bulk of the NaOH solution to be collected in the base of the tube.

[24] C. B. Marsh, G. V. Merola, and M. E. Simpson, *Text. Res. J.* **23,** 831 (1953).

The sample is transferred to a weighing bottle and weighed immediately. Controls should be included in each experiment. The increase in swollen weight relative to the control is calculated.

A graph of increase in swollen weight against enzyme concentration has been found to be linear for the cellulases of *Trichoderma koningii* and *Fusarium solani*.[25]

Unit of Activity. The unit of activity can be chosen arbitrarily from the linear relationship obtained by plotting increase in swollen weight against enzyme concentration.

Comments. The assay is extremely sensitive and, indeed, probably the most sensitive assay for measuring attack on cotton fiber. It is speculated that attack is confined to the winding layer of cellulose found in cotton fiber, but this has not been proven. It is our opinion that the value of this assay has been underrated in view of the uncertainty that still persists on the question of which enzyme(s) initiate attack on crystalline cellulose. Some enzymes classified as endo-1,4-β-glucanases show S-factor activity, but highly purified cellobiohydrolases do not. Cellobiohydrolases do, however, act synergistically with endo-1,4-β-glucanases to enhance the S-factor activity shown by certain endoglucanases. For this reason the assay could also be used to measure total cellulase activity.

Measurement of Endo-1,4-β-glucanase by Release of Short Fibers from Cotton[26]

Total cellulase preparations, and some endo-1,4-β-glucanases isolated from these cellulase preparations, fragment cotton fiber into short insoluble fragments in a few hours on shaking. It seems likely that the sites of attack are the areas of weakness which are reported to traverse the whole of the fiber. However, in view of the prevailing uncertainty as to which of the enzymes of the cellulase system actually initiate the attack, the possibility that there is some special significance of the formation of short fiber in this context cannot be ruled out.

Procedure. The reaction mixture which contains, in a screw-capped bottle, 25 mg cotton fiber, 5 ml of 0.2 M sodium acetate buffer, pH 4.8, and enzyme and water to give a total volume of 9 ml, is shaken vigorously by hand a few times to wet the fibers and then mechanically for 2 hr at 37°. The mass of apparently unhydrolyzed fibers is removed with forceps and the suspension of short fibers is poured into a measuring cylinder. Further amounts of short fibers are extracted from the mass of unhydrolyzed fibers by shaking vigorously by hand approximately 30 times in the original bottle with 5 ml of water. This extraction is repeated with a further 5

[25] T. M. Wood, *Biochem. J.* **109**, 217 (1968).
[26] G. Halliwell, *Biochem. J.* **95**, 270 (1965).

ml of water. The aqueous extracts are added to the short fiber suspension and the volume made up to 20 ml. The short fiber suspension is sufficiently stable to permit extinction measurements to be made at 600 nm using cuvettes with a 4 cm light path.

Unit of Activity. This is arbitrarily selected from a linear plot of extinction against enzyme concentration.

Comments. (1) Not all of the many endo-1,4-β-glucanases normally found in a cellulase system can generate short fibers. (2) Some purified cellobiohydrolases have been reported to be unable to produce short fibers from cotton; others, however, can apparently completely fragment filter paper which has been prepared from cotton fiber. The reason for this apparent contradiction is not clear. (3) Purified cellobiohydrolase can act in synergism with the endo-1,4-β-glucanase that do release short fiber to increase the amount produced. The assay can therefore also be used to measure total cellulase activity. (4) There does not appear to be any correlation between endo-1,4-β-glucanases that contain Swelling-factor activity (see above) and the ability to produce short fibers. However, a correlation exists between short fiber formation and losses in tensile strength of cotton yarn.

Measurement of β-Glucosidase/Cellobiase Activity

Cellobiose is the substrate of choice as aryl-β-glucosidases exist which do not hydrolyze cellobiose. However, in practice the assays involving aryl-β-glucosides are more frequently used because of their simplicity: both types of assay are included here.

Initial rates of reaction are used in calculation of units of activity, which are expressed in international units. However, the cellobiase assay recommended by the Commission on Biotechnology is based on significant and equal conversion of cellulose to glucose: this method is also listed.

Measurement of Aryl-β-D-glucosidase Using
p-Nitrophenyl-β-D-glucoside

The nitrophenol liberated is measured colorimetrically.

Reagents

    0.1 *M* sodium acetate buffer, pH 4.8
    5 m*M* p-nitrophenyl-β-D-glucoside in the acetate buffer
    0.4 *M* glycine buffer, pH 10.8 [prepared by dissolving 60 g glycine in
        1500 ml distilled water and adding 50% (w/v) NaOH until the pH is
        10.8 and then diluting to 2 liters]

p-Nitrophenol standard (20 mg/100 ml)

Procedure. p-Nitrophenyl-β-D-glucoside substrate (1.0 ml) is pipetted into a test tube along with 1.8 ml of acetate buffer and equilibrated to 50° in a water bath. Diluted enzyme solution (200 µl) is added, the contents of the tubes mixed, and then incubated at 50° for 30 min. Glycine buffer (4.0 ml) is added to stop the reaction and the liberated p-nitrophenol is measured at 430 nm. The usual enzyme and reagent blanks are included. The absorbance values obtained (less the enzyme and substrate controls) are translated to micromoles of nitrophenol using a standard graph relating micromoles of nitrophenol to absorbance.

Units. The unit of activity is the amount of enzyme required to release 1 µmol p-nitrophenol/min under the conditions of the assay.

Comments. (1) The p-nitrophenyl-β-D-glucoside can be replaced by the ortho isomer, but the activity of most fungal β-glucosidases is much lower on this substrate. 4-Methylumbelliferyl-β-glucoside is favored in some circumstances because of the ease of use and the sensitivity of the assay.[27] (2) Glycosides such as salicin[28] and esculin[29] are commercially available and have been used as substrates.

Measurement of Cellobiase

The cellobiose is hydrolyzed to glucose which is estimated using the glucose oxidase/peroxidase reagent.[30]

Reagents

0.1 M sodium acetate buffer, pH 5.0

0.4% (w/v) cellobiose solution

Tris–phosphate–glycerol buffer pH 7.0 (dissolve 36.3 g Tris and 50 g $NaH_2PO_4 \cdot 2H_2O$ in 400 ml water; add 400 ml glycerol and make up to 1 liter with water. Adjust to pH 7.0 by adding solid $NaH_2PO_4 \cdot 2H_2O$)

Glucose oxidase reagent [dissolve the following in 100 ml of Tris–phosphate–glycerol buffer: glucose oxidase (Boehringer, *Aspergillus niger* Grade II), 10 mg; horseradish peroxidase (Boehringer, Grade II), 3 mg; o-dianisidine dihydrochloride (Sigma), 10 mg]

Procedure. Cellobiose substrate solution (0.1 ml) is added to 0.5 ml acetate buffer followed by enzyme and water to give a total volume of 1 ml. The reaction mixture is incubated at 50° for 30 min and then 2 ml of

[27] R. Mullings, Ph.D. thesis, University of Leeds, England.
[28] M. Mandels and E. T. Reese, *Dev. Ind. Microbiol.* **5,** 5 (1964).
[29] B. M. Eberhart, R. S. Beck, and K. M. Goolsby, *J. Bacteriol.* **130,** 181 (1977).
[30] J. B. Lloyd and W. J. Whelan, *Anal. Biochem.* **30,** 467 (1969).

glucose oxidase reagent is added and the reaction allowed to proceed at 37° for 30 min. The reaction is stopped by the addition of 4 ml of 5 N HCl. Finally, the reaction mixture is mixed well and the absorbance is read at 520 nm. All the necessary enzyme and reagent blanks are included and subtracted from the absorbance obtained in the enzyme–substrate reaction mixture. The absorbance is then translated into micromoles glucose from a linear calibration graph obtained by plotting enzyme concentration against micromoles glucose.

Unit of Activity. The unit of activity is defined as the amount of enzyme required to release 2 μmol of glucose/min under the conditions of the assay. Alternatively, units can be expressed in microkatals, i.e., the amount of enzyme required to hydrolyze 1 μmol of cellobiose/sec under the assay conditions.

Measurement of Cellobiase—IUPAC Biotechnology Commission Recommended Method[1]

Reagents

15 mM cellobiose in 0.05 M citrate buffer, pH 4.8 (see under measurement of total cellulase using filter paper for preparation of buffer). This solution should be prepared fresh daily

0.05 M citrate buffer, pH 4.8

Glucose oxidase reagent kit

Procedure. Diluted enzyme (1.0 ml) in citrate buffer is pipetted into small test tubes. At least two dilutions should be made for each sample investigated such that one of these dilutions should release slightly more than 1 mg of glucose in the reaction, the other slightly less. The solution is equilibrated at 50° in a water bath, 1 ml of substrate solution added, and the reaction mixture incubated at 50° for 30 min. The reaction is terminated by immersing the tubes in boiling water for 5 min exactly. The tubes are transferred to a cold water bath and the glucose released determined by a standard procedure recommended by the manufacturer of a kit based on the glucose oxidase reaction. It is advisable to terminate the reaction by the addition of 0.2 ml 72% H_2SO_4 whether or not this is recommended by the manufacturer. All the necessary enzyme and reagent blanks are included and subtracted from the absorbance obtained in the glucose oxidase reaction.

Units. The unit of activity is the amount of enzyme required to release exactly 1.0 mg of glucose from cellobiose under the conditions of the assay.

The unit is calculated as follows. (1) The glucose concentrations (mg/ml) in the cellobiase reaction mixture is determined using at least two different enyzme dilutions. (2) The glucose concentrations obtained in (1) are multiplied by 2 to convert glucose concentrations into absolute amounts (mg). (3) The enzyme dilutions are converted into concentrations:

$$\text{Concentration} = 1/\text{dilution}$$

(4) The concentration of enzyme which would have released exactly 1.0 mg of glucose is obtained by plotting glucose liberated (as in 2, above) against enzyme concentrations (as in 3, above). (5) Finally, the cellobiase activity is calculated:

$$\text{Units/ml} = \frac{0.0926}{\text{enzyme concentration to release 1.0 ml glucose}}$$

The value 0.0926 is derived as follows:

$$1 \text{ IU} = 1 \text{ } \mu\text{mol/min of cellobiose converted}$$
$$= 2 \text{ } \mu\text{mol/min of glucose formed}$$

At the critical dilution 1.0 mg of glucose is released and 1.0 mg glucose = μmol of glucose or 0.5/0.18 μmol cellobiose converted. As this amount of cellobiose was converted by 1.0 ml of enzyme in 30 min (i.e., 1.0 mg glucose = 0.5/0.18 × 1.0 × 30 μmol/min/ml, the cellobiose converted is 0.0926 μmol/min/ml).

[10] Cellulase Assay Based on Cellobiose Dehydrogenase

By GIORGIO CANEVASCINI

$$\text{Cellulose} \xrightarrow{\text{cellulase(s)}} \text{cellobiose}$$

$$\text{Cellobiose} + 2 \text{ ferricyanide} \xrightarrow[\text{dehydrogenase}]{\text{cellobiose}} \text{cellobionic acid} + 2 \text{ ferrocyanide}$$

Most cellulolytic microorganisms (e.g., fungi, aerobic and anaerobic bacteria) bring about the degradation of cellulose by means of different kinds of hydrolytic enzymes which are thought to act in concert on the insoluble substrate producing, at least transiently, different cellodextrins and finally cellobiose. Cellulase assays are thus based on the determination of the increase in reducing power upon action on various cellulosic substrates, e.g., cotton fiber, filter paper, Avicel, carboxymethylcellulose. The discontinuous test described here is based on the specific determination of the cellobiose produced during the cellulase reaction with an ancillary cellobiose dehydrogenase.[1,2]

Assay Method

Principle

The cellulase reaction is carried out in the presence of ferricyanide and cellobiose dehydrogenase thus allowing the continuous oxidation of cellobiose produced (or alternatively the cellodextrins which may or may not be transiently formed) to cellobionic acid. The reduced ferrocyanide is subsequently estimated by the classic colorimetric determination of colloidal Prussian blue.[3]

Reagents

Phosphoric acid-swollen cellulose (referred to as amorphous cellulose): 0.5% (w/v) suspension in 0.1 M sodium acetate buffer, pH 5.0, containing 1% Triton X-100 and 10 mM potassium ferricyanide. Amorphous cellulose may be prepared from Avicel as described elsewhere[4]

SDS (sodium dodecyl sulfate) solution in 0.8 M phosphoric acid (0.3 g SDS, 9.5 ml of 85% phosphoric acid and water to 100 ml)

Ferric sulfate 2.5% (w/v) aqueous solution

Cellobiose dehydrogenase: a suitable preparation should contain between 1500 and 2500 units (nanomole) per ml (cf. 1)

Procedure

One milliliter of amorphous cellulose suspension in buffer is placed in a small glass centrifuge tube (3 ml volume) and equilibrated in a water bath at 30° (agitation is superfluous). The (thick) cellulose suspension is conveniently distributed with an automated sampler (Eppendorf multipet). Then 5 μl of cellobiose dehydrogenase is added followed immediately by 50 μl of the cellulase preparation to be analyzed (or a dilution thereof). Two other tubes are run in parallel: one is a reagent blank and the other is a control without added cellulase. After incubation, the reaction is terminated by adding 0.4 ml of the SDS–phosphoric acid reagent,

[1] G. Canevascini, this volume [52].
[2] G. Canevascini, *Anal. Biochem.* **147**, 419 (1985).
[3] M. Ameyama, this series, Vol. 89, p. 20.
[4] G. Canevascini, M. R. Coudray, J.-P. Rey, R. J. G. Southgate, and H. Meier, *J. Gen. Microbiol.* **110**, 291 (1979).

the residual cellulose being then centrifuged off in a bench centrifuge (3000 g, 10 min). The supernatant is carefully decanted and mixed with 0.1 ml of ferric sulfate solution followed by 2.5 ml of water and left at 30° for 20 min to allow for full color development. The absorbance is read at 660 nm against the reagent blank. The activity is the difference between the test sample and control tube (with cellobiose dehydrogenase alone). This is more conveniently expressed in nanomole cellobiose produced (viz. oxidized) per minute. Specific activity is subsequently given in μmole min^{-1} mg protein^{-1}. The relation between the Prussian blue color formed and cellobiose concentration is easily determined with a standard curve in which the (same) substrate suspension without cellulase is mixed with increasing amounts of cellobiose (up to 150 × nanomole per reaction sample). By adding 12–15 cellobiose dehydrogenase units (nanomole) the reaction should be complete within 10 min.

Linearity of the Activity Response

The assay as it is described, i.e., with a colloidal suspension of amorphous cellulose as substrate, gives a linear response between activity and enzyme concentration or, respectively, between activity and reaction time as long as the amount of the cellulase added does not produce more than 3 nmol of cellobiose per min. A typical response curve is given in Fig. 1 which represents an assay carried out with a crude cellulase preparation obtained from *Sporotrichum thermophile*. Similar responses were obtained with cellulases from various sources, e.g., from *Trichoderma viride* (Cellulase SP 122, Novo Industry; Sigma Type V cellulase) or from *Penicillium funiculosum* (Sigma Type VII cellulase). The amount of crude enzyme added in the test was between 0.5 and 2.5 μg protein per sample. Incubation was for 20 or 30 min. Table I shows the reproducibility of the assay. The assay correlates well with the same test, but where the reducing sugar was determined with a classical colorimetric reaction.

Sensitivity and Specificity

As can be seen from Fig. 1, this cellulase assay detects concentrations of cellobiose in the nanomole range. In comparison, other standard reducing sugar reagents like copper-molybdate or dinitrosalicylic acid are less sensitive (micromole range). In addition, cellobiose dehydrogenase is absolutely specific for cellobiose and cellodextrins and does not react with glucose. The only other sugar which has been found to be oxidized and can actually be detected by means of this enzyme is lactose,[5] which,

---

[5] G. Canevascini, K. Etienne, and H. Meier, Z. *Lebensm. Unters. Forsch.* **175,** 125 (1982).

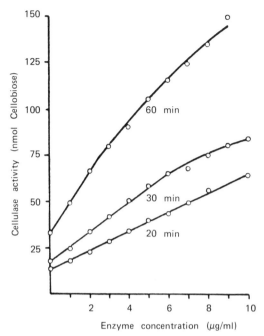

FIG. 1. Relation between activity, enzyme concentration, and time of incubation. The test was performed as described in the text. Conditions were as follows: cellulase (from *Sporotrichum thermophile*), from 1 to 10 μg of a freeze-dried culture filtrate; cellobiose dehydrogenase, 7 U; incubation time, 20, 30, 60 min, respectively. Values are the means of duplicate determinations. From Canevascini.[2]

TABLE I
REPRODUCIBILITY OF THE CELLULASE ASSAY

Cellulase source	Cellulase specific activity[a] (nmol min^{-1} mg^{-1} protein)				Mean (%)
Trichoderma viride (SP 122 Novo Industry)	666 (42)	598 (52)	589 (43)	609 (34)	615 ± 8
Penicillium funiculosum (Type VII Sigma)	960 (85)	986 (54)	956 (23)	996 (54)	974 ± 2
Sporotrichum pulverulentum	667 (81)	714 (42)	635 (45)	725 (31)	685 ± 6
Sporotrichum thermophile	1134 (89)	1148 (96)	1060 (71)	1100 (45)	1107 ± 4

[a] Each value represents the mean of 10 estimations (the corresponding standard deviation is given in parentheses) performed with different diluted enzyme samples and the values are given for four different enzyme dilutions. The reaction mixtures contained, in a volume of 1 ml, from 0.5 μg (left) to 2.5 μg (right) protein as determined by the method of Lowry. From Canevascini.[2]

however, is not likely to be present in a cellulase test. An excess of a contaminating β-glucosidase, however, as this is very often the case, would obviously compete in this assay with the ancillary cellobiose dehydrogenase for cellobiose (hydrolysis versus oxidation) and consequently decrease the value for the cellulase activity. This is contrary to a classic cellulase test (reducing sugar estimation) where the presence of β-glucosidase leads to an overestimation of the cellulase activity as shown by Galas et al.[6] The negative effect of β-glucosidase can easily be suppressed by inhibiting this enzyme with a suitable amount of D-glucono-δ-lactone: 0.5 mg per reaction sample has been found to restore the original cellulase activity even in the presence of a very large excess of β-glucosidase.[2]

As the cellobiose formed by the cellulase action is continuously removed from the reaction mixture by the cellobiose dehydrogenase there is virtually no back-inhibition of this substance on the whole cellulase reaction.

This assay has no specificity in terms of discriminating between endo- and exo-acting cellulases but is suitable for the estimation of any kind of cellulolytic enzyme, individually or in a mixture, producing cellobiose. A critical point is the actual pH at which the cellulase test is performed, which of course has to match the enzymatic oxidation of cellobiose with ferricyanide. The optimum of this reaction is close to pH 3.5–4.0 but a 75% response is still obtained at pH 3.0 or 5.0, while no oxidation takes place at pH 7.0. The assay is therefore unsuitable for cellulase enzymes working preferentially in a neutral or alkaline environment. It is suggested that for such enzymes, the ancillary cellobiose dehydrogenase should be used in conjunction with phenazine methosulfate.[7] This redox substance being autoxidizable, the cellulase reaction is directly measured by oxygen consumption either manometrically or by means of an oxygen electrode.

Concerning the cellulosic substrate, although the procedure described here makes use of amorphous cellulose (which forms a rather stable colloidal suspension easy to handle) any other kind of unsubstituted cellulose (Avicel, cotton fiber, filter paper) is suitable. The cellulase activity is then very strongly modulated by the concentration of the substrate as well as by some physical heterogeneity of its suspensions. It may be that with crystalline forms of cellulose, the cellulase action is initially restricted to some regions of the microfibrils more accessible to enzymatic attack. On the other hand, agitation is necessary to favor enzyme access to the fibrous substrate. Optimal conditions for a cellulase assay with native forms of cellulose should therefore be carefully worked out in each particular case.

[6] E. Galas, R. Pyc, K. Pyc, and K. Siwinska, *Microbiol. Biotechnol.* **11**, 229 (1981).
[7] M. R. Coudray, G. Canevascini, and H. Meier, *Biochem. J.* **203**, 277 (1982).

[11] Nephelometric and Turbidometric Assay for Cellulase

By T.-M. ENARI and M.-L. NIKU-PAAVOLA

Introduction

For an efficient and complete hydrolysis of crystalline cellulose to glucose the cooperation of exo- and endoenzymes (1,4-β-D-glucan cellobiohydrolase, EC 3.2.1.91; and 1,4-β-D-glucan 4-glucanohydrolase, EC 3.2.1.4) as well as β-glucosidase (β-D-glucoside glucohydrolase, EC 3.2.1.21) is necessary.

The specificity of a cellulolytic enzyme is difficult to determine because the substrates available and used for this purpose are heterogeneous and ill defined. They include insoluble crystalline and amorphous cellulose preparations containing impurities and showing structural differences. Dewaxed crystalline cotton, microcrystalline Avicel,[1] and the recently introduced[2] highly crystalline algal cellulose are used as representatives of highly ordered native cellulose. The degree of crystallinity of different preparations varies and the results are accordingly different.

Methods where insoluble cellulose preparations are used require long hydrolysis times, from 30 min to several days to bring about a measurable change in the substrate. They do not give figures for the initial reaction rate of individual enzymes but a value for an overall hydrolysis rate. The longer the hydrolysis time, the better the total effect of all enzymes present, including end-product inhibition, is revealed. A particular problem here is the change in the nature of the substrate as hydrolysis proceeds.

The hydrolysis of insoluble cellulose is usually analyzed only by the soluble hydrolysis products. Thus, insoluble reaction products are normally not studied. Often the cellulolytic activities are analyzed with substrates where cellulose has been rendered soluble by substitution or physical treatments. Also small molecular size, soluble cellodextrins or their derivatives have been used. Conclusions based on such hydrolysis results are not necessarily valid for insoluble cellulose.

An attempt to coordinate the activity determinations for cellulolytic enzymes was recently made by IUPAC Commission on Biotechnology.

[1] T.-M. Enari, in "Microbial Enzymes and Biotechnology" (W. M. Fogarty, ed.), p. 183. Applied Science Publishers, London, 1983.
[2] H. Chanzy, B. Henrissat, R. Vuong, and M. Schuelein, FEBS Lett. 153, 113 (1983).

Recommendations for universal methods were given[3] but these, however, list only the most commonly used methods. Unfortunately no improvement or real standardization was achieved.

It is known that cellulolytic enzymes do not hydrolyze efficiently well-ordered areas of cellulose, but attack only amorphous regions. In the method presented here, a more refined amorphous substrate has been prepared by ball milling. This substrate is completely dissolved by cellulases and can be used to measure the initial velocity of the reaction. Initial hydrolysis is followed by means of a nephelometer and the slopes of nephelometric plots are used to calculate the cellulase activity. The initial reaction rate can also be estimated with a photometer where the activity is measured as the decrease in absorbance of a suspension of amorphous cellulose.

A good correlation has been found between the initial rate of reaction and the amount of sugar released at given times.

Nephelometric and Turbidometric Methods for Insoluble Particles

The nephelometric method measures the right angle scattering of light at 600–1000 nm, from a particle in suspension, while the turbidometric method gives an estimate for the light absorption by the particle suspension. Nephelometry is suitable for diluted suspensions whereas turbidometry is better for thick suspensions of particles. Thus, slight changes in particle size are more dependably detected by nephelometry than by turbidometry. These methods are useful for a wide variety of purposes where insoluble particles are formed or degraded. Nephelometric method have been used, e.g., to detect the haze formation in beer[4] and the precipitation of antigen by antibody.[5] Turbidometry is a common method in analysis of microbial growth.[6] Nephelometric methods have been used also to follow the degradation of insoluble macromolecules. Examples are the measurement of amylolytic activity from barley extract[7] and body fluids.[8] Here the naturally soluble substrates, β-limit dextrin and amylo-

---

[3] Commission on Biotechnology, International Union of Pure and Applied Chemistry, in "Measurement of Cellulase Activities" (T. Ghose, ed.). Indian Institute of Technology, New Delhi, India, 1984.

[4] M. Nummi, M. Loisa, M. Chemardin, and L. Chapon, Brauwissenschaft 25, 229 (1972).

[5] E. A. Kabat and M. M. Mayer, "Experimental Immunochemistry." Thomas, Springfield, Illinois, 1961.

[6] G. G. Meynell and E. Meynell, "Theory and Practice in Experimental Bacteriology." University Press, Cambridge, England, 1970.

[7] B. T. O'Connell, G. L. Rubenthaler, and N. L. Murbach, Cereal Chem. 57, 411 (1980).

[8] L. Zinterhofer, S. Wardlaw, P. Jatlow, and D. Seligson, Clin. Chim. Acta 43, 5 (1973).

pectin, were rendered insoluble by gelatination or by treatment with organic solvent, respectively.

Amorphous cellulose is insoluble and can be standardized to a homogeneous substrate for nephelometry and turbidometry by exhaustive ball milling and filtering through fritted glass filter.

Preparation of Amorphous Cellulose Substrate

The crystallinity of cellulose is due to the highly ordered tight packages of parallel polyglucose chains held together by intramolecular hydrogen bonding. When these bonds are ruptured cellulose turns into a disordered amorphous form. In natural cellulose there are amorphous and crystalline areas as can be seen by X-ray diffractometric analysis. The crystalline areas can be rendered amorphous artificially by chemical or physical treatment. When the crystallinity of cellulose decreases the X-ray scattering curve as a function of the glancing angle changes in a typical manner. By comparing the relative heights in the maximum and in the minimum area an arbitrary estimate for the crystallinity is obtained. The estimate, index of order, correlates with the relative crystallinity of different cellulose preparations.[9]

In preparation of the amorphous cellulose substrate for the cellulase assay an index of order smaller than 0.5 was used as a criterion for loss of crystallinity in the substrate (Table I).

The substrate for hydrolysis was prepared[10] by milling 25 g Whatman CF11 medium length fiber cellulose powder (Whatman, Maidstone, England) in 300 ml 94% ethanol solution at 20° for 24 to 72 hr with a ball mill (Schwingmühle Vibrator, Siebtechnik, Mülheim, FRG) using steel balls. The substrate obtained was amorphous according to the index of order after 28 hr milling. For the standard procedure a milling time of 48 hr was chosen. The suspension was removed from the mill and the residue washed out with ethanol. The suspension was then transferred to 0.5 M sodium citrate. The iron dust released from the steel balls during milling was removed from the suspension by a magnetic stirring rod during a short period of stirring. The slurry was centrifuged and transferred to 0.05 M citrate buffer, pH 5.0, the optimal pH for the enzyme studied.

To obtain a homogeneous substrate for stable suspensions, which is necessary in nephelometry, cellulose was further filtered through fritted glass filters (Schott no. 1 and 2) and a fraction with fine particles (diameter

[9] O. Ant-Wuorinen and A. Visapää, *Paper Timber* **47**, 311 (1965).
[10] M. Nummi, P. C. Cox, M.-L. Niku-Paavola, and T.-M. Enari, *Anal. Biochem.* **116**, 133 (1981).

TABLE I
EFFECT OF MILLING TIME ON THE INDEX OF
ORDER OF CELLULOSE

Milling time (hr)	Index of order
2	0.81
8	0.76
24	0.54
48	0.40

90–150 μm) was collected into 300 ml 0.05 M citrate buffer, pH 5.0. This solution was used as the stock solution and diluted (1 : 100) for subsequent studies. The diluted suspension was further filtered through serum filters (Seraclear, Technicon Instruments Co., NY) just before use.

The amorphous cellulose formed a stable suspension in citrate buffer, pH 5.0. No settling or aggregation occurred within a few days as shown by constancy of nephelometer readings and reproducibility of cellulase assays. The amorphous cellulose was completely (95%) hydrolyzed by a cellulase preparation in the assay conditions during 6 hr.

Cellulase Preparation

Trichoderma reesei cellulases[11] precipitated with ammonium sulfate between 20 and 40% saturation were used to test the assay. The precipitate was dissolved and dialyzed against a 0.05 M citrate buffer, pH 5.0. The protein concentration in the preparation was 130 mg/ml by the Biuret method.[12]

Sugar Estimation

The carbohydrate contents of cellulose solutions were estimated with anthrone–sulfuric acid.[13] Glucose was used as standard.

Nephelometric Measurement

Nephelometric estimations were performed with Perkin-Elmer amylase–lipase Analyzer model 91 (Perkin-Elmer, IL). The temperature of the

[11] M. Nummi, M.-L. Niku-Paavola, T.-M. Enari, and V. Raunio, *FEBS Lett.* **113**, 164 (1980).
[12] H. U. Bergmeyer, E. Bernt, K. Gawehn, and G. Michal, *Methods Enzymatic Anal.* **1**, 174 (1974).
[13] R. L. Whistler and M. L. Wolfram, *Methods Carbohydr. Chem.* **1**, 390 (1962).

incubation chamber was 40°. The apparatus measured the light reflected at a right angle to the axis of the exciting light. The apparatus was connected to a Honeywell Electronic 194 chart recorder (Honeywell Corp.).

The nephelometer was calibrated to zero using a nonreflecting black block and adjusted to 50 nephelos units by using a standard supplied by the manufacturer. This provides a method of calibrating the machine continually and gives internal standardization. The cellulose stock suspension was diluted 1 : 100, to give a total reflection of about 80 units. A scaling of 10 mV on the chart recorder was sufficient to give a reading in the middle of the chart that could be accurately measured.

Turbidometric Measurement

Turbidity was measured as the absorbance at 620 nm by means of a Hitachi-Perkin-Elmer UV–VIS spectrophotometer.

Nephelometric Assay for Cellulase[11]

Samples of 20 μl of appropriately diluted cellulase solutions are incubated with 3 ml of cellulose suspension in 0.05 M citrate buffer, pH 5.0. The concentration of cellulose is 0.04% calculated from total sugar estimation and should give a nephelometric value of 80. Figure 1 presents a typical plot for a cellulase activity determination. The slope is calculated for the initial velocity from the linear part of the nephelometric trace as demonstrated in the picture. It was observed that if the substrate was completely hydrolyzed during a shorter period than 10 min the cellulase concentration was too high for measurement of the initial velocity of the reaction. When the cellulase concentrations are such that 10–20% of the substrate is hydrolyzed during 10 min the slopes represent the initial velocity: the slopes are linearly proportional to the enzyme concentration (Fig. 2).

The unit of activity is defined as the decrease in nephelometric value per minute.

By calculating the Δ nephelometric units as the amount of product liberated the cellulase activity can be converted to katals (Fig. 3).

Turbidometric Assay for Cellulase[11]

The cellulase activity as the initial velocity of the hydrolysis can also be determined with a photometer by measuring the change in absorbance at 620 nm.

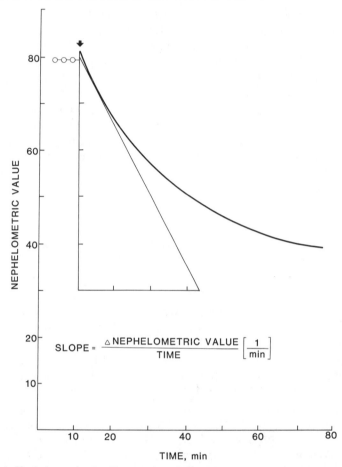

FIG. 1. Nephelometric plot for a crude cellulase preparation from *Trichoderma reesei*. The zero line (O—O—O) represents the light reflection of cellulose suspension alone. The arrow indicates the moment for enzyme addition. The calculation of slope is demonstrated in the picture.

Cellulase solution (200 μl) is incubated with 3 ml cellulose suspension (0.04%, A_{620} = 0.400) in 0.05 M citrate buffer, pH 5.0, at 50° for 10 min. This time is short enough to measure the initial activity for a broad range of enzyme concentrations and at the same time long enough to make the measurement of a series of samples possible. The absorbance of the solutions is measured after exactly 10 min against a citrate buffer blank. The unit of activity is defined as the change in absorbance per minute in the reaction conditions used. By converting the decrease in absorbance into

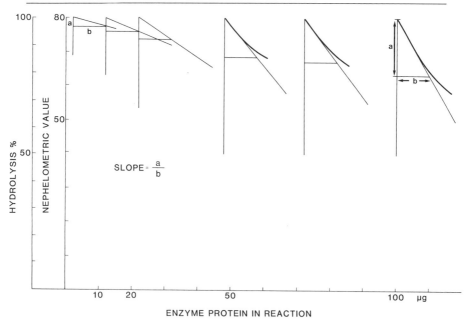

FIG. 2. Dependence between slopes and enzyme concentrations.

the product liberated the cellulase activity can be expressed as katals (Fig. 4).

The activity determinations should be made with a series of enzyme dilutions to find a linear relationship between the activity and the enzyme concentration for the cellulase preparation concerned.

The nephelometric and turbidometric assays described here measure the initial reaction rate in hydrolysis of cellulose. Thus, the methods are well suited for the determination of cellulase activities.

Turbidometric Microassay for Cellulase

Labsystems Multiskan MCC spectrophotometer (Labsystems, Helsinki, Finland) operating by fiber optics can be used to carry out the turbidometric method in microscale. On serum titer plates (Flow Laboratories, Inc., McLean, VA) 150 μl of amorphous cellulose (corresponding to 50 μg of substrate) is incubated with 0.5–5.0 μg of cellulase preparation at 50°. After 10 min, 30 min, 1 hr, and 20 hr the absorption was measured at 620 nm. The decrease in turbidity was plotted against the enzyme amount used (Fig. 5). It is seen that as little as 0.5 μg of enzyme can be detected by this method.

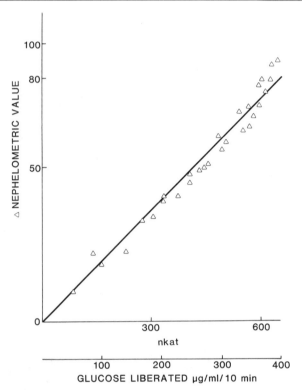

FIG. 3. Expression of Δ nephelometric units as katals.

Measurement of Xylanase Activity[14]

The nephelometric method can also be used to measure xylanase (EC 3.2.1.8, 1,4-β-D-xylan xylanohydrolase and EC 3.2.1.37, 1,4-β-D-xylan xylohydrolase) activity of fungal culture liquids.[14]

Insoluble xylan is prepared from birch pulp. Birch wood is pulped using a laboratory grinding machine equipped with a polishing stone (diameter 0.3 m, breadth 0.05 m) and rotated at 25 m/sec. Pulp consisting of 40% particles smaller than 200 mesh is used to prepare insoluble xylan. A 100-g portion of pulp is washed with 2.5 liters of 0.5 M sodium citrate buffer, pH 5.0, containing 0.1 M EDTA (Merck, Darmstadt, FRG) at room temperature for 20 hr. The washed pulp is then extracted with 1.7

[14] M. Nummi, J. M. Perrin, M.-L. Niku-Paavola, and T.-M. Enari, *Biochem. J.* **226,** 617 (1985).

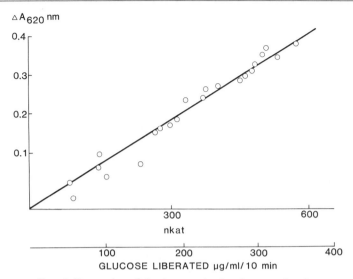

FIG. 4. Expression of the decrease in absorbance as katals.

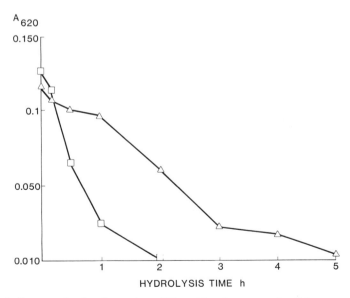

FIG. 5. Decrease in absorbancy in turbidometric microassay of cellulase preparation. Protein in reaction mixture: (□) 5 μg; (△) 0.5 μg.

liters of 1 M NaOH for 20 hr to get xylan into solution. After centrifugation the xylan extract is filtered through a Whatman (Maidstone, Kent, U.K.) GF/C 11 filter. The pH of the filtrate is adjusted to pH 4.5 by adding 10 M acetic acid to precipitate the xylan. The xylan suspension is stirred for 2 hr, after which the liquid is decanted and the xylan precipitate suspended in 0.1 M sodium acetate buffer, pH 4.0. After 20 hr the suspension is centrifuged and the xylan pellet dissolved in 400 ml of 0.6 M NaOH. After 20 hr the insoluble impurities are removed by centrifugation and filtration of the supernatant through a Whatman GF/C 11 filter. The xylan from the clear filtrate is thereafter precipitated with 800 ml of ethanol. The xylan precipitate is finally washed with 0.1 M sodium acetate buffer, pH 4.0, and then resuspended in the same buffer. The stability of the xylan substrate has been checked during long-term storage.

The hydrolysis mixture for nephelometric measurements contains 1 mg of xylan and 5–500 μg of enzyme protein in 3.2 ml of 0.05 M sodium acetate buffer, pH 4.0. A Perkin-Elmer (Oak Brook, IL) nephelometric amylase–lipase analyzer (model 91) is used to monitor the hydrolysis.[10] The temperature of the incubation chamber is 40°. The decrease in nephelometric values is measured and plotted as a function of the amount of enzyme protein used.

Acknowledgment

Dr. Martti Nummi believed that natural substrates should be used for the determination of enzymes hydrolyzing macromolecular substrates. His untimely death May 16, 1985 ended his pursuance of this concept. This chapter is largely based on his ideas.

[12] Selective Assay for Exo-1,4-β-glucanases

By M. V. DESHPANDE, L. G. PETTERSSON, and K.-E. ERIKSSON

The ability to estimate in absolute terms the activity of the individual enzymatic components of the cellulase complex, even in the presence of the other constituents, is important for several reasons. Whereas endo-1,4-β-glucanase (EC 3.2.1.4, cellulase) and β-1,4-glucosidase (EC

3.2.1.21) can be estimated by known reliable methods,[1] the third component, exo-1,4-β-glucanase [cellobiohydrolase (CBH), EC 3.2.1.91, cellulose 1,4-β-cellobiosidase], is difficult to measure unless it is obtained in pure form due to interference from the other enzymes and to the lack of specific substrates for this enzyme.

Two types of exoglucanases are documented, i.e., enzymes which split off a glucose unit from the nonreducing end of a cellulose chain and enzymes which release either glucose and/or cellobiose.[2-4] Amorphous cellulose, such as phosphoric acid-swollen cellulose, has been used for the determination of exoglucanases.[5] However, this technique has certain limitations. Crystalline cellulose is efficiently hydrolyzed only by a mixture of endo- and exoglucanases. Frequently, the exoglucanase activity is expressed in a semiquantitative manner as endo- and exoglucanase act in synergism.[5]

Rabinovich et al.[6] have suggested new substrates, carboxymethyl-substituted cellodextrins, for the selective determination of exoglucanase activity. However, this method is limited to the determination of those exoglucanases, also named exoglucosidases[7] which cleave off only glucose units from the nonreducing end of the cellulose chain and is not applicable to the cellobiohydrolases which release cellobiose units. Bartley et al.[8] have suggested a method to detect and differentiate between endo- and exoglucanases involving activity staining after polyacrylamide gel electrophoresis. Here, endo- and exoglucanases are efficiently separated before determination by using biospecific affinity chromatography according to van Tilbeurgh et al.[9] The procedure we describe

[1] T. K. Ghose, B. S. Montenecourt, and D. E. Eveleigh, "Measurement of Cellulase Activity: Substrates, Assays, Activities and Recommendations." Prepared for the Commission for Biotechnology, International Union of Pure and Applied Chemistry, July 1980.

[2] K.-E. Eriksson and B. Pettersson, Eur. J. Biochem. 51, 213 (1975).

[3] T. M. Wood and S. I. McCrae, Adv. Chem. Ser. 181, 181 (1979).

[4] International Union of Biochemistry, "Enzyme Nomenclature: Recommendations of the International Union of Biochemistry," p. 327. Academic Press, New York, 1984.

[5] K.-E. Eriksson and S. C. Johnsrud, "Experimental Microbial Ecology" (R. G. Burns and J. H. Slater, eds.), p. 134. Blackwell Scientific, London, 1982.

[6] M. L. Rabinovich, V. A. Mart'yanov, G. A. Chumak, and A. A. Klesov, Bioorg. Khim. 8, 396 (1982).

[7] N. W. Lützen, M. H. Nielsen, K. M. Oxenboell, M. Schülein, and B. Stentebjerg-Olesen, Philos. Trans. R. Soc. London, Ser. B 300, 283 (1983).

[8] T. D. Bartley, K. Murphy-Holland, and D. E. Eveleigh, Anal. Biochem. 138, 481 (1984).

[9] H. van Tilbeurgh, R. Bhikhabhai, L. G. Pettersson, and M. Claeyssens, FEBS Lett. 169, 215 (1984).

here uses synthetic substrates for the selective measurement of exoglucanases in the presence of other celluloytic components.[10]

Assay Method

Principle. The assay method is based on the finding by Nishizawa[11] that p-nitrophenyl-β-D-cellobioside is hydrolyzed by a β-glucosidase-free preparation of *Irpex* cellulase complex. Thus, the estimation of exoglucanases in a mixture containing other cellulolytic components utilizes the heterobiosides, p-nitrophenyl-β-D-cellobioside (pNPC) and/or p-nitrophenyl-β-D-lactoside (pNPL). Substrate linkages cleaved by the respective enzymes are

p-Nitrophenyl-β-D-cellobioside (pNPC):

(i)   pNP–Glu–Glu $\xrightarrow{\text{exoglucanase (CBH)}}$ pNP + Glu–Glu
      Action on the agluconic bond only

(ii) pNP–Glu–Glu $\xrightarrow[\beta\text{-glucosidase}]{\text{endoglucanase and}}$ pNP + Glu

      Both enzymes exhibit activity toward the agluconic bond and the holosidic bond

p-Nitrophenyl-β-D-lactoside (pNPL):

(i)   pNP–Glu–Gal $\xrightarrow[\text{endoglucanase}]{\text{exoglucanase (CBH) and}}$ pNP + Glu–Gal

      Both enzymes are active toward the agluconic bond only

(ii) pNP–Glu–Gal $\xrightarrow{\beta\text{-glucosidase}}$ no action

Reagents

Sodium acetate buffer, 50 mM, pH 5.0
p-Nitrophenyl-β-D-cellobioside   (or   p-nitrophenyl-β-D-lactoside),
    1 mg/ml in acetate buffer
Sodium carbonate, 2%
Glox glucose reagent (CAB Kabi Diagnostica, Stockholm, Sweden)
Potassium phosphate buffer, 50 mM, pH 7.0
Sulfuric acid, 41%
D-Glucono-1,5-δ-lactone, 0.5 mg/ml in acetate buffer

Procedure. The reaction mixture contains 1.8 ml of a 1 mg/ml solution of pNPC (or pNPL) in 50 mM sodium acetate buffer, pH 5.0, 0.5 mg/ml of D-glucono-δ-lactone,[10] and 0.2 ml of suitably diluted enzyme. The reaction mixture is incubated at 50° for 30 min and divided into two equal parts. One milliliter is used for spectrophotometric estimation of the p-

---

[10] M. V. Deshpande, K.-E. Eriksson, and L. G. Pettersson, *Anal. Biochem.* **138**, 481 (1984).
[11] K. Nishizawa, *J. Biochem.* (*Tokyo*) **42**, 825 (1955).

nitrophenol produced by the addition of 1 ml 2% Na_2CO_3. The amount of p-nitrophenol released is calculated from a molar extinction coefficient of 18.5 ml/μmol/cm for p-nitrophenol at 410 nm.

The remaining 1 ml of the assay sample is used for the estimation of glucose resulting from holosidic (in pNPC)/heterosidic (in pNPL) bond cleavage activity. After stopping the reaction by heating in a boiling water bath for 5 min, the glucose (and/or galactose) produced is estimated using the glucose oxidase–peroxidase test.[12,13]

Definition of an Enzyme Unit. One unit of agluconic bond cleavage activity is defined as the amount of enzyme liberating 1 μmol of p-nitrophenol per min under the assay conditions.[14] Linearity is obtained by diluting the enzyme to give a value of about 0.03–0.07 μmol of p-nitrophenol.

One unit of holosidic/heterosidic bond cleavage activity is defined as the amount of enzyme liberating 1 μmol of glucose (and/or galactose) per min under the assay conditions. Glucose production is linear to about 0.07–0.1 μmol.

Specific Inhibition of β-Glucosidases. D-Glucono-1,5-δ-lactone is a powerful inhibitor of β-glucosidases, but a poor inhibitor of exoglucanases, endoglucanases, and related enzymes.[15,16] It is important to determine the concentration of D-glucono-δ-lactone in the assay mixture which is necessary to eliminate interfering effect of β-glucosidases. D-Glucono-δ-lactone levels should be sufficient to inhibit β-glucosidase activity completely without affecting exo- and endoglucanase activities. Where enzyme mixtures from *Trichoderma reesei* QM 9414 and *Sporotrichum pulverulentum* are concerned, β-glucosidase activity is suppressed by 0.5 mg/ml D-glucono-δ-lactone.[10]

Influence of Endoglucanases on the Assay Procedure. To eliminate the interfering effect of endoglucanases, an initial standardization of the assay procedure is required. Using pure endoglucanase from the culture solution of mixed cellulases, it is necessary to determine endoglucanase (CMCase)[1] : agluconic bond cleavage activity and endoglucanase (CMCase) : holosidic bond cleavage activity ratios. In the case of endoglucanases from *T. reesei* and *S. pulverulentum* with pNPC as substrate, the endoglucanase (CMCase) : agluconic bond cleavage ratio is 50 : 1 for both Endo-1 and Endo-2 and the agluconic : holosidic ratio is 1 : 1 for Endo-1 and 3 : 1 for Endo-2.[10]

[12] H. U. Bergmeyer, K. Gawehn, and M. Grassl, "Methods of Enzymatic Analysis" (H. U. Bergemeyer, ed.), p. 457. Academic Press, New York, 1974.
[13] R. Bentley, this series, Vol. 9, p. 86.
[14] D. Herr, F. Baumer, and H. Dellweg, *Eur. J. Appl. Microbiol. Biotechnol.* **5,** 29 (1978).
[15] E. T. Reese and F. W. Parrish, *Carbohydr. Res.* **18,** 381 (1971).
[16] T. M. Wood and S. I. McCrae, *J. Gen. Microbiol.* **128,** 2973 (1982).

The exoglucanase activity is then obtained by deducting the estimated endoglucanase activity toward the agluconic bond from the total agluconic bond cleavage activity.

The limitation of the assay procedure is that purification of the endoglucanases is necessary for the standardization.

Acknowledgment

Financial support to one of us (M.V.D.) from the United Nations Industrial Development Organization (UNIDO) is gratefully acknowledged.

[13] Viscosimetric Determination of Carboxymethylcellulase Activity

By Michael A. Hulme

Cellulase (EC 3.2.1.4, 1,4-β-D-glucan 4-glucanohydrolase) cleaves β-1,4-glucosidic bonds randomly,[1] but this activity is difficult to measure with cellulose because the polymer is not soluble in water. Hence water-soluble derivatives such as carboxymethylcellulose are used where the degree of substitution is just sufficient to confer water solubility. Enzymatic hydrolysis of β-1,4-glucosidic bonds in this water-soluble substrate is thus termed carboxymethylcellulase activity.

Assay Method

Principle. Enzymatic hydrolysis of the β-1,4-glucosidic bond in carboxymethylcellulose generates one extra molecule for each bond that is hydrolyzed. The increased number of molecules generated in a given time can be measured as changes in the number–average molecular weight of the polymer and this weight can be calculated from the viscosity–average molecular weight obtained by viscometry. Viscosimetric methods can be applied only when enzymes cleave their substrate randomly because the assumption is made that the ratio of viscosity–average molecular weight to number–average molecular weight remains constant during the period

[1] International Union of Biochemistry, "Enzyme Nomenclature: Recommendations of the Commission on Biochemical Nomenclature." Academic Press, New York, 1984.

of assay, i.e., the initial period of hydrolysis. Most carboxymethylcellu-lases are considered to cleave their substrate randomly.[2,3]

A number of viscosimetric methods have been suggested for measuring carboxymethylcellulase activity. Some devise arbitrary units of activity[4]; some make substantial approximations of recognized equations relating viscosity measurements to number–average molecular weight[5]; some devise new empirical relationships involving constants that vary unsystematically[6,7]; and some choose assay conditions that require extensive corrections for shearing rates and kinetic energies of the solutions.[4,8] The following method was developed by Hulme[9] for calculating enzyme activity in standard international units from recognized equations without introducing approximations; it also allows the calculation to be simplified in defined conditions with little loss of accuracy; finally it can provide a simple method of comparing enzyme activities from rates of decrease of efflux times. The method depends on measuring reaction rates in a viscosimeter using defined concentration ranges of substrate and enzyme in conditions that need negligible correction of viscosity measurements for the kinetic energy and shearing characteristics of the liquid.

Reagents

Sodium carboxymethylcellulose with known degree of substitution between 0.5 and 0.7 and known number–average molecular weight
Acetate buffer (acetic acid plus sodium hydroxide) 50 mM, pH 5.0

Procedure. A 0.5% (w/v) solution of sodium carboxymethylcellulose in 50 mM acetate buffer (pH 5.0) is prepared by dissolving the polymer overnight at room temperature, filtering the solution through M porosity sintered glass, and freezing the filtrate at $-70°$ for 5 min. The solution is then thawed and boiled under reflux for 10 min. Efflux times in a No. 1 Ubbelöhde viscosimeter should be between 150 and 200 sec reproducible within 0.1 sec using a water bath set between 25 and 40° with the temperature maintained within 0.04°. If the efflux time is outside this range the polymer concentration should be adjusted. The efflux time for 50 mM acetate buffer (pH 5.0) is determined within 0.1 sec at the set temperature

[2] E. T. Reese (ed.), "Enzymic Hydrolysis of Cellulose and Related Materials." Macmillan (Pergamon), New York, 1963.
[3] G. J. Hajny and E. T. Reese (eds.), *Adv. Chem. Ser.* **95** (1969).
[4] R. Thomas, *Aust. J. Biol. Sci.* **9**, 159 (1956).
[5] R. Werner, *J. Polym. Sci., Part C* **16**, 4429 (1969).
[6] K. E. Almin and K.-E. Eriksson, *Biochim. Biophys. Acta* **139**, 238 (1967).
[7] K. E. Almin, K.-E. Eriksson, and C. Jansson, *Biochim. Biophys. Acta* **139**, 248 (1967).
[8] K. Manning, *J. Biochem. Biophys. Methods* **5**, 189 (1981).
[9] M. A. Hulme, *Arch. Biochem. Biophys.* **147**, 49 (1971).

(it will be near 70 sec). Stock solutions of sodium carboxymethylcellulose should be stored frozen to prevent depolymerization by microbial contaminants.

A suitable concentration of carboxymethylcellulase for assay is selected as follows. Sodium carboxymethylcellulose solution (10 ml) is pipetted into a 50-ml conical flask half immersed in the water bath and enzyme solution (3 ml) is added by pipet. The solutions are briefly mixed by swirling and poured into a No. 1 Ubbelöhde viscosimeter. The efflux time of the mixture is measured about 5 min after the start of pipetting the enzyme solution and again after 10 min. The difference in efflux times should be between 1 and 2 sec; suitable alteration of the enzyme concentration in the pipetted 3 ml may be needed. The enzyme activity is then assayed by repeating the above procedure with a suitable concentration of enzyme, recording the elapsed time to the nearest second between the start of enzyme addition and the start of measuring efflux time, and taking readings of elapsed time and corresponding efflux time every 5 min for 20 min. Units of enzyme activity are then calculated by one of three methods (see below).

Definition of Enzyme Activity. One standard international unit of carboxymethylcellulase is the amount of enzyme that catalyzes the initial hydrolysis of one microequivalent of β-1,4-glucosidic bonds per minute of a defined carboxymethylcellulose (known degree of substitution and number–average molecular weight) in defined conditions of temperature and pH.[1]

Since each bond broken generates one extra polymer molecule, the number of enzyme units per assay, A, that initially cleave P μg of sodium carboxymethylcellulose with number–average molecular weight M_n is

$$A = P \left[\frac{d}{dt} \left(\frac{1}{M_n} \right) \right]_{t=0} \tag{1}$$

The derivative is evaluated from a plot of $1/M_n$ against elapsed time extrapolated to zero time: in present assay conditions at least the first three readings should lie on a straight line with the required slope. Note that the intercept on this graph has no physical meaning: the measurement of elapsed time is set for simplicity at the start of the efflux period because only differences between successive readings are needed. The approximate reaction time is elapsed time plus half the efflux time.

Three progressively simpler (and progressively less accurate) ways of calculating A follow: the first requires calculation of each 5-min value of M_n; the second evaluates A in terms of the 5-min values of calculated specific viscosities; and the third compares enzyme activities in terms of the 5-min change in efflux times.

1. The specific viscosity, η_{sp} of the solution is calculated from the efflux time of the solution, t_s (sec), and the efflux time of the buffer, t_0 (sec):

$$\eta_{sp} = (t_s/t_0) - 1$$

Then using Huggins' equation[10] and the Mark–Houwink equation,[11] specific viscosity is related to the viscosity–average molecular weight of the polymer which in turn is related by a constant to the number–average molecular weight[12] (see Hulme[9] for details of equation manipulations). Hence

$$M_n = \frac{1}{\kappa} \left(\frac{(1 + 4K\eta_{sp})^{1/2} - 1}{2HKc} \right)^{1/x}$$

where κ is the constant relating number–average molecular weight to viscosity–average molecular weight, K is the constant in Huggins' equation, H is the constant multiplier, x is the constant exponent of the Mark–Houwink equation, and c is the polymer concentration in grams per deciliter. M_n is thus calculated for each reading using literature or experimentally determined values for the constants (reasonable values are given below). Hence A is calculated from Eq. (1).

2. To a good approximation, when specific viscosities are in the range recommended here, namely 0.5–1.5, and the Mark–Houwink exponent is within the common range for carboxymethylcellulose of 0.8–0.9,[13] then specific viscosity will be linearly related to the number–average molecular weight calculated above. Figure 1 shows an example using reasonable literature values for the constants of $\kappa = 2$,[13] $K = 0.24$,[9] $x = 0.86$,[13] and $H = 1.33 \times 10^{-4}$ dl/g.[7] The cited references also describe methods of measuring the constants if the reader chooses not to work with literature values. The value of c is 0.385 g/dl. Pursuing this example we find $M_n = 3.94 \times 10^4 \eta_{sp}$ and since $P = 5 \times 10^4 \mu g$ per assay then from Eq. (1) A can be evaluated as

$$A = 1.27 \times \frac{d}{dt} \left(\frac{1}{\eta_{sp}} \right)_{t=0}$$

The derivative is evaluated from a plot of $1/\eta_{sp}$ against elapsed time extrapolated to zero time: in present assay conditions at least the first three readings should lie on a straight line with the required slope. Again, the intercept on this graph has no physical meaning.

[10] M. L. Huggins, *J. Am. Chem. Soc.* **64,** 2716 (1942).
[11] R. Houwink, *J. Prakt. Chem.* **157,** 5 (1940).
[12] M. L. Miller, "The Structure of Polymers." Reinhold, New York, 1966.
[13] K. E. Almin and K.-E. Eriksson, *Arch. Biochem. Biophys.* **124,** 129 (1968).

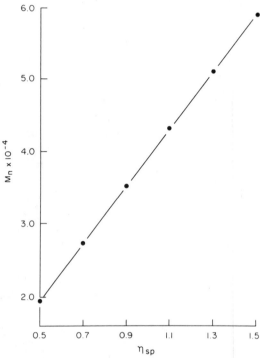

FIG. 1. The almost linear fit between M_n and η_{sp} for small ranges of η_{sp} when the Mark–Houwink exponent is 0.86.

3. If only ratios of enzyme activity are required and the Mark–Houwink exponent is between 0.8 and 0.9 then ratios of the rate of decrease in efflux time can be compared directly. These ratios will generally be within 1% of the ratios obtained when calculating international enzyme units by method 2. The concurrence is easily checked. If the specific viscosity falls from η_{sp1} to η_{sp2} [corresponding to efflux times t_1 (sec) and t_2 (sec)] in a short time t (min) then

$$\frac{d}{dt}\left(\frac{1}{\eta_{sp}}\right) = \frac{\eta_{sp1} - \eta_{sp2}}{\eta_{sp1}\eta_{sp2}}\frac{1}{t}$$

$$= \frac{(t_1 - t_2)}{(t_1 - t_0)(t_2 - t_0)}\frac{t_0}{t}$$

where t_0 (sec) is the efflux time for 50 mM acetate buffer. The part of the denominator describing efflux times should remain essentially constant in assay conditions specified here. Hence enzyme activity is directly proportional to the drop in efflux time per elapsed minute of incubation, i.e.,

$$A \propto \frac{t_1 - t_2}{t}$$

If the value of the Mark–Houwink exponent is outside the range of 0.8–0.9 then enzyme activity estimated above becomes less accurate. This lack of accuracy can be largely corrected by introducing a constant exponent, y, into the above relationship of the form

$$A \propto \left(\frac{t_1 - t_2}{t}\right)^y$$

However, evaluation of this exponent requires knowledge of all parameter values listed in method 1: it is calculated from a logarithmic plot of M_n against η_{sp} because M_n is found empirically to be almost exactly proportional to $(\eta_{sp})^y$.

Application of Assay Method to Crude Enzyme Preparations. The assay can be carried out on crude or purified enzyme preparations provided that suspended particles are removed from the enzyme solution either by filtering through M porosity sintered glass or by centrifuging at 38,000 g for 20 min. Occasionally, especially with partially purified enzymes, proteinase contaminants may slowly denature the enzyme,[14,15] resulting in irreproducible assays if the proteinase is not removed or inhibited.

[14] K. Selby and C. C. Maitland, *Arch. Biochem. Biophys.* **118**, 254 (1967).
[15] M. A. Hulme and D. W. Stranks, *Aust. J. Biol. Sci.* **27**, 457 (1974).

[14] Staining Techniques for the Detection of the Individual Components of Cellulolytic Enzyme Systems

By Michael P. Coughlan

Complete cellulolytic systems regardless of origin are multienzymatic in nature. Those produced by fungi, and possibly by bacteria, include endocellulases (cellulase or 1,4-β-D-glucan 4-glucanohydrolase, EC 3.2.1.4), exocellulases (cellulose 1,4-β-D-cellobiosidase or 1,4-β-D-glucan cellobiohydrolase, EC 3.2.1.91), and β-glucosidases (β-D-glucoside glucohydrolase, EC 3.2.1.21).[1] Moreover, in most such systems, multiple

[1] M. P. Coughlan, *Biotechnol. Genet. Eng. Rev.* **3**, 39 (1985).

forms of each of these enzymes are present. Whether or not these are true isoenzymes, each component of a system must be isolated and characterized before its possible role in cellulolysis *in vivo* may be established. To this end, one must be able to separate the various enzymes and to detect the relevant activities on the electrophoretic gels used in separation or in testing for homogeneity. In Sherlock Holmes' words,[2] "detection is, or ought to be, an exact science." Unfortunately, while endocellulase and β-glucosidase activities may readily be detected, there is currently no direct assay available for exocellulase activity.[1,3] Thus, it must be detected by indirect means (see below). One is reminded of Oliver's answer to Rosalind in "As You Like It,"[4] "and how, and why, and where this handkercher was stain'd." Indeed, a "handkercher," if made of cotton, may be a useful matrix for the separation of these enzymes.

Principles of Enzyme Detection

In the text that follows, it is assumed that the electrophoretic conditions necessary for separation of the enzymic components of the system under investigation have been established. Thus, we concentrate on the techniques used to locate a particular activity. In principle, activity on gels may be detected by several different means. The most obvious method is to cut the gel into thin slices, elute the protein from each slice, and assay for activity. An alternative is to run replicate gels. One is stained for protein. The zones corresponding to the protein bands on the first gel are then excised from the second gel, the protein content eluted, and assayed for activity. However, these procedures are time consuming, not particularly sensitive, and may lead to erroneous results if bands are closely spaced or protein concentration in any section is low.

More useful procedures include the incorporation of the substrate in the gels prior to electrophoresis. Subsequently, the gel is sprayed with, or immersed in, a solution containing reagents to detect the original substrate or the products of its hydrolysis. The location of the activity in question is then indicated by the position of the colored bands in an otherwise clear gel or, when appropriate, by that of the clear bands in an otherwise colored gel. Alternatively, the gel is sprayed with or immersed in a solution containing the appropriate buffered substrate and a suitable chromogenic reagent and the location of activity are determined as above.

[2] A. C. Doyle, "The Sign of Four." Spencer Blackett, London, 1890.
[3] R. Mullings, *Enzyme Microbiol. Technol.* **7**, 586 (1985).
[4] W. Shakespeare, "As You Like It," Act IV, Scene 3, Line 98. "The Yale Shakespeare," revised edition (S. C. Burchell, ed.). Yale University Press, New Haven, Connecticut, 1954.

In another procedure the gel containing the separated proteins is over-layed with a second gel containing buffered substrate and, when neces-sary, auxiliary enzymes and chromogenic reagents. Once again, the activ-ity in question is located by either clear or colored bands, as appropriate, on the original or on the overlaid gel. One advantage of this technique is that when activity has been located on the overlay, the original gel can then be stained for protein or for another activity and the results com-pared. Immunological detection of cellulolytic components, although more time consuming initially, is becoming more common because of its potential specificity. For example, proteins separated in one dimension by electrophoresis are "run" in a second dimension, reacted with specific antibody preparations (labeled with an enzyme such as peroxidase), and detected using a chromogenic substrate appropriate to the label.

It should be stressed that in dealing with a new cellulolytic enzyme system the investigator will have to determine the most suitable pH, buffer composition, substrate type, concentration, incubation time, and/or staining time to use for enzyme detection since these are likely to vary from one case to the next. It must also be stressed that because substrate specificities of the separate components of these systems frequently over-lap one should run at least two gels and stain each one for activity with a different substrate.

Reagents

Most of the chemicals mentioned in the text may be obtained from Sigma Chemical Co. (Poole, Dorset, UK and St. Louis, MO). These include, o-aminoazotoluene diazonium salt (Fast Garnet GBC), 6-bromo-2-naphthyl-β-D-glucoside, various carboxymethylcelluloses yielding so-lutions of different viscosities, cellobiose, cellulose azure (type II), cetylpyridinium chloride, Congo Red, 2,6-dichlorophenolindophenol, 1,5-glucono-D-lactone, hexadecyltrimethylammonium bromide, 4-meth-ylumbelliferyl-β-D-glucoside, 4-methylumbelliferyl-β-D-cellobioside, p-nitrophenyl-β-D-glucoside, p-nitrophenyl-β-D-cellobioside, and a kit for the colorimetric determination of glucose. The kit includes hexokinase, glucose-6-phosphate dehydrogenase, ATP, NADP$^+$, and iodonitrotetra-zolium. Another kit for the colorimetric determination of glucose, and hence of detecting enzymes that produce glucose, contains glucose oxi-dase, horseradish peroxidase, and ABTS, i.e., diammonium-2,2'-azinodi-(3-ethylbenzenethiazolinesulfonic acid), and is available from Boehringer-Mannheim GmbH (W. Germany). Hydroxyethylcellulose containing covalently linked Ostazin Brilliant Red H-3B may be obtained from Che-mapol Ltd., Division 15 (Prague, Czechoslovakia), agar from Difco (Detroit, MI), and GelBond (Marine Colloids Division, FMC Corp.) from Bio-Rad (Richmond, CA). Cellooligosaccharides and phosphoric acid-swollen cellulose may be prepared as described elsewhere.[5,6]

Detection of β-Glucosidases (EC 3.2.1.21)

1. Following electrophoresis, immerse gels in buffer, appropriate to the β-glucosidases under investigation, containing 20 mM cellobiose and incubate for up to 1 hr at 37°. Wash lightly with distilled water and then cover with clear agar containing glucose oxidase/peroxidase/ABTS reagent. [This agar overlay is made by heating 10 ml of a 1% (w/v) solution of agar in distilled water; the solution is then cooled to 50° and 10 ml of glucose oxidase reagent is added; the gel is then allowed to cool further and set.] The location of β-glucosidase (cellobiase) activity, i.e., the site at which cellobiose is cleaved to glucose, is indicated by a dark green band against a clear background.

2. As above except that the location of cellobiase activity is detected by the use of an agar overlay in which is incorporated 10 ml of a reagent comprised of hexokinase, glucose-6-phosphate dehydrogenase, ATP, NADP$^+$, and iodonitrotetrazolium in the appropriate buffer containing 20 mM cellobiose. The tetrazolium is converted to an insoluble colored formazan at the site of β-glucosidase activity.

3. Immerse gels in appropriate buffer containing 20 mM p-nitrophenyl-β-D-glucoside and incubate for up to 1 hr at 37°. Wash lightly in distilled water and then immerse in 0.4 M glycine–NaOH buffer, pH 10.8. The yellow color characteristic of the p-nitrophenolate ion develops at the site(s) of activity.

4. As (3) above except that 0.1% (w/v) 6-bromo-2-naphthyl-β-D-glucoside is used as substrate.[7] The brownish color characteristic of the bromonaphtholate ion develops at the site(s) of activity. Alternatively, the bromonaphthol released by enzyme action may be detected by coupling with diazonium salt to form a red complex.[8,9] Thus, as outlined in Ref. 8, the staining solution should contain 6-bromo-2-naphthyl-β-D-glucoside (10 mg), o-aminotoluene diazonium salt (150 mg), sucrose (3.24 g), methanol (10 ml), and 90 ml of a suitable buffer. Diffusion of the band, indicating the location of β-glucosidase activity, should not occur because

[5] T. D. Bartley, K. Murphy-Holland, and D. E. Eveleigh, *Anal. Biochem.* **140,** 157 (1984).
[6] A. McHale and M. P. Coughlan, *Biochem. J.* **199,** 267 (1981).
[7] V. Raunio, Ph.D. thesis. University of Turku–Finland, Kaleva, Oulu, Finland (1968).
[8] J. E. Rissler, *Appl. Environ. Microbiol.* **45,** 315 (1973).
[9] M. E. Mace, *Phytopathology* **63,** 243 (1973).

of "the affinity of bromonaphthol for protein and the water-insolubility of the red complex."[8]

5. Gels are sprayed with an appropriate buffer containing 10 mM 4-methylumbelliferyl-β-D-glucoside and, following a suitable period of incubation, are observed under near-UV light (e.g., 350 nm).[10] The location of activity is indicated by the fluorescence due to the methylumbelliferol released by enzyme action.

Detection of Endocellulases (EC 3.2.1.4)

Substrate Incorporated in Electrophoretic Gel

1. Following electrophoresis, gels containing carboxymethylcellulose (e.g., 0.5%, w/v) are washed with the appropriate buffer and incubated for 30 min at a suitable temperature.[11] They are then immersed in dinitrosalicylate reagent[12] (see Ref. 13 for method of preparation) and incubated at 100° for 15 min. The excess of dinitrosalicylate reagent is then decanted. Activity is located at the dark brown bands resulting from the reaction of dinitrosalicylate with the products of carboxymethylcellulose hydrolysis.

2. Following electrophoresis, gels containing 0.05% (w/v) carboxymethylcellulose are immersed in a suitable buffer and incubated for 10 min at 37°.[14] At this time the buffer is decanted and, if necessary, incubation is carried out for a further period ranging from 5 min to 18 hr. Following incubation, the enzymatic reaction is stopped by immersion of the gels in 60% H_2SO_4 for 5–10 min. The acid is then decanted, the gels are washed with distilled water, then immersed in solution containing 2% (w/v) KI and 0.2% (v/v) I_2, and incubated at room temperature for 30–60 min (or for shorter periods at 37°). The reaction of iodine with carboxymethylcellulose is analogous to the starch–iodine reaction. Consequently, the clear zones, at which substrate hydrolysis has taken place, indicate the location of endocellulase (CMCase) activity. Gels should be examined immediately in diffuse light as the color fades in a few minutes in direct light.[15] One should be aware of a possible artifact in cellulase screening by iodine staining recently reported by Zitomer and Eveleigh.[15] Small amounts of

[10] W. Hösel, E. Surholt, and E. Borgmann, *Eur. J. Biochem.* **84**, 487 (1978).
[11] J. N. Saddler and A. W. Khan, *Can. J. Microbiol.* **27**, 288 (1981).
[12] G. L. Miller, *Anal. Chem.* **31**, 426 (1959).
[13] M. P. Coughlan, this volume [40].
[14] R. Goren and M. Huberman, *Anal. Biochem.* **75**, 1 (1976).
[15] S. W. Zitomer and D. E. Eveleigh, *Enzyme Microb. Technol.* **9**, 214 (1987).

starch contaminating commercial agars gave false positive results in cellu-lase-negative, amylase-positive culture fluids. Accordingly, they recom-mended that amylase-digested agar, or better still, Gelrite be used as gelling agent in the preparation of agars for culture screening. It is un-likely that the above would be a problem in the case of amylase-free culture fluids or purified cellulase preparations or when polyacrylamide gels are being used. However, one should always be on the lookout for possible artifacts. In the iodine staining procedure described by Zitomer and Eveleigh[15] hydrolysis zones were visualized by flooding plates with KI/I_2 solution (KI, 1%; I_2, 0.5%) for 1–2 min (3–5 min for media contain-ing Gelrite) followed by a distilled water rinse and immediate examination in diffuse light.

3. It should also be possible to detect the location of endoglucanase activity by adapting the screening procedure of Farkas et al.[16] Thus, hydroxyethylcellulose containing 13.5% (w/v) of covalently linked Ostazin Brilliant Red H-3B should be incorporated in gels to a final con-centration of 0.2%. Following electrophoresis the gels should be incu-bated under conditions (buffer, pH, temperature, and time) appropriate to the endoglucanases in question. Activity is indicated by the appearance of clear zones against a colored background.

Substrate Incorporated in Overlay Gel

1. Following electrophoresis, gels are immersed in an appropriate buffer for a short period.[5,17–25] The buffer is then decanted and the gels are overlaid with a thin film (0.5–1.5 mm) of agarose (usually 1–2%, w/v in appropriate buffer) containing 0.5% (w/v) amorphous cellulose or, de-pending on the type, 0.1–1.4% (w/v) CM-cellulose and incubated under moist conditions for periods ranging from 15 min to 24 hr depending on the substrate used, the temperature of incubation, and the amount of activity present.[17,18] The above procedure is facilitated[18] by preparing the agarose film on GelBond. Whichever of the above substrates is used, the

[16] V. Farkas, M. Liskova, and P. Biely, *FEMS Microbiol. Lett.* **28,** 137 (1985).
[17] M. Nummi, M.-L. Niku-Paavola, T.-M. Enari, and V. Raunio, *FEBS Lett.* **113,** 164 (1980).
[18] B. Sprey and C. Lambert, *FEMS Microbiol. Lett.* **18,** 217 (1983).
[19] R. Lamed, E. Setter, and E. A. Bayer, *J. Bacteriol.* **156,** 828 (1983).
[20] L. Hankin and S. L. Anagnostakis, *J. Gen. Microbiol.* **98,** 109 (1977).
[21] B. S. Montenecourt and D. E. Eveleigh, *Adv. Chem. Ser.,* **181,** 289 (1979).
[22] P. J. Wood, *Carbohydr. Res.* **94,** C19 (1981).
[23] R. M. Teather and P. J. Wood, *Appl. Environ. Microbiol.* **43,** 777 (1982).
[24] P. Beguin, *Anal. Biochem.* **131,** 333 (1983).
[25] C. R. McKenzie and R. E. Williams, *Can. J. Microbiol.* **30,** 1522 (1984).

location of endocellulase activity in the original gel is indicated by the position of the clear zones in the otherwise turbid overlay. The fact that exocellulases are active against amorphous cellulose but inactive against CM-cellulose[1] provides an indirect method for locating these enzymes.

2. As above except that visualization of the clear zones formed after incubation of original and overlay is facilitated by precipitating the undigested substrate. This may be accomplished by flooding the overlay with 2-propanol[19] or with 1% (w/v) aqueous hexadecyltrimethylammonium bromide[20] or with 1% (w/v) cetylpyridinium chloride[21] and incubating for a short time (2–3 hr).

3. The fact that certain polysaccharides including β-D-glucans and substituted celluloses react with dyes such as Congo Red to form colored dye–polysaccharide complexes whereas cellooligosaccharides do not provide another means of detecting β-D-glucanase action.[22] Thus, visualization of the clear zones formed as in (3) above is facilitated by washing the overlay briefly in 1 M NaCl, then staining with an aqueous solution of Congo Red (1 mg/ml) for 30 min at room temperature. The excess of dye is then decanted and the overlay is immersed in a solution of 1 M NaCl until the hydrolysis zones become visible (5–10 min).[5,23–25] The overlay is then rinsed with 1 N HCl or with 5% (w/v) trichloroacetic acid or with 1 N NaOH and the sites of endocellulase activity indicated by the clear zones against a blue (acid) or red (alkaline) background. The procedure is facilitated by preparing the agarose overlay on GelBond plastic film. Indeed, it is claimed that as little as 1 ng of enzyme can be detected using the technique described.[25] Staining with Congo Red may also be used to facilitate visualization of zones of endoglucanase action in those procedures in which CM-cellulose is incorporated into the electrophoretic gel.

4. In a variation[6] of the technique described in (2), the gels containing the separated proteins are immersed in buffer containing 2% (w/v) agar and 5% (w/v) phosphoric acid-swollen cellulose azure. The agar is allowed to solidify and, when solidified, the mixture is incubated for 1 hr at a suitable temperature. During incubation, dye released from the cellulose-azure diffuses into the original gel at the locations corresponding to the endo- and exocellulases that effect its release. The purple bands indicating the location of these enzymes are readily observed on rinsing the original gel with distilled water. However, for photographic or scanning purposes it is better to immerse the gels in 50 mM NaOH so that bands are colored blue against a clear background. Once again one should note that both endo- and exocellulases will be detected by this technique whereas only the former are detected when CM-cellulose is used as substrate (as, e.g., in 3 above).

Gels are Immersed in Buffered Substrate Solution

The fact that cellooligosaccharides are active substrates for β-D-glu-canases is the principle underlying another detection technique.[5] Following electrophoresis, gels are washed for 15 min in two changes of appropriate buffer. They are then rinsed with distilled water and immersed in an appropriately buffered solution containing a sodium borohydride-reduced mixture of cellohexaose and cellopentaose and incubated for 15 min at a suitable temperature. After incubation the gel is rinsed with distilled water, then with 0.1 M iodoacetamide for 5 min, and then again with distilled water. The gels are then placed in a 2-liter beaker containing 200 ml of 0.1% (w/v) triphenyltetrazolium chloride in 0.5 M NaOH. The solution is then heated over a gas burner with gentle agitation until red bands, indicative of the location of reaction products and hence of β-glucanase activity, appear (usually several minutes). The gel is immediately rinsed in distilled water and soaked in several changes of 7.5% (w/v) acetic acid. In this procedure it is important that heating and initial storage of the gel be carried out in the dark and that heating be terminated before the appearance of a general red background. The use of replicate lanes in a single slab gel, one of which is overlaid with agarose containing CM-cellulose and the other treated as here, serves to distinguish between endo- and exocellulases.

Detection of Exocellobiohydrolases (EC 3.2.1.91)

As mentioned above, exocellobiohydrolases can be distinguished from endocellulases by virtue of the fact that they are active against cellooligo-saccharides and acid-swollen cellulose but inactive against CM-cellulose whereas endocellulases are active against all three substrates.[1] Exocello-biohydrolases may also be detected as follows.

1. Following electrophoresis, gels are washed briefly in an appropriate buffer and then soaked in the same buffer containing 10 mM 4-methyl umbelliferyl-β-D-cellobioside (MUC) and 1 mM gluconolactone. After a suitable period of incubation the gels are rinsed in distilled water and observed under near-UV light (e.g., 350 nm). The location of exocello-biohydrolase activity is indicated by fluorescence due to methylumbelli-ferol released during the reaction. Gluconolactone is included in the incubation mixture so as to inhibit β-glucosidase activity since these enzymes may also be active with the substrate used (B. Sprey, personal communication). Some endoglucanases are also active against MUC.[26] Such en-

[26] M. Claeyssens, P. Tomme, and H. van Tilbeurgh, *Proc. Int. Conf. Biotechnol. Pulp Pap. Ind., 3rd*, p. 183 (1986).

zymes may be distinguished from exocellobiohydrolases by virtue of the fact that they, i.e., the endoglucanases, would also stain for activity with CM-cellulose as substrate (see above). Exocellobiohydrolases may also be distinguished from endoglucanases by virtue of the fact that the latter may be inhibited by cellobiose (A. McHale, personal communication). Thus, by including appropriate concentrations of cellobiose in MUC staining solutions only the exocellobiohydrolase bands should fluoresce. However, it must be stated that differential inhibition by cellobiose has been demonstrated for a relatively small number of cases as yet (A. McHale, personal communication). Cellobiose dehydrogenases may also utilize MUC as substrate but do not release free methylumbelliferol. In any event they should be inactive under the staining conditions used because of the lack of a suitable electron acceptor (see below).

2. Following electrophoresis, immerse gels in buffer containing 10 mM p-nitrophenyl-β-D-cellobioside and 1 mM gluconolactone and incubate for up to 1 hr at 37°. Wash lightly with distilled water and immerse in 0.4 M glycine–NaOH buffer, pH 10.8. The yellow bands, characteristic of the p-nitrophenolate ion, indicate the site of exocellobiohydrolase action.

3. Since exocellobiohydrolases by definition catalyze the release of cellobiose from susceptible substrates one should be able to detect such activity as follows. Following electrophoresis, the gel is overlaid with agarose containing cellobiose dehydrogenase and suitable chromogenic reagents (see below). The location of exocellobiohydrolase activity is indicated by the clear or colored bands, as appropriate.

Detection of Exoglucohydrolases (EC 3.2.1.74)

Exoglucohydrolases (1,4-β-D-glucan glucohydrolases, EC 3.2.1.74) catalyze the successive removal of glucose units from 1,4-β-D-glucans.[1] Whereas all such enzymes are active against cellooligosaccharides, activity against amorphous cellulose and CM-cellulose depends on the source of the enzyme as does inhibition by gluconolactone.[27,28] If inactive against CM-cellulose they may be distinguished from endocellulases. The fact that they effect the removal of successive glucose units (detectable using the glucose oxidase/peroxidase/ABTS reagent) rather than cellobiose units (detectable as in section 2 above) from amorphous cellulose or cellooligosaccharides would distinguish them from exocellobiohydrolases. Moreover, if they are not inhibited by gluconolactone they may be differentiated from β-glucosidases.

[27] D. R. Barras, A. E. Moore, and B. A. Stone, *Adv. Chem. Ser.*, **95,** 105 (1969).
[28] T. M. Wood and S. I. McCrae, *Carbohydr. Res.* **110,** 291 (1982).

Detection of Cellobiose Dehydrogenase (EC 1.1.99.18)

The cellobiose dehydrogenase (cellobiose : acceptor 1-dehydrogenase, EC 1.1.99.18) produced by species of *Monilia* catalyzes the oxidation of cellobiose to cellobiono-1,5-lactone using dichlorophenol indophenol as an electron acceptor.[29] The location of this activity on gels may be determined by immersion in 100 mM citrate–phosphate buffer, pH 6.6, containing 2 mM cellobiose, 1 mM dichlorophenol indophenol (DCI), and 1 mM glucono-1,5-lactone.[29] After 10–15 min incubation at 37° the gel, due to absorption of DCI, is blue except for a colorless band(s) where the dye has been reduced as a result of cellobiose dehydrogenase activity. Glucono-1,5-lactone is included in the reaction mixture to inhibit β-glucosidases which, if present, would catalyze cleavage of the cellobiose.

Immunochemical Methods

Cellulolytic enzymes retain their activity after immunoprecipitation. Nummi *et al.*[17] made use of that fact to effect partial characterization of the system produced by *Trichoderma reesei*. Thus, following immunoelectrophoresis, the locations of endocellulase and β-glucosidase activities on the gel were determined as described above. Lützen *et al.*[30] identified more than 20 antigenic components in the culture filtrate of this organism. More recently, Riske *et al.*[31] have initiated the characterization of this system using mono- and polyclonal antibody preparations to identify the various components after electrophoresis. The principles of the Western blot detection techniques used in such procedures have been well documented.[32] Single immunodiffusion and ELISA techniques using appropriate antisera and affinity-purified IgG, respectively, have been used to determine the specific amounts of exocellobiohydrolase in culture filtrates of *Trichoderma viride*.[33] The results obtained by these immunochemical methods agreed well with those obtained by traditional assays of activity against filter paper.

[29] R. F. H. Dekker, *J. Gen. Microbiol.* **120,** 309 (1980).
[30] N. W. Lützen, M. H. Nielsen, K. M. Oxenboell, M. Schülein, and B. Stentebjerg-Olesen, *Philos. Trans. R. Soc. London, Ser. B* **300,** 283 (1983).
[31] F. J. Riske, I. Labudova, L. Miller, D. E. Eveleigh, and J. D. Macmillan, *Annu. Meet. Am. Soc. Microbiol.* Abstr. O1 (1985).
[32] W. N. Burnette, *Anal. Biochem.* **112,** 195 (1981).
[33] T. K. Oh, S. H. Kim, and K. H. Park, *Biotechnol. Lett.* **8,** 403 (1986).

[15] Liquid Chromatography of Carbohydrate Monomers and Oligomers

By J. K. LIN, B. J. JACOBSON, A. N. PEREIRA, and M. R. LADISCH

Introduction

Liquid chromatography is in wide use for separating monosaccharides and oligosaccharides. There are four basic modes of separation involved: (1) gel-permeation chromatography (GPC), (2) ion-exchange chromatography (IEC), (3) adsorption chromatography, and (4) partition chromatography. Numerous papers have indicated use of these four modes in carbohydrate analysis by classical liquid chromatography as well as high-performance liquid chromatography (HPLC).[1-3]

GPC is based on size exclusion, and has been extensively employed to fractionate homologous series of oligosaccharides, including cellodextrins, cyclodextrins, xylodextrins, and maltodextrins. Packings capable of separating various carbohydrate oligomers include BioGel P-2, 4, and 6 (polyacrylamide gel),[4-6] Bio-Glas (granular porous glass),[7] Sephadex (dextran gels),[8,9] and μBondagel and μPorasil GPC 60 Å columns.[10] Maltodextrins and xylodextrins with a degree of polymerization (DP) of 13–15 are fractionated in 11–20 hr.[4,5]

Water or aqueous alcohols are generally used as the eluent in GPC. GPC is capable of fractionating water-soluble oligosaccharides but not monosaccharides which have similar sizes. The main problem associated with GPC gels are their poor mechanical strength and inability to withstand high pressures associated with higher flow rates. Thus analysis time in GPC is also long requiring up to 24 hr per analysis. In comparison, adsorption chromatography, based on the affinity of a solute for the adsorbents, has been shown to give good separation of monosaccharides

[1] A. Heyraud and M. Rinaudo, *J. Liq. Chromatogr.* **4** (Suppl. 2), 175 (1981).
[2] M. R. Ladisch and G. T. Tsao, *J. Chromatogr.* **166**, 85 (1978).
[3] M. R. Ladisch, A. L. Huebner, and G. T. Tsao, *J. Chromatogr.* **147**, 185 (1978).
[4] N. K. Sabbagh and I. S. Fagerson, *J. Chromatogr.* **120**, 55 (1976).
[5] M. John, J. Schmidt, C. Wandrey, and H. Sahm, *J. Chromatogr.* **247**, 281 (1982).
[6] K. Hamacher, G. Schmid, H. Sahm, and C. Wandrey, *J. Chromatogr.* **319**, 311 (1985).
[7] G. Belue and G. D. McGinnis, *J. Chromatogr.* **97**, 25 (1974).
[8] W. Brown, *J. Chromatogr.* **52**, 273 (1970).
[9] W. Brown and Ö. Anderson, *J. Chromatogr.* **57**, 255 (1971).
[10] P. J. Kundsen, P. B. Eriksen, M. Fenger, and K. J. Florentz, *J. Chromatogr.* **187**, 373 (1980).

and oligosaccharides using carbon,[11,12] silica,[13] and polysaccharide[14] based supports in classical liquid chromatography using nonaqueous or partially aqueous solvents. Separation of carbohydrate monomers and oligomers reported for cation-exchange and anion-exchange resins include glucose, ribose, xylose, sucrose, raffinose, galactose, fructose, lyxose, mannose, arabinose, uronic acids, aldonic acids, aldobionic acids, xylooligosaccharides, and maltodextrins using water as eluent (refer to Table 2-9 in Ref. 1).

Partition chromatography is based on the relative solubilities of the solute in the mobile phase versus the support. Partition chromatography falls into two categories, normal-phase and reversed-phase chromatography. Normal-phase chromatography can be done on underivatized silica packing or bonded phases such as cyanopropyl, aminopropyl, and polyfunctional amine to silica.[15] Separation of carbohydrate monomers and oligomers by chemical bonded phases have been achieved by using aminopropyl-bonded phase[16,17] (for starch oligomers up to DP 7), polyfunctional amine-bonded[18,19] (for starch oligomers and maltodextrins up to DP 20) in 1 hr with acetonitrile/water as the eluent. Furthermore, the carbohydrate columns such as μBondapak and Partisil-10 PAC with amino- and cyano-bonded phases are also commercially available for separating mono- and oligosaccharides.[20-24] This system appears to be a rapid method for sugar analysis. However, an aqueous organic eluent is not suitable for oligosaccharides such as cellodextrins which have low solubility or samples which have noncarbohydrate components which precipitate in a semiaqueous environment. Thus, this form of partition chromatography has limitations.

In reversed-phase chromatography, the silica with nonpolar functional groups such as C_{18}, phenyl, C_8, cyano, and amine is less polar than the mobile phase. Thus, the compounds having a low polarity are eluted later

[11] G. L. Miller, J. Dean, and R. Blum, *Arch. Biochem. Biophys.* **91**, 21 (1960).
[12] D. French, J. F. Robyt, M. Weintraub, and P. Knock, *J. Chromatogr.* **24**, 68 (1966).
[13] E. E. Dickey and M. L. Wolfrom, *J. Am. Chem. Soc.* **71**, 825 (1949).
[14] S. Gardell, *Acta Chem. Scand.* **7**, 201 (1953).
[15] Millipore Corporation, "Waters Sourcebook for Chromatography." Waters Chromatography Division, Millipore Corporation, Massachusetts, 1985.
[16] R. Schwarzenbach, *J. Chromatogr.* **117**, 206 (1976).
[17] B. B. Wheals and P. C. White, *J. Chromatogr.* **176**, 421 (1979).
[18] K. Aitzetmüller, *J. Chromatogr.* **156**, 354 (1978).
[19] C. A. White, P. H. Corran, and J. F. Kennedy, *Carbohydr. Res.* **87**, 165 (1980).
[20] J. C. Linden and C. L. Lawhead, *J. Chromatogr.* **105**, 125 (1975).
[21] B. Zsadon, K. H. Otta, F. Tudös, and J. Szejtli, *J. Chromatogr.* **172**, 490 (1979).
[22] F. M. Rabel, A. G. Caputo, and E. T. Butts, *J. Chromatogr.* **126**, 731 (1976).
[23] G. J. L. Lee and H. Tieckelmann, *Anal. Biochem.* **94**, 231 (1979).
[24] G. J. L. Lee and H. Tieckelmann, *J. Chromatogr.* **195**, 402 (1980).

than polar compounds. In comparison with normal phase, these packings are faster to equilibrate, and use less organic solvent. Carbohydrate separations on reversed-phase HPLC have been achieved by using amino-bonded phase[25,26] (for starch oligomer up to DP 8, cyclosophoraose up to DP 35, curdlan up to DP 15, dextrans up to DP 25, inulin up to DP 20, and cellodextrins up to DP 5) and C_{18}-bonded phase[27-30] (for starch oligomers up to DP 14, maltodextrins up to DP 13, cellodextrins up to DP 6, amylose up to DP 30). In most cases, water or acetonitrile–water mixture is used as the eluent and analyses take less than 60 min. For rapid analysis of oligosaccharides, the Dextro-Pak and Sugar-Pak columns are also suitable.[15] These columns are available in cartridge form for use in radial compression devices.

Ion-exchange resins form a major class of chromatographic supports in sugar separations. Two types of ion-exchange resins, cation- and anion-exchange resins, have been shown to be capable of separating saccharides.[2,3,31,32] Cation-exchange resins separate mono- and oligosaccharides in 30 min using water as the sole eluent.[2,3] Anion-exchange resins also give good separation of monosaccharides using aqueous ethanol[31] or acetonitrile/water.[32] However, anion-exchange resins tend to promote sugar conversion reactions. Thus, application of this type of resin to carbohydrate separation is limited.[33]

Cation-exchange chromatography has been successfully applied to rapid separation of mono- and oligosaccharides. Ladisch et al.[2,3] developed a low pressure LC system designed for rapid analysis of carbohydrate. Cation-exchange resins such as Aminex Q15S and Aminex 50W-X4 in the calcium form give excellent separation of monosaccharides in 30 min and oligosaccharide, e.g., cellodextrins (up to DP 7) in 30 min using water as the sole eluent. Several counterions on cation exchange resins are applicable for different separations. These counterions include silver, calcium, lead, hydrogen, and cadmium.[2,3,34-38] Of these counterions, the

[25] B. Porsch, J. Chromatogr. **320**, 408 (1985).
[26] M. D'Amboise, D. Noël, and T. Hanai, Carbohydr. Res. **79**, 1 (1980).
[27] N. W. H. Cheetham and P. Sirimanne, J. Chromatogr. **207**, 439 (1981).
[28] N. W. H. Cheetham and G. Teng, J. Chromatogr. **336**, 161 (1984).
[29] L. A. T. Verhaar and B. F. M. Kuster, J. Chromatogr. **284**, 1 (1984).
[30] P. Vratny, J. Coupek, S. Vozka, and Z. Hostomska, J. Chromatogr. **254**, 143 (1983).
[31] R. Oshima, N. Takai, and J. Kumanotani, J. Chromatogr. **192**, 452 (1980).
[32] D. Noël, T. Hanai, and M. D'Amboise, J. Liq. Chromatogr. **2**, 1325 (1979).
[33] P. Jandera and J. Churáček, J. Chromatogr. **98**, 55 (1974).
[34] H. D. Scobell and K. M. Brobst, J. Chromatogr. **212**, 51(1981).
[35] L. E. Fitt, W. Hassler, and D. E. Just, J. Chromatogr. **187**, 381 (1980).
[36] Bio-Rad Laboratories, "Bio-Rad HPLC Column for Carbohydrate Analysis." Bio-Rad Laboratories, Richmond, California, 1986.
[37] B. J. Jacobson, M.S. thesis. Purdue University, West Lafayette, Indiana, 1982.
[38] J. Schmidt, M. John, and C. Wandrey, J. Chromatogr. **213**, 151 (1981).

calcium form with water as eluent is most suitable for the separations of most monosaccharides, oligosaccharides, cellodextrins (up to DP 7), and starch oligomers (up to DP 8).[2,3,35-37] The lead form with water as eluent gives separation of monosaccharides derived from hemicellulose hydrolysis (i.e., xylose, arabinose, mannose, glucose, and galactose).[36,37] The silver forms of Aminex 50W-X4, Aminex Q15S, and Aminex A-7 separate oligosaccharides, found in corn syrup, to a greater extent than the calcium form of the same resins.[34] However, the columns packed with resin in the silver form are less stable than those in the Ca^{2+} form.[37]

The principles of separation using cation-exchange resins have been reviewed by Jacobson.[37] The separation of carbohydrates may result from size exclusion, complexing with counterions, and/or adsorption effects. Of these effects, a key factor for saccharide separation in an aqueous system is the choice of the appropriate counterion since the cations adsorbed on a resin significantly influence the separation.

The advantages of using cation-exchange resins for carbohydrate monomers and oligomers are readily apparent. The system is capable of speedy analysis at low pressure. Furthermore, water is used as the sole eluent. This enhances the stability of the column used. The column may be continuously used for over 2600 hr and some columns in our laboratory have lasted for over 18 months of intermittent use.[2] Perhaps most important is the stability of properly packed cation-exchange resins when used with chemically "dirty" samples containing noncarbohydrate components, which otherwise foul normal- and reversed-phase supports. Liquid chromatography of carbohydrates, cellulose and biomass hydrolyzates, cellodextrins, maltodextrins, sugar alcohols, volatile fatty acids, fermentation broths, and plant sugars is almost exclusively carried out using cation-exchange resin columns packed in our laboratory. Methods which have evolved in our laboratory over a period of 10 years are described below. Thus, in this chapter, emphasis is placed upon appropriate techniques for using cation-exchange resins for separation and quantification of carbohydrate monomers and oligomers.

Methods

Instrumentation

Basic instrumentation for liquid chromatography is supplied by many manufacturers and consists of an eluent reservoir, pump, injector, column, detector, and connecting tubing. An example is given in Fig. 1a. All components between the injector, where the sample is introduced, and the detector cause dispersion which can interfere with the column's performance. This is referred to as band spreading. A typical instrument can be checked for band spreading by removing the column and inserting a "zero" dead volume union in its place. A sample containing a solute detectable by the detector is then injected. For a differential refractometer, glucose at 1 mg/ml concentration will be convenient. For a normal instrument, solute contained in a 10 μl sample injection should elute in a volume of 150 μl or less (as measured from the peak width at the baseline). If the volume (peak width or instrument band width) is significantly greater, the instrument components should be checked for an improperly functioning injector, poor tubing connections between various instrument components, a detector cell which has a dead volume greater than 5–10 μl, or tubing connections having improperly seated ferrules.

A variety of detectors are available for detection of sugars and oligosaccharides. Our experience has shown the differential refractometer type of detector to be the best suited for separations involving carbohydrates.

The component in the liquid chromatograph responsible for fractionating the solubles is, of course, the column. The remainder of this chapter describes preparation of packed columns for separating carbohydrate monomers and oligomers using aqueous eluents at isocratic conditions.

Packing Materials

The column packing support is usually either Aminex 50W-X4 (Bio-Rad, Richmond, CA; 4% cross-linking, 20–30 μm particle size) or Aminex Q15S (Bio-Rad, 8% cross-linking 22 μm particle size) in either the calcium or hydrogen forms. The higher degree of cross-linking provides the mechanical stability to the resin matrix, although this resin excludes oligomers of DP 3 or higher. A column packed with Aminex Q15S may be operated at higher flow rates, making it suitable for separation of monosaccharides and oligosaccharides (corn syrup up to DP 3).[39,40] For a resin with a low degree of cross-linking, the resin matrix is more open, and therefore, more sensitive to compaction and plugging. The advantage of the Aminex 50W-X4 is that it will separate oligosaccharides (cellodextrins up to DP 7 and corn syrup up to DP 8).[2,40]

Resin Preparation

The cation-exchange resins in the sodium form were purchased from Bio-Rad (Rockville Center, NY) (Aminex Q15S, 22 ± 3 μm in diameter and Aminex 50W-X4, 20–30 μm in diameter), or from Calbiochem (San Diego, CA) and then converted to calcium or other counterion form be-

[39] J. K. Palmer and W. B. Brandes, J. Agric. Food Chem. **22**, 709 (1974).
[40] H. D. Scobell, K. M. Brobst, and E. M. Steele, Cereal Chem. **54**, 905 (1977).

fore use. The procedure used in our laboratory for preparing cation-exchange resin prior to packing is given below. Calcium chloride is used for converting to the calcium form. In the case that the resin is to be prepared in another counterion form, the chloride salt of that ion will typically be used in place of $CaCl_2$. Otherwise the procedure is the same for $CaCl_2$.

1. Place 25–50 g (wet) resin into a 1-liter graduated cylinder, and fill with 1000 ml of deionized water in 200 ml increments while swirling. Let

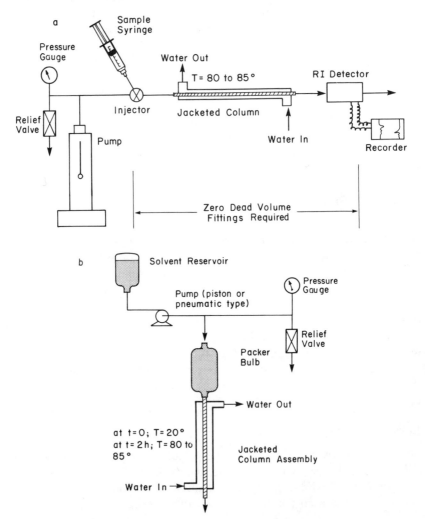

FIG. 1. Schematic diagram of key steps in packing appropriately treated resin. (a) Instrument configuration; (b) packing configuration; (c) filling procedure.

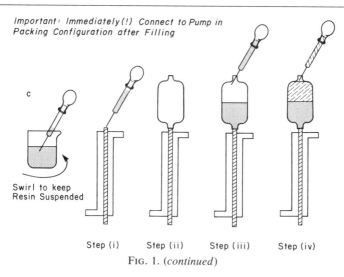

Step (i)   Step (ii)   Step (iii)   Step (iv)

FIG. 1. (*continued*)

the resin settle overnight and decant the water. This procedure is repeated 10 times or more until the supernatant above the settled resin is devoid of fines, with subsequent resin settling times of about 1–2 hr. Several days are usually needed to complete step 1.

2. Remove the excess water and rinse the resin with 1 liter of the acidic solutions, respectively, in the following order: 0.5 N HCl, 2.0 N HCl, and 6.0 N HCl. The slurrying/settling procedure is the same as step 1.

3. If the H^+ form is desired, leave the resin in 6 N HCl solution and go to step 5. Otherwise, rinse with deionized water using the procedure in step 1, until the pH is between 5 and 7 and then go to step 4.

4. Rinse the resin with 1.5–2 liters of $CaCl_2$ solutions, respectively, in the following order: 0.5, 2.5, 5.0, and 10% $CaCl_2$. A gradual increase in salt concentration is used here to minimize fracturing of the ion-exchange resins due to osmotic pressure shock otherwise induced by a large change in ion concentration. Leave the resin in the 10% $CaCl_2$ solution.

5. The resin is heated to boiling under reflux for 30 min and then allowed to cool to room temperature.

6. Rinse with 10 volumes of deionized water three to five times.

The resin slurry is then degassed at room temperature before packing.

Column Preparation

Column diameters may vary from 2, 3.2, and 4 mm i.d. (0.25 in. o.d.) to 6 and 8 mm i.d. (0.375 in. o.d.). Column lengths may vary from 5 to greater than 60 cm. Column diameters of 3.2 to 4 mm i.d. (0.25 in. o.d.) or

6 mm i.d. (0.375 in. o.d.) are recommended, based on the experience in this laboratory. The 2-mm-i.d. columns give both excellent resolution and sensitivity by minimizing solute dispersion. However, these columns are difficult to reproducibly pack unless one has had significant experience in slurry packing of LC columns. The 3.2- and 4-mm-i.d. (0.25 in o.d.) columns in a 30 to 60 cm length give good results, particularly for resolution of oligomers of DP $\leq$ 3, monosaccharides, alcohols, selected ketones, and furfurals when packed with Aminex 50W-X4 in the Ca^{2+} or H^+ form. In some cases a larger diameter column (6 mm i.d., 0.375 in. o.d.) of the same length packed with the same resin will give resolution of oligomers of DP up to 8. A thorough study on the impact of column diameter and length, for the same chromatographic support, is given by Jacobson[37] who packed over 50 columns on an internally consistent basis and then compared their performance.

The 8-mm-i.d. column can also give excellent results particularly when packed with Q15S or other resins of relatively high cross-linking. However, when Aminex 50W-X4 (or equivalent resin) is used, the packing characteristics of a 6-mm-i.d. column are dramatically different from an 8-mm-i.d. column.[2,3] The 8-mm-i.d. column gives high pressure drops (up to 3000 psig)[3] while a 6-mm-i.d. column gives lower pressures (100–200 psig) with little loss in resolving capabilities.[2,3] Consequently, the 6-mm-i.d. column is recommended for Aminex 50W-X4.

The column constructed of 316 stainless-steel tubing must be cut evenly, preferably by an experienced machinist. The ends are then smoothed and deburred. Before packing, the interior surface of the column should be treated as follows[41]:

1. rinse with deionized water at 1 ml/min for 10 min and connect standard liquid chromatography end fittings with either 2-, 5-, or 10-μm cut-off sintered disks;
2. pump 50% nitric through the column acid for 10 min at room temperature;
3. pump deionized water through the column until a pH of 5 is obtained;
4. pump 10–100 ml (i.e., 3 to 4 column volumes) of acetone through the column; and
5. then disconnect column from pump, remove end fittings, and dry thoroughly with air.

[41] J. K. Lin, S. J. Karn, and M. R. Ladisch, *Biotechnol. Bioeng.*, in press (1986).

This procedure is particularly important for resin in the H^+ form, where contact of the resin with the wall can cause corrosion and pitting, resulting in the formation of gas. The gas may only partially dissolve in the eluent, thus passing as bubbles into the detector cell.[41] This causes spikes in the resulting chromatogram which render the chromatogram useless.

Column Packing and Operation

The column can be packed by either a pneumatic amplifier pump[2,37] (constant pressure) or by a constant-flow pump.[3,41] However for a 2-mm-diameter column, the constant-flow pump is more suitable for column packing.[37,41] Details of the column packing are given elsewhere,[2,3,37,41] with the key steps summarized in Fig. 1b and c.

The resin, prepared as described above, was finally slurried in 150 ml of degassed water in a beaker. The slurry of resin was loaded, by pipetting, into an empty column assembly (capped with a 10-μm end fitting) at ambient temperature (Fig. 1c). The column (3/8 in. o.d., 6 mm i.d., by 50 cm long, with 10-μm outlet end-fitting) was then connected to the prepacker assembly which was subsequently also filled with resin slurry. It is important to keep the resin slurry evenly suspended in the beaker during the loading procedure. Intermittent swirling of the beaker during pipetting is usually sufficient.

Water from a feed reservoir attached to the pump inlet was then pumped into the prepacker column assembly (Fig. 1b). During packing, the column was slowly heated to 85° over a 2-hr period with a circulating water bath. The pressure was gradually increased to 120 psi, and the column was allowed to pack for 16 hr. The pressure was kept constant, with flow rate decreasing as the resin packed. The type of pump used in our laboratory is either a Milton Roy positive displacement pump or equivalent, or a pneumatic amplifier pump. For laboratories where only a few columns are to be packed, a positive displacement pump is recommended. In this case the pump must be given constant attention, and the flow rate controlled by manual adjustment to keep the pressure constant. A pressure gauge and relief valve as indicated in Fig. 1b is thus required for packing as well as for instrument operation (see Fig. 1a). When a large number of columns are to be packed a pneumatic amplifier pump is recommended.[2,3] This pump automatically adjusts volumetric flow rate to maintain a constant pressure at changing flow resistance such as occurs when a column is packed. This type of pump may require a greater degree of mechanical maintenance than a constant displacement pump. Microprocessor controllable constant displacement pumps which maintain a steady pressure can also be purchased and are suitable for both packing (constant pressure mode) as well as for normal instrument operation (constant volume mode).

After column packing, the column is connected to the liquid chromatograph and maintained at 80°. The liquid chromatograph includes a degassed, distilled water reservoir, an injector system, a pump, and a differential refractometer connected to an integrator/chart recorder. All analyses are carried out at a constant flow rate using water as the eluent. The typical flow rates for the 4-mm and 6-mm-i.d. column × 60 cm long are 0.25 and 0.5 ml/min at pressure drops of 200 psig or less, respectively.[37] The chromatograms in Figs. 2 to 7 illustrate a variety of separations.

Troubleshooting of LC System

Troubleshooting a liquid chromatograph can be made easier, if one knows the symptoms produced by an instrument malfunction, leaks, or

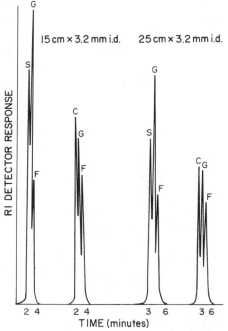

FIG. 2. Separation of sucrose (S), glucose (G), fructose (F); cellobiose (C), glucose (G), fructose (F) at 2 mg/ml, each. Disproportionate glucose peak probably caused by sample overlap. Packing: Calbiochem resin (Lot No. 002156), Ca²⁺ form; 10 μl sample; 8X attn; chart speed of 20 cm/hr; column temperature of 85°.

column problems. The most common problems and possible remedies in the LC laboratory are summarized below. The manufacturer's instructions should, of course, always be consulted and followed with respect to details on correctional procedures.

1. Bubbles in detector cells cause spikes on the chart paper output. The bubbles can be flushed out by disconnecting the detector, connecting a 10-ml hypodermic syringe with tubing or a swagelock fitting, and pushing the degassed eluent through. Then confirm that the eluent in the solvent reservoir is also properly degassed.

2. Contamination in a detector cell appears as a noisy baseline. The cells should be flushed with appropriate acid, e.g., 6 N HNO_3 and/or methanol/water as given in manufacturer's specification.

3. Baseline drift may occur due to room temperature changes (baseline drifts upward by 2–5 cm/hr). Check working environment for constant temperature and cooling, where appropriate. Note that as little as 0.1° fluctuation can cause full scale deflection of the recorder pen.

4. An increase in column pressure can occur when the column is plugged. This is normally caused by a "dirty" sample. The sample should be filtered before injection. The column may be regenerated by disconnecting from instrument, connecting column outlet to pump, and slowly flushing overnight, with the inlet fitting draining directly into a small

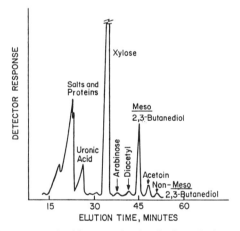

FIG. 3. Liquid chromatograph of fermentation broth after 8-hr fermentation of xylose by *Klebsiella pneumoniae*. Column dimensions: 6 mm i.d. × 60 cm long. Temperature at 85°, water as eluent at 0.5 ml/min. Column packed with Aminex 50W-X4, Ca^{2+} form, 20–30 μm (Bio-Rad). [Reprinted with permission from M. Voloch, M. R. Ladisch, M. Cantarella, and G. T. Tsao, *Biotechnol. Bioeng.* **26,** 557 (1984). Copyright © 1984 by John Wiley & Sons.]

beaker. The column is then turned around again and reconnected. If the pressure remains high, a void volume will likely form at the column inlet, and cause band broadening and loss of resolution. The column should then be repacked.

5. Poor peak response can occur due to leaks in the system, especially the injector or in a dirty cell (see item 2 above). In case of injector problems, the rotor seal in the injector may be worn and needs replacing. In this case, repeated injections of the same volume of sample will give peaks of widely different heights and baseline widths and/or shoulders.

6. Loss of resolution may occur after the column has been used for a long period of time. This may be due to loss of the counterion or fouling. The resin needs to be unloaded, and if a cation-exchange resin, boiled in 6 N HCl under reflux. Then, repeat resin preparation and column packing procedures as described previously.

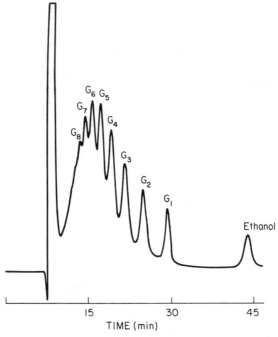

FIG. 4. Chromatogram of cellodextrins with water adjusted to pH 2 (using H_2SO_4) as eluent. Column 6 mm i.d. × 60 cm long packed with Aminex 50W-X4; H^+ form, 20–30 μm particle size (Bio-Rad). [Reprinted with permission from M. Voloch, M. R. Ladisch, V. W. Rodwell, and G. T. Tsao, *Biotechnol. Bioeng.* **23**, 1289 (1981). Copyright © 1981 by John Wiley & Sons.]

7. When a differential refractometer is used as the detector, pre- or postpeak negative dips can occur due to improper mask adjustment. The mask needs to be properly adjusted.[41]

8. No response on chart recorder or a slow drifting of the base-line can also occur if the detector cell is broken due to back pressure changes exceeding the tolerance limit (usually at 50 psig or less). A new detector cell would then be required. A more likely cause is a dirty cell surface due to microbial growth or protein fouling. In this case the cell needs to be cleaned based on the manufacturer's instructions.

Column performance should be routinely checked by injecting standards on a daily basis. Major shifts in retention time and increase in peak width may indicate void volume formation in the column. The column would then need to be repacked. Decreases in peak height indicate a dirty detector cell.

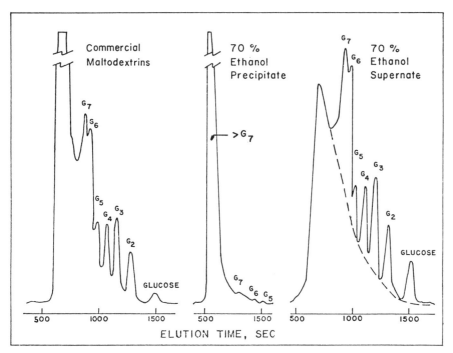

FIG. 5. Chromatogram of maltodextrins. Column size, type of packing, and conditions the same as given in Fig. 4.

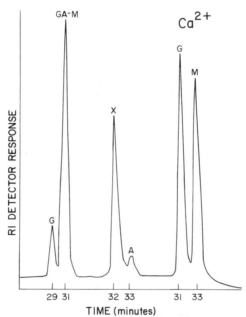

FIG. 6. Separation of glucose (G), galactose (GA), and mannose (M); xylose (X) and arabinose (A); and glucose (G) and mannose (M) for 50-cm × 3.2-mm-i.d. column using Aminex 50W-X4 resin (control No. 20654, Lot No. 15970) in the Ca^{2+} form; 10 μl sample; 8X attn.; chart speed of 20 cm/hr; 85° column temperature.

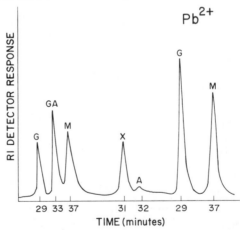

FIG. 7. Separation of glucose (G), galactose (GA), and mannose; xylose (X) and arabinose (A); and glucose (G) and mannose (M) for 50-cm × 3.2-mm-i.d. column using Aminex 50W-X4 resin (control No. 2654, Lot No. 15970) in the Pb^{2+} form; 10 μl sample; 8X attn.; chart speed of 20 cm/hr; 85° column temperature.

Conclusions

Liquid chromatography, as described in this chapter, is a useful tool for quantitating carbohydrate monomers and oligomers in a short analysis time using aqueous based eluents. The key is proper choice of sorbent and eluent, together with appropriate packing techniques. An attempt was made in this chapter to describe the salient techniques.

Acknowledgments

The procedures in this chapter were independently confirmed by Mandy Mobedshahi, K. Kohlmann, B. Woodruff, R. Hendrickson, and Jill Porter who repeated procedures written in this chapter in order to confirm clarity of presentation. Their helpful suggestions in preparing this manuscript are greatly appreciated. The material in this work was supported by NSF Grant CPE8351916 and USDA 83-CRSR-2-2250.

[16] Quantitative Thin-Layer Chromatography of Sugars, Sugar Acids, and Polyalcohols

By R. KLAUS and W. FISCHER

Introduction

The main procedures used for the quantitative determination of sugars (and sugar substitutes[1]) are polarimetry, chromatography, and enzymatic methods. Though offering extremely high selectivity, enzymatic analysis is based on a specific reaction mechanism for each sugar.[2,3] The other two procedures adopt a fundamentally different approach and permit the simultaneous analysis of multiple component mixtures. For the analysis of pure substances polarimetry is ideal both in terms of effort involved and accuracy. With multiple component mixtures, however, polarimetry very quickly reaches the limit of its capabilities. For instance, in a two-component mixture (presupposing a difference in rotatory dispersions)

---

[1] H. Förster, *Med. Monatsschr. Pharm.* **2**, 42 (1978).
[2] H. O. Beutler and J. Becker, *Dtsch. Lebensm. Rundsch.* **73**, 182 (1977).
[3] A. Lomard and M. L. Tourn, *J. Chromatgr.* **134**, 242 (1977).

rotation measurements have to be performed at two different wavelengths, λ_1 and λ_2. Progressing from these measurements, the amounts of the components are calculated from a system of equations with two unknowns. Chromatography is a much more expedient means of determining multiple component or unknown mixtures. The following chromatographic techniques are available: (1) gas chromatography (GC), (2) thin-layer (TLC) and high-performance thin-layer chromatography (HPTLC), and (3) low- and high-pressure liquid chromatography (HPLC).

The chemical constitution of the substances in question is such that none of the above methods directly produces a qualitative or quantitative result which is totally satisfactory. Whether or not gas chromatography can be applied for a separation depends on a prior chemical derivatization. In cases where separation is by one of the liquid chromatographic methods, suitable detection reactions have to be performed during or after the separation (except where detection is by refractometry in column chromatography or HPLC). Readily implementable detection reactions have so far been few and far between in HPLC. Far more numerous in the literature are publications dealing with *in situ* reactions[3-11] on thin layers, derived in part from paper chromatography.[12,13] The works describe various coloring methods and solvent systems which can be used for visual evaluations.[7,14-17] Judicious combination of the different TLC parameters enables a selectivity to be achieved which is satisfactory for virtually any qualitative sugar analysis.

Not quite so satisfactory is the performance of most color reactions in producing a quantitative result. This chapter describes an *in situ* reaction

[4] H. Scherz, G. Stehlik, E. Bancher, and K. Kaindl, *Chromatogr. Rev.* **10**, 1 (1968).
[5] G. Avigad, *J. Chromatogr.* **139**, 343 (1977).
[6] J. Cerbulis, *J. Chromatogr.* **155**, 226 (1978).
[7] M. Ghbregzabher, S. Rufini, B. Monaldi, and M. Lato, *J. Chromatogr.* **127**, 133 (1976).
[8] F. Hsu, D. Nurok, and A. Zlatkis, *J. Chromatogr.* **158**, 411 (1978).
[9] R. Gausch, U. Leuenberger, and E. Baumgartner, *J. Chromatogr.* **174**, 195 (1979).
[10] K. Y. Lee, D. Nurok, and A. Zlatkis, *J. Chromatogr.* **174**, 187 (1979).
[11] J. Iwakawa, H. Kobataka, I. Suzuki, and H. Kushida, *J. Chromatogr.* **193**, 333 (1980).
[12] G. Puey, *Ann. Falsif. Expert. Chim.* **69**, 605 (1976).
[13] T. Momose and M. Nakamuva, *Talanta* **10**, 115 (1963).
[14] G. K. Munshi and P. R. Bhattacharya, *J. Inst. Chem.* (*India*) **49**, 103 (1977).
[15] U. Siegenthaler and W. Ritter, *Mitt. Geb. Lebensmittelunters. Hyg.* **68**, 448 (1977).
[16] V. Prey, H. Scherz, and E. Bancher, *Mikrochim. Acta* p. 567 (1963).
[17] B. P. Kremer, *J. Chromatogr.* **110**, 171 (1975).

which yields products amenable to *in situ* fluorimetric analysis. These products play a vital part in quantitative analysis.[18]

Materials and Methods

Chromatography

All the separations described are carried out on HPTLC layers, which are generally superior to normal TLC plates in that they have better resolution, lower detection limits, shorter analysis times, and better band capacity.[19]

From the range of options available, silica gel 60 in the form of precoated plates and aluminum sheets was selected as adsorbent. Not only are these plates and sheets ideal for carrying out *in situ* reactions but water-resistant silica gel layers[20] have also been introduced, which unlike the frequently recommended cellulose layers[7] have practically no constraints on the composition of the eluant. Aluminum sheets have an additional advantage in that once evaluated, they can be filed together with the analysis reports and data on which the calculations are based.

There is an abundance of literature[21] to assist in the selection of suitable solvent systems for specific separations.

Application of Samples

Ordinary commercial HPTLC applicators are used to apply samples. Two devices in particular[22] can be used for applying nanoliter quantities: (1) the Camag nanoapplicator for applying variable volumes of between 0 and 230 nl, and (2) the Camag microapplicator for applying variable volumes of between 0 and 2.3 μl. Microliter capillaries can also be used for applying microliter quantities of sample. Not only do these capillaries permit sample spotting, but they can also be used for streakwise application, a feature which is particularly useful in *in situ* fluorimetric analysis. In view of the important advantages offered by streakwise application, including better separation at full measuring sensitivity, we draw attention to a semiautomatic streak applicator described in the literature.[23]

[18] R. Klaus and J. Ripphahn, *J. Chromatogr.* **244**, 99 (1982).
[19] J. Ripphahn and H. Halpaap, *J. Chromatogr.* **112**, 81 (1975).
[20] W. Jost, H. E. Hauck, and F. Eisenbeiss, *J. Chromatogr.* **256**, 182 (1983).
[21] E. Stahl, "Dünnschicht-Chromatographie, Laboratoriums-Handbuch. 2. Springer-Verlag, Auflage, 1967.
[22] Available from Camag, CH-4132 Muttenz, Switzerland.
[23] R. Klaus, *J. Chromatogr.* **333**, 276 (1985).

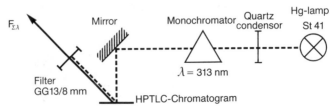

FIG. 1. Light beam arrangement for fluorimetric measurements in a chromatogram spectrophotometer.

In Situ Reaction

Tanner[24,25] first proposed the *in situ* reaction employed, i.e., chemical transformation into visually or photometrically evaluable substances, for substances with vicinal diol groups, e.g., sorbitol and other polyhydroxy compounds.

Activation. Following chromatographic development the HPTLC plate or aluminum sheet is dried with a hair drier, then dipped in the reaction solution for about 10 sec. The solution is prepared anew for each immersion, which is best performed in a vertical narrow tank of suitable dimensions. The dipping solution is prepared from two separate solutions: a saturated solution of lead tetraacetate in glacial acetic acid and a 0.1–1% solution of 2,7-dichlorofluorescein. Volumes of 5 ml of each solution are mixed and made up to 200 ml with toluene. The plate is then activated by heating with a hair drier or in a drying cabinet for approximately 3 min at approximately 100°. When replenished with lead tetraacetate and dichlorofluorescein solutions, the reaction solution can be used repeatedly for up to a month, although the fluorescence diminishes with time.

Fluorimetry and Photometry

The first step to be taken in quantitative *in situ* determination is a photometric scan of the individual bands of the HPTLC plate using a chromatogram scanner. Fluorimetric data such as the fluorescence and excitation spectra of fluorescent components produced by the activation process[26] are also important at this stage. The beam path arrangement in Fig. 1 has been shown to be effective. A more intense Hg 366 beam can be used instead of the Hg 313 excitation beam, to the slight detriment of the signal-to-noise ratio. The photometric signal produced is the wavelength

[24] H. Tanner, Z. Obst.-Weinbau **103**, 610 (1967).
[25] H. Tanner and M. Duperrex, Fruchtsaft-Ind. **13**, 98 (1968).
[26] G. Gübitz, R. W. Frei, and H. Bethke, J. Chromatogr. **117**, 337 (1976).

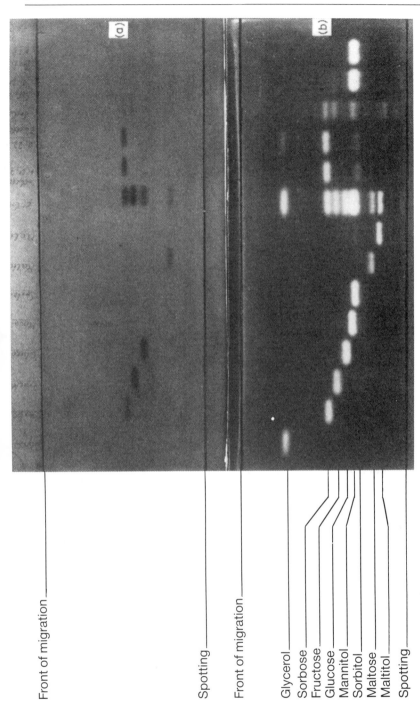

Fig. 2. Chromatic separation of sugars and sugar alcohols. (a) Derivatization: *in situ* reaction. Dipping solution: naphthoresorcinol solution, 0.2% in ethanol, 100 ml; phosphoric acid (85%), 10 ml. Further treatment: 5–10 min, ~100°. Detection: absorption in the visible range. (b) Derivatization: *in situ* reaction. Dipping solution: dichlorofluorescein/lead tetraacetate/toluene. Detection: fluorescence.

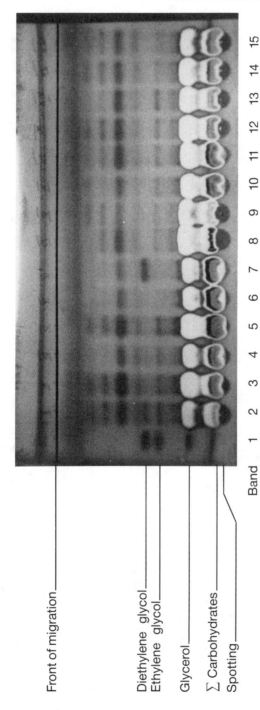

Front of migration

Diethylene glycol
Ethylene glycol

Glycerol

Σ Carbohydrates
Spotting

Band   1   2   3   4   5   6   7   8   9   10   11   12   13   14   15

Fig. 3. Chromatographic separation in the R_f region of glycols (notably EG and DEG) in various wine samples. Band 1, glycerol–EG–DEG 1.6–1.4–2.8 μg; bands 2–6, wine sample II–VI; band 7, wine sample VI + 4.5 μg DEG; band 8 (VII); band 9 (IX); band 10 (X), 11 (XI), 12 (XIII), 13 (XIII), 14 (XV), 15 (XVI). Applied volumes 4 × 5 μl, with drying (hair drier) of each 5 μl following application. Thereafter concentration with 2-propanol/water (2 : 5) to ∼8 mm above the starting line. Adsorbent: HPTLC precoated plate silica gel 60 without fluorescent indicator. Solvent: 2-propanol/chloroform/methanol/10% ammonia solution (45 : 45 : 10 : 10). Normal chamber with chamber saturation. Derivatization (plate must be free of ammonia, test for odor): spray the chromatogram (vanadium pentoxide spray) until almost transparent, heat at ∼100° until spots are visible. Variation: leave to cool, respray briefly until the background is decolored.

integral of the fluorescent radiation $F_{\Sigma\lambda}$ in the visible spectral range. The plate can be scanned in a direction parallel to or perpendicular to the direction of solvent flow. The direction will be governed by the particular separation being performed and by R_f constancy. Scanning in the perpendicular direction is simple and quick. According to the separation being performed and the spot configuration, the variable on which evaluation is based may either be the recorded peak height or the electronically integrated fluorescent radiation $\Sigma F_{\Sigma\lambda}$ over the spot.

It is interesting to note that for the group of substances in question the fluorescence spectra and excitation spectra are largely identical. Consequently, with this *in situ* reaction, components inadequately resolved by the TLC separation cannot be further separated by chromatic means.

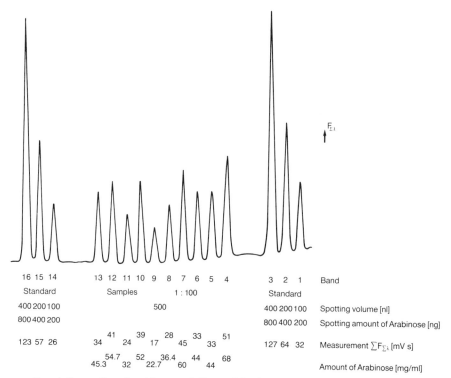

16 15 14		13 12 11 10 9	8 7 6	5 4		3 2 1	Band
Standard		Samples	1 : 100			Standard	
400 200 100			500			400 200 100	Spotting volume [nl]
800 400 200						800 400 200	Spotting amount of Arabinose [ng]
		41 39 28 33 51					
123 57 26		34 24 17 45 33				127 64 32	Measurement $\Sigma F_{\Sigma\lambda}$ [mV s]
		54.7 52 36.4 44 68					
		45.3 32 22.7 60 44					Amount of Arabinose [mg/ml]

FIG. 4. Separation of arabinose and ribose. Adsorbent: HPTLC aluminum sheet silica gel 60 without fluorescent indicator. Solvent: ethyl acetate/pyridine/tetrahydrofuran/water/glacial acetic acid (50 : 22 : 15 : 15 : 4). Normal chamber with chamber saturation. Migration distance: 1 × 7 cm. Running time: 45 min. Fluorimetric scan perpendicular to the direction of solvent flow, and table of values for the arabinose component.

Alternative in Situ Derivatizations

More substance-specific derivatizations can be performed to yield better qualitative results. Of the vast number of potential procedures a few typical cases are shown below.

Reagent for Differentiating, Inter Alia, Aldoses and Ketoses

Spray solution
   Diphenylamine, 4 g
   Aniline, 4 ml
   85% phosphoric acid, 20 ml
   Dissolved in 200 ml of acetone
Further treatment: heat the chromatogram at ~85° for ~10 min

Similar Differentiations Are Possible with the Following Solution

Spray solution: dissolve 0.5 ml of anisaldehyde in 50 ml of glacial acetic acid plus 1 ml of concentrated sulfuric acid
Further treatment: heat the chromatogram at ~100° until maximum color development is achieved

It is of great benefit in qualitative carbohydrate analysis to distinguish between sugars and sugar alcohols. The in situ derivatization with lead tetraacetate/dichlorofluorescein/toluene, described above, can differentiate the two groups of substances only through their specific fluorescences. Additional information can be obtained by performing the derivatization below, inasmuch as only sugars are detectable as absorbing spots.

Chromatic Separation of Sugars/Sugar Alcohols

Dipping solution
   Naphthoresorcinol 0.2% in ethanol, 100 ml
   85% phosphoric acid, 10 ml
Further treatment: heat the chromatogram at ~100° for 5–10 min

Figure 2 shows two identical chromatograms derivatized using two different systems.

A further reaction which is widely used in qualitative carbohydrate analysis emphasizes the importance of derivatization techniques.

In the analysis of glycols (notably diethylene glycol) in alcoholic drinks the dichlorofluorescein reaction is seen to be totally inadequate. Only following in situ derivatization below can the substance be detected within the concentration range in question. Figure 3 shows a chromatogram obtained in the qualitative detection of ethylene glycol (EG) and diethylene glycol (DEG) in a variety of wine samples. DEG concentrations of 10 mg/liter of wine can be detected without pretreatment of the sample.

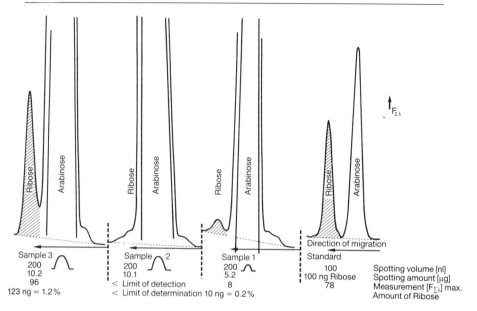

FIG. 5. Separation of arabinose and ribose, trace analysis of ribose. Adsorbent: HPTLC aluminum sheet silica gel 60 without fluorescent indicator. Solvent: ethyl acetate/pyridine/tetrahydrofuran/water/glacial acetic acid (50 : 22 : 15 : 15 : 4). Normal chamber with chamber saturation. Migration distance: 1 × 7 cm. Running time: 45 min. Fluorimetric scans of one reference and three sample bands, recorded parallel to the direction of solvent flow, and table of values.

General Detection Reaction for Carbohydrates/Absorbing Reaction Compounds[27]

Spray solution: dissolve 18.2 g of vanadium pentoxide with heating in 300 ml of 1 M sodium carbonate solution, leave to cool, mix with 460 ml of 2.5 M sulfuric acid, and make up to 1 liter with distilled water

Further treatment: heat the chromatogram at ~100° until the colors have developed

General Points Concerning Quantitative Findings

Accuracy and precision are important criteria in the evaluation of analytical findings.[28] Since quantitative HPTLC analysis makes use of reference substances, the accuracy achieved is dependent among other

[27] K. M. Haldorsen, *J. Chromatogr.* **134**, 467 (1977).
[28] R. Klaus, *J. Chromatogr.* **34**, 539 (1968).

things on the purity of the standard substances employed; corrections may need to be made. Also, the rules of statistics must be observed if precise results are to be obtained. Frequently, however, single or duplicate determinations are sufficient. This is especially true in trace analysis, a field of work in which the tremendous advantages of HPTLC are very apparent. Considering the fact that statistical analyses with more than five determinations gain virtually nothing in precision (expressed in terms of

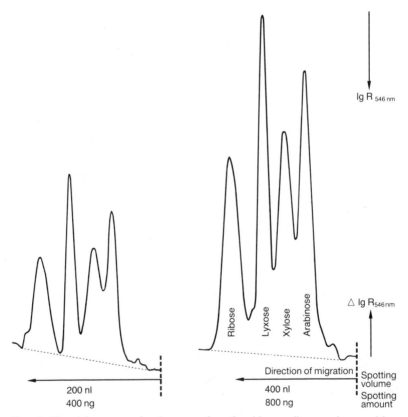

FIG. 6. Absorption curves for the separation of arabinose, ribose, xylose, and lyxose. Adsorbent: HPTLC precoated plate silica gel 60 $F_{254\,s}$ (water resistant). Impregnant: phosphate buffer pH 8, plate immersed for ~15 sec, then dried in a drying cabinet (30 min, 120°). Preparation of the dipping solution. Adjust phosphate buffer from pH 6.88 to pH 8 by dropwise addition of 1 N NaOH (indicator paper). Solvent: methyl ethyl ketone/acetic acid/ saturated boric acid solution (90 : 10 : 10). Normal chamber with chamber saturation. Migration distance: 2 × 7 cm. Running time: 2 × 25 min. Derivatization: *in situ* reaction. Dipping solution: 1,3-naphthalenediol (0.2% in ethanol)/H_2SO_4 (20%) (1 : 1). Immersion time: 15 sec. Further treatment: drying cabinet (10 min, 110°).

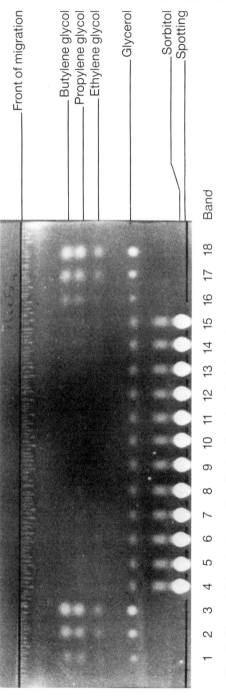

FIG. 7. Glycerol determination and qualitative detection of ethylene glycol, propylene glycol, and butylene glycol. Adsorbent: HPTLC aluminum sheet silica gel 60 without fluorescent indicator. Solvent: 2-propanol/chloroform/methanol/10% ammonia solution (45:45:20:10). Normal chamber with chamber saturation. Migration distance: 1 × 7 cm. Running time: 50 min.

standard deviation), the 20–30 bands *in toto* which can be accommodated on a single plate mean that several samples can be statistically evaluated at a time.

Analytical Applications

The analyses outlined below are just a few examples of the wide range of uses of the chromatographic procedure featured in this chapter. All

TABLE I
COMPUTER-GENERATED REPORT ON THE STATISTICAL
DETERMINATION OF GLYCEROL IN SORBITOL[a]

Spotted amounts	Band	Computed results	
		Time (sec)	Fluorescence integral $\Sigma F_{\Sigma\lambda}$ (mV × sec)
			260.547
		9.7	0.446
Standards			
658 ng	18	23.9	35.411
329 ng	17	43.9	17.524
		55.4	0.040
165 ng	16	64.2	6.076
	15	83.6	13.744
	14	103.0	12.753
	13	123.5	11.816
	12	143.2	12.223
Sample			
~120 μg	11	163.7	11.782
	10	181.8	11.328
	9	201.2	11.566
	8	223.1	12.814
	7	243.1	9.957
	6	262.5	12.631
	5	283.3	11.470
	4	301.6	11.575
Standards			
658 ng	3	321.9	32.592
329 ng	2	341.9	17.787
165 ng	1	363.2	7.009

[a] Based on the chromatogram in Fig. 7.

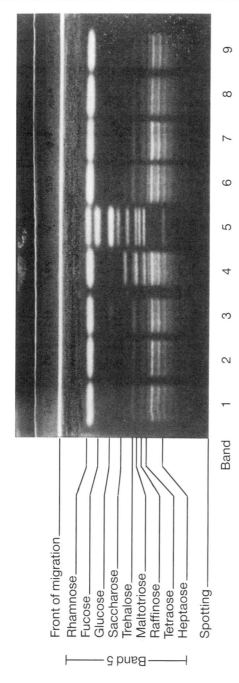

Band   1   2   3   4   5   6   7   8   9

Front of migration
Rhamnose
Fucose
Glucose
Saccharose
Trehalose
Maltotriose
Raffinose
Tetraose
Heptaose
Spotting

|— Band 5 —|

FIG. 8. Separation in the region of the higher oligosaccharides. Adsorbent: HPTLC precoated plate silica gel 60 without fluorescent indicator, HPTLC aluminum sheet silica gel 60 without fluorescent indicator. Solvent: 1-propanol/nitromethane/water/glacial acetic acid (50 : 30 : 20 : 1). Normal chamber without chamber saturation. Migration distance: 2 × 7 cm. Running time: 2 × 45 min. Streakwise application with 5 μl microcapillary. Applied quantities: bands 1–4 and 6–9, 20 μg of sample to each; band 5, reference mono-ditrisaccharides (1.25 μg of each), tetraose and heptaose (5 μg of each).

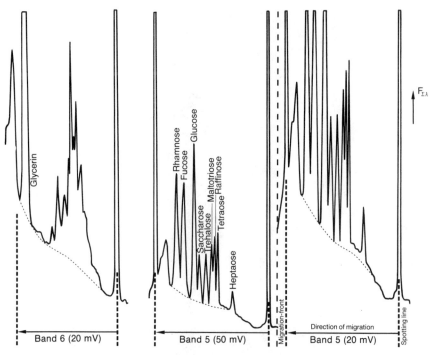

FIG. 9. Fluorimetric scans of bands 5 and 6 from the chromatogram in Fig. 8.

examples are taken from routine work; they have been selected to cover a wide variety of separations and quantification techniques based on fluorimetric detection.

Quantitative TLC of Arabinose/Ribose/Xylose/Lyxose

Figures 4 and 5 are fluorimetric scans obtained from two arabinose/ ribose separations. The TLC separation featured in Fig. 4 was devised for the quantitative determination of arabinose in 10 different ribose-containing samples. Figures 5, on the other hand, depicts the trace analysis of ribose in an arabinose sample. Details of chromatographic conditions and the quantification are given in the respective legends.

With the advent of water-resistant silica gel layers a number of new applications can be expected, especially in carbohydrate analysis. One such application is shown in Fig. 6, in which four components are completely separated. Again, procedural details are contained in the legend.

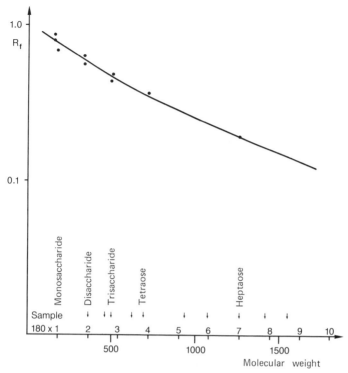

FIG. 10. Attempt made at roughly correlating the molecular weights of oligosaccharides with their R_f dependence.

Quantitative Statistical Glycerol Determination and Qualitative Detection of Ethylene Glycol, Propylene Glycol, and Butylene Glycol in a Sorbitol Sample

The solvent used for the chromatogram depicted in Fig. 7 enables a satisfactory spread of R_f values to be achieved for the glycols (C_2, C_3, C_4) while the carbohydrates, with R_f values of less than 0.1, are excluded from assessment. Statistical analysis is based on a series of reference bands with only a single sample repeatedly applied. Bivalent alcohols are not detectable. For the qualitative determination of glycerol a photometric scan is performed perpendicular to the line of solvent flow. Table I is a computer-generated report of the results, listing the amounts of sample and reference applied together with the recording parameter-dependent times for the individual bands, and the fluorescence integral $\Sigma F_{\Sigma\lambda}$ (mV ×

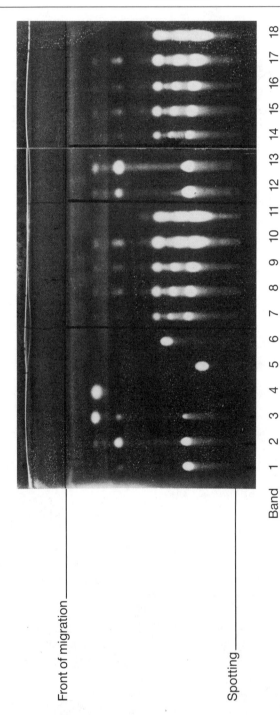

FIG. 11. Fluorimetric detection of various sugar acids in an exchanger eluate. Adsorbent: HPTLC precoated plate silica gel 60 without fluorescent indicator. Solvent: 1-propanol/nitromethane/water (50:30:20). Normal chamber without chamber saturation. Migration distance: 1 × 7 cm. Application: band 1 gluconic acid, band 2 gluconic acid, band 3 gluconic acid δ-lactone, band 4 glucuronolactone, band 5 galacturonic acid, band 6 ketogulonic acid (500 nl of each, equiv. to ~2.5 µg); bands 7–11 and 14–18 sample (exchanger eluate), various dilutions and volumes; band 12 gluconic acid, band 13 gluconic acid δ-lactone (1 µl of each, equiv. to ~5 µg).

sec). Subsequent calculation gives a glycerol content in the sorbitol sample of 242.5 ng/2 μl of applied solution, equivalent to 0.22%. Error calculations gave the following values: standard deviation for individual values $S_{abs} \sim 10$ ng, $S_{rel} \approx 4.3\%$. The expected standard deviation for the mean is $(S_{\bar{x}})_{rel} \approx 1.3\%$.

Analysis of Oligosaccharides in α-Amylases

The solvent proposed in this section is suitable for analyzing oligosaccharides. Figure 8 shows a chromatogram devised for the qualitative and semiquantitative analysis of oligosaccharides in α-amylase. The reference solution contains three monosaccharides, two disaccharides, two trisaccharides, one tetraose, and one heptaose. The strongly fluorescing band in the amylase samples close to the solvent front was shown to be glycerol. Figure 9 shows the scans obtained for one sample band and one reference band.

Evaluation. The R_f dependence of oligosaccharides seen in the reference band offers a way of roughly identifying components in the sample bands. Figure 10 depicts the relationship. The additional scale printed on the *x* axis represents multiples of the molecular weight of glucose and gives the approximate oligo values in the sample bands.

Measurement, additionally, of the specific fluorescences of the various oligo regions of the reference bands makes possible a quantitative estimation of the identified sample components. The factors on which semiquantitative analysis is based are derived from the tendency shown by these specific values.

Separation and Detection of Sugar Acids

As well as being suitable for detecting sugars and sugar alcohols, the lead tetraacetate/dichlorofluorescein/toluene reaction solution can also be used for fluorimetric detection of sugar acids. Figure 11 shows a chromatogram for the qualitative detection, in an exchanger eluate, of the sugar acids listed in the legend. Gluconic acid and its δ- and γ-lactones are detectable qualitatively in the sample.

While quantitative analyses of galacturonic acid, ketogulonic acid, and glucuronolactone can be performed with virtually no problem, in aqueous solution there is an equilibrium, dependent on several parameters, between free gluconic acid and its δ- and γ-lactones. The use of the latter as standard substances therefore requires that certain corrections be made.

[17] High-Performance Thin-Layer Chromatography of Starch, Cellulose, Xylan, and Chitin Hydrolyzates

By Landis W. Doner

The development of small-particle ($<10 \mu$m) silica and bonded-phase silica stationary phases with narrow particle size distributions has led to the mode of planar chromatography known as high-performance thin-layer chromatography (HPTLC). As a result, TLC and HPTLC remain the most widely used chromatographic methods. Combined with appropriate mobile phases and sensitive visualization techniques, TLC is very useful for monitoring reactions of carbohydrates and for examining their structure. Qualitative information is derived very rapidly, since multiple samples and standards can be run simultaneously, and unlike other chromatographic methods, extensive sample cleanup is unnecessary.

Several reports describe effective TLC separations of monosaccharides and lower oligosaccharides on silica gel. Resolution is often enhanced by impregnating the stationary phase with NaH_2PO_4,[1,2] or boric acid.[3] A quantitative HPTLC method used boric or phenylboronic acid in the mobile phase[4]; this mode of separation had been developed earlier using conventional silica gel TLC plates.[5] These methods all include an acidic component either in the mobile phase or impregnated in the stationary phase, and the flow rates of the mobile phases are quite slow.

Sugar separations by HPLC often utilize aminopropyl-bonded silica stationary phases and acetonitrile–water mobile phases. HPTLC plates with this stationary phase have been prepared,[6,7] and they are commercially available. Effective and rapid sugar separation are accomplished on these plates when covalent interactions between reducing sugars and aminopropyl functionalities are inhibited by impregnating these plates with NaH_2PO_4.[8]

This chapter will describe applications of aminopropyl-bonded silica HPTLC plates to sugar separations, with emphasis on separation of ho-

[1] S. A. Hansen, J. Chromatogr. 107, 224 (1975).
[2] M. Lato, B. Brunelli, G. Ciuffini, and T. Mezzetti, J. Chromatogr. 39, 407 (1969).
[3] P. G. Pifferi, Anal. Chem. 37, 925 (1985).
[4] R. Klaus and J. Ripphahn, J. Chromatogr. 224, 99 (1982).
[5] M. Ghebregzabher, J. Chromatogr. 180, 1 (1979).
[6] M. Okamoto, F. Yamada, and T. Omori, Chromatographia 16, 159 (1982).
[7] L. W. Doner, C. L. Fogel, and L. M. Biller, Carbohydr. Res. 125, 1 (1984).
[8] L. W. Doner and L. M. Biller, J. Chromatogr. 287, 391 (1984).

mologous series of oligosaccharides produced by partial hydrolysis of the abundant plant (cellulose, starch, xylan) and animal (chitin) biomass materials. Also, the application of recently developed underivatized silica (Si 50,000) HPTLC plates to these separations will be described, since glucans to degree of polymerization (DP) 30 have been resolved using this stationary phase.[9] Separation of homologs derived from polygalacturonic acid will also be discussed, extending a procedure described[10] from silica TLC to silica HPTLC plates.

Procedures

HPTLC Plate Preparation and Irrigation

For separation of galacturonic acid oligomers, Whatman type HP-KF (10 × 10 cm) precoated silica gel HPTLC plates were activated at 110° for 30 min and stored in a vacuum desiccator before use. These plates were irrigated twice with *n*-butanol–formic acid–water (4 : 6 : 1 v/v/v).

The HP-KF silica gel HPTLC plates were used after derivatization with aminopropyl groups for separation of partial hydrolyzates of starch, cellulose, xylan, and chitin. Details of reaction with 3-aminopropyltriethoxysilane (3-APTS) and subsequent impregnation with NaH_2PO_4 were described previously.[7,8] The plates were irrigated three times with acetonitrile–water (72 : 28 v/v).

Oligosaccharides derived from starch, cellulose, and xylan were also separated by HPTLC on Merck Si 50,000 plates (10 × 10 cm). The plates were irrigated three times with *n*-butanol–pyridine–water (8 : 5 : 4 v/v/v) after activation at 110° for 30 min and storage in a vacuum desiccator.

Between irrigations and after the final irrigation, the three types of HPTLC plates were all dried by first placing them horizontally in a fume hood for 10 min and then in a vacuum oven at 40° for 15 min.

Chromatographic Procedures

HPTLC was conducted in closed glass tanks to which freshly prepared mobile phases had been added no earlier than the previous day. Generally, 1.0 μl of 5% solutions of oligosaccharide mixtures or 0.5 μl of 1% solutions of a given standard sugar was applied to the plates. Spots on the plates were detected by spraying with a reagent consisting of a mixture of

[9] K. Koizumi, T. Utamura, and Y. Okada, *J. Chromatogr.* **321,** 145 (1985).
[10] A. Koller and H. Neukom, *Biochim. Biophys. Acta* **83,** 366 (1964).

aniline (4 ml), diphenylamine (4 g), 85% H_3PO_4 (30 ml), and acetone (200 ml). This reagent is stable for several weeks when stored in the dark at $-5°$.

Preparation of Oligosaccharide Mixtures

Xylan (75 mg, Pfanstiehl Laboratories, Inc.) was added to 15 ml 0.16 N trifluoroacetic acid in a screw-top vial, and after 10 min, the vial was placed in an oven at 100°. After 1 hr, the reaction mixture was cooled to room temperature, unreacted xylan was removed by filtration, and the filtrate was neutralized by stirring 15 min with Dowex 2-X8 (OH⁻) ion-exchange resin. After filtration, the xylooligosaccharide mixture was evaporated to a syrup under reduced pressure.

Starch-derived maltooligosaccharide mixture M-250 was obtained from Grain Processing Corporation, and cellulose and chitin partial hydrolyzates were provided by Dr. K. B. Hicks.

Oligogalacturonic acids were prepared by adding 0.1 ml pectinase (Sigma P-5146) to a solution of 100 mg polygalacturonic acid (Sigma P-3889) in 4 ml 0.1 M phosphate buffer (pH 7.4). After 5 min, the reaction mixture was placed in boiling water bath for 5 min to denature and flocculate the enzyme. After filtration the solution was evaporated to dryness.

Applications

HPTLC analyses of partial hydrolyzates of the important biomass resources cellulose, starch, xylan, and chitin are shown in Fig. 1. The homologous series of glycans derived from these materials are readily separated from one another on both aminopropyl bonded-phase silica (APS, Fig. 1A) and Si 50,000 HPTLC plates (Fig. 1B). The latter plates are superior when one wishes to monitor fragments of starch or cellulose up to a DP range of 25–35.[9] APS-HPTLC plates are more suitable for more extensively hydrolyzed polysaccharides, when oligomers ranging only to DP 5 to 10 are encountered.[8] Also, Si 50,000 plates are not useful for resolving chitooligosaccharides. Irrigation of APS plates is much more rapid, requiring just 17 min for a 10-cm plate as compared to 55 min for Si 50,000 plates. With appropriate mobile phase adjustments,[9] TLC plates such as silica gel 60 (Merck) give results comparable to Si 50,000.

In Fig. 1C is shown the separation on HP-KF silica HPTLC plates of oligogalacturonic acids produced by the action of limited pectinase digestion on polygalacturonic acid. These plates effectively resolve the oligomers of less than DP 8. The separation of oligomers to DP 6 on conventional silica TLC plates has been described earlier.[10] Separations of oligogalacturonic acids comparable to those reported here have been

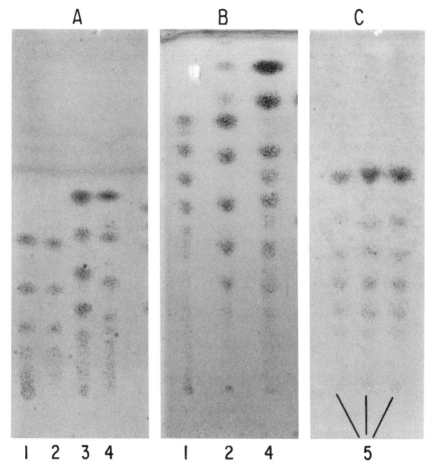

FIG. 1. HPTLC of carbohydrates on (A) aminopropyl-bonded silica impregnated with monosodium phosphate; (B) type Si 50,000; (C) type HP-KF silica. Sample identities: 1, maltooligosaccharide mixture M-250; 2, cellooligosaccharide mixture; 3, chitooligosaccharide mixture; 4, xylooligosaccharide mixture; and 5, oligogalacturonic acid mixture.

achieved on microcrystalline cellulose plates,[11] but are somewhat more difficult to visualize.

In Table I are listed R_f values of several frequently encountered sugars on APS and Si 50,000 HPTLC plates, after a single irrigation. APS plates are recommended for applications such as resolution of fructose, glucose,

[11] O. Markovic and A. Slezarik, *J. Chromatogr.* **312**, 492 (1984).

TABLE I
COMPARISON OF R_f VALUES FOR SUGARS ON AMINOPROPYL-BONDED PHASE
SILICA[a] AND Si 50,000 HPTLC[b] PLATES

	R_f			R_f	
Sample	APS	Si 50,000	Sample	APS	Si 50,000
Stachyose	0.03	0.07	Cellobiose	0.10	0.32
Raffinose	0.06	0.19	Fructose	0.20	0.41
Melezitose	0.08	0.31	Glucose	0.16	0.39
Maltotriose	0.06	0.28	Galactose	0.16	0.32
Maltose	0.10	0.33	Mannose	0.20	0.46
Isomaltose	0.08	0.21	Rhamnose	0.39	0.77
Lactose	0.10	0.24	Arabinose	0.25	0.44
Sucrose	0.13	0.37	Xylose	0.30	0.57

[a] Single irrigation with acetonitrile–water (72 : 28 v/v).
[b] Single irrigation with n-butanol–pyridine–water (8 : 5 : 4 v/v).

and sucrose, while Si 50,000 plates are more effective in resolving individual di and trisaccharides. Multiple irrigation of plates further separates the sugars, and on Si 50,000 plates, R_f values after several irrigations can be predicted using the formula

$$R_{f_n} = 1 - (1 - R_f)^n$$

where R_f is the value after a single irrigation and R_{f_n} is the predicted value after n irrigations. By applying this formula one can determine the number of mobile phase irrigations required for a monosaccharide to possess an R_f value approaching unity. Then more distance on the plate is available to separate larger oligosaccharides with lower R_f values.

[18] Screening of Prokaryotes for Cellulose- and Hemicellulose-Degrading Enzymes

By DIETER KLUEPFEL

Introduction

Various methods have been developed for the detection of microbial hydrolysis of lignocellulose. The screening for such activities is complicated by the fact that the biodegradation of each of the main components of this substrate, i.e., cellulose, hemicellulose, and lignin, requires the

METHODS IN ENZYMOLOGY, VOL. 160

action of two or more enzymes which is generally coordinated by a delicate mechanism of induction and repression by both the substrate and the hydrolysis products. Screening methods have been developed for many of the enzyme activities either individually or for the overall action of a group of enzymes.

Several methods that have been successfully applied in the screening of cellulolytic or xylanolytic prokaryotes and mutants thereof are described in this chapter. They are derived from procedures used in the detection of cellulases or xylanases produced by different microorganisms, both prokaryotes and eukaryotes, aerobes as well as anaerobes. Although adapted mainly for screening of actinomycetes, the methods outlined herein should be applicable, with only minor adjustments to other microorganisms and their enzymes mainly for their specific growth requirements.

General Principle of the Screening

Enzyme activity is demonstrated by the hydrolysis of substrate incorporated, generally as the main carbon source, in a solid agar medium distributed in petri dishes. For direct screening of the microorganisms or their mutants, the medium has to contain all the necessary ingredients (minimal medium) to afford sufficient initial growth of the inoculum in either spore or cell suspensions.

When screening culture filtrates, microbial extracts, or purification fractions for enzyme activity the substrate can be incorporated in a suitably buffered agar gel. The test solutions are placed into small wells previously punched with a cork borer of appropriate diameter. After an appropriate incubation period, the activity can be detected around the bacterial colonies or wells by the appearance of zones revealed either by substrate clearances, coloration, decoloration, or fluorescence depending on the type of technique used.

Methods for Specific Enzymes

Detection of Overall Cellulolytic Activity

Complete biodegradation of crystalline cellulose requires the concerted action of three types of enzymes, i.e., β-1,4-glucan cellobiohydrolase (EC 3.2.1.91, cellulose 1,4-β-cellobiosidase), β-1,4-glucanglucanohydrolase (EC 3.2.1.4, cellulase), and cellobiase (EC 3.2.1.21, β-glucosidase).[1] The synthesis of all these enzymes by a cellulolytic bac-

---

[1] R. Mullings, *Enzyme Microbiol. Technol.* **7**, 586 (1985).

terium can be best detected by inoculating the cells or spores on a solid agar medium containing the minimal ingredients for growth and suitably ground filter paper as main carbon source.

Since the sensitivity of the method depends to a large extent on the substrate particle size and its homogeneous distribution within the agar layer, special attention has to be given to these two factors.

Satisfactory grinding of the paper can be achieved in a jar mill (Norton, Chem. Process Div., Akron, OH, Mod. 764 AW) as follows: 6 g of shredded Whatman No. 1 filter paper suspended in 300 ml of distilled water is placed in a 2-liter grinding jar together with 100 carborundum cylinders as grinding medium; for the prevention of microbial contamination, 2 ml of chloroform is added and the jar is rolled on the mill at a steady speed of 50 rpm for 72–96 hr in a cold room at 4°. The longer the milling, the finer the paper slurry obtained, thus increasing its degradability. The slurry is recovered, sterilized in an autoclave at 121°, and stored in suitable portions for use in the medium preparation.

Homogeneous distribution of the paper particles in the agar plates is best achieved by incorporation of the slurry into the liquefied medium which is cooled under constant agitation to about 50° and distributed in 20-ml portions into standard size petri dishes. Pouring the plates on a cool surface helps to solidify the agar rapidly. As an alternative method, the double-layer technique, with 15 ml of basal medium without substrate as base, overlayed with 5 ml of the same containing the ground filter paper, has been used successfully.

The final concentration of the paper suspension can be varied from 1 to 5 g/liter of medium. The lower the substrate concentration, the greater the sensitivity for cellulase activity.[2] However, the decreasing contrast between substrate and clearing zone limits this possibility. Figure 1 shows a good example for the detection of cellulolytic activity of *Cellulomonas fimi*.[3]

Screening for β-1,4-Glucan Glucanohydrolase (Endoglucanase, CM-Cellulase)

One of the most effective methods for the detection of endoglucanase activity is based on the specific interaction of direct dyes such as Congo Red with polysaccharides. This colorant reacts with β-1,4-glucans causing a visible red shift.[4,5] The test was originally developed for the enumer-

[2] N. Daigneault-Sylvestre and D. Kluepfel, *Can. J. Microbiol.* **25,** 858 (1979).
[3] B. J. Stewart and J. M. Leatherwood, *J. Bacteriol.* **128,** 609 (1976).
[4] P. J. Wood, *Carbohydr. Res.* **85,** 271 (1980).
[5] P. J. Wood, *Carbohydr. Res.* **94,** c19 (1981).

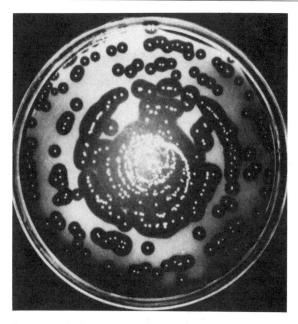

FIG. 1. Clearing zones obtained by cellulolytic activity of *Cellulomonas fimi* on minimal agar medium containing 0.1% ball-milled filter paper. From Stewart and Leatherwood.[3]

ation and characterization of cellulolytic anaerobic rumen bacteria as shown in Fig. 2.[6] This method has been adapted for the detection of endocellulase activity in actinomycetes and other aerobic bacteria as well as for the screening of their mutants.

The preferred substrate for the test is carboxymethylcellulose (CMC) which can be obtained from different sources (Hercules Inc., Wilmington DE, type 4M6F or Sigma Chem. Co., St-Louis, MO, medium viscosity). We found that satisfactory results depended largely on the degree of substitution and the level of viscosity of the CMC used and, based on our experience, the two types indicated above gave the best results.

The CMC is incorporated as the main carbon source into a minimal agar medium in quantities of 0.1–1.0%. The microorganisms can be screened directly on these plates, but the replica plating technique from a master plate is preferable since the visualization of the activity requires successive floodings with the reagents, which would render the reisolation of active colonies difficult. Such endoglucanase-producing colonies are detected after a suitable incubation time (1–3 days depending on the

---

[6] R. M. Teather and P. J. Wood, *Appl. Environ. Microbiol.* **43**, 777 (1982).

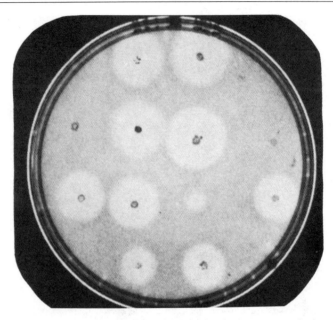

FIG. 2. Zones obtained by Congo Red staining of endoglucanase-producing colonies of anaerobic rumen bacteria, incubated with an overlay of soft agar gel containing 0.1% CMC. From Teather and Wood.[6]

growth), by flooding the plates with 10 ml of a 0.1% aqueous solution of Congo Red. The coloration is terminated after 20 min by pouring off the dye and flooding the plates with 10 ml of a 5 M NaCl solution (commercial salt can be used). After an additional 20 min, the salt solution is discarded and endoglucanase activity is revealed by a pale orange zone around the active microorganisms. In some cases, these zones can be enhanced by treating the plates with 1 N acetic acid, causing the dye to change its color to blue.

The same technique can be used as a cup-plate diffusion assay with excellent sensitivity for the determination of CM-cellulase activity in culture filtrates or during enzyme purification steps.[7]

Screening for Cellobiase (β-Glucosidase)

This β-glucosidase plays a crucial role in the regulation of the whole cellulase enzyme system. The fact that it is generally located and retained inside the cells renders the direct screening for cellobiase activity very

[7] J. H. Carger, *Anal. Biochem.* **153,** 75 (1986).

difficult. A net replication method, developed originally for the screening of β-glucosidase production in fungal spores and based on an azo dye staining technique,[8] can be adapted to prokaryotes as well.[9]

The visible staining is based on the histochemical reaction of β-glycosides with naphthol azo dyes.[10] 6-Bromo-2-naphthyl-β-D-glucoside (BNG, Sigma Chem. Co., St-Louis, MO), serving as substrate, is hydrolyzed by β-glucosidase to bromonaphthol and glucose. The bromonaphthol, when treated with o-aminoazotoluene diazonium salt (Fast Garnet, GBC, Sigma Chem. Co., St-Louis, MO), forms a water-insoluble red precipitate which binds to proteins.

The assay we developed is carried out as follows: a sterile circular Whatman No. 1 filter paper (9 cm diameter) is impregnated with a 0.1% methanolic solution of BNG and the methanol evaporated. The filter disk is placed carefully, in a marked position, on the surface of a petri plate containing the microorganisms in well-separated colonies, assuring good contact. After incubating plate and filter overnight at the appropriate growth temperature, the filter is removed carefully and sprayed twice until well covered with a freshly prepared 0.15% aqueous solution of GBC. Those colonies that have transferred to the filter paper and which produce β-glucosidase will appear as reddish pink spots. The method allows screening for β-glucosidase-positive or -negative microorganisms only; quantification of the enzyme, however, is impossible.

This azo dye method can also be adapted to the detection of cellobiase activity in cell extracts using a cup-plate technique. Petri dishes are filled with 20 ml of a medium containing 2% agar, 0.5% bovine serum albumin dissolved in 1 M sodium phosphate buffer, pH 7.0, and 0.1% BNG which is filter sterilized apart in distilled water and added just before pouring the plates. The test solutions are placed into wells and incubated overnight at the optimal temperature for enzyme action. The β-glucosidase activity is revealed after vaporizing (in a fume hood!) the surface with a solution of 0.15% GBC in water. The enzyme activity is detected by the pink zones that appear around the wells.

Screening for β-1,4-Xylan Xylanohydrolase (Endo-1,4-β-xylanase, EC 3.2.18, Xylanase)

The method most useful for the detection of xylanase activity is based on the hydrolysis of a covalently dyed xylan with Remazol Brilliant Blue R incorporated in an agar medium. The technique gives excellent results

[8] J. F. Rissler, *Appl. Environ. Microbiol.* **45,** 515 (1983).
[9] D. Kluepfel, unpublished results (1985).
[10] A. E. Ashford, *Protoplasma* **71,** 281 (1970).

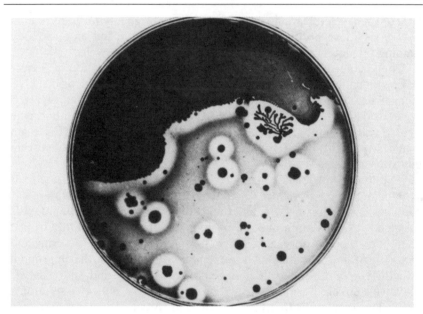

FIG. 3. Clearing zones obtained by xylanase activity of bacteria on a Czapek–Dox agar medium containing 0.2% of RBB-xylan. From Farkas *et al.*[11]

(Fig. 3) and is directly applicable to the screening of prokaryotes.[11] The dye is prepared with Remazol Brilliant Blue (RBB; Aldrich Chemical Co., Milwaukee, WI) and oat speltxylan (Sigma Chemical Co., St. Louis, MO) according to the method of Biely *et al.*,[12] with only minor adaptations.

Screening for Xylanase Activity. Xylanolytic microorganisms can be detected by spreading appropriately a diluted spore or cell suspension on petri dishes containing 20 ml of a suitable minimal agar medium including 0.1% of RBB-xylan as the main carbon source. This substrate should be sterilized separately in an aqueous solution and added just before pouring the medium. After incubation, xylanase-positive colonies can be detected, often after only 18–20 hr, by the appearance of a clearing zone around active colonies.

RBB-xylan is also useful for the determination of xylanase by the cup-plate method during enzyme purification steps and for the detection of activity during analysis in polyacrylamide or agarose gels.[13]

[11] V. Farkas, M. Liskova, and P. Biely, *FEMS Microbiol. Lett.* **28,** 137 (1985).
[12] P. Biely, D. Mislovicova, and R. Toman, *Anal. Biochem.* **144,** 142 (1985).
[13] P. Biely, O. Markovic, and D. Mislovicova, *Anal. Biochem.* **144,** 147 (1985).

[19] Chromatographic Separation of Cellulolytic Enzymes

By PETER TOMME, SHEILA MCCRAE, THOMAS M. WOOD,
and MARC CLAEYSSENS

Numerous fractionation problems have been encountered with cellulolytic complexes from microorganisms. Ion-exchange chromatography and isoelectric focusing (chromatofocusing) are the methods used most frequently.[1-3] Affinity chromatography on microcrystalline cellulose (Avicel) or on amorphous cellulose has also been attempted by several authors.[4-8] The resulting difficulties in eluting adsorbed enzymes from these supports were fully recognized.[9]

We reported earlier on the purification of a β-xylosidase by affinity chromatography on a Sepharose 4B carrier substituted with *p*-aminobenzyl-1-thio-β-D-xylopyranoside.[10,11] Using the parent cellobioside ligand a simple method for the purification of endo- and exocellulases from different sources was obtained. In the case of the enzymes from *Trichoderma reesei*[12] and *Penicillium pinophilum*[13] specific elution allows group separation of functionally related enzymes which are known to appear as isoenzymes in the culture filtrates.

The relative affinity for the ligand *p*-aminobenzyl-1-thiocellobioside is shown to determine also the applicability of the method in the isolation of an endocellulase from *Clostridium thermocellum*.

[1] L. G. Fägerstam and L. G. Pettersson, *FEBS Lett.* **98,** 363 (1979).

[2] M. Hayn and H. Esterbauer, *J. Chromatogr.* **329,** 379 (1985).

[3] R. Bhikhabhai, G. Johansson, and G. Pettersson, *J. Appl. Biochem.* **6,** 336 (1984).

[4] K. W. King, *J. Ferment. Technol.* **43,** 79 (1965).

[5] E. K. Gum and R. D. Brown, Jr., *Biochim. Biophys. Acta* **466,** 371 (1976).

[6] G. Halliwell and M. Griffin, *Biochem. J.* **169,** 713 (1978).

[7] M. Weber, M. J. Foglietti, and F. Percheron, *J. Chromatogr.* **188,** 377 (1980).

[8] M. Nummi, M.-L. Niku-Paavola, T. M. Enari, and V. Raunio, *Anal. Biochem.* **116,** 137 (1981).

[9] E. T. Reese, *Process Biochem.* **17,** 1 (1982).

[10] M. Claeyssens, H. Kersters-Hilderson, J. P. Van Wauwe, and C. K. De Bruyne, *FEBS Lett.* **11,** 336 (1970).

[11] H. Kersters-Hilderson, M. Claeyssens, E. Van Doorslaer, E. Saman, and C. K. De Bruyne, this series, Vol. 83, p. 631.

[12] H. van Tilbeurgh, R. Bhikhabhai, L. G. Pettersson, and M. Claeyssens, *FEBS Lett.* **169,** 215 (1984).

[13] M. Claeyssens and T. Wood, unpublished results (1987).

Preparation of the Affinity Column

To 25 ml Affi-Gel 10 (Bio-Rad) or CNBr-activated Sepharose 4B (Pharmacia)[14] 0.5 g p-aminobenzyl-1-thio-β-D-cellobioside,[12] dissolved in 25 ml 0.1 M sodium bicarbonate, was added. The gel after standing at room temperature (24 hr) was loaded into a cooled (4°) chromatography column (1 × 10 cm) and washed extensively with sodium acetate buffer (0.1 M, pH 5.0, 0.01% NaN$_3$).

The capacity of the gels as measured with CBH I from *Trichoderma* (see below) was considerably higher for the CNBr-activated carrier (~15 mg ml^{-1}) than for the Affi-Gel 10-derived gel (~5 mg ml^{-1}).

Taking the necessary precautions during the affinity chromatographic runs (see further) and storage (4°, dark, 0.01% NaN$_3$) gels could be reused typically 10–50 times, depending on the enzyme samples. In the long run some leakage of the ligand could be observed.

Purification of the Cellobiohydrolases I and II from *Trichoderma reesei* and *Penicillium pinophilum*

Considerable affinity of the cellobiohydrolases from *Trichoderma* sp. for aryl 1-oxygen and 1-thio-β-D-cellobiosides is reflected by the K_m and K_i values for these soluble substrates and ligands (association constants 10^4–10^5 M^{-1} at 25°).[15] This contrasts with the lower K_m or K_i values (typically 10^3 M^{-1}) for the endoenzymes (e.g., EGI) present in the cellulolytic complex of *Trichoderma* sp.[16]

Coupling of an aryl-1-thio-β-D-cellobioside to an affinity carrier was therefore expected to be useful in the fractionation of endo- and exoenzymes from *Trichoderma*. Preliminary tests indicated that CBH I purified by ion-exchange chromatography from *T. reesei*[3] was completely retained by the affinity support (Affi-Gel 10 based) (pH 5.0, 0.1 M sodium acetate, 4°). Desorption was achieved either by 0.01 M cellobiose or 0.1 M lactose solutions (pH 5.0), whereas attempts to elute the enzyme with KCl (1 M), ethylene glycol (1 M), or glucose (1 M) were unsuccessful. Partially purified CBH II samples[3] were similarly bound to the thiocellobioside column but could be eluted only by cellobiose (0.01 M). Thus, clearly a biospecific adsorbant had been prepared and selective desorption should permit successive elution of both enzymes from a complex mixture.

Figure 1 illustrates the purification procedure used for a crude cellulase preparation. D-Glucose (0.1 M) and/or gluconolactone (10 mM) were

[14] J. Porath, this series, Vol. 34, p. 123.
[15] H. van Tilbeurgh, M. Claeyssens, and C. K. De Bruyne, *FEBS Lett.* **149**, 152 (1982).
[16] M. Claeyssens, unpublished results (1985).

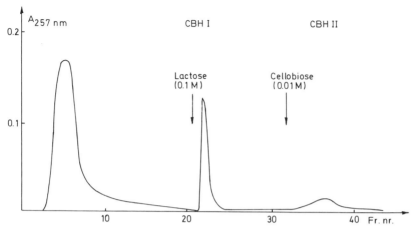

FIG. 1. Affinity chromatography of crude cellulase from *Trichoderma reesei*. Crude cellulase (celluclast, NOVO Industri, Denmark) (50 mg), dissolved in 5 ml of 0.1 M sodium acetate buffer, pH 5.0 (0.01% NaN$_3$) containing 0.1 M glucose and 10 mM gluconolactone, was applied to the affinity carrier packed in a column (1 × 10 cm) (4°) and equilibrated with the same buffer solution. Elution (30 ml/hr) was continued first with the same buffer, then with buffer supplemented with 0.1 M lactose or 0.01 M cellobiose as indicated. (Fr. nr., fraction number.)

added to all equilibrating and eluting buffer solutions. The presence of these compounds effectively suppresses the β-glucosidase (and some endocellulase) activities and in fact 0.1 M glucose is expected to enhance the affinity of the affinity ligand for CBH II considerably.[17] Elution of the adsorbed CBH I and CBH II was then performed successively with 0.1 M lactose and 0.01 M cellobiose. The isoelectric focusing patterns of both eluted fractions are shown in Fig. 2. The characteristic isoenzyme pattern of CBH I is observed for the lactose fraction, whereas the cellobiose fraction shows a series of bands, representing the CBH II isoenzyme pattern, characteristic of the present cellulase preparation. The specific activities of several fractions using standard substrates are compared in Table I.

For the bulk preparation of CBH I and CBH II from crude *Trichoderma reesei* culture filtrates (e.g., from NOVO Industri, Denmark) the following routine procedure was adapted, since it prevents nonspecific binding of proteins and other (colored) compounds present in crude extracts, which can entail considerable deterioration of the affinity carrier. The fractionating procedure is summarized in Scheme 1.

[17] H. van Tilbeurgh, G. Pettersson, R. Bhikhabhai, H. De Boeck, and M. Claeyssens, *Eur. J. Biochem.* **148**, 329 (1985).

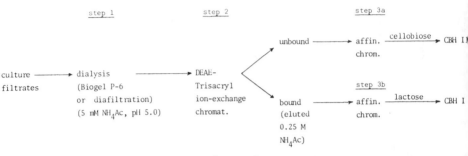

SCHEME 1.

After gel filtration (step 1) (pH 5.0, 5 mM ammonium acetate, NH$_4$Ac) of the crude extract (5 ml) on BioGel P-6 (2.5 × 50 cm) or diafiltration (PM10, Diaflo), pooled protein fractions (50 ml) are applied (step 2) on a DEAE-Trisacryl ion-exchange column (LKB-IBF, France) (1 × 15 cm) equilibrated with 5 mM NH$_4$Ac, pH 5.0. The unretarded fraction (5 mM NH$_4$Ac) (50–100 ml) is separated from the fraction (100 ml) eluted by 250 mM NH$_4$Ac.

The first fraction, containing the bulk of the β-glucosidase, CBH II, and some endocellulase activities, is concentrated by diafiltration (PM10) and adsorbed (step 3a) onto the affinity column (2 × 15 cm, pH 5.0, 10 mM glucono-1,5-lactone, 0.1 M lactose, 0.1 M glucose). When the absorbance (257 nm) of the eluates returns to the baseline, the CBH II fraction is eluted with 0.01 M cellobiose, dialyzed (pH 5.0) and concentrated (PM10, Diaflo). Solutions of 2–5 mg ml^{-1} of the CBH II isoenzyme mixture are obtained.

TABLE I

ACTIVITIES IN FRACTIONS ELUTED FROM THE AFFINITY COLUMN

Step	Specific activities (μmol glucose min^{-1} mg^{-1})		Ratio of CM-cellulase/avicelase
	Avicelase[a]	CM-cellulase[b]	
Buffer (unretarded)	0.0059	0.2828	48
Lactose (CBH I)	0.0175	0.0099	0.56
Cellobiose (CBH II)	0.0391	0.0165	0.42

[a] Activities at 50° were determined with a 1% suspension of Avicel in 0.1 M sodium acetate buffer, pH 5.0. One unit of activity was defined as the amount of enzyme needed to liberate reducing sugars corresponding to 1 μmol of glucose per min under these conditions. Specific activities are expressed per mg protein.

[b] Activities were similarly assayed with 1% solutions of carboxymethylcellulase (CM-cellulase, sodium salt, low viscosity grade) in 0.1 M sodium acetate buffer, pH 5.0 (50°).

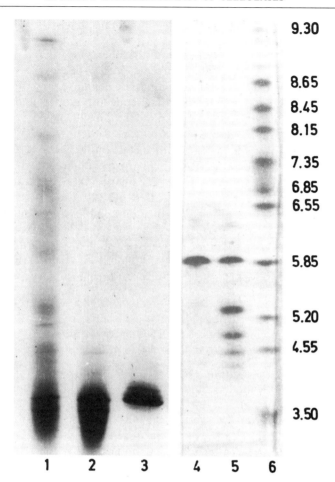

FIG. 2. Isoelectric focusing of CBH I and CBH II samples purified by affinity chromatography. The pH gradient (ampholine PAG plates, LKB) was 3.5 (bottom) to 9.5 (top). Samples contained 10–50 μg protein in distilled water. Lane 1: crude cellulase from *Trichoderma reesei*. Lane 2: CBH I–isoenzyme mixture, purified by affinity chromatography. Lane 3: CBH I (IP 3.9), purified by ion-exchange chromatography[3] and affinity chromatography. Lane 4: CBH II (IP 5.9), purified by ion-exchange chromatography[3] and affinity chromatography. Lane 5: CBH II–isoenzyme mixture, purified by affinity chromatography. Lane 6: IP marker proteins (Pharmacia).

In step 3b the fraction bound on the DEAE column and eluted at higher ionic strength (250 mM NH$_4$Ac) is run on the affinity column (pH 5.0, 0.1 M glucose). The unretarded fractions contain the endocellulase activities, whereas the CBH I components are desorbed by 0.1 M lactose.

The dialyzed and concentrated solution contain the CBH I–isoenzyme mixture (10–20 mg ml^{-1}).

Separation of the isoenzymes from these enzymes (CBH I–CBH II) can be obtained by ion-exchange chromatography or chromatofocusing.[3]

Similar results can be obtained with crude cellulase preparations from *Penicillium pinophilum* (87 160iii).[13] The concentrated (20-fold) cellulase [precipitated with $(NH_4)_2SO_4$ between the limiting 20 and 85% saturation] is desalted (5 ml) on a column (2.5 × 65 cm) of BioGel P-2, freeze dried, and redissolved in 5 ml 0.06 M sodium acetate buffer, pH 4.8, and applied to a column (2.5 × 45 cm) of DEAE-Sepharose equilibrated with the same buffer. The column is eluted first with 250 ml 0.1 M acetate buffer, pH 4.8, and then with 100 ml 0.1 M acetate buffer, pH 3.8. The colorless enzyme is finally prepared for affinity chromatography by precipitation with $(NH_4)_2SO_4$ (85% saturation), desalting on BioGel P-2, and freeze drying. As in the case of *Trichoderma reesei* two cellobiohydrolases are separated by affinity chromatography: CBH I (eluted with lactose) and CBH II (eluted with cellobiose). These fractions can further be identified by IEF and SDS–PAGE.

Purification of the Endocellulase C (EGC) from *Clostridium thermocellum* Cloned in *E. coli*

The anaerobe *Clostridium thermocellum*, secretes a highly active cellulolytic complex called cellulosome.[18] Fractionation of this complex poses a major problem and this has been approached by cloning experiments. The *celA, celB, celC,* and *celD* genes coding for the corresponding endoglucanases from *C. thermocellum* were expressed in *E. coli*.[19]

K_m values for aryl-β-D-cellobiosides indicate that the affinity of the endocellulase C (EGC) for these ligands is appreciably high.[20] The general applicability of the present biospecific chromatography is illustrated by the following fractionation of the endoglucanase C from a crude extract of frozen *Escherichia coli* JM101 (pCT 301) cells, treated for cell lysis (gift from Dr. J. Millet, Institut Pasteur, Paris).

Nucleic acid precipitation (streptomycin), heat treatment (60°), and ammonium sulfate precipitations (0–70%, 70–100%) are carried out as described.[20] The final solution is dialyzed against 50 mM sodium acetate, pH 5.0, centrifuged, and the supernatant (50 ml), containing ~50 mg

[18] R. Lamed, E. Setter, and E. A. Bayer, *J. Bacteriol.* **156,** 828 (1983).

[19] J. Millet, D. Pétré, P. Béguin, O. Raynaud, and J.-P. Aubert, *FEMS Microbiol. Lett.* **29,** 145 (1985).

[20] D. Pétré, J. Millet, R. Longin, P. Béguin, H. Girard, and J. B. Aubert, *Biochimie* **68,** 687 (1986).

protein, is applied on the affinity column as described above. The bulk of the protein is unretarded and when the absorbancy at 280 nm of the eluates returns to zero, desorption with 0.01 M cellobiose (pH 5.0) is started. Approximately 5 mg protein is obtained containing the bulk of the enzyme activity as measured with 4-methylumbelliferyl-β-D-cellobioside.[15] This fraction proves to be pure by SDS–PAGE and IEF–PAG (not shown).

This purification procedure could readily replace the classical methods used in the protocol described previously.[20]

[20] Morphological Aspects of Wood Degradation

By Robert A. Blanchette

Many species of white rot fungi can selectively degrade lignin from coniferous and deciduous wood without extensive loss of cellulose.[1,2] The patterns of cell wall degradation are distinct from the decay patterns caused by other white rot fungi that simultaneously degrade all cell wall components, as well as from other types of decay. Advanced stages of decayed wood that is lignin free can easily be identified in the field and laboratory using various histological reagents. These reagents result in color changes that aid in macroscopic observations and identification of delignified wood. Micromorphological and ultrastructural techniques are needed, however, for more precise determinations and to identify the sequential changes that occur during the process of lignin degradation within woody cell walls. The spatial relationships between fungal enzymes and sites of cell wall attack can be elucidated using these techniques. Quantitative data on lignin distribution in different morphological regions of the cell wall can be obtained by coupling scanning electron microscopy with energy-dispersive X-ray analysis.

Macroscopic Observations of Lignin-Free Wood

Several histological reagents can be used to rapidly test for lignin-free cellulose in wood with advanced stages of decay. Although these reagents have primary use in microscopic histochemistry,[3] they can be applied to

[1] R. A. Blanchette, L. Otjen, M. J. Effland, and W. E. Eslyn, *Wood Sci. Technol.* **19**, 35 (1985).

[2] R. A. Blanchette, *Appl. Environ. Microbiol.* **48**, 647 (1984).

[3] W. A. Jensen, "Botanical Histochemistry." Freeman, San Francisco, California, 1962.

METHODS IN ENZYMOLOGY, VOL. 160

cut surfaces of decayed wood for macroscopic observations. Two tests, phloroglucinol–HCl and the zinc chloroiodide reaction, have been used to identify decayed wood free of lignin. The phloroglucinol–HCl reaction consists of applying a saturated aqueous solution of phloroglucinol in 20% HCl. If lignin is present the wood will turn red-violet. The removal of lignin can also be detected by the use of reagents that will react with the unbound cellulose. The zinc chloroiodide reaction causes cellulose in delignified wood to appear dark blue. When lignin is still bound to the cellulose of the cell walls, the wood will appear yellow-brown or only faint blue. Several milliliters of a solution consisting of 50 g zinc chloride and 16 g potassium iodide dissolved in 17 ml of water (the solution is allowed to stand for several days and the supernatant used) is applied directly to the wood surface. These reagents are most successful when extensive amounts of lignin have been removed from large areas of the substrate.[4,5]

Decayed wood that is free of lignin will frequently contain black deposits of manganese dioxide.[5] These deposits can be differentiated from other black substances produced by fungi (e.g., interaction zone lines, pseudosclerotial plates, discolored wood) by applying an acidified solution (pH 3–4) of 0.2% leucoberbelin blue. This reagent is available from Dr. H. J. Altmann, Gehlberge Str. 9, D-1000 Berlin 20, West Germany. The colorless leucoberbelin reagent reacts with manganese oxides to produce a bright greenish-blue color within the wood.

Scanning Electron Microscopy Techniques

Micromorphological characteristics of delignified wood can be used to identify specifically cells void of lignin. Direct observations of decayed wood with a scanning electron microscope can select for fungi with the capacity to preferentially degrade lignin.[2] When sectioned, delignified cells readily separate from adjacent cells due to the extensive loss of middle lamellae (Fig. 1). The cells that remain are composed primarily of secondary wall layers (Fig. 2). Lignin and wood sugar analyses of decayed wood, using other techniques,[6,7] confirm the extensive loss of lignin and preferential attack of hemicellulose (Table I).

Cutting small samples of decayed wood by hand with a razor blade will result in a separation of delignified cells. The cell lumina will not be exposed. In contrast, lignified cells of sound (Fig. 3) or decayed wood

[4] R. A. Blanchette, *Phytopathology* **74**, 153 (1984).
[5] R. A. Blanchette, *Phytopathology* **74**, 725 (1984).
[6] M. J. Effland, *Tappi* **60**, 143 (1977).
[7] R. C. Pettersen, V. H. Schwandt, and M. J. Effland, *J. Chromatogr. Sci.* **22**, 478 (1985).

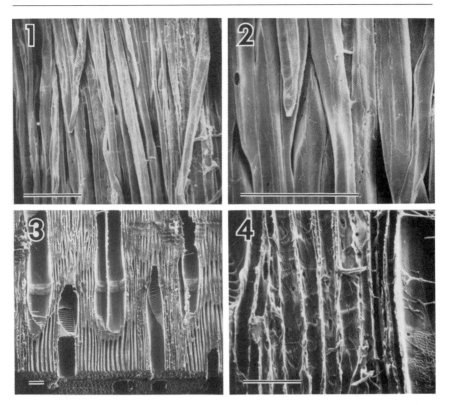

FIGS. 1 to 4. Scanning electron micrographs of sound and laboratory-decayed birch wood. Bar = 50 μm. (1 and 2) Wood delignified by *Poria medulla-panis* showing extensive loss of middle lamellae. Wood separated between cells revealing a defibration of the wood. (3) Sound wood with cells cut exposing the cell lumina. (4) Erosion troughs and holes in cell walls after decay by *Coriolus versicolor*. Sectioned wood exposed the cell lumina showing the simultaneous removal of all cell wall components from localized areas of the cell walls.

(Fig. 4) will maintain integrity during sectioning and the cell walls will be cut exposing the cell lumina. Cell wall degradation by *Coriolus versicolor*, a white rot fungus causing a simultaneous decay of all cell wall components (Table I), is presented in Fig. 4 for comparison to cells delignified by *Poria medulla-panis* (Figs. 1 and 2). Holes and erosion troughs are evident in the cell walls (Fig. 4). The localized degradation does not result in extensive degradation of middle lamella and cells remain intact.

To prepare the wood for scanning electron microscopy, specimens are mounted directly on aluminum specimen holders and dried in a desiccator or fixed in 2.0% OsO_4 for 1 hr and dehydrated through a graded ethanol series and critical point dried with CO_2 as a transitional fluid. Specimens

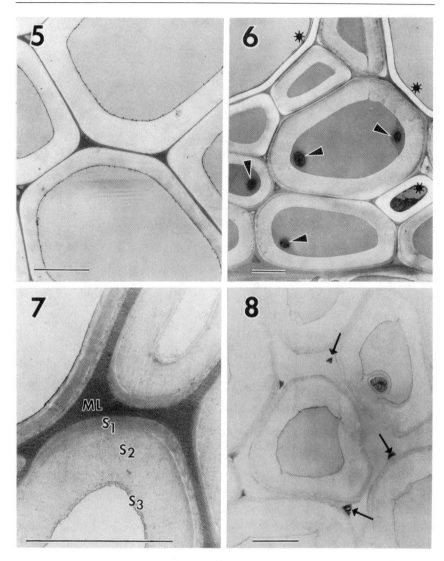

FIGS. 5 to 8. Transmission electron micrographs from transverse sections of sound aspen wood and wood decayed in the laboratory by *Phlebia tremellosus*. Bar = 5 μm. (5) Sound wood fixed in glutaraldehyde–OsO_4 and poststained with uranyl acetate had cells with electron-dense middle lamella and secondary walls with little contrast. (6) Wood delignified by *P. tremellosus* fixed in glutaraldehyde–OsO_4 and poststained with uranyl acetate. Cells with hyphae (arrowheads) in lumina had secondary walls that reacted with uranyl acetate and stained more intensely. The middle lamella between these cells was degraded. Cells with lignin remaining (*) appeared similar to cells from sound wood. (7) Sound wood fixed in $KMnO_4$ with good contrast and definition of lignified middle lamella region (ML) and sec-

TABLE I
PERCENTAGE LOSS OF WEIGHT, LIGNIN, AND WOOD SUGARS AFTER
12 WEEKS OF DECAY

Degradation specimen	Loss %				
	Weight	Lignin	Glucose	Xylose	Mannose
Phellinus pini	17	53.9	5.2	12.9	11.4
Phlebia tremellosus	34	75.2	4.1	39.4	28.5
Poria medulla-panis	34	82.7	1.0	39.3	23.4
Coriolus versicolor	53	59.2	47.9	59.2	65.1

are then coated with approximately 250 Å of 40% gold–60% palladium in a vacuum evaporator and observed with a scanning electron microscope.

Areas of delignified wood may not be uniformly distributed throughout the wood.[4,5] Often, delignified zones are surrounded by a white-rotted wood where a simultaneous erosion of all cell wall components has occurred (Fig. 4). Since two types of decay can be produced by some white rot fungi, observations should be made in several representative areas of the substrate.

Ultrastructural Techniques

The distribution of lignin within cell walls of sound wood and wood decayed to various extents by fungi can be illustrated in micrographs from transmission electron microscopy.

Samples of wood are cut into small pieces ($1 \times 1 \times 4$ mm) and processed using the following methods.

1. Specimens are fixed in 2.5% glutaraldehyde in 0.05 M phosphate buffer for 24 hr at 4° followed by a phosphate buffer wash and fixation in 2.0% OsO_4 in 0.05 M phosphate buffer for an additional 24 hr at 4°. During fixation, samples are placed in a chamber under low vacuum to aid infiltration. Samples are dehydrated through a graded acetone series and embedded in a modified Spurr's hard consistency embedding medium[8] consisting of 10.0 g vinylcyclohexene, 4.5 g diglycidyl ether of polypro-

[8] A. R. Spurr, J. Ultrastruct. Res. 26, 31 (1969).

ondary wall layers (S_1, S_2, and S_3). (8) Wood delignified in the laboratory by P. tremellosus showing complete degradation of middle lamella between cells. Cell corner regions (arrows) were not degraded. The secondary walls were less electron dense than the secondary walls of sound wood.

pylene glycol (DER 736), 26.0 g nonenylsuccinic anhydride, and 0.2 g dimethylaminoethanol. Polymerization is done at 70° for 16 hr. Sections can be obtained with a diamond knife and stained with 0.5% uranyl acetate. This method will give poor contrast of the compound middle lamella and secondary wall regions in sound wood (Fig. 5). After lignin removal, however, the secondary wall layers stain intensely in areas of unbound cellulose (Fig. 6). The removal of lignin apparently results in better penetration and reaction of uranyl acetate.

2. Samples are fixed in 2.0% $KMnO_4$ in distilled water for 3 hr at 4°. During fixation, samples were placed under low vacuum to ensure infiltration. Samples are dehydrated and embedded as stated above. No poststaining is used.

The compound middle lamella and secondary wall layers became clearly evident with good contrast after $KMnO_4$ fixation (Fig. 7). The primary wall is not distinguishable from the middle lamella but the S_1, S_2, and S_3 layers of the secondary wall are discernible. In delignified wood samples, the wall layers become less electron dense. Samples with large lignin losses show the middle lamella between cells to be completely degraded and cell corner regions persist (Fig. 8).

Energy-Dispersive X-Ray Microanalysis (STEM-EDXA)

Lignin distribution within wood cell walls can be accurately quantified by bromination of the wood in a nonaqueous system and subsequent analysis of bromine concentration using scanning electron microscopy coupled with energy dispersive X-ray analysis.[9] Under the proper conditions bromine will react with lignin and become incorporated into the various wall layers in proportion to the lignin content. This technique has been used to observe the lignification sequence in developing pine[10] and can be adapted to monitor the sequential degradation of lignin from the various wall layers by wood decomposing fungi.

Samples of wood, $1 \times 1 \times 7$ mm, are brominated as described by Saka et al.[10] Wood samples are dehydrated in a graded ethanol series and the ethanol replaced with $CHCl_3$. Bromine, 0.3 ml in 20 ml $CHCl_3$, is added to 70 ml of $CHCl_3$ containing the wood samples and stirred for 3 hr at room temperature. The reaction mixture is refluxed for an additional 3 hr and samples repeatedly washed with $CHCl_3$. The specimens were then placed in a Soxhlet apparatus and extracted with $CHCl_3$ until all unreacted bromine is removed (7–10 days). Wood samples are embedded in Spurr's embedding medium[8] and 0.2-μm sections cut with a diamond knife.

[9] S. Saka and R. J. Thomas, *Wood Sci. Technol.* **16**, 1 (1982).
[10] S. Saka and R. J. Thomas, *Wood Sci. Technol.* **16**, 167 (1982).

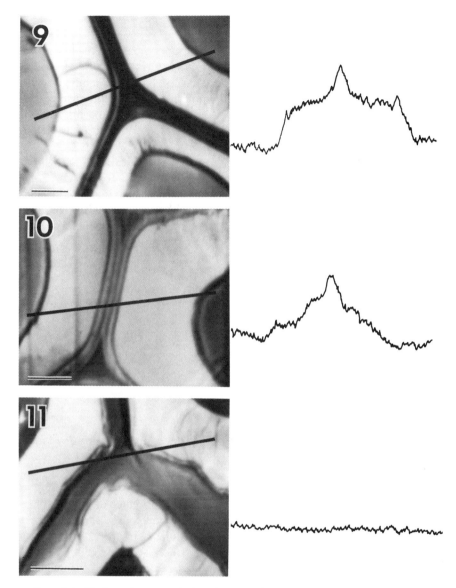

Figs. 9 to 11. Scanning transmission electron micrographs from electrons transmitted through 0.2-μm sections of sound birch wood and wood decayed by *Phellinus pini* in the laboratory and intensity distribution profiles of bromine (L-series) X rays obtained from the scan line indicated on micrograph. Bar = 2 μm. (9) Sound wood and intensity profile representing lignin distribution within the various cell wall layers. (10) Incipient stage of decay by *P. pini* with reduced Br intensity in the secondary walls nearest the cell lumina. Concentrations of Br increase toward the middle lamella demonstrating the degradation of lignin from secondary wall layers before loss from the middle lamella. (11) Advanced decay by *P. pini* with low-intensity Br distribution throughout the cell walls indicating extensive loss of lignin.

Sections are placed on Formvar-coated copper grids and coated with a thin layer of nickel in a vacuum evaporator. Grids are attached to a carbon post using spectroscopically pure carbon paint and analyzed with an energy dispersive spectrometer in conjunction with a scanning electron microscope and transmitted electron detector (STEM). The distance of the X-ray detector, accelerating voltage, condenser aperature, and electron beam spot size are all held constant.

Bromine distribution profiles along a scan line, corresponding to lignin concentration, are presented for sound wood and wood decayed by *Phellinus pini,* a white rot fungus that preferentially removed lignin. Although the image quality of micrographs from electrons transmitted through thick sections is not excellent, it illustrates the exact location of the scan line (Figs. 9–11). The bromine (L-series) X-ray distribution profile provides the relative concentration of bromine among the different morphological regions. Lignin is most concentrated in the middle lamella region and secondary wall layers have a reduced concentration. *Phellinus pini* removes the lignin from the secondary wall layer from the lumen toward the middle lamella region. In advanced stages of decay, lignin is removed from throughout the secondary wall and middle lamella. SEM-EDXA can also provide X-ray distribution mapping and point analysis in addition to the line scan analysis.

[21] Cellulases of *Pseudomonas fluorescens* var. *cellulosa*

By Kunio Yamane and Hiroshi Suzuki

Cellulose is an insoluble and highly ordered polymer of glucose. Microorganisms growing on cellulose as a sole carbon source should secrete cellulases to the outside of the cells. The cellulases hydrolyze the highly polymerized structure of cellulose to yield soluble cellooligosaccharides.[1,2] Most studies on cellulases have been performed using extracellular cellulases. However, considerable amounts of cellulases have also been found inside cells.[3,4]

Cellulolytic bacteria are suitable for studies not only on the nature of the cellulolytic enzyme system but also on the regulation mechanism of

[1] E. T. Reese, R. G. H. Siu, and H. S. Levinson, *J. Bacteriol.* **59,** 485 (1950).
[2] K. Selby, *Adv. Chem. Ser.* **95,** 34 (1969).
[3] K. W. King, *J. Dairy Sci.* **42,** 1848 (1959).
[4] B. Norkrans, *Adv. Appl. Microbiol.* **9,** 97 (1967).

cellulase formation in microorganisms, because the growth rate of bacterial cells is higher and experimental conditions are easier to control than in fungal cells. *Pseudomonas fluorescens* var. *cellulosa* produces two extracellular cellulases (A and B), and one cell-bound cellulase (C). Cellulase C differs from cellulases A and B in enzymatic properties and also in physiological responses to culture conditions.[5-7]

Assay Method

Principle. The activity is measured by following the increase in reducing sugars due to enzymatic hydrolysis of cellulose. The enzyme activities for a soluble cellulose derivative, carboxymethylcellulose (CMC), and for an insoluble crystalline cellulose, Avicel, are called CMCase and Avicelase, respectively.

Reagents

McIlvaine's buffer (0.2 M $Na_2HPO_4 \cdot 2H_2O$–0.1 M citric acid), pH 7.0
CMC solution, 1 g CMC (degree of substitution = 0.6, Daiichi Kogyo Seiyaku Co., Kyoto, Japan) in 100 ml of H_2O
Avicel suspension, 1 g Avicel (American Viscose Co. Ltd.) in 100 ml of H_2O
Somogyi's copper reagent[8] and Nelson's arsenomolybdate reagent[9] for determination of reducing sugars[10]

Procedure. The reaction mixture consisting of CMC solution or Avicel suspension, McIlvaine's buffer, and enzyme solution (1 : 2 : 1, v/v/v) is incubated at 30°. After an appropriate period (1 hr for CMCase and 24 hr for Avicelase activity), 1.0 ml of the mixture is withdrawn and the reducing power increase is determined by the method of Somogyi and Nelson.[10] Color development is measured by the absorbance at 660 nm with a spectrophotometer.

Definition of Unit. One unit of CMCase activity is defined as the amount of enzyme which produces 1.0 μmol of reducing sugars as glucose/min/ml of enzyme solution and 1 unit of Avicelase activity is defined as for CMCase except that the rate is in μmol/hr.

[5] K. Yamane, H. Suzuki, M. Hirotani, H. Ozawa, and K. Nisizawa, *J. Biochem.* (*Tokyo*) **67,** 9 (1970).
[6] K. Yamane, H. Suzuki, and K. Nisizawa, *J. Biochem.* (*Tokyo*) **67,** 19 (1970).
[7] H. Suzuki, K. Yamane, and K. Nisizawa, *Adv. Chem. Ser.* **95,** 60 (1969).
[8] M. Somogyi, *J. Biol. Chem.* **195,** 19 (1952).
[9] N. Nelson, *J. Biol. Chem.* **153,** 375 (1944).
[10] R. G. Spiro, this series, Vol. 8, p. 7.

Origin and Culture Medium of Organism

Pseudomonas fluorescens var. *cellulosa* (IAM 12622, the culture collection of Institute of Applied Microbiology, The University of Tokyo, Bunkyo-ku, Tokyo, 113 Japan) isolated in 1948 by Ueda *et al.*[11] from soil was provided by Dr. Ueda. This cellulolytic bacterium is a gram-negative straight rod (0.5–0.7 × 1.2–3.0 μm) with a polar flagellum. Its taxonomic position is being recharacterized. Bacteria are maintained on Dubos minimal medium[12] consisting of 0.1 g of K_2HPO_4, 0.05 g of $MgSO_4 \cdot 7H_2O$, 0.1 g of $NaNO_3$, 0.05 g of NaCl, and a trace of $FeCl_3$ in 100 ml of tap water (pH 7.0–7.2). A strip of filter paper is the sole carbon source. NaCl is generally omitted and 0.02–0.05% $FeCl_3 \cdot 6H_2O$ provides optimal growth.

Purification of *Pseudomonas* Cellulases

Pseudomonas fluorescens produces two extracellular components (cellulase A which moves rapidly to the cathode and cellulase B which moves slowly), and one cell-bound component (cellulase C), upon zone electrophoresis on cellulose acetate film (Fig. 1). Cellulases A and B are obtained from the supernatants of cultures grown on cellulose as the carbon source. In contrast, cellobiose (G_2), one of the major products of cellulolysis, stimulates only the formation of cellulase C, which seems to be located at the wall or periplasmic space of the cells.[5,7] Purification of cellulases A, B, and C is summarized in Table I.[6,7]

Preparation of Cellulases A and B

The organism is grown on Dubos minimal medium containing 0.5% (w/v) Avicel in 20-liter jar fermentors for 4 days at 37°. After centrifugation at 10,000 g, the supernatant (approximately 120 liters in all from 6 batches of the culture) is concentrated to about 1/10 volume using a flash evaporator under reduced pressure at 30–35°. Cellulases were precipitated therefrom by ammonium sulfate (65 g/100 ml) at pH 7, collected by filtration through Celite 535, and extracted with 3 liters of distilled water. The dissolved material was dialyzed against tap water and then distilled water. The dialyzed solution was concentrated under reduced pressure and lyophilized (yield, about 12 g). All following procedures are carried out at 2–5° unless otherwise stated using Na_2HPO_4–KH_2PO_4 buffers.

Chromatography on Sephadex G-100 (Separation of Cellulase A from B). Half a gram of the lyophilized preparation was dissolved in 50 ml of 20

[11] K. Ueda, S. Ishikawa, T. Itami, and T. Asai, *J. Agric. Chem. Soc. Jpn.* **26**, 35 (1952).
[12] R. J. Dubos, *J. Bacteriol.* **15**, 223 (1928).

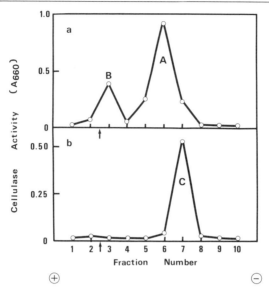

FIG. 1. Zone electrophoretic patterns of an extracellular cellulase preparation from a 0.5% cellulose culture (a) and a cell-bound cellulase preparation from a 0.5% cellobiose culture (b) on cellulose acetate films (2 × 11 cm). Arrows indicate the origins. A, B, and C represent cellulases. Electrophoresis was conducted at about 2° for 2 hr with a constant current of 0.6 mA/cm in Veronal buffer, pH 8.6. After the run, the films were cut crosswise into 1-cm segments and the cellulase (CMCase) activity in each was assayed.

mM buffer at pH 7.0. After dialysis against the same buffer overnight, the cellulase solution was passed through a Sephadex G-100 column (42 × 650 mm), and eluted with the same buffer; fractions of 100 ml were collected. Most of the cellulase activity was found in peak fractions 3–6, but some activity tailed into fractions 7–15 or 17. The former and latter combined fraction (AI and BI) consisted mainly of cellulase A and cellulase B, respectively. AI and BI from 22 chromatographic runs were concentrated.

Chromatography on DEAE-Sephadex A-50. AI and BI were dialyzed against 20 mM buffer, pH 6.0. The solution was divided into five portions and each was applied to a DEAE-Sephadex A-50 column (42 × 650 mm) equilibrated with the same buffer. Upon stepwise treatment with 1.5 liter each of the buffers of 20 mM, pH 6.0, 50 mM pH 7.0, 100 mM, pH 8.0, and 100 mM, pH 9.0 with 0.5 M NaCl, cellulase AI elutes as a major peak with 100 mM buffer at pH 8.0; cellulase BI elutes as a major peak with 50 mM buffer, pH 7.0. The fractions in each major peak were combined, concentrated, and dialyzed against 20 mM buffer, pH 7.0. The combined

TABLE I
PURIFICATION OF EXTRACELLULAR AND CELL-BOUND CELLULASES AND
THEIR RECOVERIES

Purification step	Recovery	
	Total activity[a] $(\times 10^{-2})$	Specific activity[b]
Cellulase A		
Crude cellulase preparation	1030	80
Sephadex G-100 (cellulase AI)	760	(390)
DEAE-Sephadex A-50 (cellulase AII)	—	(640)
Sephadex G-100 (cellulase AIII)	—	(1260)
DEAE-Sephadex A-50 (cellulase A)	344	1150
Cellulase B		
Crude cellulase preparation	1030	80
Sephadex G-100 (cellulase BI)	96	(10)
DEAE-Sephadex A-50 (cellulase BII)	—	(100)
Starch zone electrophoresis (cellulase B)	12	90
Cellulase C		
Cells	380	5
EDTA–lysozyme–sucrose treatment	—	(30)
Sephadex G-150 (cellulase CI)	210	(70)
DEAE-Sephadex A-50 (cellulase CII)	—	(1520)
Sephadex G-150 (cellulase C)	59	1860

[a] CMCase activity units.
[b] Units per Lowry protein as bovine serum albumin. Figures in parentheses are calculated from the protein measured by the absorbance at 280 nm, providing that a 0.1% solution of protein gives an absorbance of 1.0.

preparations, named cellulase AII and BII, were almost free of amylase and β-glucosidase activities.

Gel Filtration of Cellulase AII with Sephadex G-100. AII (40 ml) was applied to a Sephadex G-100 column (42 × 700 mm) and eluted with 20 mM buffer, pH 7.0. A peak of cellulase activity, fractions 7–11 (250 ml), was concentrated (AIII).

Rechromatography of AIII on DEAE Sephadex A-50 Column. AIII was subjected to DEAE-Sephadex A-50 column chromatography under the same conditions as above. Eluates with 100 mM buffer, pH 8.0, were combined, concentrated, dialyzed against 2 mM buffer, pH 7.0, and lyophilized. The preparation was called cellulase A.

Starch Zone Electrophoresis of Cellulase BII. BII was divided into four portions of 10 ml each and each portion was subjected to starch zone electrophoresis in a plastic tray (35 × 10 × 1.5 cm) for 40 hr at 0–2° with a

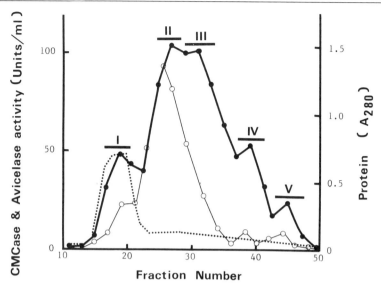

Fɪɢ. 2. Gel filtration pattern of a crude extracellular cellulase preparation obtained from a 0.5% Avicel culture on a BioGel P-150 column (4 × 70 cm). Eluates were collected every 10 ml. (●) CMCase activity; (○) Avicelase activity; dotted line, absorbance at 280 nm.

constant current, 2.3 mA/cm, in Veronal buffer of pH 8.6 (0.05 μ). After the run, the starch block was cut crosswise into 1 cm widths, each block is extracted with 10 ml of 50 mM buffer, pH 7.0, and measured for CMCase activity. The major peak, which behaved as component B on cellulose acetate film electrophoresis, was combined, concentrated, dialyzed against 2 mM buffer, pH 7.0, and lyophilized. This was called cellulase B.

Alternative Procedure for Fractionation and Purification of Extracellular Cellulases A and B. Figure 2 shows the gel filtration pattern on a BioGel P-150 of CMCase and Avicelase activities of the crude extracellular cellulase preparation obtained from the 0.5% Avicel culture. Five peaks (I–V) were detected, whose molecular weights are estimated to be >10 × 10⁴, 6.5 × 10⁴, 6 × 10⁴, 5 × 10⁴, and 4 × 10⁴, respectively. Peak II showed the highest Avicelase activity. Peak I mainly consists of cellulase B and peak V contains only cellulase A, whereas peaks II–IV are composed of both cellulase A and B. Cellulase A in peak V was purified by DEAE-Sephadex A-50 column chromatography. Cellulase B in peak I can be further purified by gel filtration on BioGel P-300 column and DEAE-Sephadex A-50 column chromatography.[13]

[13] T. Yoshikawa, H. Suzuki, and K. Nisizawa, *Sci. Rep. Tokyo Kyoiku Daigaku* **16,** 87 (1975).

Preparation of Cellulase C

Bacteria are grown at 37° for 8 hr by shaking in 150 500-ml flasks each containing 200 ml Dubos minimal medium with 0.5% (w/v) G_2. Cells, harvested by centrifugation at 10,000 g, were suspended in 4 liters of 0.32 M sucrose in 25 mM Tris–HCl buffer, pH 8.0, containing 2 g of lysozyme (Seikagaku Kogyo Co. Ltd., Tokyo, Japan) and 8 g of EDTA, and incubated at 30° for 30 min under gentle stirring. After removing the resultant spheroplasts by centrifugation at 10,000 g for 1 hr, cellulase in the supernatant was precipitated by ammonium sulfate (65 g/100 ml), cellected by centrifugation, and dissolved in 20 mM buffer at pH 7.0.

Gel Filtration on Sephadex G-150. After desalting by passing through a Sephadex G-25 column the crude preparation of cell-bound cellulase was divided into 20 portions of 25 ml each and each portion was subjected to gel filtration on Sephadex G-150 columns (42 × 650 mm) equilibrated with 20 mM buffer, pH 7.0. The eluate with the same buffer was collected every 50 ml. The peak of cellulase activity, fractions 13–17, was pooled and concentrated (CI).

Chromatography on DEAE-Sephadex A-50. CI was dialyzed against 20 mM buffer, pH 6.0, and divided into four portions. Each was applied to a DEAE Sephadex A-50 column (42 × 650 mm) under conditions similar to those described for extracellular cellulases. The cellulase activity was recovered in the eluate with 20 mM buffer, pH 6.0, and was completely separated from amylase and β-glucosidase activity. The peak fractions were pooled and concentrated (CII).

Second Filtration on Sephadex G-150. CII was dialyzed against 20 mM buffer, pH 7.0, and divided into six portions of 4.5 ml each. Each was applied to a Sephadex G-150 column (24 × 1,000 mm) and eluted with the same buffer. The peak of cellulase activity, fractions 18–21 (60 ml), were combined, concentrated, dialyzed against 2 mM buffer, pH 7.0, and lyophylized. This was designated cellulase C.

Purity of Cellulases A, B, and C

The three preparations showed a single cellulase peak in zone electrophoreses on starch and on cellulose acetate film, and in analytical ultracentrifugation, although a small amount of heavier component was detected in the ultracentrifugal pattern of cellulase C.

Properties of Cellulases A, B, and C

Chemical Analyses. Cellulases A, B, and C showed no significant differences in their amino acid composition.[6] All three contain carbohydrate, amounting to about 13, 36, and 33%, respectively.[6] The main constituent is galactose in cellulase A and glucose in cellulases B and C.

pH Optima and Heat Stability. The optimal pH of CMCase activity is 8.0 for cellulase A and B and 7.0 for cellulase C. All three are most stable at pH 7.0–8.0, and completely inactivated by heating at 60° for 10 min, although cellulase B is slightly activated by treatment at 40° for 10 min.

Activities toward Cellooligosaccharides, Celluloses, and Their Derivatives. Cellulases A, B, and C are similarly incapable of attacking G_2 and *p*-nitrophenyl-β-glucoside, but they hydrolyze various substrates such as cellooligosaccharides, amorphous celluloses, and Avicel. The activities of each cellulase toward these substrates are markedly different from one another. Cellulase B generally shows much lower activity toward soluble substrates including CMC than cellulase A and C (approximately 1/10 to 1/20), whereas their activity toward highly polymerized substrates such as insoluble swollen cellulose, Avicel, and DEAE-cellulose is not so different.

The most conspicuous difference is found in the activity toward cellotriose (G_3) and cellotriosylsorbitol (G_{4H}). Only cellulase C hydrolyzes them easily. This indicates that at least three consecutive glucosyl residues seem to be necessary for the hydrolysis of cellooligosaccharides by cellulase C. In contrast, four consecutive glucosyl residues are necessary for hydrolysis by cellulases A and B, since they hydrolyze cellotetraose (G_4) and cellotetraosylsorbitol (G_{5H}) but not G_{4H}. Table II is a summary of relative hydrolysis rates of each bond of nonreduced and reduced cellooligosaccharides by the three cellulases. Cellulase C mostly binds the G_2 residue of the nonreducing end to cleave it when it forms an enzyme–substrate complex. Cellulases A and B similarly bind a G_2 residue at the nonreducing end of lower cellooligosaccharides, but they more easily bind the G_3 residue of the nonreducing end of higher cellooligosaccharides.

There is no significant difference in the degree of randomness in CMC hydrolysis by the three cellulases, as shown by the relationship between the decrease in viscosity and the increase in reducing power during incubation. Therefore, these cellulases hydrolyze CMC in a similarly random mechanism.

Effect of Culture Conditions on Cellulase Formation

Bacteria can grow on various carbohydrates, amino acids, and organic acids.[14] However, cellulase production is remarkably affected by the kind

---

[14] K. Yamane, H. Suzuki, K. Yamaguchi, M. Tsukada, and K. Nisizawa, *J. Ferment. Technol.* **43**, 721 (1965).

TABLE II
RELATIVE HYDROLYSIS RATES OF THE GLUCOSYL BONDS IN NONREDUCED AND REDUCED CELLOOLIGOSACCHARIDES

Substrates with molecular model[a]	Bond attacked preferentially[b]		
	Cellulase A	Cellulase B	Cellulase C
G_3 O—O—O< (1–2–3)	c	c	1 or 2
G_4 O—O—O—O< (1–2–3–4)	2 > 1 or 3	2 > 1 or 3	2 > 1 or 3
G_{4H} O—O—●—O (1–2–3–4)	c	c	1 > 2 >>> 3
G_5 O—O—O—O—O< (1–2–3–4)	2 or 3 >>> 1 or 4	2 or 3 >>> 1 or 4	2 or 3 >>> 1 or 4
G_{5H} O—O—O—●—O (1–2–3–4)	2 > 3 >>> 1 or 4	2 > 3 >>> 1 or 4	2 >>> 1, 3 or 4
G_6 O—O—O—O—O—O< (1–2–3–4–5)	3 ≧ 4 or 2 >>> 1 or 5	3 ≧ 4 or 2 >>> 1 or 5	2 or 4 > 3 >>> 1 or 5
G_{6H} O—O—O—O—●—O (1–2–3–4–5)	3 >>> 4 = 2 >>> 1 or 5	3 >>> 4 > 2 >>> 1 or 5	3 = 4 >>> 1, 2 or 5

[a] O, O<, and ● represent the nonreducing and reducing glucosyl residues, and the terminal sorbotol residues, respectively. Numbers from 1 to 5 indicate the position of bonds from the nonreducing end.

[b] The symbols >, >>, and >>> represent the relative rates of bond hydrolysis in the factors of 1.5–3 times, 5–10 times, and extremely high, respectively. The rates are based on quantitative analysis of each product by paper chromatography.

[c] Not hydrolyzed.

of carbohydrates (0.5% w/v) in batch cultures at 37°. Glucose supports excellent growth, but does not stimulate cellulase formation. In contrast, cellulose and sophorose (2-O-β-glucosyl-D-glucose) increase the production of cellulases, especially the extracellular form. Sophorose is an inducer of cellulase in the cellulolytic fungus, *Trichoderma viride*.[15,16] G_2 stimulates the formation of a cell-bound cellulase but not the extracellular enzyme. Growth rates (increase in turbidity at 610 nm/hr) in the logarithmic phase of growth on glucose, G_2, sophorose, and cellulose were 0.183, 0.174, 0.183, and 0.03, respectively. Sophorose supported excellent growth (as did glucose) and hyperproduction of the extracellular cellulase (as did cellulose). Effects of starch, lactose, galactose, and xylose on cell growth and cellulase formation were similar to that of glucose, while those of cellooligosaccharides (G_3, G_4, G_5, G_6, and G_{25}) were similar to that of G_2.[5]

In the cellulose culture supporting the enhanced cellulase production, the concentration of soluble sugars in the medium should not be high. Slow feeding of cultures was then attempted with glucose, G_2, and sophorose. The growth rate in these cultures was suppressed to about one-third of that found in the batch cultures. Cellulase formation, particularly extracellular cellulase formation, in all the slowly fed cultures was stimulated to the level comparable to the cellulose batch culture. The extracellular cellulase was also composed of cellulase A and B.[5,17] The level of the cell-bound cellulase in these cultures was almost equal to the levels in batch cultures on G_2 and cellooligosaccharides. The cell-bound cellulase C can be solubilized during spheroplasting with the treatment of lysozyme and EDTA in an isotonic solution. Therefore, this component seems to be located at the wall or periplasmic space of the cells.[17]

Sophorose highly stimulates the production of extracellular cellulase not only in the batch culture but also in the slowly fed culture. To investigate the sophorose effect further, the effects of various concentrations of carbohydrates on cellulase and amylase formation were analyzed using washed cell suspensions. The results suggest that sophorose does not act as an actual inducer for cellulase formation in this bacterium, and that the observed stimulation of cellulase formation was rather closely related to release from so-called catabolite repression, although we have not yet clarified why sophorose supports such excellent growth and hyperproduc-

[15] M. Mandels, F. W. Parrish, and E. T. Reese, *J. Bacteriol.* **83,** 400 (1962).
[16] T. Nisizawa, H. Suzuki, M. Nakayama, and K. Nisizawa, *J. Biochem.* (*Tokyo*) **70,** 375 (1971).
[17] K. Yamane, T. Yoshikawa, H. Suzuki, and K. Nisizawa, *J. Biochem.* (*Tokyo*) **69,** 771 (1971).

tion of cellulase. Since the formation of cell-bound cellulase is only slightly influenced by culture conditions, it seems to be less sensitive to catabolite repression than that of extracellular cellulases.[18]

In Vitro Conversion of Cellulase B to Cellulase A

Cellulases A and B are secreted into the culture medium and are similar in their substrate specificity; but they differ in their specific activity, electrophoretic mobility, and carbohydrate content. Characteristics of cellulase B such as high adsorbability on Sephadex and slow electrophoretic mobility, possibly due to a high adherence to cellulose acetate film, may be due to its high glucose content. The proportion of cellulase B to A decreases with the culture period not only in the Avicel culture but also in the culture slowly fed with glucose.[17,19] An attempt was made to convert cellulase B to A in vitro using Trichoderma viride β-glucosidase as a modifying agent. The β-glucosidase preparation, free from protease activity for casein as well as from CMCase and other glycanases, was obtained from a commercial cellulase preparation of Trichoderma viride, cellulase Onozuka (Yakult Co. Ltd., Tokyo Japan) by consecutive column chromatography on Amberlite CG-50 and DEAE-Sephadex A-50.[20] When a purified cellulase B preparation from peak I (Fig. 2) was incubated with the β-glucosidase at 30° for 72 hr at pH 7.0, cellulase B was completely lost and an electrophoretically rapid-moving cellulase A-like component was detected. Similar changes in cellulase B by β-glucosidase treatment were observed in peaks II–IV (Fig. 2).[13] Although we have not yet obtained evidence other than the electrophoretic behavior for identification of the conversion product with cellulase A, the results strongly indicate the possibility of an enzyme-catalyzed conversion of cellulase B to A. In contrast, no changes in CMCase activity and electrophoretic mobility occurred upon the β-glucosidase treatment of cellulase A and C, and peak V. These results suggest that cellulase B is an aggregated form of cellulase A, which differs from cellulase C.

[18] H. Suzuki, in "Symposium on Enzymatic Hydrolysis of Cellulose" (M. Bailey, T. M. Enari, and M. Linko, eds.), p. 155. Technical Research Centre of Finland, Helsinki, Finland, 1975.

[19] T. Yoshikawa, H. Suzuki, and K. Nisizawa, J. Biochem. (Tokyo) 75, 531 (1974).

[20] G. Okada, K. Nisizawa, and H. Suzuki, J. Biochem. (Tokyo) 63, 591 (1968).

[22] Cellulases of *Cellulomonas uda*

By KATSUMI NAKAMURA and KUMPEI KITAMURA

Cellulomonas uda secretes several types of cellulases. The isolation of a cellulase active on crystalline cellulose, β-1,4-glucan cellobiohydrolase (EC 3.2.1.91, cellulose 1,4-β-cellobiosidase) is described. Other components of cellulases are also reported.

Assay Method

Cellulase activity is determined by measuring the reducing sugar released from crystalline cellulose and CM-cellulose.

Reagents

Crystalline cellulose (Avicel, E. Merck AG. no. 2330)
CM-cellulose (CMC, Sigma Co. no. C-8758)
Acetate buffer, 0.2 M, pH 5.5
Reagents for Somogyi–Nelson method[1]
Reagents for dinitrosalicylic acid method[2]

Procedure

Hydrolysis of Crystalline Cellulose. The reaction mixture consisting of 2.0 ml of 1% crystalline cellulose suspension in 0.2 M acetate buffer (pH 5.5) and 0.2 ml enzyme solution is incubated at 45° for 1 hr. The reaction is stopped by heating the mixture at 100° for 5 min and then the mixture is filtered. Reducing sugar in the filtrate is determined with the Somogyi–Nelson method.[1] The enzyme solution is diluted to give a value of less than 0.5 mg of reducing sugar.

Hydrolysis of CM-Cellulose. The reaction mixture consisting of 2.0 ml of 1% CM-cellulose in 0.2 M acetate buffer (pH 5.5) and 0.1 ml of enzyme solution is incubated at 45° for 10 min and the reducing sugar liberated is determined with the dinitrosalicylic acid method[2] or Somogyi–Nelson method.[1] The enzyme solution is diluted to give a value of less than 0.5 mg of reducing sugar.

[1] M. Somogyi, *J. Biol. Chem.* **195**, 19 (1952).
[2] G. L. Miller, *Anal. Chem.* **31**, 42 (1959).

Definition of Unit and Specific Activity

One unit of activity is defined as the amount of enzyme liberating 1 μmol of reducing sugar as glucose per min under the assay condition. Specific activity is defined as units per milligram of protein. Protein content is determined by the method of Lowry *et al.*[3] using bovine serum albumin as a standard.

Cell Growth

Cellulomonas uda CB4 (FERM BP-199, Fermentation Research Institute, Tsukubagun, Ibaraki 305, Japan) is cultured at 30° for 48 hr with shaking in the Avicel-peptone medium[4] composed (per liter) of 20 g crystalline cellulose (Avicel, Asahi Kasei Co.), 20 g Polypepton (Daigo Eiyo Co.), 2 g K_2HPO_4, 0.5 g $MgSO_4 \cdot 7H_2O$, 0.5 g yeast extract (Difco Co.), and 1 g Tween 80 (pH 8.0).

Enzyme Purification

All operations[5] are performed at 4°.

Ammonium Sulfate Precipitation

Culture of *C. uda* CB4 is very viscous because of the polysaccharide produced, and centrifugation at high speed (20,000 g for 10 min) is needed to remove the cells. The culture (6 liter) is brought directly to 10% saturation with ammonium sulfate, and then the solution is centrifuged at low speed (8000 g for 10 min), which removes the cells, and the resulting supernatant containing no cells was further fractionated with 50% saturation with ammonium sulfate. The resulting precipitate is dissolved in 30 ml of 0.05 M Tris buffer, pH 7.2, and desalted by gel filtration with Sephadex G-25 column (3 × 90 cm) equilibrated with 0.05 M Tris buffer (pH 7.2), and eluted with the same buffer. Fractions with larger molecular size proteins were collected.

DEAE-Sepharose Chromatography

The G-25 eluate is applied to a DEAE-Sepharose column (2.6 × 34 cm) equilibrated with 0.05 M Tris buffer, pH 7.2. The column is washed with the same buffer and enzyme protein is eluted with a linear gradient com-

[3] O. H. Lowry, N. J. Rosenbrough, A. L. Farr, and R. J. Randall, *J. Biol. Chem.* **193,** 265 (1951).
[4] K. Nakamura and K. Kitamura, *J. Ferment. Technol.* **60,** 343 (1982).
[5] K. Nakamura and K. Kitamura, *J. Ferment. Technol.* **61,** 379 (1983).

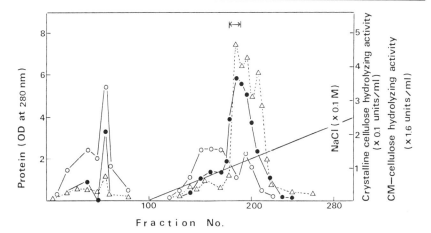

FIG. 1. Elution pattern of cellulolytic enzymes on the first DEAE-Sepharose column (2.6 × 34 cm) chromatography. Fractions of 9.5 ml were collected. (△) Protein (absorbance at 280 nm), (●) activity toward crystalline cellulose, (○) activity toward CM-cellulose.

posed of 1 liter 0.05 M Tris buffer (pH 7.2) and 1 liter 0.3 M NaCl in the same buffer, as shown in Fig. 1. Fractions 178–189, which show high activity on crystalline cellulose and low activity on CM-cellulose, are collected and concentrated by ammonium sulfate precipitation, desalted as above, and applied to a DEAE-Sepharose column with dimensions and equilibration buffer as above, and eluted with a linear gradient of 0–0.2 M NaCl constructed as above.

Toyopearl HW-55F Gel Filtration

Active fractions from DEAE rechromatography were precipitated with 50% saturation of ammonium sulfate. The precipitate was dissolved in 4 ml 0.05 M Tris buffer, and applied on Toyopearl HW-55F column (3 × 63 cm) equilibrated with 0.05 M Tris buffer containing 0.2 M NaCl. As shown in Fig. 2, two protein peaks were eluted. Fractions 60–65, which showed high activity on crystalline cellulose, were pooled and precipitated with ammonium sulfate (50% saturation). The resulting precipitate was dissolved in 4 ml 0.05 M Tris buffer (pH 7.2), applied on a Toyopearl HW-55F column (3 × 63 cm) equilibrated with 0.05 M Tris buffer (pH 7.2) containing 0.2 M NaCl, and eluted by the same buffer. The cellulase was eluted as a single peak. The purification steps are summarized in Table I. Recovery of the activity on crystalline cellulose is 0.4%. The purified enzyme appears homogeneous on polyacrylamide disc gel electrophoresis.[6]

[6] K. Weber and M. Osborn, *J. Biol. Chem.* **244**, 4406 (1969).

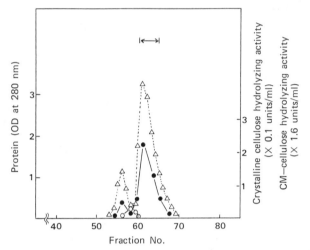

FIG. 2. Elution pattern of cellulolytic enzymes on a Toyopearl HW-55F column (3 × 63 cm). Fractions of 4.5 ml were collected. (△) Protein (absorbance at 280 nm), (●) activity toward crystalline cellulose, (○) activity toward CM-cellulose.

TABLE I

PURIFICATION OF A CELLULASE FROM *Cellulomonas uda* CB4[a]

Step	Total protein (mg)	Total activity (units)	Specific activity (units/mg)	Yield (%)
Culture filtrate	31,100	1,410	0.045	100
Ammonium sulfate	5,560	428	0.077	30.4
DEAE-Sepharose (second)	203	20	0.099	1.4
Toyopearl HW-55F (second)	50	6	0.12	0.4

[a] Cellulase activity was measured with Avicel as a substrate.

Properties

Molecular Weight. The molecular weight of purified enzyme is estimated to be 81,000 by SDS–polyacrylamide gel electrophoresis.[7]

Isoelectric Point. The isoelectric point of purified enzyme is estimated to be pH 4.4 by electrofocusing on ampholine polyacrylamide gel plate (LKB Co., Stockholm, Sweden).

Optimum pH and Temperature. The optimum pH and temperature for the hydrolysis of Avicel are pH 5.5–6.5 and 45–50°, respectively.

[7] U. K. Laemmli, *Nature (London)* **227,** 689 (1970).

TABLE II
MULTIPLICITY IN CELLULASES OF *Cellulomonas uda* CB4

Enzyme	Derived fractions in Fig. 1	Specific activity (units/mg) on Avicel	CMC	Activity ratio CMC/Avicel	Molecular weight[a]	Glyco-sylation[b]
Cellulase I	178–189	0.12	0.016	0.13	81,000	+
Cellulase II	30–50	<0.01	60.9	>6090	48,000	+
Cellulase III	50–70	0.10	10.6	106	55,500	+
Cellulase IV	130–170	0.04	62.3	1558	48,000	+
Cellulase V	130–170	0.06	48.0	800	55,000	+
Cellulase VI	190–210	0.03	4.7	157	56,000	+

[a] Molecular weight was estimated by SDS–polyacrylamide gel electrophoresis.
[b] Glycoprotein was stained with Schiff's reagent [G. Dubray and G. Bezard, *Anal. Biochem.* **119**, 325 (1982)].

Stability. The enzyme is stable between pH 5.5 and 8.0 at 25° for 24 hr. The activity of enzyme is completely lost on incubation at 60° for 10 min in 0.2 M acetate buffer, pH 5.5, without substrate.

Activation. The enzyme is activated about 2-fold by addition of 1 mM Co^{2+} or Mn^{2+}.

Substrate Specificity. Avicel is the best substrate (specific activity 0.12 units/mg protein) and the rate of the hydrolysis of cotton, CM-cellulose and filter paper are 16, 13, and 1% of that of Avicel, respectively. The apparent K_m value for Avicel is about 2.9 mg/ml. The main product from Avicel is cellobiose, with a trace of cellotriose. Cellotetraose is readily hydrolyzed to cellobiose, but cellotriose is not hydrolyzed.

Comments

The purified enzyme hydrolyzed Avicel to release cellobiose and a trace of cellotriose, indicating that this enzyme is β-1,4-glucan cellobiohydrolase (EC 3.2.1.91, cellulose 1,4-β-cellobiosidase). Cellotriose is not hydrolyzed by the purified enzyme, which is different from cellobiohydrolase from *Trichoderma reesei*.[8] As shown in Table I, purification (-fold) and yield are very low, indicating that Avicel hydrolysis requires the participation of various cellulases. Multiplicity in cellulases of *Cellulomonas uda* CB4 appears in Fig. 1. Fractions 30–50, 50–70, 130–170, and 190–210 in Fig. 1 are chromatographed on a Toyopearl HW-55F column

[8] M. L. Ladisch, J. Hong, M. Voloch, and G. T. Tsao, *Basic Life Sci.* **18**, 55 (1981).

(3 × 63 cm) equilibrated with water. Five main cellulase peaks are detected on these chromatograms. Each cellulase peak is rechromatographed on a Toyopearl HW-55F column under the same conditions and eluted as a single peak. Five cellulases obtained appear homogeneous on polyacrylamide gel electrophoresis. Table II shows some properties of these cellulases. Cellulase I is a cellobiohydrolase, whose purification is described here in detail. Cellulase II is active on CMC alone and is considered to be typical β-1,4-glucan glucanohydrolase (EC 3.2.1.4, cellulase). Cellulases IV and V are also considered to be a glucanohydrolase-type cellulase. Cellulases III and VI show substrate specificity intermediate between that of cellobiohydrolase and glucanohydrolase. The major components in the culture filtrate of *Cellulomonas uda* CB4 are cellulase I and III.

[23] Cellulase of *Ruminococcus albus*

By THOMAS M. WOOD

Bacterium

The American Type Culture Collection holds two cultures (ATCC No. 27210; 27211) of *Ruminococcus albus*. The particular strain used by the author to study the cellulases (strain SY3) was isolated by Dr. C. Stewart, Rowett Institute, Aberdeen, U.K. in March 1976 from a clearing in an anaerobic cellulose roll-tube[1] that had been inoculated with a 10^6-fold dilution of rumen contents from a sheep. *R. albus* SY3 is maintained at the Rowett Institute.

Culture Methods

Anaerobic Culture Methods. Media for stock cultures and for growth experiments are prepared and maintained in culture tubes and flasks under 100% CO_2 by using the strictly anaerobic technique of Bryant.[2] Larger cultures for enzyme preparations are prepared under CO_2 by the method of Allison *et al.*[3]

Culture Maintenance. The bacterium can be maintained by transfer of stock cultures at approximately monthly intervals. Stock cultures are 10-

[1] N. O. van Gylswyk, *J. Agric. Sci.* **74**, 169 (1970).
[2] M. P. Bryant, *Am. J. Clin. Nutr.* **25**, 1324 (1964).
[3] M. J. Allison, M. P. Bryant, I. Katz, and M. Keeney, *J. Bacteriol.* **83**, 1084 (1962).

TABLE I
COMPOSITION OF MEDIA USED FOR 100-ml CULTURES OF
Ruminococcus albus

Components	Habitat-simulating medium[a]		Defined medium
	A	B	
Bacto-casitone	1.0 g	1.0 g	1.0 g
Bacto-yeast extract	0.25 g	0.25 g	0.25 g
NaHCO$_3$	0.4 g	0.4 g	0.4 g
Mineral solution I[b]	15 ml	15 ml	15 ml
Mineral solution II[c]	15 ml	15 ml	15 ml
Rumen fluid	30 ml	30 ml	—
Vitamin solution[d]	—	—	1.0 ml
Cellobiose	0.2 g	—	—
Glucose	0.2 g	—	—
Maltose	0.2 g	—	—
Sodium lactate	1.0 ml	—	—
Volatile fatty acid soln.[e]	—	—	7 ml
Resazurin (0.1 % w/v)	0.1 ml	0.1 ml	0.1 ml
Water	40 ml	40 ml	60 ml

[a] To prepare habitat-simulating medium A or B, the carbohydrate constituent(s) are added to the media and the flask and its contents are heated to boiling to remove dissolved oxygen. The flask is removed from the source of heat and 0.1 g L-cystine–HCl added. CO$_2$ is bubbled through the solution continuously before the solution is dispensed into screw-capped tubes (first culture) or flasks equipped with rubber bungs (third culture). The tubes/flasks are autoclaved. The preparation of media for enzyme preparation is described separately in the text.

[b] Mineral solution I consists of (per liter) 3.0 g K$_2$HPO$_4$.

[c] Mineral solution II consists of (per liter) 3.0 g KH$_2$PO$_4$, 6.0 g (NH$_4$)$_2$SO$_4$, 6.0 g NaCl, 0.09 g MgSO$_4 \cdot$ 7H$_2$O, 0.69 g CaCl$_2 \cdot$ H$_2$O.

[d] Vitamin solution consists of (per 100 ml) 20 mg pyridoxine–2HCl, 20 mg riboflavin, 20 mg thiamin–HCl, 20 mg nicotinamide, 200 mg calcium D-pantothenate, 1 mg *p*-aminobenzoic acid, 0.5 mg folic acid, 0.5 mg biotin, 0.5 mg vitamin B$_{12}$.

[e] Volatile fatty acid solution contains (per liter) 20 ml glacial acetic acid, 1 ml isobutyric acid, 1.2 ml isovaleric acid, 1.2 ml *n*-valeric acid, 1.2 ml 2-methylbutyric acid.

ml broths of habitat-simulating medium (see Table I) containing 0.5% w/v cellobiose in anaerobic Hungate tubes[4] with butyl-rubber-septum stoppers (Bellco, Vineland, NJ). Alternatively, cultures (0.5 ml, habitat-simulating

[4] R. E. Hungate, *Methods Microbiol.* **3B,** 117 (1969).

medium A) can be preserved either by freeze-drying in 10-ml ampoules in an Edwards BSA Freeze-drier (Edwards High Vacuum, Crawley, Sussex, U.K.) with a vacuum during secondary drying of between 7 and 3 Pa, or in liquid nitrogen. Neither method is particularly satisfactory and generally marked changes in the morphology (and in biochemical characteristics) of the cell are observed with storage.[5]

Composition of Culture Media. Both defined and habitat-simulating (HS) media are used for the preparation of cultures (Table I) of *R. albus* S43.

First Culture. A 0.5-ml inoculum is added to 7.5 ml habitat-simulating medium A (Table I) in a screw-capped tube (15 ml): CO_2 gas is bubbled into the tubes during the dispensing. The culture is incubated at 37°.

Second Culture. A 0.5 ml-inoculum of the first culture is added to 7.5 ml of fresh habit-simulating medium A (Table I) using exactly the same conditions used for the preparation of the first culture. This second culture is necessary to ensure rapid growth of the bacterium after it has been preserved in liquid nitrogen or in the freeze-dried state.

Third Culture. A 2-ml inoculum of the second culture is added for 50 ml of habitat-simulating medium (medium B, Table I) under CO_2 and incubated at 37° for 24 hr.

Fourth Culture. Medium B (750 ml) (Table I) in a 1-ml conical flask is heated to boiling and autoclaved with a gauze plug. On removal from the autoclave a sterile rubber bung carrying a tube and vent is fitted and CO_2 gas is bubbled in. A solution of 2.7 g cellobiose and 0.9 g cysteine–HCl in 45 ml water is filter sterilized and 40 ml of this solution is added to the conical flask containing medium B. CO_2 is bubbled through the medium until it is cool. Finally, a 50-ml inoculum from the third culture is added and the flask placed in an incubator at 37° for 106 hr. Cellulose swollen in H_3PO_4 (see chapter [3] in this volume) can be used in place of cellobiose as an energy source. Defined medium (750 ml) can also be used in place of 750 ml of habitat-simulating medium B.

Enzyme Preparation

Enzyme can be prepared from the cultures of *R. albus* SY3 which have reached the end of the exponential growth phase (6 hr for cultures containing cellobiose and 12 hr for cultures containing cellulose) or during the stationary phase (24 hr for cultures containing cellobiose and 7 days

[5] T. M. Wood, C. A. Wilson, and C. S. Stewart, *Biochem. J.* **205**, 129 (1982).

for cultures containing cellulose).[5] Cultures containing cellulose yield about 7-fold more enzyme than cultures containing cellobiose. Cellulase (CM-cellulase; EC 3.1.2.4, endo-1,4-β-glucanase) is assayed with CM-cellulose as substrate using the reducing sugar method discussed in chapter [9] of this volume.

Extracellular Enzyme. Cultures are centrifuged (75,000 g for 20 min) and the enzyme in the supernatant is precipitated immediately by the addition of solid $(NH_4)SO_4$ (39 g/liter) at 1° until the solution is 80% saturated. The precipitate is collected by centrifugation (75,000 g for 20 min), dissolved in 0.1 M KH_2PO_4 (NaOH buffer, pH 6.5, which is 0.02% with respect to NaN_3), to give a 5- to 20-fold concentrate, and stored at $-18°$.

The $(NH_4)_2SO_4$ precipitation must be performed as soon as possible after removal of the cells as the crude enzyme is relatively unstable. Concentrated enzyme is much more stable and can be stored at 10° for several weeks without significant loss of activity. At $-18°$ there is no detectable loss of activity. About 85% of the original cell-free cellulase (i.e., CM-cellulase) is recovered by precipitation with $(NH_4)_2SO_4$: freeze-drying is equally effective.

Cell-Bound Cellulase. The cells from 800-ml cultures incorporating cellobiose as an energy source are isolated by centrifugation (75,000 g for 20 min), suspended in 50 ml of KH_2PO_4–NaOH buffer, pH 6.67, ionic strength 0.1, or, alternatively, water, and kept at room temperature for 30 min before centrifugation (75,000 g for 20 min). Much of the cell-bound enzyme (80–88%) can be released simply by repeated buffer or water treatment: 90% of the remaining bound cellulose can be released by a 5-min ultrasonic treatment.

The enzymes are concentrated by precipitation with solid $(NH_4)_2SO_4$ (80% saturation) followed by centrifugation (75,000 g for 20 min) and dissolution of the pellet in 0.1 M KH_2PO_4–NaOH buffer, pH 6.5.

Enzyme Purification

Gel filtration on a column of Sephacryl S-300 (Pharmacia) reveals a pattern of high-molecular-weight (1 million) and low-molecular-weight cellulases (CM-cellulase) which varies according to the culture conditions and the conditions used to prepare the enzyme.

The enzyme would appear to exist on the bacterial cell wall as high-molecular-weight enzyme but to exist in solution as either high-molecular-weight or low-molecular-weight enzyme (25,000–30,000) depending on the culture conditions. It appears that the high-molecular-weight enzyme may dissociate into fragments of molecular weight 25,000–30,000. For example, when the bacterium is cultured in defined medium with cellulose as the energy source, very little high-molecular-weight material is detected at any stage of the fermentation. However, when cellobiose is the energy source, small quantities of high-molecular-weight component appear in solution along with the low-molecular-weight enzyme.

When rumen fluid is present in the medium (habitat-simulating medium) the stability of the high-molecular-weight material appears to be enhanced. Indeed, in those cultures where cellobiose is the energy source the high-molecular-weight enzyme is the major component in late stationary phase cultures.

Only the low-molecular-weight enzyme found in cultures of cellulose and defined medium has been purified to homogeneity: only the low-molecular-weight enzyme is discussed further.

Step 1. Gel Filtration. Concentrated (20-fold) extracellular cellulase (0.5 ml) from a culture of *R. albus* SY3 prepared in defined medium with cellulose as the energy source is applied to a column (85 × 1.5 cm) of Sephacryl S-300 equilibrated with 0.1 M KH_2PO_4–NaOH buffer, pH 6.5, and eluted with that buffer. Fractions (1 ml) are collected and assayed for cellulase (CM-cellulase): those fractions containing protein are identified from the absorption at 280 nm. Fractions 82–120, which contain the CM-cellulase activity, are pooled, cooled to 1°, and the enzyme precipitated by the addition of solid $(NH_4)_2SO_4$ to 80% saturation. The pellet collected by centrifugation (75,000 g for 20 min) is dissolved in 3 ml 0.1 M KH_2PO_4–NaOH buffer, pH 6.5.

Step 2. Isoelectric Focusing. Enzyme from Step 1 is desalted on a column of BioGel P-2, freeze-dried, dissolved in 2 ml of a 10% (w/v) sucrose solution, and applied in 100-μl aliquots to the tops of 20 (7.5 × 0.5 cm) polyacrylamide gels (7.5%) containing 1% (v/v) of LKB carrier ampholyte solution covering the pH range 6–8. The samples (approximately 300 μg of protein as determined by the Lowry method) are protected from the electrode solution by a layer of ampholyte solution (1% v/v) in 5% (w/v) sucrose. A current of 2 mA/gel is applied for 2 hr. The cathode (top of gel) and anode solutions consist of ethanolamine (0.4%, v/v) and H_2SO_4 (0.2%, v/v), respectively. After focusing, the gels are removed from the glass tubes, washed with distilled water, and sliced into disks 2 mm thick. The slices are added to 1 ml of water (0.02% with respect to NaN_3), and left overnight at 4°. The contents of the tubes are mixed, centrifuged, and the supernatants tested for pH and assayed for cellulase (CM-cellulase) activity. Fractions (45–50) containing the major CM-cellulase peak of activity (p*I* 6.0) are pooled.

A summary of the purification is shown in Table II.

TABLE II
PURIFICATION OF ENDO-1,4-β-GLUCANASE

Step and fraction	Total activity (units)	Total protein (mg)	Specific activity (units/mg)	Yield (%)
Crude enzyme	20.3	53.71	0.38	100
Step 1				
After gel filtration	10.7	5.82	1.84	53
After (NH₄)₂SO₄ pptn. and desalting	9.8	5.60	1.75	48
Step 2				
After isoelectric focusing in 20 tube gels	5.8	2.9	2.0	28

Properties

The purified cellulase component has a molecular weight of approximately 30,000 (gel filtration) and an isoelectric pH of 6.0–6.1 (4°). In addition to its ability to hydrolyze CM-cellulose, the purified cellulase component can hydrolyze H_3PO_4-swollen cellulose causing a very rapid fall in degree of polymerization while effecting only a small degree (0.5%) of hydrolysis.[6] These properties, along with the observation that it releases large amounts of cellotriose and small amounts of cellotetraose from H_3PO_4-swollen cellulose, would suggest that the enzyme is an endo-1,4-β-glucanase.

[6] T. M. Wood and C. A. Wilson, *Can. J. Microbiol.* **30**, 316 (1984).

[24] Cellulase of *Trichoderma koningii*

By THOMAS M. WOOD

Maintenance of Strain

Trichoderma koningii Oudemans (IMI 73 022) is available from the Commonwealth Mycological Institute, Ferry Lane, Kew, Surrey, U.K. It can be maintained for periods of 6 months without subculturing if grown on potato–carrot–agar slants in screw-capped bottles and stored at 1°. Growth on potato–carrot–agar slants is very weak and repeated subculturing on this medium does not appear to affect the properties of the cellulase produced by the fungus. Repeated subculturing on PDA slants (39 g per liter potato–dextrose–agar–Oxoid), in contrast, has a deleterious effect on the capacity of the fungus to produce extracellular cellulase in submerged cultures. Mycelial growth on PDA is profuse.

Potato–carrot–agar can be prepared by adding 20 g of grated carrot and 20 g of grated potatoes to 1 liter of tap water and then boiling for 1 hr. After cooling the suspension is made up to 1 liter with tap water and sieved through a domestic sieve. Twenty grams of agar (Oxoid No. 3) is added and the mixture steamed for 1 hr. The clear solution is dispensed into 10-ml screw-capped bottles, autoclaved, and allowed to solidify to form slants.

When the potato–carrot–agar slants are covered with liquid paraffin (specific gravity 0.86) and stored at 1°, the viability of the fungus and the biochemical properties of the extracellular cellulase are retained even after several years without subculturing.

Production of Enzyme

Stationary Cultures. Cotton fiber (4 g) is shaken until saturated with 150 ml of the salt medium of Saunders *et al.*,[1] which consists of (in grams per liter) KH_2PO_4, 0.2; K_2HPO_4, 0.15; $NaH_2PO_4 \cdot 5H_2O$, 2.3; Na_2HPO_4, 1.5; NH_4NO_3, 0.6; $MgSO_4 \cdot 7H_2O$, 0.3; $NaNO_3$, 3.8; yeast, 0.02; and 0.1 ml of salt solution A [A: $ZnSO_4$, 50 mg; $Fe_2(SO_4)_3$, 54 mg; $MnSO_4 \cdot 4H_2O$, 6 mg dissolved separately in water, mixed, 0.8 ml 12% (v/v) HCl added, and the volume made up to 100 ml] and 0.1 ml of salt solution B (B: $CuSO_4 \cdot 5H_2O$, 2.5 mg; H_3BO_3, 57 mg dissolved in 100 ml water). The cotton is teased out and placed inside a 1-liter Roux flask together with the solution which is not absorbed by the fiber. This method results in an even distribution of the saturated fiber over the available surface area of the flask. The flasks are then inoculated with spores/mycelium from a 5-day PDA slant and incubated in the dark at 27°. Mycelial growth appears in 36 hr, and in 7 days the whole surface area of the cotton exposed to the air is covered with a thick mat of mycelium and spores which is green in some areas and white or bright yellow in others.

After 20 days incubation, the culture medium is filtered (Whatman GFA), centrifuged at 66,000 g for 20 min at 3°, made 5 mM with respect to NaN_3 to prevent bacterial growth, and stored at 1°.

Enzyme Concentration. The clarified enzyme preparation is normally partially purified by the addition of solid $(NH_4)_2SO_4$ (11 g/liter) until 20% saturation is achieved, centrifuged at 75,000 g for 20 min, and the sedi-

---

[1] P. R. Saunders, R. G. H. Siu, and R. N. Genest, *J. Biol. Chem.* **174,** 697 (1948).

mented material discarded. Further additions of solid $(NH_4)_2SO_4$ to the supernatant are made until the solution is 80% saturated (49 g/liter), and this results in the quantitative precipitation of the cellulase and β-glucosidase enzymes. The suspension is stirred for at least 1 hr before the precipitate is collected by centrifugation (75,000 g for 30 min). The pellet is dissolved in 0.1 M acetate buffer, pH 5.0, which is 5 mM with respect to NaN$_3$, to give an approximately 50-fold concentration of the original culture filtrate. Any undissolved material is removed by centrifugation (75,000 g for 20 min). Finally, the residual sulfate ions are removed by desalting in 5-ml aliquots on a column of Sephadex G-25 (62.5 × 2.5 cm) equilibrated with 0.01 M ammonium acetate buffer, pH 5.0, and freeze-dried. The enzyme is stable indefinitely if kept in the powdered state at 5°, but, in fact, even in the liquid form enzyme solutions have been kept in the refrigerator in this laboratory for 10 years with no apparent loss of cellulase or β-glucosidase activity. The risk of minor modification of the enzymes by proteases present in the solution, or by protein–protein aggregation or dissociation, is clearly real, however, and if fractionation, purification, and characterization of the enzymes are contemplated it is advisable to keep the enzyme in powdered form. Recovery of activity after $(NH_4)_2SO_4$ precipitation is 110% for endo-1,4-β-glucanase, 108% for β-glucosidase, and 65% for protein.

An attractive alternative method of concentrating the enzyme solution is to use the hollow fiber diafiltration technique with an Amicon HCP10-8 cartridge with a molecular weight cut-off of 10,000. Any residual salts can be removed by desalting on Sephadex G-25 or by dialysis using the same hollow fiber cartridge.

Assay Methods

Endo-1,4-β-glucanase activity is measured according to method given in chapter [9] of this volume which involves the use of carboxymethylcellulose as substrate and measurement of the reducing sugar released by the Somogyi–Nelson reagents. β-Glucosidase is measured using o- or p-nitrophenyl-β-D-glucoside according to the method given in chapter [9] of this volume. Protein is estimated using the method of Lowry[2] or by absorbance at 280 nm.

Units of activity are expressed, in the case of endo-1,4-β-glucanase, as the release of 1 μmol of reducing sugar (glucose equivalent) per minute, or, in the case of β-glucosidase, as the release of 1 μmol of o-nitrophenol per minute.

[2] O. H. Lowry, N. J. Rosebrough, A. L. Farr, and R. J. Randall, *J. Biol. Chem.* **193,** 376 (1951).

Separation and Purification of the Enzyme System

The cellulase system of *T. koningii* contains two cellobiohydrolases,[3] four major endo-1,4-β-glucanases,[4] and two β-glucosidases[4]; these can be separated and purified by a series of chromatographic procedures involving ion-exchange chromatography, gel filtration, isoelectric focusing, chromatofocusing, and affinity chromatography (Fig. 1).

The purification effected by these procedures is monitored and measured using assays for β-glucosidase with nitrophenyl-β-glucoside as substrate and for endo-1,4-β-glucanase with CM-cellulose as substrate.[3] Cellobiohydrolase purification has been followed by the author by first identifying the protein that acts synergistically with the endo-1,4-β-glucanase to solubilize cotton fiber, and then purifying that protein to homogeneity. Avicel is frequently used as a substrate for cellobiohydrolase,[5] but Avicel is also attacked to some extent by other cellulase components. The reader is referred to chapter [9] in this volume and the review by Wood and McCrae[6] for further understanding of the problems of assaying for cellulase components.

Separation and Purification of the β-Glucosidases

Step 1. Affinity Chromatography on Concanavalin A-Sepharose. A column (1.5 × 27 cm) of concanavalin A-Sepharose is equilibrated with 0.1 *M* sodium acetate buffer, pH 6.0, which contains other salts at the following concentrations: 1 *M* NaCl, 1 m*M* MnCl$_2$, 1 m*M* CaCl$_2$, and 0.003 m*M* NaN$_3$. An enzyme solution (2 ml) containing approximately 130 mg of protein (determined by Folin–Lowry method), and which has been equilibrated with the above buffer by dialysis in a collodion tube (Sartorius), is applied to the column in the cold room. When the column is washed with the same buffer solution at 20 ml/hr, the β-glucosidase is eluted along with only 3% of the endo-1,4-β-glucanase applied to the column. After washing the column with a further 300 ml of buffer, the adsorbed fraction (endo-1,4-β-glucanase and cellobiohydrolase) is eluted with the above buffer made 0.5% with respect to α-methyl-D-mannoside.

Step 2. Isoelectric Focusing. The pooled β-glucosidase component from the concanavalin A-Sepharose column is concentrated by ultrafiltration through an Amicon PM10 membrane, desalted (5-ml aliquots) on a column (43.6 × 2.6 cm) of Sephadex G-25, and freeze-dried. The powder is redissolved in 10 ml of an aqueous solution of LKB 40% (w/v) ampholine carrier ampholyte (125 μl diluted to 10 ml with distilled water) cover-

[3] T. M. Wood and S. I. McCrae, *Biochem. J.* **128**, 1183 (1972).
[4] T. M. Wood and S. I. McCrae, *J. Gen. Microbiol.* **128**, 2973 (1982).
[5] T. M. Wood and S. I. McCrae, *Carbohydr. Res.* **148**, 331 (1986).
[6] T. M. Wood and S. I. McCrae, *Adv. Chem. Ser.* **181**, 181 (1979).

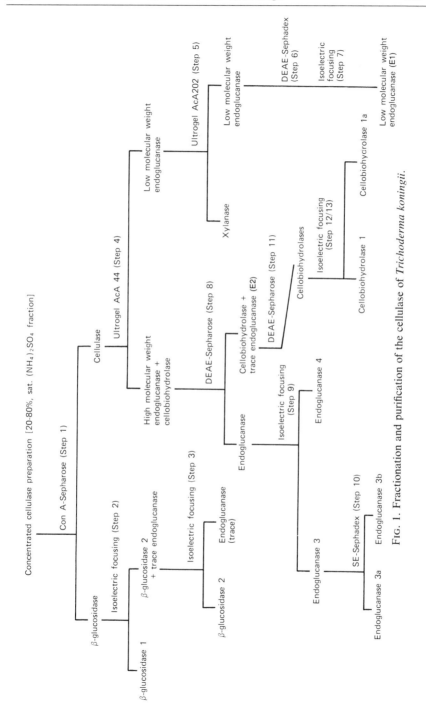

FIG. 1. Fractionation and purification of the cellulase of *Trichoderma koningii*.

ing the pH range 4–6, and purified by isoelectric focusing using an LKB isoelectric focusing column of 110 ml capacity containing 1% (v/v) ampholine carrier ampholyte (pH 4–6) in a sucrose density gradient. The experiment is set up and the reagents prepared as follows.

Preparation of Dense Solution. LKB ampholine carrier ampholyte 40% (w/v) covering the pH range 4–6 (1.87 ml) is diluted to 36 ml with distilled water and 28 g of sucrose dissolved in the solution.

Preparation of Light Solution. LKB ampholine carrier ampholyte 40% (w/v) covering the pH range 4–6 (0.62 ml) is diluted to 50 ml with distilled water.

Preparation of Dense Electrode Solution. Phosphoric acid (85%) (0.2 ml) is diluted with 14 ml of distilled water and 12 g sucrose dissolved in the solution.

Preparation of Light Electrode Solution. NaOH (0.1 g) is dissolved in 10 ml of distilled water.

The dense and light solutions are used as shown below to make up 24 fractions each containing a total of 4.6 ml. During this procedure, the

Fraction number	Dense solution added (ml)	Light solution added (ml)
1	4.6	Nil
2	4.4	0.2
3	4.2	0.4
4	4.0	0.6
5	3.8	0.8
6	3.6	1.0
7	3.4	1.2
8	3.2	1.4
9	3.0	1.6
10	2.8	1.8
11	2.6	2.0
12	2.4	2.2
13	2.2	2.4
14	2.0	2.6
15	1.8	2.8
16	1.6	3.0
17	1.4	3.2
18	1.2	3.4
19	1.0	3.6
20	0.8	3.8
21	0.6	4.0
22	0.4	4.2
23	0.2	4.4
24	0	4.6

solution containing the β-glucosidase is used to replace the light solution added to some of the middle fractions.

The dense electrode solution is poured into the isoelectric focusing column and each of the fractions is then pumped into the column at a rate not exceeding 4 ml/min, followed by the light electrode solution. The electrodes are placed so that the anode is at the bottom of the column. After focusing at 5° for 70 hr (600 V, 3 mA at the start; 700 V, 1 mA at the end of the run) the column is emptied at the rate of 120 ml/hr and the fractions (2 ml) which are collected are tested for pH, β-glucosidase, and endo-1,4-β-glucanase (CM-cellulase). β-Glucosidase 1 is isoelectric at pH 5.53 (5°) and the peak of activity is found at this pH: β-glucosidase 2 is isoelectric at pH 5.85. Fractions containing β-glucosidase 1 and β-glucosidase 2 are pooled separately. β-Glucosidase 1 is obtained pure at this stage and can be freed from sucrose and carrier ampholyte by gel filtration (5-ml aliquots) on a column of BioGel P-4 (90 × 2.5 cm). β-Glucosidase 2 is associated with a trace of CM-cellulase activity and requires further purification.

Note. The sucrose density gradient can also be set up using the LKB Gradient Mixer constructed for the purpose. In this case, the whole of the light solution is placed in the nonmixing chamber and the whole of the dense solution is placed in the mixing chamber. The enzyme sample is dissolved in the dense solution after approximately one-third of the gradient has been formed in the column.

Step 3. Further Isoelectric Focusing. The pooled fractions containing β-glucosidase 2 are concentrated to 10 ml on an Amicon ultrafiltration cell using a PM10 membrane, diluted with 10 ml of 2% (v/v) carrier ampholyte solution (pH 5–7) and refocused using the same column and conditions used in Step 2 except that the carrier ampholyte used covers the pH range 5–7. The contents of the column are emptied, collected in 2-ml fractions, and assayed for β-glucosidase. The relevant fractions are pooled, then passed in 5-ml aliquots down a column of BioGel P-4 (90 × 2.5 cm) to obtain β-glucosidase free from sucrose and ampholyte.

A summary of a typical purification procedure is given in Table I.

Isolation of Low-Molecular-Weight Endo-1,4-β-glucanase

Step 4. Gel filtration on Ultrogel AcA 44. The fraction adsorbed on Concanavalin A-Sepharose and eluted with α-methylmannoside (Step 1) is cooled (1°) and the enzyme precipitated by the addition of solid $(NH_4)_2SO_4$ (68 g/100 ml) with stirring until 80% saturation is reached. After stirring for 1 hr to ensure equilibrium, the suspension is centrifuged (75,000 g for 20 min) and the pellet dissolved in 0.01 M ammonium acetate buffer, pH 5.0. The solution is desalted in 5-ml aliquots on a column of

TABLE I
PURIFICATION OF β-GLUCOSIDASES

Step and fraction	Total β-glucosidase (units)	Total protein (mg)	Specific activity[a] (units/mg)	Yield (%)
Concentrated [(NH$_4$)$_2$SO$_4$)] cell-free culture filtrate	63.6	130	0.49	100
Step 1. After concanavalin A-Sepharose	62.2	2.2	28.3	98
Step 2. After isoelectric focusing				
β-Glucosidase 1	8.6	0.18	47.8	13
β-Glucosidase 2	49.8	—	—	78
Step 3. After isoelectric focusing				
β-Glucosidase 2	47.4	0.51	92.9	74

[a] Enzyme activity was measured at 37° using o-nitrophenyl-β-D-glucoside as substrate.

Sephadex G-25 (46.3 × 2.6 cm), freeze-dried, and dissolved in 5 ml 0.01 M ammonium acetate buffer, pH 5.0. The sample is applied to a column (82.0 × 2.5 cm) of LKB Ultrogel AcA 44 gel filtration medium previously equilibrated with 0.01 M ammonium acetate buffer, pH 5.0, and eluted with the same buffer at 15 ml/hr. Fractions (5 ml) are collected and assayed for endo-1,4-β-glucanase (CM-cellulase). Fractions 50–75 contain the high-molecular-weight endo-1,4-β-glucanase and cellobiohydrolase enzymes: the low-molecular-weight enzyme (E1) is eluted in fractions 76–91 approximately. Fractions 50–75 and 76–91 are pooled separately and freeze-dried. Freeze-dried powder from fractions 76–91 is redissolved in 5 ml 0.01 M ammonium acetate buffer, pH 5.0, and purified by rechromatography on the same Ultrogel column. Fractions (76–91) are again pooled, freeze-dried, and the enzyme redissolved in 5 ml 0.01 M ammonium acetate buffer, pH 5.0.

Step 5. Gel filtration on Ultrogel AcA 202. The low-molecular-weight endo-1,4-β-glucanase fraction from Step 4 is applied (6 ml) to a column (82.5 × 2.6 cm) of Ultrogel AcA 202 previously equilibrated in 0.01 M ammonium acetate buffer, pH 5.0, and eluted with the same buffer at 20 ml/hr. Fractions (5 ml) are collected and assayed for CM-cellulase and xylanase[7] activity. Fractions 30–41 contain the further purified low-molecular-weight endo-1,4-β-glucanase: fractions 51–70 contain a low-molecular-weight xylanase. Fractions 30–41 are pooled and freeze-dried.

[7] T. M. Wood and S. I. McCrae, *Carbohydr. Res.* **148**, 331 (1986).

Step 6. Ion-Exchange Chromatography on DEAE-Sephadex. The freeze-dried powder from Step 5 is dissolved in 2 ml of 0.05 M acetic acid–NaOH buffer, pH 5.6, added to a column (28 × 1.5 cm) of DEAE-Sephadex equilibrated with the same buffer, and eluted under the starting conditions. Fractions containing CM-cellulase activity are pooled, concentrated to 5 ml in a collodion tube, dialyzed against 0.01 M ammonium acetate buffer, pH 5.0, and freeze-dried.

Step 7. Isoelectric Focusing. The freeze-dried residue from Step 6 is dissolved in a mixture of sucrose and LKB carrier ampholyte solution (pH 3–5) as previously described (Step 2), and subjected to isoelectric focusing in a 110-ml capacity column using the same conditions described in Step 2. The purified low-molecular-weight endo-1,4-β-glucanase is isoelectric at pH 4.73 (4°). The relevant fractions are pooled, concentrated to 5 ml, precipitated by the addition of solid $(NH_4)_2SO_4$ (80% saturation), centrifuged (75,000 g for 20 min), and the pellet dissolved in 0.01 M ammonium acetate buffer, pH 5.0. After dialysis in a collodion tube against 0.01 M ammonium acetate buffer, pH 5.0, until free from sulfate ions, the solution is freeze-dried. Purified in this way, the low-molecular-weight endo-1,4-β-glucanase shows a single band on SDS–gel electrophoresis.

Separation and Purification of the High-Molecular-Weight Endo-1,4-β-glucanases

Step 8. Ion-Exchange Chromatography on DEAE-Sepharose. The high-molecular-weight endo-1,4-β-glucanase from Step 4 is dissolved in 5 ml 0.1 M acetic acid–NaOH buffer, pH 5.0, and applied to a column (26.0 × 1.5 cm) of DEAE-Sepharose CL-6B. The column is eluted at a rate of 20 ml/hr first with start buffer and then, after 80 fractions (10 ml) have been collected, with a pH gradient (pH 5.0–3.8) constructed from 500 ml start buffer and 500 ml of 0.1 M acetic acid–NaOH buffer, pH 3.8, with a LKB Gradient Former using a linear gradient template (0–100%). The pH gradient obtained is linear between pH 4.0 and 3.8. Fractions are assayed for CM-cellulase activity and for protein content: at least 95% of the endo-1,4-β-glucanase (CM-cellulase) is eluted under the start conditions. The cellobiohydrolase which separates as a single peak (pH 3.95 at 4°) after the application of the pH gradient is the major protein component. The different types of activity are pooled (fractions 3–80 and 105–120, approximately) separately, cooled (1°), precipitated by the addition of solid $(NH_4)_2SO_4$ to 80% saturation, desalted on a column of Sephadex G-25, and freeze-dried. Fractions 3–80 contain the endo-1,4-β-glucanases and fractions 105–120 (approximately) the cellobiohydrolases.

Note. The LKB Gradient Former can be replaced by two small beakers of equal dimensions and connected by a siphon. The liquid in the mixing beaker is agitated with a magnetic stirrer.

Step 9. Isoelectric Focusing. The freeze-dried enzyme (fractions 3–80 from Step 8) is dissolved in LKB carrier ampholyte solution (pH 4–6) and added to, and focused in, a LKB column (110 ml) as detailed in Step 2. After 40 hr, the contents of the column are emptied and collected in 2-ml fractions. Fractions are assayed for CM-cellulase and protein and measured (5°) for pH. Two peaks of CM-cellulase activity are obtained (fractions 15–20, and 27–35) with peaks of activity at pH 4.32 (endo-1,4-β-glucanase 3) and 5.09 (endo-1,4-β-glucanase 4), respectively. The pooled fractions are cooled (1°), precipitated separately with $(NH_4)_2SO_4$ (80% saturation), and the precipitates collected by centrifugation (75,000 g for 20 min). The pellets are dissolved separately in 5 ml 0.01 M ammonium acetate buffer, pH 5.0, and the solutions desalted on a column (43.6 × 2.6

TABLE II
PURIFICATION OF ENDO-1,4-β-GLUCANASES (CM-CELLULASES)

Step and fraction	Total endoglucanase (units)	Total protein (mg)	Specific activity[a] (units/mg)	Yield (%)
Concentrated [$(NH_4)_2SO_4$)] cell-free culture filtrate	540	120	4.5	100
Step 4. E1				
Ultrogel AcA44	167	10.1	16.5	29
Step 5. E1				
Ultrogel AcA 202	154	8.2	18.8	28
Step 6. E1				
DEAE-Sephadex	162	7.8	20.8	30
Step 7. E1				
Isoelectric focusing	146	5.1	28.6	27
Step 8. E3 + E4[b]				
DEAE-Sepharose	371	41	9.0	69
Step 9. E3				
Isoelectric focusing	166	18.2	9.1	31
E4				
Isoelectric focusing	168	17.1	9.2	31
Step 10. E3a				
SE-Sephadex	158	4.8	32.9	29
E3b				
SE-Sephadex	17	5.2	3.2	3

[a] Enzyme activity was measured at 37°.
[b] Endoglucanase E2 is a minor component and is discarded.

cm) of Sephadex G-25 equilibrated with ammonium acetate buffer. The enzymes are finally recovered by freeze-drying. Endoglucanase 4 shows a single band (Coomassie Blue stain) on polyacrylamide gel electrophoresis with and without SDS. Endoglucanase 3 requires further purification.

Step 10. Ion-Exchange Chromatography on Sulfoethyl-Sephadex. Freeze-dried endoglucanase 3 from Step 9 is dissolved in 0.01 M succinic acid–NaOH buffer, pH 4.4 (1.5 ml), and applied to a column (70 × 1.6 cm) of SE-Sephadex equilibrated with the same buffer. The column is eluted with the start buffer at 9 ml/hr. Fractions (2 ml) are collected and assayed for CM-cellulase activity and for protein. The two peaks of CM-cellulase activity (endoglucanase 3a and 3b) are well separated. Fractions 16–31 (endoglucanase 3a) and 33–47 (endoglucanase 3b) are pooled separately and concentrated to approximately 5 ml using an Amicon ultrafiltration cell equipped with a PM10 membrane. These two endo-1,4-β-glucanases (3a and 3b), along with those isolated in Steps 9 (endoglucanase 4) and 7 (low-molecular-weight endoglucanase E1), account for approximately 90% of the endo-1,4-β-glucanase (CM-cellulase) activity found in the original culture filtrate. Minor endo-1,4-β-glucanase components are discarded.

A summary of a typical purification procedure for the endo-1,4-β-glucanases is given in Table II.

Comments. Sulfoethyl-Sephadex is no longer available and has been replaced by sulfopropyl-Sephadex: the separation of the enzymes appears to be similar on this new ion-exchange material.

Purification of the Cellobiohydrolases

Step 11. Ion-Exchange Chromatography on DEAE-Sephadex. The freeze-dried sample of partially purified cellobiohydrolase (Step 8) is dissolved in 5 ml 0.1 M formic acid–NaOH buffer, pH 4.8, and applied to a column (21.6 × 1.5 cm) of DEAE-Sephadex (formate form). The column is eluted at a rate of 15 ml/hr with a pH gradient which is approximately linear between pH 4.0 and 3.6. The gradient is constructed from 150 ml of 0.1 M formate buffer, pH 3.5, and 150 ml of 0.1 M formate buffer, pH 4.8, by a LKB Gradient Former using a linear template (0–100%). Fractions (4.2 ml) are collected and assayed for protein and CM-cellulase activity. The elution profile shows a major protein component (cellobiohydrolase) eluted at approximately pH 3.9, with a minor protein peak associated with CM-cellulase activity eluted at pH 3.7. The fractions containing the cellobiohydrolases are pooled, and the enzyme precipitated with $(NH_4)_2SO_4$ (80% saturation) and collected by centrifugation (75,000 g for 20 min). The pellet is redissolved in 0.01 M ammonium acetate buffer, pH 5.0, and the solution desalted on a column (46.3 × 2.6 cm) of Sephadex G-25: the enzyme is isolated by freeze-drying. Isoelectric focusing in polyacrylamide gels using ampholyte covering the pH range 3–5 at this stage reveals two protein bands on staining with Coomassie Blue: both are cellobiohydrolases (1 and 1a, Fig. 1).

Step 12. Preparation of Narrow pH Range Ampholyte for Isoelectric Focusing. A sucrose density gradient is prepared manually for an LKB 100-ml isoelectric focusing column as detailed in Step 2 except that an 8% v/v solution of LKB carrier ampholyte is used in the isoelectric focusing column instead of a 1% v/v solution. Thus, the dense solution is prepared by diluting 14.96 ml of carrier ampholyte (pH 3–5) to 36 ml with distilled and dissolving 28 g of sucrose in the solution, and the light solution by diluting 4.96 ml of carrier ampholyte solution (pH 3–5) to 50 ml with distilled water. The dense electrode solution is composed of phosphoric acid (0.2 ml), water (14.0 ml), and sucrose (15.0 g); the light electrode solution comprises NaOH (0.1 g) and water (10.0 ml). The electrodes are arranged so that the anode is at the bottom of the column. A voltage is applied (initially 520 V; 2 mA) for 142 hr, when it should be steady at 720 V with a current of 3 mA. The contents are removed by pumping them through the top of the column into fraction collector tubes by way of a capillary tube and a pump. The capillary tube is kept just below the surface of the liquid during the operation. Fractions (2 ml) covering the pH range 3.72–4.25 are combined to give an approximately 8% ampholyte–sucrose solution (Solution A).

Step 13. Isoelectric Focusing in a pH Gradient Covering Only 0.5 pH Unit. The sucrose density gradient and the electrode solutions are constructed as in Step 2, except that Solution A (Step 12) is used instead of the 40% (w/v) LKB ampholine carrier ampholyte solution to prepare the dense and light solutions.

A sample (20 mg) of the enzyme, prepared for electrofocusing as detailed in Step 11, is dissolved in 10 ml narrow-range ampholyte solution (Solution A Step 12) and added to the gradient in the isoelectric focusing column as detailed in Step 2. After focusing for 90 hr, approximately (700 V; 3 mA at the end of the run), fractions (2 ml) are removed from the bottom of the column at a rate of 120 ml/hr. The pH of each fraction is measured at 5°. The enzyme in each fraction is then precipitated at 1° by the addition of solid $(NH_4)_2SO_4$ (80% saturation), centrifuged (75,000 g for 20 min), redissolved in 0.01 M ammonium acetate buffer, pH 5.0, dialyzed in collodion tubes (Sartorius) against the same buffer at 4° for 6 days, and assayed for protein. The two protein components with peaks isoelectric at pH 3.8 (fractions 17–23, approximately) and pH 3.95 (fractions 27–33, approximately), respectively, are pooled separately, and the purified enzymes isolated by freeze-drying. Recovery of protein applied to the iso-electric focusing columns is of the order of 90%: 24% of the protein is associated with the minor cellobiohydrolase (cellobiohydrolase 1a, Fig. 1).

Comment. Recoveries of cellobiohydrolase activity in the various steps are not given in tabular form. The reader is referred to the discussion under Separation and Purification of the Enzyme System (above) for an explanation.

Characteristic Properties of the Enzymes

Cellobiohydrolase (CBH). Both CBH1 and CBH1a have molecular weights of 62,000 (by gel filtration). CBH1 is isoelectric at pH 3.95 and CBH1a at pH 3.8. CBH1 is associated with 9% w/w carbohydrate; CBH1a, 33%. Substrates hydrolyzed are H_3PO_4-swollen cellulose, cello-tetraose, cellohexaose, and on prolonged incubation, dewaxed cotton and Avicel: cellobiose is the major product in all cases. CM-cellulose is not hydrolyzed significantly by either CBH. An important property of both cellobiohydrolases is that they act synergistically with some purified endo-1,4-β-glucanases to effect the rapid dissolution of crystalline cellulose in the form of the cotton fiber.

Endo-1,4-β-glucanases. The characteristic properties are as follows: endoglucanase 1, 3a, 3b, and 4 (Fig. 1) have molecular weights (gel filtration) of 13,000, 48,000, 48,000, and 31,000, respectively: pIs are 4.7, 4.3, 4.3, and 5.09, respectively. Only endoglucanase 3a and 4 act synergistically with CBH1 and CBH1a to solubilize cotton fiber. Acting alone, neither can solubilize cotton significantly, but all hydrolyze CM-cellulose, H_3PO_4-swollen cellulose, and cellooligosaccharides.

β-Glucosidases. Both β-glucosidases have molecular weights of 39,800. The respective pI values are 5.53 and 5.85 for β-glucosidase 1 and 2. Affinities for o-nitrophenyl-β-glucoside differ [K_m 0.37 (β-glucosidase 1), 0.85 (β-glucosidase 2)] as does the effect of the inhibitor gluconolactone [K_i 1.8 (β-glucosidase 1) and 1.17 (β-glucosidase 2)]. Neither β-glucosidase can hydrolyze crystalline cellulose but both act synergistically with the other components of the cellulase system to solubilize crystalline cellulose.

[25] Cellulases of *Trichoderma reesei*

By MARTIN SCHÜLEIN

Introduction

Trichoderma reesei (formerly *Trichoderma viride*) QM6a and a mutant from this strain produce very effective cellulases. Many publications have described the purification and characterization of the cellulases from this strain. The synergy between exocellobiohydrolase (CBH 1) (EC 3.2.1.91, cellulose 1,4-β-cellobiosidase) and the endoglucanase (EC 3.2.1.4, cellulase) has to be revised. The protein CBH 1 is the most abundant of the enzymes from *Trichoderma reesei*. This protein can be purified in high yields and in an active and stable form. The endoglucanases lose their stability when they are highly purified.[1]

Cellulolytic enzymes form a multienzyme complex which can degrade cellulose. To describe the individual enzymes one needs a well-defined substrate. But the substrate, cellulose, is not well defined; it is insoluble in water and has different regions which are amorphous or crystalline. The basic building block is glucose units which are linked in a β-1,4 configuration. The chain length has to be less than 6 glucose units for the polymer to be soluble in water. However by replacing the hydroxyl group of the glucose carbon C-2, C-3, or C-6 with a carboxymethyl group (CMC), the polymer becomes much more soluble. The polymer can be used to determine β-1,4-glucanase activity. Another polymer, barley β-glucan, is also soluble in water. It consists of β-1,4 and β-1,3 linkages. There are 2 or 3 β-1,4 linkages for every β-1,3. This polymer can also be used to determine β-glucanase activities. The insoluble substrate, filter paper, is only suitable for measurement of the efficiencies of the whole enzyme complex; the activity toward insoluble cellulose decreases dramatically when the different enzymes are separated.

Trichoderma reesei also produces cellobiase activity (EC 3.2.1.21, β-glucosidase). In the product Celluclast (Celluclast is the enzyme product supplied from Novo Industri A/S, Bagsvaerd, Denmark), this activity is very low and the addition of *Aspergillus niger* cellobiase (EC 3.2.1.21) is recommended for total saccharification of cellulose.

The enzyme complex is commercially available from different companies. It is produced by a selected strain of *Trichoderma reesei* which is

---

[1] M.-L. Niku-Paavola, A. Lappalainen, T.-M. Enari, and M. Nummi, *Biochem. J.* **231,** 75 (1985).

grown under optimum submerged fermentation conditions for production of this product. The extracellular enzyme product is partially purified, concentrated, and stabilized.

Enzyme Analysis

Measurement of Cellulolytic Activity

Many different methods have been used to determine cellulolytic activities. The following method is used because it is reproducible, and no synergy or cooperation between the different enzymes exists. This is important during purification. The substrate is carboxymethyl cellulose (CMC) 7 LFD from Hercules Powder Co. The enzyme is incubated for 20 min with 0.5% CMC in a sodium acetate buffer, 50 mM, pH 4.8. The reducing capacity equivalent to glucose is measured and one cellulase unit is defined as the amount of enzyme which degrades CMC to reducing carbohydrate corresponding to 1 μmol glucose per min at 37°.

For fast analysis the following method has been developed. It is carried out at pH 7.0 which is not the optimal pH value, but it is fast and very convenient for following cellulolytic activities eluted from chromatographic columns. The substrate is CMC 7 LFD and the buffer is 20 mM Tris–HCl, pH 7.0. The glucose is detected by glucose oxidase using the GOD-Perid kit from Boehringer-Mannheim GmbH. Substrate (CMC), enzymes and glucose oxidase, and the chromogen, 2,2'-azinobis(3-ethyl-benzthiazoline sulfonic) acid (ABTS) are mixed in microtiter plates and incubated at 37° for 20 min. The development of a green color indicates degradation of CMC. The intensity is measured by means of a Titertek spectrophotometer at 690 nm.

Protein Characterization Using Immunomethods

The enzyme complex in Celluclast was used for raising antiserum. A freeze-dried powder contained 8000 cellulase units/g was used. Four rabbits were immunized every second week with 1 ml containing Freund's adjuvant and 10 mg Celluclast (80 cellulase units). After 3 months the rabbits were bled every second week giving about 20 ml of serum per animal. This serum was pooled and the γ-globulin purified by dialysis and ammonium sulfate precipitation. The precipitate was dialyzed and sodium azide added; it was 3-fold more concentrated than the serum and was kept as a clear solution at 4°.

Analysis of the protein components of Celluclast was performed by crossed immunoelectrophoresis. Figure 1 shows the result. The first di-

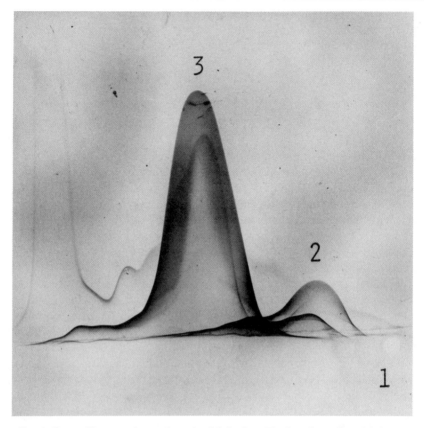

FIG. 1. Crossed immunoelectrophoresis of Celluclast. The first dimension with the anode at the left. The second dimension with the anode at the top. Staining: Coomassie Brilliant Blue R.

mension is 1.5 hr electrophoresis with 10 V/cm. The buffer is Tris-maleic acid, pH 7.0. The agarose is a 1% solution of equivalent amounts of Litex HSA and HSB. The temperature is 15°. In the second dimension, 0.3 ml purified γ-globulin is added per plate (10 × 10 cm). The time is 18 hr with 1 V/cm. All proteins with a pI lower than 7.0 are visualized with Coomassie Blue staining of the γ-globulin. They form characteristic precipitates with the antigens.

The enzymes can be visualized by incubation with substrate. The plate is incubated with 1% CMC in pH 5.0 buffer instead of staining with Coomassie Blue. After incubation for 1 hr at 45° the plate is incubated with GOD-Perid reagent. Development of a green color indicates formation of glucose. The endoglucanase 1 arch is the first to appear, but after 1 hr the cellobiohydrolase 1 also produces a green color. Around the application well a third green area is formed, indicating the presence of endoglucanase 2. The degradation of CMC is now visualized by precipitation of CMC with CTAB (*N*-cetyl-*N,N,N*-trimethylammonium bromide). Five percent CTAB solubilized in water and heated to 100° was added to the plate after cooling. After about 1 hr the three enzyme active zones were still clear and the rest of the plate was white from the precipitate of CMC–CTAB complex. The immunoprecipitate showed other protein components, but these did not degrade CMC. This immunological method is used for keeping track of the cellulolytic enzymes during purification. If the immunoprecipitation arch configuration changes, this indicates that the properties of the purified enzyme are not identical with the properties of that protein in the total complex before purification.

Purification

Celluclast has a very high dry matter content. For purification and characterization of the different enzyme components, the product is diluted twice with water and then dialyzed using an Amicon hollow fiber cartridge (Amicon Corp., Danvers, MA) with a molecular weight cut off below 10,000. For dialysis of cellulolytic enzymes cellulose acetate dialysis bags cannot be used as they are degraded too rapidly. Instead the product is dialyzed using a hollow fiber concentrator until the conductivity is below 2.5 mS. It can then be freeze-dried or used directly for further characterization.

Anion-Exchange Chromatography

The development of fast methods for purification of proteins has made it possible to purify the enzyme-active proteins in milligram amount within a few days.

Mono Q is a fast liquid chromatography column from Pharmacia Inc. It is a anion-exchange column with a total volume of 2 ml. Figure 2 shows the separation of components from Celluclast using Mono Q chromatography. The sample volume was 0.5 ml containing 12.5 mg of the freeze-dried powder or 100 cellulase units of Celluclast adjusted to pH 7.0. The buffer was 20 mM Tris–HCl, pH 7.0, and the flow rate was 1 ml/min. After applying the sample the column was washed with 10 ml of buffer. Then a linear gradient of 0–1 M NaCl was started and run for 20 min. The fraction size was 0.3 ml. Figure 2 shows the elution pattern of the protein measured as the absorbance at 280 nm and the CMC activity measured with the fast method and measured as the absorbance at 690 nm. The

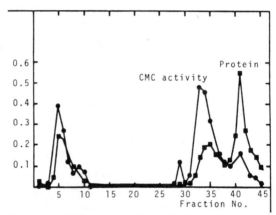

FIG. 2. Mono Q elution of Celluclast at pH 7.0. Protein was measured as absorbance at 280 nm, and CMC activity was measured as described in the text. Fraction size was 0.6 ml. Flow rate was 1 ml/min.

nonbound protein with maximum CMC activity (endoglucanase 2) was in fraction 5 which was further analyzed by HPLC. The next CMC-active enzyme eluting at 0.2 M NaCl (fraction 33) contained the maximum bound CMC activity (endoglucanase 1). At 0.3 M NaCl the main protein eluted and fraction 41 which had maximum protein concentration (cellobiohydrolase 1) was further analyzed on a HPLC sizing column TSK 3000W.

This method can be scaled up using DEAE-Sephadex. Table I shows the result of a purification of Celluclast in larger scale. The anion-exchange resin DEAE-Sepharose can also be used. The total volume is large, 100 g freeze-dried desalted Celluclast was solubilized in 5000 ml buffer; the flow rate was 500 ml/hr and the column diameter was 25 cm.

TABLE I
LARGE-SCALE PURIFICATION OF CELLUCLAST ON
DEAE-SEPHADEX[a]

Component	Weight (g)	Concentration (NCU/g)
Fraction 1 (Endo 2)	25.6	6670
Fraction 2 (Endo 1)	9.4	8060
Fraction 3 (CBH 1, Endo 1)	39.4	10700

[a] Freeze-dried Celluclast (100 g) was partly purified on 5 liters DEAE-Sephadex. Buffer 20 mM Tris–HCl, pH 7.0. Fraction 1 is nonbound protein, fraction 2 elutes with low NaCl, and fraction 3 elutes with high NaCl. The fractions were concentrated on hollow fiber and freeze-dried.

The unbound enzyme was concentrated and freeze-dried and used for purification of endoglucanase 2. The endoglucanase 1 is eluted with 50 mM NaCl in buffer in a total volume of 10 liters. Fraction 3 is eluted with 500 mM NaCl in a total volume of 12 liters.

Size Chromatography

Sephacryl S-200 is used for further purification of the different components. The column (5 × 90 cm) has a flow rate of 50 ml/hr and the sample size is 40 ml. The buffer was 20 mM Tris–HCl, pH 7.0. Cellobiohydrolase 1 has a higher molecular weight than endoglucanases and can be purified free of these. One gram of cellobiohydrolase is purified with this column at a time.

Analytical and Preparative HPLC Chromatography

Figure 3 shows the elution pattern of the crude 12.5 mg freeze-dried Celluclast applied to a 250 ml TSK 3000W column. A 1-ml sample was applied and fractionated in 5-ml fractions; the flow rate was 5 ml/min. The CMC activity and the protein content were measured. In the first peak, high-molecular-weight proteins eluted with very low CMC activity. This includes the cellobiase enzymes. Then CBH 1 elutes at 25 min showing some CMC activity. The main endoglucanases eluted between 28 and 32 min, and at last the low-molecular-weight CMC activity components eluted. When the elution patterns of Figs. 2 and 3 are compared, the contri-

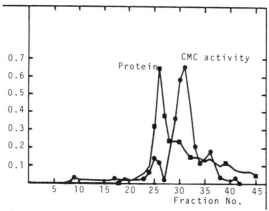

FIG. 3. HPLC TSK 3000W chromatography of Celluclast at pH 7.0. Protein was measured as absorbance at 280 nm. CMC activity was measured as described in the text. Fraction size was 5 ml. Flow rate was 5 ml/min.

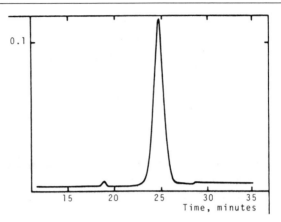

FIG. 4. HPLC TSK 3000W chromatography of fraction 33 from Mono Q at pH 7.0. Protein was measured with a flow cell as the absorbance at 280 nm. Flow rate was 5 ml/min.

bution of the different components to the CMC activity can be calculated. The CBH 1 contributes less than 10% of the activity, endoglucanases 1 and 2 contribute more than 80% of the CMC activity, and the low-molecular-weight endoglucanase contributes less than 10% of the activity.

Figure 4 shows the elution pattern of fraction 33 from the Mono Q column. It is an almost pure endoglucanase 1 with a retention time of 29 min. This purified enzyme was used for immunization of rabbits.

Figure 5 shows the elution pattern of fraction 41 from the Mono Q

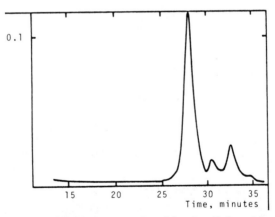

FIG. 5. HPLC TSK 3000W chromatography of fraction 41 from Mono Q at pH 7.0. Protein was measured with a flow cell as the absorbance at 280 nm. Flow rate was 5 ml/min.

column. It is an almost pure cellobiohydrolase 1 with a retention time of 26 min.

Properties of the Purified Components

The cellobiohydrolase 1 (CBH 1) (EC 3.2.1.91, cellulose 1,4-β-cellobiosidase) is the major protein of the cellulase system of *Trichoderma reesei*. The enzymatic activity is very difficult to measure because CBH 1 binds very strongly to its substrate. In 1973 the enzyme was first purified by Berghem and Pettersson[2] who called it exocellobiohydrolase. The enzyme is immunochemically identical to CBH 1 purified from Celluclast. The enzyme was sequenced by Fägerstam[3] and it was cloned and sequenced by Shoemaker *et al.*[4] and Teeri *et al.*[5] Some sequence homology exists with the active site of egg white lysozyme (an endo-β-glucanase where the glucose unit contains *N*-acetylglucosamine).[6]

The CBH 1 protein behaves as a specific β-1,4-glucanase with very high inhibition by cellobiose and binding to insoluble cellulose. The degradation of soluble barley β-glucan indicates that the action pattern is endo.[7] The enzyme is strongly absorbed on the insoluble substrate until the substrate is degraded, and the main soluble product is cellobiose. CBH 1 does not bind to the reducing end of the cellulose crystal, but to the side of the crystal. Electron microscopy shows that the degradation takes place because the enzyme can hydrolyze its way through the crystal by breaking β-1,4 linkages in endo fashion.[8] Addition of endoglucanase 1 increases the degradation. Highly purified endoglucanase 1 cannot degrade insoluble crystalline cellulose.[7]

Table II shows some properties of the highly purified components. The highly purified CBH 1 has only one immunoprecipitate using the polyspecific antiserum, and this precipitate is identical with the precipitate before purification. It forms a single sharp peak in HPLC TSK 3000W chromatography and a single band in SDS gradient gel electrophoresis.

[2] L. E. R. Berghem, L. G. Pettersson, and U. B. Axio-Fredriksson, *Eur. J. Biochem.* **53,** 55 (1975).

[3] L. G. Fägerstam, L. G. Pettersson, and J. A. Engstrom, *FEBS Lett.* **167,** 309 (1984).

[4] S. P. Shoemaker, V. Schweilkart, M. Ladner, D. Gelfand, S. Kwok, K. Myambo, and M. Innis, *Bio/Technology* **1,** 691 (1983).

[5] T. Teeri, I. Salovuori, and J. Knowles, *Bio/Technology* **1,** 696 (1983).

[6] M. G. Paice, M. Desrochers, D. Rho, L. Jurasek, C. Roy, C. F. Rollin, E. De Miguel, and M. Yaguchi, *Biotechnology* **2,** 535 (1984).

[7] B. Henrissat, H. Driguez, C. Viet, and M. Schülein, *Biotechnology* **3,** 722 (1985).

[8] H. Chanzy, B. Henrissat, R. Vuong, and M. Schülein, *FEBS Lett.* **153,** 113 (1983).

TABLE II
PROPERTIES OF HIGHLY PURIFIED COMPONENTS[a]

Component	NCU/mg	MW	HPLC (min)	pI
CBH 1	0.6	65,000	26	3.8–4.0
Endo 1	60	52,000	29	4.0–5.0
Endo 2	40	48,000	32	7.0

[a] Molecular weight was determined by means of 5–20% gradient SDS gel electrophoresis. HPLC retention time was measured using a TSK 3000W column. Bovine serum albumin has a retention time of 25 min. pI, isoelectric focusing using LKB ampholine gel 3.5–9.5.

After isoelectric focusing, two sharp protein bands are formed, indicating isoenzymes.

Properties of the Endoglucanases

Two endoglucanases can be separated. The most alkaline does not bind to anion-exchange columns at pH 7.0. The properties of purified endoglucanase 2 can be seen in Table II. The pI is 7.0 so it does not form a distinct immunoprecipitate in immunoelectrophoresis because γ-globulin has the same pI value.

Endoglucanase 1 binds to the anion-exchange column. After purification into a monocomponent using size chromatography, antiserum was raised against this component, but it shows reactivity against a range of isoenzymes with pI around 4–5. It forms a characteristic immunoprecipitate, even when partially purified. It can be purified free of CBH 1, and the highest specific activity is 60 cellulase units/mg protein. The enzyme, when rechromatographed on a TSK 3000W column, forms a sharp protein peak (absorbance at 280 nm) at 29 min, and a single band in SDS gradient gel. It cannot degrade crystalline cellulose.

Suggested Mode of Action

CBH 1 binds and hydrolyzes the β-1,4 linkage in the polymer. It will then disturb the hydrogen binding between the polymers and allow water to move into the crystal. Then the endoglucanase can attack the polymer and render the cellobiose and glucose soluble.

[26] Cellulases of a Mutant Strain of *Trichoderma viride* QM 9414

By A. G. J. VORAGEN, G. BELDMAN, and F. M. ROMBOUTS

For the complete hydrolysis of crystalline cellulose a multicomponent enzyme system, including 1,4-β-D-glucan glucanohydrolase (endoglucanase, EC 3.2.1.4, cellulase), 1,4-β-D-glucanocellobiohydrolase (exoglucanase, EC 3.2.1.91, cellulose 1,4-β-cellobiosidase), and β-D-glucoside glucohydrolase (β-glucosidase, EC 3.2.1.21) is necessary. Cellulases of fungal origin are known to be most powerful in cellulose hydrolysis. The methodology for the integral fractionation of a commercial cellulase preparation derived from a mutant *Trichoderma viride* QM 9414 strain[1] and chemical and physical properties of all endoglucanases, exoglucanases, and β-glucosidases which could be detected are described below.

Enzyme Preparation

Maxazym C1, a commercial cellulase preparation, produced by a hyperproducing mutant strain of *Trichoderma viride* QM 9414 grown in submerged culture, was kindly provided by Gist-Brocades (Delft, The Netherlands).

Assay Methods

Principles. Enzymes are incubated with various substrates for which the enzymes have a certain specificity. Hydrolysis of a substrate results in an increase in reducing groups which is measured by the copper reduction method. Alternatively, with *p*-nitrophenylglucoside, the concentration of released *p*-nitrophenol is measured spectrophotometrically, and with cellobiose, glucose is measured enzymatically using hexokinase/glucose-6-phosphate dehydrogenase. Endoglucanase activity is also measured viscosimetrically as the rate of change in specific fluidity. Reaction products from Avicel are also separated and quantified by HPLC.

---

[1] The interrelationship of various mutants of *Trichoderma viride/reesei,* in use for cellulase production, is described by B. S. Montenecourt, *Trends Biotechnol.* **1,** 156 (1983).

Reagents

Avicel (Type SF, Serva, Heidelberg, FRG)

CM-cellulose (Type Akucell AF 0305, DS 0.80–0.85, AKZO, Arnhem, The Netherlands)

H_3PO_4-swollen cellulose[2]

p-Nitrophenyl-β-D-glucopyranoside   (Koch-Light,   Colnbrook, Bucks, England)

Cellobiose (BDH Chemicals, Ltd., Poole, England)

Reagents for the Somogyi–Nelson copper reduction method[3]

Test-combination glucose/fructose (Cat. no. 139106, Boehringer-Mannheim, FRG)

Buffers: 0.05 and 0.1 *M* sodium acetate, pH 5; 0.5 *M* glycine–NaOH, pH 9.0 in 0.002 *M* Na_2EDTA

Avicelase. This activity is measured in a reaction mixture consisting of 0.5 ml of a 1% suspension of Avicel in 0.05 *M* sodium acetate buffer, pH 5, and 0.5 ml of enzyme solution. Incubation takes place in a shaking incubator at 30° for 20 hr. After centrifugation, 0.5 ml of the reaction mixture is analyzed for reducing sugars by the method of Somogyi–Nelson.[3] Specific activities are measured at a protein concentration of 25 µg/ml.

CM-Cellulase. Activity is measured in a mixture containing 0.4 ml of a 0.5% solution of CM-cellulose in 0.05 *M* sodium acetate, pH 5.0, and 0.1 ml of enzyme solution. After incubation at 30° for 1 hr, the reaction is stopped and reducing sugars are measured by the Somogyi–Nelson method.[3] Samples are centrifuged before reading absorbance at 520 nm. CM-cellulase activity is also measured viscosimetrically at 30° according to Almin and Eriksson,[4] using a 0.25% substrate concentration. Activity is expressed as rate of change in specific fluidity, $\Delta\phi_{sp}$ (min^{-1}). A protein concentration of 1 µg/ml is used for specific activity measurements.

β-Glucosidase. Activity is estimated by addition of 0.1 ml of enzyme solution to 0.9 ml of 0.1% *p*-nitrophenyl-β-D-glucopyranoside in 0.05 *M* sodium acetate buffer, pH 5.0, and incubation for 1 hr at 30°. The reaction is stopped by addition of 1 ml of 0.5 *M* glycine buffer, pH 9.0, containing 0.002 *M* disodium EDTA. The concentration of released *p*-nitrophenol is measured at 400 nm, using an extinction coefficient of 13,700 M^{-1} cm^{-1}. Specific activities are measured using a protein concentration of 5 µg/ml.

Cellobiase. Activity is measured in enzyme fractions containing β-glucosidase activity by incubation of 0.4 ml enzyme solution in 0.1 *M*

[2] T. M. Wood, *Biochem. J.* **121,** 353 (1971).

[3] R. G. Spiro, this series, Vol. 8, p. 3.

[4] K. E. Almin and K. E. Eriksson, *Biochim. Biophys. Acta* **139,** 238 (1967).

sodium acetate, pH 5.0, containing 20 μg protein with 0.1 ml 0.2% cellobiose in the same buffer. After incubation at 30° for 1 hr, the enzymes are inactivated by placing the tubes in a boiling water bath for 10 min. Glucose concentration is measured enzymatically by the test combination glucose/fructose.

Activity toward H_3PO_4-Swollen Cellulose. This is measured by addition of 0.1 ml enzyme solution to 0.9 ml 0.55% H_3PO_4-swollen cellulose in 0.05 M sodium acetate, pH 5.0. Incubation takes place for 20 hr at 30°. After centrifugation 0.5 ml of the supernatant is analyzed for reducing sugars. Specific activities are measured at a protein concentration of 1 μg/ml.

Enzyme activities are expressed in units, one unit of enzyme activity producing one microequivalent of reducing groups per minute under the above specified conditions.

Reaction products released from Avicel are analyzed by HPLC using an Aminex 87-P column and water as eluent. The column is thermostatted at 85° and compounds are detected by refractive index. Enzyme protein (60 μg/ml) is incubated with a 0.5 ml 2% suspension of Avicel at 30° in 0.05 M sodium acetate, pH 5.0, for 6 hr. After centrifugation to remove residual cellulose, the enzymes in the supernatant are inactivated by heating for 5 min at 100°. The samples are further cleaned by treatment with lead nitrate prior to injection.

Fractionation and Purification

The crude cellulase preparation is fractionated and the various enzymes are purified according to the scheme[5] in Fig. 1. Ten grams of the preparation is dissolved in 30 ml 0.01 M sodium acetate buffer, pH 5.0. After centrifugation for 20 min at 50,000 g, the clear supernatant is desalted on a BioGel P-10 column (3 × 95 cm), previously washed with the same buffer. All of the cellulase activity appears in the void volume. This peak is pooled and chromatographed on a DEAE-BioGel A column, as shown in Fig. 2. Enzyme activity peaks II_1, II_3, and II_4 are pooled (Fig. 2).

Peak II_1, containing most of the β-glucosidase and CM-cellulase activity, is lyophilized, dissolved in a small volume, and applied to a BioGel P-100 column (2 × 90 cm), equilibrated in 0.02 M sodium citrate, pH 3.5. After elution with this buffer three peaks are obtained, two of which (III_1 and III_2) show cellulase activity.

[5] G. Beldman, M. F. Searle-van Leeuwen, F. M. Rombouts, and A. G. J. Voragen, *Eur. J. Biochem.* **146,** 301 (1985).

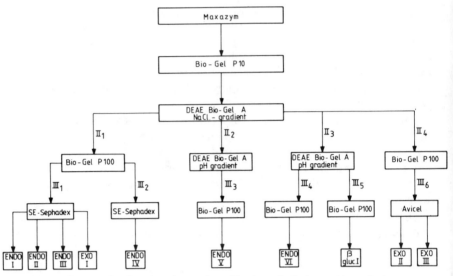

FIG. 1. Scheme for fractionation and purification of all detectable endoglucanases, exoglucanases, and β-glucosidases from a cellulase preparation of *Trichoderma viride*. The roman numerals with arabic subscripts refer to pools of enzyme activity peaks. From Beldman *et al.*[5]

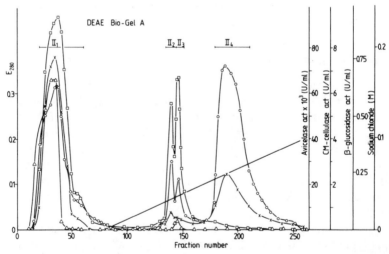

FIG. 2. DEAE-BioGel A chromatography of the cellulase preparation. Fractions from the BioGel P-10 column (140 ml) are applied to a DEAE-BioGel A column (3 × 20 cm), previously equilibrated with 0.01 *M* sodium acetate, pH 5.0. The column is washed with this buffer, and then eluted with a sodium chloride gradient in the same buffer, at a flow rate of 25 ml/hr and fractions of 5.5 ml are collected. Absorbance at 280 nm (○); avicelase (×); CM-cellulase (□); β-glucosidase (△); sodium chloride concentration (——). From Beldman *et al.*[5]

Peak III$_1$ is dialyzed and concentrated in 0.01 M sodium citrate, pH 3.5, in an ultrafiltration cell (Diaflo type 50, Amicon, Danvers, MA) using the PM10 membrane. It is then chromatographed on an SE-Sephadex C-50 column (Fig. 3). Four enzyme peaks emerge (endo I, endo II, endo III, and exo I), two of which (endo II and endo III) are further purified by rechromatography on the same SE-Sephadex column. At this stage these four enzymes appear as single bands in SDS–polyacrylamide gel electrophoresis.

Pool III$_2$ from the BioGel P-100 column (Fig. 1) is concentrated by ultrafiltration in an Amicon filter cell and applied to an SE-Sephadex column (2 × 20 cm) equilibrated in 0.02 M sodium citrate, pH 3.5. After washing with 475 ml of the same buffer, the column is eluted with a linear gradient to 0.1 M sodium citrate, pH 3.5 (300 ml). The peak containing CM-cellulase activity is pooled, dialyzed, and concentrated in 0.05 M sodium citrate, pH 5.0, by ultrafiltration. This enzyme is the electrophoretically homogeneous endo IV glucanase.

Peak II$_2$ from the DEAE-BioGel A column (Fig. 1) is rechromatographed on the same column but now equilibrated with 0.02 M sodium acetate buffer, pH 5.6. For this purpose the enzyme is diluted with water to the same conductivity as the equilibration buffer. The column is eluted with a pH gradient in 0.02 M sodium citrate from pH 5.8 to pH 3.8.

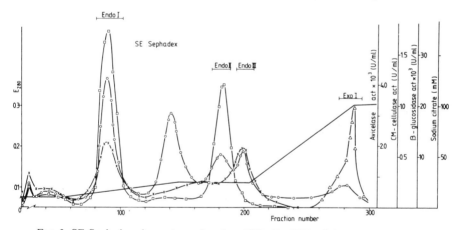

FIG. 3. SE-Sephadex chromatography of pool III$_1$. Pool III$_1$, dialyzed and concentrated in an Amicon filter is applied to a SE-Sephadex column (2.5 × 30 cm) equilibrated in 0.01 M sodium citrate, pH 3.5. The column is washed with 200 ml of the same buffer and then eluted with a linear gradient from 0.01 to 0.025 M sodium citrate, pH 3.5. Next the column is washed with 275 ml 0.025 M sodium citrate, pH 3.5, and finally elution is finished with a linear gradient to 0.1 M sodium citrate, pH 3.5. The flow rate is 6 ml/hr and fractions of 5 ml are collected. Absorbance at 280 nm (○); avicelase (×); CM-cellulase (□); β-glucosidase (△); sodium citrate concentration (——). From Beldman *et al.*[5]

The peak III_3, containing CM-cellulase activity, is concentrated by lyophilization and chromatographed on a BioGel P-100 column (2 × 90 cm), equilibrated in 0.05 M sodium citrate, pH 4.0. The peak containing CM-cellulase activity (endo V) is dialyzed and concentrated in 0.05 M sodium acetate, pH 5.0, by ultrafiltration.

By the same pH gradient purification step on DEAE-BioGel A the enzyme pool II_3 is fractionated into two CM-cellulase peaks and one β-glucosidase peak. The first CM-cellulase peak has been identified as the same material present in pool III_3. The second CM-cellulase peak (pool III_4) and the β-glucosidase peak (pool III_5) are further purified by BioGel P-100 chromatography, carried out as described for pool III_3, to yield pure enzymes endo VI and β-glucosidase I, respectively.

Peak II_4 (Fig. 2), containing essentially only avicelase activity, is concentrated by lyophilization and chromatographed on BioGel P-100, as described for pool III_3. A peak III_6 (Fig. 1) containing most of the avicelase activity is dialyzed by ultrafiltration to 0.1 M potassium/sodium phosphate, pH 6.0, and applied to a column (2.2 × 10 cm) of Avicel cellulose equilibrated in the same buffer. The column is washed with 500 ml buffer and then eluted with a pH gradient in 0.1 M potassium/sodium phosphate from pH 6.0 to pH 11.8 (1500 ml), at a flow rate of 50 ml/hr. The fractions containing protein are immediately neutralized by addition of an equal volume of 0.1 M KH_2PO_4. The two completely separated and pure peaks of avicelase activity (exo II, eluted at pH 10.3 and exo III, eluted at pH 11.8) are pooled, dialyzed, and concentrated by ultrafiltration in 0.05 M sodium acetate, pH 5.0.

Properties

Molecular properties, specific activities, and degree of random attack are summarized in Table I; adsorption characteristics, reaction kinetic parameters, reaction products formed, and degree of synergistic effect on Avicel are summarized in Table II. Most of the purified enzymes show great similarities with enzymes already described in the literature.[6-12] β-Glucosidase I differs however from other β-glucosidases in that it is also

[6] S. P. Schoemaker and R. D. Brown, *Biochim. Biophys. Acta* **523**, 133 (1978).

[7] S. P. Schoemaker and R. D. Brown, *Biochim. Biophys. Acta* **523**, 147 (1978).

[8] L. Fägerstam, U. Häkanssen, G. Pettersson, and L. Andersson, *Proc. Bioconversion Symp. Ind. Inst. Technol., New Delhi*, p. 165 (1977).

[9] G. Okada, *J. Biochem. (Tokyo)* **77**, 33 (1975).

[10] L. E. R. Berghem and L. G. Pettersson, *Eur. J. Biochem.* **46**, 295 (1974).

[11] C. S. Gong, M. R. Ladisch, and G. T. Tsao, *Biotechnol. Bioeng.* **19**, 959 (1977).

[12] M. Inglin, B. A. Feinberg, and J. R. Loewenberg, *Biochem. J.* **185**, 515 (1980).

TABLE I

MOLECULAR PROPERTIES, SPECIFIC ACTIVITIES, AND DEGREE OF RANDOMNESS OF THE CELLULASES OF *Trichoderma viride*

| Enzyme | Molecular properties | | | pH optimum | Specific activity toward | | | Degree of randomness[c] $\Delta\phi_{sp}$/reducing power |
	MW[a]	pI[b]	Glycoprotein[a]		CMC	Avicel	H$_3$PO$_4$-swollen cellulose	
Endo I	50,000	5.3	+	5.0	13.1	0.0130	0.64	1
Endo II	45,000	6.9	+	4.0	20.1	0.0068	0.46	0.77
Endo III	58,500	6.5	+	5.0	3.2	0.0156	0.45	0.40
Endo IV	23,500	7.7	−	5.5	9.6	0.0026	0.33	0.88
Endo V	57,000	4.4	+	4.5	14.7	0.0073	0.61	0.50
Endo VI	52,000	3.5	+	4.5	15.8	0.0044	0.36	0.49
Exo I[d]	53,000	5.3	+	5.5–5.7	1.8	0.0160	0.63	
Exo II	60,500	3.5	+		0.26	0.0064	0.053	
Exo III	62,000	3.8	+	4.5–7	0.033	0.0078	0.030	
β-Glucosidase I[e]	76,000	3.9	+	4.7	0.29	0.0018	0.45	

[a] By SDS–polyacrylamide slab gel electrophoresis according to U. K. Laemmli [*Nature (London)* **227**, 680 (1970)]; stained for protein with Coomassie Brilliant Blue and for carbohydrate with periodic acid–Schiff reagent (J. P. Seqrest and R. L. Jackson, this series, Vol. 28 [5]).

[b] By analytical thin-layer gel isoelectric focusing on, for example, LKB 2117 Multiphor.

[c] Ratio of specific activity on CM-cellulose measured viscosimetrically and the activity calculated from increase in reducing power.

[d] Activity toward PNP-β-D-glucopyranoside and cellobiose, 0.79 and 0.19 U/mg.

[e] Activity toward PNP-β-D-glucopyranoside, 0.40 U/mg.

· TABLE II

ADSORPTION CHARACTERISTICS, REACTION KINETIC PARAMETERS, REACTION PRODUCTS, AND MOLAR ENDOGLUCANASE TO EXO III RATIOS FOR OPTIMAL SYNERGISTIC EFFECTS[a]

Enzyme	K_p^b (mg/ml)$^{-1}$	$P_{ads,m}^b$ (mg protein/ g cellulose)	R^b (ml/g cellulose)	K_m (mg/ml)	V_{max} (μmol/mg min)	Products from Avicel	Endo/exo III for maximal DSE[c]
Endo I	0.88	125.6	110.5	46.3	0.196	$G_2 > G_1 = G_3$	1.9[d]
Endo II	0.28	89.9	25.1	90.6	0.099	$G_2 \gg G_1 \sim G_3$	5.2
Endo III	11.67	25.5	297.6	14.4	0.085	$G_2 \gg G_3 > G_1$	2.7[d]
Endo IV	2.50	2.8	7.0	130.7	0.152	$G_3 > G_2 > G_1$	13.2
Endo V	0.89	105.3	93.7	64.3	0.133	$G_2 \sim G_1$	2.2[d]
Endo VI	3.44	4.1	14.1	122.9	0.122	$G_2 > G_1$	3.3
Exo II	4.96	6.6	32.7	44.1	0.060	G_2	
Exo III	6.92	62.5	435.5	12.0	0.019	G_2	

[a] Substrate is Avicel.

[b] K_p is adsorption equilibrium constant (mg/ml)$^{-1}$, $P_{ads,m}$ is maximum adsorbed enzyme (mg/g cellulose), $R = K_p P_{ads,m}$ is initial rate of adsorption (ml/g cellulose). From Beldman et al.[13]

[c] Degree of synergistic effect (DSE) is the ratio between the activity of the combination of enzymes and the sum of the activities of the individual enzymes. From Beldman et al.[14]

[d] Adsorb strongly on Avicel cellulase.

highly active toward PNP-β-xyloside, unable to hydrolyze cellobiose, and moderately active toward CM-cellulose, Avicel, and xylan. Endoglucanase III, in contrast to other endoglucanases, is marked by its low specific activity and low degree of randomness toward CM-cellulose and its relatively high specific activity toward Avicel. It shows, however, no synergistic action with other endoglucanases but acts synergistically with exo III.

Based on their initial rate of adsorption the glucanases can be divided in two groups: endo I, III, and V and exo III adsorbing strongly on Avicel and endo II, IV, and VI and exo II adsorbing only moderately or slightly.[13]

The exoglucanases act synergistically with the endoglucanases on Avicel but not on H_3PO_4-swollen cellulose on which they only have a low synergistic activity. The degree of synergistic effect on Avicel appears largely dependent on the initial mass ratio of endoglucanases to exoglucanases and reaches an optimum at a specific value.[14] This optimum ratio can be qualitatively related to the adsorption behavior of the endoglucanases indicating that only those endoglucanase and exoglucanase molecules which are adsorbed on the cellulose chains as a complex in a specific ratio are involved in synergistic effects on insoluble crystalline cellulose. This supports the hypothesis of Wood and McCrae[15] that endo- and exoglucanases work in very close cooperation, maybe as an enzyme–enzyme complex on the surface of the cellulose chains.

[13] G. Beldman, A. G. J. Voragen, F. M. Rombouts, M. F. Searle-van Leeuwen, and W. Pilnik, *Biotechnol. Bioeng.* **30,** 251 (1987).
[14] G. Beldman, A. G. J. Voragen, F. M. Rombouts, and W. Pilnik, *Biotechnol. Bioeng.* **31,** 173 (1988).
[15] T. M. Wood and S. I. McCrae, *Biochem. J.* **171,** 61 (1978).

[27] Cellulases from *Eupenicillium javanicum*

By Mitsuo Tanaka, Masayuki Taniguchi, Ryuichi Matsuno, and Tadashi Kamikubo

Extracellular cellulases which a fungus produces usually consist of several components with different substrate specificities. These cellulase components act synergistically in the hydrolysis of crystalline cellulose. *Eupenicillium javanicum* produces several cellulase components in growth medium. The cellulase activity from *E. javanicum* is comparable to that from *Trichoderma reesei* QM 9414.[1]

Assay Methods

Reagents

Microcrystalline cellulose (MCC, No. 2330, E. Merck AG., FRG) suspension, 1.1% in sodium acetate buffer, 0.05 M, pH 5.0

Carboxymethylcellulose (CMC, Wako Pure Chemical Industries, Ltd., Japan) solution, 0.44% in sodium acetate buffer, 0.05 M, pH 5.0

p-Nitrophenyl-β-D-glucopyranoside (*p*-NPG, Nakarai Chemicals, Ltd., Japan) solution, 0.03% in sodium acetate buffer, 0.2 M, pH 5.0

Sodium hydroxide solution, 0.25 N

Sodium carbonate solution, 1 M

Procedure

MCC-Solubilizing Activity. The MCC suspension is incubated for more than 20 hr at 40° before the enzyme reaction is started. Reaction mixture containing 1.0 ml of the MCC suspension and 0.1 ml of an enzyme solution is incubated at 40° with shaking. After 1 or 20 hr of reaction, 1.1 ml of the sodium hydroxide solution is added to the reaction mixture, and the whole mixture is centrifuged at 2000 g for 10 min to remove residual cellulose. Total sugar produced in the supernatant is assayed by the phenol–sulfuric acid method of Dubois *et al.*[2] with glucose as the standard. Absorbance is read at 490 nm. One unit of enzyme activity is defined as the amount of enzyme releasing 1 μmol of total sugar from the substrate used per minute.

CMC-Saccharifying Activity. Reaction mixture containing 0.5 ml of the CMC solution and 0.05 ml of an enzyme solution is incubated at 40° for 20 min without shaking. To the reaction mixture is added the same amount of the sodium hydroxide solution after the enzyme reaction. The reducing sugar produced is assayed by the Somogyi–Nelson method[3,4]

[1] M. Tanaka, M. Taniguchi, T. Morinaga, R. Matsuno, and T. Kamikubo, *J. Ferment. Technol.* **58,** 149 (1980).

[2] M. Dubois, K. A. Gilles, J. K. Hamilton, P. A. Rebers, and F. Smith, *Anal. Chem.* **28,** 350 (1956).

[3] M. Somogyi, *J. Biol. Chem.* **195,** 19 (1952).

[4] N. Nelson, *J. Biol. Chem.* **153,** 375 (1944).

with glucose as the standard. Absorbance is read at 650 nm. One unit of enzyme activity is defined as the amount of enzyme releasing 1 μmol of reducing sugar from the substrate used per minute.

CMC-Liquefying Activity. Reaction mixture containing 14 ml of the CMC solution and 1 ml of an enzyme solution is incubated in an Ubbelohde viscometer at 40°. The viscosity of the reaction mixture is measured after appropriate intervals.

β-Glucosidase Activity. Reaction mixture containing 0.1 ml of an enzyme solution and 1.0 ml of the *p*-NPG solution is incubated at 40° for 10 min. Then 1.0 ml of the sodium carbonate solution is added to the reaction mixture. The *p*-nitrophenol liberated is estimated from the absorbance at 400 nm with *p*-nitrophenol as the standard. One unit of enzyme activity is defined as the amount of enzyme releasing 1 μmol of *p*-nitrophenol from the substrate used per minute.

Assay of Protein. Protein is assayed by the method of Lowry *et al.*[5] using crystalline bovine serum albumin as the standard or by measuring absorbance at 280 nm.

Enzyme Source

A mycelial mat from a stock culture of *E. javanicum* (Beyma) Stolk *et* Scott which was isolated from soil on a malt extract-agar slant is inoculated into a test tube (24 × 200 mm) containing 10 ml of a medium of the following composition[1]: 2 g of KH_2PO_4, 1.4 g of $(NH_4)_2SO_4$, 0.3 g of $MgSO_4 \cdot 7H_2O$, 0.3 g of $CaCl_2$, 5 mg of $FeSO_4 \cdot 7H_2O$, 1.6 mg of $MnSO_4 \cdot H_2O$, 1.4 mg of $ZnSO_4 \cdot 7H_2O$, 2.0 mg of $CoCl_2$, 0.3 g of urea, 1.0 g of proteose peptone (Difco), 20 g of MCC, and 1 liter of distilled water, pH 4.5. The test tube is incubated at 28° for 5 days with vigorous shaking. The entire contents of the test tube are then poured into a 500-ml shake flask containing 100 ml of the same medium, and the flask is incubated at 28° for 5 days on a reciprocal shaker at 105 strokes/min. The entire contents of the shake flask are poured into a 2-liter shake flask containing 400 ml of the same medium, and this flask is incubated at 28° for 32 days on the reciprocal shaker. The contents of several flasks are pooled and about 8.2 liters of culture broth is centrifuged continuously at 7000 *g* with cooling to remove mycelia and residual cellulose. The supernatant is concentrated with a rotary evaporator under reduced pressure at 30°. The whole concentrated solution (about 600 ml) is centrifuged at 6000 *g* for 15 min with cooling to remove precipitate. Ammonium sulfate is added slowly to

[5] O. H. Lowry, N. J. Rosebrough, A. L. Farr, and R. J. Randall, *J. Biol. Chem.* **193**, 265 (1951).

the supernatant to 3.1 M and the mixture is left to stand overnight at 5°. The precipitate formed is recovered by centrifugation at 15,000 g for 15 min at 4°, and dissolved in 80 ml of 0.1 M sodium acetate buffer, pH 5.0. This solution is used as the crude cellulases.

Purification Procedure

All procedures but electrophoresis are performed at 5°. The activity toward CMC measured after each purification step is saccharifying activity.

Step 1. Column Chromatography on CM-Sephadex C-50. To remove salts, 45 ml of the crude cellulase solution described above is put on a BioGel P-6 column (3.8 × 35 cm) equilibrated with 0.005 M sodium acetate buffer, pH 5.0, and eluted with the same buffer at a flow rate of 5.5 ml/15 min. Fractions of 5.5 ml each are collected. Fractions 17–42, which contain cellulase activity, are combined and stored at −20°. A 45-ml portion of the combined fractions is put on a CM-Sephadex C-50 column (3.8 × 35 cm) equilibrated with 0.005 M sodium acetate buffer, pH 5.0, and eluted stepwise with the same buffer of increasing concentrations from 0.005 to 0.25 M at a flow rate of 5.5 ml/15 min. Fractions of 5.5 ml each are collected. The volume of each buffer used for elution is as follows: 660 ml of 0.005 M buffer, 660 ml of 0.015 M buffer, 710 ml of 0.03 M buffer, 740 ml of 0.1 M buffer, and 470 ml of 0.25 M buffer. Six components, tentatively named A-1 to A-6, have different CMC-saccharifying and MCC-solubilizing activities.[6] Only two components, A-1 and A-4, obtained by elution with the buffer concentrations of 0.005 and 0.1 M, respectively, have high specific activities toward MCC and CMC. Component A-3, obtained by elution with the buffer concentration of 0.1 M, has a much higher β-glucosidase activity than the other components. The components not mentioned have little cellulase activity.

Step 2. Column Chromatography on DEAE-Sephadex A-50. The total volume (286 ml) of component A-1 (fractions 20–51) is concentrated to 30 ml by ultrafiltration under reduced pressure with a collodian bag (a nitrocellulose membrane, MW cutoff, 12,000, Sartorious GmBH, FRG) arranged in a suction flask.[6] The concentrate is put on a DEAE-Sephadex A-50 column (3.8 × 35 cm) equilibrated with 0.005 M sodium acetate buffer, pH 5.0, and eluted stepwise with the same buffer with concentration increasing from 0.005 to 0.25 M at a flow rate of 5.5 ml/15 min. Fractions of 5.5 ml each are collected. The volume of each buffer used for

[6] M. Tanaka, M. Taniguchi, R. Matsuno, and T. Kamikubo, *J. Ferment. Technol.* **59**, 177 (1981).

elution is as follows: 400 ml of 0.005 M buffer, 630 ml of 0.03 M buffer, 720 ml of 0.1 M buffer, 550 ml of 0.2 M buffer, and 470 ml of 0.25 M buffer. Nine components, tentatively named B-1 to B-9, have different cellulase activities. Of these components, only three components, B-1, B-5, and B-7, obtained by elution with the buffer concentrations of 0.03, 0.1, and 0.2 M, respectively, have high specific activities toward MCC and CMC. Although the specific activities toward MCC and CMC of component B-4, obtained by elution with the buffer concentration of 0.1 M, are low, the ratio of the specific activity toward MCC to that toward CMC is high.

Step 3. Fractionation by Electrophoresis. Only five components, A-4, B-1, B-4, B-5, and B-7, which have higher specific activity toward MCC or CMC than the other components, are further purified by polyacrylamide gel electrophoresis. The electrophoresis is done by a modification of the method of Ornstein[7] and Davis[8] using 0.005 M Tris– 0.038 M glycine buffer, pH 8.3, as the reservoir buffer. The resolving gel, 7.7% polyacrylamide gel, is prepared using 10 ml of 1.5 M Tris–HCl buffer, pH 8.9, 10 ml of 30% acrylamide including 0.8% N,N'-methylenebisacrylamide, 0.4 ml of 10% ammonium persulfate, 19.6 ml of distilled water, and 0.024 ml of N,N,N',N'-tetramethylethylenediamine. The stacking gel, 3.1% polyacrylamide gel, is prepared using 1.25 ml of 1.0 M Tris–HCl buffer, pH 6.8, 1.00 ml of the above acrylamide solution, 0.50 ml of 10% ammonium persulfate, 7.25 ml of distilled water, and 0.010 ml of N,N,N',N'-tetramethylethylenediamine. Sample containing 50–80 μg of protein and a small volume of glycerin and bromphenol blue is put on the top of the stacking gel, and then electrophoresis is done at 2 mA/gel. After electrophoresis, the gels are stained with 0.25% Coomassie Brilliant Blue G 250 for about 20 min and destained with 7.0% acetic acid solution in a boiling water bath by occasionally changing the acetic acid solution. On the basis of the mobility of the bands of each component obtained from the destained gels, the main band of each component in the other gels without staining is cut out, extracted with a small volume of 0.05 M sodium acetate buffer, pH 5.0, in a homogenizer under cooling, and dialyzed against the same buffer in a collodion bag.[6] The cellulase components are tentatively named A-4-E, B-1-E, B-4-E, B-5-E, and B-7-E. Figure 1 shows the electrophoresis of the cellulase components obtained by the treatment described above. Each cellulase component has a single band at a different migration distance. The purification of cellulases from *E. javanicum* is summarized in Table I.

[7] L. Ornstein, *Ann. N.Y. Acad. Sci.* **121,** 321 (1964).
[8] B. J. Davis, *Ann. N.Y. Acad. Sci.* **121,** 404 (1964).

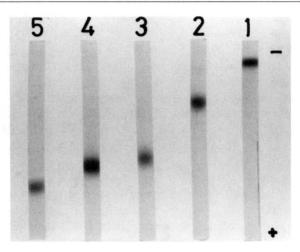

FIG. 1. Polyacrylamide gel electrophoresis of cellulase components after fractionation by electrophoresis. 1, A-4-E; 2, B-1-E; 3, B-4-E; 4, B-5-E; 5, B-7-E. Reproduced from Tanaka *et al.*[6] with permission of the publisher.

Properties

CMC-Saccharifying and MCC-Solubilizing Activities. As shown in Table I, each cellulase component fractionated by electrophoresis has different specific activities toward MCC, CMC, and *p*-NPG. The low recovery of protein in the cellulase components after electrophoresis might be due to extraction in a homogenizer being insufficient. The specific activities toward MCC and CMC of cellulase component B-4-E are low, whereas those of cellulase component B-5-E are high. However, the former component has the highest ratio of specific activity toward MCC to that toward CMC. Cellulase component A-4-E has no detectable activity toward MCC, though it has higher activity toward *p*-NPG than the other cellulase components. Component A-3-E, obtained by fractionation by electrophoresis, and showing a single band by electrophoresis when checked for purity, has extremely high activity toward *p*-NPG, as shown in footnote *b* of Table I. This component is probably a β-glucosidase.

Molecular Weight. The molecular weight of each purified cellulase component is estimated by sodium dodecyl sulfate–polyacrylamide gel electrophoresis with 0.025 M Tris–0.192 M glycine buffer, pH 8.3, containing 0.1% sodium dodecyl sulfate as the reservoir buffer by a modification of the method of Weber and Osborn.[9] The resolving gel, 11% polyacrylamide, is prepared using 9.0 ml of 1.5 M Tris–HCl buffer, pH 8.9, 0.36

[9] K. Weber and M. Osborn, *J. Biol. Chem.* **244**, 4406 (1969).

TABLE I
PURIFICATION OF CELLULASES FROM *Eupenicillium javanicum*[a]

Purification steps and components	Total protein (mg)	Total activity[b] (units)	Specific activity (units/mg) toward		
			MCC	CMC[b]	p-NPG
Crude cellulases	1,380	13,520	0.27	9.8	
BioGel P-6	990	11,580	0.36	11.7	0.14
CM-Sephadex C-50					
A-1	760	5,930	0.29	7.8	
A-3	18	34	0.04	1.9	1.07
A-4	51	1,280	0.20 (0.040)	25.1	0.02
DEAE-Sephadex A-50					
B-1	16	352	0.52 (0.085)	22.0	0.03
B-4	35	98	0.22 (0.042)	2.8	0.01
B-5	15	617	0.23 (0.056)	41.4	0.04
B-7	10	223	0.48 (0.055)	22.3	0.09
Electrophoresis					
A-4-E	0.37	16	($\approx$0)	42.2	2.95
B-1-E	0.26	14	(0.094)	52.4	0.41
B-4-E	0.35	2	(0.069)	5.8	0.34
B-5-E	0.23	16	(0.130)	70.8	0.45
B-7-E	0.16	7	(0.056)	43.2	0.93

[a] Reproduced from Tanaka *et al.*[6] with permission of the publisher.

[b] Based on CMC-saccharifying activity. The values in parentheses are MCC-solubilizing activity on the basis of total sugar produced during 20 hr of reaction. The activity toward p-NPG of component A-3-E fractionated by electrophoresis is 136 units/mg.

ml of 10% sodium dodecyl sulfate, 9.0 ml of 30% acrylamide including 0.8% N,N'-methylenebisacrylamide, 0.36 ml of 1% ammonium persulfate, 17.26 ml of distilled water, and 0.018 ml of N,N,N',N'-tetramethylethylenediamine. The stacking gel, 3.1% polyacrylamide gel, is prepared using 1.25 ml of 1.0 M Tris–HCl buffer, pH 6.8, 0.1 ml of 10% sodium dodecyl sulfate, 1.0 ml of the above acrylamide solution, 0.5 ml of 1% ammonium persulfate, 7.14 ml of distilled water, and 0.010 ml of N,N,N',N'-tetramethylethylenediamine. Sample containing about 10 μg of protein, Tris–HCl buffer, pH 6.8, 3% sodium dodecyl sulfate, and a small volume of glycerin and bromphenol blue is boiled for 3 min, and then electrophoresis is done at 2 mA/gel. The following standard proteins are used: bovine serum albumin (MW 67,000), ovalbumin (45,000), chymotrypsinogen (25,000), and cytochrome c (12,400). The gels are stained and destained as described in Step 3 of the Purification Procedure. The molecular weights of cellulase components A-4-E, B-1-E, B-4-E, B-5-E, and B-7-E are about 21,000, 65,000, 47,000, 41,000, and 30,000, respectively.

Optimum pH and Optimum Temperature. The effect of pH on the cellulolytic activities of the cellulase components is studied at 40° using CMC and MCC as substrates. Substrate solutions are adjusted to the appropriate pH values with 0.04 M Britton–Robinson's wide-range buffer. The buffer solution is prepared by mixing 0.2 N NaOH and a solution composed of 3.92 g H_3PO_4, 2.40 g of acetic acid, and 2.47 g of H_3BO_3 per liter. The substrate concentration and the volume ratio of cellulase solution to substrate solution are the same as in Assay Methods. The optimum pH values for CMC-saccharifying activity and MCC-solubilizing activity are 5.0 and 5.5, respectively, for all cellulase components after fractionation by electrophoresis. The optimum temperature is about 55° for all cellulase components.

Heat Stability and pH Stability. To test the heat stability of CMC-saccharifying activity, the cellulase components are incubated in 0.05 M sodium acetate buffer, pH 5.0, without substrate at various temperatures. All cellulase components except A-4-E are stable for 60 min at 30°, and lose their activities in 60 min at 80°. The cellulase component A-4-E loses about 30 and 50% of the original activity in 60 min at 30 and 40°, respectively; other cellulase components lose about 5% of the original activity in 60 min at 40°. The pH stability of CMC-saccharifying activity is also investigated. Solutions of cellulase components are adjusted to the appropriate pH values with 0.04 M Britton–Robinson's wide-range buffer, and kept at 40° for 2 hr or 3° for 24 hr. The reaction mixture consists of 0.5 ml of the solution of the cellulase component adjusted to the appropriate pH value and 1.0 ml of 0.62% CMC solution in 0.5 M sodium acetate buffer, pH 5.0. All cellulase components are stable between pH 4 and pH 7 at 3° for 24 hr and between pH 4 and pH 6 at 40° for 2 hr.

Randomness in CMC Hydrolysis. The degree of randomness of cellulase components in CMC hydrolysis is examined by measuring the decrease in viscosity and increase in reducing sugar during the reaction. The degree of randomness increases as the ratio of specific fluidity to reducing sugar produced increases. Specific fluidity is expressed as the reciprocal of specific viscosity. All cellulase components give straight lines of different slopes (specific fluidity/reducing sugar), and can be grouped into more-random (A-4-E and B-7-E) and less-random (B-1-E, B-4-E, and B-5-E) types.

Synergistic Effect. Synergistic effects are expressed as the ratio of total sugar produced by the combination of cellulase components to the sum of total sugar produced by each cellulase component after 100 hr of reaction with 1.0% MCC as a substrate. The combination of cellulase components B-4-E and B-5-E, which gives the maximum difference in CMC-saccharifying activity, has stronger synergistic action as the proportion of the cellulase component B-4-E increases. The combination of cellulase components B-5-E and B-1-E also has strong synergistic action. The combination of cellulase components A-4-E and B-4-E, with the maximum difference in ratio of specific activity toward MCC to that toward CMC, and the combination of cellulase components B-4-E and B-7-E, with the maximum difference in randomness in CMC hydrolysis, also have weak synergistic action. The combination of all cellulase components has the strongest synergistic action.

[28] Cellulase of *Aspergillus niger*

By GENTARO OKADA

Cellulases [EC 3.2.1.4; 1,4-(1,3;1,4)-β-D-glucan 4-glucanohydrolase] play an important role in maintaining the carbon balance in nature. In most cellulolytic microorganisms, several cellulase components secreted into the culture filtrate together constitute a "cellulase system," and insoluble cellulosic materials are converted to soluble sugar by their synergistic action. Recently, a cellulolytic enzyme was extensively purified from a commercial crude cellulase preparation from *Aspergillus niger*, and characterized.[1]

Assay Method

Principle. Cellulase activity is assayed spectrophotometrically by measuring the release of reducing sugar from sodium carboxymethylcellu-

[1] G. Okada, *Agric. Biol. Chem.* **49**, 1257 (1985).

lose (CMC) using the method of Somogyi[2] and Nelson.[3] Absorbance at 660 nm is measured.

Reagents

Sodium carboxymethylcellulose (Cellogen BS, a product of Daiichi Industrial Pharmaceutical Co., Ltd., Kyoto, Japan; D.S. 0.62–0.64), 1%, in distilled water

Sodium acetate buffer, 50 mM, pH 4.0

Enzyme, Clear supernatant of crude extract or product of subsequent steps. Dilute in 5 mM acetate buffer (pH 4.0), so that 0.25 ml contains about 0.03 unit as defined below

Solution A (copper reagent)[2] consists of 4 g of $CuSO_4 \cdot 5H_2O$, 24 g of anhydrous Na_2CO_3, 16 g of $NaHCO_3$, 12 g of Rochelle salt, and 180 g of anhydrous Na_2SO_4 in 1 liter of distilled water. This reagent should be stored at 37°

Solution B (arsenomolybdate color reagent)[3] consists of 50 g of $(NH_4)_6Mo_7O_{24} \cdot 4H_2O$, 42 ml of concentrated H_2SO_4, and 6 g of $Na_2HAsO_4 \cdot 7H_2O$ in 1 liter of distilled water. This reagent should be stored in a brown bottle

Glucose standard solution

Procedure. CMC saccharification activity is used as the standard assay of cellulase. The reaction mixture contains 0.25 ml of CMC, 0.5 ml of sodium acetate buffer, and 0.25 ml of enzyme solution in a total volume of 1.0 ml, and the reaction is initiated by the addition of CMC. After incubating at 30° for an appropriate period, the reaction is stopped by adding 1.0 ml of solution A, and mixed well. The mixture is heated for 20 min in a boiling water bath, and cooled for 5 min in running tap water. After adding 1.0 ml of solution B the mixture is left at room temperature for 10 min. The mixture is diluted with 10 ml of distilled water and the resulting blue color is measured at 660 nm.

For each set of measurements a CMC blank, an enzyme blank, and three glucose standards (20–80 μg per ml) should be included.

Definition of Unit and Specific Activity. One unit of the enzyme activity is defined as the amount of enzyme that hydrolyzes 0.25% CMC to produce reducing sugar equivalent to 1 μmol of D-glucose per minute under standard assay conditions. Specific activity is expressed as units per milligram of protein as measured by the method of Lowry et al.[4] using

[2] M. Somogyi, *J. Biol. Chem.* **195**, 19 (1952).
[3] N. Nelson, *J. Biol. Chem.* **153**, 375 (1944).
[4] O. H. Lowry, N. J. Rosebrough, A. L. Farr, and R. J. Randall, *J. Biol. Chem.* **193**, 265 (1951).

crystalline bovine serum albumin (Miles Laboratories, Inc., Kankakee, IL) as the reference standard.

Purification Procedure[1]

Crude Enzyme Material. The ethanol-precipitated powder (lot No. C6600N) from a water extract of a wheat bran culture of *A. niger*, kindly supplied by Amano Pharmaceutical Co., Ltd., Nagoya, Japan, is used as a starting material for the purification of cellulase.

The following purification steps are carried out in a cold room (4°). The enzyme preparations at various stages of purification are concentrated by ultrafiltration with a Diaflo UM10 membrane (Amicon Corp., Danvers, MA).

Step 1. Chromatography on Amberlite CG-50. The crude cellulase powder (12 g) is dissolved in 400 ml of distilled water, and fractionated with solid ammonium sulfate. The precipitate sedimenting between 60% saturation (39 g/100 ml) and 80% saturation (an additional 14.3 g/100 ml) of ammonium sulfate is dissolved in 20 ml of 0.2 M acetate buffer (pH 3.5), and is then dialyzed extensively against the same buffer. The dialyzed solution is applied to a column (4.4 × 70 cm) of Amberlite CG-50 previously equilibrated with the same buffer used for dialysis. The enzyme is then eluted stepwise with three acetate buffers (0.2 M, pH 3.5; 0.3 M, pH 4.5; 0.4 M, pH 5.5). Fractions (60 ml) are collected at a flow rate of 10 ml/8 min, and monitored for protein and enzyme activity. The major bulk of cellulase is eluted with 0.3 M acetate buffer (pH 4.5). Fractions 70–91 (1327 ml), which contain most of the cellulase activity, are pooled and concentrated to about 4.7 ml.

Step 2. First Gel Filtration on BioGel P-150. The concentrated enzyme solution is layered on a column (2.2 × 67 cm) of BioGel P-150 previously equilibrated with 50 mM acetate buffer (pH 5.0). The column is developed with the same buffer. Fractions (4 ml) are collected at a flow rate of 1 ml/10 min and analyzed for enzyme activity and protein concentration. Fractions 40–51 containing most of the cellulase activity are pooled and concentrated to about 3.3 ml. The concentrated sample is used directly for the next step.

Step 3. Second Gel Filtration on BioGel P-150. The same chromatography procedure as in step 2 is performed again. The fractions containing cellulase activity are pooled and concentrated to about 3.4 ml.

Step 4. First Gel Filtration on Sephadex G-50. The concentrated enzyme solution is passed through a column (2.2 × 62 cm) of Sephadex G-50 previously equilibrated with 50 mM acetate buffer (pH 5.0). The column is developed with the same buffer. Fractions (4 ml) are collected at a flow

TABLE I
PURIFICATION OF CELLULASE FROM *Aspergillus niger*

Fraction	Volume (ml)	Total protein (mg)	Total activity (units)	Specific activity[a] (units/mg protein)	Yield (%)
1. Amberlite CG-50	1327	724.4	20657.9	28.5	100
2. First BioGel P-150	48	566.8	19950.6	35.2	96.6
3. Second BioGel P-150	47	473.8	17842.0	37.7	86.4
4. First Sephadex G-50	43	179.7	14181.9	78.9	68.7
5. Second Sephadex G-50	44	148.4	13984.7	94.2	67.7
6. Third BioGel P-150	32	108.6	12687.4	116.8	61.4

[a] Specific activities are determined at 30°.

rate of 1 ml/10 min and analyzed for enzyme activity and protein concentration. The active fractions (30–40) are pooled and concentrated to about 3.7 ml. The concentrated sample is used directly for the next step.

Step 5. Second Gel Filtration on Sephadex G-50. The same chromatography procedure as in step 4 is performed again. The fractions containing cellulase activity are pooled and concentrated to about 3 ml.

Step 6. Third Gel Filtration on BioGel P-150. The concentrated enzyme solution is passed through a column (2.2 × 40 cm) of BioGel P-150 previously equilibrated with 50 mM acetate buffer (pH 5.0). The column developed with the same buffer. Fractions (4 ml) are collected at a flow rate of 1 ml/10 min and analyzed for enzyme activity and protein concentration. A protein peak containing only cellulase activity is obtained in this stage. The active fractions (20–27) are then combined.

A summary of a typical purification is given in Table I.

Properties[1]

Homogeneity. The purified enzyme gives a single protein band on 7.5% (w/v) polyacrylamide gel electrophoresis at pH 8.0. The locations of the enzyme activity extracted from the gel and the stained band coincide. Electrophoresis of the purified enzyme on SDS–polyacrylamide gel yields a single protein band. The enzyme is also homogeneous on ampholine electrophoresis.

Molecular Properties. From the results of isoelectric focusing,[5] the purified enzyme is an acidic protein with an isoelectric point at pH 3.67.

[5] O. Vesterberg and H. Svensson, *Acta Chem. Scand.* **20,** 820 (1966).

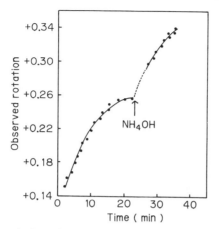

FIG. 1. Changes in optical rotation during the hydrolysis of cellopentaose by the purified cellulase and after base-catalyzed mutarotation. Cellopentaose (22.5 mg) was dissolved in 1.5 ml of distilled water (1.5%, w/v), and centrifuged to remove undissolved materials. The supernatant was left at room temperature overnight to ensure mutarotation. An aliquot of 0.5 ml of buffered enzyme solution at pH 4.0 (11.0 units) was added to 1.0 ml of the above cellopentaose solution, and thoroughly mixed. Then, a part of this mixture was quickly transferred to 1-dm polarimeter tube, and changes in the rotation were measured at 25° at intervals using a polarized D line of a sodium lamp as a light source. After incubation for 23 min, one drop of concentrated ammonium hydroxide was added to the mixture, and the rotation was again measured after thorough mixing. From Okada[1] with permission.

The molecular weight of the enzyme is estimated to be about 31,000 by SDS–polyacrylamide gel electrophoresis.[6] No carbohydrate moiety associated with the enzyme protein is detected by the phenol–sulfuric acid method.[7]

Catalytic Properties. CMC saccharification is catalyzed by the purified enzyme with a specific activity of 116.8 units/mg of protein. The catalytic activity of cellulase on CMC is maximal at pH 4.0. The K_m value of the enzyme for CMC is 0.086%.

The enzyme split cellopentaose, retaining the β-configuration of the anomeric carbon atoms in the hydrolysis products (Fig. 1). It is active on CMC and cellooligosaccharides (cellotriose to cellohexaose), but not on either cellobiose or *p*-nitrophenyl-β-D-glucoside. The enzyme is characterized as a typical endo-type cellulase (EC 3.2.1.4) on the basis of its action on CMC and cellooligosaccharides.

[6] K. Weber and M. Osborn, *J. Biol. Chem.* **244,** 4406 (1969).
[7] M. Dubois, K. A. Gilles, J. K. Hamilton, P. A. Rebers, and F. Smith, *Anal. Chem.* **28,** 350 (1956).

The purified enzyme hydrates the enolic bond of cellobial to form 2-deoxycellobiose. By carrying out the incubation in deuterated buffer and ^{1}H NMR spectra for product analysis, the enzyme is found to protonate (deuterate) cellobial from below the double bond, i.e., from a direction opposite that assumed for protonation of the β-D-glycosidic linkages of cellulose and cellodextrin.[8]

Stability. The enzyme is completely stable over the range of pH 5.0–8.0 at 4° for 24 hr, and retains about 50% of its original activity after heating at 70° for 10 min. The enzyme is completely inactivated by heating at 80° for 10 min.

Inhibitors. The enzyme is partly inactivated by 1 mM Ag$^+$, Hg^{2+}, and Fe^{2+}, corresponding to about 75, 67, and 55% inhibition, respectively. EDTA and sulfhydryl reagents have no effect on the activity of the enzyme.

[8] T. Kanda, C. F. Brewer, G. Okada, and E. J. Hehre, *Biochemistry* **25**, 1159 (1986).

[29] Cellulase and Hemicellulase from *Aspergillus fumigatus* Fresenius

By JOHN C. STEWART and JOHN HEPTINSTALL

As early as 1948, *Aspergillus fumigatus* was reported to make full use of lignocellulosic materials[1] and Reese and Levinson[2] compared the cellulolytic ability with many bacteria and fungi from all the major groups. The mold is, however, a mild pathogen and this has restricted its potential exploitation. More recently there have been a number of reports of *A. fumigatus* as a cellulase and xylanase producer which may be linked to the fact that it is now claimed[3] that there is far less risk attached to the organism than was originally thought.

Trivedi and Rao[4] investigated factors affecting cellulase production in a mesophilic strain and there have been recent reports[5-7] on thermophilic

[1] S. N. Basu, *J. Text. Inst.* **34**, T232 (1948).

[2] E. T. Reese and H. S. Levinson, *Physiol. Plant.* **5**, 345 (1966).

[3] W. M. Olver, *J. Waste Recycl.* **March/April,** 36 (1979).

[4] L. S. Trivedi and K. K. Rao, *Indian J. Exp. Biol.* **17**, 671 (1979).

[5] E. J. Vandamme, J. M. Logghe, and H. A. M. Geeraerts, *J. Chem. Technol. Biotechnol.* **32**, 968 (1982).

[6] J. M. Logghe and E. J. Vandamme, *Proc. Eur. Cong. Biotechnol., 1st,* p. 1. (1978).

[7] H. M. Shaker, M. A. Farid, and A. I. El-Diwany, *Enzyme Microb. Technol.* **6**, 212 (1984).

strains capable of producing a wide range of cellulases according to the cellulosic substrate provided. Stewart and Parry[8] also investigated cellulase production by a mesophilic strain. Reese and Levinson[2] reported that *A. fumigatus* reduced the tensile strength of cotton fibers and thus indicated that the organism produced a complete cellulase complex,[9] i.e., it contained the three enzyme activities exocellobiohydrolase (1,4-β-D-glucan cellobiohydrolase, EC 3.2.1.91, cellulose 1,4-β-cellobiosidase), endoglucanase [1,4-(1,3; 1,4)-β-D-glucan 4-glucanohydrolase, EC 3.2.1.4, cellulase], and β-glucosidase (β-D-glucoside glucohydrolase, EC 3.2.1.21). This was confirmed by Stewart and Parry[8] who also purified the major endoglucanase.[10] Several reports state that the organism is a good producer of β-glucosidase under certain culture conditions. Stewart *et al.*,[11] have shown the organism to produce a number of xylanase (EC 3.2.1.8, endo-1,4-β-xylanase) activities when grown on cellulose and lignocellulose substrates. Recently[12] attempts have been made to increase enzyme titers with the organism growing on wheat straw.

The organism is a known lignin degrader[13] and is a common contaminant of hay and straw. This ligninolytic ability, coupled with the broad spectrum of cellulase and hemicellulase activities produced, means that although the cotton-solubilizing activity of *A. fumigatus* is not as great as that found in *Trichoderma reesei*, for example, its lignocellulolytic ability is equal or superior to many other fungi.

Enzyme Assays

Exoglucanase

This enzyme can be measured in crude extracts without interference by endoglucanase either absorptiometrically or fluorimetrically by determination of MU (4-methylumbelliferone) released from MU-cellobiose (β-D-Glc$p$-(1,4)-β-D-Glc$p$-7-O-4-methylumbelliferone; obtained from Koch-Light, Haverhill, Suffolk, UK) in the presence of 2.5 mM D-glucono-1,5-lactone.[14] The latter is included to inhibit β-glucosidase activity which otherwise interferes with the assay. At 2.5 mM, the glucono-1,5-lactone inhibits exoglucanase activity by about 8% and β-glucosidase activity by

[8] J. C. Stewart and J. B. Parry, *J. Gen. Microbiol.* **125**, 33 (1981).
[9] T. M. Wood, S. I. McCrae, and C. C. MacFarlane, *Biochem. J.* **189**, 51 (1980).
[10] J. B. Parry, J. C. Stewart, and J. Heptinstall, *Biochem. J.* **213**, 437 (1983).
[11] J. C. Stewart, A. Lester, B. Milburn, and J. Heptinstall, *Biotechnol. Lett.* **7**, 581 (1985).
[12] D. A. J. Wase, A. K. Vaid, and C. McDermott, *Enzyme Microb. Technol.* **7**, 134 (1985).
[13] V. S. Bisaria and T. K. Ghosh, *Enzyme Microb. Technol.* **3**, 91 (1981).
[14] J. Heptinstall, J. C. Stewart, and M. Seras, *Enzyme Microb. Technol.* **8**, 70 (1986).

about 90%. The assay mixture consisted of 60 μl of 10 mM MU-cellobiose, 180–420 μl of 50 mM sodium acetate buffer, pH 5.2, 20 μl of 50 mM D-glucono-1,5-lactone and enzyme to give a total volume of 500 μl. The reaction was started by the addition of substrate which had been preincubated at 50° to ensure dissolution and was terminated after 20 min by the addition of 3.5 ml of 0.5 M glycine/NaOH buffer, pH 10.4. MU was measured absorptiometrically at 365 nm or fluorimetrically by emission at 450 nm following excitation at 365 nm. Standard curves were prepared using MU up to 20 μM for the spectrophotometer and in the ranges 0–2 and 0–20 μM for the fluorimeter. Stable standards for fluorimetry in these two ranges are, respectively, 1 and 10 μM quinine sulfate in 0.05 M H_2SO_4.[15] A continuous assay was run fluorimetrically by scaling up the reaction mixture to 3.0 ml with λ_{ex} 355 nm and λ_{em} 450 nm. The sensitivity of this assay was found to be less than 7% that of the discontinuous (stopped) assay.

Exoglucanase activity was qualitatively determined in column factions by incubating aliquots of up to 20 μl with 18 μl of 0.1 M sodium acetate buffer, pH 5.2, 2 μl of 50 mM D-glucono-1,5-lactone, and 10 μl of 10 mM MU-cellobiose in wells of microtiter trays. The trays were incubated at 50° for 20 min when 50 μl of 0.5 M glycine/NaOH buffer was added to each well. The release of MU was visualized by placing the tray on an ultraviolet illuminator.

β-Glucosidase

This enzyme can be assayed quantitatively, and its presence determined qualitatively, using the same procedures as described for exoglucanase except the substrate is MU-Glc(β-D-Glc$p$-7-O-4-methylumbelliferone) and D-glucono-1,5-lactone is omitted. The reaction is not susceptible to interference by either exoglucanase or endoglucanase activities.[14]

Endoglucanase

This assay is dependent on the release of reducing sugar from carboxymethylcellulose (CMC).[16] Enzyme (1 ml) was incubated with 1 ml of 1% (w/v) CMC (sodium salt medium viscosity) in 0.1 M sodium acetate buffer, pH 4.8, at 50° for 15 min. Dinitrosalicylic acid reagent[17] (1.5 ml) was added and the mixture heated at 100° for 10 min, cooled, and 1.5 ml

[15] D. H. Leaback, "An Introduction to the Fluorimetric Estimation of Enzyme Activities," 2nd Ed. Koch-Light Laboratories, Colnbrook, England, 1975.
[16] K. Matsumoto, *J. Biochem. (Tokyo)* **76,** 563 (1974).
[17] J. B. Sumner and V. A. Graham, *J. Biol. Chem.* **65,** 393 (1925).

water added. The absorbance was measured at 640 nm, dilutions of the enzyme were chosen to give readings less than 0.15 so that the reaction was not limited by substrate concentration. The activity was expressed as μmol glucose equivalents released min^{-1}, based on a standard curve of 0–300 μg ml^{-1} glucose.

Xylanase

The procedure was the same as that for endoglucanase except that the substrate was 1% (w/v) oat xylan (Sigma) and the activity was expressed as μmol xylose equivalents released min^{-1}.

Polyacrylamide Gel Electrophoresis

SDS–PAGE was carried out in vertical gels (130 mm × 130 mm × 1.5 mm) at 10% (w/v) acrylamide concentration, by the method of Laemmli.[18] They were developed at 50 V for 1 hr to concentrate the proteins in the stacking gel and separation was achieved at 150 V for approximately 3 hr. After fixing in 25% (v/v) 2-propanol/10% (v/v) acetic acid to remove the detergent, they were stained in Coomassie Blue for 1 hr and destained in 10% (v/v) acetic acid until the background gel was no longer blue. Protein samples of 50–150 μg were used together with a molecular weight marker mixture of BSA (68,000), egg albumin (45,000), DNase (32,000), α-chymotrypsinogen (23,000), and cytochrome c (12,500). Some gels were stained with periodic acid–Schiff reagent[19] to distinguish between glycoproteins and nonglycoproteins.

Organism and Culture Conditions

A. fumigatus is a common organism in composts of various types. The strain used here was isolated from garden compost by enrichment culturing and identified by the Commonwealth Micrological Institute, Kew, U.K. and designated IMI 246651 (ATCC 46324).

For producing inocula it was maintained on malt extract agar at 37° but there must be free access of air otherwise it will not produce spores. Longer term maintenance was achieved on a mineral salts/agar medium with sterile filter paper strips as substrate.

The basic mineral salts medium had the following composition (per liter): $(NH_4)_2SO_4$, 2.5 g; KH_2PO_4, 1.0 g; $MgSO_4 \cdot 7H_2O$, 0.5 g; KCl, 0.5 g; $FeSO_4 \cdot 7H_2O$, 5 mg; $MnSO_4 \cdot H_2O$, 1.5 mg; $ZnSO_4 \cdot 7H_2O$, 1.4 mg;

[18] U. K. Laemmli, Nature (London) 227, 680 (1970).
[19] J. P. Segrest and R. L. Jackson, this series, Vol. 28, p. 54.

$CoCl_2 \cdot 6H_2O$, 2 mg. Carbon source (1% w/v) CF11 cellulose, for cellulase and hay (mixed sward) or straw, for xylanase, was added and the medium sterilized at 121° for 15 min. Spore suspensions in water were added to 1 liter medium in 2-liter Erlenmeyer flasks and shaken at 37° and 120 rpm for 7 days. Applying simple statistical methods[12] has led to a maximum increase of 28% in the endoglucanase activity.

Purification of Enzymes

The purification procedures used for the cellulase enzymes of *A. fumigatus* and the recoveries from each procedure are as shown in Table I.

Concentration and Desalting of Culture Filtrates

Mycelium and unused cellulose/hay was removed by filtration through a glass sinter (porosity 1) and the filtrate is concentrated by lyophilization. The dried material was dissolved in water (30 ml water/liter lyophilized culture filtrate) and applied to a BioGel P-6 column (350 × 32 mm) equilibrated in 100 mM citrate buffer, pH 5, and eluted with the same buffer at 10 ml/hr.

Attempts were made to desalt the concentrated culture filtrates on Sephadex G-25 but this led to a 25% loss in column bed volume and poor recovery of activity due to the crude concentrate containing dextranase activity. No such problems were encountered with the polyacrylamide gel filtration medium. However, absorption effects were noticed at buffer

TABLE I

RECOVERY OF ENZYME ACTIVITIES DURING PURIFICATION OF *A. fumigatus* CELLULASES[a]

Treatment	Endo-glucanase	Exo glucanase[b]	β-Glucos-idase[c]	Protein
Original filtrate	100 (4500 U/liter)	100 (80 U/liter)	100 (300 U/liter)	100 (100 mg/liter)
Lyophilization	90	33	97	100
BioGel P-6 Column	38	12	65	52
Dialysis	38	6	35	55
BioGel P-60 Column				
Peak I	0.1		11	6.7
Peak II	1.5	2.2	18	32
Peak V	33	1.3	0	6.4

[a] Enzyme activities and protein concentrations are expressed as percentages of the values found in the crude culture filtrate.
[b] Expressed as activity against Avicel.
[c] Expressed as activity with *p*-nitrophenyl-β-D-glucopyranoside as substrate.

concentrations below 100 mM and this led to peak broadening and loss of resolution.

Fractions containing protein (A_{280}) were combined and dialyzed in Visking tubing at 4° against distilled water overnight. Nondiffusible material was lyophilized and dissolved in distilled water (2 ml) for chromatography on BioGel P-60.

Chromatography on BioGel P-60

The solution from the BioGel P-6 column (2 ml) was chromatographed on a BioGel P-60 column (750 × 25 mm) equilibrated with 5 mM citrate buffer, pH 5, and eluted with the same buffer at 10 ml/hr. The column was calibrated with proteins of known relative molecular mass (RMM) by the method of Andrews.[20] Cellulose grown filtrates gave two major and three minor peaks (Fig. 1, peaks I–V) but with hay a sixth peak was obtained eluting at three void volumes (Fig. 2).

β-Glucosidase was present in peaks I and II, peak I being totally excluded from the column. Two β-glucosidase activities from A. *fumigatus* have been purified[21,22] with RMMs of 340,000 and 40,000. The peak II components have RMMs of the order of 66,000 by SDS–PAGE clearly higher than would be expected, but the RMMs were determined by different methods. Cotton solubilizing activity and 5% of the total endoglucanase activity was in peak II and some xylanase activity was also present in this peak. Xylanase activity was also present in peaks IV and VI; peak IV appears to be a constitutive xylanase being produced on both cellulosic and hemicellulosic substrates whereas peak VI is produced only with grasses as carbon source. Peak V contained the major endoglucanase with 95% of the total activity. The known characteristics of peaks I–VI are shown in Table II.

The RMMs of peaks IV–VI as determined from the BioGel P-60 column are at variance with those determined by SDS–PAGE, the former always being lower, the elution volume being larger than expected. While it is known that some of the proteins are glycosylated, as determined by periodic acid–Schiff staining, and that glycoproteins run more slowly than would be expected on SDS–PAGE,[23] nevertheless peak V endoglucanase is not glycosylated but yet elutes later in gel filtration than would be expected from its mobility in SDS–PAGE. At the pH of the column buffer

[20] P. Andrews, *Biochem. J.* **96,** 595 (1965).
[21] M. J. Rudick and E. D. Elbein, *J. Biol. Chem.* **248,** 6506 (1973).
[22] M. J. Rudick and E. D. Elbein, *J. Bacteriol.* **124,** 534 (1975).
[23] J. P. Segrest, R. L. Jackson, E. P. Andrews, and V. T. Marchesi, *Biochem. Biophys. Res. Commun.* **44,** 390 (1971).

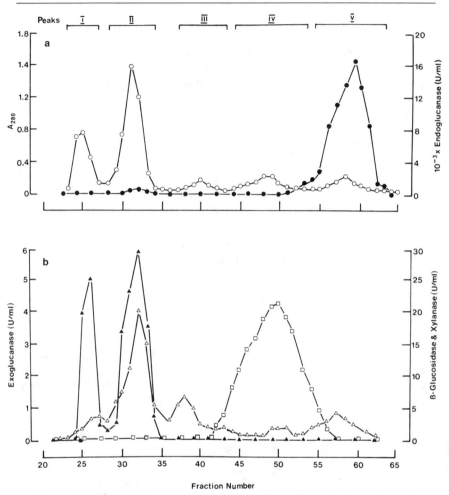

FIG. 1. Gel filtration of concentrated, desalted culture filtrate from *A. fumigatus* on BioGel P-60. (a) Protein distribution (○) was measured as A_{280} and endoglucanase (●) measured by the release of reducing sugars from CM-cellulose. (b) Exoglucanase (△) was measured by the release of soluble carbohydrates from Avicel, β-glucosidase (▲) was determined with *p*-nitrophenyl-β-D-glucopyranoside as substrate, and xylanase (□) was measured as reducing sugars released from xylan. (Fraction volume 3.5 ml.)

(pH 5) peaks IV–VI are all below their isoelectric points and must be positively charged, peak IV being nearest its pI and peak VI being far removed from its pI. We suggest that the BioGel P-60 column is negatively charged at pH 5 and the peaks are being delayed due to the ion-exchange properties of the column.

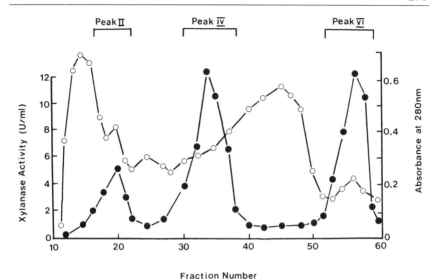

FIG. 2. Gel filtration of concentrated, desalted culture filtrate from A. *fumigatus* grown on hay, on BioGel P-60. Absorbance at 280 nm (○) and xylanase (●) measured as reducing sugar released from xylan. (Fraction volume 5 ml.)

Purification of the Exoglucanase (Peak II)

Samples of peak II from the BioGel P-60 chromatography (40–50 mg protein) were chromatographed on DEAE-Sephadex A-50 (200 × 25 mm) in 10 mM sodium phosphate buffer, pH 6.5. Over 80% of the exoglucanase activity was recovered in a single peak upon application of a 300 ml linear gradient of 10–500 mM sodium phosphate, pH 6.5. A number of minor peaks were also found but these varied from sample to sample.

TABLE II
KNOWN PROPERTIES OF THE BIOGEL P-60
PROTEIN PEAKS

Peak	Relative elution volume ($V_0 = 1$)	pI	RMM[a]
I	1.0	<5	
II	1.3	4.02	66,000
IV	1.8	5–6	16,000–31,000
V	2.2	7.1	12,500
VI	3.0	>10	

[a] Determined by sodium dodecyl sulfate–polyacrylamide gel electrophoresis.

The exoglucanase was further fractionated by isoelectric focusing in polyacrylamide gels, in the pH range 3.5–6, prepared according to the LKB manual. Activity was demonstrated in the gels by application of 2 mM MU-cellobiose in water followed by visualization on an ultraviolet illuminator. Bands were excised from the gel and eluted with 5 ml boiled distilled water. After measurement of pH, dialysis against distilled water and lyophilization, the major band of activity, with a pI of 4.02, was found to represent 80% of the activity recovered from the gel. Minor bands were also found, varying in number from 1 to 4 depending on the sample, with pI values in the range 3.8–4.5.

The major exoglucanase was found not to be synergistic in the breakdown of cotton with the purified major endoglucanase from peak V off the BioGel P-60 column. The composition of the cellulase complex required to hydrolyze cotton remains to be established.

Partial Purification of the Xylanase (Peak IV)

Chromatography on BioGel P-60 revealed that production of the three xylanase activities (peaks II, IV, and VI) by *A. fumigatus* was dependent on the lignocellulosic substrate used as shown in Table III. High xylanase activities in peak II were produced only on hay and dried grasses as substrate. Xylanase was not produced after delignification of the substrate by the chlorite method or when isolated xylans were used. However, the cellulase complex associated with peak II was still produced on these substrates, the profile from the BioGel P-60 column being similar to that given with CF11 cellulose as substrate. High xylanase activities in peak II were mainly produced with hay or dried grasses, but most straws and CF11 cellulose gave low, but detectable enzyme activity. With delignified grasses or hay, or xylan extracted from hay as the growth substrates, no xylanase activity was detectable. It is possible that this enzyme is produced in response to an as yet unidentified hemicellulose present in these untreated substrates, the hemicellulose being labile under the conditions used for delignification or xylan extraction. The xylanase associated with peak IV was produced on all the materials tested in the medium, including CF11 cellulose, partially purified xylans, and delignified materials. This enzyme is probably constitutive or at least induced along with the cellulase complex. SDS–PAGE of the BioGel P-60 peak IV samples from *A. fumigatus* grown on a variety of substrates revealed that, although the materials eluted as a single peak from the gel filtration column, in general a number of different proteins were present. From the electrophoresis it was possible to divide the proteins into three groups.

TABLE III
EFFECT OF SUBSTRATE SOURCE ON EXTRACELLULAR
XYLANASE LEVELS

Substrate	Peaks from BioGel P-60 Column[a]		
	Peak 2	Peak 4	Peak 6
Hay mixed grasses	+++	+++	+++
Hay xylan	−	+++	−
Hay delignified	−	+++	−
Grass Fescue 1	+++	+	+
Grass Fescue 2	+++	+++	+
Grass mixed sward	++	++	+++
Barley straw	+	+++	+
Barley straw delignified	−	+++	+
Barley straw xylan	−	+++	−
Oat straw	+	+++	+
Oat xylan (Sigma)	−	+++	−
Wheat straw	−	+++	−
Wheat straw xylan	−	+++	−
Rice straw	+	+++	+
Larch xylan	−	+++	+
CF11 cellulose	+	+++	+

[a] (−) Activity not detectable, (+) activity detectable, (+++) highest activity detectable within a sample.

Group 1. (RMM 31K) was produced by all the materials tested, though in some cases only trace amounts were present. Larch xylan gave only this band.

Group 2. (RMM 28K–20K). These proteins were produced on all the substrates except oat and larch xylan. Within this group three bands could be distinguished, but the pattern was not constant from one substrate to another, hay gave a band at 23K RMM but upon delignification, a band at 25K RMM was produced.

Group 3. (RMM 16K) was produced on all substrates except for larch xylan. Oat xylan as substrate gave only this band.

Purification of the Major Endoglucanase (Peak V)

SDS–PAGE and staining with periodic acid–Schiff reagent revealed that peak V was not a glycoprotein but that peaks I–IV were. Thus peak

V from the BioGel column was chromatographed on a concanavalin A-Sepharose column (40 × 16 mm) equilibrated with 50 mM sodium phosphate buffer, pH 7, containing 0.15 M NaCl, 1 mM MnCl$_2$, 1 mM MgCl$_2$, and 1 mM CaCl$_2$. The unbound endoglucanase was eluted with this buffer, bound glycoproteins being removed by elution with 20 mM α-methylmannoside in the same buffer. The endoglucanase was further fractionated by isoelectric focusing in polyacrylamide gels, in the pH range 3–10. After focusing, the gels were cut into 5-mm strips, each strip being eluted with 5 ml boiled distilled water and the pH and endoglucanase activity measured. The major endoglucanase, pI 7.1, was dialyzed against distilled water and lyophilized.

More work remains to be done on the synergistic relationships between the various cellulase and hemicellulase isoenzymes in order to determine the degree of significance of each in the overall hydrolytic process. However, the fact that *A. fumigatus* produces a wide spectrum of enzymes to break down all the components of lignocellulosic material may make it a useful organism to incorporate into experimental strategies for the development of commercial processes.

[30] Cellulases of *Aspergillus aculeatus*

By SAWAO MURAO, REIICHIRO SAKAMOTO, and MOTOO ARAI

Cellulase assays involve a multiplicity of enzymes and substrates.[1-5] The assays should be standardized; however, this is difficult because cellulose is not one substrate but varies depending on the source and pretreatment, and because cellulase is not one enzyme but a complex of enzymes containing mainly exo- and endoglucanases, and β-glucosidases (cellobiases). It is not known whether there are regions of less ordered chains in the cellulose microfibril. The reaction of cellulases with highly ordered cellulose proceeds differently from that with less ordered cellulose. The rate of hydrolysis declines rapidly and the initial rate is not necessarily significant. Carboxymethylcellulose (CM-cellulose) and Avi-

[1] M. R. Ladisch, K. W. Lin, M. Voloch, and G. T. Tsao, *Enzyme Microb. Technol.* **5,** 82 (1983).
[2] M. Mandels, *Annu. Rep. Ferment. Processes* **5,** 35 (1982).
[3] T. M. Wood and S. I. McCrae, *Carbohydr. Res.* **110,** 291 (1982).
[4] V. S. Bisaria and T. K. Ghose, *Enzyme Microb. Technol.* **3,** 90 (1981).
[5] Y. H. Lee and L. T. Fan, *Adv. Biochem. Eng.* **17,** 101 (1980).

cel have been used as substrates by many workers. Cellulases which are highly inhibited by carboxymethyl substituents cannot be detected using this substrate. Also, crystalline cellulose such as Avicel is not readily degraded by most purified cellulases and hence requires a long reaction time, or the hydrolysis of cellulose is slight. Efforts have been made to find a substrate with which any cellulase will react satisfactorily and rapidly. The insoluble cellooligosaccharide[6] with an average polymerization degree of 20 formed by hydrolyzing cellulose powder with HCl seems to be suitable for this purpose. In any event, to discover all of the different cellulase components of a fungus, several different substrates are required. An array of enzymes, substrates, and products for cellulase assays is shown in Table I.

Assay Method

Principle. The nine cellulase components of *Aspergillus aculeatus* No. F-50[7–11] are assayed by measuring reducing sugars released in each reaction mixture using four substrates: Avicel, CM-cellulose, insoluble cellooligosaccharide, and salicin.

Reagents

Sodium acetate buffer, 0.1 M, pH 5.0

Avicel (microcrystalline cellulose). Avicel SF is obtained from Asahi Kasei Kogyo Co., Ltd., Tokyo.

Carboxymethylcellulose. CM-cellulose (degree of substitution, DS, is 0.65, degree of polymerization, $\overline{DP}$, is 500) is obtained from Wako Pure Chemical Industries, Osaka.

Salicin. Salicin is obtained from Wako.

Insoluble cellooligosaccharide.[6] Thirty grams of Cellulose Powder D (Toyo Roshi Co., Ltd., Tokyo) or of Avicel SF is suspended in 400 ml of 36% HCl at 0°, and is dissolved thoroughly by adding gaseous HCl. This is left at 25° for 2 hr, and the material is poured into 1 liter of water at 0° and brought to pH 3 by the addition of sodium carbonate. The gelatinous material that forms is collected on filter paper, washed until free of chloride, and then dispersed in water ultrasonically. The total sugar content of the precipitated cellulose

[6] R. Sakamoto, M. Arai, and S. Murao, *J. Ferment. Technol.* **62,** 561 (1984).

[7] S. Murao, J. Kanamoto, and M. Arai, *J. Ferment. Technol.* **57,** 151 (1979).

[8] S. Murao, J. Kanamoto, R. Sakamoto, and M. Arai, *J. Ferment. Technol.* **57,** 157 (1979).

[9] S. Murao and R. Sakamoto, *Agric. Biol. Chem.* **43,** 1789 (1979).

[10] R. Sakamoto, J. Kanamoto, M. Arai, and S. Murao, *Agric. Biol. Chem.* **49,** 1275 (1985).

[11] S. Murao, R. Sakamoto, and M. Arai, *Agric. Biol. Chem.* **49,** 3511 (1985).

TABLE I
CELLULASE ASSAYS

Enzyme	Substrate	Product measured
1,4-β-D-Glucan 4-glucanohydro-lase (EC 3.2.1.4, cellulase)		
C_x	CM-cellulose[a]	Reducing sugar[a-c]
CM-cellulase (CMCase)	HE-cellulose[b]	Reduction in viscosity[a]
Endo-β-1,4-Glucanase	Phosphate swollen cellulose[c]	
Endoglucanase	Soluble cellooligosaccharide[c]	
1,4-β-D-Glucan cellobiohydro-lase (EC 3.2.1.91, cellulose 1,4-β-cellobiosidase)		
C_1	Cotton[d]	Reducing sugar[e-g]
Cellobiohydrolase	Avicel[e,f,h]	Loss of weight[a,d]
Avicelase	Phosphate-swollen cellulose[a,i]	Reduction in absorbance[h]
Exoglucanase	Bacteria cellulose[g]	Soluble sugar[i]
Hydrocellulase	Soluble cellooligosaccharide[e]	
1,4-β-D-Glucan glucohydrolase (EC 3.2.1.74, glucan 1,4-β-glucosidase)		
Exo-β-1,4-glucosidase	CM-cellulose[j]	Reducing sugar[j]
Exoglucanase	Phosphate-swollen cellulose[k]	Glucose[k]
	Soluble cellooligosaccharide[k]	o-Nitrophenol[k]
	p-Nitrophenyl-β-glucoside[j]	
	o-Nitrophenyl-β-glucoside[k]	
β-D-Glucoside glucohydrolase (EC 3.2.1.21, β-glucosidase)		
β-Glucosidase	p-Nitrophenyl-β-glucoside[e]	p-Nitrophenol[e,l]
Cellobiase	o-Nitrophenyl-β-glucoside[m]	o-Nitrophenol[m]
	Salicin[l,n]	Glucose[d,n]
	Cellobiose[d]	Reducing sugar[l,n,o]
	Insoluble cellooligosaccharide[o]	
Others		
Swelling factor	Cotton[m,p]	Uptake of alkali[m,p]
FPase	Filter paper[q]	Reducing sugar[q]
Degradation of filter paper	Filter paper[r]	Time required for complete disintegration[r]
Cellulase	Dyed cellulose[s]	Release of dye[s]
	Cotton thread[t]	Breaking strength[t]
Hydrocellulase	Insoluble cellooligosaccharide[o]	Reducing sugar[o]
Synergistic Avicelase activity	Avicel[u]	Reducing sugar[u]

[a] T. M. Wood and S. I. McCrae, *Biochem. J.* **128,** 1183 (1972).
[b] T. Iwasaki, H. Suzuki, and M. Funatsu, *J. Biochem.* (*Tokyo*) **55,** 30 (1964).
[c] S. P. Shoemaker and R. D. Brown, *Biochim. Biophys. Acta* **523,** 133 (1978).
[d] T. M. Wood, *Biochem. J.* **115,** 457 (1969).

is assayed by the phenol–sulfuric acid method[12] and the reducing sugar content by the Somogyi–Nelson method.[13,14] The average degree of polymerization is 20. The yield of product based on the original cellulose is 47%.

Copper reagent.[13] Rochelle salt (12 g) and anhydrous sodium carbonate (24 g) are dissolved in about 240 ml of water. To this, a solution of 4.0 g of cupric sulfate pentahydrate in 40 ml of water is added with stirring, followed by 16 g of sodium hydrogen carbonate. A solution of 180 g of anhydrous sodium sulfate in 500 ml of water is boiled to expel air, then the two solutions are combined and diluted to 1 liter. This is left at 37° for 1 week, then the clear solution filtered through a paper filter is stored in a glass-stoppered brown bottle.

Arsenomolybdate reagent.[14] To 50 g of ammonium molybdate in 500 ml of water is added 46 ml of 95% sulfuric acid, followed by 6.0 g of disodium hydrogen arsenate heptahydrate dissolved in 25 ml of water. This solution is diluted to 1 liter and incubated for 24 hr at 37°. The clear solution is stored in a glass-stoppered brown bottle.

Procedure

Avicelase Activity.[7] First, 0.1 ml of the substrate solution prepared by suspending 1 g of Avicel in 100 ml of the acetate buffer is put in a test tube

---

[12] M. Dubios, K. A. Gilles, J. K. Hamilton, P. A. Rebers, and F. Smith, *Anal. Chem.* **28,** 350 (1956).
[13] M. Somogyi, *J. Biol. Chem.* **195,** 19 (1952).
[14] N. Nelson, *J. Biol. Chem.* **153,** 375 (1944).

---

[e] L. E. R. Berghem and L. G. Pettersson, *Eur. J. Biochem.* **37,** 21 (1973).
[f] K. Wakabayashi, T. Kanda, and K. Nishizawa, *J. Ferment. Technol.* **43,** 739 (1965).
[g] G. Halliwell and M. Griffin, *Biochem. J.* **135,** 587 (1973).
[h] L. H. Li, R. M. Flora, and K. W. King, *J. Ferment. Technol.* **41,** 98 (1963).
[i] T. M. Wood and S. I. McCrae, *Carbohydr. Res.* **57,** 117 (1977).
[j] L. H. Li, R. M. Flora, and K. W. King, *Arch. Biochem. Biophys.* **111,** 439 (1965).
[k] T. M. Wood and S. I. McCrae, *Carbohydr. Res.* **110,** 291 (1982).
[l] J. H. Hash and K. W. King, *J. Biol. Chem.* **232,** 395 (1958).
[m] T. M. Wood, *Biochem. J.* **109,** 217 (1968).
[n] T. Hirayama, S. Horie, H. Nagayama, and K. Matsuda, *J. Biochem. (Tokyo)* **84,** 27 (1978).
[o] R. Sakamoto, M. Arai, and S. Murao, *J. Ferment. Technol.* **62,** 561 (1984).
[p] T. Nishizawa, H. Suzuki, and K. Nishizawa, *J. Ferment. Technol.* **44,** 659 (1966).
[q] M. Mandels, R. Andreotti, and C. Roche, *Biotechnol. Bioeng. Symp.* **6,** 21 (1976).
[r] T. Kitamikado and N. Toyama, *J. Ferment. Technol.* **40,** 85 (1962).
[s] R. P. Poincelot and P. R. Day, *Appl. Microbiol.* **22,** 875 (1972).
[t] K. Ogawa and N. Toyama, *J. Ferment. Technol.* **41,** 282 (1963).
[u] S. Murao, R. Sakamoto, and M. Arai, *Agric. Biol. Chem.* **49,** 3511 (1985).

and preincubated at 37° for 10 min. Then, 0.1 ml of the enzyme solution diluted appropriately with the acetate buffer is added to initiate the reaction. After 100 min at 37°, the reaction is terminated by adding 1.0 ml of copper reagent. The tubes, together with a blank in which copper reagent was added before the enzyme, are closed with glass beads, and heated on a vigorously boiling water bath for 15 min. The samples are cooled with tap water, then 1.0 ml of arsenomolybdate reagent is added, and the mixture left for 15 min. The samples are diluted to 10 ml with water, and the absorbance of samples at 500 nm is measured.

Synergistic Avicelase Activity.[11] Samples are assayed as for Avicelase activity except that 1.8 μg of FI CM-cellulase and 0.8 μg of β-glucosidase 1 prepared as described below are added to the reaction mixtures.

CM-Cellulase Activity.[7] Samples are assayed as for Avicelase activity except that 0.5 ml of 1% CM-cellulose is used as a substrate, that 0.5 ml of enzyme solution is added to the reaction mixtures, and that the reaction time is 10 min.

Hydrocellulase Activity.[6] Samples are assayed as for CM-cellulase activity except for using 0.1 ml of 1% insoluble cellooligosaccharide as a substrate and 0.1 ml of enzyme solution.

β-Glucosidase Activity.[10] Samples are assayed as for hydrocellulase activity except for using 1% salicin as a substrate.

Evaluation of the Assay. When β-glucosidase released less than 50 μg of reducing sugar as glucose under the above assay conditions, the activity gives a lower value for the enzyme concentration. However, this problem has been solved by adding 20 μg of egg albumin. If the enzyme releases 20–30 μg of reducing sugar as glucose from the substrate under these assay conditions, the reaction rate of other cellulase activities is constant with time and proportional to the enzyme concentration. When measuring activities for insoluble cellulosic substrates, care must be taken before pipetting to stir vigorously to obtain a uniform suspension.

Definition of Unit and Specific Activity. One unit of each enzyme activity is defined as the amount of enzyme that releases 1 μmol of reducing sugar as glucose per minute under the assay conditions. The specific activity is expressed in units of cellulolytic activity per milligram of protein based on absorbancy of the enzyme solutions at 280 nm.

Source

Aspergillus aculeatus No. F-50. Murao *et al.*[7,15,16] found that *A. aculeatus* No. F-50 produced a potent cellulolytic enzyme which was able to

[15] R. Sakamoto, H. Hayashi, K. Moriyama, M. Arai, and S. Murao, *Hakkokogaku* **60,** 333 (1982).
[16] H. Hayashi, M. Arai, R. Sakamoto, and S. Murao, *Hakkokogaku* **61,** 413 (1983).

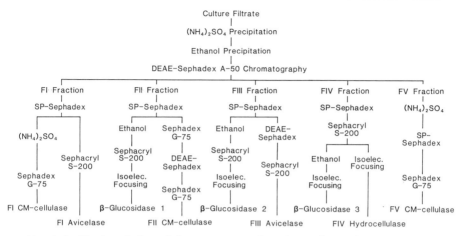

FIG. 1. Purification of nine cellulase components from *Aspergillus aculeatus* No. F-50.

hydrolyze synergistically cellulosic substances in combination with *Trichoderma* cellulases, and which yielded only such monosaccharides as glucose and xylose.

Purification Procedure

The nine cellulase components of *A. aculeatus* No. F-50 were separated and purified using the steps shown in Fig. 1.

Culture of Fungi. The fungus was cultured in a 100-liter jar fermentor.[17] The medium was 3% wheat bran, 0.5% bleached kraft pulp (hardwood), 0.5% sodium glutamate, 0.2% corn steep liquor, 0.5% K_2HPO_4, 0.1% $MgSO_4 \cdot 7H_2O$, 0.2% NaCl, and 0.1% Tween 80 in 60 liters of tap water (pH 5.5). The culture was carried out with an air flow of 60 liters/min and an agitator speed of 150–300 revolutions per minute at 30°C. After 4 days of culture, the mycelia were removed by filtration through a cloth, and were washed with tap water. The 70 liters of culture filtrate obtained contained 4290 units of Avicelase activity, 249,000 units of CM-cellulase activity, 168,000 units of hydrocellulase activity, and 120,000 units of β-glucosidase activity.

Preparation of Crude Cellulase (Acucelase). First, to 70 liters of the culture filtrate, 42.6 kg of ammonium sulfate was added to give 80% saturation, and then the sample was left at 5° for 1 week. The supernatant was removed by decantation, and the precipitate was first collected by centrifugation, and then dissolved in water. The insoluble material formed

---

[17] R. Sakamoto, I. Shooro, M. Arai, and S. Murao, *Hakkokogaku* **60,** 327 (1982).

was separated by centrifugation and washed in water. The 7 liters of the clear supernatant collected was dialyzed with PVA-Hollow Fiber Dialyzer (KL-2-50, A membrane, Kuraray Co., Ltd., Osaka) against water. The sediment formed during dialysis was centrifuged off. The enzyme solution was condensed to yield 2020 ml in a rotary evaporator under reduced pressure at below 30°, and then was cooled with ice bath containing NaCl. To this, 8080 ml of ethanol ($-20°$) was added to yield 80% (v/v), the precipitate that formed was collected by centrifugation and dissolved in water. Then the precipitate was dissolved to yield 2025 ml containing 0.02 M sodium acetate buffer, pH 5.5.

DEAE-Sephadex A-50 Chromatography. The 1 liter of crude cellulase was applied to a DEAE-Sephadex A-50 column (9.6 × 40 cm) equilibrated with 20 mM acetate buffer, pH 5.5. The protein was eluted with a five-stage pH gradient[10] as follows. The first stage was formed by adding 5 liters of 50 mM acetate buffer, pH 5.0, to 5 liters of 20 mM acetate buffer, pH 5.5. The second stage was formed by adding 5 liters of 0.2 M acetate buffer, pH 4.5, to 5 liters of 50 mM acetate buffer, pH 5.0. The third stage was formed by adding 5 liters of 0.5 M acetate buffer, pH 4.0, to 5 liters of 0.2 M acetate buffer, pH 4.5. The fourth stage was formed by adding 5 liters of 0.5 M acetate buffer containing 20 mM NaCl, pH 3.5, to 5 liters of 0.5 M acetate buffer, pH 4.0. The fifth stage was formed by adding 5 liters of 0.5 M acetate buffer containing 20 mM NaCl, pH 2.7, to 5 liters of the same buffer, pH 3.5. One hundred-milliliter fractions were collected, and each fraction was separated into five fractions, FI to FV. The elution pattern is shown in Fig. 2. The fractionation of the crude cellulase with DEAE-Sephadex A-50 was repeated once more. The FI fraction emerged at pH 5.5 without being absorbed on the DEAE-Sephadex, and had CM-cellulase, hydrocellulase, and synergistic Avicelase activity. The FII fraction eluted at around pH 4.8 and had CM-cellulase, hydrocellulase, and β-glucosidase activity. The FIII fraction eluted at around pH 4.5 and had hydrocellulase, synergistic Avicelase, and β-glucosidase activity. The FIV fraction eluted at around pH 4.1 and had hydrocellulase, Avicelase, and β-glucosidase activity. The FV fraction eluted at around pH 3.7 and had CM-cellulase and hydrocellulase activity.

Purification of FI Fraction

SP-Sephadex C-50 Chromatography. To 35 liters of the FI fraction, 22 kg of ammonium sulfate was added to give 83% saturation. The resulting precipitate was dissolved in 20 mM acetate buffer, pH 5.0. The clear enzyme solution was dialyzed with PVA-Hollow Fiber Dialyzer (KL-2-50, A membrane) against 20 mM acetate buffer, pH 5.0. The enzyme solution (1.7 liters) was diluted to 3 liters with 50 mM acetate buffer

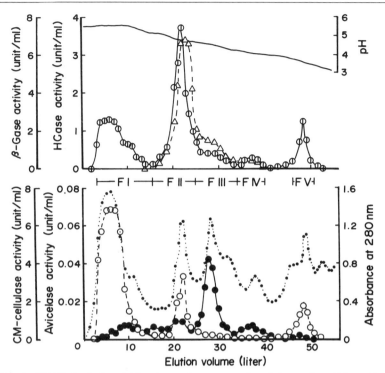

Fɪɢ. 2. DEAE-Sephadex A-50 column chromatography of the crude cellulase (Acuce-lase). Fractions FI to FV were pooled as shown (⊢——⊣). (····) Absorbance at 280 nm; (——) pH; (△) β-glucosidase activity (β-Gase activity); (Φ) activity toward insoluble cellooligosac-charide (HCase activity); (●) activity toward Avicel (Avicelase activity); (○) activity toward CM-cellulose (CM-cellulase activity).

containing 20 mM NaCl, pH 3.5. The sample was adjusted to pH 3.5 with acetic acid and applied to a SP-Sephadex C-50 column (8.0 × 30 cm) equilibrated with the same buffer (pH 3.5). The protein was eluted with a two-stage pH gradient[11] as follows. The first stage was formed by adding 3 liters of 50 mM acetate buffer, pH 4.5, to 3 liters of 50 mM acetate buffer containing 20 mM NaCl, pH 3.5. The second stage was formed by adding 3 liters of 50 mM acetate buffer, pH 5.5, to 3 liters of the same buffer, pH 4.5. One hundred-milliliter fractions were collected. The two protein peaks were emerged at pH 3.5–4.0 and pH 4.2–4.6. The first peak had synergistic Avicelase and hydrocellulase activity, and the second peak had CM-cellulase and hydrocellulase activity. The peaks were referred to as the FI Avicelase and FI CM-cellulase fraction, respectively.

Ammonium Sulfate Fractionation of FI CM-Cellulase Fraction. To 2.5 liters of the FI CM-cellulase fraction, 1050 g of ammonium sulfate was added to yield 55% saturation. The resulting precipitate was dissolved with 20 mM acetate buffer, pH 5.0, and desalted by filtration through a BioGel P-2 column (5.0 × 30 cm) equilibrated with the same buffer, pH 5.0.

Sephadex G-75 Chromatography of FI CM-Cellulase Fraction. The 150 ml of enzyme solution obtained by desalting were condensed to yield 15 ml under reduced pressure as described above. The sample (7 ml) was applied to a Sephadex G-75 column (4.0 × 80 cm) equilibrated with 50 mM acetate buffer containing 0.5 M NaCl, pH 5.0. Elution was carried out at 30 ml/hr with the same buffer, and 10-ml fractions were collected. The gel filtration with Sephadex G-75 was repeated once more. The peak of protein emerged at around 780 ml, and coincided well with the peak of CM-cellulase and hydrocellulase activity. The fractions which had the constant specific activity of CM-cellulase and hydrocellulase were collected. The enzyme solution was condensed under reduced pressure as described above, and desalted by filtration through a BioGel P-2 as described above. This purified enzyme was referred to as FI CM-cellulase.

Sephacryl S-200 Chromatography of FI Avicelase Fraction. To 2480 ml of the FI Avicelase fraction, 1.56 kg of ammonium sulfate was added to give 83% saturation. The resulting precipitate was dissolved and desalted as for the FI CM-cellulase fraction. The enzyme solution was condensed to yield 10 ml under reduced pressure as described above, and applied to a Sephacryl S-200 (superfine) column (3.0 × 100 cm) equilibrated with 50 mM acetate buffer containing 0.5 M NaCl, pH 5.0. Elution was carried out at 40 ml/hr with the same buffer. The peak of protein emerged at around 425 ml, and coincided well with the peak of synergistic Avicelase and hydrocellulase activity. The fractions which had the constant specific activity of synergistic Avicelase and hydrocellulase were collected. The enzyme solution was condensed and desalted as for the FI CM-cellulase fraction. This purified enzyme was referred to as FI Avicelase.

Purification of FII Fraction

SP-Sephadex C-50 Chromatography. To the 19.2 liters of FII fraction, 12.1 kg of ammonium sulfate was added to yield 83% saturation. The resulting precipitate was dissolved with 20 mM acetate buffer, pH 5.0, and desalted by filtration through a BioGel P-2 column (5.0 × 30 cm) equilibrated with the same buffer, pH 5.0. The enzyme solution (525 ml) was diluted to 1.2 liters with 50 mM acetate buffer containing 20 mM NaCl, pH 3.5. The sample was adjusted to pH 3.5 with acetic acid and applied to a SP-Sephadex C-50 column (6.0 × 23 cm) equilibrated with the same buffer, pH 3.5. The column was washed with 600 ml of the same buffer, and then the protein was eluted with a three-stage pH gradient[10] as follows. The first stage was formed by adding 1.3 liters of 50 mM acetate buffer, pH 4.5, to 1.3 liters of 50 mM acetate buffer containing 20 mM NaCl, pH 3.5. The second stage was formed by adding 1.3 liters of 50 mM acetate buffer, pH 6.0, to 1.3 liters of the same buffer, pH 4.5. The third stage was only 1 liter of the same buffer, pH 6.0. Twenty-milliliter fractions were collected, and 1 ml of 2 M sodium acetate was immediately added into each fraction of less than pH 4. The protein peak that had CM-cellulase and hydrocellulase activity emerged at pH 3.3–3.9, and the later peak at pH 5.2–5.8 had hydrocellulase and β-glucosidase activity. The peaks were referred to as the FII CM-cellulase and β-glucosidase 1 fraction, respectively.

First Sephadex G-75 Chromatography of FII CM-Cellulase Fraction. To 680 ml of the FII CM-cellulase fraction, 430 g of ammonium sulfate was added to give 83% saturation. The resulting precipitate was dissolved and desalted by filtration through a BioGel P-2 column as described above. The enzyme solution was condensed to yield 14 ml under reduced pressure as described above, and fractionated with Sephadex G-75 column (4.0 × 80 cm) as for the FI CM-cellulase fraction. The protein peak that had CM-cellulase and hydrocellulase activity emerged at around 430 ml. The active fractions were collected.

DEAE-Sephadex A-50 Chromatography of FII CM-Cellulase Fraction. The enzyme solution was condensed to yield 20 ml under reduced pressure as described above, and desalted by filtration through a BioGel P-2 column (5.0 × 30 cm) equilibrated with 50 mM acetate buffer, pH 5.5. The sample (90 ml) was applied to a DEAE-Sephadex A-50 column (2.5 × 19 cm) equilibrated with the same buffer, pH 5.5. The protein was eluted with 1.4 liters of the same buffer, pH 5.5. The protein peak that emerged at around 530 ml had CM-cellulase and hydrocellulase activity. The active fractions (250 ml) were collected, and to this, 160 g of ammonium sulfate was added. The precipitate was dissolved in 20 mM acetate buffer, pH 5.0.

Second Sephadex G-75 Chromatography of FII CM-Cellulase Fraction. The enzyme solution (3 ml) was fractionated again with Sephadex G-75 as described above. The protein peak that emerged at around 420 ml had the constant specific activity of CM-cellulase and hydrocellulase. The active fractions were collected and condensed under reduced pressure as described above. The enzyme solution was desalted as for the FI CM-cellulase. This purified enzyme was referred to as FII CM-cellulase.

Ethanol Fractionation of β-Glucosidase 1 Fraction. The β-glucosidase 1 fraction (700 ml) was condensed to yield 250 ml under reduced pressure as described above, and was cooled in an ice bath containing

NaCl. Then 220 ml of ethanol ($-20°$) was added to give 47% (v/v). The precipitate obtained by centrifugation was dissolved in 20 mM acetate buffer, pH 5.0. The insoluble material was centrifuged off, and the clear supernatant was condensed to yield 3 ml under reduced pressure.

Sephacryl S-200 Chromatography of β-Glucosidase 1 Fraction. The enzyme solution was put through a Sephacryl S-200 (superfine) column (3.0×94 cm) and fractionated as for the FI Avicelase fraction. The protein peak with highest specific activity of both β-glucosidase and hydrocellulase emerged at 320–370 ml. The active fractions were collected and condensed under reduced pressure as described above, followed by desalting by filtration through a BioGel P-2 column (5.0×30 cm) equilibrated with 20 mM acetate buffer, pH 5.0.

Isoelectric Focusing of β-Glucosidase 1 Fraction. The enzyme solution (36 ml) was applied to an LKB electrofocusing column (440 ml) with a sucrose density gradient (0–50% as sucrose concentration) containing 1% carrier ampholites (pH 4 ~ 6). Focusing was done for 73 hr at 700 V, then 4-ml fractions were removed from the bottom of the column. The β-glucosidase activity was found in the fractions from pH 4.5 to 4.8, and the peak coincided with those of protein and hydrocellulase activity. The fractions, which had the constant specific activities of both enzymes, were collected, and the enzyme solution was condensed and desalted as for the FI CM-cellulase. This purified enzyme was referred to as β-glucosidase 1.

Purification of FIII Fraction

SP-Sephadex C-50 Chromatography. To the 18.2 liters of FIII fraction, 11.5 kg of ammonium sulfate was added to yield 83% saturation. The resulting precipitate was dissolved with 20 mM acetate buffer, pH 5.0, and desalted by filtration through a BioGel P-2 column (5.0×30 cm) equilibrated with the same buffer, pH 5.0. The enzyme solution (535 ml) was diluted to 1.5 liters with 50 mM acetate buffer containing 20 mM NaCl, pH 3.5. The sample was adjusted to pH 3.5 with acetic acid, and applied to a SP-Sephadex C-50 column (6.0×31 cm) equilibrated with the same buffer, pH 3.5. The column was washed with 500 ml of the same buffer, and then the protein was eluted with a three-stage pH gradient[11] as follows. The first stage was formed by adding 1.8 liters of 50 mM acetate buffer, pH 4.5, to 1.8 liters of 50 mM acetate buffer containing 20 mM NaCl, pH 3.5. The second stage was formed by adding 1.5 liters of 50 mM acetate buffer, pH 6.0, to 1.5 liters of the same buffer, pH 4.5. The third stage was only 1 liter of the same buffer, pH 8.0. Twenty-milliliter fractions were collected, and 1 ml of 2 M sodium acetate was immediately added into each fraction of less than pH 4. The protein peak that had synergistic Avicelase and hydrocellulase activity emerged at pH 3.6–3.9, and was referred to as FIII Avicelase fraction. The two peaks that had hydrocellulase and β-glucosidase activity emerged at pH 4.8–5.5 and pH 5.5–6.7. The middle peak which had by far most of both enzyme activities was referred to as the β-Glucosidase 2 fraction. The third peak with minor activity was discarded.

DEAE-Sephadex A-50 Chromatography of FIII Avicelase Fraction. To 715 ml of the FIII Avicelase fraction, 450 g of ammonium sulfate was added to give 83% saturation. The resulting precipitate was dissolved and desalted by filtration through a BioGel P-2 column equilibrated with 50 mM acetate buffer, pH 5.0. The enzyme solution was diluted to 400 ml with the same buffer, and was applied to a DEAE-Sephadex A-50 column (5.0 × 15 cm) equilibrated with the same buffer. The protein was eluted with a two-stage pH gradient[11] as follows. The first stage was formed by adding 600 ml of 0.2 M acetate buffer, pH 4.0, to 600 ml of 50 mM acetate buffer, pH 4.5. The second stage was formed by adding 450 ml of 0.2 M acetate buffer containing 30 mM NaCl, pH 3.5, to 450 ml of 0.2 M acetate buffer, pH 4.0. Ten-milliliter fractions were collected. A large peak of protein appeared from pH 4.3–4.1, and at the front of this peak the synergistic avicelase activity coincided approximately with hydrocellulase activity. At rear of the peak, however, the synergistic Avicelase activity was slightly less than that of hydrocellulase. Therefore, these fractions were discarded. The fractions which were constant as to both specific activities were collected. The enzyme solution was condensed under reduced pressure as described above and desalted by filtration through a BioGel P-2 column equilibrated with 20 mM acetate buffer, pH 5.0.

Sephacryl S-200 Chromatography of FIII Avicelase Fraction. The enzyme solution was condensed to yield 15 ml under reduced pressure, and applied to a Sephacryl S-200 (superfine) column and fractionated as for the FI-Avicelase fraction. Both peaks of synergistic Avicelase and hydrocellulase activities appeared in the same place as the peak of protein, which emerged at around 425 ml, near the elution position of FI Avicelase. The fractions with the constant specific activities of both enzymes were collected. The enzyme solution was condensed and desalted as for the FI CM-cellulase. This purified enzyme was referred to as FIII Avicelase.

Ethanol Fractionation of β-Glucosidase 2 Fraction. The β-glucosidase 2 fraction (1640 ml) was condensed under reduced pressure and desalted by filtration through a BioGel P-2 column equilibrated with 20 mM acetate buffer, pH 5.0. The enzyme solution (100 ml) was cooled in an ice bath containing NaCl, and then 100 ml of ethanol ($-20°$) was added to yield 50% (v/v). The precipitate obtained by centrifugation was dissolved in 20 mM acetate buffer, pH 5.0. The insoluble material was centrifuged off, and the clear supernatant was condensed to yield 3 ml under reduced pressure.

Sephacryl S-200 Chromatography of β-Glucosidase 2 Fraction. The enzyme solution was applied to a Sephacryl S-200 (superfine) column (3.0 × 100 cm) and fractionated as for the β-glucosidase 1. The protein peak with the highest specific activity of both β-glucosidase and hydrocellulase emerged at 345 ml, near the elution position of β-glucosidase 1. The active fractions were collected and condensed under reduced pressure as described above, followed by desalting by filtration through a BioGel P-2 as described above.

Isoelectric Focusing of β-Glucosidase 2 Fraction. The enzyme solution was applied to an LKB electrofocusing column (110 ml) with a sucrose density gradient (0–50% as sucrose concentration) containing 1% carrier ampholites (pH 2.5 ~ 4 : pH 4 ~ 6 = 1 : 4), and electrofocused as for the β-glucosidase 1. β-Glucosidase and hydrocellulase activity and by far most of the protein was found in the fractions from pH 4.2 to 4.4, and the peaks all emerged together at pH 4.3. The fractions with the both specific activities constant were collected. The enzyme solution was condensed and desalted as for the FI CM-cellulase. This purified enzyme was referred to as β-glucosidase 2.

Purification of FIV Fraction

SP-Sephadex C-50 Chromatography. To the 13 liters of FIII fraction, 8.2 kg of ammonium sulfate was added to yield 83% saturation. The resulting precipitate was dissolved with 20 mM acetate buffer, pH 5.0, and desalted by filtration through a BioGel P-2 column (5.0 × 30 cm) equilibrated with the same buffer, pH 5.0. The enzyme solution was adjusted to pH 3.5 with acetic acid, and was applied to an SP-Sephadex C-50 column (4.0 × 40 cm) equilibrated with 50 mM acetate buffer (ionic strength 0.02, pH 3.5). The column was washed with 1 liter of the same buffer. Under these conditions, most of the hydrocellulase and β-glucosidase activities were eluted without being absorbed on the SP-Sephadex. The active fractions were collected, and to 340 ml of the enzyme solution, 215 g of ammonium sulfate was added to give 83% saturation. The resulting precipitate was dissolved with 20 mM acetate buffer, pH 5.0.

Sephacryl S-200 Chromatography. Gel filtration was done on a Sephacryl (superfine) column (3.0 × 100 cm) as for the β-glucosidase 1. Two protein peaks emerged, one at around 345 ml and the other at around 475 ml. The first which was near the elution position of β-glucosidase 1 had β-glucosidase activity. The second had hydrocellulase activity and slight CM-cellulase activity. These were referred to as the β-glucosidase 3 and FIV hydrocellulase fractions, respectively.

Ethanol Fractionation of β-Glucosidase 3 Fraction. The β-glucosidase 3 fraction (180 ml) was condensed under reduced pressure and desalted as for the β-glucosidase 2. The ethnol fractionation of the enzyme solution was carried out as for the β-glucosidase 2.

Isoelectric Focusing of β-Glucosidase 3 Fraction. The enzyme solution was electrofocused as for β-glucosidase 1 except that carrier ampholites (pH 2.5 ~ 4 : pH 3.5 ~ 10 = 4 : 1) were used. The protein and β-glucosidase activity was at pH 3.4. The fractions with the constant specific activity were collected. The enzyme solution was desalted as described above. This purified enzyme was referred to as β-glucosidase 3.

Isoelectric Focusing of FIV Hydrocellulase Fraction. The 260 ml of FIV hydrocellulase fraction was condensed under reduced pressure as described above and desalted by filtration through a BioGel P-2 column as described above. The enzyme solution was electrofocused as for the β-glucosidase 1 except for using carrier ampholites (pH 2.5 ~ 4). The protein and hydrocellulase activity emerged at pH 3.5, and separated from most of the CM-cellulase activity. The fractions with the constant specific activity were collected and desalted by filtration through a BioGel P-2 column equilibrated with 20 mM acetate buffer, pH 5.0. This purified enzyme with only hydrocellulase activity was referred to as FIV hydrocellulase.

Purification of FV Fraction

Ammonium Sulfate Fractionation. To 9.4 liters of the FV fraction, 4.2 kg of ammonium sulfate was added to give 60% saturation. The resulting precipitate was desalted by filtration through a BioGel P-2 column equilibrated with 20 mM acetate buffer, pH 5.0.

SP-Sephadex C-50 Chromatography. The enzyme solution (432 ml) was diluted to 600 ml with 0.1 M acetate buffer containing 20 mM NaCl, pH 3.0. The sample was adjusted to pH 3.0 with 6 N HCl and was applied to a SP-Sephadex C-50 column (5.0 × 20 cm) equilibrated with the same buffer, pH 3.0. The column was washed with 100 ml of the same buffer, and then the protein was eluted with a two-stage pH gradient as follows. The first stage was formed by adding 500 ml of 50 mM acetate buffer containing 20 mM NaCl, pH 3.5, to 500 ml of 0.1 M acetate buffer containing 20 mM NaCl, pH 3.0. The second stage was formed by adding 500 ml of 50 mM acetate buffer containing 20 mM NaCl, pH 4.0, to 500 ml of the same buffer, pH 3.5. Ten-milliliter fractions were collected, and 1 ml of 2

M sodium acetate was immediately added to each fraction. The protein peak that emerged at around pH 3.4 had CM-cellulase and hydrocellulase activities. The active fractions were collected, and to 140 ml of the enzyme solution, 64 g of ammonium sulfate was added to give 60% saturation. The resulting precipitate was dissolved with 20 mM acetate buffer, pH 5.0.

Sephadex G-75 Chromatography. The enzyme solution was put through a Sephadex G-75 column and fractionated as for the FI CM-cellulase fraction. The peak of protein emerged at around 540 ml, and coincided well with the peaks of CM-cellulase and hydrocellulase activity. The fractions that had the constant specific activities of both enzymes were collected. The enzyme solution (140 ml) was salted out by ammonium sulfate as described above. The resulting precipitate was dissolved and desalted by filtration through a BioGel P-2 column equilibrated with 20 mM acetate buffer, pH 5.0. This purified enzyme was referred to as FV CM-cellulase.

Purification of Nine Cellulolytic Enzymes

The purification of the three CM-cellulases is summarized in Table II, and that of the hydrocellulase[6,9] and the two Avicelases,[11] and the three β-glucosidases[10] is in Tables III and IV, respectively.

Overall Recovery of Cellulolytic Enzymes

So far many cellulolytic enzymes have been purified from fungi and bacteria. But reports presenting enzyme recovery relating to the purification steps are few in cellulase research. Although the enzyme activity for crystalline cellulose decreased significantly by separating endoglucanase and cellobiohydrolase, it seems that the overall recovery for both enzymes has not been reported yet. As the nine cellulase components were purified from one culture filtrate of *A. aculeatus* No. F-50, the overall recovery of cellulolytic enzymes was investigated and is summarized in Table V. The total yield of protein was 6.4% of the crude cellulase estimated by the Lowry method,[18] and the total yields of CM-cellulase and β-glucosidase activity against the culture filtrate were 31 and 34%, respectively. Although the total yield of Avicelase activity was very low, in recombination assays of the nine cellulase components, the recovery of Avicelase activity increased markedly to 33%. Similarly the recovery of hydrocellulase activity increased from 12 to 35% by reconstituting all

[18] O. H. Lowry, N. J. Rowebrough, A. L. Farr, and R. J. Randall, *J. Biol. Chem.* **193,** 265 (1951).

TABLE II
PURIFICATION OF THREE CM-CELLULASES FROM *A. aculeatus* NO. F-50

Step	Total protein (A_{280})	Hydrocellulase activity			CM-Cellulase activity		
		Total (units)	Specific (units/A_{280})	Yield (%)	Total (units)	Specific (units/A_{280})	Yield (%)
Culture filtrate	1,701,000	168,000	0.16	100	249,000	0.23	100
Crude cellulase	118,000	118,000	1.00	70.4	164,000	1.39	65.7
FI CM-Cellulase							
DEAE-Sephadex A-50	15,000	23,600	1.57	14.0	92,800	6.19	37.2
SP-Sephadex C-50	3,580	12,200	3.41	7.26	79,600	22.2	32.0
(NH$_4$)$_2$SO$_4$ fractionation	3,180	11,300	3.55	6.73	76,200	24.0	30.6
Sephadex G-75	2,640	10,200	3.86	6.07	68,900	26.1	27.7
FII CM-Cellulase							
DEAE-Sephadex A-50	7,200	25,200	3.51	15.0	19,600	2.72	7.87
SP-Sephadex C-50	2,040	1,220	0.60	0.73	11,800	5.78	4.74
Sephadex G-75	327	1,100	3.36	0.65	9,970	30.5	4.00
Second DEAE-Sephadex A-50	175	692	3.95	0.41	6,680	38.2	2.68
Second Sephadex G-75	118	493	4.18	0.29	4,710	39.8	1.89
FV CM-Cellulase							
DEAE-Sephadex A-50	7,820	4,140	0.53	2.46	11,600	1.48	4.64
(NH$_4$)$_2$SO$_4$ fractionation	2,940	2,900	0.99	1.73	7,630	2.60	3.06
SP-Sephadex C-50	269	1,450	5.39	0.86	4,080	15.2	1.64
Sephadex G-75	169	1,220	7.20	0.72	3,420	20.2	1.37

TABLE III
PURIFICATION OF HYDROCELLULASE AND AVICELASES FROM *A. aculeatus* NO. F-50

Step	Total protein (A_{280})	Hydrocellulase activity			Synergistic Avicelase activity		
		Total (units)	Specific (units/A_{280})	Yield (%)	Total (units)	Specific (units/A_{280})	Yield (%)
Culture filtrate	1,701,000	168,000	0.16	100	249,000	0.23	100
Crude cellulase	118,000	118,000	1.00	70.4	164,000	1.39	65.7
FI Avicelase							
DEAE-Sephadex A-50	15,000	23,600	1.57	14.0	289	0.019	6.7
SP-Sephadex C-50	2,190	1,370	0.63	0.82	140	0.064	3.3
Sephacryl S-200	1,380	1,140	0.83	0.68	115	0.084	2.7
FIII Avicelase							
DEAE-Sephadex A-50	10,600	5,200	0.49	3.06	581	0.055	13.5
SP-Sephadex C-50	2,070	535	0.26	0.32	323	0.156	7.5
Second DEAE-Sephadex A-50	1,004	214	0.21	0.11	199	0.198	4.6
Sephacryl S-200	803	186	0.23	0.11	151	0.188	3.5
FIV Hydrocellulase							
DEAE-Sephadex A-50	5,060	1,920	0.38	1.14			
SP-Sephadex C-50	2,500	1,760	0.70	1.05			
Sephacryl S-200	980	770	0.78	0.46			
Isoelectric focusing	790	500	0.63	0.30			

TABLE IV

PURIFICATION OF THREE β-GLUCOSIDASES FROM *A. aculeatus* No. F-50

Step	Total protein (A_{280})	β-Glucosidase activity			Hydrocellulase activity		
		Total (units)	Specific (units/A_{280})	Yield (%)	Total (units)	Specific (units/A_{280})	Yield (%)
Culture filtrate	1,701,000	168,000	0.16	100	249,000	0.23	100
Crude cellulase	118,000	118,000	1.00	70.4	164,000	1.39	65.7
β-Glucosidase 1							
DEAE-Sephadex A-50 (FII)	7,200	63,800	8.87	53.2	25,200	3.51	15.0
SP-Sephadex C-50	1,230	48,200	39.2	40.2	8,860	7.20	5.27
Ethanol fractionation	580	44,900	77.4	37.4	8,240	14.2	4.90
Sephacryl S-200	422	40,400	95.7	33.7	7,420	17.6	4.42
Isoelectric focusing	289	30,600	105.7	25.5	5,460	18.9	3.25
β-Glucosidase 2							
DEAE-Sephadex A-50 (FIII)	10,600	16,000	1.59	13.3	5,200	0.49	3.10
SP-Sephadex C-50	938	12,400	13.2	10.3	2,230	2.38	1.33
Ethanol fractionation	138	11,700	86.7	9.75	2,100	15.6	1.25
Sephacryl S-200	103	10,600	102.9	8.83	1,890	18.3	1.13
Isoelectric focusing	74	8,400	114.0	7.00	1,520	20.6	0.90
β-Glucosidase 3							
DEAE-Sephadex A-50 (FIV)	5,060	3,000	0.59	2.5	1,920	0.38	1.14
SP-Sephadex C-50	1,700	2,340	1.37	1.95	1,570	0.92	0.93
Sephacryl S-200	157	1,900	12.1	1.58	7.9	0.050	0.005
Ethanol fractionation	98	1,750	17.9	1.46	4.0	0.041	0.002
Isoelectric focusing	84	1,600	19.0	1.33	3.0	0.036	0.002

TABLE V

Overall Recovery of the Nine Cellulase Components from A. aculeatus No. F-50[a]

| Cellulase component | Protein | | Cellulase activity | | | | | | | |
| | | | CM-Cellulase | | Hydrocellulase | | Avicelase | | β-Glucosidase | |
	mg	%	Units	%	Units	%	Units	%	Units	%
Crude cellulase	59,000	100	164,000	65.7	118,000	70.4	2,940	68.4	100,000	83.7
FI CM-Cellulase	877	1.49	68,900	27.7	10,200	6.07	4.22	0.10	—	0.0
FII CM-Cellulase	71	0.12	4,710	1.89	493	0.29	0.45	0.01	—	0.0
FV CM-Cellulase	61	0.10	3,420	1.37	1,220	0.72	0.41	0.01	—	0.0
FIV Hydrocellulase	658	1.12	12	0.00	498	0.30	2.37	0.05	—	0.0
FI Avicelase	1,179	2.00	97	0.04	1,140	0.68	10.6	0.25	—	0.0
FIII Avicelase	613	1.04	14	0.01	186	0.11	5.78	0.13	—	0.0
β-Glucosidase 1	191	0.32	12	0.00	5,460	3.25	0.00	0.00	30,600	25.5
β-Glucosidase 2	47	0.08	3	0.00	1,520	0.90	0.00	0.00	8,400	7.0
β-Glucosidase 3	64	0.11	4	0.00	3	0.00	0.00	0.00	1,600	1.3
Total	3,761	6.37	77,170	31.0	20,720	12.3	23.86	0.55	40,600	33.8
Recombination			85,160	34.2	58,130	34.6	1,400	32.6	40,900	34.1

[a] The protein of crude cellulase was estimated by the Lowry method, and that of the purified enzymes was assayed by the absorbance at 280 nm. Recombination was composed of each of the purified enzymes corresponding to the amount proportional to yield against the crude cellulase, and its activity was assayed similarly.

enzymes. Therefore, the overall recovery of the four enzyme activities amounted to 33 ~ 35%. It seemed that the main cellulolytic enzymes of this fungus are three CM-cellulases, one hydrocellulase, two Avicelases, and three β-glucosidases.

Properties[6–11,19,20]

Physicochemical Properties. The physicochemical properties of FI, FII, and FV CM-cellulase, FI and FIII Avicelase, FIV hydrocellulase, and β-glucosidase 1, 2, and 3 are summarized in Table VI. The molecular weights of the enzymes were estimated by SDS–polyacrylamide gel disc electrophoresis. Both Avicelases had a higher molecular weight (over 100,000) than cellulases from other fungi.[21–31] In addition, both Avicelases contained more sugar residues than do most cellulases. FII CM-cellulase, which also had a high sugar content, was an endoglucanase.

Enzymatic Properties. To study the enzymatic properties of purified cellulase components, insoluble cellooligosaccharide seems to be the best substrate, because all cellulase components except β-glucosidase 3 are more active toward it and because the reaction time is very short (10 min). This cellulose is, furthermore, insoluble in water, like native cellulose, and has no substituents, unlike CM-cellulose. The enzymatic properties of the nine cellulase components are summarized in Table VI. The optimum pH values of FIV hydrocellulase and FIII Avicelase are as low as that of cellobiohydrolase[21,27] and of acid cellulase.[32] The pH activity profile of FIII Avicelase plateaus at pH 4–5, as observed by Wood.[27] Each cellulase component of *A. aculeatus* was stable at acid pH, and labile at alkaline pH and to heat.

Substrate Specificity. The activities of CM-cellulases, hydrocellulase, and Avicelases were assayed with 0.5% (w/v) solutions and suspensions

[19] J. Kanamoto, R. Sakamoto, M. Arai, and S. Murao, *J. Ferment. Technol.* **57**, 163 (1979).
[20] R. Sakamoto, M. Arai, and S. Murao, *Agric. Biol. Chem.* **49**, 1283 (1985).
[21] T. M. Wood and S. I. McCrae, *Biochem. J.* **128**, 1183 (1972).
[22] L. E. R. Berghem and L. G. Pettersson, *Eur. J. Biochem.* **37**, 21 (1973).
[23] T. M. Wood and S. I. McCrae, *Carbohydr. Res.* **57**, 117 (1977).
[24] Y. Tomita, H. Suzuki, and K. Nishizawa, *J. Ferment. Technol.* **52**, 233 (1974).
[25] S. Shikata and K. Nishizawa, *J. Biochem. (Tokyo)* **78**, 499 (1975).
[26] C. S. Gong, M. R. Ladisch, and G. T. Tsao, *Adv. Chem. Ser.* **181**, 261 (1979).
[27] T. M. Wood, S. I. McCrae, and C. C. Macfarlane, *Biochem. J.* **189**, 51 (1980).
[28] M. Gritzall and R. D. Brown, Jr., *Adv. Chem. Ser.* **181**, 237 (1979).
[29] K. Nishizawa, *J. Ferment. Technol.* **51**, 267 (1973).
[30] T. Kanda, S. Nakakubo, K. Wakabayashi, and K. Nishizawa, *J. Biochem.* **84**, 1217 (1978).
[31] K. E. Eriksson and B. Pettersson, *Eur. J. Biochem.* **51**, 213 (1975).
[32] R. Ikeda, T. Yamamoto, and M. Funatsu, *Agric. Biol. Chem.* **37**, 1153 (1973).

TABLE VI
PHYSICOCHEMICAL AND ENZYMATIC PROPERTIES OF THE NINE CELLULASE COMPONENTS OF *A. aculeatus* No. F-50

	CM-Cellulase			Hydrocellulase	Avicelase		β-Glucosidase		
	FI	FII	FV	FIV	FI	FIII	1	2	3
$E_{1\ cm}^{1\%}$ at 280 nm	30.1	16.7	27.6	12.0	11.7	13.1	15.1	15.2	13.1
Molecular weight	25,000	66,000	38,000	68,000	109,000	112,000	133,000	132,000	136,000
Isoelectric point	4.8	4.0	3.4	3.5	4.7	4.0	4.7	4.3	3.6
Neutral sugar content (%)	—	25.4	5.3	6.7	32.1	21.8	22.7	22.3	15.1
Aminosugar content (%)	—	1.6	1.7	1.6	0.4	1.1	4.3	4.2	3.5
Optimum pH	4.5	5.0	4.0	2.5	5.5	2.5	4.5	4.5	3.0[a]
Optimum temperature (°C)	50	70	65	60	65	65	55	60	65
pH stability	2.0 ~ 9.0	3.0 ~ 8.0	3.5 ~ 9.0	3.5 ~ 6.0	2.7 ~ 7.0	2.5 ~ 5.0	3.5 ~ 7.0	3.0 ~ 7.0	2.5 ~ 6.0
Thermal stability (°C)	45	30	50	50	50	50	50	50	60

[a] For salicin.

TABLE VII

SUBSTRATE SPECIFICITY OF CELLULASE COMPONENTS OF *A. aculeatus* NO. F-50[a]

Substrate	CM-Cellulase			Hydrocellulase	Avicelase	
	FI	FII	FV	FIV	FI	FIII
CM-cellulose (DS = 0.65, $\overline{DP}$ = 500)	78.6	66.5	55.8	0.018	0.082	0.022
Avicel	0.005	0.006	0.007	0.004	0.009	0.009
Alkali-swollen cellulose ($\overline{DP}$ = 150)	6.44	5.51	7.07	0.29	0.58	0.17
Phosphate-swollen cellulose ($\overline{DP}$ = 100)	7.98	6.18	9.00	0.41	0.94	0.24
Insoluble cellooligosaccharide ($\overline{DP}$ = 20)	11.6	6.98	19.9	0.76	0.97	0.26
Cellohexaose (DP = 6)	88.5	44.7	75.9	9.1	7.91	3.72
Cellopentaose (DP = 5)	66.2	10.7	25.8	3.0	5.10	1.00
Cellotetraose (DP = 4)	6.08	0.64	9.56	1.9	35.1	1.05
Cellotriose (DP = 3)	0.081	0.007	0.047	0.082	0.033	0.027
Cellobiose (DP = 2)	<0.001	<0.001	<0.001	<0.001	<0.001	<0.001
Salicin	<0.001	<0.001	<0.001	<0.001	<0.001	<0.001

[a] Data given in units/mg.

TABLE VIII
SUBSTRATE SPECIFICITY OF β-GLUCOSIDASES FROM *A. aculeatus* No. F-50

Substrate	β-Glucosidase (units/mg)		
	1	2	3
CM-cellulose (DS = 0.65, $\overline{\mathrm{DP}}$ = 500)	0.060	0.068	0.069
Avicel	<0.001	<0.001	<0.001
Alkali-swollen cellulose ($\overline{\mathrm{DP}}$ = 150)	0.004	0.005	<0.001
Phosphate-swollen cellulose ($\overline{\mathrm{DP}}$ = 100)	0.007	0.008	0.002
Insoluble cellooligosaccharide ($\overline{\mathrm{DP}}$ = 20)	28.5	32.8	0.047
Cellohexaose (DP = 6)	174	197	0.217
Cellopentaose (DP = 5)	182	203	0.594
Cellotetraose (DP = 4)	181	201	1.11
Cellotriose (DP = 3)	178	192	1.54
Cellobiose (DP = 2)	132	157	2.29
Salicin	160	181	24.9
p-Nitrophenyl-β-D-glucoside	28.9	33.5	34.2
β-Methyl-D-glucoside	44.5	49.2	7.02

of various β-1,4-glucans; results are summarized in Table VII. That of β-glucosidases are similarly summarized in Table VIII.[20] The three CM-cellulases are more active toward amorphous cellulose and soluble cellooligosaccharides, but less active toward Avicel. On the other hand, FIV hydrocellulase and both Avicelases are less active not only on Avicel but on CM-cellulose, and among the amorphous celluloses, insoluble cellooligosaccharide is the substrate for which these cellulases are most active. FI Avicelase has a very high activity toward cellotetraose. It is not recognized that any cellulases of *A. aculeatus* except for the three β-glucosidases have activity toward cellobiose and salicin.

Hydrolysis of Cellulose. Avicel and insoluble cellooligosaccharide are the preferred substrates to study the action of purified cellulase components on cellulose.

Each cellulase component is incubated with 4 mg of Avicel, 0.08 mg of NaN_3, 4 mg of egg albumin, and 40 μmol of acetate buffer, pH 5.0, in a total volume of 0.4 ml at 37° for 6 days. The soluble sugars released in the reaction mixture are separated using a Lichrosorb Si 60 (size 5 μm) column (4.0 × 250 mm, Merck, Darmstadt, G.F.R.) with acetonitrile : water (3 : 2, v/v) containing 0.01% HPLC Amine Modifier[33] (NATEC, Hamburg, G.F.R.) at 1.1 ml/min of flow rate. The area of sugar peaks detected by their refractive index was computed on a Chromatogram Processor

[33] K. Aitzelmüller, *J. Chromatgr.* **156**, 354 (1978).

TABLE IX

DEGRADATION OF AVICEL AND ITS SUGAR PRODUCTS BY CELLULASE COMPONENTS OF *A. aculeatus* No. F-50

Cellulase component (amount added)	Reaction time	Hydrolysis (%)	Degradation (%)	Sugars yielded						$\overline{DP}$ of sugars yielded
				Weight %			Molar %			
				G_1	G_2	G_3	G_1	G_2	G_3	
FI CM-Cellulase	2 hr	1.4	3.4	6	13	80	16	17	67	2.4
(150 µg)	6 day	3.7	5.8	30	64	6	47	50	3	1.6
FII CM-Cellulase	2 hr	2.1	4.7	4	50	46	9	56	35	2.3
(150 µg)	6 day	6.3	12	9	90	1	16	83	1	1.8
FV CM-Cellulase	2 hr	0.9	2.1	8	31	61	18	36	46	2.3
(150 µg)	6 day	2.6	3.7	45	55	—	62	38	—	1.4
FIV Hydrocellulase	2 hr	1.1	2.3	5	69	26	11	71	18	2.1
(300 µg)	6 day	3.7	6.4	14	86	—	24	76	—	1.8
FI Avicelase	2 hr	2.2	4.3	5	85	10	10	83	7	2.0
(200 µg)	6 day	11	20	4	96	—	8	92	—	1.9
FIII Avicelase	2 hr	1.5	2.7	7	93	—	13	87	—	1.9
(200 µg)	6 day	11	19	16	84	—	28	72	—	1.7

TABLE X

DEGRADATION OF INSOLUBLE CELLOOLIGOSACCHARIDE AND ITS SUGAR PRODUCTS BY CELLULASE COMPONENTS OF A. *aculeatus* No. F-50

Cellulase component (amount added)	Reaction time	Hydrolysis (%)	Degradation (%)	Sugars yielded										$\overline{\mathrm{DP}}$ of sugars yielded
				Weight %					Molar %					
				G_1	G_2	G_3	G_4	G_5	G_1	G_2	G_3	G_4	G_5	
FI CM-Cellulase (0.20 mg)	10 min	5	12	8	18	50	24	—	20	23	42	15	—	2.5
	18 hr	43	90	15	28	57	—	21	31	30	39	—	12	2.1
FII CM-Cellulase (0.17 mg)	10 min	7	19	1	32	21	25	21	2	47	20	18	12	2.9
	18 hr	45	95	2	81	17	—	—	5	83	11	—	—	2.0
FV CM-Cellulase (0.05 mg)	10 min	13	31	5	37	37	21	—	12	45	30	13	—	2.5
	18 hr	48	95	9	62	28	1	—	19	62	19	0	—	2.0
FIV Hydrocellulase (1.6 mg)	10 min	11	21	11	68	21	—	—	21	66	13	—	—	1.9
	18 hr	51	96	11	78	12	—	—	20	73	7	—	—	1.9
FI Avicelase (1.0 mg)	10 min	6	14	—	69	31	—	—	—	77	23	—	—	2.2
	18 hr	40	75	8	90	2	—	—	14	85	1	—	—	1.9
FIII Avicelase (3.5 mg)	10 min	9	17	11	72	17	—	—	22	67	11	—	—	1.9
	18 hr	54	96	15	81	4	—	—	26	72	2	—	—	1.8

Model 7000AS (System Instruments Co., Ltd., Tokyo). The results are in Table IX. The hydrolysis and degradation rates are expressed as the ratio of reducing sugars formed/Avicel and that of total sugars formed/Avicel, respectively. The degree of polymerization was estimated on the basis of the ratio of total sugars formed/reducing sugars formed. Soluble cellooligosaccharides larger than cellotriose were not detected in any reaction mixture. At the initial stage of the reaction, the cellulases, except for FIII Avicelase, yielded cellotriose, but this is nearly undetectable in the reaction mixtures at the final stage. The molar ratio of cellobiose of FI Avicelase at the final stage is over 90%, but that of FIV hydrocellulase and FIII Avicelase is about 70%. Both Avicelases seem to act on the crystalline region of cellulose, because the enzymes degrade Avicel to the extent of 31 ∼ 47% as the amount of enzyme against the substrate is increased. The β-glucosidases were not able to hydrolyze Avicel.

Next, 20 mg of insoluble cellooligosaccharide was incubated with the enzyme, 0.4 mg of NaN_3, and 200 μmol of acetate buffer, pH 5.0, in a total volume of 2.0 ml at 37° for 18 hr. The soluble sugars released in the reaction mixture were separated with using a Unicil C_{18} (size 10 μm) column (4.6 × 300 mm, Gasukuro Kogyo Inc., Tokyo) with water at the flow rate of 0.5 ml/min.[6] Sugars were assayed as described above, and results are summarized in Table X. All cellulases degraded the cellulose more than 70% under these conditions. FII CM-Cellulase yielded cellopentaose at the initial stage. Every CM-cellulase produced cellotetraose, but this is not detected in any reaction mixture of FIV hydrocellulase or either Avicelase. FI Avicelase released cellotriose but no glucose at the initial stage, and at the final stage made the highest molar ratio of cellobiose among these cellulases. The molar ratio of cellobiose on FIV hydrocellulase and FIII Avicelase at the final stage was about 70%, close to that of Avicel. β-Glucosidase 1 and 2 yielded glucose only from insoluble cellooligosaccharide, and degraded it to the extent of 46% when 1 mg of egg albumin was added to the reaction mixture.[20] Both of those enzymes also hydrolyzed cellobiose to glucose completely. β-Glucosidase 3 was, however, less active on insoluble cellooligosaccharide and decomposed cellobiose only to the extent of 62%.

[31] Cellulases from *Thermoascus aurantiacus*

By MAXWELL G. SHEPHERD, ANTHONY L. COLE, and CHOW CHIN TONG

The cellulases from thermophiles with the ability to operate at temperatures of 55° and higher offer the advantages of an increased rate of reaction and a stable enzyme system, compared with the cellulases from mesophiles. Furthermore, the high operating temperature and acid pH required by the cellulases from thermophilic fungi restrict the growth of contaminating organisms and for many industrial processes obviate the removal of heat from the system.

Methods

Preparation of Culture Filtrate. Thermoascus aurantiacus can be isolated from compost heaps or obtained from the American Type Culture Collection. Hyphal isolates of *T. aurantiacus* were subcultured on agar medium containing (per liter) 5 g yeast extract, 10 g glucose, and grown at 50°. An agar/mycelium disk (0.8 cm diameter) was cut from a yeast extract/glucose agar plate and transferred to a Wheaton medical flask (C-16, 500 ml) containing 60 ml of Fergus medium[1] with filter paper (Whatman No. 1) as a carbon source. This contained (per liter): filter paper, 30 g; K_2HPO_4, 1.0 g; $MgSO_4 \cdot 7H_2O$, 0.3 g; bactopeptone, 1.0 g; yeast extract, 0.1 g; $Fe(NH_4)_2(SO_4)_2 \cdot 6H_2O$, 1 mg; $ZnSO_4$, 1 mg; $MnSO_4 \cdot 4H_2O$, 0.5 mg; $CaSO_4 \cdot 5H_2O$, 0.08 mg; $CoCl_2$, 0.07 mg; H_3BO_3, 0.1 mg; final pH 6.5. After 6 days incubation at 50° the culture was filtered through glass-fiber paper (Whatman GF/C) to remove hyphal fragments and residual insoluble cellulose. The culture filtrate (~1.5 liters) was freeze-dried and then resuspended in 100 ml of 0.1 *M* citrate/phosphate buffer, pH 5.0. The mixture was stirred overnight and insoluble material was removed by centrifugation (30 min at 10,000 *g*). The cellulases were then purified as described below.

Determination of Protein. Protein was determined by a modification[2] of the method of Lowry *et al.*[3] with sodium citrate being used instead of

---

[1] C. L. Fergus, *Mycologia* **61**, 120 (1969).
[2] M. Eggstein and F. H. Kreutz, *in* "Techniques in Protein Chemistry" (J. L. Bailey, ed.), p. 340. Elsevier, London, 1967.
[3] O. H. Lowry, N. J. Rosebrough, A. L. Farr, and R. J. Randall, *J. Biol. Chem.* **193**, 265 (1951).

sodium tartrate. Crystalline bovine serum albumin was used as a standard. The A_{280} was used for monitoring protein in column effluents.

Determination of Reducing Sugars. The number of reducing sugar groups created by hydrolysis of the cellulosic substrates was measured spectrophotometrically by using the Somogyi–Nelson procedure.[4]

Assay of Filter Paper-Degrading Activity. An indication of total cellulolytic activity was obtained by the determination of filter paper-degrading activity. The standard reaction mixture, containing 20 mg of filter paper (Whatman No. 1), 0.9 ml of citrate/phosphate buffer, pH 5.0, 0.1 ml of enzyme solution of appropriate dilution, and one drop (10 μl) of toluene, was incubated at 60° for 24 hr. The mixture was then analyzed for the production of reducing sugar. The toluene added to prevent bacterial growth was found to have no effect on enzyme activity. Reaction mixtures were checked for contamination by withdrawing samples and streaking them onto nutrient agar plates which were incubated at 37 and 50°.

Assay of CM-Cellulase Activity. Cellulase activity toward CM-cellulose (carboxymethylcellulose 7HF, degree of substitution 0.75, Hercules, Wilmington, DE) was also measured by the appearance of reducing end groups in a solution of CM-cellulose. The assay conditions were 0.9 ml of 0.75% (w/v) CM-cellulose in citrate/phosphate buffer, pH 4.5, and 0.1 ml of enzyme solution incubated for 30 min at 70°, and the rate of production of reducing sugars was determined.

An absolute definition of a unit of cellulase activity is difficult. This is because, in a substrate such as CM-cellulose, the glucose molecules are substituted with carboxymethyl groups, and the products of the enzyme reaction on filter paper and CM-cellulose are heterogeneous polymers; the effect of this on the absorption coefficient of reducing end groups is not known. It is not, therefore, valid to use a glucose standard to determine the quantity (micromoles) of reducing end groups. In addition, there is little to be gained by expressing the activity in terms of glucose equivalents, since glucose is not the only product of the enzyme reaction. A unit of filter paper-degrading activity or CM-cellulase activity is defined as that amount of enzyme that produces an increase in absorbance of 0.10 at 560 nm under the conditions defined. A change of 0.1 A is equivalent to 96 μg of glucose under the conditions given, and thus cellulase preparations with units quoted in glucose equivalents can be compared.

Assay of β-Glucosidase. β-Glucosidase activity was assayed using *p*-nitrophenyl-β-D-glucoside as substrate. Enzyme solution (0.1 ml) and 0.4 ml of 1 mM *p*-nitrophenyl-glucoside in citrate/phosphate buffer, pH

---

[4] M. Somogyi, *J. Biol. Chem.* **195**, 19 (1952).

5.0, were incubated for 30 min at 70°. After incubation, 1.0 ml of 1 M sodium carbonate solution was added to 0.5 ml of the assay mixture, diluted with 10 ml of distilled water, and the p-nitrophenol released was measured from the A_{420}. One unit of β-glucosidase activity is defined as that amount of enzyme needed to liberate 1 μmol of p-nitrophenol per minute under the conditions of the assay.

Determination of Carbohydrate. Total carbohydrate was measured by the anthrone/H_2SO_4[5] method with glucose as standard.

Column Chromatography. Unless otherwise stated, column chromatography was carried out at room temperature. Each gel was swollen in the appropriate buffer and active fractions corresponding to the enzyme activity were combined and concentrated by freeze-drying before the next purification step.

For the determination of molecular weights, a column of BioGel P-60 (2.2 × 47 cm) was calibrated with marker proteins in 0.05 M Tris–HCl, pH 7.5, containing 0.1 M KCl. Samples (2–6 mg) were applied in a volume of 1 ml and a constant flow rate of 15 ml/hr was used for elution. Fractions of 2.5 ml were collected for analysis. Marker proteins were located by their A_{280}.

Polyacrylamide Gel Electrophoresis. Disc gel electrophoresis was performed at 4° by the method of Ornstein and Davis[6] in 7.5% (w/v) polyacrylamide gel with bromphenol blue as tracking dye. The gels were prepared in glass tubes (0.5 × 9.0 cm) with citrate/phosphate buffer, pH 5.5. The citrate/phosphate buffer in both the upper and the lower tanks was five times more concentrated than that used for gel preparation. The gels were run at 3.0 mA/tube for 30 min and then at 4.5 mA/tube for 8 hr. The staining technique of Reinsner *et al.*[7] was used. For preparative purposes, several disc gels (0.7 × 12.0 cm) were run at 4.0 mA/tube for 1 hr and then at 7.0 mA/tube for a further 23 hr. SDS–polyacrylamide gels were run and stained by the method of Weber *et al.*[8]

Kinetic Analysis. The enzyme kinetic parameters K_m and V_{max} were derived from data analyzed by the direct linear plot[9] using a program developed for an Apple computer. Velocities were expressed as units per μg of protein and K_m values as mg of substrate per ml of μmol of p-nitrophenol in the case of β-glucosidase.

Buffers. The citrate/phosphate buffers from pH 3.0 to 8.0 were prepared from 0.1 M citric acid and 0.2 M dibasic sodium phosphate. The

[5] D. Herbert, P. J. Phipps, and R. E. Strange, *Methods Microbiol.* **5B**, 265 (1971).
[6] L. Ornstein and B. J. Davis, *Ann. N.Y. Acad. Sci.* **121**,321 (1964).
[7] H. H. Reinsner, P. Nemes, and C. Bucholtz, *Anal. Biochem.* **64**, 509 (1975).
[8] J. R. Weber, J. R. Pringle, and M. Osborn, this series, Vol. 26, p. 3.
[9] R. Eisenthal and A. Cornish-Bowden, *Biochem. J.* **139**, 715 (1974).

TABLE I
$(NH_4)_2SO_4$ FRACTIONATION OF *T. aurantiacus* CELLULASES[a]

$(NH_4)_2SO_4$		Total protein (mg)	Total activity		
Saturation (%)	Addition (g/100 ml)		Filter paper-degrading activity (units $\times$ 10^{-3})	CM cellulase activity (units $\times$ 10^{-3})	β-Glucosidase (units)
10–20	5.3	36	0	2.75	66.6
20–30	5.5	51	5.50	5.75	266.6
30–40	5.6	86	15.75	200.00	300.0
40–50	5.8	120	37.80	136.00	31.6
50–60	6.0	76	10.00	2.75	8.2

[a] For details see the text.

buffer used for column chromatography was 0.05 M ammonium formate, pH 5.0; 0.2 M Tris–HCl buffer was used for pH optimum studies.

Purification Procedure

The method given here is essentially that described by Tong *et al.*[10] Apart from the column chromatography, which was carried out at room temperature, all operations were conducted at 0–4°. Concentration of the enzyme fractions was achieved by freeze-drying; in this procedure the salt concentration did not increase because of the volatile nature of the buffer.

Step 1. Solid $(NH4)_2SO_4$ was slowly added to 100 ml of stirred crude culture filtrate to give 10% increments in $(NH_4)_2SO_4$ saturation as indicated in Table I. Stirring was continued for a further 30 min after the addition of the $(NH_4)_2SO_4$. The solution was centrifuged (10,000 g for 30 min) and the precipitate resuspended in 20 ml of formate buffer. Insoluble material was removed and discarded by centrifugation (10,000 g for 30 min). Fractions containing the various $(NH_4)_2SO_4$ concentrations were treated separately by using the following purification steps.

Step 2. Fractions from the ammonium sulfate precipitation were desalted on a column (2.7 $\times$ 30 cm) of BioGel P-2 matrix (100–200 mesh) equilibrated with the ammonium formate buffer (flow rate ~100 ml/hr) with a void volume of 50 ml. Fractions (5 ml) were collected. The eluate (~60 ml) was freeze-dried and the residue resuspended in 5 ml of the

[10] C. C. Tong, A. L. Cole, and M. G. Shepherd, *Biochem. J.* **191,** 83 (1980).

formate buffer. Any undissolved material was removed and discarded by centrifugation (10,000 g for 30 min).

Step 3. The desalted samples were layered onto a column (2.75 × 91.8 cm) of Sephadex G-100 equilibrated with ammonium formate buffer. At a flow rate of ~45 ml/hr, fractions (3.5 ml) were collected, and those fractions with more than 30% of the activity of the peak fraction were pooled and freeze-dried. The residue was suspended in formate buffer and centrifuged (10,000 g for 30 min) to remove undissolved material. Recycling through the same column of Sephadex G-100 was necessary in order to separate completely the β-glucosidase from the cellulases. The purified β-glucosidase was desalted on a BioGel P-2 column.

Step 4. The final purification of the cellulases was by disc gel electrophoresis. A 0.1 ml sample of the partially purified cellulase enzyme solution was mixed with 40 μl of 80% (v/v) glycerol and 10 μl of bromphenol blue (0.2%, w/v) before it was layered on top of each gel tube. After electrophoresis, unstained gels were scanned at 280 nm on a Joyce Loebl UV scanner and then sliced. Protein was eluted in formate buffer, pH 5.0, by grinding the gels in uniform poly(tetrafluoroethylene) (PTFE) pestle/glass body homogenizers with repeated washings. The gelatinous material was removed by centrifuging at 10,000 g for 30 min. The supernatant obtained was filtered through a Millipore filter (0.45 μm pore size) and the filtrate concentrated by freeze-drying. The dried material was dissolved in 3.0 ml of formate buffer and desalted on a BioGel P-2 column. The active fractions were pooled and the resulting purified cellulases were used for subsequent studies.

An important feature of the purification scheme was the $(NH4)_2SO_4$ fractionation. At 10–60% $(NH_4)_2SO_4$ saturation, 95% of the enzyme activity was recovered, whereas 61% of the protein was removed. Table I shows that 96% of the CM cellulase activity was associated with the 30–50% saturated $(NH_4)_4SO_4$ pellet; 80% of the β-glucosidase was with the 20–40% pellet, whereas the filter paper-degrading enzyme was spread between 30 and 60% $(NH_4)_2SO_4$ saturation. Pure β-glucosidase was obtained from chromatography of the 10–40% pellet on Sephadex G-100. A typical purification scheme for cellulase III is given in Table II.

Properties

Purity. The purified cellulases and the β-glucosidase all migrated as single bands on a polyacrylamide gel in citrate/phosphate buffer, pH 5.5. The cellulases are referred to as cellulase I, cellulase II, and cellulase III according to the positions they occupied in the gel in descending sequence.

TABLE II
PURIFICATION OF A CELLULASE (CELLULASE III) FROM *T. aurantiacus*

Fraction	Volume (ml)	Total activity (units × 10⁻³)	Total protein (mg)	Specific activity (units × 10⁻³/mg)	Purification (-fold)
Crude extract	1510.0	195.0	1087.2	0.17	—
30–40% saturated (NH₄)₂SO₄ precipitate	20.0	160.0	86.0	1.86	10.94
BioGel P-2 eluate	57.3	148.5	67.0	2.21	13.0
Sephadex G-100 eluate second chromatography	130.0	104.0	41.6	2.50	14.7
Disc gel electrophoresis eluate	20.0	150.0	23.6	11.02	64.8

Stability. The purified β-glucosidase and cellulases were stable for at least a year when stored as frozen solutions at $-20°$. In addition, dilute solutions (5 μg/ml) could be kept at 4° for several days without significant loss of activities.

Molecular Weights. The molecular weights of the enzymes were determined from a calibrated BioGel P-60 column and were 85,000 for β-glucosidase, 78,000 for cellulase I, 48,000 for cellulase II, and 34,000 for cellulase III. Similar molecular weights were found with SDS–polyacrylamide gel electrophoresis except that cellulase I dissociated into multiple bands. These data indicate that the β-glucosidase and cellulases II and III exist as monomers.

Carbohydrate Content. The carbohydrate content for each of the enzymes was (% w/w): 33, β-glucosidase; 5.5, cellulase I; 2.6, cellulase II; and 1.8, cellulase III.

Kinetics. The K_m and V_{max} for the β-glucosidase and the three cellulases were determined from saturation curves by the direct linear plot[8] and these data are summarized in Table III. The V_{max} for cellulases I, II, and III acting on filter paper were 0.12, 0.34, and 0.18, respectively.

Temperature and pH Optima. The effects of both temperature and pH on enzyme activity were studied with the substrate giving the highest activity for each enzyme: β-glucosidase on *p*-nitrophenyl-β-D-glucoside, cellulase I on CM cellulose and yeast glucan, cellulase II on filter paper, and cellulase III on CM cellulose. Optimum temperatures for the β-glucosidase and cellulolytic activities fell within the range 60–70°. The highest optimum temperature was observed with cellulase I acting on CM cellulose (75°), whereas hydrolysis of yeast glucan by the same enzyme was most efficient at 65°. It should be noted that the incubation time for the enzyme assay was 24 hr with yeast glucan compared with 0.5 hr on CM-

TABLE III
K_m AND V_{max} VALUES FOR β-GLUCOSIDASE AND
CELLULASES I, II, AND III FROM *T. aurantiacus*

Component	$K_m{}^a$	$V_{max}{}^b$
β-Glucosidase (on p-nitrophenyl-D-glucoside)	0.52	6.5×10^4
Cellulase I (on CM-cellulose)	3.9	6.3
Cellulase I (on yeast glucan)	1.2	1.1
Cellulase II (on filter paper)	34.4	0.34
Cellulase III (on CM-cellulose)	1.9	33

[a] The K_m values for the cellulases are expressed as mg of substrate/ml; for β-glucosidase the units of K_m are micromoles of p-nitrophenyl/ml.

[b] The V_{max} for the cellulases have been determined after converting the ΔA_{560} to μg of reducing sugar and the units are moles of glucose equivalents produced/second per mole of cellulase. For the β-glucosidase the units are mole of p-nitrophenol/second per mole of β-glucosidase.

cellulose. The optimum temperature for cellulase II acting on filter paper decreased from 68 to 60° when the incubation period of the assay was increased from 2 to 24 hr. Activity of β-glucosidase on p-nitrophenyl-β-D-glucoside was optimal at 70°. Cellulase III showed a temperature optimum at 65° with CM-cellulose. The enzymes all showed a sharp optimum on the pH activity profile between pH 4.5 and 5.0.

Temperature and pH Stability. The three cellulases were completely stable at temperatures up to 65°, but above this temperature they were very quickly denatured. At 75°, less than 10% of any of the cellulolytic activities remained after a 1-hr incubation in 0.1 M citrate/phosphate buffer, pH 4.5. β-Glucosidase was more thermostable than the cellulases. The enzyme was totally stable at 70° and still retained 70% of its activity after a 1-hr incubation at 75°. The pH stability of the cellulase enzymes differed: cellulase II exhibited a remarkably broad range of stability from pH 2 to 12, and cellulases I and III a narrower range of pH 5–9. β-Glucosidase had a pH stability range of pH 6–8.

Substrate Specificity. All three cellulases partially degraded native cellulose. Cellulase I, but not cellulases II and III, readily hydrolyzed the mixed β-1,3/β-1,6-polysaccharides such as carboxymethylpachyman, yeast glucan, and laminarin. Both cellulase I and the β-glucosidase degraded xylan, and it is proposed that the xylanase activity is an inherent feature of these two enzymes. Lichenan (β-1,4; β-1,3) was degraded by all three cellulases. Cellulase II could not degrade CM cellulose, and with filter paper as substrate the end product was cellobiose, which indicates that cellulase II is an exo-β-1,4-glucan cellobiosylhydrolase. Degradation of cellulose (filter paper) can be catalyzed independently by each of the three cellulases; there was no synergistic effect between any of the cellulases, and cellobiose was the principal product of degradation.[11]

Mode of Action. The mode of action of one cellulase (cellulase III) was examined by using reduced cellulodextrins.[11] The central linkages of the cellulodextrins were the preferred points of cleavage, which, with the rapid decrease in viscosity of CM-cellulose, confirmed that cellulase III was an endocellulase. The rate of hydrolysis increased with chain length of the reduced cellulodextrins, and these kinetic data indicated that the specificity region of cellulase III was five or six glucose units in length.

[11] M. G. Shepherd, C. C. Tong, and A. L. Cole, *Biochem. J.* **193**, 67 (1981).

[32] 1,4-β-D-Glucan Cellobiohydrolase from Sclerotium rolfsii

By JAI C. SADANA and RAJKUMAR V. PATIL

The fungus *Sclerotium rolfsii*, when grown on cellulose as the sole carbon source, produces an oxidative enzyme, cellobiose dehydrogenase,[1] and three different types of hydrolytic enzymes,[2-5] (1) a 1,4-β-D-glucan cellobiohydrolase (EC 3.2.1.91, cellulose 1,4-β-cellobiosidase)[2] (cellobiohydrolase), (2) three endo-β-glucanases,[3] and (3) four β-D-glucosidases.[4,5] All these enzymes have been obtained in a homogeneous state. The hydrolytic enzymes form the cellulase complex of *S. rolfsii*. All of the

[1] J. C. Sadana and R. V. Patil, *J. Gen. Microbiol.* **131**, 1917 (1985).
[2] R. V. Patil and J. C. Sadana, *Can. J. Biochem. Cell Biol.* **62**, 920 (1984).
[3] J. C. Sadana, A. H. Lachke, and R. V. Patil, *Carbohydr. Res.* **133**, 297 (1984).
[4] J. G. Shewale and J. C. Sadana, *Arch. Biochem. Biophys.* **207**, 185 (1981).
[5] J. C. Sadana, J. G. Shewale, and R. V. Patil, *Carbohydr. Res.* **118**, 205 (1983).

hydrolytic enzymes are required for the efficient hydrolysis of crystalline cellulose. The cellobiose dehydrogenase plays only a minor role in the overall degradation of crystalline cellulose to glucose when *S. rolfsii* is grown on cellulose.[1]

1,4-β-D-Glucan cellobiohydrolase shows high activity toward phosphoric acid-swollen cellulose but no viscosity-lowering activity toward carboxymethylcellulose. It functions in the hydrolysis of crystalline cellulose.[2] The enzyme acts in an endwise fashion by removing successive cellobiosyl units (93–96%) from the nonreducing end of the β-1,4-glucan chains; glucose (4–7%) is also detected. In addition, it shows an initial endo-type mode of action on cotton sliver (dewaxed deltapine cotton) producing short fibers.[6] A similar type of enzyme which shows an endo-type mode of action on crystalline cellulose, besides its endwise action of removing cellobiosyl units from the nonreducing chain ends of cellulose, has been reported in *Trichoderma reesei*.[7]

Assay Method

Principle. The enzyme is assayed by determining the soluble sugars released in the supernatant fluid from phosphoric acid-swollen cellulose by the Somogyi–Nelson[8] or *p*-hydroxybenzoic acid hydrazide[9] method or by determining *p*-nitrophenol from *p*-nitrophenyl-β-D-cellobioside. The range of the Somogyi–Nelson method is 5–100 μg, and that of *p*-hydroxybenzoic acid hydrazide method 1–50 μg of reducing sugars as glucose.

The method of estimation of cellobiohydrolase activity, using phosphoric acid-swollen cellulose as substrate, has certain limitations and is applicable to preparations which are free of endo-β-glucanase and β-D-glucosidase activities. The *S. rolfsii* cellobiohydrolase is much more active against phosphoric acid-swollen cellulose than is β-glucosidase, but is much less active than endo-β-glucanase. The interfering effect of β-glucosidase is overcome by the addition of 2 mM D-glucono-1,5-δ-lactone. The *Sclerotium* cellobiohydrolase is not inhibited by 10 mM D-glucono-1,5-δ-lactone. In the presence of endo-β-glucanase, the assay will give erroneous values of cellobiohydrolase activity.

A general assay technique for the determination of cellobiohydrolase activity does not exist as no specific substrate is available. However, in cases where the sample solution does not contain either β-glucosidase or

[6] J. C. Sadana and R. V. Patil, *Can. J. Biochem. Cell Biol.* **63,** 1250 (1985).
[7] H. Chanzy, B. Henrissat, and R. Vuong, *FEBS Lett.* **172,** 193 (1984).
[8] M. Somogyi, *J. Biol. Chem.* **195,** 19 (1952).
[9] P. L. Hurst, J. Nielsen, P. A. Sullivan, and M. G. Shepherd, *Biochem. J.* **165,** 33 (1977).

endo-β-glucanase, or the endo-β-glucanase does not hydrolyze cellobiose, p-nitrophenyl β-D-cellobioside can be used as a specific substrate for the determination of cellobiohydrolase activity. The cellobiohydrolase from *S. rolfsii* hydrolyzes p-nitrophenyl β-D-cellobioside to p-nitrophenol and cellobiose but it has no action on cellobiose or p-nitrophenyl β-D-glucopyranoside.[2] The three endo-β-glucanases, A, B, and C from *S. rolfsii*, which have been obtained in a homogeneous state, do not hydrolyze cellobiose, p-nitrophenyl β-D-glucopyranoside or p-nitrophenyl β-D-cellobioside.[3]

Reagents

Phosphoric acid-swollen cellulose suspension in 50 mM citrate buffer, pH 4.5, 20 mg/ml

Citric acid–trisodium citrate buffer, 50 mM, pH 4.5

Sodium azide, 10 mM in 50 mM citrate buffer, pH 4.5

D-Glucono-1,5-δ-lactone, 20 and also 60 mM in 50 mM citrate buffer, pH 4.5

p-Nitrophenyl β-D-cellobioside, 2 mM in 50 mM citrate buffer, pH 4.5

Sodium carbonate, 2%

Preparation of Phosphoric Acid-Swollen Cellulose. Avicel P.H. 101 (Honeywill and Stein Ltd., U.K.) (10 g) is suspended in o-phosphoric acid (88%, w/v) and left for 1 hr with occasional stirring at 4°. The mixture is poured into ice-cold water (4 liters) and left for 3 min. The swollen cellulose is washed several times with cold water by decantation. After washing with 1% (w/v) NaHCO$_3$ solution, the thick suspension of swollen Avicel is dialyzed against water in a cold room. After 1 min treatment in a Waring blender, 50 mM citrate buffer, pH 4.8, is added to the suspension so that 1 ml of suspension contains 20 mg of cellulose. Cellulose content is estimated by the anthrone–sulfuric acid method.[10] Phosphoric acid-swollen cellulose prepared in this way is in the form of an even suspension. Phosphoric acid-swollen cellulose is stored at 4–5°.

Procedure

With Phosphoric Acid-Swollen Cellulose as Substrate. A 1-ml suspension of phosphoric acid-swollen cellulose is mixed with 0.3 ml of 10 mM sodium azide, 0.1 ml of D-glucono-1,5-δ-lactone (if the sample solution contains β-glucosidase), and 1.5 ml of 50 mM citrate buffer, pH 4.5, in a Pyrex test tube. The enzyme solution (0.1 ml) containing 10 to 15 μg

[10] D. M. Updegraff, *Anal. Biochem.* **32**, 420 (1969).

enzyme protein (0.6–1 mU) is added to the reaction mixture to give a final volume of 3 ml, mixed thoroughly, and incubated at 50° for 4 hr. A control is run with boiled enzyme. The reaction is stopped by immersing the tubes in a boiling water bath for 5 min and is centrifuged after cooling. The released soluble sugars are determined in a 1-ml aliquot of the supernatant fluid by the Somogyi–Nelson[8] or hydroxybenzoic acid hydrazide[9] method.

The crude filtrate from *S. rolfsii* shows high endo-β-glucanase and β-glucosidase activities. The interference due to β-glucosidase is overcome by adding D-glucono-1,5-δ-lactone at a final concentration of 3 mM which inhibits β-glucosidase. The interference due to endo-β-glucanase cannot be overcome in the above assay procedure.

With p-Nitrophenyl-β-D-Cellobioside as Substrate. In this assay the activity is measured as the amount of *p*-nitrophenol released from *p*-nitrophenyl-β-D-cellobioside. One mole of *p*-nitrophenol is released from 1 mol of *p*-nitrophenyl-β-D-cellobioside. A suitably diluted enzyme (0.1 ml) containing 50–100 μg enzyme protein is added to a small Pyrex test tube containing 0.8 ml 2 mM *p*-nitrophenyl-β-D-cellobioside in 50 mM citrate buffer, pH 4.5, and 0.1 ml of 20 mM D-glucono-1,5-δ-lactone which has been previously preincubated at 37°. After incubation at 37° for 30 min, the reaction is stopped by adding 1 ml 2% sodium carbonate. The absorbancy of *p*-nitrophenol released is determined spectrophotometrically at 410 mm in a 1-cm light path cuvette after 10–15 min. The amount of *p*-nitrophenol released is calculated from the absorbancy index of 18.5 cm^2/μmol for *p*-nitrophenol at 410 nm.[11] Linearity is obtained when a value of 0.02–0.05 μmol of *p*-nitrophenol is released per minute.

Definition of Unit of Activity. One unit of cellobiohydrolase activity is defined as the amount of enzyme that releases 180 μg reducing sugars (1 μmol as D-glucose equivalent and corresponds to the production of one reducing end group) per minute at pH 4.5 and 50°, or 1 μmol of *p*-nitrophenol from *p*-nitrophenyl-β-D-cellobioside per minute at 37°, pH 4.5.

Purification

Preparation of Crude Filtrate. Sclerotium rolfsii CPC 142,[12] used as the enzyme source, is grown as given in chapter [10] of this volume for cellobiose dehydrogenase.

The purification of cellobiohydrolase follows the procedure given in chapter [10] of this volume for cellobiose dehydrogenase. When the concentrated enzyme solution from ultrafiltration step (step 3) of the cellobi-

[11] D. Herr, F. Baumer, and H. Dellwig, *Eur. J. Appl. Microbiol. Biotechnol.* **5,** 29 (1978).
[12] *Sclerotium rolfsii* CPC 142 culture (NCIM No. 1084) is available from the National Collection of Industrial Microorganisms, National Chemical Laboratory, Poona 411 008, India.

TABLE I
PURIFICATION OF CELLOBIOHYDROLASE FROM S. *rolfsii*[a]

| | | Activity toward (total units) | | |
Fraction	Protein (mg)	H₃PO₄-swollen cellulose	CMC	PNPG
Culture filtrate	17,202	27,523	520,109	45,284
Ammonium sulfate, 0–3.4 *M*	13,937	26,480	462,620	42,785
Sephadex G-75 chromatography fraction A	10,774	22,324	170,210	28,370
Ultrafiltration of fraction A (Amicon PM10)	8,470	21,260	122,300	21,430
DEAE-Sephadex A-50 chromatography	329	206	4.8	57
Preparative poly-acrylamide gel electrophoresis	52	3.4	0.018	0

[a] From Ref. 2. One unit of activity towards H₃PO₄-swollen cellulose or CMC (carboxymethylcellulose) is defined as the amount of enzyme that releases 180 μg reducing sugars (1 μmol as D-glucose equivalent) per minute at pH 4.5 and 50°, using a 4-hr incubation for H₃PO₄-swollen cellulose and a 30-min incubation for CMC.[6] β-D-Glucosidase is determined as described in chapter [10] of this volume. PNPG, *p*-Nitrophenyl-β-D-glucopyranoside.

ose dehydrogenase procedure is chromatographed on a DEAE-Sephadex A-50 column, endo-β-glucanase and β-glucosidase activities are not adsorbed on the column, whereas a major amount of protein containing cellobiohydrolase is adsorbed. This is eluted by 0.1 *M* citrate buffer, pH 4.5 (fraction numbers 50–70, see Fig. 1 in chapter [10] of this volume for cellobiose dehydrogenase). This fraction is concentrated by lyophilization and dialyzed against 5 m*M* Tris–glycine buffer, pH 8.5. Further purification of cellobiohydrolase is obtained by preparative polyacrylamide gel electrophoresis as described for cellobiose dehydrogenase in this volume [10]. The results of a typical purification of cellobiohydrolase from S. *rolfsii* are presented in Table I.

Properties of the Purified Enzyme

Purity. The purified enzyme shows one protein band on polyacrylamide gel electrophoresis at pH 2.9 and 8.9, with or without SDS treatment, and in analytical isoelectric focusing in 7.5% polyacrylamide gel over the pH range 3.5–10.0.

Stability. The enzyme is stable when stored in 50 mM citrate buffer, pH 4.5, at −15°. The enzyme is most stable at pH 4.5–5.0. The enzyme is stable to repeated freezing and thawing.

Physical Properties. The relative molecular weight, M_r, estimated by gel filtration on BioGel P-150,[13] electrophoresis by slope method,[14] and by its migration in SDS–polyacrylamide gel[15,16] is 41,500, 41,700, and 42,000, respectively. The pI of the pure *S. rolfsii* enzyme is 4.32.

Chemical Properties. The enzyme is composed of a single polypeptide chain as the carboxyamidomethylated derivative of the reduced form of cellobiohydrolase gives one protein band with a relative molecular weight (M_r 42,000) corresponding to the native protein.

The enzyme is a glycoprotein containing 7.0% total carbohydrate; it contains 3.6 residues of glucosamine per molecule of enzyme but no galactosamine. The enzyme is high in acidic and low in basic amino acids and contains no cystine or half-cystine.[2]

Enzymatic Properties. With phosphoric acid-swollen cellulose, the pH and temperature optima are 4.5 and 50° (4 hr assay), and with Avicel (24 hr assay) they are 4.2 and 37°. The rate of phosphoric acid-swollen cellulose hydrolysis is linear up to 15 μg enzyme protein (4 hr assay), and with 10 μg enzyme, the hydrolysis is linear up to 8 hr under standard assay conditions. The purified enzyme (10 μg) produces 40.8 μg reducing sugars as glucose equivalent from phosphoric acid-swollen cellulose in 4 hr, 50°, pH 4.5. With phosphoric acid-swollen cellulose, the activation energy, calculated from Arrhenius plot, is 8.55 cal/mol (1 cal = 4.1868 J). The cellobiohydrolase shows no S-factor activity,[17] though it potentiates S-factor activity of endo-β-glucanases.[3]

Kinetics. The $[S]_{0.5}$ and V_{max} values for phosphoric acid-swollen cellulose at pH 4.5, 50°, calculated from Lineweaver–Burk plot, are 1.66 mg/ml and 42 μg reducing sugars/mg protein per minute. The $[S]_{0.5}$ for cellotriose is 2.2 mM (glucose liberated is measured by glucose oxidase–peroxidase method[18]) and for cellotriose it is 1.81 mM (cellobiose liberated is measured by p-hydroxybenzoic acid hydrazide method[9]). The V_{max} for cellotriose is 46 μg glucose/mg protein per minute, and for cellotetraose 208 μg reducing sugar/mg protein per minute.

[13] P. Andrews, *Biochem. J.* **91**, 222 (1964).
[14] J. L. Hedrick and A. J. Smith, *Arch. Biochem. Biophys.* **126**, 155 (1968).
[15] K. Weber and M. Osborn, *J. Biol. Chem.* **244**, 4406 (1969).
[16] A. L. Shapiro, E. Vinuela, and J. V. Maizel, *Biochem. Biophys. Res. Commun.* **28**, 815 (1967).
[17] C. B. Marsh, G. V. Merola, and M. E. Simpson, *Text. Res.* **23**, 831 (1953).
[18] H. U. Bergmeyer, K. Gawein, and M. Grassl, *Methods Enzymatic Anal.* **1**, 457 (1974).

Specificity. The enzyme is specific for β-D-(1→4) linkage.

Action of Cellobiohydrolase on Insoluble and Soluble Substrates. The cellobiohydrolase hydrolyzes Cellulose-123 (Carl Schliecher and Schull Co., W. Germany), Avicel, α-cellulose, phosphoric acid-swollen cellulose, cellooligosaccharide G_{37} (i.e., degree of polymerization 37), and cotton. The principal product from each substrate is cellobiose (93–96%); glucose (4–7%) is also detected. The enzyme hydrolyzes carboxymethylcellulose to a small extent which is not due to its contamination with endo-β-glucanase; cellobiose is the only product detected. The cellobiohydrolase has no action on cellobiose or *p*-nitrophenyl-β-D-glucopyranoside; however, it hydrolyzes *p*-nitrophenyl β-D-cellobioside to *p*-nitrophenol and cellobiose. The enzyme hydrolyzes cellotriose and higher molecular weight cellodextrins; the rate of hydrolysis increases with increasing polymerization of the cellodextrin. Cellotriose is hydrolyzed to cellobiose and glucose, cellotetraose to cellobiose, cellopentaose to cellobiose and cellotriose initially and finally to cellobiose and glucose, and cellohexaose to cellobiose and cellotetraose initially and finally to only cellobiose.

Inhibitors. D-Glucono-1,5-δ-lactone at 10 mM is not inhibitory. *N*-Bromosuccinimide, a tryptophan-specific reagent, at 0.1 mM inhibits the enzyme activity completely.

Endo-type Mode of Action of Cellobiohydrolase. The enzyme forms insoluble short fibers from native dewaxed cotton initially.[6] The formation of insoluble short fibers from cotton by *S. rolfsii* cellobiohydrolase suggests an endo-type mode of action, besides its endwise action of removing cellobiosyl units from the nonreducing end of cellulose chains.

Synergism between Cellobiohydrolase and Endo-β-Glucanase. The pure cellobiohydrolase and endo-β-glucanases from *S. rolfsii*, when acting in concert, show active synergistic effects in reconstitution experiments in the solubilization of Avicel and cotton. With phosphoric acid-swollen cellulose, the synergistic effect is much smaller.[19]

Function

Initiation of Enzymatic Degradation of Crystalline Cellulose. The greater solubilization of Avicel by cellobiohydrolase and of phosphoric acid-swollen cellulose by endo-β-glucanases when acting alone, and the beneficial effect of pretreatment of Avicel with cellobiohydrolase, and of phosphoric acid-swollen cellulose by endo-β-glucanases, prior to the addition of the alternative type of enzyme,[19] suggest that endo-β-glucanase initiates the attack on amorphous cellulose (creating more ends for the

---

[19] J. C. Sadana and R. V. Patil, *Carbohydr. Res.* **140,** 111 (1985).

cellobiohydrolase to act), and cellobiohydrolase initiates the attack on crystalline cellulose (thereby making the substrate more accessible for hydrolysis[19]). The observation that cellobiohydrolase also displays an initial endo-type mode of action besides its endwise action of removing cellobiosyl units from the nonreducing end of the cellulose chains lends support to this concept.[6]

[33] Cellulases of *Thermomonospora fusca*

By DAVID B. WILSON

Introduction

Thermomonospora fusca YX is a filamentous actinomycete that has a doubling time of about 4 hr when grown at 55° in defined medium with cellulose (Solka-Floc) as a carbon source. The supernatant from a culture of cellulose-grown *T. fusca* contains a high level of cellulase activity, xylanase activity, and an active protease which modifies the different cellulases without altering their cellulase activity.[1] Even though the protease activity can be inhibited by phenylmethylsulfonyl fluoride (PMSF), treated supernatant still contains proteolytically derived isozymes and is extensively degraded when it is heated in SDS. We have isolated a mutant strain of *T. fusca,* ER-1[2] that produces no detectable extracellular protease activity and the supernatant from this strain has been used to purify the different cellulase activities produced by *T. fusca.* Polyacrylamide gel electrophoresis of ER-1 supernatant in the presence of SDS separates 16 protein bands of varying intensity, with none of the bands containing more than 10% of the total protein present in the culture supernatant. We have isolated five different β-1,4-endoglucanases and one xylanase from the culture supernatant of ER-1 cells grown on cellulose. At the present time we have not detected any enzyme that is a true exocellulase. However, one enzyme, E_3, appears to resemble the fungal exocellulases in that it has relatively little activity on CMC and it shows synergism when it is assayed on filter paper with two of the other four enzymes.

[1] R. E. Calza, D. C. Irwin, and D. B. Wilson, *Biochemistry* **24,** 7797 (1985).
[2] E. Lin, unpublished observations (1985).

METHODS IN ENZYMOLOGY, VOL. 160

Assay Method—CMCase Activity

Principle. The amount of reducing sugar released from carboxymethylcellulose (CMC) is measured by the DNS procedure of Miller.[3]

Reagents. Carboxymethylcellulose (Hercules type 4M6F-CMC) 1% solution in KP_i buffer, 0.05 M, pH 6.5 and sodium azide 0.004%, sonicated to give a uniform suspension.

DNS Reagent

Sodium potassium tartrate, 200 g
Dinitrosalicylic acid, 10 g
Phenol, 2 g
Na_2SO_3, 0.5 g
NaOH, 10 g
To 1 liter with H_2O
To achieve a linear standard curve, add 26 μl of 0.5 M glucose to 100
  ml DNS reagent within 8 hr of its use.

Procedure. The sample to be assayed is brought to a volume of 0.30 ml with 0.05 M KP_i, pH 6.5, and 0.10 ml of CMC is added. The mixture is incubated at 55° for the appropriate time (15–60 min) and the reaction is stopped by adding 0.75 ml of the DNS reagent. The samples are boiled for 15 min, cooled, and the optical density at 600 nm is recorded. If a tube is turbid it is centrifuged before the OD_{600} is read. The OD_{600} values are converted to micromoles of glucose equivalent reducing sugar released, using a glucose standard curve. One unit produces 1 μmol of glucose equivalent reducing sugar per minute in the above assay. When purified samples need to be diluted before assay, it is important to dilute them in 0.05 M KP_i buffer, pH 6.5, that contains 0.10 mg/ml bovine serum albumin (BSA). Protein is measured by the procedure of Bradford.[4]

Method—Filter Paper Assay

The sample to be assayed is added to a disk of Whatman No. 1 filter paper that is cut out with a hand paper hole punch. A disk has a diameter of 7 mm and weighs 3 mg. The sample is brought to a final volume of 0.4 ml with 0.05 M potassium phosphate buffer, incubated at 55° for 4–24 hr and the reaction is stopped and reducing sugars are determined by the procedure described for the CMCase assay.

[3] G. L. Miller, R. Blum, W. E. Glennin, and A. L. Benton, *Anal. Biochem.* **21,** 127 (1960).
[4] M. Bradford, *Anal. Biochem.* **72,** 248 (1976).

Strains. T. fusca YX and mutant ER-1 are available from the author for use in basic research.

Growth of Cells

Although *T. fusca* can be grown in a shake flask, the highest yield of cellulase is obtained by growing the organism in a fermentor (New Brunswick Scientific Corp., New Brunswick, NJ). A 250-ml, 18-hr shake culture of *T. fusca* mutant Er-1 grown in Hagerdal medium plus 0.2% cellobiose is added to 10 liters of Hagerdal medium[5] containing 1% Solka-Floc as the carbon source (Brown Paper products, Concord, NH).

Hagerdal medium	10× Macro salts	50× Micro salts
10× macro salts, 1 liter	NaCl, 15 g	EDTA, 0.50 g
50× micro salts, 200 ml	$(NH_4)_2SO_4$, 31 g	$MgSO_4 \cdot H_2O$, 2.00 g
Yeast extract, 10 g	Na_2HPO_4, 91 g	$ZnSO_4 \cdot H_2O$, 0.08 g
Thiamin, 10 mg	KH_2PO_4, 9 g	$FeSO_4 \cdot H_2O$, 0.20 g
Biotin, 10 mg	H_2O to 1 liter	$MnSO_4 \cdot H_2O$, 0.15 g
H_2O to 10 liters		$CaCl_2$, 0.20 g
		H_2O to 200 ml

The culture is grown with vigorous aeration and agitation for 26 hr and then PMSF is added to a final concentration of 0.1 mM. The culture is cooled to 10° and solids are removed with a refrigerated Sharples continuous flow steam-driven centrifuge at a flow rate of 400 ml/min.

Purification Procedure

All steps are carried out at 0–4°. Millipore immersible CX30 ultrafiltration units are used for concentration and desalting the various fractions.

Ammonium Sulfate Precipitation. The proteins present in the culture supernatant are concentrated by adding 351 g/liter of ammonium sulfate, allowing the solution to sit for 50 min and centrifuging at 14,000 g for 15 min. The pellets are resuspended in a total volume of 350 ml of 0.005 M KP$_i$ buffer, pH 6.5, allowed to sit overnight, and centrifuged for 10 min at 9000 g. The supernatant is then concentrated to approximately 10 mg/ml protein.

Additional cellulase activity (20% more) can be obtained by washing the solids removed in the Sharples centrifugation with distilled water. This material was not used in this purification scheme.

Gel Filtration. The first step in the purification is gel filtration which

---

[5] B. G. R. Hagerdal, J. D. Ferchak, and E. K. Pye, *Microbiology* **36**, 606 (1978).

also desalts the sample. Aliquots (15 ml) of Fraction I are chromatographed on a 2.5 × 100 cm column of Ultragel AcA 34 (LKB, Bromma, Sweden) that is equilibrated and eluted with 0.005 M KP$_i$ buffer, pH 6.5. The column is run at a flow rate of 1.4 ml/min and 12-ml fractions are collected. The eluant is monitored at 280 nm and the first peak to elute, which contains more than 95% of the CMCase and filter paper activity present in Fraction I, is combined, concentrated, and washed with 0.005 M KP$_i$, pH 6.5 (Fraction II).

Hydroxylapatite Chromatography. Fraction II (200 mg) is loaded on a 2.5 × 37 cm column of hydroxylapatite (HAP) (Clarkson Chemical Co., Williamsport, PA) and eluted with a 2 liter gradient from 0.005 to 0.110 M KP$_i$ buffer, pH 6.5, at a flow rate of 0.5 ml/min. Fractions of 7 ml are collected and assayed for OD$_{280}$, CMCase, and filter paper activity. Five peaks elute from this column and each peak is combined and concentrated (Fig. 1). The results of the purification through this step are summarized in Table I.

Enzyme E$_1$. The last peak to elute (HAP-5) contains endocellulase E$_1$ and this material has a purity of about 90% when tested either by SDS–polyacrylamide gel electrophoresis using a 10% gel or by isoelectric focusing.

Enzyme E$_2$. HAP-1 contains endocellulase E$_2$ and this material has a purity greater than 90% when tested by the above procedures. Attempts to purify E$_1$ and E$_2$ further result in large losses of activity, particularly of filter paper activity.

Enzyme E$_3$. E$_3$ is prepared by fractionating HAP-4 on a Canalco Disc Prep 20 apparatus (Canalco, Fort Lee, NJ) using a 7.5% acrylamide gel. The recipe for this gel is as follows:

Separating gel	Stacking gel
Acrylamide : bisacrylamide, 29.2 : 0.8%, 4.6 ml	Acrylamide : bisacrylamide, 10 : 2.5%, 2 ml
Separating buffer (Tris base 18.1 g, 1 N HCl 24 ml, H$_2$O to 100 ml), 4.6 ml; H$_2$O, 9.2 ml	Stacking buffer (Tris base 2.23 g, 1 M H$_3$PO$_4$ 12.8 ml, H$_2$O to 100 ml), 2 ml; H$_2$O, 4 ml
Ammonium persulfate, 10% 150 μl	Ammonium persulfate, 10% 125 μl
TEMED, 13 μl	TEMED, 8 μl

	Upper buffer	Lower buffer
	Tris base, 5.16 g	Tris base, 14.5 g
	Glycine, 3.48 g	4 N HCl, 15 ml
	H$_2$O to 1 liter	Glycerol, 100 ml
		H$_2$O to 1 liter

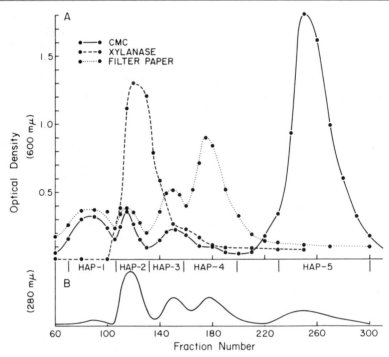

FIG. 1. Hydroxylapatite column chromatography of Ultragel AcA 34 cellulase peak. (A) Enzymatic activities. CMC, 1.5 μl samples, 10-min incubation ($\times$); filter paper, 10 μl samples, 18-hr incubation ($\bigcirc$); xylanase, 2 μl samples, 10-min incubation ($\triangle$). (B) UV monitor protein trace.

TABLE I
PURIFICATION OF CELLULASES FROM *T. fusca*[a]

Major cellulase component	Protein (mg)	CMC		Filter paper		Xylan	
		U	SA	U	SA	U	SA
Ammonium sulfate	384	20,000	52	214	0.56	34,850	91
Ultragel AcA 34	233	21,042	90	195	0.84	15,750	68
Hap 1 E_2	7.1	730	103	43	6.1	8	1
Hap 2 E_5	39.4	1,360	35	19	0.48	12,172	309
Hap 3 E_4	30.8	820	27	19	0.62	1,042	34
Hap 4 E_3	54	558	10	31	0.58	162	3
Hap 5 E_1	14	15,144	1125	1	0.07	182	13
Recovery (%)		96		53		39	

[a] SA, Specific activity; U, units.

The gel is run at a constant current of 10 mA, and the elution buffer is pumped at 1 ml/min and collected in 3 ml fractions. The eluant is monitored at 280 nm and the fractions containing protein are assayed for CM-Case and filter paper activity. There is a single broad peak of protein and activity eluted from this gel. The fractions containing pure E_3 are identified by SDS–polyacrylamide gel electrophoresis of the samples from the prep gel. E_3 has a molecular weight of 65,000 and is the major cellulase present in HAP-4.

Enzyme E_4. E_4 is prepared by preparative gel electrophoresis of HAP-3. The sample is fractionated as described for enzyme E_3. In this case two protein peaks elute from the gel, both of which have activity, but E_4 has a specific activity that is several times higher than that of the other peak. The final step in the purification of E_4 is chromatography on a BioGel P-100 column. The fractions from the prep gel are concentrated to a volume of 2 ml and loaded on a 1.5 g 70 cm column of BioGel P-100 that is equilibrated and eluted with 0.005 M KP$_i$ buffer, pH 6.5. The flow rate is 5 ml/hr and 2 ml fractions are collected. The eluate is monitored at 280 nm and samples from the first peak to elute are tested by SDS–polyacrylamide gel electrophoresis using a 10% gel. The fractions in which a 106,000 MW band is the major component are combined and concentrated.

Enzyme E_5. HAP-2 contains endocellulase E_5 as well as the xylanase produced by cellulose grown *T. fusca*. These can both be isolated in nearly pure form by chromatography of HAP-2 on a 28 ml DEAE-Sephadex column. HAP-2 (36 mg) is loaded on a 1.5 × 14 cm column of DEAE-Sephadex A50 that is equilibrated with 0.10 M sodium acetate buffer, pH 5.5. The column is washed with 100 ml of the above buffer and then eluted with a 500 ml linear gradient from 0 to 0.3 M sodium chloride in the same buffer. The flow rate is 16 ml/hr and fractions of 4 ml are collected. Enzyme E_5 eluted just before the peak of the xylanase activity.

From the yields of these pure enzymes and the amounts of activity in the HAP peaks it appears that E_5 (2.1 mg) is present in the smallest amount while enzymes E_1 (7 mg), E_2 (14 mg), E_3 (8 mg), and E_4 (3 mg) are present in nearly equal amounts. The purified enzymes are very stable when stored frozen at $-70°$.

Purity

Duplicate samples of each purified protein are run on an 8.5% "native" acrylamide slab gel (1 mm thick). Three microgram samples of each protein are used in the lanes to be Coomassie Blue stained and 0.004 units of CMC activity are used in lanes to be activity stained.[6] The gel is

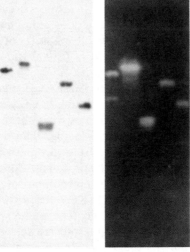

FIG. 2. Polyacrylamide "native" gel electrophoresis of purified enzymes; lane 1, E_1; lane 2, E_2; lane 3, E_3; lane 4, E_4; lane 5, E_5.

electrophoresed at a constant current of 20 mA until the bromphenol blue dye reaches the bottom.

A CMC agarose overlay is prepared by heating to boiling a solution of 1% agarose and 0.05% CMC in 0.01 M Tris–HCl, pH 6.5, buffer. The hot agarose (15 ml) is pipetted on a 110 × 125 mm sheet of Gel Bond film agarose gel support medium (FMC Corp., Rockland, ME).

After electrophoresis the gel is cut to remove the portion to be stained. The CMC agarose overlay is placed on top of the remaining gel, wrapped in plastic and incubated for 30 min at 55°. The overlay is then removed and stained with Congo Red. Figure 2 shows the five purified cellulases tested by this method. For each enzyme the single Coomassie blue-stained band corresponds to the major activity band. In addition, each enzyme was tested for purity by SDS–gel electrophoresis in a 10% gel and by isoelectric focusing. Each enzyme gave one major band containing at least 90% of the total protein by both procedures.

Properties

Some of the properties of the purified enzymes are summarized in Table II. It is clear that these enzymes differ greatly in their activities on

TABLE II
PROPERTIES OF PURIFIED ENZYMES

Enzyme	Molecular weight	Isoelectric point	Specific activity (units/mg)			PNP cellobiose hydrolysis (activity ratios)
			CMC	Amorphous cellulase	Filter paper	
E_1	108,000	3.2	1100	47	0.07	1.0
E_2	42,000	4.7	100	55	6.1	0.006
E_3	71,000	3.1	3.9	2.3	0.68	0.037
E_4	106,000	3.6	6.0	4.0	0.13	0.002
E_5	45,000	4.5	150	31	0.14	1.0

the different substrates, with E_1 having a specific activity on CMC that is 10 times higher than any other enzyme and E_2 having a specific activity on filter paper that is 10 times higher than any other enzyme.

All of these enzymes are endocellulases since every one reduces the viscosity of CMC at a much greater rate than it releases reducing sugars from CMC. E_2 and E_5 reduce the viscosity of CMC twice as rapidly for a given release of reducing sugar than E_1 while E_3 and E_4 initially reduce the viscosity at the high rate but after a short time the viscosity reduction slows to the lower rate.

E_1, E_2, and E_5 are distinct antigenically, as rabbit antisera prepared against each enzyme specifically inhibit and react only with the enzyme they are prepared against. E_3 and E_4 do react with antisera prepared against each other but are antigenically distinct from the other enzymes. Some of the enzymatic properties of E_3 and E_4 are similar, but these enzymes have very different molecular weights. In addition the N-terminal sequences of all five enzymes have been determined on a gas phase sequencer (Table III) and are all different from each other.

Enzymes E_3 and E_2 are glycoproteins while enzymes E_1, E_4, and E_5 are not. This was shown by staining SDS gels run on the purified enzymes by both the periodic acid–Schiff reagent procedure[7] and the more sensitive blot transfer avidin–biotin technique of Bayer et al.[8] E_3 appears to contain more carbohydrate than E_2.

Because E_1 and E_2 were isolated first a number of additional properties

[6] P. Béguin, *Anal. Biochem.* **131**, 333 (1983).
[7] J. G. Beeley, "Glycoprotein and Proteoglycan Techniques: Laboratory Techniques in Biochemistry and Molecular Biology" (R. H. Burden and P. H. van Knippenberg, eds.). Elsevier, Amsterdam, 1985.
[8] E. Bayer, H. Ben-Hur, and M. Wilchek, *Anal. Biochem.* **161**, 123 (1987).

TABLE III
N-Terminal Sequence

Residue number	E_1	E_2	E_3	E_4	E_5
1	X[a]	Asn	Ala	Glu	Ala
2	Glu	Asp	Gly	Pro	Gly
3	Val	Ala (Ser)	Cys	Ala	Leu
4	Asp	Pro	Ser	Phe	Thr
5	Glu	Phe	Val	Tyr	Ala
6	Phe, Leu	Tyr	Asp	Asn	Thr
7	Cys, Arg, His	Val	Tyr	Ala	Val
8	Asn	Asn	Thr	Ala	Thr
9	Gly	Pro	Val	Glu	Pro
10	Asp	Asn	Asn	Leu	Val
11		Met	Ser	Gln	Thr
12		Thr (Asp, Asn)	Trp	Lys	Lys
13		Ser	Gly	Ser	Glu
14		Ala	Thr	Met	Ser
15		Gln	Gly	Phe	Ser
16		Trp	Phe	Phe	Trp
17			X	Tyr	
18			Ala	Glu	
19			Asn	Glu, Ala	
20			Val	Gln	
21			X	Arg	
22			Thr		
23			Asn		

[a] X, Residue destroyed.

TABLE IV
Amino Acid Composition

Amino acid	mol%		Amino acid	mol%	
	E_1	E_2		E_1	E_2
Asp	10.8	11.9	Val	8.5	6.0
Glu	8.7	7.6	Leu	7.1	3.9
Ser	6.6	8.4	Ile	3.6	5.8
Thr	9.1	8.6	Phe	3.4	2.7
Gly	12.5	9.9	Tyr	3.6	2.4
Ala	9.5	13.8	Lys	2.5	1.7
Pro	7.5	8.5	His	1.8	2.4
Cys	1.0	1.1	Arg	2.7	3.4
Met	1.1	2.0			

of these enzymes have been determined. Their amino acid compositions are given in Table IV. The K_m for CMC of E_1 is 360 μg/ml while the value for E_2 is 120 μg/ml. Both enzymes have broad pH optimums centered around pH 6.5 and both are inhibited completely by 10 ppm $HgCl_2$ and 50% by 100 ppm $AgNO_3$. E_1 readily hydrolyzes cellotriose to glucose and cellobiose while E_2 does not hydrolyze this substrate. Consistent with this finding, limit digests of amorphorous cellulose produced by E_1 contain mainly cellobiose and a trace of glucose while limit digests produced by E_2 contained mainly cellotriose and higher oligomers with some cellobiose and a trace of glucose.

There are two major differences between the cellulases produced by *T. fusca* and those produced by cellulolytic fungi. One is that *T. fusca* does not appear to produce an exocellulase. The other is that we have been unable to detect any evidence for complex formation between the different *T. fusca* cellulase components. The cellulase activity present in the *T. fusca* culture supernatant sediments at the same rate during glycerol gradient centrifugation as do the individual purified enzymes. However, we have been able to detect some synergism when we mix the purified enzymes. Enzyme E_3 appears to be essential for synergism since a mixture of all five enzymes shows twice the activity on filter paper predicted from the sum of the individual activities. Mixtures of E_3 plus E_2 or E_5 also show synergism while other pairs do not.

Acknowledgments

I would like to thank Dr. W. Dexter Bellamy for the culture of *T. fusca* YX and for his advice and encouragement. I would also like to acknowledge the efforts of Dr. Roger Calza and Diana Irwin in the work reported in this chapter. This research was supported by Grant PCM-931432 from the National Science Foundation and Grant DE-FG02-84 ER13233 from the Department of Energy.

[34] Cellulases of *Humicola insolens* and *Humicola grisea*

By SHINSAKU HAYASHIDA, KAZUYOSHI OHTA, and KAIGUO MO

The enzymatic hydrolysis of native cellulose is a complex process requiring the participation of several enzymes. There are three major types of cellulolytic enzymes produced by fungi,[1] namely, cellobiohydro-

---

[1] T. M. Enari, *in* "Microbial Enzymes and Biotechnology" (W. M. Fogarty, ed.), p. 183. Applied Science Publishers, London, 1983.

lase (Avicelase, 1,4-β-D-glucan cellobiohydrolase; EC 3.2.1.91, cellulose 1,4-β-cellobiosidase), endoglucanase (CMCase, 1,4-β-D-glucan 4-glucanohydrolase; EC 3.2.1.4, cellulase), and β-glucosidase (β-D-glucoside glucohydrolase; EC 3.2.1.21, cellulose 1,4-β-glucosidase). Cellobiohydrolase is the enzyme with the highest affinity for cellulose and is capable of degrading crystalline cellulose (Avicel). Endoglucanase usually hydrolyzes only soluble substrates like CMC (carboxymethylcellulose). β-Glucosidase is the enzyme that hydrolyzes cellotriose and cellobiose to glucose. The thermostable cellobiohydrolase, endoglucanase, β-glucosidase from *Humicola insolens* and the Avicel-disintegrating endoglucanase from *Humicola grisea* are described here.

Assay Method

Principle. The assay is based on the release of reducing sugars or glucose following the incubation of Avicel, CMC, or cellobiose with the enzyme.

Reagents

Sodium acetate buffer, 0.1 M, pH 5.0
Avicel, 10 mg/ml in buffer
CMC (carboxymethylcellulose), 10 mg/ml in buffer
Cellobiose, 10 mg/ml in buffer
Bertrand reagents, solution A: $CuSO_4$, 0.16 M; solution B: potassium sodium tartrate, 0.7 M, NaOH, 3.8 M; solution C: $Fe_2(SO_4)_3$, 0.13 M, H_2SO_4, 2 M
$KMnO_4$, 3.2 mM
Glucose-oxidase colorimetric reagents (kits from Sigma Chemical Co.)

Procedure. (1) Cellobiohydrolase activity.[2] The reaction mixture, containing 100 mg of Avicel in 5 ml of buffer and 5 ml of enzyme solution, is shaken in an L-shaped tube on a Monod shaker at 90 strokes per min at 50° for 24 hr. Liberated reducing sugar is determined by the Bertrand micromethod.[3] One milliliter of the supernatant of reaction mixture, 2 ml of deionized water, 3 ml of solution A, and 3 ml of solution B are mixed and boiled for 6 min. It is immediately cooled by running water and centrifuged. The precipitate is washed twice with deionized water and dissolved in 2 ml of solution C. The solution is titrated with $KMnO_4$ (3.2 mM). The amount of reducing sugar is obtained from the conversion table.[3] (2) Endoglucanase activity. The reaction mixture, containing 100

[2] S. Hayashida and H. Yoshioka, *Agric. Biol. Chem.* **44**, 1721 (1980).
[3] G. Klein, *in* "Handbuch der Pflanzenanalyse, Spezielle Analyse I," p. 786. Wien Verlag von Julius Springer, Vienna, Austria, 1932.

mg of CMC in 5 ml of buffer and 5 ml of enzyme solution, is shaken in an L-shaped tube on a Monod shaker at 90 strokes per min at 50° for 1 hr. Liberated reducing sugar is determined by the Bertrand micromethod,[3] as described above. Viscometric assay is performed in size 150 Cannon-Fenske viscometers in a water bath at 50°. (3) Filter paper-disintegrating activity.[4] The reaction mixture, containing 10 strips of filter paper (1 × 1 cm) in 5 ml of buffer and 5 ml of enzyme solution, is shaken in an L-shaped tube on a Monod shaker at 90 strokes per min at 50°. The time for complete disintegration of filter paper is determined. (4) β-Glucosidase activity.[5] The reaction mixture, containing 100 mg of cellobiose in 5 ml of buffer and 5 ml of enzyme solution, is incubated at 50° for 1 hr. The glucose formed is determined colorimetrically with glucose oxidase.

Definition of Unit and Specific Activity. One unit of cellulase activity is defined as the amount of enzyme releasing 1 μmol of reducing sugar from the substrate per minute. Specific activity is expressed as units per milligram of protein. The protein content is determined by the method of Lowry *et al.*[6]

Organisms and Culture Conditions

Humicola insolens YH-8[2] and *Humicola grisea* var. *thermoidea* YH-78[7] were isolated from manure and compost heaps in our laboratory (these organisms can be obtained from Department of Agricultural Chemistry, Kyushu University, Fukuoka 812, Japan). A protease-negative mutant of *H. grisea* var. *thermoidea* YH-78 is obtained by ultraviolet and *N*-methyl-*N'*-nitro-*N*-nitrosoguanidine combined treatment.[4] The stock culture is maintained on slants of carrot juice agar medium (carrot 2000 g, chloramphenicol 50 mg, agar 20 g, tap water 1000 ml). The organism from a fresh slant culture is inoculated into the wheat bran medium (wheat bran 100 g, Avicel 10 g, tap water 100 ml) in a 2-liter Erlenmeyer flask, and the 20 flasks are incubated at 50° for 4 days.

Purification Procedure

All purification steps are carried out at 0–4° unless otherwise stated. Centrifugation at each step is performed at 13,000 g for 20 min. Diluted enzyme solution from each chromatography is concentrated by lyophilization, followed by dissolving in a minimum volume of deionized water.

[4] S. Hayashida and K. Mo, *Appl. Environ. Microbiol.* **51,** 1041 (1986).
[5] H. Yoshioka and S. Hayashida, *Agric. Biol. Chem.* **44,** 1729 (1980).
[6] O. H. Lowry, N. J. Rosebrough, A. L. Farr, and R. J. Randall, *J. Biol. Chem.* **193,** 265 (1951).
[7] H. Yoshioka, S. Anraku, and S. Hayashida, *Agric. Biol. Chem.* **46,** 75 (1982).

Preparation of Crude Extract. To the fungal culture in each flask, 600 ml of tap water is added and mixed well. The mixture is left to stand for 3 hr and clarified by centrifugation. The supernatant is designated the crude extract.

Ammonium Sulfate Fractionation. Solid ammonium sulfate is added to the crude extract (60 g/100 ml initial volume) with stirring and left to stand for 24 hr. The precipitate is collected by centrifugation and then dissolved in a small volume of deionized water. The supernatant is dialyzed against deionized water by PVA-Hollow Fiber (Kuraray Co., Osaka, Japan) and used for purification of cellulase.

Purification of Cellobiohydrolase[2]

Step 1. Avicel Adsorption. Avicel is added to the enzyme solution from *H. insolens* YH-8 (10 g/100 ml initial volume), and the mixture is stirred for 10 min in order to adsorb cellobiohydrolase onto Avicel. The Avicel is collected by centrifugation and washed three times with 0.05 M acetate buffer (pH 5.0). Ammonium hydroxide–ammonium chloride buffer (0.2 M, pH 9.0; 50 ml buffer/g Avicel) is added for elution of cellobiohydrolase from Avicel. The mixture is kept overnight and the Avicel is then removed by centrifugation. The pH of the supernatants is adjusted to 3.5 and left to stand for 24 hr. The pH is then adjusted to 5.5. Any insoluble material formed is removed by centrifugation. Solid ammonium sulfate is added to the supernatant (20–50 g/100 ml initial volume).

Step 2. Chromatography on Sephadex G-50 and G-100. The precipitate is dissolved in a minimum volume of deionized water and applied to a Sephadex G-50 column (4.5 × 103 cm). Filtration is carried out with deionized water at a rate of 24 ml/hr. The fractions containing cellobiohydrolase activity are pooled and concentrated. The concentrated enzyme solution (8 ml) is applied to a Sephadex G-100 column (2.3 × 57 cm) and filtration is performed with deionized water at a rate of 22 ml/hr. The fractions containing cellobiohydrolase activity are pooled and concentrated.

Step 3. Chromatography on DEAE-Sephadex A-50. The concentrate is placed on a DEAE-Sephadex A-50 column (2.3 × 57 cm) equilibrated with 0.05 M phosphate buffer (pH 5.5) and is eluted with a linear gradient of 0.05–1.0 M phosphate buffer (pH 5.5) in 500 ml. The fractions containing cellobiohydrolase activity are collected and concentrated.

Step 4. Rechromatography on Sephadex G-100 and DEAE-Sephadex A-50. The concentrate is rechromatographed on Sephadex G-100 and DEAE-Sephadex A-50 columns as described above. The fractions containing cellobiohydrolase activity are collected, concentrated, and desalted by filtration through a Sephadex G-50 column, and then lyophi-

TABLE I
SUMMARY OF CELLOBIOHYDROLASE-PURIFICATION

Step	Volume (ml)	Protein (mg)	Activity (units)	Specific activity (units/mg)	Recovery (%)
Crude extract	10,000	2,550	142	0.056	100
Ammonium sulfate	1,400	1,750	141	0.083	99
Sephadex G-50	170	274	48	0.18	34
First Sephadex G-100	51	150	43	0.29	31
First DEAE-Sephadex A-50	57	24	23	0.95	16
Second Sephadex G-100	36	13	20	1.55	14
Second DEAE-Sephadex A-50	42	11	17	1.60	12

lized. The lyophilized preparation shows a single protein band on disc electrophoresis in 7.5% polyacrylamide gel at pH 8.3 in Tris–glycine buffer by the method of Davis.[8] A summary of the purification procedure of cellobiohydrolase from *H. insolens* YH-8 is given in the Table I.

Purification of Endoglucanase[2]

Step 1. Avicel and Heat Treatments. Avicel is added to the enzyme solution from *H. insolens* YH-8 (20 g/100 ml initial volume), and the mixture is stirred for 10 min. After removal of Avicel by centrifugation, the supernatant is heated in a water bath maintained at 75° for 5 min. The pH is adjusted to 2.5 and left to stand for 24 hr. The pH of the solution is then adjusted to 5.5. Any insoluble material formed is removed by centrifugation. Solid ammonium sulfate is added to the enzyme solution (20–50 g/100 ml initial volume) with stirring.

Step 2. Chromatography on Sephadex G-50 and G-100. The above precipitate is dissolved in a minimum volume of deionized water and applied sequentially to Sephadex G-50 and G-100 columns. Filtration is carried out as described above. The fractions containing endoglucanase activity are pooled and concentrated.

Step 3. Chromatography on DEAE-Sephadex A-50 and Rechromatography on Sephadex G-100. The concentrate is applied to a DEAE-Sephadex A-50 column equilibrated with 0.05 *M* phosphate buffer (pH 5.5). The chromatography is carried out as described above. The fractions containing endoglucanase activity are collected, concentrated, and rechromatographed on a column of Sephadex G-100. The fractions containing endoglucanase activity are collected and concentrated.

[8] B. J. Davis, *Ann. N.Y. Acad. Sci.* **121**, 404 (1964).

Step 4. Chromatography on Hydroxyapatite. The concentrate is applied to a hydroxyapatite column (2.3 × 57 cm) equilibrated with 0.05 M phosphate buffer (pH 5.5). Elution is with a linear gradient of 0.05–1.0 M phosphate buffer (pH 5.5) in 500 ml at a rate of 24 ml/hr. The fractions containing endoglucanase activity are pooled and concentrated.

Step 5. Chromatography on Sephadex G-200 and Rechromatography on DEAE-Sephadex A-50. The concentrate is chromatographed on a Sephadex G-200 column (2.3 × 57 cm). Elution is performed with deionized water at a rate of 22 ml/hr. The fractions containing endoglucanase activity are concentrated and again placed on a DEAE-Sephadex A-50 column equilibrated with 0.5 M acetate buffer (pH 3.2). Elution is performed with a pH gradient of 3.2–6.0 in the 0.5 M acetate buffer. The fractions containing endoglucanase activity are pooled and desalted by filtration through a Sephadex G-50 column, and then lyophilized. The lyophilized preparation shows a single protein band on disc electrophoresis.[8] The purification procedure of endoglucanase from *H. insolens* YH-8 is summarized in Table II.

Purification of β-Glucosidase[5]

Step 1. Ammonium Sulfate Fractionation, and Sephadex G-50 and G-100 Chromatography. Solid ammonium sulfate is added to the enzyme solution from *H. insolens* (20–50 g/100 ml initial volume) with stirring. After being kept overnight, the resulting precipitate is dissolved in a minimum volume of deionized water and desalted by filtration through a Sephadex G-50 column. The concentrate is applied to a Sephadex G-100 column. The fractions containing β-glucosidase activity are pooled and concentrated.

TABLE II
Summary of Endoglucanase Purification

Step	Volume (ml)	Protein (mg)	Activity (units)	Specific activity (units/mg)	Recovery (%)
Crude extract	10,000	2,550	18,404	7.23	100
Ammonium sulfate	1,400	1,730	18,348	10.56	100
Sephadex G-50	230	564	10,731	18.9	58
First Sephadex G-100	63	281	9,174	32.8	50
First DEAE-Sephadex A-50	87	68	5,838	86.18	32
Hydroxyapatite	40	32	3,653	112.87	20
Sephadex G-200	48	29	3,508	122.32	19
Second DEAE-Sephadex A-50	24	9	2,007	217.95	11

Step 2. Chromatography on DEAE-Sephadex A-50. The concentrate is chromatographed on a DEAE-Sephadex A-50 column (2.3 × 57 cm) equilibrated with 0.05 *M* phosphate buffer (pH 5.5), and is eluted with a linear gradient from 0.05 to 1.0 *M* phosphate buffer (pH 5.5) in 500 ml. The fractions containing β-glucosidase activity are pooled and concentrated.

Step 3. Chromatography on Sephadex G-200. The concentrate is applied to a Sephadex G-200 column. Filtration is carried out with deionized water. The fractions containing β-glucosidase activity are pooled and then lyophilized. The lyophilized preparation is designated the purified β-glucosidase. The purification procedure of β-glucosidase from *H. insolens* YH-8 is summarized in Table III.

Purification of Avicel-Adsorbable, Disintegrating Endoglucanase[4]

Avicel is added to the enzyme solution from a protease-negative *H. grisea* var. *thermoidea* mutant (20 g/100 ml initial volume), and the mixture is stirred for 10 min in order to adsorb Avicel-adsorbable endoglucanase. The Avicel is collected by centrifugation and washed three times with 0.05 *M* acetate buffer (pH 5.5). Ammonium hydroxide–ammonium chloride buffer (0.2 *M*, pH 9.0; 50 ml buffer/g Avicel) is added for elution of endoglucanase from Avicel. The mixture is kept overnight and the Avicel is then removed by centrifugation. The pH of the supernatant is adjusted to 3.0 and left to stand for 24 hr. The pH is then adjusted to 5.5. Any insoluble material formed is removed by centrifugation. Solid ammonium sulfate is added to the supernatant (60 g/100 ml initial volume) with stirring. The precipitate is dissolved in a minimum volume of deionized water and desalted by filtration through a Sephadex G-50 column. The

TABLE III
SUMMARY OF β-GLUCOSIDASE PURIFICATION

Step	Volume (ml)	Protein (mg)	Activity (units)	Specific activity (units/mg)	Recovery (%)
Crude extract	10,000	2,550	21,823	8.56	100
Ammonium sulfate (60 g/100 ml)	1,400	1,730	21,801	12.62	100
Ammonium sulfate (20–50 g/100 ml)	500	928	15,835	17.07	73
Sephadex G-50	120	211	11,014	52.15	51
Sephadex G-100	65	120	8,518	71.17	39
DEAE-Sephadex A-50	40	18	4,710	266.32	22
Sephadex G-200	25	18	4,700	266.88	22

TABLE IV

SUMMARY OF AVICEL-ADSORBABLE, DISINTEGRATING ENDOGLUCANASE PURIFICATION

Step	Protein (mg)	Activity (units)	Specific activity (units/mg)	Recovery (%)
Crude extract	2100	7245	3.5	100
First DEAE-Sephadex A-50	540	5670	10.5	78
Sephadex G-100	310	3798	12.3	52
Second DEAE-Sephadex A-50	89	1925	21.8	27
Sephadex G-200	48	1369	28.7	19

concentrated sample is chromatographed sequentially on DEAE-Sephadex A-50, Sephadex G-100, DEAE-Sephadex A-50, and Sephadex G-200 columns as described previously for *H. insolens* endoglucanase. The fractions containing Avicel-adsorbable endoglucanase activity at each step are collected and lyophilized. The final lyophilized preparation is designated the purified Avicel-adsorbable, disintegrating endoglucanase. The purification procedure of Avicel-adsorbable, disintegrating endoglucanase from a protease-negative *H. grisea* var. *thermoidea* mutant is summarized in Table IV. For comparison, the Avicel-nonadsorbable, nondisintegrating endoglucanase from a protease-positive *H. grisea* mutant is also purified by the procedure described for *H. insolens* endoglucanase.

Properties

Molecular Properties. The molecular weights of cellobiohydrolase, endoglucanase, and β-glucosidase from *H. insolens* YH-8 are estimated as 72,000, 57,000, and 250,000, respectively, by SDS–gel electrophoresis.[2,5] The molecular weight of Avicel-adsorbable, disintegrating endoglucanase from the protease-negative *H. grisea* var. *thermoidea* mutant is determined as 128,000,[4] whereas that of Avicel-nonadsorbable, nondisintegrating endoglucanase from the protease-positive mutant is estimated as 63,000.[4] The total content of carbohydrate of cellobiohydrolase and endoglucanase is 26.1 and 39.0%, respectively. The carbohydrate residues of both enzymes are abundant in *N*-acetylglucosamine, contain 14.0 and 20.0% of mannose, and 12.1 and 19.0% of *N*-acetylglucosamine, respectively.[9] The carbohydrate content in β-glucosidase is 2.5% and the constituent sugars are 1.7% of mannose and 0.8% of glucose.[5] The isoelectric

[9] S. Hayashida and H. Yoshioka, *Agric. Biol. Chem.* **44,** 481 (1980).

point of β-glucosidase is pH 4.23. The amino acid composition of β-glucosidase shows that aspartic acid, glycine, and alanine are characteristically abundant.

Thermal Stability and Optimal Temperature. There is no significant loss in activity when cellobiohydrolase, endoglucanase, and β-glucosidase are stored at 0–4° in 0.05 M acetate buffer (pH 5.0) for at least 2 weeks. Cellobiohydrolase, endoglucanase, and Avicel-adsorbable endoglucanase are stable with the heat treatment at 65° for 5 min. β-Glucosidase is stable to heating at 60° for 5 min.[5] The endoglucanase retains 45% of the original activity after heating at 95° for 5 min.[2] When 90% of the carbohydrate residues of cellobiohydrolase and endoglucanase is liberated, cellobiohydrolase is inactivated after heating at 65° for 5 min, and endoglucanase is inactivated after heating at 75° for 5 min. The thermal stabilities of the enzymes thus have a correlation with their carbohydrate moieties.[9] The optimum temperature of the enzymes for activity is 50°.

pH Stability and pH Optimum. The enzyme solution is incubated at various pH values at 4° for 24 hr, and residual activity is assayed under optimum conditions. Cellobiohydrolase and endoglucanase are stable at pH 3.5–9.5 and pH 3.0–11.0, respectively.[2] Avicel-adsorbable endoglucanase and β-glucosidase are stable at pH 3.0–11.0 and pH 4.0–8.0, respectively.[4,9] The optimum pH of the enzymes for activity is 5.0. The pH stability behaviors of the enzymes have a correlation with their carbohydrate moieties.

Adsorbabilities onto Cellulose. The enzyme solution (10 μg/ml, pH 5.0) is mixed with Avicel (20 g/100 ml initial volume), and the mixture is stirred at 4° for 10 min. After centrifugation, the enzyme activity in the supernatant fluid is assayed and the adsorbabilities are calculated. Cellobiohydrolase can be completely adsorbed onto Avicel. Avicel-adsorbable endoglucanase is adsorbed 72% onto Avicel and 70% onto cellulose powder.[4]

Action on Cellulosic Substrate. Cellobiohydrolase hydrolyzes Avicel to 40% of the original amount in the prolonged incubation at 50°.[2] The conventional endoglucanase from *H. insolens* and Avicel-adsorbable endoglucanase from *H. grisea* can rapidly decrease the viscosity of a 0.5% solution of CMC in 0.05 M acetate buffer (pH 5.0).[2,4] Avicel-adsorbable endoglucanase disintegrates completely filter paper within 30 min and Avicel within 24 hr at 50°. However, endoglucanase shows little disintegrating activity on filter paper and Avicel under the same conditions.[4] Avicel-adsorbable endoglucanase is observed via scanning electron microscopy to disintegrate Avicel fibrils layer by layer from the surface yielding thin sections with exposed chains; endoglucanase only attacks the surface of Avicel. A mixed preparation of cellulases exhibits synergistic action in the hydrolysis of native cellulose. The Avicel-adsorbable, disintegrating endoglucanase exhibits a higher synergistic effect with cellobiohydrolase than with the conventional endoglucanase, because it disintegrates not only the surface of cellulose fibrils, but also the inner part of fibrils, removing thin sections layer by layer, newly exposing glucan chains, and making the fibrils more accessible to cellobiohydrolase. Avicel-adsorbable, disintegrating endoglucanase plays an important role in the initial step of the hydrolysis of native cellulose. The endoglucanases can be designated Avicel-disintegrating endoglucanase C_1 and Avicel-nondisintegrating endoglucanase C_x according to Reese's hypothesis. On incubation with subtilisin, Avicel-adsorbable, disintegrating endoglucanase C_1 is converted to Avicel-unadsorbable, nondisintegrating endoglucanase C_x. It is suggested that Avicel-disintegrating endoglucanase C_1 have "cellulose-affinity site" that is different from cellulase active site.

β-Glucosidase has no action on cellulosic substrates, such as Avicel and CMC, but the enzyme attacks not only cellobiose but also β-glucoside such as p-nitrophenyl-β-D-glucopyranoside and salicin.[5]

Effect of Various Chemicals.[5] None of the tested metal ions markedly stimulates the β-glucosidase activity. Cu^{2+} inhibits 55% of the original activity. However, β-glucosidase activity is significantly resistant to urea and organic solvents, such as dimethyl sulfoxide and ethanol at high concentration. The enzyme activity shows low activity even at a low concentration of NaCl.

[35] Cellulase–Hemicellulase Complex of *Phoma hibernica*

By HENRYK URBANEK and JADWIGA ZALEWSKA-SOBCZAK

The walls of plant cells contain a variety of polysaccharide polymers. These polymers are cross-linked and form compact structures. A number of substrate-specific hydrolases which degrade cellulose, hemicelluloses, and pectins have been described from fungi and bacteria.[1,2] Because of the complexity of the cell wall its degradation requires coordinated and sequential attack by enzymes. For this reason stable enzymatic complexes capable of degrading cell wall constituents may be expected to be formed by microorganisms. Enzyme preparation and cellulolytic and hemicellulo-

[1] R. F. H. Dekker and G. N. Richards, *Adv. Carbohydr. Chem. Biochem.* **32,** 277 (1976).
[2] R. M. Cooper, *in* "Biochemical Plant Pathology" (J. A. Callow, ed.), p. 101. Wiley, New York, 1983.

lytic activities obtained from the culture supernatant of *Phoma hibernica* are studied in this chapter.

Assay

The cellulolytic and hemicellulolytic activities were determined by estimating the increment of reducing groups according to Nelson.[3] The reaction mixture contained 0.25 ml of 0.5% CM-cellulose (Koch-Light, Colubrook, United Kingdom) or 0.5% of one of the hemicellulose substrates, 0.25 ml of 0.01 M sodium barbital–sodium acetate buffer, pH 5, and 0.5 ml of enzyme solution appropriately diluted. Hemicellulose substrates used were xylan (Koch-Light), poly-D-galactomannan (Serva, Heidelberg, FRG), glucomannan from sulfite hemlock, and galactoglucomannan from kraft hemlock (ITT Rayonier Inc., NY). One unit of cellulase or hemicellulase activity was defined as the amount of enzyme liberating 1 μmol of reducing groups per 30 min. Specific activity was expressed as units per milligram of protein as measured by the method of Lowry *et al.*[4]

Purification Procedure[5]

Organism and Cultural Conditions. Phoma hibernica strain PE s5 isolated from forest soil was obtained from the Botany Department of the University of Warsaw. The fungus was cultured in a medium composed of 9 g of $NaNO_3$, 2 g of KH_2PO_4, 0.5 g of $MgSO_4 \cdot 7H_2O$, 0.3 g of $CaCl_2 \cdot 2H_2O$, 0.01 g of $FeSO_4 \cdot 7H_2O$ in 1 liter of tap water. Filter paper (type Filtrak 3 w from VEB Freiberger Zellstoff-und Papierfabrik zu Weissenborn, East Germany) was the only source of carbon added. Surface cultures were performed at 28° in 1-liter Erlenmeyer flasks containing 95 ml of medium and 5 ml of inoculum. Double-folded disks of filter paper were laid on the surface of the mineral medium. Glass beads were placed in the bottom of the Erlenmeyer flask to support the paper disks. The inoculum was prepared by extracting surface growth of 7-day-old slant cultures of *Phoma hibernica* with water. After 12 days the mycelium was centrifuged and the culture supernatant was the material used for further purification.

Purification Steps. All operations were carried out at 4°. Ammonium sulfate was added to the culture supernatant to 3.3 M (0.8 saturation).

[3] N. Nelson, *J. Biol. Chem.* **153**, 375 (1944).
[4] O. H. Lowry, N. J. Rosebrough, A. L. Farr, and R. J. Randall, *J. Biol. Chem.* **193**, 265 (1951).
[5] H. Urbanek, J. Zalewska-Sobczak, and A. Borowińska, *Arch. Microbiol.* **118**, 265 (1978).

After standing overnight, the resulting precipitate was centrifuged, dissolved in 0.1 M NaCl, and the solution dialyzed for 12 hr first against distilled water and then for 12 hr against 0.01 M Tris–HCl buffer, pH 8.0. The dialyzate was applied to a column of DEAE-Sephadex A-50 (1.5 × 30 cm) equilibrated with the same buffer. Enzymes were eluted by a linear gradient of 0–0.8 M NaCl in 0.01 M Tris–HCl buffer at pH 8.0. Three enzymatically active peaks were found in eluates from the DEAE-Sephadex column (Fig. 1). Peaks I and II possessed only xylanase activity; peak III, exhibiting activity on CM-cellulose, xylan, galacto-, gluco-, and galactoglucomannan, was purified further. The activities toward galacto-, gluco-, and galactoglucomannan were similar; therefore only the results concerning galactomannan are reported in detail. Fractions 26–33 (Fraction III) were pooled, dialyzed for 12 hr against distilled water, and concentrated to 6 ml by lyophilization. The concentrated solution was dialyzed for 12 hr against 0.01 M sodium barbital–sodium acetate buffer, pH 5.0, and the enzyme solution applied to a column of Sephadex G-100 (2 × 45 cm) and eluted with the same buffer. Only one fraction exhibiting the same enzymatic activities as fraction III from DEAE-Sephadex A-50 was

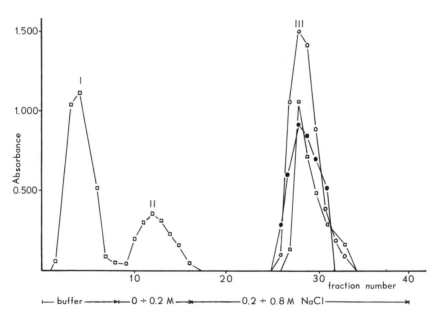

FIG. 1. Chromatography of cellulase and hemicellulase activities on DEAE-Sephadex A-50. Six-milliliter fractions were collected. Fractions 26–33 were pooled (fraction III). (○) Substrate, CM-cellulose; (●) substrate, galactomannan; (□) substrate, xylan.

obtained. The specific cellulase activity was slightly augmented following Sephadex G-100 filtration of fraction III. After the last purification step the specific cellulase activity of the obtained preparation was higher than the xylanase and galactomannase specific activities by about 50%. A summary of the purification steps is shown in Table I.

Attempts at separating cellulase, xylanase, and galactomannase activities by rechromatography on DEAE-Sephadex A-50 or by chromatography on CM-Sephadex C-50, hydroxyapatite, and Sephadex C-50 were unsuccessful. Only one fraction with all three mentioned activities was obtained in each case.

Analytical Isoelectric Focusing. The enzymatic preparation obtained by gel filtration on Sephadex G-100 was subjected to isoelectric focusing with pH 3.5–10 ampholine carrier ampholytes according to Gronov and Griffiths.[6] One distinct protein band corresponding to an isoelectric point of about 5.0 was obtained with activity toward CM-cellulose, xylan, and galactomannan. Only very weak xylanase activity was found in the segment of gel corresponding to the pH range 7.5–7.8. Application of electrofocusing over a pH range of 4.0–6.0 yielded only one protein band with the three investigated activities corresponding to an isoelectric point of 5.3.

Chromatofocusing. The enzymatic preparation from Sephadex G-100 was also subjected to chromatofocusing according to the procedure given in the handbook of Pharmacia Fine Chemicals (Uppsala, Sweden). Chromatofocusing was performed over a pH range of 7–4 using polybuffer exchanger PBE 94 and polybuffer 74. Only one adsorbed fraction corresponding to a pI of 5.3 exhibiting activity toward CM-cellulose, xylan, and galactomannan was obtained. This fraction had specific activity and hydrolytic ability similar to those from Sephadex G-100.

Properties of the Enzyme Complex[5]

To study enzymatic properties the fraction obtained after gel filtration on Sephadex G-100 was used.

pH Optimum. The enzyme complex had an identical pH optimum of 4.5 for degradation of CM-cellulose and galactomannan, while the xylanase activity exhibited a pH optimum in the range of 4.5–5.0 and it was only slightly lower at pH 6.0.

Effect of Some Chemicals. Tetranitromethane (10^{-3} M), urea (5 M), and Fe^{3+} (2×10^{-3} M) significantly inhibited all activities. *p*-Chloromer-

---

[6] M. Gronov and B. Griffiths, *FEBS Lett.* **15**, 340 (1971).

TABLE I
PURIFICATION OF CELLULASE AND HEMICELLULASE ACTIVITIES

Purification steps	Total volume (ml)	Activity with CM-cellulose			Activity with xylan			Activity with galactomannan		
		Units	Units/mg protein	Yield (%)	Units	Units/mg protein	Yield (%)	Units	Units/mg protein	Yield (%)
Culture supernatant	450	2070	12.1	100	2700	15.9	100	1620	9.5	100
Ammonium sulfate	45	1772	23.3	86	2196	28.8	81	1350	17.7	83
DEAE-Sephadex A-50 (fraction III)	84	1260	28.6	61	980	22.2	36	875	19.8	54
Sephadex G-100	36	1080	31.7	52	785	23.0	29	686	20.1	42

TABLE II
HYDROLYSIS OF NITROPHENYL DERIVATIVES AND HIGHLY POLYMERIZED SUBSTRATES
BY PURIFIED ENZYME PREPARATION

Substrate	p-Nitrophenol liberated ($A_{420\ nm}$)	Aldehyde groups liberated (μmol/hr $\times$ mg protein)
p-Nitrophenyl-β-D-glucopyranoside	0.600	3.5
p-Nitrophenyl-β-D-xylopyranoside	0.085	0.4
p-Nitrophenyl-β-D-mannopyranoside	0.810	4.7
CM-cellulose	—	61.0
Xylan	—	42.0
Galactomannan	—	39.2

curibenzoate, acetylimidazole, diisopropyl fluorophosphate, and divalent metal ions such as Fe^{2+}, Ca^{2+}, Mg^{2+}, and Cu^{2+} did not affect the enzymatic activities.

Reaction Products. Purified enzyme preparations attacked the different carbohydrate polymers in different ways. In the case of xylan degradation, a large amount of xylobiose and lesser amounts of higher oligosaccharides were found. Traces of xylose were detectable only after several hours of hydrolysis. In contrast, glucose constituted the main product released during degradation of CM-cellulose and insoluble cellulose. Cellobiose and some higher oligosaccharides were detected only in small amounts. Considerable amounts of mannose, dimers, and more polymerized products were detected on degradation of galactomannan.

Ability to Degrade p-Nitrophenyl Carbohydrate Derivatives. The enzyme complex showed a capacity for decomposition of nitrophenyl carbohydrate derivatives as measured according to Kanda *et al.*[7] *p*-Nitrophenyl-β-D-mannopyranoside was degraded approximately 8-fold more slowly than galactomannan, *p*-nitrophenyl-β-D-glucopyranoside about 17-fold more slowly than CM-cellulose, and *p*-nitrophenyl-β-D-xylopyranoside about 100-fold more slowly than xylan (Table II). The optimum pH for the hydrolysis of nitrophenyl carbohydrate derivatives was near 6.0.

Competition among CM-cellulose, Xylan, and Galactomannan. When the enzymatic preparation was incubated with either CM-cellulose and xylan or CM-cellulose and galactomannan at their respective K_m values

[7] T. Kanda, K. Wakabayashi, and K. Nisizawa, *J. Biochem. (Tokyo)* **79**, 977 (1976).

then the rate of reducing groups released was considerably higher than the value obtained with only one substrate, though it was lower than those calculated theoretically on the basis of individual substrate hydrolysis. This suggests that in the hydrolysis of the polysaccharides used more than one active center is involved. Values of K_m obtained from Lineweaver–Burk plots were 0.057% for CM-cellulose, 0.045% for xylan, and 0.068% for galactomannan. However, when the enzymatic preparation was incubated with a mixture of the same substrates but at much higher concentrations (0.125% each) the rate of reducing groups released was equal to the value obtained for one substrate. At high concentration the substrates could mutually inhibit their hydrolysis in the presence of different active centers.

Results indicate that the purified enzymatic preparation from *Phoma hibernica* forms a distinct complex with different enzymatic activities. It is capable of hydrolysis of the β-1,4-glycosidic linkages in polysaccharides formed by either glucose, xylose, or mannose residues. Detection of mannose but not galactose monomer during hydrolysis of galactomannan and the ability of the complex to decompose p-nitrophenyl-β-D-mannopyranoside suggest that the enzyme preparation should split only the linkages between mannose residues that are β-1,4-linked in the substrate.

[36] Cellulases from *Sporocytophaga myxococcoides*

By JOSTEIN GOKSØYR

Cellulose → β-1,4-glucan + β-1,4-glucan (randomly)

Sporocytophaga myxococcoides belongs to the group of flexible bacteria having gliding movement on solid surfaces. Vegetative cells are rod shaped, 5–8 μm long, and 0.3–0.5 μm in diameter. Resting cells (cysts) are coccoidal, 1.5 μm in diameter. Colonies are yellow. When attacking cellulose, the bacteria grow in close contact with the microfibrils.[1] The bacteria are frequently found in soil, and can be enriched by various techniques. Obtaining a pure culture requires some skill, and for isolation of the cellulases, it is advisable to acquire a stock culture from a culture collection.

---

[1] B. Berg, B. V. Hofsten, and G. Pettersson, *J. Appl. Microbiol.* **35**, 215 (1972).

The cellulases (β-1,4-glucan 4-glucohydrolase; EC 3.2.1.4) are inducible, and probably also are repressed by catabolic repression. *Sporocytophaga myxococcoides* will grow on glucose, although impurities, especially caramelization during autoclaving, may inhibit growth. Only very small amounts of cellulase will be formed in the presence of glucose. In order to obtain high cellulase activity, a medium containing cellulose or a cellulose derivative has to be used.

Culture and Harvest of Cells

Sporocytophaga myxococcoides QMB 482 (ATCC 10011) was kept on stock as shake cultures in 500-ml conical flasks with 100 ml mineral nutrient solution, containing 1.0 g $(NH_4)_2SO_4$, 0.2 g $MgSO_4 \cdot 7H_2O$, and 0.01 g $CaCl_2 \cdot 2H_2O$ in 1000 ml 0.02 M phosphate buffer, pH 7.2, to which had been added 1 ml of a trace element solution, containing 0.5 g $ZnSO_4 \cdot 7H_2O$, 0.5 g $MnSO_4 \cdot 7H_2O$, 0.1 g $CuSO_4 \cdot 5H_2O$, 0.1 g $Co(NO_3)_3 \cdot 6H_2O$, 2.0 g $Na_2MoO_4 \cdot 2H_2O$, and 2.5 g EDTA-iron salt per liter.[2] Whatman CF 11 cellulose (0.5%) was added as carbon source. The inoculum was 1–3 ml of a previous culture, and the incubation temperature 30°. The incubation lasted until the cultures had a weak yellow color. They could then be stored in a cold room for periods up to a month.

For enzyme production, the bacteria were grown in a 14-liter fermenter tank containing 10–12 liters mineral nutrient solution with 0.25% Whatman CC 41 cellulose. As inoculum 100 ml stock culture is used. Incubation temperature is 30°, rate of aeration is 250 liters/min, and stirring velocity is 100 rpm. The cultures were harvested after 2 days, when the bacterial density, measured as the protein content of the culture, was close to its maximum. Before harvesting, the culture was cooled down while aerated, and ammonium sulfate to 10% (w/v) was added to simplify the centrifugation.[2] The bacteria were separated from the culture supernatant in a high-speed continuous centrifuge (Cepa Schnell-Zentrifuge, type LE with cylinder type SK; C. Padberg, BRD) at 48,000 g. After addition of 0.02% sodium azide (as a preservative), the supernatant was concentrated in an Amicon ultrafiltration unit (Amicon, The Netherlands), consisting of a Diaflo model 402 ultrafiltration cell with membrane PM10. The slime formed by the bacteria and present in the supernatant, slows down the filtration, but it is possible to concentrate 10 liters supernatant to 15 ml in 2 weeks. The concentration is done batchwise: the concentrate after 700–800 ml supernatant is collected and the filter is washed before a new batch is added.

[2] J. P. Verma and H. H. Martin, *Arch. Microbiol.* **59,** 355 (1967).

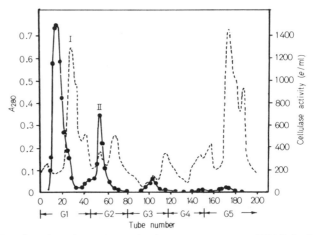

FIG. 1. Fractionation of concentrated culture supernatant on DEAE-Sephadex A-50. Column diameter 19 mm, gel height at start 29 cm. Elution with phosphate buffer, pH 7.0, and stepwise increasing ionic strength: G1, 0.02 M; G2, 0.04 M; G3, 0.1 M; G4, 0.1 M + 0.1 M NaCl; G5, 0.1 M + 0.5 M NaCl. Fraction size, 5 ml. (---) Protein measured as A_{280}; (●) cellulase activity (e, defined in text). Reprinted with permission from Osmundsvåg and Goksøyr.[4]

Measurement of Enzymatic Activity

Viscosimetry. Carboxymethylcellulose (CMC) (Koch-Light, England), degree of substitution between 0.5 and 0.6, and degree of polymerization about 500 is dissolved to 0.55% in 0.02 M phosphate buffer, pH 7.0, with 0.02% sodium azide. The solution was placed on a shaker for at least 24 hr before use. Of this solution 10 ml was mixed with 0.1–0.2 ml buffered enzyme solution in a Cannon-Fenske viscosimeter, kept in a water bath at 30°. The fall time in the viscosimeter was normally measured after incubation periods of 1, 3, 6, and 10 min. For calculation of enzyme activity, the following terms were used: η_{rel}, relative viscosity (fall time for CMC solution/fall time for buffer), and η_{sp}, specific viscosity ($\eta_{rel} - 1$). The inverse specific viscosity (η_{sp}^{-1}) is proportional to the incubation time and the amount of enzyme used.[3] The amount of enzyme (e) required to change the inverse specific viscosity by 0.001/min under the conditions described was chosen as the enzyme unit.

Reducing End Groups. The method of Nelson and Somogyi with glucose as reference standard, or other methods of comparable sensitivity can be used. Samples were removed every 10 min from an incubation mixture with CMC, and assayed for reducing end groups.

[3] R. Thomas, *Aust. J. Biol. Sci.* **9**, 159 (1956).

Purification[4]

A sample of the supernatant concentrate was put on a DEAE-Sephadex A-50 column. The column was eluted with phosphate buffer, pH 7.0, with stepwise or continuously increasing ionic strength (Fig. 1). Owing to the acidic nature of the slime, it was retained on the column, while at the same time a good separation of the proteins was obtained. Cellulase was eluted from this column in two major fractions, called cellulases I and II, and one or two minor fractions. Both major fractions were further purified on a column of Sephadex G-100 by elution with 0.02 M phosphate buffer, pH 7.0. Final purification of cellulase I was obtained by fractionation on SE-Sephadex C-50 with 0.01 M phosphate buffer, pH 5.9, to which was added a linear gradient of NaCl, starting at fraction 18 and reaching 0.2 M at fraction 61. Cellulase II was purified on DEAE-Sephadex A-50 by eluting with 0.01 M phosphate buffer, pH 7.15, to which was added a linear gradient of NaCl, starting at fraction 26 and reaching 0.2 M at fraction 94.

Comments

For testing the fractionation procedures, columns of about 20 mm diameter and 30 cm gel height are used. The first purification step is also run, using the total supernatant concentrate from a 10 liter culture, on a column 54 mm in diameter and with a gel height of 27 cm. The starting buffer was 0.01 M phosphate buffer, pH 7.4. The elution was carried out in a continuous NaCl gradient, starting with 0.1 M. The result was similar to that in Fig. 1. For step 2, Sephadex G-75 (column diameter, 19 mm, gel height 101.5 cm) was tried. It gave an elution picture very similar to that for Sephadex G-100, but seemed to give a better purification of cellulase II. Fraction volumes were in all cases 5 ml.

Characterization

As proteins, the cellulases can be characterized using standard procedures for determination of properties such as molecular weight, isoelectric point, carbohydrate content, and amino acid composition. It should be noted that molecular weight determinations of cellulases, based on Sephadex gel filtration, may be erroneous, due to affinity absorption. Thus, we found the molecular weight using this method to be about one-half of that found by sodium dodecyl sulfate–polyacrylamide gel electrophoresis. The latter is believed to be correct, and gave a molecular weight

[4] K. Osmundsvåg and J. Goksøyr, *Eur. J. Biochem.* **57**, 405 (1975).

of 46,000 for cellulase I and 52,000 for cellulase II. Using the orcinol method, the carbohydrate content was found to be low in both cellulases, leading to the conclusion that they are not glycoproteins. By isoelectric focusing, the isoelectric point (pI) was found to be 7.5 for cellulase I and 4.75 for cellulase II. pH optimum for cellulase I was about 7.0, and for cellulase II a flat optimum range between 5.5 and 7.5 was found.

Manner of Action

An impression of how the cellulases attack cellulose can be obtained by relating viscosity decrease to the formation of reducing end groups. An endoglucanase attacking glucosidic bonds randomly will cause a rapid decrease in viscosity compared to the rate of formation of reducing end groups, while an exoglucanase attacking the cellulose chain from the end will give a small reduction in viscosity compared to the rate of formation of reducing end groups. Both cellulases from *S. myxococcoides* behaved like typical endoglucanases.[4]

[37] Cellulases in *Phaseolus vulgaris*

By MARY L. DURBIN and LOWELL N. LEWIS

Introduction

Research on cellulase (β-1,4-glucan 4-glucanohydrolase, EC 3.2.1.4) in the bean plant, *Phaseolus vulgaris* L. cv. Red Kidney, has led to the discovery of a family of cellulases with differing forms and functions. The presence of these various forms had initially caused some confusion as to the role of cellulase in bean. For example, since cellulase levels often did not appear to increase significantly during abscission, the enzyme was not thought to be involved in this process. Isoelectric focusing techniques, however, revealed several forms of cellulase[1,2]: at least two acidic forms with isoelectric points (pI) of 4.5 and 4.8, respectively, and one basic form with a pI of 9.5 (9.5 cellulase). Cellulase activity also has been found to be associated with the plasma membrane.[3]

[1] L. N. Lewis and J. E. Varner, *Plant Physiol.* **46,** 194 (1970).
[2] F. T. Lew and L. N. Lewis, *Phytochemistry* **13,** 1359 (1974).
[3] D. E. Koehler, R. T. Leonard, W. Vanderwoude, A. E. Linkins, and L. N. Lewis, *Plant Physiol.* **58,** 324 (1976).

The development of methods for differentiating cellulase activities made it possible to begin sorting out the functions of the various cellulases. With the purification of the 9.5 cellulase[4] and subsequent production of antibodies to it, the amount of 9.5 cellulase activity vs acidic p*I* cellulases could be quantitated. It was found that 9.5 cellulase is synthesized *de novo* in the abscission zone of bean leaves in response to ethylene.[5] Further, synthesis of 9.5 cellulase is confined to a very narrow band of cells in the abscission zone.[6] The acidic forms of cellulase occur throughout the plant, particularly in young rapidly expanding tissue.[5,7] These forms appear to be auxin regulated and involved in growth and differentiation.[8,9]

A question still remains about the specific substrate for the cellulases in bean, since the enzymes do not readily degrade crystalline cellulose. A modified cellulose, carboxymethylcellulose (CMC), is the substrate used for the viscometric assay of cellulase activity. For this reason, some researchers feel it is more accurate to term the enzymes CM-cellulases. However, we have demonstrated that the 9.5 cellulase hydrolyzes tobacco callus cell walls,[8] and we have further tested it for activity on many substrates. In addition, we have exploited the affinity of 9.5 cellulase for its substrate and purified the enzyme on a cellulose column. Hydrolysis of the cellulose does occur to some extent, as reducing sugars are found in the peak fractions containing the purified enzyme. Since the acidic cellulases have only been partially purified, their substrate specificity has not been firmly established. It is likely that the insoluble, crystalline nature of purified cellulose inhibits hydrolysis of the substrate by the enzyme.

Assay

The assay for cellulase measures the reduction in viscosity of carboxymethylcellulose. Viscosity data are converted to intrinsic viscosity and relative units of activity. This provides a linear relationship between viscosity and enzyme activity.[10]

[4] D. E. Koehler, L. N. Lewis, L. Shannon, and M. L. Durbin, *Phytochemistry* **20,** 409 (1981).

[5] M. L. Durbin, R. Sexton, and L. N. Lewis, *Plant Cell Environ.* **4,** 67 (1981).

[6] R. Sexton, M. L. Durbin, L. N. Lewis, and W. W. Thomson, *Nature (London)* **283,** 873 (1980).

[7] L. N. Lewis and D. E. Koehler, *Planta* **146,** 1 (1979).

[8] L. N. Lewis, A. E. Linkins, S. O'Sullivan, and P. D. Reid, *Proc. Int. Conf. Plant Growth Substances, 8th* p. 708 (1975).

[9] A. E. Linkins, L. N. Lewis, and R. L. Palmer, *Plant Physiol.* **52,** 554 (1973).

[10] K. E. Almin, K. E. Eriksson, and C. Jansson, *Biochim. Biophys. Acta* **139,** 238 (1967).

Reagents and Materials

CMC. Type 7H35F (Hercules, Inc., Wilmington, DE) 1.30% (w/v) in 0.02 M Tris–HCl, pH 8.1, sterilized.

We have found the type 7H35F CMC to be a particularly good CMC substrate because of its high degree of substitution and polymerization. It also has the advantage of being readily water soluble. Even so, it can take 1–2 days of constant stirring to produce a completely homogeneous solution. As a result, it is convenient to prepare a large batch at one time and divide it into 250-ml aliquots. These are autoclaved for 30 min and stored at −20°. Repeated autoclaving and freeze-thawing should be avoided, because this reduces the viscosity. Sterility must be maintained, since CMC is a good substrate for many organisms. A 1.0-ml repipetter is a convenient way of dispensing the substrate.

Calibrated Pipet. Our pipets are calibrated to drain water in 0.6 sec. However, the pipet can be calibrated to any convenient drainage time as long as this figure is used in the calculations for intrinsic viscosity (N). The tip of a 0.1-ml pipet is flamed and drawn out to slightly constrict its flow. A length of rubber tubing is attached to the mouth of the pipet. The pipet is then clamped vertically in a vise. A pipetman tip can be used as a disposable mouthpiece on the free end of the rubber tubing. The pipet is calibrated by measuring the distance water falls in 0.6 sec. This is done by drawing water up into the pipet past the zero mark. Suction is released, and the distance the meniscus of the water falls from the zero mark in 0.6 sec is measured, timed with a stopwatch. The CMC solution, with buffer added in place of sample, should have a drainage time of 30–40 sec in that distance. This initial viscosity varies slightly with each new batch of substrate. This does not significantly affect the reproducibility of the assay.

Procedure. First, 0.2 ml of enzyme sample is added to 0.4 ml of CMC solution and thoroughly mixed by vortexing. Then, initial viscosity is measured: the liquid is drawn up into the pipet past the zero mark, suction is released, and a stopwatch is used to measure the drainage time (T_1), in seconds, for the meniscus to travel the distance previously calibrated for water. Duplicate readings should be taken and the time recorded. It is important that the tip of the pipet not touch the side of the test tube so that the liquid will drop freely. A second reading (T_2) is taken a few minutes to 20 hr later, depending on the amount of enzyme activity. The time is again recorded. The assay can be done at room temperature, provided there are no temperature fluctuations. All viscometric measurements should be performed at the same temperature.

Calculation of Activity

Relative activity (B) is calculated from the following equation:

$$B = \frac{[(N)_{T_2}^{-a} - (N)_{T_1}^{-a}]}{t} C_s$$

where $(N)_{T_2}$ is the intrinsic viscosity at any given reaction time, $(N)_{T_1}$ is the initial intrinsic viscosity, C_s is substrate concentration in g/liter (after dilution with sample or buffer), t is reaction time in hours, and a is an empirical constant determined by Almin et al.[10] to be 3.66.

Intrinsic viscosity (N) can be calculated from the following equation:

$$(N)_T = \frac{8}{C_s} \left[\left(\frac{T}{T_0} \right)^{1/8} - 1 \right]$$

where 8 is the number establishing the relationship between viscosities and substrate concentration.[10] C_s is the substrate concentration in g/liter (after dilution with sample or buffer), T is the drainage time in seconds for substrate solution in the viscometer, and T_0 is the drainage time of solvent in seconds in the viscometer (0.6 sec for water).

Example. A sample has an initial viscometric measurement (T_1) of 33.0 sec. In 30 min, the viscometric measurement is 18.3 sec (T_2). First, calculate intrinsic viscosity for both drainage times:

$$(N)_{T_1} = \frac{8}{8.8} \left[\left(\frac{33.0}{0.6} \right)^{1/8} - 1 \right] \qquad (N)_{T_2} = \frac{8}{8.8} \left[\left(\frac{18.3}{0.6} \right)^{1/8} - 1 \right]$$

$$(N)_{T_1} = 0.5911 \qquad\qquad (N)_{T_2} = 0.4845$$

Note. 8.8 is used for the substrate concentration to account for dilution by the sample.

Next relative activity (B) is calculated:

$$B = \frac{(0.4845)^{-3.66} \, 8.8 - (0.5911)^{-3.66} \, 8.8}{0.5 \text{ hr}}$$

$$B = 64.54 \text{ relative activity/0.5 hr}$$
$$B = 129.08 \text{ relative activity/hr}$$

Table I, derived from these equations, provides a quick and easy method for calculating enzyme activity from the viscometric readings. The table is a computation of B values for given drainage times. The

TABLE I
VALUES[a] FOR B

Drainage time (sec)	Drainage time (tenth of seconds)									
	0.0	0.1	0.2	0.3	0.4	0.5	0.6	0.7	0.8	0.9
45.0	42.48	42.38	42.17	42.17	42.07	41.97	41.87	41.77	41.67	41.57
44.0	43.54	43.43	43.22	43.22	43.11	43.00	42.90	42.79	42.69	42.58
43.0	44.66	44.55	44.43	44.32	44.21	44.09	43.98	43.87	43.76	43.65
42.0	45.84	45.72	45.60	45.48	45.36	45.24	45.12	45.01	44.89	44.78
41.0	47.09	46.96	46.83	46.71	46.58	46.45	46.33	46.21	46.08	45.96
40.0	48.40	48.27	48.13	48.00	47.87	47.74	47.60	47.47	47.34	47.21
39.0	49.80	49.66	49.51	49.37	49.23	49.09	48.95	48.81	48.68	48.54
38.0	51.28	51.13	50.98	50.83	50.68	50.53	50.38	50.23	50.09	49.94
37.0	52.85	52.69	52.53	52.37	52.21	52.05	51.90	51.74	51.59	51.43
36.0	54.35	54.36	54.19	54.02	53.85	53.68	53.51	53.35	53.18	53.02
35.0	56.32	56.13	55.95	55.77	55.59	55.41	55.23	55.05	54.88	54.70
34.0	58.23	58.03	57.84	57.64	57.45	57.26	57.07	56.88	56.69	56.50
33.0	60.28	60.07	59.86	59.65	59.44	59.23	59.03	58.83	58.63	58.43
32.0	62.48	62.25	62.02	61.80	61.58	61.36	61.14	60.92	60.70	60.49
31.0	64.84	64.60	64.36	64.11	63.87	63.64	63.40	63.17	62.93	62.70
30.0	67.40	67.13	66.87	66.61	66.35	66.10	65.84	65.59	65.34	65.09
29.0	70.17	69.88	69.59	69.31	69.03	68.75	68.48	68.21	67.94	67.67
28.0	73.17	72.86	72.55	72.24	71.94	71.64	71.34	71.04	70.75	70.46
27.0	76.45	76.11	75.77	75.43	75.10	74.77	74.45	74.12	73.80	73.49
26.0	80.03	79.65	79.28	78.92	78.56	78.20	77.84	77.49	77.14	76.79
25.0	83.96	83.55	83.14	82.74	82.34	81.95	81.56	81.17	80.78	80.40
24.0	88.29	87.84	87.39	86.95	86.51	86.07	85.64	85.21	84.79	84.37
23.0	93.09	92.59	92.09	91.60	91.11	90.63	90.15	89.68	89.21	88.75
22.0	98.44	97.88	97.32	96.77	96.23	95.69	95.16	94.64	94.12	93.60
21.0	104.42	103.79	103.17	102.55	101.95	101.34	100.75	100.16	99.58	99.01
20.0	111.16	110.45	109.75	109.05	108.37	107.69	107.02	106.36	105.71	105.06
19.0	118.81	118.00	117.20	116.41	115.63	114.86	114.10	113.35	112.61	111.88
18.0	127.55	126.62	125.70	124.80	123.91	123.03	122.16	121.31	120.46	119.63
17.0	137.62	136.55	135.49	134.44	133.41	132.40	131.40	130.42	129.45	128.49
16.0	149.35	148.09	146.85	145.64	144.44	143.26	142.09	140.95	139.82	138.71
15.0	163.16	161.67	160.20	158.77	157.35	155.96	154.60	153.25	151.93	150.63
14.0	179.62	177.83	176.08	174.36	172.67	171.01	169.38	167.79	166.22	164.67
13.0	199.54	197.36	193.23	193.14	191.09	189.09	187.12	185.19	183.30	181.44
12.0	224.08	221.37	218.73	216.15	213.62	211.14	208.72	206.35	204.04	201.77
11.0	254.93	251.50	248.15	244.88	241.69	238.58	235.54	232.58	229.68	226.85
10.0	294.72	290.24	285.88	281.64	277.51	273.50	269.59	265.78	262.07	258.46
9.0	347.66	341.60	335.74	330.05	324.54	319.19	314.01	308.97	304.08	299.34
8.0	420.90	412.36	404.12	396.18	388.52	381.12	373.97	367.06	360.38	353.91
7.0	527.56	514.80	502.58	490.87	479.64	468.86	458.50	448.55	438.98	429.77
6.0	693.98	673.37	653.82	635.24	617.57	600.75	584.72	569.43	554.84	540.89
5.0	980.43	943.16	908.26	875.54	844.80	815.88	788.64	762.94	738.67	715.72

[a] $(N_T^{-3.66} C_s)$ at various drainage times in seconds using a substrate concentration of 1.33% and a viscometer calibrated to drain water in 0.6 sec. $C_s = 8.8$ to account for dilution of the substrate by the sample. See text for an explanation of how to use this table.

calculations are based on a substrate concentration of 1.33% CMC and a drainage time of 0.6 sec for water. To use the table, one finds the drainage time in seconds in integers in the left column and the value in tenths of a second across the top column. Where these numbers intersect in the table is the *B* value for that drainage time. The value from the table for the initial drainage time is subtracted from the value for the final drainage time. This figure is divided by the time in hours.

Example:

	Time	Drainage time (sec)	Value from table
First reading	10:05	33.0	60.28
Second reading	10:35	18.3	124.84

Relative units cellulase/hr = (124.83 − 60.28)/0.5 = 129.08

The assay is linear over 20 hr at low enzyme concentrations. With high levels of enzyme activity, the sample should be diluted for greater accuracy. Linearity of activity vs time should be determined for a cellulase not previously assayed. Samples are often assayed in the presence of sodium chloride and Triton X-100. Neither affects the accuracy of the assay, although Triton X-100 reduces initial viscosity.[3] The pH of 8.1 for the substrate is used as an example. The pH for optimum activity of the cellulase to be assayed should be determined and always used thereafter.

Purification of 9.5 Cellulase

Preparation of Plant Material. Seedlings of *Phaseolus vulgaris* L. cv. Red Kidney, 12–14 days old, are induced for 9.5 cellulase biosynthesis, first by deblading at the distal end of the petiole with the abscission zone left intact. The plants are then sprayed with Ethrel (Amchem Products, Inc.) at a concentration of 7 mM active ingredient (2-chloroethylphosphonic acid). They can also be placed in ethylene chambers at a concentration of 20–50 ppm. After 48 hr, the laminar abscission zones are harvested.

Enzyme Extraction. All procedures are performed at 4°. The tissue is homogenized in a blender with 3 ml buffer (0.02 M Tris–HCl, pH 8.1, 3 mM EDTA) per gram of tissue. For tissue containing high levels of phenolics, it may be necessary to add insoluble polyvinylpyrrolidone (PVP). Antioxidants and protease inhibitors may also be included, if necessary. The homogenate is filtered through 40-μm mesh nylon cloth and the residue reextracted with one-half the volume of the same buffer. The residue is then extracted twice more with high salt buffer (0.02 M Tris–HCl, pH 8.1, 3 mM EDTA, 0.5 M NaCl), again at one-half the volume. The two high-salt filtrates are combined and centrifuged at 13,000 g for 20 min.

Ammonium Sulfate Fractionation. Solid ammonium sulfate is slowly added to the high-salt supernatant to a concentration of 40% ammonium sulfate (28.2 g/100 ml). This is then centrifuged at 10,000 g and the pellet discarded. The supernatant is then taken to 65% ammonium sulfate concentration by the addition of 17.4 g/100 ml ammonium sulfate. This is centrifuged and the pellet collected. The cellulase is quite stable in the 65% ammonium sulfate pellet and can be stored at 4° for several days.

Cellulose Column. CF-11 cellulose (Whatman, Inc., Clifton, NJ 07014) is thoroughly washed, first with distilled water, then with column buffer (0.02 M Tris–HCl, pH 8.1, 0.1 M NaCl). A column is poured (5 × 18 cm) and equilibrated with column buffer. The 40–65% ammonium sulfate pellet is reconstituted in column buffer and applied to the column. Care must be taken not to overload the column. When large amounts of cellulase are applied, a hard, impermeable mat can form at the surface, presumably caused by hydrolysis of the cellulose. This can cause the remaining sample and buffer to channel down the sides of the column, resulting in a poor yield. When applying a very active enzyme sample, formation of the mat can be avoided by stirring the sample into the top one-third of the column and allowing the slurry to resettle before starting the buffer wash. The column is then washed with column buffer until the OD_{280} of the eluate returns to baseline. The cellulase is then eluted by washing the column with column buffer containing 0.1 M cellobiose. The enzyme peak is concentrated by adding ammonium sulfate (59.9 g/100 ml) to an 85% concentration. We have found that cellulase has a partial affinity for many different matrixes such as dialysis tubing and membranes used in concentrating devices. Techniques conventionally employed with other proteins therefore cannot be routinely applied to cellulases. Thus, we use ammonium sulfate precipitation as a means of concentrating the enzymes. The purified enzyme also stores well as an ammonium sulfate pellet.

Purity

The procedure just described results in about a 600-fold purification.[4] The cellulase is essentially pure after one column, but a second column can be used to ensure absolute purity. The affinity of cellulase for its substrate was exploited for this purification procedure. The buffers chosen were at the optimum pH (8.1) of the enzyme and, indeed, the enzyme exhibited the best affinity for the column at this pH. Cellobiose was chosen as an elutant because it is one of the end products of cellulose hydrolysis. It has the advantage of being highly specific for eluting only cellulase from the column. The cellulase used for making antiserum was further purified on a Reisfeld gel[11] to separate the protein from the carbohydrates that resulted from hydrolysis of the column.

This procedure also proved successful for purifying avocado cellulase,[12] but not acidic forms of bean cellulase. We found some absorption to the column at a lower pH (the optima of the acidic forms), but were not able to release the enzyme from the column.

Quantitation of Cellulase Using Antibody to 9.5 Cellulase. Antibody produced in rabbits against the purified 9.5 cellulase does not cross-react with the other cellulases in bean. The cellulases thus do not appear to be structurally related, a situation that could be exploited to differentiate between 9.5 cellulase and the other cellulases. Antiserum in sufficient quantity to precipitate all the 9.5 cellulase activity is added to the sample to be assayed. Preimmune serum is added to a duplicate sample. The samples are incubated 30 min at room temperature and then for several hours at 4°. Next, 10 μl of IgGSORB (Enzyme Center, Inc., Boston, MA) is added to aid precipitation of antibody complexes. The samples are incubated for 30 min at 4° and then centrifuged at 5000 g for 10 min. The supernatants are then assayed for cellulase activity. The enzyme activity that is precipitated from the supernatant by the antibody is 9.5 cellulase. The activity remaining in the supernatant is acidic cellulase. By subtracting the amount remaining in the antibody-treated sample from the preimmune sample, one can calculate the amount of 9.5 cellulase activity.

Substrate Specificity

Purified 9.5 cellulase showed no activity against the following substrates: apple or orange pectin, tamarind seed xyloglucan, laminarin, xylan, β-1,3- or β-1,4-glucan.

The 9.5 cellulase antibody and isoelectric focusing was used to establish that acidic cellulase also had no affinity for xyloglucan. Xyloglucanase is a basic protein with a pI similar to 9.5 cellulase. Use of the antibody to 9.5 cellulase as described above removed the basic cellulase activity from the isoelectric focusing column fractions, while there was no reduction in either acidic cellulase activity or xyloglucanase activity (see Fig. 1).

[11] R. A. Reisfeld, U. J. Lewis, and D. E. William, *Nature (London)* **195,** 281 (1962).
[12] M. Awad and L. N. Lewis, *J. Food Sci.* **45,** 1625 (1980).

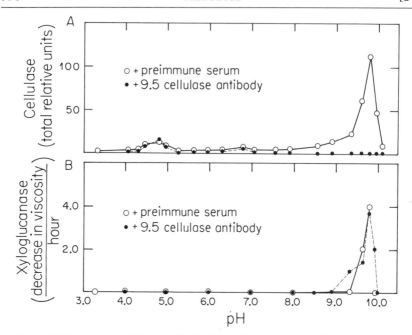

FIG. 1. (A) Immunoprecipitation of 9.5 cellulase from isoelectric focusing column fractions. Unprecipitated activity is 4.5 cellulase. (B) Xyloglucanase activity is unaffected by 9.5 cellulase antibody. The sample for the column was prepared from ethylene-treated bean abscission zones.

Summary of Properties of Cellulases of *Phaseolus vulgaris*

9.5 Cellulase. 9.5 cellulase is a monomer[4] with a molecular weight of approximately 51,000. Previously published estimates were lower[13] based on elution profiles from columns such as Sephadex. A partial affinity for the Sephadex probably retarded the elution of the cellulase, thus lowering its estimated molecular weight. We have been unable to detect any carbohydrate moieties on the protein. The isoelectric point is 9.5. The enzyme has a broad pH activity curve spanning 6.0–8.5. We found, however, that the best affinity for cellulose occurred at 8.1. There are no apparent cofactor requirements for its activity. This basic cellulase shows no structural homology with the acidic forms of cellulase, as judged by lack of cross-reactivity with the antibody to 9.5 cellulase. The 9.5 cellulase is synthesized *de novo* in response to ethylene and is inhibited by auxin. Synthesis of this form of cellulase is limited to a narrow band of cells in

[13] L. N. Lewis, F. T. Lew, P. D. Reid, and J. E. Barnes, *in* "Plant Growth Substances" (D. J. Carr, ed.), p. 234. Springer-Verlag, Berlin, Federal Republic of Germany, 1970.

the abscission zone. In bean, it appears to function in the shedding of various organs such as fruits, flowers, and leaves.

Acidic Cellulases. Less is known about the acidic forms of cellulase, since they have not been purified sufficiently. Their pH optima are in the 5.0–6.0 range. Their isoelectric points are 4.5 and 4.8. They are induced by auxin[9] and appear in high levels in young, rapidly expanding tissue.[5] These cellulases may function in loosening the cellulose fibrils of the cell wall to allow expansion and growth.

Membrane-Bound Cellulase. Cellulase is also found to be associated with the plasma membrane.[3] It is present before abscission, but declines in the abscission zone during the abscission process. The membrane cellulase has a pI of 4.5 and a pH optimum of 5.1. It is believed to be a glycoprotein, because it binds to Sepharose-concanavalin A and can be eluted with 0.2 M methyl α-D-mannopyranoside. The enzyme is partially inhibited by dextran,[14] and is strongly inhibited by D-gluconic acid lactone. The function of this form of cellulase and its relation to the soluble forms of acidic cellulase are not yet known.

[14] E. del Campillo and L. N. Lewis, *J. Cell. Biochem., Suppl.* **10B,** 20 (1986).

[38] Endoglucanase from *Clostridium thermocellum*

By THOMAS K. NG and J. G. ZEIKUS

Cellulose + H$_2$O → cellodextrins

Clostridium thermocellum produces several endoglucanases which are part of a cellulase complex termed cellulosome.[1] Five of the endoglucanases have been purified and characterized.[2–6] Most interestingly, two of these endoglucanases from *C. thermocellum* are produced by *Escherichia coli* strains carrying the genes that encode for the proteins.[5,6] The nucleotide sequence of the genes has also been determined.[7,8] Although

[1] R. Lamed and E. A. Bayer, this volume [57].
[2] T. K. Ng and J. G. Zeikus, *Biochem. J.* **199,** 341 (1981).
[3] N. Creuzet and C. Frixon, *Biochimie* **65,** 149 (1983).
[4] J. Pètre, R. Longin, and J. Millet, *Biochimie* **63,** 629 (1981).
[5] P. Béguin, P. Cornet, and J. Millet, *Biochimie* **65,** 495 (1983).
[6] D. Petre, J. Millet, R. Longin, P. Béguin, H. Girard, and J.-P. Aubert, *Biochimie* **68,** 687 (1986).
[7] P. Béguin, P. Cornet, and J.-P. Aubert, *J. Bacteriol.* **162,** 102 (1985).
[8] O. Grepinet and P. Béguin, *Nucleic Acids Res.* **14,** 1791 (1986).

an exoglucanase from *C. thermocellum* has not been identified, all the endoglucanases reported so far contain exoglucanase activity.[2-6] One of these five endoglucanases, termed endoglucanase I, will be described here.

Assay Method

Principle. The formation of free, reducing C-1 hydroxyl groups generated by internal cleavage of β-1,4 bonds is measured with dinitrosalicylic acid (DNS) using glucose as the standard.

Reagents

Carboxymethylcellulose 7HS (Hercules)

DNS (3,5-dinitrosalicylic acid) reagent: 10 g NaOH, 200 g potassium sodium tartrate, 2 g phenol, 0.5 g sodium sulfite, and 10 g DNS dissolved in the given order in 1 liter of distilled water. The reagent is stored under nitrogen gas in the dark at 4° for up to 3 months

0.05 M sodium acetate, pH 5.0, with 0.03% NaN$_3$

Procedure. Before the assay, a 1% solution of carboxymethylcellulose in acetate buffer is prepared by stirring the suspension at room temperature for 24 hr. Aliquots of 1 ml are dispensed into 13 × 100-mm tubes and kept frozen at −20°. The reaction is started by the addition of 1 ml enzyme solution to test tubes with the frozen substrate. Tubes with enzyme are vortexed and incubated at 60° for 5, 10, and 15 min. The reaction is stopped by immersion in an ice-water bath and the addition of 3 ml DNS reagent. The tubes are then heated in a boiling water bath for 15 min and cooled immediately to 4° in an ice-water bath. The intensity of the color is measured spectrophotometrically at 640 nm. The amount of reducing sugars is determined using glucose as the standard.

Definition of Unit and Specific Activity. One unit of endoglucanase is defined as the micromoles of reducing sugars as glucose equivalent released per minute under the conditions described. Specific activity is expressed as units per milligram of protein determined by the Lowry procedure.[9]

Purification Procedure

All operations except where otherwise specified are performed at 0–4°. All buffers contain 3 μM of dithiothreitol.

---

[9] O. H. Lowry, N. J. Rosebrough, A. L. Farr, and R. J. Randall, *J. Biol. Chem.* **193,** 265 (1951).

Step 1. Growth of Bacteria. *Clostridium thermocellum* strain LQRI (available as strain DSM 2360, German Culture Collection, Göttingen, FRG) is cultured under strict anaerobic condition as described by Ng *et al.*[10] For the purification of endoglucanase I, *C. thermocellum* is grown in a 14-liter fermenter with 12 liters of medium for 24 hr at 60° under constant stirring at 100 rpm and gassing with oxygen-free nitrogen at 100 ml/min. The culture medium contains per liter of distilled water: K_2HPO_4, 0.5 g; $KH_2PO_4 \cdot 3H_2O$, 1.0 g; urea, 2 g; $MgCl_2 \cdot 6H_2O$, 0.5 g; $CaCl_2 \cdot 2H_2O$, 0.05 g; $FeSO_4 \cdot 5H_2O$, 0.5 mg; 4-morpholinepropanesulfonic acid, 10 g; cysteine hydrochloride hydrate, 1 g; cellobiose, 5 g; and yeast extract, 2 g. Yeast extract used in the medium is pretreated by diafiltration to remove proteins with molecular weight greater than 10,000.[2]

Step 2. Preparation of Crude Enzyme. After 24 hr of growth, the culture broth is centrifuged at 10,000 *g* for 30 min and the supernatant is concentrated 10-fold by ultrafiltration with a 10,000 molecular weight cutoff membrane. Ammonium sulfate is added to the concentrated supernatant with gentle stirring at 80% of saturation (0.516 g/ml). After 30 min, the precipitate is collected by centrifugation, redissolved in minimal volume of 20 m*M* ammonium acetate buffer, pH 5.0, and dialyzed exhaustively by ultrafiltration against the same buffer.

Step 3. Ion-Exchange Chromatography. Dialyzed crude enzyme (400 ml with 3.34 g of protein) is applied to a 4.5 × 50 cm column of DEAE-Sephadex A-50 preequilibrated in 50 m*M* ammonium acetate buffer, pH 5.0. The column is eluted in steps with 750 ml of 50 m*M*, 600 ml of 100 m*M*, 500 ml of 600 m*M*, and 750 ml of 800 m*M* of the same buffer. Fractions eluted at 50 m*M* (270–830 ml) which contain the highest specific and total endoglucanase activity are pooled, concentrated by ultrafiltration, and reapplied to a 2.6 × 30 cm column of DEAE-Sephadex A-50 in 50 m*M* Tris–HCl buffer, pH 7.2. The column is eluted with a stepwise increase of NaCl (600 ml of 0 m*M*, 800 ml of 50 m*M*, 600 ml of 100 m*M*, and 600 ml of 150 m*M*) in the same buffer. The protein eluted at 50 m*M* NaCl contains 60% of the total endoglucanase activity. Fractions are pooled and refractionated on a 2.6 × 30 cm SP-Sephadex C-50 column in 20 m*M* sodium citrate buffer, pH 3.5. The column is eluted with a continuous NaCl gradient (0–300 m*M*) in 400 ml of citrate buffer and protein with the highest activity elutes at approximately 200 m*M* NaCl. The fractions are pooled, dialyzed against 50 m*M* sodium acetate buffer, pH 5.0, and concentrated 20-fold by ultrafiltration.

Step 4. Column Electrophoresis. Electrophoretic fractionation is performed with a preparative electrophoresis apparatus described else-

[10] T. K. Ng, P. J. Weimer, and J. G. Zeikus, *Arch. Microbiol.* **114**, 1 (1977).

TABLE I

PURIFICATION OF ENDOGLUCANASE I FROM *C. thermocellum*[a]

Procedure	Volume (ml)	Total activity (units)	Total protein (mg)	Specific activity (units/mg)	Yield (%)
Cell-free supernatant	12,000	20,300	6,720	3.0	—
Ultrafiltrate	1,000	17,800	4,750	3.7	87.6
(NH₄)₂SO₄ ppt.	400	13,800	3,370	4.2	68.3
DEAE-Sephadex A50-I	400	6,400	624	10.3	31.6
DEAE-Sephadex A50-II	60	2,750	119	23.0	13.6
SP-Sephadex C-50	25	975	21	45.9	4.8
Preparative gel electrophoresis	10	514	7.9	65.1	2.5

[a] From Ng and Zeikus.[2] Reprinted by permission of the Biochemical Society, London.

where.[11,12] A 4 cm length of 8% separating gel with a 0.5 cm section of 4% stacking gel is used. The polyacrylamide gels are prepared by mixing solutions of 0.3 M Tris–HCl, pH 7.5, buffer with 0.23% of N,N,N',N'-tetramethylenediamine, 40% acrylamide with 0.8% of N,N'-methylenebisacrylamide, water, and 0.14% ammonium persulfate in the ratio of 1 : 2 : 1 : 4 for 8% gel and 1 : 1 : 2 : 4 for 4% gel. The upper reservoir buffer consists of 0.8 M Tris-sulfate, pH 8.0, and the lower reservoir buffer is twice that strength. Both reservoirs are fed constantly with buffer at 1.0 ml/min, while the elution chamber is swept with 0.36 M Tris–HCl buffer, pH 7.6, at 1.0 ml/min. A 2-ml sample with 43 mg of protein in 10% glycerol is applied to the column and the potential across the electrodes is set at 400 V (~34 mA); 2.5-ml fractions are collected. After 16 hr, endoglucanase I is eluted between fractions 384 and 389.

The purification procedure is summarized in Table I. A 22-fold increase in specific activity with 2.5% recovery in total activity is achieved.

Properties

Physical Properties. The endoglucanase I is a glycoprotein with ~11% carbohydrate. The molecular weights determined by sedimentation equilibrium analysis, amino acid composition analysis, and polyacrylamide gel electrophoresis are 83,000, 88,000, and 94,000, respectively. The molar absorption coefficient at 280 nm is 53,750 M^{-1} cm^{-1} and the isoelectric

[11] T. Jovin, A. Chambach, and M. A. Naughton, *Anal. Biochem.* **9**, 351 (1964).

[12] A commercial apparatus is available as PolyPrep 200 from Buchler, Fort Lee, New Jersey.

point is 6.72. The pH and temperature optima are 5.2 and 62°. The enzyme contains 2.5% methionine but no cysteine.

Specificity. The enzyme catalyzes most actively the hydrolysis of carboxymethylcellulose. Activity toward cellodextrins increases with the degree of polymerization, with cellopentaose having the highest activity. Cellotriose and cellobiose are not hydrolyzed. The enzyme–substrate complex formation prefers cellotriosyl units over cellobiosyl units and that from glucosyl units is rare. Avicel and acid-swollen cellulose are attacked but the activity is very low.

Kinetics Constants. The $[S]_{0.5v}$ and V_{max} for cellopentaose and cellohexaose are 2.30×10^{-3} M, 39.3 μmol/min/mg protein (54.5 S^{-1} assuming a molecular weight of 83,000), and 0.56×10^{-3} M, 58.7 μmol/min/mg protein (81.2 S^{-1}), respectively.

[39] Crystalline Endoglucanase D of *Clostridium thermocellum* Overproduced in *Escherichia coli*

By Pierre Béguin, Gwennaël Joliff, Michel Juy,
Adolfo G. Amit, Jacqueline Millet, Roberto J. Poljak,
and Jean-Paul Aubert

Introduction

Clostridium thermocellum, a thermophilic anaerobe, has been shown to produce a very active cellulase complex[1] composed of a large number of individual enzymes.[2] Attempts to resolve the complex into its active subunits have not been successful and only two endoglucanases (EG) (EC 3.2.1.4, cellulase) have been purified from culture supernatant.[3,4] Construction of *C. thermocellum* genomic libraries in *Escherichia coli*[5–7] showed that, as a general rule, *C. thermocellum* genes were readily expressed in *E. coli*. Consequently, it appeared that it was easier to purify

[1] E. A. Johnson, M. Sakajoh, G. Halliwell, A. Madia, and A. Demain, *Appl. Environ. Microbiol.* **43**, 1125 (1982).

[2] R. Lamed, E. Setter, and E. A. Bayer, *J. Bacteriol.* **156**, 828 (1983).

[3] J. Pètre, R. Longin, and J. Millet, *Biochimie* **63**, 629 (1981).

[4] T. K. Ng and J. G. Zeikus, *Biochem. J.* **199**, 341 (1981).

[5] P. Cornet, D. Tronik, J. Millet, and J.-P. Aubert, *FEMS Microbiol. Lett.* **16**, 137 (1983).

[6] J. Millet, D. Pétré, P. Béguin, O. Raynaud, and J.-P. Aubert, *FEMS Microbiol. Lett.* **29**, 145 (1985).

[7] W. Schwarz, K. Bronnenmeier, and W. L. Staudenbauer, *Biotechnol. Lett.* **7**, 859 (1985).

C. thermocellum cellulases from *E. coli* strains carrying recombinant DNA than from *C. thermocellum* culture supernatant. Four endoglucanases, EGA,[8] EGB,[9] EGC,[10] and EGD,[11] have already been isolated and studied. We report here on purification, crystallization, and properties of EGD.[11,12]

Assay Methods

Principle. Endoglucanases readily hydrolyze carboxymethylcellulose (CMC), a soluble derivative of cellulose. The reaction can be monitored by assaying for the appearance of reducing ends which are determined by the Somogyi–Nelson method[13] (Method A). Alternatively, the reduction of the average chain length of the substrate molecules can be followed viscometrically (Method B).[14]

Method A

Reagents

Buffer: 50 mM K$_2$HPO$_4$, 12.5 mM citric acid, pH 6.3 (PC buffer)

Substrate: 1.5% (w/v) CMC (Sigma, D.S. 0.7, medium viscosity) in PC buffer

Somogyi reagent: Dissolve 70.5 g Na$_2$HPO$_4$ · 12H$_2$O and 40 g sodium potassium tartrate · 4H$_2$O in 350 ml water. Slowly add 100 ml 1 N NaOH, followed by 80 ml 10% (w/v) CuSO$_4$ · 5H$_2$O and 180 g Na$_2$SO$_4$ · 10H$_2$O. Bring the volume up to 1 liter with water. Leave for 48 hr at 37° and filter through Whatman No. 1

Nelson reagent: Dissolve 25 g (NH$_4$)$_6$Mo$_7$O$_{24}$ · 4H$_2$O (ammonium heptamolybdate) in 450 ml warm water. After cooling, add 25 ml concentrated H$_2$SO$_4$ and let cool. Add 25 ml 12% (w/v) Na$_2$HAsO$_4$ · 7H$_2$O, leave at 37° for 48 hr, and filter on Whatman No. 1. The reagent should be yellow, but occasionally develops a

[8] W. H. Schwarz, F. Gräbnitz, and W. L. Staudenbauer, *Appl. Environ. Microbiol.* **51**, 1293 (1986).

[9] P. Béguin, P. Cornet, and J. Millet, *Biochimie* **65**, 495 (1983).

[10] D. Pétré, J. Millet, R. Longin, P. Béguin, H. Girard, and J.-P. Aubert, *Biochimie* **68**, 687 (1986).

[11] G. Joliff, P. Béguin, M. Juy, J. Millet, A. Ryter, R. Poljak, and J.-P. Aubert, *Bio/Technology* **4**, 896 (1986).

[12] G. Joliff, P. Béguin, J. Millet, J.-P. Aubert, P. Alzari, M. Juy, and R. Poljak, *J. Mol. Biol.* **89**, 249 (1986).

[13] N. Nelson, *J. Biol. Chem.* **153**, 375 (1944).

[14] R. Thomas, *Aust. J. Biol. Sci.* **9**, 159 (1956).

greenish tinge upon storage. The original color can be restored by adding a few drops 0.02 N KMnO$_4$.

Procedure. One milliliter of CMC solution is incubated at 60° with 0.2 ml of appropriately diluted enzyme. At defined times, 50-μl aliquots are withdrawn and transferred to 0.95 ml Somogyi reagent diluted with water 1 : 2.8 (v/v). Samples are heated for 20 min at 100°, cooled to room temperature and 0.25 ml Nelson reagent is added, followed by 3.75 ml water. After 10 min at room temperature, the absorbance is read at 650 nm. A blank corresponding to zero time incubation is subtracted. Since the kinetics of the assay are not linear, values of ΔOD_{650} should not exceed 0.150. A standard curve is determined with 0–100 nmol glucose.

Definition of Unit and Specific Activity. One unit of activity corresponds to the amount of enzyme that releases 1 μmol glucose equivalent per minute. Specific activity is expressed in units per milligram protein. Protein is determined by the Coomassie Blue binding assay,[15] using bovine serum albumin as a standard.

Method B

Substrate. CMC (1.5%) in PC buffer, as in method A.

Procedure. The assay mixture is composed of 10 ml 1.5% CMC solution, 2–x ml PC buffer, and x ml enzyme appropriately diluted in PC buffer. The mixture of CMC solution and PC buffer is equilibrated at 60° before adding the enzyme. At defined times, 1.2-ml aliquots are withdrawn and the reaction is stopped by heating for 5 min at 100°. After equilibration at room temperature (23°), the fluidity of the sample is determined by measuring the efflux time of a 0.2 ml sample in a 0.2-ml Kimax 51 No. 37022 pipet. The increase in specific fluidity $\Delta\phi_{sp}$ is calculated according to the equation:

$$\Delta\phi_{sp} = T_{PC}/(T - T_{PC}) - T_{PC}/(T_0 - T_{PC})$$

where T_{PC} is efflux time of PC buffer, T is efflux time of incubated sample, and T_0 is efflux time of nonincubated sample. The change in specific fluidity is proportional to the increase in reducing sugar equivalent with a coefficient specific for each enzyme.[14]

Purification Procedure

The procedure has been designed for the purification of EGD from *E. coli* JM101(pCT603), in which the enzyme is overproduced.[11] EGD can be

[15] J. J. Sedmak and S. E. Grossberg, *Anal. Biochem.* **79,** 544 (1977).

TABLE I

PURIFICATION OF ENDOGLUCANASE D FROM GRANULES PRODUCED BY *E. coli* JM101
(pCT603)[a]

Step	Volume (ml)	Total protein (mg)	Total activity (units)	Specific activity (units/mg)	Yield (%)
Crude extract after solubilization of granules in 5 *M* urea and dialysis	7	154	8800	57	100
Streptomycin sulfate supernatant	5.5	44	8600	195	98
Heat-treated extract	4	20	7900	395	89
Ammonium sulfate precipitate after dialysis	1	14	6000	428	68

[a] Purification is from 1 liter of an overnight culture at 37°.

recovered either from the soluble cytoplasmic fraction or from insoluble granules which form inside the cytoplasm. Although the enzyme present in the granule fraction amounts to only 22% of the total activity, it is more easily purified to homogeneity. A slightly less pure preparation can be obtained by the procedure outlined below starting from the soluble cytoplasmic fraction. Unless otherwise stated, all operations are performed at 4°. The recovery and the specific activity of the preparation after each step are summarized in Table I.

Growth Conditions. *E. coli* cells are grown overnight at 37° in Luria broth[16] containing 100 μg/ml carbenicillin. Induction of the culture with isopropyl-β-thiogalactoside (IPTG) is not required provided that the cells are harvested in stationary phase.

Step 1. Preparation of Granule Fraction. Bacteria are harvested by centrifugation (13,500 g_{max} for 20 min), resuspended in 100 ml PC buffer, and disrupted by sonication for 2 × 4 min in a Branson B-12 sonifier. The lysate is centrifuged at 1000 g_{max} for 5 min to remove cell debris. Granules present in the supernatant are pelleted at 25,000 g_{max} for 15 min, washed twice with 0.154 *M* NaCl, and dissolved in 5 ml 0.1 *M* Tris–HCl, pH 8.5, containing 5 *M* urea. After 5 min any remaining insoluble material is removed by centrifugation at 27,000 g_{max} for 15 min. The supernatant is dialyzed four times for a total of 30 hr against 500 ml 20 m*M* Tris–HCl, pH 7.7.

[16] T. Maniatis, E. F. Frisch, and J. Sambrook, "Molecular Cloning: A Laboratory Manual." Cold Spring Harbor Lab., Cold Spring Harbor, New York, 1982.

Step 2. Streptomycin Sulfate Precipitation. Ten milliliters 10% (w/v) streptomycin sulfate in PC buffer is added per gram protein to the solubilized granule fraction. After stirring for 1 hr, the precipitate is removed by centrifugation at 30,000 g_{max} for 30 min and discarded.

Step 3. Heat Treatment. The supernatant is heated with gentle swirling in a 60° water bath until the temperature of the extract reaches 60°. The preparation is then kept at 60° for a further 10 min and chilled on ice prior to centrifugation at 30,000 g_{max} for 30 min. Traces of contaminating pellet material are removed from the supernatant by filtration through a Whatman GF/C filter.

Step 4. Ammonium Sulfate Precipitation. Solid ammonium sulfate (0.226 g) is added per ml of heat-treated supernatant (40% saturation). After stirring for 45 min, the precipitate is removed by centrifugation at 30,000 g_{max} for 30 min, dissolved in 40 mM Tris–HCl, pH 7.7, and dialyzed four times for a total of 30 hr against 500 ml of the same buffer. Preparations meant to be crystallized should not be frozen, but kept at 4° with 0.06 M NaN$_3$ added as a preservative.

Purification of EGD from the Soluble Cytoplasmic Fraction. The enzyme present in the supernatant from the granule preparation (step 1) can be purified according to the same procedure (steps 2–4). Recoveries and specific activities are given in Table II.

Occasional problems due to the degradation and leakage of the dialysis bag have been encountered during the dialysis of the redissolved, concentrated enzyme after the ammonium sulfate precipitation. Therefore, unless a fully desalted preparation is required, the last dialysis is probably best omitted or replaced by another step such as gel filtration.

TABLE II
PURIFICATION OF ENDOGLUCANASE D FROM THE CYTOPLASMIC SUPERNATANT OF *E. coli* JM101 (pCT603)[a]

Step	Volume (ml)	Total protein (mg)	Total activity (units)	Specific activity (units/mg)	Yield (%)
Crude cytoplasmic supernatant	92	460	31,000	67	100
Streptomycin sulfate supernatant	90	420	31,000	74	100
Heat-treated extract	82	85	24,500	288	79
Ammonium sulfate precipitate after dialysis	5	70	22,000	315	71

[a] Purification is from 1 liter of an overnight culture at 37°.

Crystallization of EGD. Crystallization buffer: 0.1 M potassium phosphate, pH 6.2, containing 0.8 M $(NH_4)_2SO_4$ [prepared from stock solutions of potassium phosphate adjusted to pH 6.6 and nonneutralized $(NH_4)_2SO_4$]. Crystallization of EGD is performed by vapor diffusion using the hanging drop method.[17] Wells of disposable tissue culture trays (preferably with flat rims, e.g., LINBRO FB 16-24 TC, with 24 17 × 18-mm wells) are filled with 2 ml crystallization buffer, and the rim of each well is coated with vacuum grease using a glass rod. Two microliter drops of EGD solution (about 15 mg/ml protein in 40 mM Tris–HCl, pH 7.7, plus 0.06 M NaN_3) are placed on siliconized cover slips. The drops are mixed with 2 μl crystallization buffer and the cover slips are placed on top of the wells with the drops facing downward. The sides of the slips are pressed with a glass rod against the grease to ensure an air-tight seal and the tray is kept free from vibrations for 24 hr at room temperature. The X-ray intensities of native protein crystals are recorded on the flat film cassettes of a Nonius Arndt–Wonacott oscillation camera,[18] to 0.25 nm resolution, using synchrotron radiation. The crystals are rotated about the c axis with oscillation angles of 1.2°.

Properties

Purity. SDS–polyacrylamide gel electrophoresis[19] of 7 μg EGD purified from cytoplasmic granules shows no impurity detectable by staining with Coomassie Brilliant Blue R 250. Crystallizing and redissolving the enzyme did not affect its specific activity. The preparation obtained from the soluble cytoplasmic fraction displays a minor band (<10%) with M_r 63,000.

Heat Stability. Upon assaying EGD at various temperatures for 20 min, no deviation from Arrhenius' law is observed up to 60°. However, the enzyme is denatured above 65°.

Physical Properties. The migration of the denatured and reduced EGD on SDS–polyacrylamide gels corresponds to a molecular weight of 65,000. The same value was found by sedimentation equilibrium of the native protein,[11] showing that the enzyme is monomeric. The isoelectric point of EGD determined by isoelectric focusing on polyacrylamide gels is 5.4.

Crystals. EGD crystals appear as hexagonal prisms terminated by

[17] A. Wlodawer and K. O. Hodgson, *Proc. Natl. Acad. Sci. U.S.A.* **72**, 398 (1975).
[18] V. W. Arndt and A. J. Wonacott, "The Rotation Method in Crystallography." North-Holland, Amsterdam, 1972.
[19] U. K. Laemmli, *Nature (London)* **227**, 680 (1970).

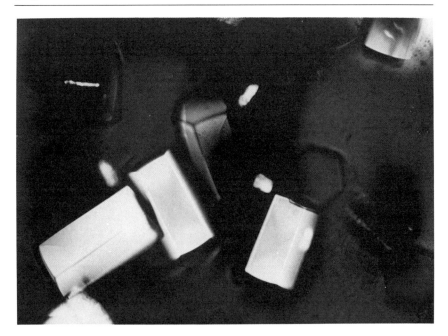

FIG. 1. Crystals of endoglucanase D.

bipyramidal faces, with a size up to $1.2 \times 0.2 \times 0.2$ mm (Fig. 1). The crystals diffract X rays to a resolution of about 0.25 nm using synchrotron radiation, $\lambda = 0.141$ nm. They are suitable for a high resolution structure determination. X-Ray diffraction analysis on precession cameras shows the space group to be trigonal $P3_1 12$ (or $P3_2 12$) with $a = b = 9.91$ (± 0.01) nm and $c = 19.26$ (± 0.02) nm. The unit cell dimensions and symmetry, compared to those of other crystalline proteins, suggest that there is one molecule of endoglucanase D in the asymmetric unit. The assumption of two molecules per asymmetric unit would imply a solvent content of about 40%, an unusually small value. With one molecule per asymmetric unit, the solvent content would be about 70%.[20] A search for multiple heavy atom isomorphous substitutions has been started. Intensity changes have been detected in heavy atom derivatives obtained using CH_3HgCl, K_2PtCl_4, and $Nb_6Cl_{14} \cdot 8H_2O$.

Homology with Other Cellulases. The amino acid sequence of EGD derived from the nucleotide sequence of the corresponding gene[21] shows no homology with cellulases or β-glucanases from other organisms whose

[20] B. W. Matthews, *J. Mol. Biol.* **33**, 491 (1968).
[21] G. Joliff, P. Béguin, and J.-P. Aubert, *Nucleic Acids Res.* **14**, 8605 (1986).

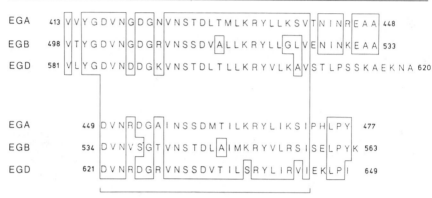

Reiterated region

FIG. 2. Homology between the carboxy-terminal regions of *C. thermocellum* endoglucanases A, B, and D. Numbers refer to the position of residues within the sequence of each protein. In each case, the last displayed amino acid is the COOH-terminal residue of the protein. The six homologous, reiterated segments of 24 amino acids each are lined up with each other. Boxes include amino acids which are identical or have closely related chemical properties (V, L, I, M; K, R; S, T). Reprinted from Joliff *et al.*[21] with permission.

sequence has been determined so far.[22–26] However, comparison with *C. thermocellum* EGA and EGB shows that a region of 65–69 amino acids located at the COOH-end of the three enzymes is highly conserved.[21] This region is itself subdivided into two homologous regions of 24 amino acids each (Fig. 2).

Catalytic Properties. The hydrolysis of CMC by EGD shows an optimum pH of 6, more than 50% maximum activity being observed between pH 4.8 and 8.2. The activation energy is 26 kJ mol^{-1}. Under the assay conditions described above, the ratio between the increase of the specific fluidity of CMC and the increase in reducing sugar equivalents is equal to 0.6 mM^{-1}. Rates of hydrolysis of Walseth cellulose, *p*-nitrophenyl-β-cellobioside, and lichenan are 4.7, 2.9, and 26 units/mg protein, respectively. Cellulose MN 300, Avicel, xylan, *p*-nitrophenyl-β-glucoside, cellobiose, and cellotriose are not hydrolyzed.

[22] G. O'Neill, S. H. Goh, R. A. J. Warren, D. G. Kilburn, and R. C. Miller, Jr., *Gene* **44**, 325 (1986).

[23] W. K. R. Wong, B. Gerhard, Z. M. Guo, D. G. Kilburn, R. A. J. Warren, and R. C. Miller, Jr., *Gene* **44**, 315 (1986).

[24] S. Shoemaker, V. Schweickaert, M. Ladner, D. Gelfand, S. Kwok, K. Myambo, and M. Innis, *Bio/Technology* **1**, 691 (1983).

[25] S. Shoemaker, D. H. Gelfand, M. A. Innis, S. Y. Kwok, M. B. Ladner, and V. Schweickaert, European Patent Application 137,280 (1984).

[26] N. Murphy, D. J. McConnell, and B. A. Cantwell, *Nucleic Acids Res.* **12**, 5355 (1984).

[40] Isolation of 1,4-β-D-Glucan 4-Glucanohydrolases of *Talaromyces emersonii*

By MICHAEL P. COUGHLAN and AIDAN P. MOLONEY

Various strains of the moderately thermophilic fungus, *Talaromyces emersonii*, produce a complete cellulase system when grown on media containing cellulose as carbon source.[1-6] *Talaromyces emersonii* CBS814.70, the most active strain in this context, was obtained from Centraal Bureau voor Schimmelcultures, Baarn, The Netherlands. The extracellular system it produces is comprised of four endocellulases (1,4-β-D-glucan 4-glucanohydrolase, EC 3.2.1.4, cellulase), at least an equal number of exocellulases (1,4-β-D-glucan cellobiohydrolase, EC 3.2.1.91, cellulose 1,4-β-cellobiosidase), and three β-glucosidases (β-D-glucoside glucohydrolase, EC 3.2.1.21).[4,7-11] Procedures for the preparation of the endocellulases, published in detail elsewhere,[10] are given below, while those for isolation of the β-glucosidases are outlined later in this volume.[12]

Growth of the Organism

Talaromyces emersonii CBS814.70 was routinely cultured on Sabouraud dextrose agar medium at 45° for 3–5 days. The plates were then sealed and stored at 30° until required for use. Under these conditions the cultures remained viable for at least 1 month. For longer term storage freeze-drying was shown to be effective. However, it should be noted that the latter procedure is not feasible with cellulase-hyperproducing mutants of this strain.[6] An alternative procedure for long-term storage of both wild

[1] M. A. Folan and M. P. Coughlan, *Int. J. Biochem.* **9**, 659 (1978).
[2] M. A. Folan and M. P. Coughlan, *Int. J. Biochem.* **10**, 505 (1979).
[3] M. A. Folan and M. P. Coughlan, *Int. J. Biochem.* **13**, 243 (1981).
[4] A. McHale and M. P. Coughlan, *Biochim. Biophys. Acta* **662**, 145 (1981).
[5] A. P. Moloney, P. J. Considine, and M. P. Coughlan, *Biotechnol. Bioeng.* **25**, 1169 (1983).
[6] A. P. Moloney, T. J. Hackett, P. J. Considine, and M. P. Coughlan, *Enzyme Microbiol. Technol.* **5**, 260 (1983).
[7] A. McHale and M. P. Coughlan, *FEBS Lett.* **117**, 319 (1980).
[8] A. McHale and M. P. Coughlan, *Biochim. Biophys. Acta* **662**, 152 (1981).
[9] A. McHale and M. P. Coughlan, *J. Gen. Microbiol.* **128**, 2327 (1982).
[10] A. P. Moloney, S. I. McCrae, T. M. Wood, and M. P. Coughlan, *Biochem. J.* **225**, 365 (1985).
[11] International Union of Biochemistry, "Enzyme Nomenclature." Academic Press, New York, 1984.
[12] M. P. Coughlan and A. McHale, this volume [51].

type and mutant strains is to suspend agar squares containing mycelial mat in glycerol and keep at $-70°$. For the preparation of inocula and for larger scale fermentations the following medium was used: crystalline cellulose (Solka-Floc or Avicel) 20 g; corn steep liquor, 5 g; yeast extract, 0.5 g; KH_2PO_4, 5 g; $MgCl_2$, 0.5 g; $FeSO_4 \cdot 7H_2O$, 12.5 mg; $CaCl_2 \cdot 2H_2O$, 5 mg; and 2.5 mg each of $MnSO_4 \cdot 4H_2O$, $ZnSO_4 \cdot 7H_2O$, Na_2MoO_4, and $Co(NO_3) \cdot 6H_2O$. The pH was adjusted to 4.5 with NaOH and the volume made to 1.0 liter with tap water before autoclaving. During autoclaving of the 10-liter fermenter, the sparger was covered with aluminum foil to prevent clogging by the insoluble cellulose. For the preparation of fermentation inocula, two 250-ml Erlenmeyer flasks each containing 100 ml of the above medium were inoculated with a piece of mycelial mat and incubated at 45° on a rotary shaker at 250 rpm for 48 hr. The contents of both flasks were then transferred aseptically to 10 liters of the above medium in the fermenter. During fermentation the temperature of the medium was maintained at 45°, aeration at 10 liters/min, and agitation at 400 rpm. The course of fermentation was monitored by removing samples of medium at intervals and measuring pH, cell wet weight, cell protein concentration, and the total cellulase activity of the filtrate.[1,13] Culture fluids were harvested at 3–5 days and filtered through Celite. When necessary the filtrate was freeze-dried and resolubilized in 0.1 M sodium acetate buffer, pH 5, for use.

Reactions Catalyzed by Endocellulases

Endocellulases hydrolyze internal 1,4-β-D-glycosidic linkages in cellulose, soluble derivatives thereof (e.g., carboxymethyl cellulose), lichenin, and cereal β-D-glucans.

Endocellulase Assays

Endocellulase activity was determined by measuring the release of reducing equivalents after incubation of 5 μl of culture filtrate with 6% (w/v) low-viscosity carboxymethylcellulose in 0.1 M sodium acetate buffer, pH 5, at 60° for 10 min or less, depending on linearity. The reducing equivalents were measured by the dinitrosalicylate method[14] or by the Somogyi–Nelson method.[15] Endocellulase activity was also measured by determining the decrease in viscosity of a 1.0% (w/v) solution of Cellofas

[13] M. Mandels, R. E. Andreotti, and C. Roche, *Biotechnol. Bioeng. Symp.* **6**, 21 (1976).
[14] G. L. Miller, *Anal. Biochem.* **31**, 426 (1959).
[15] N. Nelson, *J. Biol. Chem.* **153**, 376 (1952).

B in 0.1 M sodium acetate buffer, pH 5.4, containing 0.01% (w/v) Merthiolate (thimerosal) after incubation with a sample of enzyme at 30° for
30 min.[16]

Reagents

Sabouraud dextrose agar and yeast extract (Oxoid, Basingstoke,
    Hants, UK)
Solka-Floc, type SW40 (James River Corp., Berlin, NH)
Avicel, type 105 (FMC Corp., Philadelphia, PA)
Corn steep liquor (Biocon Ltd., Carrigaline, Co., Cork, Ireland)
Sephadex G-150 and DEAE-Sephadex A-50 (Phamacia Fine Chemi-
    cals, Milton Keynes, Bucks, UK)
0.1 M sodium acetate buffer, pH 5
6% (w/v) CM-cellulose (type C8758; viscosity of 2% aqueous solu-
    tion at 25° is 10–20 cP; Sigma Chemical Co., Poole, Dorset, UK) in
    above buffer
1% (w/v) CM-cellulose (Cellofas B; degree of substitution 0.5; I.C.I.,
    Stevenston, Ayrshire, Scotland) in above buffer containing 0.01%
    (w/v) Merthiolate
0.1% (w/v) Coomassie Blue in 50% (w/v) trichloroacetic acid
Dinitrosalicylate reagent,[14] prepared as follows: to 10 g of dinitrosa-
    licylate in 200 ml of 2 N NaOH add 300 g of sodium potassium
    tartarate in 500 ml distilled water. Heat to dissolve reagents and
    make to 1 liter with distilled water

Purification of the Endocellulases

Step 1. Ammonium Sulfate Precipitation. Ammonium sulfate was
added to culture filtrate at 4° to a concentration of 3.9 M. The mixture was
stirred for 1 hr and the precipitated protein was collected by centrifugation at 2500 g, resuspended in distilled water, dialyzed overnight against
distilled water, and then freeze-dried. The dialysis tubing was not degraded by the filtrate at 4°.

Step 2. Gel Filtration. Approximately 1 g of freeze-dried material was
resolubilized in a minimal volume of 0.1 M potassium acetate buffer, pH
3.5, containing 0.1 M NaCl and subjected to gel filtration on a column
(4.5 × 72 cm) of Sephadex G-150. The column was irrigated at 4° with the
same buffer at a flow rate of 50 ml/hr. Endocellulase activity eluted as a
single symmetrical peak at an elution volume centered at 800 ml. The
appropriate fractions were pooled, concentrated approximately 10-fold by

[16] T. M. Wood and S.I. McCrae, *Biochem. J.* **128**, 1183 (1972).

using an Amicon ultrafiltration device with a PM10 membrane, and de-salted by passage of several volumes of distilled water through the appa-ratus. The concentrated material, which contains most of the original exocellulase in addition to endocellulase activity but which is substan-tially free from β-glucosidase activity, was then freeze-dried.

Step 3. First Ion-Exchange Chromatography. A 50 mg portion of the freeze-dried material above was redissolved in a minimal volume of 50 mM sodium acetate buffer, pH 5, containing 0.15 M NaCl and applied to a DEAE-Sephadex A-50 ion-exchange column (1.7 × 30 cm). The column was irrigated at 4° with 3–4 column volumes of the above buffer and then with 500 ml of a convex exponential gradient, 0.15–0.35 M, of NaCl in 50 mM sodium acetate buffer, pH 5. The gradient was set up using two reservoirs. One, a cylindrical vessel, contained 250 ml of 0.35 M NaCl in 50 mM sodium acetate buffer, pH 5. This was connected to a conical vessel containing 0.15 M NaCl in 50 mM sodium acetate buffer, pH 5. The levels of the solutions in both flasks were the same. The conical flask, which was the mixing flask, was in turn connected to the column. En-doglucanase and exoglucanase activities eluted separately as single sym-metrical peaks at ionic strength values corresponding to 0.0175 and 0.23 M NaCl, respectively. The endoglucanase fractions were pooled, dialyzed against distilled water, and freeze-dried.

Step 4. Second Ion-Exchange Chromatography. The above material was resolubilized in 50 mM sodium acetate buffer, pH 4.3, containing 0.1 M NaCl and applied to a DEAE-Sephadex ion-exchange column (1.7 × 45 cm) equilibrated with the same buffer. Two column volumes of starting buffer were run through at 4° followed by 600 ml of a linear gradient (0.11–0.17 M) of NaCl in 50 mM sodium acetate buffer, pH 4.3, and then by 200 ml of a linear gradient (0.17–0.22 M) of NaCl in the same buffer. Four overlapping peaks of endocellulase activity were eluted at volumes from 300 to 700 ml. Fractions corresponding to the best cut from each peak were pooled separately, dialyzed against distilled water, and freeze-dried.

Step 5. Electrophoresis (Analytical and Preparative). Endocellulase preparations at all stages of purification were examined for homogeneity by vertical slab gel electrophoresis in homogeneous 5% (w/v) polyacryl-amide gels using the buffer system described by Reisfeld et al.[17] Samples were applied in 10% (w/v) sucrose in the electrode buffer. Bromocresol Purple (0.04%, w/v) was used as the marker dye, and representative portions (see below) of the gels were stained for protein with 0.1% (w/v) Coomassie Blue in 50% (w/v) trichloroacetic acid. The above electropho-retic procedure was also used for preparative purposes. Each of the four

[17] R. A. Reisfeld, U. J. Lewis, and D. E. Williams, Nature (London) 195, 281 (1962).

freeze-dried endocellulase preparations from Step 4 above were individually dissolved in electrode buffer and electrophoresed for 1200 V-h. Thin strips cut from each edge of the developed gel were stained for protein and realigned with the unstained gel. Transverse strips corresponding to each of the major protein bands so located were then cut from the gel. These were homogenized, using a Potter–Elvehjem homogenizer, in 5 ml of 0.1 M sodium acetate buffer, pH 5. The homogenates were left at 4° for 4 hr so as to allow protein to leach into solution, and then centrifuged at 13,000 g for 1 hr at 4°. The pellet was resuspended in buffer, homogenized as before, and stored overnight at 4°, and then again centrifuged. After a total of four washes enzyme activity could no longer be extracted. Typically, at least 60% of the protein applied in the electrophoresis step could be recovered. The pooled supernatants corresponding to each of the endocellulases were dialyzed against distilled water and freeze-dried if necessary.

Purity of the Endocellulase Preparations

Samples of each of the freeze-dried preparations above were dissolved in electrode buffer and subjected to electrophoresis as described. Staining for protein indicated the presence of a single band in each preparation. The four endocellulases were designated I to IV, the distance of migration of each toward the anode under the conditions used being I, 4.8 cm; II, 5.35 cm; III, 5.9 cm; IV, 6.2 cm. Each preparation was also found to be homogeneous as judged by reductive sodium dodecyl sulfate–polyacrylamide gradient gel electrophoresis and by isoelectric focusing in polyacrylamide gel. The M_r value in each case was 35,000. The pI values were as follows: I, 3.19; II, 3.08; III, 2.93–3.0; IV, 2.86.

Properties of the Endocellulases

The properties of these enzymes, which have been described in detail elsewhere,[10,18] may be summarized as follows: Activity in each case is optimal at pH 5.5–5.8 and at 75–80°. Half-lives at pH 5 and 75° were from 2 to 4 hr. As judged by the release of product, none of the purified forms acted on cotton, Avicel, filter paper, phosphocellulose, DEAE-cellulose, dextran, DEAE-Sephadex, inulin, laminarin, xylan, p-nitrophenyl-β-D-xyloside, cellobiose, or p-nitrophenyl-β-D-glucoside even with long incubation times. By contrast, phosphoric acid-swollen cellulose, CM-cellulose of various degrees of substitution, methylcellulose, and cellooli-

---

[18] A. P. Moloney and M. P. Coughlan, *Biochem. Soc. Trans.* **13**, 457 (1985).

gosaccharides (from cellohexaose to cellotriose) were active substrates. The specific activity with any individual substrate was the same for each enzyme as was the ratio of activity from one substrate to the next. After 52-hr incubation of any one of the endocellulases with phosphoric acid-swollen filter paper the products were shown to be glucose and cellobiose with smaller amounts of higher cellooligosaccharides. The ratio of reducing sugar to glucose in hydrolyzates was much the same (3.5–4.2) for each enzyme. Similarly, the products of hydrolysis of cellooligosaccharides after 4 hr at pH 5 and 60° were mainly glucose and cellobiose. Since cellobiose is not a substrate for any of the enzyme forms, such findings are consistent with these enzymes being true endocellulases, i.e., they participate in cellulose hydrolysis by cleavage of internal glycosidic linkages at random. This being so, one should expect the rate of release of reducing sugars on hydrolysis of CM-cellulose to be linearly related to the rate of decrease in viscosity of the solution. Such was found to be the case. Moreover, the ratio of reducing equivalents released to increase in fluidity of solution, i.e., the slope of the line, was the same for each endocellulase. It is our opinion (although further work will be required to substantiate this) that the finding of four endocellulases in the culture fluid of T. emersonii reflects differential glycosylation of a single enzyme rather than genetically determined differences in primary structure.

Acknowledgment

The author thanks the Biochemical Society for permission to quote from material published in Biochem. J., 225, 365 (1985).[10]

[41] Endo-1,4-β-glucanases of Sporotrichum pulverulentum

By K.-E. Eriksson and B. Pettersson

Introduction

Endo-1,4-β-glucanases are found in varying concentrations in cell-free culture filtrates of all wood-rotting fungi.[1] Endoglucanases are also produced by aerobic and anaerobic bacteria.[2]

Endoglucanases hydrolyze cellulose chains resulting in a rapid reduction in the degree of polymerization. Normally, three different types of

[1] K.-E. Eriksson and T. M. Wood, in "Biosynthesis and Biodegradation of Wood Components" (T. Higuchi, ed.), pp. 469–503. Academic Press, London, 1985.

[2] L. G. Ljungdahl and K.-E. Eriksson, Adv. Microb. Ecol. 8, 237 (1985).

hydrolytic enzymes are involved in the degradation of cellulose, at least in the case of crystalline cellulose. These enzymes are endo-1,4-β-glucanases, exo-1,4-β-glucanases, and 1,4-β-glucosidases. Brown rot fungi, which can degrade cellulose extensively, produce endoglucanases but appear to lack the exoglucanases. In spite of this, brown-rot fungi rapidly depolymerize cellulose but clearly by a mechanism other than a synergistic action between endo- and exoglucanases.[1]

Assay Methods

Assay methods for the determination of cellulase activity are based on several principles including weight loss of insoluble substrates, changes in turbidity of cellulose suspensions, increase in reducing end groups, decrease in the viscosity of cellulose derivatives, colorimetric determinations, measurements of clearance zones in cellulose agar, and polarography.[3]

Soluble cellulose derivatives are convenient substrates for the determination of endoglucanase activity since they are not hydrolyzed by exoglucanases. Carboxymethylcellulose (CMC) substituted to different degrees is most frequently used. The method developed by Almin and Eriksson[4] and Almin *et al.*,[5] whereby viscosity changes can be converted to absolute units, has been used in our studies of endoglucanases from *S. pulverulentum*.[6,7]

Reagents

Sodium acetate buffer 50 mM, pH 5.0. Carboxymethylcellulose (CMC), 2.62 g/liter buffer (type 7H3SXF, Hercules Inc., Wilmington, DE). Degree of substitution (DS) = 0.88, number average molecular weight (M_n) = 142,000, viscosity average molecular weight (M_w) = 229,000.

Procedure

CMC solutions are prepared by stirring a weighed quantity of CMC in a known volume of 50 mM sodium acetate buffer, pH 5.0, for 2 hr at room

---

[3] K.-E. Eriksson and S. C. Johnsrud, *in* "Experimental Microbial Ecology" (R. G. Burns and J. H. Slater, eds.), pp. 134–153. Blackwell Scientific, London, 1982.

[4] K. E. Almin and K.-E. Eriksson, *Biochim. Biophys. Acta* **139**, 238 (1967).

[5] K. E. Almin, K.-E. Eriksson, and C. Jansson, *Biochim. Biophys. Acta* **139**, 248 (1967).

[6] K.-E. Eriksson and B. Pettersson, *Eur. J. Biochem.* **51**, 193 (1975).

[7] K. E. Almin, K.-E. Eriksson, and B. Pettersson, *Eur. J. Biochem.* **51**, 207 (1975).

temperature and then overnight at 6°. The resultant solution is either used directly or stored at −24° and thawed before use. Solutions are equilibrated at 25° prior to assay.

A Cannon-Fenske type viscometer is used for the viscometric measurements and the constants of this viscometer are given in Ref. 7. The assay is carried out as follows: 10 ml of the CMC solution is incubated with 0.1 ml of enzyme solution 25 ± 0.1°. After 15 min the enzyme–substrate solution is quickly transferred to the viscometer and the efflux time is estimated.

Calculation of Molecular Activity

Viscosimetric data can be converted into absolute enzyme units according to Almin et al.[4,5,7] The molecular activity of an enzyme, defined as the number of bonds in the substrate which are cleaved per unit time by one molecule of enzyme at a substrate concentration giving maximum degradation velocity for a given enzyme concentration, is calculated according to Almin et al.[5]

Purification Procedure

Growth of the Organism

S. pulverulentum (ATCC 32629, anamorph of Phanerochaete chrysosporium) is cultivated on a modified Norkrans medium the composition of which is given in Table I. The carbon source, Munktell's powder cellulose for chromatography (No. 400), is obtained from Grycksbo Pappersbruk (STORA, Grycksbo, Sweden). Cultures are grown at 30 ± 1° in 1-liter conical flasks on a rotary shaker operated at 150 rpm with a shaking diameter of 40 mm.

Flasks of culture medium may be inoculated either with fungal mycelium or spores. To obtain reproducible results it is important that the inoculation is carefully standardized. A recommended spore density for start up of growth cultures is 10^6–10^7 spores/ml.

After 6 days growth the solids are filtered off using a sintered glass funnel and the culture solution, normally 30 liters/batch, concentrated approximately 30-fold with a Diaflo membrane (UM10) ultrafiltration system (Amicon Corp., Cambridge, MA). Proteins are then precipitated with solid ammonium sulfate, added to 90% of saturation, and collected by centrifugation. The precipitate is resuspended in cold distilled water and dialyzed in collodion tubes (Membrane filter Gesellschaft, Göttingen, Germany) for 24 hr against 20 mM ammonium acetate buffer.

TABLE I

COMPOSITION OF THE CULTURE MEDIUM FOR
GROWTH OF *Sporotrichum pulverulentum*

Component	Amount (g/liter)
$NH_4H_2PO_4$	2.0
KH_2PO_4	0.6
$MgSO_4 \cdot 7H_2O$	0.5
K_2HPO_4	0.4
$CaCl_2 \cdot 2H_2O$	0.074
Ferricitrate	0.012
$ZnSO_4 \cdot 7H_2O$	0.0066
$MnSO_4 \cdot 4H_2O$	0.005
$CoCl_2 \cdot 6H_2O$	0.001
Thiamin	0.0001
Carbon source	5.0
(Munktell's powder cellulose)	

Step 1: Fractionation of DEAE-Sephadex. The initial fractionation step is carried out on a 45 × 400 mm Sephadex A-50 column (Pharmacia Fine Chemicals, Uppsala, Sweden) using 20–100 mM ammonium acetate buffer, pH 5.0. Component separation is achieved by discontinuously increasing the buffer strength as follows: fractions 0 to end of first absorbancy peak at 280 nm (approx. fraction 40), 20 mM buffer; from end of first absorbancy peak to the end of last absorbancy peak (approx. fraction 120), 100 mM buffer. An even flow rate of 25–30 ml/hr is obtained using an LKB peristaltic pump (type 4912 A) and 12–15 ml/per fractions are collected.

This fractionation step yields three endoglucanase peaks designated T_1, T_2, and T_3. Peaks T_2 and T_3 are each resolved further, into peaks T_{2a}, T_{2b}, and T_{3a}, T_{3b}, respectively on a 15 × 350 mm Sephadex A-50 column using sodium phosphate buffer, pH 6.0, with discontinuous increases in buffer concentration over the range 20–100 mM as follows: T_2, fractions 0 ~ 35, 20 mM buffer; fractions 35 ~ 60 (end of absorbancy 280 nm), 100 mM buffer. T_3, fractions 0 ~ 52, 50 mM buffer; fractions 52 to end of absorbancy 280 (approx. fraction 80), 100 mM buffer. The flow rate is maintained at 16 ml/hr using the same pump system as above, and 8-ml fractions are collected.

Step 2: Fractionation by Gel Filtration. Further purification of all five previously obtained endoglucanase fractions is carried out on a 25 × 900 mm polyacrylamide gel P-60 column (Bio-Rad Lab., Richmond, CA) with sodium citrate buffer 50 mM, pH 4.0, serving as eluant. The column is

TABLE II

SEPARATION OF ENDO-1,4-β-GLUCANASE ACTIVITY INTO FIVE ISOENZYME FRACTIONS FROM CULTURE SOLUTION OF THE FUNGUS *Sporotrichum pulverulentum*[a]

Fractionation	Volume (ml)	A_{280}	Endo-1,4-β-glucanase				Protein total (A_{280} units)
			Activity (U/ml)	Total activity (U)	Specific activity	Yield (%)	
T_1, after DEAE-Sephadex A-50, ammonium acetate, pH 5.0	465	0.900	63	29295	70	32.1 (100)	419.0
T_1, after P-60, sodium citrate, pH 4.0	170	1.100	167	28390	152	30.7 (95.5)	187.0
T_1, after concanavalin A, 0.10 M sodium phosphate, pH 6.0	100	0.350	204	20440	584	22.4 (73)	35.0
T_1, after SP-Sephadex C-50, sodium citrate, pH 3.5	120	0.200	148	17780	740	19.5 (87)	24.0
T_2, after DEAE-Sephadex A-50, ammonium acetate, pH 5.0	34	2.200	478	16252	217	17.8 (100)	75.0
T_{2a}, after DEAE-Sephadex A-50, sodium phosphate, pH 6.0	80	0.300	91	7313	304	8.0 (45)	24.0
T_{2a}, after P-60 sodium citrate, pH 4	25	0.440	246	6150	559	6.7 (84)	11.0
T_{2a}, after SP-Sephadex C-50, sodium citrate, pH 3.5	12	0.200	247	2960	1235	3.3 (48)	2.4
T_{2b}, after DEAE-Sephadex A-50, sodium phosphate, pH 6.0	70	0.310	93	6500	300	7.1 (40)	21.7
T_{2b}, after P-60, sodium citrate, pH 4.0	20	0.500	296	5900	592	6.5 (91)	10.0
T_{2b}, after SP-Sephadex C-50, sodium citrate, pH 3.5	10	0.220	287	2870	1300	3.2 (49)	2.2
T_3, after DEAE-Sephadex A-50, ammonium acetate, pH 5.0	37	2.105	370	13700	176	15.0 (100)	78.0

TABLE II (*continued*)

Fractionation	Volume (ml)	A_{280}	Activity (U/ml)	Total activity (U)	Specific activity	Yield (%)	Protein total (A_{280} units)
				Endo-1,4-β-glucanase			
T_{3a}, after DEAE-Sephadex A-50, sodium phosphate, pH 6.0	20	1.402	288	5750	206	6.3 (42)	28.0
T_{3a}, after P-60, sodium citrate, pH 4.0	12	1.007	425	5100	425	5.6 (89)	12.0
T_{3a}, after SP-Sephadex C-50, sodium citrate, pH 3.5	5	0.300	408	2040	1360	2.3 (40)	1.5
T_{3b}, after DEAE-Sephadex A-50, sodium phosphate, pH 6.0	19	1.500	310	5890	207	6.5 (43)	28.5
T_{3b}, after P-60, sodium citrate, pH 4.0	13	1.152	420	5450	365	6.0 (93)	15.0
T_{3b}, after SP-Sephadex C-50, sodium citrate, pH 3.5	6	0.272	365	2190	1350	2.4 (40)	1.6

[a] T_1, $T_2 \rightarrow T_{2a}$ and T_{2b}; $T_3 \rightarrow T_{3a}$ and T_{3b}. The order of the purification steps is given in Scheme 1. Specific activity is calculated as the ratio of total units of endo-1,4,-β-glucanase activity to total A_{280}. The first figure in the yield column, 32.1, 30.7, etc. indicates activity yield compared with the total amount of endo-1,4-β-glucanase activity in the original culture solution, 91200 units. The second figure in the column (within parentheses) indicates activity yield in that particular purification step. From Eriksson and Pettersson.[6]

operated at an upward flow rate of 24 ml/hr and 12-ml fractions are collected.

Step 3: Fractionation of the Endoglucanase Peak T_1 on Concanavalin A-Sepharose. Of the five endoglucanase fractions obtained only fraction T_1 is further purified on a concanavalin A-Sepharose column (concanavalin A bound to Sepharose 4B by the cyanogen bromide method, Pharmacia Fine Chemicals AB, Uppsala, Sweden). The fractionation is performed on a 10 × 80 mm column using 100 mM sodium phosphate buffer, pH 6.0. The column is operated with this buffer until the absorbancy at 280 nm decreases to almost zero (approx. after 10 fractions). The elution of the cellulase activity peak is then carried out with 0.5% α-methyl-D-mannoside in the same buffer. The column is operated at a flow rate of 6 ml/hr and 2-ml fractions are collected.

Step 4: Fractionation on SP-Sephadex C-50. The final step in the purification of all five endoglucanase fractions is carried out on a 15 × 300 mm SP-Sephadex C-50 cation exchanger (Pharmacia Fine Chemicals AB, Uppsala, Sweden) using 20–100 mM sodium citrate buffer, pH 3.5. The column is operated at a flow rate of 8 ml/hr and 4-ml fractions are collected. The column elution conditions for the five different endoglucanase activity peaks are as follows: T_1, fractions 0 ~ 18 (when $A_{280\,nm}$ decreased to almost 0), 20 mM buffer; fractions 18–27 ($A_{280\,nm}$ ~ 0), 100 mM buffer. T_{2a}, fractions 0 ~ 33 (when $A_{280\,nm}$ decreased to almost 0), 20 mM buffer; fractions 33 ~ 55 ($A_{280\,nm}$ = 0), 100 mM buffer. T_{2b}, fractions 0 ~ 25 ($A_{280\,nm}$ = 0), 20 mM buffer; fractions 25–40 ($A_{280\,nm}$ = 0), 100 mM buffer. T_{3a}, fractions 0 ~ 30 ($A_{280\,nm}$ = 0), 20 mM buffer; fractions 30 ~ 55 ($A_{280\,nm}$ almost 0), 100 mM buffer. T_{3b}, fractions 0 ~ 25 ($A_{280\,nm}$ = 0), 20 mM buffer; fractions 25–50 ($A_{280\,nm}$ = 0), 100 mM buffer.

A typical purification program is described in Scheme 1. Increases in specific activity and yields are given in Table II.

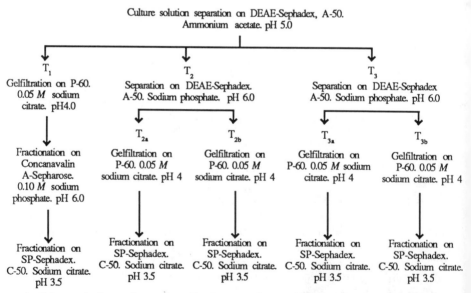

SCHEME 1. Separation and purification procedures for five endoglucanases isolated from culture solutions of the fungus *Sporotrichum pulverulentum*. From Eriksson and Pettersson.[6]

TABLE III

AMINO ACID AND CARBOHYDRATE COMPOSITIONS OF THE FIVE
ENDO-1,4-β-GLUCANASE ISOENZYMES FROM CULTURE SOLUTION
OF THE FUNGUS *Sporotrichum pulverulentum*[a]

Amino acid	Number of residues in				
	T_1	T_{2a}	T_{2b}	T_{3a}	T_{3b}
Lysine	5	7	5	7	7
Histidine	4	5	2	4	5
Arginine	5	6	2	6	5
Aspartic acid	29	36	25	36	37
Threonine	27	33	28	35	32
Serine	31	42	32	39	34
Glutamic acid	22	29	21	29	31
Proline	21	26	16	24	23
Glycine	28	36	31	35	39
Alanine	27	35	20	31	32
Half-cysteine	1	2	3	—	—
Valine	20	25	16	18	18
Methionine	1	2	1	6	5
Isoleucine	11	15	8	18	19
Leucine	21	27	12	23	25
Tyrosine	10	11	15	10	10
Phenylalanine	11	13	9	17	19
Tryptophan	4	5	6	4	4
Σ	278	355	251	342	345
Monosaccharides	21[b]	0	14[c]	11[d]	5[d]
Σ	299	355	265	353	350
Carbohydrate (%)	10.5	0	7.8	4.7	2.2

[a] Number of amino acid determinations: $T_1 = 2$; $T_{2a} = 3$; $T_{2b} = 1$;
$T_{3a} = 3$; $T_{3b} = 1$.
[b] Two glucose and 19 mannose per enzyme molecule; number of
determinations = 6.
[c] Five mannose, 7 galactose, 1 glucose, and 1 arabinose per enzyme molecule; number of determinations = 2.
[d] Carbohydrate composition has not been determined with sufficient accuracy.

Properties of the Five Isolated Endoglucanases

General. The purity of the enzyme preparations was confirmed both
by analytical isoelectric focusing on flat-bed polyacrylamide gel electrophoresis and on dodecyl sulfate gels. Purification is not considered complete until the respective enzyme preparations show up as single protein

bands with both techniques. Identification of the endoglucanase bands on polyacrylamide isoelectric focusing gels is achieved using the zymogram technique developed by Eriksson and Pettersson.[8] The isoelectric points of the five endoglucanases vary between pH 4.2 and 5.3. Molecular weights lie within the range of 28,300–37,500. Molar absorption coefficients vary between 4.0×10^4–$5.7 \times 10^4 \, M^{-1} \, cm^{-1}$. The ratio of activity distributed in the five different enzyme peaks is calculated to be $4:1:1:1:1$ with peak T_1 as the main peak.

Amino Acid and Carbohydrate Composition of the Five Endoglucanases. Samples (2–3 mg) of the different enzymes are hydrolyzed separately in 6 *M* HCl at 100° for 24 and 72 hr, respectively. Hydrolyzates are analyzed in a Biochrom automatic amino acid analyzer equipped with an Infrotronichs CRS-12 integrator. Results of these analyses are presented in Table III where average values are used for all amino acids except serine, threonine, and tyrosine where "zero time" values are calculated by first-order extrapolation, and histidine, proline, half-cysteine, and isoleucine where only the higher 72-hr values are used. The tryptophan content is determined spectrophotometrically according to Bencze and Schmid.[9]

Carbohydrate analyses are carried out using the orcinol method according to Vasseur.[10] In order to identify the monosugars, between 1 and 3 mg of the individual enzymes is hydrolyzed with 2 *N* HCl on a boiling water bath for 2 hr. Monosaccharides are determined by gas–liquid chromatography of the alditol acetate derivatives according to Sawardeker *et al.*[11]

[8] K.-E. Eriksson and B. Pettersson, *Anal. Biochem.* **56**, 618 (1973).
[9] W. L. Bencze and K. Schmid, *Anal. Chem.* **29**, 1193 (1957).
[10] E. Vasseur, *Acta Chem. Scand.* **2**, 693 (1948).
[11] J. S. Sawardeker, J. H. Stoneker, and A. Jeanes, *Anal. Chem.* **37**, 1603 (1965).

[42] Carboxymethylcellulase from *Sclerotium rolfsii*

By WILLIAM A. LINDNER

Carboxymethylcellulases are predominantly endoglucanases able to catalyze the hydrolysis of glycosidic bonds in the soluble, substituted cellulose, carboxymethylcellulose (CMC). These enzymes, formerly called C_x, are important components of the cellulase complex that catalyzes the degradation of crystalline cellulose.

A method for purifying a carboxymethylcellulase from the fungus, *Sclerotium rolfsii*, has been published elsewhere.[1] An alternative method, requiring less specialized equipment than the published method is described here for purifying this enzyme, designated F2.

Assay Method

Principle. During hydrolysis of CMC, reducing sugars are generated. Timed samples are withdrawn from the reaction mixture and assayed for reducing sugar using a modification of the improved Somogyi–Nelson method.[2] Samples assayed in this way are used to construct a linear progress curve for the measurement of initial velocity.

Reagents

Substrate solution: 20 mg/ml CMC (low viscosity, degree of substitution 0.7, Sigma Chemical Co., St. Louis, MO) in acetic acid–NaOH buffer, 0.05 M, pH 4.5. The resulting pH is about 4.9
Somogyi–Nelson reagents
Reagent A: 25 g of Na_2CO_3, 25 g of potassium sodium tartrate, 20 g of $NaHCO_3$, and 200 g of Na_2SO_4 in 1.0 liter
Reagent B: 30 g of $CuSO_4 \cdot 5H_2O$ dissolved in 200 ml of water containing 0.4 ml of concentrated sulfuric acid
Reagent C: 1 part of B added to 25 parts of A, made fresh before use
Arsenomolybdate reagent: Solution A: 25 g of ammonium molybdate made up to 450 ml with water. To this is added 21 ml of concentrated sulfuric acid. Solution B: 3 g of $Na_2HAsO_4 \cdot 7H_2O$ in 25 ml of water. Solutions A and B are mixed, made to 500 ml with water, and incubated at 55° for 30 min, and stored in a dark bottle
Procedure. Dry CMC is weighed into a test tube, a small amount of buffer is added, and the mixture is ground into a smooth transparent paste with a glass rod. Thereafter more buffer is added, with stirring, to the required volume. Enzyme in 10 μl is added to 3.0 ml of substrate solution and the resulting mixture incubated at 50°. At intervals, 0.5 ml of reaction mixture is transferred to a separate tube containing 1.0 ml of reagent C and 0.5 ml of water, thereby terminating the reaction. After mixing with reagent C, samples are placed in a boiling water bath for 20 min, cooled, and mixed rapidly with 1.0 ml of arsenomolybdate reagent. After 10 min, the contents are made up to 12.5 ml and insoluble material is removed by centrifugation before reading optical density at 520 nm.

[1] W. A. Lindner, C. Dennison, and R. K. Berry, *Biochim. Biophys. Acta* **746**, 160 (1983).
[2] J. P. Marais, J. L. De Wit, and G. V. Quicke, *Anal. Biochem.* **15**, 373 (1966).

Enzyme activity in column effluents is assayed by incubating 10 μl of enzyme with 0.2 ml of substrate solution for 20 min at room temperature (22–24°) before determining reducing sugar. Although this less rigorous method of assay gives a measure of average, rather than initial velocity, it saves time and conserves enzyme.

Definition of Unit. Owing to the often ill-defined nature of the substrate and the nonstoichiometric measurement of the product, an arbitrary unit must be used. A serviceable definition of a unit is the amount of enzyme that catalyzes the formation of 1 μmol of reducing sugar, measured as glucose, per minute under the assay conditions.

Purification Procedure

Reagents

Culture medium[3] (in grams per liter): $(NH_4)_2SO_4$ (1.4), KH_2PO_4 (2.0), urea (0.3), $CaCl_2 \cdot 2H_2O$ (0.4), $MgSO_4 \cdot 7H_2O$ (0.3), proteose peptone (0.75), Tween 80 (2.2), Avicel (10). To this is added 1.0 ml of a solution containing (in grams per liter): $FeSO_4 \cdot 7H_2O$ (9.1), $CoCl_2$ (2.0), $MnSO_4 \cdot 4H_2O$ (3.0), and $ZnCl_2 \cdot 7H_2O$ (3.0)
Acetone, redistilled
Avicel PH 102 (FMC Corporation, Philadelphia, PA)
Buffer A: 1.33 g of $Na_2HPO_4 \cdot 2H_2O$ and 1.73 g of $NaH_2PO_4 \cdot H_2O$ in 1.0 liter
SP-Sephadex C-50
Formic acid–NaOH buffers, 0.05 M, pH 3.5, 3.8, 4.1, and 4.7
BioGel P-100, 100–200 mesh
Buffer B: sodium phosphate–NaOH buffer, 0.05 M, pH 6.0, containing 0.02% (w/v) NaN_3

Growth of the Fungus. *Sclerotium rolfsii*, strain T25 (PREM 45782), was obtained from the National Fungus Collection Plant Protection Research Institute in Pretoria, South Africa. The fungus is grown on potato dextrose agar, at room temperature, until an even mycelial mat covers the agar surface (about 1 week). At this stage, before marked sclerotium formation, ~1 cm^2 of agar is cut out and transferred to 1-liter Erlenmeyer flasks containing 400 ml of culture medium. Flasks are shaken at 200 rpm for 10–12 days in a shaker (New Brunswick) at 31°, during which time the pH of the medium drops to below 3. The purification procedure is outlined in Table I.

[3] E. T. Reese and A. Maguire, *Dev. Ind. Microbiol.* **12**, 212 (1971).

TABLE I

PURIFICATION OF CARBOXYMETHYLCELLULASE F2 FROM *Sclerotium rolfsii*

Fractions	Total units	Concentration of protein[a] (mg/ml)	Specific activity (units/mg protein)
1. Crude extract	10,500	0.45	17
2. 2.5 volumes of acetone	3,350	11.0	19
3. Avicel PH 102	2,500	11.7	29
4. SP-Sephadex	214	1.5	20
5. BioGel P-100	106	0.30	40

[a] From E. F. Hartree, *Anal. Biochem.* **48,** 422 (1972). Bovine serum albumin was used as protein standard.

Preparation of the Crude Extract. Mycelial mass is removed by filtration through a nylon stocking, the pH of the filtrate is adjusted to 6.5 with NaOH, and (the filtrate) frozen. After thawing, the jellylike extracellular fungal polysaccharide, scleroglucan, is removed by filtration through glass wool, to give a clear filtrate comprising the crude extract.

Acetone Precipitation. Crude extract from 2 liters of culture fluid is freeze-dried. The resulting solid material is reconstituted in 60 ml of 4-fold diluted buffer A, centrifuged to remove undissolved solids, and cooled on ice. Protein is precipitated by adding 2.5 volumes of chilled (0°) redistilled acetone at the rate of 2 ml/min. After centrifugation, the precipitate is scraped into an ice-cold petri dish with a spatula and dried under vacuum in a vacuum desiccator. To guard against losses from spurting, the lid of the petri dish is held in place with adhesive tape.

Cellulose Chromatography. A column (3.5 × 8 cm) is packed with Avicel PH 102 previously equilibrated overnight at 4° in buffer A. The acetone powder (4 g) is resuspended in 16 ml of buffer A, applied to the column, and elution commenced with the same buffer at ~1.3 ml/min. The active fraction eluting under these conditions is called fraction A1.

SP-Sephadex Chromatography. Fraction A1 is filtered through an Amicon PM10 ultrafiltration membrane and adjusted to pH 3.5 by washing with 0.05 *M* sodium formate buffer, pH 3.5. The adjusted solution, in 14 ml, is applied to a column (1.5 × 16.5 cm) of SP-Sephadex C-50 equilibrated in the same buffer. Stepwise pH gradient elution follows with 0.05 *M* sodium formate buffer at, in sequence, pH 3.5 (40 ml), 3.8 (40 ml), 4.1 (40 ml), and 4.7 (80 ml). Fractions of 7.5 ml are collected at a flow rate of 1.2 ml/min. The proteins that emerge during elution at pH 4.7 in fractions 21–23 are called collectively S5 (Fig. 1).

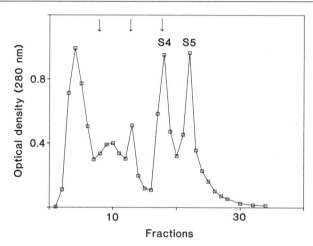

FIG. 1. SP-Sephadex chromatography of fraction A1. Stepwise pH gradient elution is performed with 0.05 M sodium formate buffer at pH 3.5, 3.8, 4.1, and 4.7 on a 1.5 × 16.5 cm column. The arrows indicate changes in the pH of the eluting buffer. The peak S5 (fractions 21–23) contains the carboxymethylcellulase, F2.

BioGel P-100 Chromatography. Fraction S5 is filtered on an Amicon PM10 ultrafiltration membrane and adjusted to pH 6.0 by washing with buffer B. The adjusted solution, in 6.6 ml, is applied to a column (2.6 × 145 cm) of BioGel P-100, 100–200 mesh, equilibrated in buffer B and eluted with the same buffer at a flow rate of 22 ml/hr. When fractions of 6.3 ml are collected, the cellulase F2 elutes as the major carboxymethylcellulase peak in fractions 47–53 (Fig. 2). At this stage, because of the proximity of the major two protein peaks, judicious pooling of fractions is necessary to obtain F2 uncontaminated by protein immediately preceding it during elution. Failing this, rechromatography on the BioGel column will achieve the same end.

Comments about the Procedure. Two further carboxymethylcellulases, present in peak S4 (Fig. 1), are readily resolved by chromatography on the BioGel column. However, their individual properties were not investigated in detail and are not reported here.

Carboxymethylcellulases usually occur as families of physicochemically distinct proteins that, nevertheless, possess the same nominal activity. Therefore, as purification proceeds, unwanted protein as well as carboxymethylcellulase activity is removed from the particular enzyme being purified. Furthermore, there is evidence of synergistic effects among carboxymethylcellulases. In view of these circumstances, the spe-

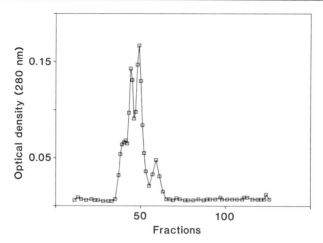

Fig. 2. BioGel chromatography of peak S5. The carboxymethylcellulase, F2, elutes in fractions 47–53. Fractions 47–53 also contain the major part of the carboxymethylcellulase activity emerging from a 2.6 × 145 cm column equilibrated and developed with 0.05 M sodium phosphate buffer, pH 6.0.

cific activity need not increase as purification proceeds and, consequently, need not reflect the degree of purity actually obtained.[4]

Properties

Homogeneity. F2 exhibits a single band on SDS–polyacrylamide gels, but has been shown to exhibit three bands after analytical electrofocusing in polyacrylamide gels.[1] All three components of F2 are glycoproteins with pI values between 4.3 and 4.5.

Specificity. In addition to catalyzing the degradation of CMC, F2 catalyzes the degradation of phosphoric acid-swollen cellulose. The enzyme fails to attack crystalline cellulose (Avicel), p-nitrophenyl-β-D-glucoside, or glucuronoxylan. Commercial locust bean galactomannan is partly degraded in the presence of F2, yielding ~30% of the reducing sugar generated from CMC under the same conditions. A commercial preparation of xylan (Sigma), that normally contains a glucan impurity, is also partially degraded in the presence of F2, releasing about 15–20% of the reducing sugar generated from CMC.

pH Optimum. The pH optimum is 4.3. The enzyme is unstable at pH values above 7.

[4] W. A. Lindner, C. Dennison, and G. V. Quicke, *Biotechnol. Bioeng.* **25**, 377 (1983).

Temperature Stability. A sharp decline in activity is observed when F2 is exposed to temperatures greater than 50° for 20 min.

Estimated Molecular Weight. The molecular weight of F2 determined by molecular exclusion chromatography on BioGel P-150 was reported to be 54,000.[1] More recent measurements using BioGel P-100 give a lower molecular weight of 45,000.

Kinetic Constants. The K_m for CMC 4H1F (degree of substitution 0.4, Hercules Powder Co., Wilmington, DE) is about 2 mg/ml.

Mode of Action on Cellooligosaccharides. F2 degrades cellooligosaccharides of degree of polymerization between 3 and 6 to a mixture of cellobiose and cellotriose, together with trace amounts of glucose. Cellobiose is not attacked. Cellooligosaccharides with a degree of polymerization between 4 and 6, reduced at C_1, are degraded in a manner suggesting that F2 acts by transferring a glucosyl unit from the nonreducing end of the oligosaccharide.[1]

[43] β-Glucanases from *Pisum sativum*

By GORDON MACLACHLAN

Introduction

Four distinct β-glucanases have been extracted and purified to near electrophoretic homogeneity from etiolated pea stems (*Pisum sativum* L. var. Alaska). Two endo-1,3-β-D-glucanases (EC 3.2.1.6) are present in buffer extracts, with one concentrated in growing regions (referred to as 1,3-β-glucanase I, apical) and the other in maturing tissues (1,3-β-glucanase II, basal).[1,2] They differ substantially in certain physical properties (e.g., electrophoretic mobility, molecular weight) and in their action patterns vs substrates of differing chain length. Two endo-1,4-β-D-glucanases (EC 3.2.1.4, cellulase) are generated in apical growing regions of pea stems following treatment with hormones of the auxin type. One is found in buffer extracts (1,4-β-glucanase I, soluble), whereas the other is buffer insoluble, but soluble in 1 *M* salt (1,4-glucanase II, insoluble.)[3,4]

[1] Y.-S. Wong and G. A. Maclachlan, *Biochim. Biophys. Acta* **571**, 244 (1979).
[2] Y.-S. Wong and G. A. Maclachlan, *Biochim. Biophys. Acta* **571**, 256 (1979).
[3] H. Byrne, N. V. Christou, D. P. S. Verma, and G. A. Maclachlan, *J. Biol. Chem.* **250**, 1012 (1975).
[4] Y.-S. Wong, G. B. Fincher, and G. A. Maclachlan, *J. Biol. Chem.* **252**, 1402 (1977).

Despite marked differences in their physical properties and amino acid composition, the pea 1,4-β-glucanases possess closely similar substrate affinities and mode of action.[5] This chapter summarizes methods used to purify these four glucanases and surveys their physical and kinetic characteristics.

The initial motivation for studying β-glucanases in young pea stems arose from the conviction that plant cell expansion must require a "relaxing" or "loosening" of the rigidity of cell walls, even while new wall materials are synthesized, i.e., wall turnover during growth.[6,7] Pea epicotyls were selected as the source tissue in which to search for β-glucanase activities because of the wealth of information available on the growth responses to hormone treatments and the cell wall composition in peas.[8]

Early studies employed carboxymethyl derivatives of appropriate insoluble polysaccharides as substrates for endo-1,3-β- and 1,4-β-glucanases. With such substrates, glucanase activity could be assayed viscometrically with great sensitivity. Crude enzymatic extracts of growing regions of epicotyls were capable of reducing the viscosity of CM-cellulose solutions[7] and this "cellulase" activity could be enhanced manyfold by pretreating the epicotyl with the hormone indoleacetic acid.[9] Similarly, epicotyl extracts reduced the viscosity of a 1,3-linked derivative, i.e., CM-pachyman solutions.[10] The overall 1,3-β-glucanase activity was only modestly enhanced by auxin treatment of the epicotyl. The present chapter concludes with brief comments on subsequent studies conducted on the localization, potential functions, and regulation of pea β-glucanases (for a review, see Ref. 11).

Materials and Basic Methods

Sources of peas, chemicals, enzymes, and chromatographic and electrophoretic materials are all given in the references listed above, as are basic methods used for preparing substrates, growing peas, and fractionating β-glucanases.

1,3-β-Glucanase activity was routinely assayed either reductometri-

[5] G. Maclachlan and Y.-S. Wong, *Adv. Chem. Ser.* **181,** 347 (1979).
[6] G. A. Maclachlan and M. Young, *Nature (London)* **195,** 1319 (1962).
[7] G. A. Maclachlan and J. Perrault, *Nature (London)* **204,** 81 (1964).
[8] G. A. Maclachlan and C. T. Duda, *Biochim. Biophys. Acta* **97,** 288 (1965).
[9] D. F. Fan and G. A. Maclachlan, *Can. J. Bot.* **44,** 1025 (1966).
[10] A. H. Datko and G. A. Maclachlan, *Plant Physiol.* **43,** 1837 (1967).
[11] D. P. S. Verma, V. Kumar, and G. A. Maclachlan, *in* "Cellulose and Other Natural Polymer Systems" (R. M. Brown Jr., ed.), p. 459. Plenum, New York, 1982.

cally or viscometrically as "laminarinase" or "CM-pachymanase." In the former, 0.05 ml enzyme was incubated with 0.45 ml 0.1% (w/v) dialyzed laminarin (from Sigma, determined by end-group analysis to have an average DP of 20) in 20 mM sodium acetate buffer (pH 5.5) at 35°, and aliquots were removed at intervals for estimation of reducing power by the Somogyi[12] method. Enzyme units were expressed as glucose equivalents generated in 2 hr during the initial stages of reaction when rates were linear. Other substrates, e.g., laminaridextrins, prepared by partial acid hydrolysis of laminarin, were assayed reductometrically in the same way. For viscometric assays, 0.1 ml enzyme was incubated with 0.9 ml 0.8% (w/v) CM-pachyman (prepared with a degree of substitution of 0.32 by the method of Stone[1] from a sample of pachyman, DP 255, donated by Dr. Bruce Stone, La Trobe University) in sodium acetate buffer (pH 5.5) containing 0.2 M NaF at 35°. Reaction was measured in Cannon-Manning semimicroviscometers.[10] The rate of viscosity loss was proportional to the amount of enzyme added and linear up to about 20% loss. One unit of CM-pachymanase was defined as the amount of enzyme required to cause 1% loss in viscosity (N_{rel}) of the above reaction mixture in 2 hr.

1,4-β-Glucanase activity was routinely assayed viscometrically[3] at 35° by adding 0.1 ml enzyme to 0.9 ml 0.1 M sodium phosphate, pH 6.0, 0.2% NaF containing 0.8% (w/v) CM-cellulose (0.6% 7LP–0.2% 7HSP, Hercules Powder Co.) in Cannon-Manning semimicroviscometers. The viscosity loss was recorded at intervals during the early (linear) stages of reaction. One unit of CMCase activity was defined as the amount of enzyme required to cause a 1% decrease in N_{rel} of the above rection mixture in 2 hr. In instances where the production of reducing power was required it was measured by the Somogyi[12] method.

β-Glucanase Purification

In early experiments,[1] 1,3-β-glucanase specific activity, as assayed reductometrically (laminarinase) in extracts of the apical 10 mm of 7-day-old epicotyls, was found to be severalfold greater than that in extracts from a basal 10-mm segment from the middle of the first internode. However, when the assay was viscometric (CM-pachymanase), the specific activity was manyfold greater in extracts of the basal segment. Thus, separate 1,3-β-glucanase activities (I and II) appeared to have developed at opposite ends of the stem. This was later confirmed[13] in studies of the

[12] M. Somogyi, *J. Biol. Chem.* **195**, 19 (1952).
[13] Y.-S. Wong and G. A. Maclachlan, *Plant Physiol.* **65**, 222 (1980).

distribution of the two activities along the length of the stem. Both activities were fully extracted by homogenizing tissue in 20 mM sodium phosphate (pH 5.5), 5% glycerol, and 0.05% sodium azide. Apical segments were used as the richest source of glucanase I and basal segments for glucanase II. Approximately 12,000 1-cm segments (200–250 g fresh wt) were homogenized in 2 vol 20 mM sodium acetate (pH 5.5), 5% glycerol, 0.05% sodium azide, centrifuged, and the supernatant was used as source of the two glucanases.

Most 1,3-β-glucanase I activity precipitated in ammonium sulfate at concentrations between 0.8 and 1.8 M salt, whereas glucanase II required 2.0 to 2.8 M. The fractions were redissolved in 70 ml buffer and eluted (35 ml/hr) from a column (30 × 2.5 cm) of DEAE-cellulose (Whatman DE-52) with a linear gradient (200 ml) of 0–2 M NaCl in buffer. Most protein eluted at salt concentrations below 1.0 M. Glucanase I eluted at concentrations between 1.0 and 1.5 M and glucanase II between 1.5 and 2.0 M. The eluates were reduced to a protein concentration of 0.1% (apical) or 0.01% (basal), by ultrafiltration (Amicon PM10 membrane) and eluted (23 ml/hr) from Sephadex G-50 columns (90 × 1.5 cm) with acetate buffer, pH 5.5, which removed most of the contaminating protein (void) from the fractionated glucanase. Fractions were pooled and concentrated by ultrafiltration. The yields of purified enzymes via these procedures were only 13 and 10% and the purifications effected were 88- and 240-fold (Table I). It was calculated that 1,3-β-glucanase I represented 0.4% (w/w) of the protein in buffer extracts of the apical tissue, and II made up 1.2% (w/w) of the soluble protein in basal tissues.

1,4-β-Glucanases were extracted from the plumules, hooks, and swollen apical pea epicotyl tissue[3] 5 days after treatment with auxin [10 ml 0.1% 2,4-dichlorophenoxyacetic acid, 0.1% Tween 80, 0.1 M NaCl spray per 500 (7-day-old) seedlings]. Approximately 500 apices (1 kg fresh wt) were homogenized in 2 vol 20 mM sodium phosphate, pH 6.2, 5% glycerol to dissolve 1,4-β-glucanase I, and glucanase II was extracted from the buffer-insoluble residue in 1 vol of the same solvent containing 1 M NaCl. The buffer-soluble glucanase precipitated in ammonium sulfate at between 1.4 and 2.4 M and the buffer-insoluble salt-soluble glucanase required 3.2 M salt to precipitate.

Salt-precipitated 1,4-β-glucanase I readily redissolved in 20 mM phosphate buffer but glucanase II required a minimum of 0.1 M buffer to solubilize. Glucanase I solution (300 ml) was applied to a column (42 × 5 cm) of DEAE-cellulose (Bio-Rad Cellex-D) to which the enzyme bound so firmly that it could not be eluted by liters of buffer containing up to 0.4 M NaCl. However, it readily eluted from the column in 1 M NaCl–phos-

TABLE I
PURIFICATION[a] OF β-GLUCANASES FROM PEA EPICOTYL EXTRACTS[b]

| Fractionation sequence | Purification | | | |
| | 1,3-β-Glucanases (fold, total units) | | 1,4-β-Glucanases (fold, total units × 10⁻³) | |
	I (apical)	II (basal)	I (soluble)	II (insoluble)
Crude extract	1 (254)	1 (1110 × 10⁻³)	1 (6600)	1 (6490)
(NH₄)₂SO₄	1 (250)	3 (585 × 10⁻³)	2 (3800)	1 (6340)
DEAE-cellulose	—	—	17 (1400)	16 (1730)
Ultrafiltration	6 (82)	181 (296 × 10⁻³)	74 (1200)	69 (1400)
Sephadex			709 (880)	—
Ultrafiltration	88 (33)	240 (106 × 10⁻³)	2500 (850)	242 (1120)

[a] The $(NH_4)_2SO_4$ concentration ranges used to precipitate the glucanases differed in each case (see text). Purity of the enzymes was demonstrated by SDS–electrophoresis with or without labeling with [³H]dimethyl sulfate.[1,3]

[b] 1,3-β-Glucanase I was purified from buffer extracts of apical (growing) regions and II from basal (nongrowing) regions of epicotyl; 1,4-β-glucanase I was extracted in 20 mM buffer from apical regions of epicotyl after treatment with 2,4-dichlorophenoxyacetic acid, and II was extracted with buffer–1 M NaCl from the residue.

phate, pH 6.2. 1,4-β-Glucanase II (50 ml) was applied to a column (35 × 2.5 cm) of DEAE-cellulose to which it bound firmly even after washing in 1 M NaCl buffer, pH 6.2, but it readily eluted at the same salt concentration when the pH was raised to 7.8. Eluates were concentrated by ultrafiltration and precipitated protein was discarded. Supernatants containing approximately 1% protein (glucanase I) or 0.3% protein (glucanase II) were fractionated on columns of Sephadex (95 × 2.5 cm of G-75 or 40 × 2.5 cm of G-100, respectively) and again concentrated by ultrafiltration, with precipitated protein removed. This procedure resulted in recoveries of 13 and 17% of 1,4-β-glucanases I and II, respectively, and purification of 2500- and 242-fold respectively (Table I). 1,4-β-Glucanase I represented only 0.04% (w/w) of the total soluble protein in crude extracts, whereas 1,4-β-glucanase II made up ~0.4% of the total buffer-insoluble salt-soluble protein.

It should be added that the two 1,4-β-glucanases are not only readily separable on the basis of solubility. If they are extracted together in salt solution, they can nevertheless be cleanly separated on columns of Sephadex (G-100) on the basis of their molecular size.[5] Crude 1,4-β-glucanase II fractionated on Sephadex with an apparent molecular weight which was about half of that of the highly purified enzyme[3] since it appears to readily dimerize in concentrated solution upon purification.

TABLE II
PHYSICAL PROPERTIES OF PURIFIED PEA β-GLUCANASES

Parameter	1,3-β-Glucanases		1,4-β-Glucanases	
	I (apical)	II (basal)	I (soluble)	II (insoluble)
pH optimum	5.5–6.0	5.5–6.0	6.0–6.5	6.0–6.5
Isoelectric point	5.4	6.8	5.2	6.9
Molecular weight × 10^{-3}				
Electrophoresis	22	37	15	70
Gel filtration	14	29	21	71
Carbohydrate (w/w) (%)	0.5	0.8	<1	<1

Physical Characteristics

Table II summarizes physical properties of the pea β-glucanases. The 1,3-β-glucanases have similar pH optima as do the 1,4-β-glucanases, all close to pH 6.0, which presumably reflects the pH of the plant cell wall and periplasmic space, where their substrates accumulate. This is a higher pH optimum by 1–2 pH units than most fungal glucanases.[14] Isoelectric points were determined by disc polyacrylamide gel electrophoresis at various gel concentrations containing ampholines in the appropriate pH range.[1,3] The 1,3- and 1,4-β-glucanases separated during isoelectric focusing, yielding p*I* values which were slightly more acidic for glucanase I than for glucanase II in each linkage group. This procedure also demonstrated the purity of the preparations. SDS–gel electrophoresis indicated that the two glucanases of each group were readily separable by such analysis and very different in molecular size. Molecular weight determination by gel filtration (Sephadex) generally yielded somewhat smaller apparent molecular sizes because these enzymes tended to bind to Sephadex. None of these glucanases appeared to be glycoproteins, unlike many of their counterparts in microorganisms, i.e., only slight reactions were obtained upon treatment of the purified products with the phenol sulfuric aid reagent (less than 1% carbohydrate).

Amino acid analyses performed on the purified 1,4-β-glucanases[3] showed that glucanase II, with a much higher molecular weight, nevertheless contained fewer of certain amino acids (histidine, arginine, valine) than glucanase I. The implication is that II could not have derived from I, unless a great deal of complex processing occurred during their genera-

---

[14] D. R. Whitaker, *in* "The Enzymes" (P. D. Boyer, ed.), p. 273. Academic Press, New York, 1970.

tion.[5] Moreover, there appeared to be no cross-reaction between antibodies to the two purified proteins, although this point should be reinvestigated because the enzymes were not denatured by detergent treatment before they were used for antibody production.

Kinetic Characteristics

The 1,3-β-glucanases[2,13] and the 1,4-β-glucanases[4,7] rapidly reduced the viscosity of solutions of carboxymethyl derivatives of 1,3- and 1,4-glucans, while generating free reducing groups at a comparatively slow rate, a reaction pattern characteristic of endohydrolysis. Even after viscosity loss was essentially complete, the enzymes continued to generate reducing chain ends at a linear rate. In the case of the 1,4-β-glucanases, this rate was identical for equal viscosity units of enzyme, but with 1,3-β-glucanases, the apical enzyme (I) generated reducing groups at a rate which was approximately 20 times that of the basal enzyme (II). This indicated that 1,3-β-glucanase I attacked internal linkages of the substrate much less effectively than 1,3-β-glucanase II. It accounts for why the former was more sensitively assayed reductometrically (as laminarinase) while the latter was best assayed viscometrically (as CM-pachymanase).

V_{max} values for the β-glucanases acting vs a variety of substrates are summarized in Table III. 1,3-β-Glucanase I (apical) hydrolyzed all substrates at a greater rate per mole of enzyme than 1,3-β-glucanase II (basal), but 1,4-β-glucanase II (insoluble) acted with a higher turnover rate than 1,4-β-glucanase I (soluble). Rates were highest with soluble β-glucans as substrates and progressively less with shorter oligosaccharides. Substitution with carboxymethyl groups to the degree employed here, or crystallization to the point of rendering a glucan insoluble, reduced the accessibility of substrates to hydrolysis. Mixed-linkage β-glucans (barley β-glucan, 68% 1,4-, 32%, 1,3-linkage, and lichenan) were hydrolyzed by both 1,3- and 1,4-β-glucanases to yield glucose laminaridextrins and 3-O-β-D-cellobiosyl-D-glucose or glucose, cellodextrins, and 4-O-β-D-laminaribosyl-D-glucose. The latter products are expected from hydrolysis adjacent to 1,3- or 1,4-linkages, according to anticipated affinities.

With respect to transitory products formed from 1,3-β-glucans during hydrolysis, both 1,3-β-glucanases generated an unbroken series of laminaridextrins (DP 2–9), but I (apical) cleaved the substrate to form lower dextrins more rapidly than II (basal).[2,13] With laminarin or *Fucus* β-glucan as substrate, 1,3-β-glucanase I very rapidly generated small laminaridextrins leaving the bulk of the original substrate intact, while 1,3-β-glucanase II, in contrast, rapidly reduced the size of the substrate and only eventually generated small dextrins. It was clear that 1,3-β-

TABLE III
V_{max} VALUES[a] AT 35° AND OPTIMAL pH WITH SUBSTRATES SATURATED

Substrate[b]	1,3-β-Glucanases		1,4-β-Glucanases	
	I (apical)	II (basal)	I (soluble)	II (insoluble)
CM-glucan	127	20	93	201
Insoluble β-glucan	95	21	13	80
Barley β-glucan	88	63	93	177
Lichenan	36	29	22	141
Soluble β-glucan	455	104	3	22
Hexa- to octasaccharide	60	45	Not determined	
Hexasaccharide	46	13	173	389
Pentasaccharide	36	2	74	206
Tetrasaccharide	34	1.3	45	78
Trisaccharide	23	0.7	21	35
Disaccharide	2	0	0	0

[a] V_{max} = micromoles reducing glucose equivalents generated/micromole enzyme/minute. Note that values are erroneously recorded as millimoles glucose in Ref. 4.

[b] The 1,3-β-glucans were pachyman (insoluble and carboxymethylated) and laminarin (soluble). The 1,4-β-glucans were Whatman cellulose (insoluble and carboxymethylated) and pea xyloglucan (soluble). The oligosaccharides were the laminaridexrin and cellodextrin series.

glucanase I preferentially hydrolyzed glucan fragments to smaller oligosaccharides while glucanase II showed highest affinity for internal linkages of the longest chains (see also relative V_{max} values, Table III). On analogy with studies of the amylase action pattern,[15] 1,3-β-glucanase I acts by a "multiple attack" mechanism while II acts by multichain attack wherein it preferentially hydrolyzes the longest available chains internally before attacking fragments of low DP.

Both 1,4-β-glucanases form glucose and cellobiose as the main final products generated from cellodextrin[4] but, with cellohexaose as substrate,[5] cellotriose is formed first and then slowly degraded. Both of these glucanases clearly show preference for the most internal linkages of substrates. Neither 1,3- nor 1,4-β-glucanases are capable of forming sorbitol from reduced dextrins,[2,4] nor can they hydrolyze disaccharides.

Apparent K_m values, shown in Table IV, indicate that the 1,4-β-glucanases have very similar apparent affinities for all of the 1,4-β-glucans tested, but the two 1,3-β-glucanases generally differ from one another in

[15] J. Thoma, J. E. Spradlin, and S. Dygert, *in* "The Enzymes" (P. D. Boyer, ed.), p. 115. Academic Press, New York, 1970.

TABLE IV
APPARENT K_m VALUES[a] AT 35° AND OPTIMUM pH

| Substrate[b] | 1,3-β-Glucanases | | 1,4-β-Glucanases | |
	I (apical)	II (basal)	I (soluble)	II (insoluble)
CM-glucan	3.3	4.5	3.5	3.6
Insoluble β-glucan	4.2	2.8	—	—
Soluble β-glucan	1.5	0.6	3.2	2.5
Hexasaccharide	2.2	7.4	3.8	3.8

[a] K_m values are expressed as milligram substrate/milliliter.

[b] The 1,3-β-glucans were CM-pachyman, curdlan (insoluble), laminarin (soluble), and laminarihexaose; the 1,4-β-glucans were CM-cellulose, pea xyloglucan (soluble), and cellohexaose.

this respect. In keeping with the observed action patterns, 1,3-β-glucanase II shows highest affinity for longer chains and 1,3-β-glucanase I has a much greater affinity for laminarihexaose than II.

Endogenous Substrates

Etiolated pea stem walls contain material that fluoresces with aniline blue,[16] a reaction characteristic of callose.[17] This fluorescence can be eliminated by treatment of tissue sections with pea 1,3-β-glucanases or by extraction with boiling water. Water-soluble material from both apical and basal regions of epicotyl is readily degraded by 1,3-β-glucanase I and II, to a series of laminaridextrins. As assayed by the production of reducing groups from water-soluble material following incubation with 1,3-β-glucanases for 12 hr, it was clear that the apical region contained an endogenous β-glucan that was far more susceptible to attack by the apical 1,3-β-glucanase (I) than by the basal enzyme (II). β-Glucan extracted from basal regions was much more rapidly hydrolyzed by the basal II than I. Evidently, natural substrates for the apical and basal 1,3-β-glucanases exist in the pea epicotyl which are preferentially susceptible to degradation by the glucanase that is concentrated in those regions. It is not known what differences exist in the structures of these substrates, but it could simply be that average glucan chain lengths are greater in basal regions, since the apical glucanase I preferentially hydrolyzes shorter chains and the basal II prefers long chains (Tables III and IV).

[16] H. B. Currier, *Am. J. Bot.* **44**, 478 (1957).
[17] M. M. Smith and M. E. McCully, *Planta* **136**, 65 (1977).

For 20 years after their discovery, it was assumed that the natural substrate for the 1,4-β-glucanases was the cellulose in growing pea primary walls. However, the finding that pea cellulose microfibrils in such tissue are coated with the 6-substituted 1,4-β-glucan, xyloglucan, and that the amount of xyloglucan is almost as great as cellulose at the time of maximum growth rate,[18] raises the question of whether this hemicellulose is the more likely substrate. Xyloglucan not only occurs bound to the surfaces of microfibrils, it also stretches between them. It thereby acts as a molecular matrix, which would be expected to be more accessible to secreted 1,4-β-glucanase than the buried cellulose microfibrils.

In fact, the buffer-insoluble 1,4-β-glucanase II, which is the glucanase that is found concentrated extracellularly on the inner surface of the young wall,[19] has a higher affinity for pea xyloglucan than for CM-cellulose or cellohexaose (Table IV), and it readily degrades xyloglucan to oligosaccharide fragments.[20,21] When cell wall "ghosts" containing the xyloglucan–cellulose macromolecular complex are treated with purified endo-1,4-β-glucanase, xyloglucan is degraded first before cellulose is attacked.[18,21] Likewise, *in vivo*, when pea tissue was treated with auxin in such a way as to increase 1,4-β-glucanase activities 85-fold, the endogenous xyloglucan suffered a marked drop in average chain length, while cellulose DP did not change.[20] Accordingly, it is proposed that xyloglucan is the preferred substrate for 1,4-β-glucanase in growing plant tissue, and that xyloglucan turnover is a clear (obligatory?) facet of wall metabolism during cell expansion.

[18] T. Hayashi and G. A. Maclachlan, *Plant Physiol.* **75,** 596 (1984).
[19] A. K. Bal, D. P. S. Verma, H. Byrne, and G. A. Maclachlan, *J. Cell Biol.* **69,** 97 (1976).
[20] T. Hayashi and G. A. Maclachlan, *Plant Physiol.* **75,** 605 (1984).
[21] T. Hayashi and G. A. Maclachlan, *in* "Cellulose: Structure Modification and Hydrolysis" (R. A. Young and R. Rowell, eds.), p. 67. Wiley, New York, 1986.

[44] Cellobiosidase from *Ruminococcus albus*

By KUNIO OHMIYA and SHOICHI SHIMIZU

p-Nitrophenyl-β-D-cellobioside + H_2O → cellobiose + p-nitrophenol
Cellooligomers + nH_2O → n cellobiose + cellooligomers
(glucose)$_m$ (glucose)$_{m-2n}$

A cellobiosidase is purified to homogeneity from the cells of *Ruminococcus albus*. The enzyme catalyzes the hydrolysis of p-nitrophenyl-β-D-cellobioside (PNPC) and cellooligomers both soluble and insoluble in water to cellobiose and smaller cellooligomers by cleaving the β-glucoside linkage from the nonreducing terminal ends. Therefore, the enzyme is 1,4-β-D-glucan glucohydrolase (EC 3.2.1.74, glucan 1,4-β-glucosidase).

Assay Method

Principle

Assay of cellobiosidase is based on measurements of the initial rates of release of p-nitrophenol from PNPC or of cellobiose from cellooligomers.

Procedure[1]

For the determination of enzyme activity, PNPC is synthesized by the method of Nishizawa and Wakabayashi[2] and used as the substrate. It is found to provide a very sensitive index of enzyme activity. PNPC solution (2 mM, 250 μl) and an enzyme sample (700 μl) are added to potassium phosphate buffer (1.0 M, pH 6.8, 50 μl) and incubated at 37° for 1 hr. p-Nitrophenyl-β-D-glucoside (PNPG, Nakarai Chemical Co.) is also used under the same conditions. In both cases, after the reaction is stopped by the addition of Na$_2$CO$_3$ (2 M, 250 μl), the absorbance of the released p-nitrophenol is measured at 405 nm. The enzyme with the activity against PNPC is denoted as PNPCase. For experiments in which cellooligosaccharides prepared by the method of Miller et al.[3] and ball-milled cellulose (BMC) are used as substrates, a reaction mixture of potassium phosphate buffer (50 mM, pH 6.8, 1.0 ml) containing 1% BMC and enzyme solution (0.5 ml) is allowed to stand at 37° for 2 hr. The reaction mixture is boiled for 10 min. The amount of reducing sugar released in the filtrate of the boiled mixture is determined by the Somogyi method.[4] The residual amount of BMC, [KC-flock W-300 (Sanyo Kokusaku Pulp Co.) which is ball-milled in a 3% suspension for 3 days] is quantified with the anthrone–sulfuric acid reagent after hydrolysis with 60% sulfuric acid overnight.

Definition of the Enzyme Unit

One unit of enzyme activity is defined as the amount of enzyme catalyzing the release of 1 μmol of p-nitrophenol per minute.

[1] K. Ohmiya, M. Shimizu, M. Taya, and S. Shimizu, J. Bacteriol. 150, 407 (1982).
[2] K. Nisizawa and K. Wakabayashi, Seikagaku 24, 41 (1952).
[3] G. L. Miller, J. Dean, and R. Blumn, Arch. Biochem. Biophys. 91, 21 (1960).
[4] M. Somogyi, J. Biol. Chem. 195, 19 (1952).

Cultivation of the Organism

Ruminococcus albus F-40 isolated from cow rumen and donated by Professor K. Ogimoto of Tohoku University, School of Agriculture, Sendai, Japan, is cultivated under strictly anaerobic conditions achieved by removing oxygen with oxygen-free carbon dioxide gas. The composition of the medium employed is K_2HPO_4, 0.045%; KH_2PO_4, 0.045%; NaCl, 0.09%; $(NH_4)_2SO_4$, 0.09%; Na_2CO_3, 0.4%; $MgSO_4 \cdot 7H_2O$, 0.009%; $CaCl_2$, 0.009%; $FeSO_4 \cdot 7H_2O$, 0.001%; $ZnSO_4 \cdot 7H_2O$, 0.001%; $MnSO_4 \cdot 4–6H_2O$, 0.001%; $CoCl_2 \cdot 6H_2O$, 0.001%; L-arginine, 0.002%; L-cysteine · HCl, 0.025%, $Na_2S \cdot 9H_2O$, 0.025%; pyridoxine · HCl, 0.0002%; *p*-aminobenzoic acid, 0.00001%; biotin, 0.000005%; resazurin, 0.0001%; acetic acid, 0.17%; propionic acid, 0.06%; *n*-butyric acid, 0.04%; isobutyric acid, 0.01%; *n*-valeric acid; 0.01%; isovaleric acid, 0.01%; DL-α-methylbutyric acid, 0.01%; D-cellobiose, 0.02%; and cellulose, 0.9–1.5%. The pH of the medium after sufficient bubbling of oxygen-free carbon dioxide gas is 6.8. This medium without rumen fluid supports the growth of *R. albus*. The cellulose source used is BMC and cultivation is performed in a jar fermenter (working volume: 1 liter) for 2 days at 37°. The pH of the culture broth is maintained at 6.5 by adding 1 *N* NaOH. During the cultivation of *R. albus* for 20–40 hr, almost all of the BMC is solubilized, and the bacterial dry cell weight increases to a maximum. The enzyme activity against PNPC increases in the supernatant obtained from the broth to a maximum after about 50 hr cultivation. This suggests that the enzyme may be synthesized and released from the cells once their growth has stopped.

Purification Procedure

All the procedures are carried out at 4°. For the evaluation of the enzymatic activity, PNPC is used as a substrate throughout the procedure.

Step 1. Preparation of Crude Enzyme. After 50 hr of cultivation of *R. albus,* the supernatant from the broth is made 10 m*M* in mercaptoethanol and dialyzed within a cellophane membrane bag against 1000 volumes of Tris–HCl buffer (0.01 *M*, pH 7.2) containing 10 m*M* mercaptoethanol. The dialyzate with 80% of original activity is used as a starting material (crude PNPCase) for the purification of the enzyme.

Step 2. DEAE-Sephadex A-25 Column Chromatography. The crude PNPCase solution (1000 ml) is loaded on a DEAE-Sephadex A-25 column (2.5 × 22 cm) equilibrated with the same buffer which is used for dialysis. Almost all of the proteins including enzyme protein are adsorbed on the column. The enzyme protein is eluted with a KCl linear gradient (0–1.0 M) in the same buffer (1.0 liter). The chromatography provides one major and two minor peaks with activity against PNPC eluting at KCl concentrations of 0.1, 0.3, and 0.5 M, respectively. The fraction that eluted with 0.5 M KCl and had major activity against BMC is denoted 1,4-β-glucanase I and is used for the determination of viscosity changes of carboxymethylcellulose (CMC; Dai-ichi Pharmaceutical Co.) solutions. The purity is increased 20-fold by this step.

Step 3. ω-Aminohexyl—Sepharose 6B Column Chromatography. The major fraction with activity against PNPC and eluted with 1 M KCl is chromatographed further on an ω-aminohexyl-Sepharose 6B column (1.6 × 2.5 cm). A broadly spread protein peak with activity is eluted with 0.3 M KCl. The fraction (15 ml) with the highest activity is dialyzed against 600 volumes of the Tris–HCl buffer containing 2 mM mercaptoethanol. The purity is increased 1.5-fold by this step.

Step 4. Isoelectric Focusing. Isoelectric focusing is carried out using Pharmalyte (Pharmacia Fine Chemicals), pH range 4–6.5, in a sucrose gradient at 4° with constant voltage (1300 V) for 85 hr. For the preparation of the sucrose gradient in Pharmalyte solution, a dense solution containing sucrose (22 g), distilled water (22 ml), and Pharmalyte (1.2 ml) and a light solution containing distilled water (30 ml) and Pharmalyte (0.375 ml) are appropriately mixed and pumped into the glass column (1.7 × 50 cm). A mixture of sucrose (30 g), phosphoric acid (0.05 ml), and distilled water (20 ml) is used as the anode (+) solution. For cathode (−) solution, distilled water (15 ml) containing 3 drops of sodium hydroxide solution (3 N) is employed. The enzyme solution is mixed with the dense solution to give the appropriate density and placed in the middle of the column. All the solutions used for isoelectric focusing contain 10 mM mercaptoethanol. After focusing, the solution is pumped out from the bottom of the column and fractionated with a fraction collector. Sucrose and Pharmalyte in the active fractions are replaced by 0.5 M KCl in a Tris–HCl buffer by gel filtration on a column (1.7 × 50 cm) containing equal volumes of Sephadex G-15 and G-50. Isoelectric focusing of fractions with enzyme activity provides a sharp protein peak coinciding with an enzyme activity at pI of 5.3. The protein in the peak fraction is homogeneous, as determined by polyacrylamide disc gel electrophoresis both in the presence and absence of sodium dodecyl sulfate (SDS). The enzyme activity corresponds to the protein band at a relative mobility of 0.1 by standard gel electrophoresis in the presence of 10 mM mercaptoethanol. This three-step purification procedure yields a 39-fold enrichment of the enzyme (Table I). The amount of purified protein obtained is about 2 mg.

TABLE I

PURIFICATION OF CELLOBIOSIDASE FROM *R. albus*[a]

Purification step	Total volume (ml)	Total protein (mg)	Total activity (OD_{405})	Specific activity (OD_{405}/mg)	Recovery of enzyme activity (%)	Purification (fold)
Culture supernatant (dialyzate)	1000	145.0	350	2.4	100	1
DEAE-Sephadex A-25	70	7.6	98	12.9	28	19
ω-Aminohexyl-Sepharose 6B	600	4.7	95	20.2	27	31
Isoelectric focusing	5	3.7	77	20.8	22	39

[a] All fractionation procedures are performed at 4°. OD_{405} is optical density at 405 nm.

Properties

Physical Characteristics

The molecular weight of the enzyme in the presence of SDS is estimated to be 100,000 by disc gel electrophoresis. The estimated molecular weight by gel filtration analysis with a BioGel P-300 column (1.7 × 60 cm; Bio-Rad Laboratories) in the absence of SDS is 200,000. Therefore, the native enzyme may be a dimer of two polypeptide chains of similar molecular weight. The molecular weight of the enzyme is larger than those from aerobes (42,000–62,000).[5,6]

Amino Acid Composition and Carbohydrate Content

The amino acid composition determined by the method of Moore *et al.*,[7] of the purified enzyme protein is given in Table II. The carbohydrate content of the enzyme evaluated by the phenol-sulfuric acid method[8] is 7.4%.

[5] G. Halliwell and M. Griffin, *Biochem. J.* **135**, 587 (1973).
[6] T. M. Wood and S. I. McCrae, *Biochem. J.* **189**, 51 (1980).
[7] S. Moore, D. H. Spackmn, and W. H. Stein, *Anal. Chem.* **30**, 1185 (1958).
[8] M. Dubois, K. A. Gilles, J. K. Hamilton, P. A. Rebers, and F. Smith, *Anal. Chem.* **28**, 350 (1956).

TABLE II
AMINO ACID COMPOSITION OF CELLOBIOSIDASE FROM R. albus

Amino acid	Composition (mol/100 mol)	Residue per molecule	
		Determined[a]	Nearest integer
Aspartic acid	13.8	98.4	98
Threonine	7.8[b]	55.5	56
Serine	8.6[b]	61.1	61
Glutamic acid	8.9	63.4	63
Proline	3.0	21.7	22
Glycine	11.8	84.1	84
Alanine	7.7	55.3	55
Cysteine	0.8[c]	6	6
Valine	5.9[d]	41.8	42
Methionine	1.8	10.8	11
Isoleucine	4.3[d]	30.7	31
Leucine	5.7	40.4	40
Tyrosine	5.3	37.8	38
Phenylalanine	3.4	24.4	24
Histidine	7.3	51.8	52
Lysine	1.5	13.1	13
Arginine	2.5	17.9	18
Total			714

[a] Values are based on a molecular weight of 100,000.
[b] Values of 22-hr hydrolyzate.
[c] Determined as carboxymethylcysteine.
[d] Values of 83 hr hydrolyzate.

Stability

The enzyme incubated in Tris–HCl buffer (10 mM, pH 7.2) for 10 min maintains almost all of the initial activity at temperatures below 30°, and about 20% of the initial activity is lost at 37°. After storage of the enzyme for 15 hr, residual activity is about 75% at 30° and about 25% at 37°. Maximal stability is noted at pH values between 5.5 and 8.0.

Activity Optimum

Maximum activity of the enzyme is observed at pH 6.8 at 37°.

Activity Changes by Chemical Reagents

Each of the reagent (1.0 mM) is added just before the reaction is initiated. Of inorganic cations, Ca^{2+}, Co^{2+}, and Mg^{2+} have little effect on enzymatic activity, whereas Zn^{2+} and Cu^{2+} depress the activity remarkably. Phenylmethylsulfonyl fluoride does not affect the activity. Reducing reagents such as mercaptoethanol, dithiothreitol, and glutathione activate the enzymatic activity slightly, whereas sulfhydryl-reacting reagents (*p*-chloromercuribenzoate and iodoacetoamide) strongly inhibit the activity, suggesting that the enzyme belongs to SH enzymes. The requirement of intact thiol groups for *R. flavefaciens* enzyme has been noted by Pettipher and Latham.[9]

Substrate Specificity

For the identification of the compounds released from PNPC and water-soluble cellooligomers, a reaction mixture of enzyme (50 μg/ml, 0.5 ml) and substrate (0.45%, 1.0 ml) is incubated at 37° for 2 hr. After the solution is boiled for 10 min, the supernatants are spotted on silica gel-precoated thin-layer plates (5 × 20 cm; Yamato replate-50 plates, Yamato Scientific Co. Ltd.). After the plates are developed for 5 hr at room temperature with a 1-butanol–water–pyridine [6 : 3 : 4 (v/v/v)] solution, *p*-nitrophenol is detected by its yellow color and the other compounds are detected as dark spots when the plates, which have been sprayed with 15% H_2SO_4 solution, are heated. Reaction products released from PNPC are identified as cellobiose and *p*-nitrophenol. Cellobiose is produced from cellotriose, cellotetraose, cellopentaose, reduced cellotriose, and reduced cellopentaose. Glucose is a product of cellotriose and cellopentaose but not of PNPC and cellotetraose. In the cases of insoluble cellooligosaccharides, BMC, CMC, and Avicel, enzymatic reactions are continued overnight with twice the usual enzyme concentration; under these conditions small amounts of cellobiose are detected but glucose is not. PNPG is not hydrolyzed even after overnight reaction.

The initial velocity of the enzymatic reaction against each substrate is determined. The value for cellotetraose (13.5 × 10^{-8} mol/min) is the largest seen for substrates having only 1,4-β-glucoside linkages. Then the value for cellopentaose is 9.3 × 10^{-8} mol/min. The velocity against cellotriose (1.9 × 10^{-8} mol/min) is the lowest of the cellooligosaccharides used. The velocity against PNPC (2.2 × 10^{-8} mol/min) is close to that observed for cellotriose. PNPG is not hydrolyzed completely. Cellooligosaccharides with modified reducing terminals such as reduced cellotriose and reduced cellopentaose are hydrolyzed to cellobiose and sorbitol and to cellobiose and reduced cellotriose, respectively, at smaller velocities than those with unmodified reducing terminal ends. Initial rates of hydrolysis against BMC, CMC, Avicel, insoluble cellooligosaccharides, and reduced cellotriose cannot be determined because of the very slow reaction.

[9] G. L. Pettipher and M. J. Latham, *J. Gen. Microbiol.* **110**, 29 (1979).

Viscosity changes in CMC solutions resulting from the action of the enzyme or from 1,4-β-glucanase I (a fraction from DEAE-Sephadex A-25 chromatography) are determined with an Ostwald viscometer (Iwaki Glass Co. Ltd.) at 37°. The changes in fluidity [$1/\eta_{sp}$ (the reciprocal of the specific viscosity)] generated by the enzyme are compared with those generated by 1,4-β-glucanase I after a reaction period during which the amounts of reducing sugar released from CMC by either of these enzymes are the same. Fluidity ($1/\eta_{sp}$) of CMC after enzyme action increases linearly with the amount of reducing sugar released. The slope of this line is smaller than that obtained after the action of 1,4-β-glucanase I on CMC. These results suggest that the purified enzyme cleaves the penultimate glucosidic linkage from nonreducing terminal ends and can be described as an exoenzyme, whereas 1,4-β-glucanase I is classified as an endoenzyme. Therefore, the former enzyme is termed a cellobiosidase.

[45] Cellobiohydrolases of *Penicillium pinophilum*

By THOMAS M. WOOD

Strain and Its Maintenance

Strain. Penicillium pinophilum, IMI 87160iii, is obtained from the Commonwealth Mycological Institute, Kew, Surrey, UK. Previously this particular strain was called *Penicillium funiculosum*, IMI 87160iii. However, in the current CMI catalog it has been assigned to a different taxonomic group (*pinophilum*).

Maintenance of Culture. A potato dextrose agar culture slope (39 g/ liter Oxoid PDA) is prepared and inoculated with spores from the original culture. After incubation at 27° for 5 days sporulation is induced by standing the culture in light for several days. Five milliliters of 10% (v/v) glycerol solution is added, and the spores suspended in this solution by gently scraping with a rod. One milliliter quantities of the suspension are dispensed into ampoules standing in ice water. The ampoules are sealed, allowed to stand at 4° for 30 min, wrapped in cotton wool, and placed in the freezer at −85°. After 24 hr the ampoules are transferred to liquid nitrogen. The cultures can be kept indefinitely under these conditions without loss of viability. Revival is easily effected by thawing, transferring 0.5 ml of the contents on to a complete sterile medium (CM agar), and incubating at 28°. The CM agar consists of (per liter) KCl, 0.5 g; $MgSO_4 \cdot 7H_2O$, 0.5 g; $FeSO_4 \cdot 7H_2O$, 0.5 g; KH_2PO_4, 1.0 g; corn steep liquor, 10 g; bacto-casitone, 2 g; yeast extract, 2 g; DL-methionine, 0.05 g; sucrose, 30 g; agar, 1.5 g. Cultures can be maintained for short periods on this medium, or on PDA slants if kept at 4°.

Enzyme Preparation

Cultures of *P. pinophilum* can be prepared by stationary or submerged culture techniques.

Stationary Culture

These are prepared using cotton as the carbon source in exactly the same way described for *Trichoderma koningii* (see chapter [24] in this volume). The enzyme is also isolated from the culture medium after 30–35 days incubation as previously described in chapter [24] in this volume.

Submerged Cultures

Inoculum. The inoculum for a submerged culture is prepared in four flasks (250 ml) with medium (125 ml) containing (per liter) $(NH_4)_2SO_4$, 1.4 g; urea, 0.3 g; proteose peptone, 1 g; KH_2PO_4, 2.0 g; $CaCl_2$, 0.3 g; $MgSO_4 \cdot 7H_2O$, 0.3 g; $FeSO_4 \cdot 7H_2O$, 5.0 mg; $MnSO_4 \cdot H_2O$, 1.56 mg; $ZnSO_4 \cdot 7H_2O$, 1.4 mg; $CoCl_2$, 2.0 mg; and ball-milled straw or Solka-Floc (1% w/v). The medium is inoculated with conidia from a fresh PDA slant to give a final concentration of 10^4 conidia/ml and incubated at 28° for 72 hr on a rotary shaker (150 rpm).

Fermenter. The inoculum is added to a 16-liter Microgen fermenter (New Brunswick Scientific, New Brunswick, NJ) containing 9.4 liters of the same medium that is used to prepare the inoculum except that the concentration of the cellulosic carbon source is 6% (w/v). The culture is stirred at 300 rpm, but is increased to 600 rpm when the dissolved oxygen level falls below 20% of saturation. Air is introduced through a sparger at 3 liters/min. The temperature is maintained at 28 or 35° and the pH controls are set so that the medium will be maintained between pH 3.5 and 5.0 by the automatic addition of HCl or NH_4OH. Foaming is controlled by the automatic addition of antifoam (A emulsion, Sigma no. A5758). After 10 days the culture is filtered through Whatman filter paper GF/A, concentrated to 250 ml by ultrafiltration in an Amicon hollow fiber cartridge (type HIP10-8), and freeze dried.

Purification Procedure

Penicillium pinophilum cellulase contains two immunologically distinct cellobiohydrolases (herein designated cellobiohydrolase I and cellobiohydrolase II).[1,2] These can be separated and purified by ion-exchange chromatography, isoelectric focusing, and chromatofocusing. In the author's work[1,2] the cellobiohydrolases were first identified as the major protein components which acted synergistically with the separated endo-1,4-β-glucanase and β-glucosidase enzymes to solubilize crystalline cellulose. These proteins were then purified to homogeneity. Other workers use Avicel as a substrate for cellobiohydrolase, and Avicelase is now regarded as being synonymous with cellobiohydrolase (but see chapter [9] in this volume for problems associated with using this substrate for measuring this activity). Other enzymes in the effluent from the columns can be monitored and measured as described in chapter [9] of this volume using *p*-nitrophenyl-β-D-glucoside (β-glucosidase) and CM-cellulose (CM-cellulase, endo-1,4-β-glucanase).

As the enzymes were purified simply as proteins in the author's work no tabulated data of the purification are presented.

Step 1. Ion-Exchange Chromatography on DEAE-Sephadex. A sample (40 mg) of concentrated enzyme is dissolved in water (5 ml), desalted on a column of BioGel P-2 (50.1 × 2.5 cm) equilibrated with 0.01 M ammonium acetate buffer, pH 4.8, freeze-dried, and redissolved in 15 ml of 60 mM sodium acetate buffer, pH 4.8. The enzyme is applied to a column (38.2 × 2.5 cm) of DEAE-Sephadex (acetate form) and eluted with start buffer at 2.5 ml/hr until 180 fractions (2.5 ml) have been collected. Fractions are assayed for endo-1,4-β-glucanase (CM-cellulase), protein (Folin–Lowry method), and β-glucosidase. A graphical presentation of the distribution of these activities will demonstrate the separation of two major protein components: cellobiohydrolase II is associated with the first (fractions 39–61); cellobiohydrolase I with the second (fractions 71–100): both must be purified further. Fractions 39–61 are also rich in CM-cellulase and β-glucosidase activity: fractions 71–100 contain only traces of these activities. The protein associated with cellobiohydrolases I and II is approximately 40–50% of that eluted from the column.

Step 2. Isoelectric Focusing of Cellobiohydrolase I. Combined fractions 71–100 from Step 1 are concentrated to 5 ml in a collodion tube (Sartorius), dialyzed against 0.01 M ammonium acetate buffer, pH 5.0, and freeze-dried. A sucrose density gradient is made using an LKB Gradi-

[1] T. M. Wood, S. I. McCrae, and C. M. Macfarlane, *Biochem. J.* **189,** 51 (1980).
[2] T. M. Wood and S. I. McCrae, *Carbohydr. Res.* **148,** 331 (1986).

ent Maker, by mixing, a light solution comprising 0.62 ml of 40% (w/v) LKB ampholine carrier ampholytes, (pH 3–5) diluted to 50 ml with water, and a dense solution consisting of a solution of the same carrier ampholytes (1.87 ml) and sucrose (28 g) diluted to 36 ml with water. The entire light solution is placed in the nonmixing chamber and the entire dense solution is placed in the mixing chamber. The freeze-dried enzyme is dissolved in the dense solution after approximately one-third of the gradient has been formed in the isoelectric focusing column. A mixture of 85% (w/v) phosphoric acid (0.2 ml), water (14.0 ml), and sucrose (12.0 g) is used for the dense electrode solution; a solution of sodium hydroxide (0.1 g) in water (10.0 ml) is used for the light electrode solution. The electrodes are placed so that the anode is at the bottom of the column. A voltage is applied for 65–70 hr until the power has stabilized (600 V; 1.6 W at the end of the run). The contents of the column are pumped out through the bottom, collected in 2 ml fractions, and assayed for β-glucosidase, protein, and CM-cellulase. The pH of each fraction is measured at 4°. β-Glucosidase in fractions 35–40 is discarded. Cellobiohydrolase I, which is identified as the major protein component (fractions 25–32) isoelectric at pH 4.4 (4°), is pooled and then purified further by a rerun of the isoelectric focusing experiment under the same conditions. The pooled enzyme is added to the isoelectric focusing column without further treatment. This is done by replacing an equivalent volume of the light solution (described above) with the pooled enzyme solution and then constructing the gradient as before.

The further purified cellobiohydrolase I is prepared free from carrier ampholytes by precipitation with $(NH_4)_2SO_4$ (80% saturation), centrifugation (75,000 g for 20 min), dialysis of the redissolved pellet in a collodion tube (Sartorius) against 0.01 M ammonium acetate buffer, pH 5, and freeze-drying. When purified in this way the enzyme should show no detectable β-glucosidase activity. Gel electrofocusing using carrier ampholyte covering the pH range 4–6 should show only one protein band (Coomassie Blue stain).

Recoveries of cellobiohydrolase protein of the order of 90% can be expected.

The final purification of cellobiohydrolase I can also be effected by chromatofocusing.[2]

Step 3. Ion-Exchange Chromatography of Cellobiohydrolase II on DEAE-Sepharose. Fractions 39–61, Step 1, are pooled and the enzyme precipitated with $(NH_4)_2SO_4$ (80% saturation). The suspension is centrifuged (75,000 g for 20 min) and the pellet redissolved in 0.01 M ammonium acetate buffer, pH 5.0. Salt-free enzyme is obtained by desalting on BioGel P-2. The enzyme is freeze-dried. After dissolution in 0.01 M sodium phosphate buffer, pH 6.9, the enzyme is applied to a column (30.0 × 2.5 cm) of DEAE-Sepharose CL-6B, equilibrated with 0.01 M phosphate buffer pH 6.9. The column is eluted at 15 ml/hr with stepwise changes in buffer. These include (1) 0.01 M phosphate buffer, pH 6.9, (2) 0.05 M phosphate buffer, pH 6.9, (3) 0.1 M phosphate buffer, pH 6.9, and (4) 0.2 M phosphate buffer, pH 5.6. Changes in eluent are made at fractions (2 ml) 150, 350, and 500. The major protein component (fractions 354–388) eluted immediately after the change in eluent to 0.2 M phosphate buffer, pH 5.6, contains cellobiohydrolase II.

Step 4. Chromatofocusing of Cellobiohydrolase II. Fractions 354–388, Step 3, are pooled and the enzyme precipitated by the addition of solid $(NH_4)_2SO_4$ (80% saturation). The suspension is centrifuged (75,000 g for 20 min) and the pellet is redissolved in 5 ml of 0.01 M ammonium acetate buffer, pH 5.0. The solution is desalted on a column of BioGel P-2 equilibrated with 0.01 M ammonium acetate buffer, pH 5.0, the enzyme isolated by freeze-drying, and then redissolved in 6 ml of 0.025 M histidine–HCl buffer, pH 6.4. This solution is applied to a column (28.5 × 0.9 cm) of PBE 94 chromatofocusing medium (Pharmacia) and eluted (15 ml/hr) with 200 ml Polybuffer 74 (Pharmacia) which has been diluted 9-fold and brought to pH 4.5 with conc. HCl. Fractions (2 ml) are collected, measured for pH, and assayed for protein, CM-cellulase and activity to Avicel. The pH gradient produced covers the pH range 4.8–6.4 and results in the elution of the cellobiohydrolase as a single protein peak (E_{280}) at pH 5.45. Another component absorbing at 280 nm is eluted at pH 4.58. This component is virtually devoid of cellulase activity. A small amount of xylanase activity is eluted at pH 5.3 (fractions 56–61, approximately). The nature of the component eluted pH 4.58 is unknown. The possibility that it has some role to play in the breakdown of cellulose cannot be excluded. Fractions 46–56 are pooled and the enzyme is recovered by precipitation with $(NH_4)_2SO_4$ (80% saturation), centrifugation (75,000 g for 20 min), desalting on BioGel P-2, freeze drying, and finally dissolution in 0.01 M ammonium acetate buffer, pH 5.0. Electrophoresis of the purified and concentrated enzyme under nondenaturing and denaturing conditions in polyacrylamide gels shows only one band on staining for protein (Coomassie Blue). Typically, 80–85% of the protein (determined by Folin–Lowry method) applied to the column is recovered.

Properties

Cellobiohydrolase I and II are similar in respect of the molecular weights (I, 46,300; II, 50,700), pI values (I, 4.36 at 4°, II, 5.0 at 4°), and in their abilities to act synergistically with a reconstituted mixture of the endo-1,4-β-glucanase components in solubilizing cotton fiber. Cellobiohydrolase I and II differ, however, in their carbohydrate contents (I, 9%; II, 19%) their heat stabilities (destroyed during 40 min at 60°: II, 97%, I, 65%), and their pH optima (I, 2.5; II, 4.5). Further, cellobiohydrolase II does not react with anticellobiohydrolase I antiserum, and cellobiohydrolase I, but not II, has appreciable activity on cellotriose. D-glucose (200 mM) stimulates the action of cellobiohydrolase I on H_3PO_4-swollen cellulose, but inhibits cellobiohydrolase II. The converse is true for cellobiose.

[46] Exocellulase of *Irpex lacteus* (*Polyporus tulipiferae*)

By TAKAHISA KANDA and KAZUTOSI NISIZAWA

The cellulase system of *Irpex lacteus* (*Polyporus tulipiferae*) includes various kinds of endocellulases of different randomness and at least one kind of exocellulase. The exocellulase may be a 1,4-β-D-glucan cellobiohydrolase (EC 3.2.1.91, cellulose 1,4-β-cellobiosidase) based on its substrate specificity. The fungus seems to attack native cellulose to obtain reducing sugars *in vivo,* using the synergistic action of the different kinds of cellulases.

Assay Method

Principle. The assays are based on the following measurements: (1) the increase in reducing power formed during reaction, and (2) the decrease in the average degree of polymerization (DP) of cellulose.[1,2] Both methods for measuring these changes should be performed in parallel to detect the exocellulase activity. The first method is conveniently carried out during the purification of exocellulase which is higher in saccharification activity, and in addition is simpler than the second procedure. In this section, therefore, we describe the reducing sugar method.

Reagents

Sodium acetate buffer, 0.05 M, pH 5.0
1% Avicel (microcrystalline cellulose) suspension in water. The suspension may be used for at least 1 month if stored in a refrigerator

[1] A. Donetzhuber, *Sven. Papperstidn.* **63,** 447 (1960).
[2] T. Kanda, K. Wakabayashi, and K. Nisizawa, *J. Biochem.* (*Tokyo*) **87,** 1635 (1980).

Copper reagent[3,4]: (a) dissolve 25 g of Na_2CO_3 (anhydrous), 25 g of Rochelle salt, 20 g of $NaHCO_3$, and 200 g of Na_2SO_4 (anhydrous) in 800 ml of distilled water in this order, and dilute up to 1 liter. (b) Dissolve 150 g of $CuSO_4 \cdot 5H_2O$ in distilled water and make up to 1 liter. Before use, reagents (a) and (b) are mixed in a ratio of 25 : 1 (v/v)

Nelson reagent[3,4]: dissolve 25 g of $(NH_4)_6Mo_7O_{24} \cdot 4H_2O$ in 500 ml of distilled water, add 42 g of concentrated H_2SO_4, 3 g of $Na_2HAsO_4 \cdot 7H_2O$, and dilute up to 1 liter

Procedure.[5] The reaction mixture consists of 1 ml of 1% Avicel suspension, 2 ml of 0.05 M sodium acetate buffer, pH 5.0, and 1 ml of enzyme solution. The mixture is incubated at 30° with shaking for 1 hr in most cases and centrifuged. To 0.5 ml of the supernatant, are added 1 ml of copper reagent and 0.5 ml of distilled water. The mixture is shaken and heated for 10 min in a boiling water bath. After cooling for 5 min, 1 ml of Nelson reagent and 2 ml of distilled water are added and mixed. The reducing power of the mixture is estimated colorimetrically by the absorbance at 660 nm.

Definition of Unit and Specific Activity. One unit of the Avicel saccharification activity is defined as the amount of enzyme which produces reducing power equivalent to 1 μmol of glucose per minute under the assay conditions described above. Specific activity is expressed as enzyme units per milligram protein. Protein is determined by the method of Lowry *et al.*[6] using bovine serum albumin as standard.

Purification Procedure[5]

All operations are performed at 4–6°.

Step 1. Ammonium Sulfate Fractionation. Driselase powder (50 g), a commercial enzyme preparation from *Irpex lacteus* by Kyowa Hakko Co., Japan, is extracted with about 300 ml of water. The suspension is centrifuged at 11,900 g for 15 min. The brown supernatant is brought to 20% saturation by addition of solid ammonium sulfate, and the precipitate which ordinarily has no cellulase activity is removed by centrifugation at 11,900 g for 30 min. The supernatant is brought to 80% saturation by

[3] M. Somogyi, *J. Biol. Chem.* **195**, 19 (1952).
[4] N. Nelson, *J. Biol. Chem.* **153**, 375 (1944).
[5] T. Kanda, S. Nakakubo, K. Wakabayashi, and K. Nisizawa, *J. Biochem.* (*Tokyo*) **84**, 1217 (1978).
[6] O. H. Lowry, N. J. Rosebrough, A. L. Farr, and R. J. Randall, *J. Biol. Chem.* **193**, 265 (1951).

further addition of solid ammonium sulfate. The resulting precipitate which contains a high cellulase activity is collected by centrifugation and dissolved in about 200 ml of water. The aqueous cellulase solution is dialyzed against running tap water at 5° for 24 hr. The dialyzed solution is concentrated by ultrafiltration with Diaflo UM10 membrane (Amicon) and lyophilized (17.5 g).

Step 2. First DEAE-Sephadex A-50 Chromatography. The crude lyophilized preparation obtained above is dissolved in 150 ml of 0.02 *M* ammonium acetate buffer, pH 5.0, and applied to a DEAE-Sephadex A-50 column (5.0 × 50 cm) preequilibrated with the same buffer. Elution is carried out stepwise with 0.02 and 0.1 *M* of the same buffer, and maintained at a flow rate of 1 ml/min. Each 20-ml fraction of the eluates is collected. Four peaks of protein are obtained, and the third fraction being eluted with 0.1 *M* acetate buffer is most active toward Avicel and comprises the greatest amount of protein among the four. This peak is pooled and concentrated by ultrafiltration and lyophilized (6.2 g).

Step 3. First BioGel P-100 Gel Filtration. The lyophilized preparation is dissolved in about 30 ml of 0.1 *M* sodium acetate buffer, pH 5.0, and applied to a BioGel P-100 column (2.7 × 130 cm) preequilibrated with the same buffer, and eluted with the same buffer. The column is operated at a flow rate of 5 ml/hr, and 2.5-ml fractions of eluate are collected. Two peaks of Avicel saccharification activity are obtained, of which the second peak is higher in cellulase activity and protein content. This fraction is concentrated and lyophilized as above (1.5 g).

Step 4. CM-Sephadex C-50 Chromatography. The lyophilized preparation is dissolved in about 10 ml of 0.01 *M* sodium acetate buffer, pH 5.0, and applied to CM-Sephadex C-50 column (3.0 × 30 cm) preequilibrated with the same buffer. The elution is carried out stepwise with 0.01 and 0.1 *M* sodium acetate buffers. The column is operated at a flow rate of 20 ml/ hr and 10-ml fractions of the eluate are collected. The Avicel saccharification activity is separated further into three peaks, and the first peak fraction shows relatively high activity and amount of protein. This cellulase fraction is pooled, concentrated, and lyophilized (228 mg).

Step 5. Second DEAE-Sephadex A-50 Chromatography. The lyophilized preparation is dissolved in about 5 ml of 0.02 *M* ammonium acetate buffer, pH 5.0 and again subjected to DEAE-Sephadex A-50 column chromatography under conditions similar to those of the first DEAE-Sephadex A-50 chromatography except for column size (3.0 × 30 cm), flow rate (26 ml/hr), and fraction volume (10 ml). Two cellulase peaks are obtained; the second peak being eluted with 0.1 *M* ammonium acetate buffer shows a typical Avicel saccharification activity as compared with CMC saccharification activity (carboxymethyl cellulose which is ordinarily used as sub-

TABLE I
PURIFICATION OF EXOCELLULASE FROM *Irpex lacteus*

Fraction	Total volume (ml)	Total protein (mg)	Total units	Specific activity (units/mg)	Recovery (%)
1. Crude extract	260	14,925	11.95	0.0008	(100)
2. Ammonium sulfate fractionation	100	2,853	5.14	0.0018	43
3. First DEAE-Sephadex A-50	430	1,011	3.24	0.0032	27
4. First BioGel P-100	50	245	1.27	0.0052	11
5. CM-Sephadex C-50	190	37	0.95	0.0256	8
6. Second DEAE-Sephadex A-50	50	14	0.61	0.0438	5
7. Second BioGel P-100	12	6.8	0.39	0.0571	3

strate for endocellulase). This peak is pooled and lyophilized after concentration (87 mg).

Step 6. Second BioGel P-100 Gel Filtration. The lyophilized preparation is dissolved in 3 ml of 0.1 M sodium acetate buffer, pH 5.0, and again applied to BioGel P-100 column under conditions similar to those of the first BioGel P-100 gel filtration except for column size (1.2 × 140 cm), flow rate (3 ml/hr), and fraction volume (3.0 ml). A protein peak containing both cellulase activities for Avicel and CMC is obtained. The peak fractions are combined and lyophilized (42 mg). The cellulase activity ratio of this enzyme preparation for Avicel to that for CMC is 9.6 when the Avicel saccharification activity is measured after incubation for 1 hr under the standard condition and CMC saccharification activity is measured after incubation for 30 min under the same conditions; this ratio shows no change upon subsequent column chromatography of the cellulase preparation. Moreover, the enzyme preparation reveals a single protein band on SDS–polyacrylamide gel electrophoresis. A summary of the purification procedure is given in Table I.

Properties

pH and Temperature Optima and Stability.[5] Exocellulase is practically stable in the pH range from 3.5 to 6.0, and the optimum pH is 5.0. The optimum activity of the enzyme is at 50°, and only 5% of its optimum activity remains after the enzyme was heated at 100° for 10 min.

Molecular Weight, Amino Acid Composition, and Carbohydrate Content.[5] The molecular weight of the exocellulase is estimated by gel filtration to be 65,000 and it contains 2.4% carbohydrate as glucose. The pattern of its amino acid content is not very different from those of other endocellulases from the same fungus, particularly the high content of acidic amino acids, glycine, serine, and threonine.[7,8]

Comparison of Randomness of Exo- and Endocellulases.[7,9] The randomness of hydrolysis by the exocellulase is compared with that by endocellulases from the same fungus on the basis of the ratio of a decrease in the degree of polymerization of substrates to an increase in the reducing power produced simultaneously. The ratio for more randomness should be larger than for less randomness. The ratios for the hydrolysis of all substrates including CMC and native cellulose by the exocellulase are the smallest. This cellulase, therefore, may be regarded as a saccharification type rather than a liquefaction type.

Substrate Specificity. The exocellulase seems to split off successively almost exclusively cellobiosyl residues from the nonreducing end of native and degraded celluloses as well as from CMC. However, the β-configuration of the anomeric carbon of the cellobiosyl residue seems not to be inversed on liberation from the substrate by the enzyme action, because an upward mutarotation is observed on addition of alkali to the reaction mixture in which cellopentaitol is used as substrate.[5] This property is entirely similar to that of endocellulase from the same fungus.[8]

The exocellulase produces glucose (G_1) and cellobiose (G_2) from cellotriose (G_3), and G_1, G_2, and G_3 from cellopentaose (G_5), but only G_2 from cellotetraose (G_4), cellohexaose (G_6), CMC, and insoluble celluloses at earlier stages of hydrolysis. G_3 and G_4 are attacked only slowly by exocellulase as compared with G_5 and G_6. G_2 is not hydrolyzed by the exocellulase even during a prolonged incubation, but p-nitrophenyl β-D-cellobioside, a synthetic derivative of cellobiose, is hydrolyzed at either its holoside or aglycon bond. The hydrolysis rate of the aglycon bond is higher than that of the holoside bond. K_m values for G_5 and G_6 are 0.190 and 0.303 mM, respectively.[5,7]

In contrast to the hydrolysis mode on $(1 \rightarrow 4)$-β-D-glucan, the exocellulase hydrolyzes internal glucosidic linkages of β-1,3;1,4-D-glucan such as barley glucan and lichenan, and it releases G_1, G_2, G_3, and two kinds of mixed oligosaccharides, 3-O-β-D-glucosylcellobiose and 3-O-β-D-cellobiosylcellobiose, but does not release laminaribiose.[10] In this respect, the enzyme is similar to an endo-type cellulase from *Streptomyces*.[11,12]

[7] T. Kanda, S. Nakakubo, K. Wakabayashi, and K. Nisizawa, *Adv. Chem. Ser.* **181**, 211 (1979).

[8] T. Kanda, K. Wakabayashi, and K. Nisizawa, *J. Biochem. (Tokyo)* **87**, 1625 (1980).

[9] M. Takai, J. Hayashi, K. Nisizawa, and T. Kanda, *J. Appl. Polym. Sci.: Apply Polym. Symp.* **37**, 345 (1983).

[10] T. Kanda, H. Yatomi, Y. Amano, and K. Nisizawa, manuscript in preparation.

[11] F. W. Parrish, A. S. Perlin, and E. T. Reese, *Can. J. Chem.* **38**, 2094 (1960).

[12] A. S. Perlin, *in* "Advances in Enzymic Hydrolysis of Cellulose and Related Materials" (E. T. Reese, ed.), p. 185. Pergamon, New York, 1963.

This enzyme seems to require a cellobiosyl residue adjacent to a $(1 \rightarrow 3)$-β-D-linked glucosyl residue, and it splits the $(1 \rightarrow 4)$-β-D-glucosidic linkage contiguous to a $(1 \rightarrow 4)$-β-D-linked glucosyl residue. This fact may be explained by the assumption that the stereochemical structure of the successive $(1 \rightarrow 4)$-β- and $(1 \rightarrow 3)$-β-glucosyl linkages is permissible in the action of this cellulase which requires two contiguous $(1 \rightarrow 4)$-β-linked sequences.[10] Furthermore, the exocellulase is able to attack $(1 \rightarrow 3)$-β-D-linkage positioned between $(1 \rightarrow 4)$-β-D-linkages of oligosaccharides such as 3-O-β-cellobiosylcellobiose to produce cellobiose almost exclusively.[10]

Other Properties. A synergistic action of exo- and endocellulase in the hydrolysis of cotton, viscose rayon and alkali cellulose has been observed. Due to this effect the hydrolysis of cotton proceeds 44% farther than the sum of the hydrolysis extents by each single cellulase of a different type. Under similar conditions, the hydrolysis effects on viscose rayon and alkali cellulose are 28 and 16%, respectively.[2]

The exocellulase transfers the cellobiosyl residue at the nonreducing end of donor substrates such as G_3, G_5, and p-nitrophenyl β-D-cellobioside to appropriate acceptors. It can, however, transfer the glucosyl residue at the nonreducing end of these substrates to a small extent. The transcellobiosylation activity of exocellulase is, however, far higher than that of endocellulase from the same fungus.[13]

[13] T. Kanda, I. Noda, K. Wakabayashi, and K. Nisizawa, *J. Biochem.* (*Tokyo*) **93**, 787 (1983).

[47] β-Glucosidase from *Ruminococcus albus*

By KUNIO OHMIYA and SHOICHI SHIMIZU

p-Nitrophenyl-β-D-glucoside + H_2O → glucose + p-nitrophenol
Cellooligomers + $n H_2O$ → n glucose + cellooligomers
(glucose)$_m$ (glucose)$_{m-n}$

A β-glucosidase is purified to homogeneity from the cells of *Ruminococcus albus*. The enzyme catalyzes not only the hydrolysis of p-nitrophenyl-β-D-glucoside (PNPG), but also the degradation of such cellooligomers as cellobiose, cellotriose, cellotetraose, and cellopentaose to glucose by cleaving β-glucoside linkages from nonreducing terminal ends. Therefore, the enzyme is 1,4-β-D-glucan glucohydrolase (EC 3.2.1.74, glucan 1,4-β-glucosidase).

Assay Method

Principle

Assay of β-glucosidase is based on measurement of the initial rate of release of *p*-nitrophenol from PNPG or of glucose from cellooligomers, respectively.

Procedure[1]

PNPG (Nakarai Chemicals Co.) is used as the substrate, because it provides a very sensitive index of the enzyme activity. PNPG solution (2 mM, 250 μl) and enzyme sample (700 μl) are added to the potassium phosphate buffer (1.0 M, pH 6.8, 50 μl), and incubated at 30° for 1 hr. *p*-Nitrophenyl-β-D-cellobioside (PNPC) is also used under the same conditions. In both cases, after the reaction is stopped by the addition of Na_2CO_3 (2 M, 250 μl), the absorbance of the released *p*-nitrophenol is measured at 405 nm. The enzyme with the activity against PNPG is denoted as PNPGase. For experiments in which cellooligosaccharides prepared by the method of Miller *et al.*[2] are used as substrates, the amount of glucose released is determined with the glucose oxidase colorimetric reagent purchased from Fujisawa Medical Supply Co. The residual amount of cellulose, KC-flock W-300 (Sanyo Kokusaku Pulp Co.), which is ball-milled in a 3% suspension for 3 days (ball-milled cellulose: BMC) is quantified with the anthrone–sulfuric acid reagent after hydrolysis with 60% sulfuric acid overnight.

Definition of the Enzyme Unit

One unit of enzyme activity is defined as the amount of enzyme catalyzing the release of 1 μmol of *p*-nitrophenol per minute.

Cultivation of Organism

Ruminococcus albus F-40 isolated from cow rumen and donated by Professor K. Ogimoto of Tohoku University, School of Agriculture, Sendai, Japan, is cultivated under strictly anaerobic conditions which are provided by removing oxygen with oxygen-free carbon dioxide gas. The composition of the medium employed is K_2HPO_4, 0.045%, KH_2PO_4, 0.045%; NaCl, 0.09%; $(NH_4)_2SO_4$, 0.09%; Na_2CO_3, 0.4%; $MgSO_4 \cdot 7H_2O$,

[1] K. Ohmiya, M. Shirai, Y. Kurachi, and S. Shimizu, *J. Bacteriol.* **161,** 432 (1985).
[2] G. L. Miller, J. Dean, and R. Blumn, *Arch. Biochem. Biophys.* **91,** 21 (1960).

0.009%; $CaCl_2$, 0.009%; $FeSO_4 \cdot 7H_2O$, 0.001%; $ZnSO_4 \cdot 7H_2O$, 0.001%; $MnSO_4 \cdot 4-6H_2O$, 0.001%; $CoCl_2 \cdot 6H_2O$, 0.001%; L-arginine, 0.002%; L-cysteine $\cdot$ HCl, 0.025%; $Na_2S \cdot 9H_2O$, 0.025%; pyridoxine $\cdot$ HCl, 0.0002%; p-aminobenzoic acid, 0.00001%; biotin, 0.000005%; resazurin, 0.0001%; acetic acid, 0.17%; propionic acid, 0.06%; n-butyric acid, 0.04%; isobutyric acid, 0.01%; n-valeric acid, 0.01%; isovaleric acid, 0.01%; DL-α-methylbutyric acid, 0.01%; D-cellobiose, 0.02%; and cellulose, 0.9–1.5%. The pH of the medium after sufficient bubbling of oxygen-free carbon dioxide gas is 6.8. This medium without rumen fluid supports the growth of *R. albus*. The cellulose source used is BMC. The cultivation is performed in a jar fermenter (working volume, 1 liter) for 2 days at 37°. The pH of the culture broth is maintained at 6.5 by adding 1 *N* NaOH. When the organism is cultivated on BMC, the highest activity against PNPG is detected in the cells of *R. albus* (17 g wet weight in 100 g of precipitate containing undigested cellulose) harvested at the middle-logarithmic growth phase after about 20 hr of cultivation, when about 50% of the initial cellulose content is consumed. A trace of PNPGase activity is detected in the supernatant when the cell concentration begins to decrease owing to autolysis. When the organism is cultivated on cellobiose, its consumption is almost complete within 20 hr. The maximum enzyme activity in the cells appears after 8–9 hr of cultivation. During growth on glucose, the activity from the cells reaches a maximum after about 15 hr of cultivation. These data suggest that the presence of cellulose is not essential for the enzyme formation.

Purification Procedure

All the procedures are carried out at 4°. For the evaluation of the enzymatic activity, PNPG is used as a substrate throughout the procedure.

Step 1. Preparation of Cell-Free Extract. Cells frozen at $-20°$ overnight are suspended in 1 liter of 10 m*M* potassium phosphate buffer (pH 6.8) containing 10 m*M* mercaptoethanol and are stirred gently for 2 hr. The supernatant of this suspension is used as the starting material (crude PNPGase) for the purification of the enzyme.

Step 2. Aminohexyl-Sepharose 6B Column Chromatography. The crude PNPGase is completely adsorbed on an aminohexyl-Sepharose 6B column (2.4 × 25 cm) previously equilibrated with the same buffer, but large amounts of other proteins are passed through the column. The enzyme protein is eluted with a KCl linear gradient (0–0.6 *M*) in the same buffer (1.2 liters). A single peak of the enzyme activity is eluted at about

0.25 *M* KCl between two large protein peaks. The purity is increased 6-fold by this step.

Step 3. Aminooctyl-Sepharose 6B Column Chromatography. Fractions (13 ml each) with higher activity in Step 2 are combined and equilibrated with 0.86 *M* $(NH_4)_2SO_4$ and chromatographed through an aminooctyl-Sepharose 6B column (1.6 × 14 cm). Almost all of the proteins in the sample loaded on the column are passed through the column. Enzyme protein having PNPG-hydrolyzing activity is adsorbed on the hydrophobic column and eluted by a dual linear gradient of $(NH_4)_2SO_4$ (0.86–0 *M*) and ethylene glycol (0–40%).[3] The purity is increased 7-fold by this step.

Step 4. DEAE-BioGel A Column Chromatography. The eluate with the activity obtained from Step 3 is loaded on a DEAE-BioGel A (Bio-Rad Laboratories) column (1.6 × 20 cm) and the enzyme protein is eluted with a KCl linear gradient (0–1.0 *M*) in the phosphate buffer (10 m*M*, pH 6.8, 1.0 liter). The fractions with activity are eluted at 0.15 *M* KCl. The purity is increased 3-fold.

Step 5. Isoelectric Focusing. Isoelectric focusing analysis with Pharmalyte, pH range 2.5–5.0, in a sucrose gradient is performed at a constant voltage (1000 V) for 90 hr. For the preparation of sucrose gradient in Pharmalyte solution, the dense solution containing sucrose (22 g), distilled water (22 ml), and Pharmalyte (1.2 ml) and the light solution containing distilled water (30 ml) and Pharmalyte (0.375 ml) are appropriately mixed and pumped into the glass column (1.7 × 50 cm). The mixture of sucrose (30 g), phosphoric acid (0.05 ml), and distilled water (20 ml) is used as the anode (+) solution. For the cathode (−) solution, distilled water (15 ml) containing three drops of sodium hydroxide solution (3 *N*) is employed. The enzyme solution mixed with the dense solution to give appropriate density is placed in the middle of the column. All solutions employed for isoelectric focusing contain 10 m*M* mercaptoethanol. After focusing, the solution is pumped out from the bottom of the column and fractionated with a fraction collector. The enzyme protein is concentrated as a sharp peak at p*I* 4.4, indicating the absence of isoenzymes. The final protein yield is 0.12 mg, with 520-fold purification and activity recovery of about 9% (Table I). This protein solution is gel filtrated on a BioGel P-100 column (1.7 × 50 cm) with phosphate buffer (10 m*M*, pH 6.8) containing 0.5 *M* KCl to remove Pharmalyte and sucrose, then desalted with ultrafiltration membrane PM10 under a nitrogen gas atmosphere and stored at 4° until used. Disc gel electrophoresis shows that the enzyme with activity against PNPG is a homogeneous protein having a relative mobility of 0.6.

---

[3] T. Horio and J. Yamashita (eds.), "Fundamental Methods of Protein and Enzyme Experiments," p. 180. Nankodo Press, Tokyo, 1981.

TABLE I
PURIFICATION OF PNPGase FROM *R. albus*

Purification step	Total activity (U)	Total protein (mg)	Specific activity (U/mg)	Recovery of activity (%)	Purification (fold)
Crude extract	17	690	0.025	100	1
Aminohexyl-Sepharose 6B	9.9	69	0.14	58	6
Aminooctyl-Sepharose 6B	7.8	7.4	1.1	46	44
DEAE-BioGel A	5.9	2.0	3.0	35	120
Isoelectric focusing	1.6	0.12	13.0	9	520

Properties

Physical Characteristics

PNPGase from *R. albus* has a molecular weight of approximately 82,000 by gradient slab polyacrylamide gel electrophoresis and by gel filtration with BioGel P-300. From the results of sodium dodecyl sulfate–polyacrylamide gel electrophoresis, this purified enzyme appears to consist of one polypeptide chain because of the appearance of a single band on the gel. The estimate of the molecular weight of the polypeptide by sodium dodecyl sulfate–polyacrylamide gel electrophoresis is 116,000, which may be overestimated due to the high content of carbohydrate (12%) in the enzyme as reported by Ng and Zeikus.[4]

Amino Acid Composition and Carbohydrate Content

The amino acid composition of the purified enzyme is given in Table II. The carbohydrate content of the enzyme determined by the phenol-sulfuric acid method[5] is 12%.

Stability

The enzyme incubated at pH 6.8 for 10 min maintains almost all of the initial activity at temperatures below 30°, and about 20% of the initial activity is lost at 37°. The enzyme is less stable at 37° than β-glucosidases

---

[4] T. K. Ng and J. G. Zeikus, *Biochem. J.* **199**, 341 (1981).
[5] M. Dubois, K. A. Gilles, J. K. Hamilton, P. A. Rebers, and F. Smith, *Anal. Chem.* **28**, 350 (1956).

TABLE II
AMINO ACID COMPOSITION OF PNPGase FROM *R. albus*

Amino acid	Composition (mol/100 mol)	Residue per molecule	
		Determined[a]	Nearest integer
Aspartic acid	8.3	74.6	75
Threonine	6.6[b]	38.6	39
Serine	5.6[b]	33.1	33
Glutamic acid	10.1	59.1	59
Proline	3.8	22.5	23
Glycine	10.1	59.1	59
Alanine	8.6	50.7	51
Cysteine	1.5[c]	9	9
Valine	6.2[d]	36.2	36
Methionine	2.8	16.6	17
Isoleucine	4.5[d]	26.2	26
Tyrosine	3.7	21.9	22
Phenylalanine	4.4	26.1	26
Histidine	1.5	8.7	9
Lysine	6.2	36.5	37
Arginine	4.4	25.9	26
Total			591

[a] Values are based on a molecular weight of 82,000.
[b] Values from 23.5-hr hydrolyzate.
[c] Determined as carboxymethylcysteine.
[d] Values from 72-hr hydrolyzate.

from other microorganisms such as *Clostridium thermocellum, Trichoderma reesei, and Thermoascus auranticus.*[6]

Kinetic Factors

Activity Optimum. Maximum activity of the enzyme is observed at pH 6.5 from 30 to 35°.

K_m *Value.* The K_m value of the enzyme against PNPG is estimated to be 2.2 mM from Lineweaver–Burk plots. When cellobiose is used, the value is 26 mM. Both values are about 10 times greater than those of β-glucosidase from fungi[6,7] such as *Trichoderma viride* cultivated on each substate, but are comparable to the values of the enzyme from *C. thermocellum.*[8]

[6] J. Woodward and A. Wiseman, *Enzyme Microb. Technol.* **4,** 73 (1982).
[7] L. E. R. Berghem and L. G. Pettersson, *Eur. J. Biochem.* **46,** 295 (1974).
[8] N. Ait, N. Creuzet, and J. Cattaneo, *J. Gen. Microbiol.* **128,** 569 (1982).

Chemical Reagents. Each of the reagents (1.0 mM) is added just before the reaction is initiated. The inorganic cations, Ca^{2+}, Co^{2+}, and Mg^{2+}, have little effect on enzymatic activity, whereas Zn^{2+}, Hg^{2+}, and Cu^{2+} depress the activity remarkably. Reducing reagents such as mercaptoethanol, dithiothreitol, glutathione, and cysteine–HCl activate the enzymatic activity slightly, whereas sulfhydryl-reacting reagents (*p*-chloromercuribenzoate and iodoacetoamide) strongly inhibit the activity, suggesting that the enzyme belongs to the class of SH enzymes.

Substrate Specificity

The rates of enzymatic reactions against PNPG and PNPC are 1.3 and 1.4 mol/min/ml, respectively. The rates against salicin and other cellooligomers (G_2, G_3, G_4, and G_5) are 0.2 and 0.04 mol/min/ml, respectively. High-molecular-weight cellulose and sugars having no 1,4-β-glucoside linkage, such as maltose, sucrose, and lactose, are not degraded. On thin-layer chromatography, and hydrolytic products of carboxymethylcellulose and BMC are not detected.

Thin-layer chromatograms of cellooligomers reveal that glucose is the product released from all susceptible substrates and that every glucose unit is released one by one from the oligomers. PNPC is hydrolyzed to glucose and PNPG and the resulting PNPG is then hydrolyzed to glucose and *p*-nitrophenol, but cellobiose is not detected. The enzyme cannot hydrolyze lactose, suggesting that the enzyme does not hydrolyze glucose at the reducing end of the substrate. Therefore, the enzyme is referred to as a β-glucosidase which can be characterized by hydrolysis of cellooligomers at the nonreducing ends.

The enzyme seems to be retained in or on the cells. The enzymes from rumen microorganisms such as *Bacteroides succinogenes*,[9] *Ruminococcus flavefaciens*,[10] and *C. thermocellum*[8] are also found to be cell associated. On the other hand, the fungal enzyme is released from the cells.[11] This localization seems to be advantageous for preventing glucose inhibition, since glucose does not accumulate outside the cells and can be converted effectively to the other compounds by the Embden–Meyerhof–Parnas pathway.[12] This β-glucosidase may be a key enzyme in generating glucose as an energy source for growth of *R. albus*.

[9] D. Groleau and C. W. Forsberg, *Can. J. Microbiol.* **27**, 517 (1981).
[10] G. L. Pettipher and M. J. Latham, *J. Gen. Microbiol.* **110**, 29 (1979).
[11] T. M. Woodward and S. I. McCrae, *J. Gen. Microbiol.* **128**, 2973 (1982).
[12] A. E. Joyer and R. L. Baldwin, *J. Bacteriol.* **92**, 1321 (1966).

[48] 1,4-β-Glucosidases of *Sporotrichum pulverulentum*

By V. DESHPANDE and K.-E. ERIKSSON

$$\beta\text{-1,4-Glucoside } (G_n) \rightarrow \beta\text{-1,4-glucoside } (G_{n-1}) + \text{D-glucose}$$
$$\text{Cellobiose} \rightarrow 2 \text{ D-glucose}$$
$$p\text{-Nitrophenyl-}\beta\text{-D-glucoside (PNPG)} \rightarrow p\text{-nitrophenol} + \text{D-glucose}$$

Introduction

Enzymes of the β-glucosidase (β-D-glucoside glucohydrolase, EC 3.2.1.21) type hydrolyze compounds containing β-D-glucoside linkages by splitting off the terminal β-D-glucose residue. Included in this class of enzymes, generally named after the substrates which they hydrolyze, are aryl-β-glucosidase, cellobiase, gentiobiase, and salicilinase. The β-glucosidases are widely distributed in nature and occur in bacteria, fungi, plants, and animals. β-Glucosidases of cellulolytic organisms are important in the conversion of cellulose to glucose and have been investigated in several laboratories over recent years.[1-3] Production, purification, and partial characterization of the 1,4-β-glucosidases from the white-rot fungus *Sporotrichum pulverulentum* was reported by Deshpande *et al.*[4] The methods used in this work are summarized here.

Location of the β-Glucosidases

In order to study the location of the β-glucosidases, i.e., whether cell bound or extracellular, the organism is grown in a modified Norkrans' medium[5] containing 0.5% glucose and 0.05% yeast extract for 20 hr (using an inoculum of 4×10^6 spores per 100 ml medium). The resulting mycelium is washed with Norkrans' medium and transferred aseptically to flasks containing the same salt medium but without yeast extract and with either 0.5% cellobiose, cellulose, or glucose as carbon source. Samples of

---

[1] K.-E. Eriksson and T. M. Wood, *in* "Biosynthesis and Biodegradation of Wood Components" (T. Higuchi, ed.), Chapter 17, pp. 469–503. Academic Press, New York, 1985.

[2] E. Sakamoto, J. Kanamoto, M. Arrai, and S. Murao, *Agric. Biol. Chem.* **49**, 1275 (1985).

[3] W. J. Chirico and R. D. Brown, *Eur. J. Biochem.* **165**, 333 (1987).

[4] V. Deshpande, K.-E. Eriksson, and B. Pettersson, *Eur. J. Biochem.* **90**, 191 (1978).

[5] A. R. Ayers, S. B. Ayers, and K.-E. Eriksson, *Eur. J. Biochem.* **90**, 171 (1978).

the broth are removed periodically and centrifuged. The β-glucosidase activity in the washed mycelium (cell bound) and in the supernatant (extracellular) is determined using p-nitrophenyl-β-D-glucoside as substrate. There is no induction of β-glucosidase activity when glucose is used as the sole carbon source indicating that the enzyme is not produced constitutively by the fungus. When cellobiose or cellulose is used as the carbon source, the enzyme is detectable on fungal cell walls 15–20 min following induction.

To investigate the conditions under which the enzyme is actively secreted into the medium, the organism is grown either on cellobiose, cellulose, or different mixtures of cellobiose and cellulose. When cellobiose serves as the carbon source, no β-glucosidase activity is detectable in the culture solution even after 24 hr, whereas with cellulose as the carbon source β-glucosidase activity appears in the culture solution after 7 hr of cultivation. Results from fungal cultivations on different mixtures of cellobiose and cellulose indicate that, with cellobiose dominating in the medium, most of the enzyme is cell bound and high extracellular activities were obtained only when cellulose serves as the sole carbon source.

Assay

Principle

β-glucosidase activity is determined using p-nitrophenyl-β-D-glucoside (PNPG) as substrate. Either of the reaction products, glucose or the aglycone, p-nitrophenyl, is measured. The methods most commonly adopted are those based on the colorimetric measurement of (1) glucose or (2) p-nitrophenol in alkaline solution.

Determination of Glucose

Glucose Oxidase/Peroxidase Method.[6] Glucose oxidase catalyzes the oxidation of glucose to gluconic acid

$$\beta\text{-D-Glucose} + H_2O + O_2 \rightarrow \text{D-gluconic acid} + H_2O_2$$

The hydrogen peroxide is decomposed by peroxidase

$$H_2O_2 \rightarrow H_2O + 1/2\ O_2$$

[6] H. Bergmeyer and E. Bernt, *in* "Methods of Enzymatic Analysis" (H. Bergmeyer, ed.), pp. 321–325. Academic Press, New York, 1963.

The liberated oxygen oxidizes a hydrogen donor DH_2 (e.g., *o*-dianisidine) to a colored derivative D

$$DH_2 + 1/2\ O_2 \rightarrow D + H_2O$$

The amount of the dye formed is a measure of glucose oxidized.

Reagents

I. Cellobiose (0.5%): dissolve 50 mg of cellobiose in 10 ml, 50 mM sodium acetate buffer, pH 5.0

II. Buffer–enzyme mixture (0.12 M phosphate buffer, pH 7.0, 40 μg peroxidase/ml, 250 μg glucose oxidase/ml): dissolve 2.07 g $Na_2HPO_4 \cdot 2H_2O$ and 1.09 g $NaH_2PO_4 \cdot 2H_2O$ in 6 mg peroxidase (Sigma, type VI) and 38 mg glucose oxidase (Sigma, type V) in double distilled water and make up to 150 ml

III. Chromogen (5 mg *o*-dianisidine hydrochloride/ml): dissolve 10 mg *o*-dianisidine hydrochloride in double distilled water and make up to 2 ml

IV. Glucose reagent: add 0.5 ml of solution III to 50 ml of solution II with vigorous stirring. (Note. Prepare the glucose reagent fresh each day)

V. Glucose standard solution (100 μg D-glucose/ml)

VI. 50% H_2SO_4

Procedure. This method is generally used for the estimation of cellobiase activity since glucose oxidase is highly specific for glucose and does not react with cellobiose unlike the other reagents which measure reducing sugars. The reaction mixture consists of 0.5 ml of the cellobiose solution (I) and a known volume of suitably diluted enzyme solution in a total volume of 1 ml. Following incubation at 40° for 20 min, the reaction is terminated by heating in a boiling water bath for 5 min using glass bulbs as stoppers. Glucose reagent (IV) (5 ml) is added to aliquots (0.1–0.2 ml) of the reaction mixture and incubated at 40° for 30 min. The color is stabilized by adding 5 ml of 50% H_2SO_4 and the absorbance of the solution read at 546 nm. The amount of glucose released in the incubation sample is proportional to the increase in absorbance at 546 nm and is calculated by comparing it with a standard glucose curve.

A commercial glucose oxidase reagent Glox (AB Kabi, Stockholm, Sweden) containing glucose oxidase, peroxidase, and *o*-dianisidine is available and 5 ml of a Glox solution (1.65 g Glox dissolved in 100 ml of 0.5 M Tris–acetate buffer, pH 7.0) may be substituted for the glucose reagent (IV).

Other Methods for Glucose Estimation

The amount of glucose released can be estimated by any of the procedures commonly used for determining reducing sugars such as the methods of Nelson,[7] Somogyi,[8] Bernfeld,[9] or the dinitrosalicylic acid (DNS) method.[10] However, these methods are less useful with substrates such as cellobiose which give high substrate blanks.

Determination of Aglycone Released

Principle. The action of β-D-glucosidase on *o*- or *p*-nitrophenyl-β-D-glucosides (PNPG) releases *o*- or *p*-nitrophenol as one of the products of the reaction. The nitrophenol gives an intense yellow color in alkaline solutions above pH 8.0 and can be quantitatively measured at 400 nm.

Reagents

Sodium acetate buffer (50 mM, pH 5.0)
PNPG (1 mM)
Dissolve 30 mg of PNPG in 100 ml of 50 mM sodium acetate buffer, pH 5.0
p-Nitrophenyl (10 mM)

Procedure. PNPG is used as the substrate for the enzymes in this method. The reaction mixture contains 1 ml, 1 mM PNPG and 100 μl enzyme solution. After incubation at 40° for 10 min, the reaction is terminated by adding 2 ml of 1 M Na$_2$CO$_3$. The mixture is diluted with 10.0 ml distilled water and the amount of *p*-nitrophenol liberated is measured spectrophotometrically by determining the absorbance at 400 nm and comparing with a standard curve for *p*-nitrophenol.

Methods for Histochemical Detection of β-Glucosidases

4-Methylumbelliferone-β-D-glucoside is widely used for the study of β-glucosidases in animal and insect tissues by measuring the fluorescence given by the aglycone in glycine buffer at pH 10.3.[11] Glucosides of naphthols and 6-bromo-2-naphthol have also been found to be useful in the detection of these enzymes in histochemical studies.[12] 6-Bro-

[7] N. Nelson, *J. Biol. Chem.* **153**, 375 (1944).
[8] M. Somogyi, *J. Biol. Chem.* **195**, 19 (1952).
[9] P. Bernfeld, this series, Vol. 1, p. 149.
[10] G. L. Miller, *Anal. Chem.* **31**, 426 (1959).
[11] D. Robinson, *Biochem. J.* **63**, 39 (1956).
[12] R. B. Cohen, *J. Biol. Chem.* **195**, 239 (1952).

monaphthyl derivatives of β-glucosides are employed for the detection of isozymes of β-glucosidase in polyacrylamide gels.[13]

Definition of the Enzyme Unit and Specific Activity

One unit of β-glucosidase activity is defined as the amount of enzyme that produces 1 μmol of glucose from cellobiose or 1 μmol of *p*-nitrophenol from PNPG per minute under the assay conditions. Specific activity is expressed as units per milligram of protein. The protein is determined by the method of Lowry *et al.*[14] with bovine serum albumin as the protein standard.

Purification Procedure

All operations are carried out at 0–4° unless otherwise stated. All solutions are prepared with double distilled water.

Growth of the Organism

S. pulverulentum (ATCC 32629 (anamorph of *Phanerochaete chrysosporium*) is grown in 1-liter conical flasks containing 300 ml of modified Norkrans's medium.[5] The fungus is cultivated while shaking for 13 days. Then the mycelium and the residual cellulose are removed by filtering the culture solution through a glass fiber filter (Sartorius 13400).

Step I: Concentration of the Enzyme. The culture filtrate, usually 15 liters per batch, is concentrated by ultrafiltration (Diaflo membrane UM10) to 1 liter. Further concentration is obtained by precipitating the protein with solid ammonium sulfate added to 90% saturation. After 40 min, the solution is centrifuged at 14,000 *g* for 60 min and the precipitated protein redissolved in water and dialyzed in collodion tubes against several changes of distilled water. This solution is used as the starting material for purification of the β-glucosidases.

Step II: Preparative Slab Gel Isoelectric Focusing I. The concentrated culture filtrate is subjected to preparative isoelectric focusing which is carried out as described by LKB (Application Laboratory, LKB Produkter AB, Stockholm, Sweden, Application Note 198). A 5% solution of Ultrodex (LKB) in 2% ampholine R (pH range 3 to 10) is poured on to the weighed plate, then 36% of the water in the slurry is evaporated, the

---

[13] V. Herducova and K. Benes, *Biol. Plant.* **19**, 436 (1977).

[14] O. H. Lowry, N. J. Rosebrough, A. L. Farr, and R. L. Randall, *J. Biol. Chem.* **193**, 265 (1951).

extent of evaporation being monitored by weighing the plate. When the gel is set, 15 ml of the concentrated and dialyzed solution containing about 300 mg of protein is mixed with the gel solution and applied to the gel bed with the aid of a sample applicator. The electrode strip, soaked in 1 M H_3PO_4 or 1 M NaOH, is placed at the anodic and the cathodic side, respectively, and the electrofocusing is run with a constant power of 8 W for a period of 16–18 hr. In order to locate the β-glucosidase activity at the end of the run, a zymogram technique is used which is similar in principle to that reported by Eriksson and Pettersson[15] for analytical polyacrylamide gels. A dry filter paper is placed carefully on top of the gel bed so as to avoid trapping air bubbles under the paper. After 3 min, the filter paper is removed and sprayed with 0.01 M PNPG solution, kept for a further 2–3 min at room temperature, and then dried in an oven at 110–120° for 5–15 min. The paper is then sprayed with 1 M Na_2CO_3 and dried as above. Yellow bands appear on the paper corresponding to the position of the β-glucosidase enzymes on the gel. A fractionating grid is pressed through the gel bed and the pH of the gel measured in the grid compartments with the aid of a surface electrode. Sections of the gel are transferred with a spatula to small disposable syringes and eluted with 5 ml of cold water. β-Glucosidase activity of those fractions shown to be positive by the zymogram investigation, and protein levels in all the fractions, are determined. The distribution of protein and β-glucosidase activity after a typical isoelectric focusing is presented in Deshpande et al.[4]

Step III: Phenyl-Sepharose Chromatography. Active fractions from step II are pooled, concentrated, and adjusted to 3 M with respect to NaCl. The sample is then applied to a column (3 × 40 cm) of phenyl-Sepharose (Pharmacia Fine Chemicals, Uppsala, Sweden) previously equilibrated with 3 M NaCl. The column is first eluted with a linear concentration gradient generated using 1 liter of 50 mM sodium acetate buffer, pH 5.0, and 1 liter of 3 M NaCl. A second elution step is carried out with a gradient formed by using 1 liter of water and 1 liter of 50% ethylene glycol. Fractions (5 ml) are collected at a flow rate of 30 ml/hr. This results in the separation of β-glucosidase activity into two distinct peaks: peak A arising from the sodium chloride gradient and peak B obtained with the ethylene glycol gradient (Fig. 1). Enzymes A and B are pooled as indicated in Fig. 1, dialyzed, concentrated, and further purified by isoelectric focusing.

Step IV: Preparative Slab Gel Isoelectric Focusing II. The procedure for the second preparative gel isoelectric focusing is the same as described in purification step II except that this time a narrow pH range of

[15] K.-E. Eriksson and B. Pettersson, *Anal. Biochem.* **56**, 618 (1973).

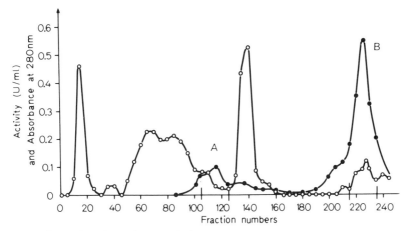

FIG. 1. Separation of protein (open circles) and 1,4-β-glucosidase activity (closed circles) after chromatography on a phenyl-Sepharose column. Elution with two different concentration gradients (A) 1 liter of 50 m*M* acetate buffer and 1 liter of 3 *M* NaCl, and (B) 1 liter of water and 1 liter of 50% ethylene glycol. β-Glucosidase activity (●); absorbance at 280 nm (○). From Deshpande *et al.*[4]

3–7 is used for the separation. This approach divides enzyme A into two peaks of activity, A_1 and A_2 (Fig. 2A), and enzyme B into three peaks of activity, namely, B_1, B_2, and B_3 (Fig. 2B). The respective active fractions are pooled, concentrated, and dialyzed to remove ampholytes. Thus, five peaks of highly purified β-glucosidases are obtained by following a purification scheme involving preparative isoelectric focusing in a wide and narrow pH range combined with a hydrophobic interaction chromatography on phenyl-Sepharose (Table I).

The purification procedure for β-glucosidases may be further simplified by using polyurethane foam strips instead of Ultrodex[16,17] as a support medium in the preparative isoelectric focusing. This obviates the need for Ultrodex, fractionating grid, and the laborious filtration using disposable syringes, and also improves the yield of proteins.

Properties of the Isolated β-Glucosidases

The five different peaks of β-glucosidase activity obtained following the purification of culture filtrates of *S. pulverulentum* may represent

[16] A. M. Bodhe, V. V. Deshpande, B. C. Lakshmikantham, and H. G. Vartak, *Anal. Biochem.* **123**, 133 (1982).

[17] V. V. Deshpande, M. Rao, S. Keskar, and C. Mishra. *Enzyme Microb. Technol.* **6**, 371 (1984).

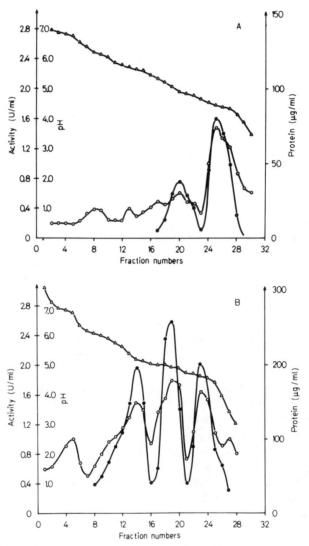

FIG. 2. Distribution of protein and 1,4-β-glucosidase activity after preparative slab gel isoelectric focusing in the pH range 3–7 of peak A (A) and peak B (B) from Fig. 1. pH values (△); β-glucosidase activity (●); protein μg/ml (○). From Deshpande et al.[4]

different isoenzymes produced by the fungus. Alternatively, they may arise through modifications of one or two original enzymes due to proteolytic activity which is considerable in the culture solution.

Both preparative and analytical isoelectric focusing indicate that the

TABLE I
PURIFICATION OF β-GLUCOSIDASES FROM CULTURE FILTRATES OF THE FUNGUS S. *pulverulentum*[a]

Step	Volume (ml)	Activity (U/ml)	Total activity (U)	Protein (mg/ml)	Total protein (mg)	Specific activity (U/mg)	Yield (%)
Culture filtrate	13,200	1.1	14,520	1.6	21,120	0.68	—
Ammonium sulfate precipitated	26	540	14,040	300	7,800	1.8	100
Preparative isoelectric focusing I	169	34	5,746	4	676	8.5	40
Phenyl-Sepharose							
A	32.5	22	715	1.4	45.5	15.7	4.9
B	32.5	60	1,950	2.6	84.5	23	13.4
Preparative isoelectric focusing II							
A₁	12	8.5	102	0.10	1.20	85.0 ⎫	78.6
A₂	23	20	460	0.22	5.06	90.9 ⎭	
B₁	20	24.5	490	0.31	6.20	79.0 ⎫	
B₂	21	29.5	620	0.29	6.09	102.0 ⎬	80.7
B₃	21	22.1	465	0.28	5.88	79.1 ⎭	

[a] From Deshpande *et al.*[4]

isoelectric points for A_1 and A_2 are pH 4.80 and 4.52, respectively, and pH 5.15, 4.87, and 4.56 for B_1, B_2, and B_3, respectively. Estimations based on electrophoresis in sodium dodecyl sulfate–polyacrylamide show the molecular weights of the enzymes to be between 165,000 and 182,000. Using preparations obtained from step III, K_m values for enzyme A and B are 4.5×10^{-3} and 3.7×10^{-3} M, respectively, for cellobiose and 1.5×10^{-4} and 2.1×10^{-4} M for PNPG. Gluconolactone is a powerful competitive inhibitor of the β-glucosidases and K_i values for enzymes A and B are 3.5×10^{-7} and 15×10^{-7} M, respectively, using PNPG as substrate.

Cell-Bound β-Glucosidases

Due to the small amount of cell-bound activity the purification of cell-bound β-glucosidase was not attempted. The K_m and K_i values for the crude preparation of cell-bound β-glucosidase were 2×10^{-3} and 1.2×10^{-4} M using p-nitrophenyl-β-glucoside as a substrate and gluconolactone as an inhibitor.

Function of β-Glucosidase in Cellulose Degradation

β-Glucosidase plays an important role in saccharification of cellulose to glucose.[1] Cellobiose, a product of the combined action of endo- and exo-1,4-β-glucanases on cellulose, is a competitive inhibitor of endo- and exoenzymes.[18] Sequential removal of cellobiose by β-glucosidase enhances the efficiency of the cellulose degradation process.[19] Adequate levels of β-glucosidase activity are essential when glucose is wanted as the major product of cellulose hydrolysis.[20]

[18] T. M. Wood and S. I. McCrae, *Proc. Symp. Bioconversion Cell. Subst. Energy, Chem. Microb. Protein, 1st* p. 111 (1978).
[19] B. S. Montenecourt and D. E. Eveleigh, *Appl. Environ. Microbiol.* **34**, 777 (1977).
[20] C. Mishra, M. Rao, R. Seeta, M. C. Srinivasan, and V. Deshpande, *Biotechnol. Bioeng.* **26**, 370 (1984).

[49] β-D-Glucosidases from *Sclerotium rolfsii*

By Jai C. Sadana, Rajkumar V. Patil, and
Jaiprakash G. Shewale

Sclerotium rolfsii produces at least four different extracellular β-glucosidases when grown on cellulose.[1,2]

Assay Method

Principle. β-D-Glucosidase (β-D-glucoside glucohydrolase, EC 3.2.1.21) activity is measured as the amount of glucose released from cellobiose or *p*-nitrophenol from *p*-nitrophenyl-β-D-glucopyranoside (PNPG). The glucose released is determined by the glucose oxidase–peroxidase test[3] using Glox glucose reagent. The amount of *p*-nitrophenol released is calculated from the molar absorbance of 18,500 for nitrophenol at 410 nm.[4]

With Cellobiose as Substrate

Reagents

Citrate buffer, 50 mM, pH 4.5
D(+)-Cellobiose, 50 mM in 50 mM citrate buffer, pH 4.5

[1] J. G. Shewale and J. C. Sadana, *Arch. Biochem. Biophys.* **207**, 185 (1981).
[2] J. C. Sadana, J. G. Shewale, and R. V. Patil, *Carbohydr. Res.* **118**, 205 (1983).
[3] H. U. Bergmeyer, K. Gawehn, and M. Grassl, *Methods Enzymatic Anal.* **1**, 457 (1974).
[4] D. Herr, F. Baumer, and H. Dellweg, *Eur. J. Appl. Microbiol. Biotechnol.* **5**, 29 (1978).

Glox glucose reagent,[5] one packet of Glox is dissolved in 500 ml of 50 mM phosphate buffer, pH 7.0. The reagent is freshly prepared to avoid high blank values

Sulfuric acid, 41%

Procedure. The enzyme is diluted just before assay in 50 mM citrate buffer, pH 4.5, to obtain a concentration of 0.05–0.3 unit of enzyme per milliliter (see definition below). The reaction is initiated by adding enzyme. The enzyme (0.1 ml) is added to 0.9 ml of cellobiose solution in a Corning test tube and the reaction mixture is incubated at 65° for 30 min. The reaction is stopped by heating the tubes in a boiling water bath for 5 min. After cooling 5 ml of Glox (glucose oxidase–peroxidase) reagent is added and the reaction mixture incubated for 1 hr at 30°. At the end of incubation, 4 ml of 41% H_2SO_4 is added and the color developed is measured at 530 nm. The color is developed immediately and is stable for 2 hr. A control is run with boiled enzyme. Under the conditions of assay, the enzyme shows linearity up to 0.4 mg of glucose produced.

With PNPG as Substrate

Reagents

Citrate buffer, 50 mM, pH 4.2

p-Nitrophenyl-β-D-glucopyranoside (PNPG), 1 mg/ml in 50 mM citrate buffer, pH 4.2

Sodium carbonate, 2% in distilled water

Procedure. The enzyme is diluted just before assay in 50 mM citrate buffer, pH 4.2, to obtain a concentration of 0.01–0.05 unit of enzyme per milliliter (see definition below). The reaction is initiated by adding 0.1 ml of enzyme to 0.9 ml of PNPG in a Corning test tube. The reaction mixture is incubated at 68° for 30 min. The reaction is stopped by adding 1 ml of 2% Na_2CO_3 and the color developed is measured at 410 nm. The color is developed immediately after addition of sodium carbonate and is stable for at least 24 hr. A control is run with boiled enzyme. The enzyme shows linearity up to 12 μg of p-nitrophenol produced.

Definitions of Unit and Specific Activity. One unit of β-D-glucosidase activity is defined as the amount of enzyme that releases 1 μmol of glucose from cellobiose at 65°, pH 4.5 or 1 μmol of p-nitrophenol from PNPG at 68°, pH 4.2, per minute. In the assay with cellobiose, 2 mol of glucose is released from 1 mol of cellobiose but the glucose values have been given as such without dividing by a factor of 2. This is according to the recom-

---

[5] Glox, glucose reagent (glucose oxidase–peroxidase reagent), was purchased from AB Kabi Diagnostica, Stockholm, Sweden.

mendations of the Commission on Biotechnology, International Union of Pure and Applied Chemistry.[6] The specific activity is defined as the number of units per milligram of protein. Protein is determined by the procedure of Lowry *et al.*[7]

Purification Procedure

Three of the purification steps, ammonium sulfate precipitation (step 1), fractionation by gel chromatography on Sephadex G-75 (step 2), and ultrafiltration of Fraction A (step 3) are identical to the steps for cellobiose dehydrogenase as described in this volume [53]. During gel filtration on Sephadex G-75 (step 2) β-glucosidase is eluted along with some high-molecular-weight cellulases, after the void volume (Fraction A) and ahead of low-molecular-weight cellulases (Fraction B). Fraction A (110–165 ml) contains about 95–98% of β-glucosidase activity and about 70% endo-β-glucanase activity, whereas Fraction B (170–265 ml) contains about 30% cellulase and 2% β-glucosidase activity.

Step 4. DEAE-Sephadex A-50 Ion-Exchange Chromatography. Fraction A after concentration on Amicon (step 3) is dialyzed in a collodion bag against 0.05 M phosphate buffer, pH 7.3, and chromatographed on DEAE-Sephadex column (1.8 × 100 cm) equilibrated with 0.05 M phosphate buffer, pH 7.3. The column is washed with the equilibration buffer. Fractions of 2 ml are collected at a flow rate of 12–15 ml/hr. β-Glucosidase and endo-β-glucanase activities are not adsorbed on the column. β-Glucosidase comes just after the void volume and forms the first peak; it is almost free of endo-β-glucanase (Fig. 1). The dark brown pigment present in the culture filtrate is removed in this step. Fractions 6–7 containing β-glucosidase of 44–50 specific activity are pooled and the pH is adjusted to 4.5 with 0.1 M citric acid. This is freeze-dried and then redissolved in 5 ml of 0.05 M citrate buffer, pH 4.5, and dialyzed against the same buffer. The β-glucosidase fraction at this stage gives one band in disc gel electrophoresis at pH 8.9. However, electrophoresis of the enzyme at pH 4.3 resolves it into four bands of proteins.

Step 5. Preparative Isoelectric Focusing. The β-glucosidase (Fractions 6–7) from step 4 is dialyzed overnight against 1 mM citrate buffer, pH 4.5, to reduce the salt concentration and is purified further by preparative isoelectric focusing. A gradient of sucrose from 50 to 5% is prepared

[6] *IUPAC, Comm. Biotechn.* **59,** 257 (1987).
[7] O. H. Lowry, N. J. Rosebrough, A. L. Farr, and R. J. Randall, *J. Biol. Chem.* **193,** 265 (1951).

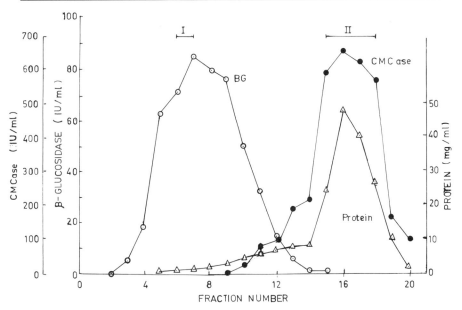

FIG. 1. Ion-exchange chromatography on DEAE-Sephadex A-50 of Fraction A from step 3. Peak I comprises fractions 6–7 and Peak II fractions 14–18. (○) β-D-Glucosidase (BG), (●) CMCase (endo-β-glucanase), (△) protein. From Shewale and Sadana.[1]

in a 110 ml LKB electrofocusing column.[8] Ampholyte solution (final concentration of 1%), pH 4–6, and β-glucosidase enzyme solution are mixed with light and dense solutions before making the gradient. The anode solution contains 0.16 M phosphoric acid in 60% sucrose, whereas the cathode solution employs 0.25 N sodium hydroxide. The cathode is at the top during the run. The voltage at the end of run (72 hr) is 500 V and the current 2 mA. Fractions of 1 ml are collected and are immediately processed for pH (5–7°), activity, and protein determination. Enzyme is freed from sucrose by extensive dialysis against 0.05 M citrate buffer, pH 4.5. β-Glucosidase activity is resolved into four separate peaks at pH 4.10, 4.55, 5.10, and 5.55 (Fig. 2) and they are designated BG-1, BG-2, BG-3, and BG-4, respectively. The observation is reproducible. The results of a typical purification of β-glucosidase enzymes are given in Table I.

Purity. The purified enzymes individually show one protein band on polyacrylamide gel electrophoresis at pH 4.3 and 8.9, in the presence or absence of SDS, and on isoelectric focusing in 7.5% polyacrylamide gel

[8] Instruction manual, LKB Preparative Isoelectric Focusing, Uppsala, Sweden.

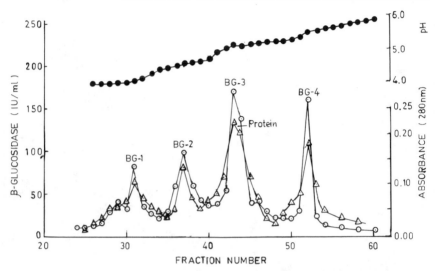

FIG. 2. Isoelectric focusing of Peak I from DEAE-Sephadex chromatography. (○) β-Glucosidase, (△) absorbance at 280 nm, (●) pH. From Shewale and Sadana.[1]

over the pH range 3.5–10. The purified β-glucosidases are free of contaminating endo-β-glucanase activity as determined viscometrically.[9]

Stability. β-Glucosidases are stable for several months when stored in 50 m*M* citrate buffer, pH 4.5, at −15°, and on repeated freezing and thawing.

Properties

Physical Properties. The relative molecular masses of the BG-1, BG-2, BG-3, and BG-4 β-glucosidases, estimated by gel filtration on BioGel P-150 according to the procedure of Andrews,[10] are 90,000, 90,000, 107,000, and 92,000, respectively, and by migration in SDS–polyacrylamide gel[11] are 95,500, 95,500, 106,000, and 95,000, respectively. The molecular weight of BG-3 β-glucosidase, determined by electrophoresis using the slope method,[12] is 100,000.

Subunit Structure. The four β-glucosidases from *S. rolfsii* are comprised of single polypeptide chains since the reduced carboxymethylated

[9] P. L. Hurst, J. Nielsen, P. A. Sullivan, and M. G. Shepherd, *Biochem. J.* **165,** 33 (1977).
[10] P. Andrews, *Biochem. J.* **91,** 222 (1964).
[11] K. Weber and M. Osborn, *J. Biol. Chem.* **244,** 4406 (1969).
[12] J. L. Hedrick and A. J. Smith, *Arch. Biochem. Biophys.* **126,** 155 (1968).

TABLE I
PURIFICATION OF β-GLUCOSIDASES FROM S. rolfsii[a]

Fraction	Total protein (mg)	β-Glucosidase			CMCase[b]		
		Total units	Specific activity (U/mg protein)	Recovery (%)	Total units	Specific activity (U/mg protein)	Recovery (%)
Culture filtrate	16,150	17,720	1.1	100	355,488	22	100
Ammonium sulfate, 0–3.4 M saturation	10,750	15,930	1.4	90	284,740	26	80
Sephadex G-75							
Fraction A	7,580	11,400	1.5	64	109,290	14	30
Fraction B	918	74	0.004	0.4	49,680	54	14
Ultrafiltration of fraction A (Amicon XM-50)	3,800	9,210	2.4	52	57,260	15	16
DEAE-Sephadex A-50							
Peak I	23	1,125	48	6.3	110	4.7	0.03
Peak II	550	127	0.23	0.71	28,359	51	7.9
Preparative isoelectric focusing of Peak I							
BG-1	0.37	10	27	0.05	0	0	0
BG-2	0.20	10	50	0.05	0	0	0
BG-3	1.80	120	57	0.57	0	0	0
BG-4	0.59	14	25	0.07	0	0	0

[a] From Shewale and Sadana.[1]
[b] CMCase, Endo-β-glucanase. One unit of endo-β-glucanase activity is defined as the amount of enzyme that releases 180 μg of reducing sugars (expressed as glucose equivalent) per minute from carboxymethylcellulose.

form of each enzyme shows one protein band on SDS–gel electrophoresis with relative molecular masses corresponding to native proteins.

Chemical Properties. All four *S. rolfsii* β-glucosidases are glycoproteins. Monitoring the values of the Michaelis constant (K_m) and maximum velocity (V_{max}) as a function of pH for BG-3 suggests involvement of a carboxylate group in the formation and dissociation of the enzyme–substrate complex.

Enzymatic Properties. The optimum pH and temperature for activity for all four β-glucosidases are pH 4.2 and 68° with PNPG as substrate, and pH 4.5 and 65° with cellobiose as substrate, respectively. The activation energies, calculated from Arrhenius plots, are 12.2, 14.9, 11.5, and 18.3 kcal/mol with PNPG and 6.5, 7.6, 6.9, and 6.4 kcal/mol with cellobiose as substrate for BG-1, BG-2, BG-3, and BG-4 enzymes, respectively.

Kinetics. The K_m values for PNPG, *p*-nitrophenyl β-D-cellobioside, cellobiose, and cellodexterins are given in Table II. The V_{max} values for cellobiose as substrate, calculated from Lineweaver–Burk plots, for BG-1, BG-2, BG-3, and BG-4 β-glucosidases are 55, 78, 175, and 51 μmol glucose/mg protein/min at 65°, pH 4.5. The K_m values of all four β-glucosidases decrease with the chain length of the cellodextrins upto cellopentaose.

Inhibitors. Glucose, glucono-1,5-δ-lactone, and nojirimycin inhibit all four *S. rolfsii* β-glucosidases. The K_i values of BG-3 β-glucosidase, with cellobiose as substrate, for glucose, glucono-1,5-δ-lactone, and nojirimycin are 0.55 mM, 0.01 mM, and 1.0 μM, respectively. The ratio of K_m to K_i for glucono-1,5-δ-lactone, 530, and for nojirimycin, 5840, suggests that these are strong inhibitors of cellobiase activity. The inhibitory effect of glucose is more marked with cellobioase as substrate when compared with higher molecular weight cellodextrins and decreases with the increase in chain length of cellodextrins.

Specificity. The specificity of the enzyme is not restricted to the β-D-(1 → 4) linkage, and all four β-glucosidases hydrolyze disaccharides having β-D-(1 → 6), β-D-(1 → 3) and β-D-(1 → 2) linkages. Laminaribiose [β-(1 → 3)] is hydrolyzed at a faster rate than gentibiose [β-(1 → 6)] and sophorose [β-(1 → 2)]. The enzymes require a strictly β-D-glucopyranosyl configuration for activity. Neither the glucosides with α-configuration nor the galactosides or xylosides are hydrolyzed. The enzymes can tolerate a

TABLE II
MICHAELIS CONSTANTS FOR *S. rolfsii* β-D-GLUCOSIDASES I–IV[a]

	β-D-Glucosidases (mM)				
Substrate	I	II	III	IV	III[b]
p-Nitrophenyl-β-D-glucopyranoside	1.07	1.38	0.89	0.79	0.51
p-Nitrophenyl-β-D-cellobioside	c	c	0.38	c	0.35
Cellobiose	3.65	3.07	5.84	4.15	3.65
Cellotriose	1.00	1.23	1.98	0.70	1.13
Cellotetraose	0.50	0.85	0.83	0.50	0.75
Cellopentaose	0.49	0.40	0.76	0.55	0.60
Cellohexaose	0.62	0.37	0.37	0.67	0.50

[a] Kinetic studies are done with the standard assay systems, at 65° and pH 4.5, and varying the substrate concentration. K_m values are calculated from Lineweaver–Burk plots, which are linear in all cases. From Sadana *et al.*[2]

[b] Assay is carried out at 65° and pH 4.5, in the presence of bovine serum albumin (0.5 mg/ml).

[c] Not determined.

wide variety of aglycon though the rate of hydrolysis depends on the nature of the aglycon moiety. Phenyl-β-D-glucopyranoside and o- or p-nitrophenyl-β-D-glucopyranosides are hydrolyzed 10–20 times faster than methyl-β-D-glucopyranoside.

Hydrolysis of Cellodextrins and β-D-Glucans. S. *rolfsii* β-glucosidases, as indicated earlier, hydrolyze cellodextrins in addition to cellobiose. The initial higher rates of hydrolysis of cellodextrins up to cellopentaose and lower K_m values for higher molecular weight cellodextrins (Table II) indicate cellopentaose as the preferred substrate for all four β-glucosidase enzymes. The reduced cellodextrins, cellotetraitol, and cellopentaitol are hydrolyzed to almost the same extent as those for the corresponding unreduced cellodextrins. Cellobiotol, however, is completely resistant. None of the four β-glucosidases acts on highly ordered substrates such as Avicel, but the enzymes slowly hydrolyze disordered substrates such as phosphoric acid-treated Avicel and carboxymethylcellulose but no measurable decrease in viscosity of a carboxymethylcellulose solution is observed in presence of these enzymes.

Products of Hydrolysis. Analysis of the products formed as a result of enzyme action on cellodextrins, reduced cellodextrins, and phosphoric acid-treated Avicel shows that the β-glucosidases from S. *rolfsii* act by cleaving D-glucosyl groups from the nonreducing end of the chain of the substrates.

Characterization of β-Glucosidases. An enzyme having the capacity to attack long-chain polymers, such as phosphoric acid-swollen cellulose, carboxymethylcellulose, and cellodextrins, is classified, according to the definition of Reese *et al.,*[13] as an exoglucanase. The β-glucosidase enzymes of S. *rolfsii,* thus, behave rather as exo-β-D-glucan glucohydrolases.

Role of β-Glucosidase in Cellulose Hydrolysis. The higher reaction rate of the four S. *rolfsii* β-glucosidases with higher molecular weight cellodextrins as compared to cellobiose, the decrease in their K_m values, and the decrease in the inhibitory effect of D-glucose on the rate of hydrolysis with the increase in chain length of cellodextrins indicate that higher molecular weight cellodextrins, and not cellobiose, are the major route of D-glucose formation from cellulose.

[13] E. T. Reese, A. H. Maguire, and F. W. Parrish, *Can. J. Biochem.* **46**, 25 (1968).

[50] Purification and Assay of β-Glucosidase from *Schizophyllum commune*

By Amy C. Lo, Gordon Willick, Roger Bernier, Jr., and Michel Desrochers

The breakdown of cellulose by fungi is catalyzed by a multiple enzyme system. The major enzymes of this system are endo-1,4-β-D-glucanase (cellulase, EC 3.2.1.4), cellobiohydrolase (exocellobiohydrolase; EC 3.2.1.91, cellulose 1,4-β-cellobiosidase), and β-1,4-D-glucosidase (β-glucosidase, EC 3.2.1.21). Although the β-glucosidase does not hydrolyze cellulose, it does play an important role in cellulose degradation by hydrolyzing soluble oligosaccharides, including cellobiose, with a concomitant release of glucose. This prevents the accumulation of cellobiose, an inhibitor of cellobiohydrolase, thus increasing the rate of cellulose hydrolysis. In order to study the kinetics and enzyme hydrolysis mechanism of the β-glucosidase of the Basidiomycete *Schizophyllum commune*, we have developed procedures for the purification and the detection of β-glucosidase. We have also utilized rapid techniques for the analysis of oligosaccharides of cellulose by HPLC chromatography.

Procedures for Testing β-Glucosidase Activity

Two procedures were adopted to measure the activity of β-glucosidase in the present work; the *p*-nitrophenyl-β-D-glucosidase (PNPGase) assay first developed by Nisizawa *et al.*[1] and a commercial glucose detection kit supplied by Sigma Co. (kit 115). The PNPGase activity is used in the enzyme purification steps and the glucose determination kit for studying enzyme kinetics. However, we have adapted both assays to microtiter dish scale so that as many as 96 samples can be assayed and scanned by the plate reader very rapidly. These microassays also offer another advantage in that only a minute amount of sample is required for accurate measurement. For example, both the glucose and the PNPGase microassays can detect up to 1.4 μg/ml of glucose or 0.8 μg/ml of *p*-nitrophenol in a microtiter well. Furthermore, one commercial glucose kit can accommodate up to 1500 microassays instead of 100 assays as suggested by the supplier.

Generally, in the PNPGase microassay, 10 μl of enzyme preparation is

---

[1] T. Nisizawa, H. Suzuki, and K. Nisizawa, *J. Biochem. (Tokyo)* **70**, 387 (1971).

METHODS IN ENZYMOLOGY, VOL. 160

mixed with 240 μl of *p*-nitrophenyl-β-D-glucopyranoside and incubated in a Titertek microplate incubator (Flow Laboratories) for 15 min at 30°. The reaction is stopped by the addition of 50 μl of 1 *M* Na_2CO_3 and the release of *p*-nitrophenol is assayed by a multiscan plate reader Titertek (Flow Laboratories) using a 414-nm filter. In the glucose microassay, each well contains 25 μl of glucose standard or enzyme reaction mixture and 50 μl of glucose color reagent. The mixture is incubated at room temperature for 20 min, whereupon 100 μl of 0.158 *N* HCl is added to stop the reaction. The plate is then scanned, using a 510-nm filter.

Methods for Analyzing the Hydrolysis of Oligosaccharides

Cellobiase activity (measured as production of glucose from cellobiose) can be monitored using high-performance liquid chromatography and a refractive index detector.[2] Figure 1 shows the chromatogram of cellodextrins of DP 1–6 (except the cellotriose) prepared by the method of Miller[3] and applied on a 3.9 × 30 cm carbohydrate analysis column (Waters Associates, Milford, MA). Water in acetonitrile (66% CH_3CN : 34% H_2O) is used as the mobile phase, at a flow rate of 1.0 ml/min, and backpressure of approximately 1000 psi.

The other procedure utilizes the Interaction CHO-620 carbohydrate column (Interaction Chemicals, Mountain View, CA) for analyzing the hydrolysis of oligosaccharides (DP 2–6) by the β-glucosidases. The amount of sugar can be detected by a refractometer with a minimum sensitivity of about 100 μg/ml oligosaccharides. This method is also very simple and fast, provided the sample is properly desalted and dissolved in water. Generally, it only takes 15 min to analyze one sample after the hydrolysis step. These two columns are complementary in the sense that the carbohydrate analysis column fully resolves oligosaccharides of DP 4–6, whereas the CHO-620 resolves those of DP 1–5.

The β-glucosidase is dissolved in 25 μl of 10 m*M* NH_4OAc, pH 5.0 (final concentration of 6 μg/ml), and mixed with equal volumes of 2 mg/ml of substrate (DP 2–6). The reaction is carried out in a test tube at 30°. Samples are withdrawn at specific time intervals and placed in a boiling water bath for at least 1 min to denature the protein. Ammonium acetate is removed by evaporation and the sample is redissolved in 50 μl of water for analysis. The Interaction column is normally operated at 90° and is eluted with HPLC purity-grade degassed water running at 0.55 ml/min and a back-pressure of 700 psi. The protein from the reaction mixture can

[2] E. K. Gum, Jr., and R. D. Brown, Jr., *Anal. Biochem.* **82,** 372 (1977).
[3] G. L. Miller, *Methods Carbohyd. Chem.* **3,** 134 (1963).

FIG. 1. Separation of cellodextrins on a carbohydrate analysis column. Each sugar was diluted in water to a final concentration of 1 mg/ml. Volumes of injection were 25 μl. The retention times (min) of each sugar are indicated above the peaks.

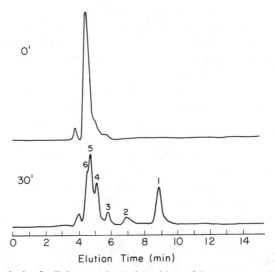

FIG. 2. Hydrolysis of cellohexaose by β-glucosidase of *S. commune* at 0 and 30 min. The samples (1 mg/ml cellohexaose) in 10 mM NH$_4$OAc, pH 5.2, plus 3 μg/ml of enzyme was eluted with water at 90° at a flow rate of 0.55 ml/min using a CHO-620 carbohydrate column. The degree of polymerization is indicated by the numbers above the peaks.

be removed either by a Prepsep C_{18} column or, in our case, by a guard cartridge attached to the column. Figure 2 shows the resolution of digestion products of cellohexaose by β-glucosidase II.

Procedure for Purification

This procedure is a modification of the original procedure of Desrochers *et al.*[4] The major modification involves the use of a high-performance anion exchanger for the final stage of the purification. This method involves three steps (A. C. Lo *et al., in preparation*). First, a two-stage ethanol precipitation of the culture filtrate is required. The first precipitation removes polysaccharides secreted into the medium by this organism in order to carry out the chromatography steps. The second precipitation concentrates the β-glucosidase. The ethanol precipitation is followed by an initial separation by conventional ion-exchange chromatography, which also serves to remove all residual polysaccharides. The last step is the final purification using a high-performance ion-exchange column.

Growth of Fungus

A 1-cm^2 piece of the growing edge of the mycelium of *Schizophyllum commune* No. 13 (ATCC 38548) culture grown on YEPD plates (2% yeast extract, 2% peptone, and 2% glucose) is used to inoculate 200 ml of 1.5% malt extract (Difco) in a polypropylene flask containing a large glass marble (20 mm diameter) to keep the mycelium well dispersed. After cultivation for 3 days (200 rpm, 30°), the suspension is transferred (5% v/v) into a polypropylene flask (200 rpm, 30°) containing a modified medium optimized for β-glucosidase production[4] (in g/100 ml) peptone (3), $CaNO_3 \cdot 4H_2O$ (1.5), $MgSO_4 \cdot 7H_2O$ (0.05), KH_2PO_4 (1.3), trace element mix [in g: $FeSO_4$, (2.5); $MNSO_4 \cdot H_2O$ (0.98), or $MnCl_2 \cdot 4H_2O$, (0.89); $ZnCl_2$ (0.83) or $ZnSO_4 \cdot H_2O$ (1.76); $CoCl_2 \cdot 6H_2O$ (1.0) or $Co(NO_3)_2 \cdot 6H_2O$ (1.25); HCl, concentrated, 5 ml; distilled water 495 ml] (Mandels *et al.*[5]) 0.1 ml, and cellulose (Solka-Floc SW-40, Brown Company, IL) (3).

Purification

When the β-glucosidase concentration reaches 15–22 IU/ml (usually about 10 days), the culture is filtered through a cheesecloth and centri-

[4] M. Desrochers, L. Jurasek, and M. G. Paice, *Appl. Environ. Microbiol.* **41,** 222 (1981).
[5] M. Mandels, D. Sternberg, and R. E. Andreotti, *in* "Symposium on Enzymatic Hydrolysis of Cellulose" (M. Bailey, T. M. Enari, and M. Linko, eds.), p. 81. The Finnish National Fund for Research and Development, Helsinki, Finland, 1975.

fuged to remove the residual mycelium. To 1 liter of filtrate (room temperature), 250 ml of ethanol ($-20°$) is added, and the precipitated polysaccharides are removed by centrifugation (6000 g for 20 min. in a GS-3 rotor). A further 500 ml of ethanol is then added, and the resulting precipitate, containing the β-glucosidase activity, is collected by centrifugation. The ethanol precipitate is dissolved in 0.1 M NH$_4$OAc, pH 5.2 (about 1.5 g total protein in 275 ml) and applied to a 4.5 × 17.5 cm column of DEAE-BioGel A (Bio-Rad Laboratories). After washing with 1 liter of 0.1 M NH$_4$OAc, pH 5.2, the proteins are eluted with a 3 liter linear gradient of NH$_4$OAC (0.2–0.4 M, pH 5.2) at 4°, and a flow rate of 60 ml/hr. Two major activity peaks are eluted and designated β-glucosidase I and β-glucosidase II. The fractions from each peak are pooled separately. Each is further purified on a Mono Q HR5/5 column (Pharmacia Fine Chemicals). The column is eluted with a 25 ml linear gradient of NH$_4$OAc (0.2–1 M, pH 5.2) at 1 ml/min. The column has a capacity of 15 A_{280} units of protein under these conditions. The run and recycle time is 30 min. Fractions (0.5 ml) are assayed for β-glucosidase activity. With the ethanol precipitate activity defined as 100%, the recovery of each β-glucosidase is about 40% after DEAE-BioGel chromatography, and 12% after the final Mono Q chromatography. The latter step requires at least four passes through the Mono Q column, and the final fractions retained are best identified by 10% SDS–gel chromatography.[6]

Comments

The Mono Q column is also available in preparative size, the use of which would considerably speed up the purification. The final purification can also be carried out on a P200 column,[4] but this method is much slower. We have found that silica-based columns, such as the TSK (Toyo Soda Co.), cannot be used for the chromatography of cellulase and β-glucosidase from *S. commune* because of large losses from irreversible adsorption. This also holds true for polysaccharide-based columns such as Sephadex. It is also advisable to avoid freeze-drying, if full enzyme activity retention is desired.

The purified enzymes have molecular weights of 110,000 (β-glucosidase I) and 96,000 (β-glucosidase II). There is some molecular weight heterogeneity with each; this has been suggested to be due to glycosylation heterogeneity.[7]

Cellodextrins suitable for use in the kinetic studies can be prepared by

[6] U. K. Laemmli, *Nature* (*London*) **227,** 680 (1970).
[7] G. E. Willick and V. L. Seligy, *Eur. J. Biochem.* **151,** 89 (1985).

cellulose hydrolysis with HCl followed by column purification.[3] We have purified the resulting crude oligosaccharide mix by further chromatography of P-2 (Bio-Rad Laboratories; 11 × 90 cm column), eluted with water at 65° at a flow rate of 15 ml/hr. Reese and Mandels[8] reported an enzymatic hydrolysis of cellulose by using cellulase preparations of high activity. Although commercial preparations of cellodextrins are available (Pfanstiehl Laboratories Inc., Waukegan, IL), we recommend that the purity of such preparations be tested by using HPLC chromatography with either or both of the carbohydrate analysis or CHO-620 columns.

Acknowledgments

We thank J. R. Barbier for technical assistance and M. G. Paice for helpful advice on HPLC chromatography. R.B. Jr is an industrial fellowship from the N.R.C. of Canada. This is publication number 25522 of the National Research Council of Canada.

[8] E. T. Reese and M. Mandels, *Methods Carbohydr. Chem.* **3**, 139 (1963).

[51] Purification of β-D-Glucoside Glucohydrolases of *Talaromyces emersonii*

By MICHAEL P. COUGHLAN and ANTHONY McHALE

Talaromyces emersonii CBS814.70 produces a complete extracellular cellulase system when grown on media containing cellulose as the carbon source.[1] Among the components of this system are three enzymes exhibiting β-glucosidase (β-D-glucoside glucohydrolase; EC 3.2.1.21) activity.[2–4] Each form exists as a single polypeptide, the M_r values, determined by electrophoresis under denaturing conditions in polyacrylamide gradient gels, being I, 135,000; II, 100,000; III, 45,700. An intracellular form, β-glucosidase IV, having an M_r value of 57,600, is also produced by the organism. Procedures for the isolation of forms I, III, and IV are given below.[2,4,5]

[1] M. P. Coughlan and A. P. Moloney, this volume [40].
[2] A. McHale and M. P. Coughlan, *Biochim. Biophys. Acta* **662**, 152 (1981).
[3] A. McHale and M. P. Coughlan, *J. Gen. Microbiol.* **128**, 2327 (1982).
[4] A. McHale and M. P. Coughlan, *Biochim. Biophys. Acta* **662**, 145 (1981).
[5] A. McHale and M. P. Coughlan, *FEBS Lett.* **117**, 319 (1980).

Growth Conditions

Talaromyces emersonii CBS814.70 was grown as described elsewhere in this volume[1] except that ammonium nitrate rather than ammonium sulfate was used as a nitrogen source. Culture filtrates and mycelia were harvested at the times appropriate to the isolation of the individual enzymes.[4] In pH-uncontrolled fermentations, β-glucosidase III is induced concurrently with the endo- and exocellulases. Its activity is maximal at 36 hr but by 45 hr it has disappeared from the medium as a result of the low pH that develops during growth. In fermentations in which the pH of the medium is maintained at 5 by the inclusion of sodium citrate, form III begins to accumulate at 20 hr and is still present at much later stages of cultivation. In pH-controlled and uncontrolled fermentations β-glucosidase I activity appears at about 50 hr and is maximal at about 75 hr. β-Glucosidase II behaves as does type III but is not readily obtained in quantities sufficient for its characterization. The intracellular form, β-glucosidase IV, is present throughout the growth cycle. However, it should be noted that under the growth conditions described cell lysis begins at approximately 45 hr.

Reactions Catalyzed by β-Glucosidases

β-Glucosidases catalyze the hydrolysis of cellobiose, removal of glucose from the nonreducing ends of cellooligosaccharides, glucosyl transfer to cellobiose, and hydrolysis of various artificial β-D-glucosides.

β-Glucosidase Assay

β-glucosidase activity is measured as follows. Reaction mixtures containing *p*-nitrophenyl-β-D-glucoside (1.25 mM in the case of β-glucosidase I and IV, 5 mM in the case of β-glucosidase III), 50 mM sodium acetate buffer, pH 5, and an aliquot of enzyme in a final volume of 4 ml were incubated at 37°. Under these conditions activity was constant for up to 1 hr for β-glucosidase I and III but for only 10 min in the case of β-glucosidase IV. Reactions were stopped by the addition of 4 ml of 0.4 M glycine/NaOH buffer, pH 10.8, and the absorbance at 430 nm was read. Activity is expressed as μmol *p*-nitrophenol released/min/ml of enzyme. With cellobiose as substrate, conditions were as described above except that the volume of the mixture was 1 ml and the reaction was stopped by boiling in sealed tubes for 3 min. The amount of glucose released was measured using the glucose oxidase method of Werner *et al.*[6] in which

[6] W. Werner, H. G. Rey, and H. Weilinger, *Z. Anal. Chem.* **252**, 224 (1970).

2,2′-azinodi(3-ethylbenzenethiazolinesulfonic acid) is oxidized to a colored complex. Zymogram stains for locating β-glucosidase activity on electrophoretic gels are described elsewhere in this volume.[7]

Reagents

1.25 and 5 mM *p*-nitrophenyl-β-D-glucoside (Sigma Chemical Company, Poole, Dorset, UK) in 50 mM sodium acetate buffer, pH 5
5 mM cellobiose (Sigma Chem. Co.) in above buffer
0.04% (w/v) bromphenol blue
Sephadex G-75 and DEAE-Sephadex A-50 (Pharmacia Fine Chemicals, Milton Keynes, Bucks, UK)
Glucose oxidase/peroxidase reagent[6] (Boehringer Corp., Ltd., Dublin, Ireland)

Purification of β-Glucosidase I

Culture filtrate was harvested at about 80 hr as described in the accompanying paper[1] at which time the predominant β-glucosidase present was type I.[4]

Step 1. Ammonium Sulfate Fractionation. The fraction of culture filtrate that precipitates at 4° at ammonium sulfate concentrations between 0.8 and 3.9 M was collected by centrifugation, redissolved in distilled water, and freeze-dried.

Step 2. Gel Filtration. The freeze-dried material was dissolved in 0.1 M potassium acetate buffer, pH 3.5, and subjected to gel filtration at 4° on a column (4.5 × 72 cm) of Sephadex G-75 equilibrated with the same buffer. Two major peaks of protein eluting between 300–800 and 1000–1600 ml, respectively, were obtained. The first, pool A, contained β-glucosidase, endocellulase, exocellulase, laminarinase, and a trace of protease activity and accounted for 80% of the total protein applied. The second, pool B, accounted for the remaining 20% of the applied protein and contained exocellulase, laminarinase, and dextranase activities. Pool A was concentrated 10-fold using an Amicon ultrafiltration device with a PM30 membrane and desalted by passage of several volumes of distilled water through the apparatus. The desalted solution was freeze-dried and, if necessary, stored at 4°. No loss of activity occurred over a period of 1 month.

Step 3. Ion-Exchange Chromatography. Aliquots (30–40 mg) of the freeze-dried material were redissolved in 2 ml of 50 mM sodium acetate buffer, pH 5, containing 0.15 M NaCl and fractionated on a column (1.7 ×

---

[7] M. P. Coughlan, this volume [14].

26 cm) of DEAE-Sephadex A-50. Three peaks of protein were eluted by irrigation with about 4 column volumes of the same buffer. The first contained β-glucosidase and accounted for approximately 1.5% of the protein of the original ammonium sulfate precipitate. The others contained endocellulase and laminarinase activities and between them accounted for 24% of the original protein. Two further peaks of protein, containing endocellulase and exocellulase activities, respectively, were eluted on application of a 500 ml convex exponential gradient, 0.15–0.35 M, of NaCl in 50 mM sodium acetate buffer, pH 5. The gradient was set up using two reservoirs. One, a cylindrical vessel, contained 250 ml of 0.35 M NaCl in 50 mM sodium acetate buffer, pH 5. This was connected to a conical vessel containing 0.15 M NaCl in 50 mM sodium acetate buffer, pH 5. The levels of the solutions in both flasks were the same. The conical flask, which was the mixing flask (solution stirred magnetically), was in turn connected to the column.

The fractions exhibiting β-glucosidase activity were pooled, dialyzed against distilled water, and freeze-dried. Polyacrylamide gradient gel electrophoresis (see below) showed the presence in this material of two protein bands, one of which displayed β-glucosidase activity.

Step 4. Gradient Gel Electrophoresis (Preparative and Analytical). Gradient gels (5–27% with respect to polyacrylamide concentration) were cast in slab gel molds as described by Margolis and Kenrick.[8] The gels (2.5 mm × 7 cm × 8 cm) were prepared in 88.7 mM Tris, 2.5 mM EDTA, 81.57 mM boric acid buffer, pH 8.3, and run in the same buffer using the GE2/4LS apparatus and the ECPS3000/150 power supply provided by Pharmacia Fine Chemicals (Uppsala, Sweden). Sample applicators inserted in the top of the gel prior to running allowed of the application of 8 × 30 μl samples to each gel slab. The freeze-dried samples from step 3 above had been redissolved in 0.5 ml of the electrophoresis buffer containing 10% (v/v) glycerol. On application of the samples, the voltage was set at 300 V and maintained at this value until entry of the samples into the gel was complete (~15 min). Electrophoresis was then carried out at 150 V for 1–2 hr after the tracker dye (bromphenol blue, 0.04%, w/v) had migrated from the gel.

After electrophoresis a representative portion of the gel slab (i.e., one of the lanes) was stained for β-glucosidase activity.[7] The band corresponding to β-glucosidase I in each of the replicate lanes was excised and homogenized in 2 ml of 0.2 M sodium acetate buffer, pH 5, by extrusion through a narrow-bore needle. The resulting mixture of gel and buffer was then subjected to further homogenization at 4° using a Potter–Elvehjem

---

[8] J. Margolis and K. G. Kenrick, *Anal. Biochem.* **25**, 347 (1968).

homogenizer. The homogenate was allowed to stand at 4° for 4 hr and then centrifuged at 10,000 g for 20 min and the supernatant containing the enzyme was collected. The remaining gel was washed with 2 ml of 0.2 *M* sodium acetate buffer, pH 5, to maximize recovery of the enzyme. Typically, about 85–90% (i.e., 7 of the 8 IU) of the β-glucosidase type I activity applied is recovered in this fashion and, as expected, is found to be homogeneous by the polyacrylamide gradient gel electrophoresis procedure described above.

Purification of β-Glucosidase III

Culture filtrate was harvested at 36 hr at which time β-glucosidase III was the predominant form of this enzyme present.[4]

Step 1. Ammonium Sulfate Fractionation. The fraction precipitating at 4° at ammonium sulfate concentrations between 0.8 and 3.9 *M* was collected as described above.

Step 2. Gel Filtration. Approximately 1 g of freeze-dried material from the previous step was dissolved in 0.1 *M* potassium acetate buffer, pH 3.5, and fractionated by gel filtration on Sephadex G-75 as described above. β-Glucosidase I, if present, eluted between 200 and 400 ml while β-glucosidase III eluted as a broad peak centered at 1100 ml. The appropriate fractions were pooled, concentrated, and dialyzed as before using an Amicon ultrafiltration device equipped with a PM10 membrane, and then freeze-dried.

Step 3. Ion-Exchange Chromatography. Samples (typically 50 mg protein) of the freeze-dried material were dissolved in 1 ml of 50 m*M* sodium acetate buffer, pH 5, containing 0.25 *M* NaCl and chromatographed on a column (1.7 × 26 cm) of DEAE-Sephadex A-50 equilibrated and irrigated with the same buffer. The column was then irrigated with a linear salt gradient (0.25–0.55 *M* NaCl) in 50 m*M* sodium acetate buffer, pH 5. A single symmetrical peak of β-glucosidase III was eluted toward the end of the gradient. The appropriate fractions were concentrated, dialyzed and freeze-dried as before.

Step 4. Polyacrylamide Gradient Gel Electrophoresis. Preparations of β-glucosidase III isolated as described in steps 1 to 3 were frequently found to migrate as a single band on gradient gel electrophoresis. When necessary, "cleaning up" could be effected by using the preparative electrophoresis as described above for β-glucosidase I.

Purification of β-Glucosidase IV

Mycelia harvested at about 50 hr were washed successively with 10 volumes of distilled water, 0.5 *M* sodium acetate buffer, pH 5, distilled water, 0.2 M sodium acetate buffer, pH 5, and the suspended in the latter. The suspended material in the presence of glass beads (0.5–0.75 mm) was homogenized in a Waring blender for 3 × 1 min. Following centrifugation at 100,000 g to remove particulate matter, the supernatant was dialyzed against distilled water and then subjected to preparative polyacrylamide gradient gel electrophoresis and elution as above.

Purity of the β-Glucosidase Preparations

Because of the purification methods used, each of the enzyme preparations was homogeneous as judged by gel filtration and by gradient gel electrophoresis. The β-glucosidase III preparation was also homogeneous as judged by isoelectrofocusing, migrating as a single band with a pI value of 3.6.[3] When subjected to isofocusing the β-glucosidase IV preparation gave three distinct bands of protein, each active, the pI values being, 4.41, 4.47, and 4.50, respectively. In the case of the β-glucosidase I preparation, no distinct bands were observed following isofocusing. Rather, the activity and protein stains, although coincident, were streaked along the gel from regions corresponding to pI values of 3.4–4.17. We have reason to believe that the microheterogeneity of the β-glucosidase I and IV preparations may result from differences in the content or composition of attached carbohydrate (50% by weight in the case of type I) rather than from differences in primary structure.

Properties of the β-Glucosidase

With respect to activity, the pH and temperature optima of the β-glucosidases were as follows: I, 4.1 and 70°; III, 5.1 and 70°; IV, 5.7 and 35°. The half-lives at 70°, pH 5, were I, 410 min; III, 175 min; IV, 2 min.[2] The time course of production, K_m value, K_i value, substrate specificity, and reaction products of each form of the enzyme provided clues as to its possible function $in\ vivo$.[2–4] Thus, β-glucosidase I operates as a true cellobiase. By cleaving the cellobiose, produced from cellulose by the action of the endo- and exocellulases, it not only provides glucose for the cell but also relieves the inhibitory effect of cellobiose on cellulase activity. By contrast, β-glucosidase III may be primarily engaged in the hydrolysis of the cellooligodextrins arising during the course of cellulose digestion. As such, this form of the enzyme may be considered to be a 1,4-β-D-glucan glucohydrolase (glucan 1,4-β-glucosidase, EC 3.2.1.74). The intracellular form, β-glucosidase IV, by virtue of its hydrolase activity, converts to glucose the cellobiose taken into the cell, and, by virtue of its transferase and hydrolase activities, may be involved in the formation of specific trisaccharides or disaccharides. There have been numerous suggestions that such compounds may act as inducers of cellulase synthesis in fungi.[9]

Acknowledgment

The author thanks Elsevier Scientific Publishers B.V. (Biomedical Division), Amsterdam for permission to quote from material published in *Biochim. Biophys. Acta* **662**, 145 and 152 (1981).

[9] M. P. Coughlan, *Biotechnol. Genet. Eng. Rev.* **3**, 39 (1985).

[52] Cellobiose Dehydrogenase from *Sporotrichum thermophile*

By Giorgio Canevascini

$$\text{Cellobiose} + \text{acceptor} \xrightarrow[\text{dehydrogenase}]{\text{cellobiose}} \text{cellobionolactone} + \text{reduced acceptor}$$

$$\downarrow \text{H}_2\text{O}$$

$$\text{cellobionic acid}$$

An enzyme capable of oxidizing cellobiose at C_1 of the reducing residue was first discovered some years ago in a few representatives of a group of lignolytic Basidiomycetes, collectively called white-rot fungi.[1,2] Similar isofunctional enzymes have subsequently been found in other organisms not taxonomically related to the former, but all sharing the common ability to degrade cellulose.[3,4] The presence of oxidative enzymes, acting on cellulose degradation products, in cellulolytic organisms raises the question of their possible function (regulatory, metabolic) in the whole process of cellulose degradation. In this chapter a simple enrichment procedure is described to obtain cellobiose dehydrogenase preparations virtually free of cellulase using the culture supernatant of *Sporotrichum thermophile,* an organism producing high yields of this enzyme during growth on cellulose.[5]

[1] U. Westermark and K.-E. Eriksson, *Acta Chem. Scand., Ser. B* **28**, 204 (1974).
[2] A. R. Ayers and K.-E. Eriksson, this series, Vol. 89, p. 129.
[3] R. F. H. Dekker, *J. Gen. Microbiol.* **120**, 309 (1980).
[4] P. Fähnrich and K. Irrgang, *Biotechnol. Lett.* **4**, 775 (1982).
[5] M. R. Coudray, G. Canevascini, and H. Meier, *Biochem. J.* **203**, 277 (1982).

Enzyme Assay: Cellobiose Dehydrogenase

Cellobiose dehydrogenase is tested photometrically with dichlorophenolindophenol as electron acceptor at pH 7.0 and 30°. The reaction mixture (in a total volume of 3 ml) contains the following: sodium potassium phosphate buffer, pH 7.0, 0.3 mmol; dichlorophenolindophenol, 0.12 μmol; cellobiose, 1 μmol; and enzyme. After temperature equilibration, the reaction is started by adding enzyme and monitoring the change in absorbancy (A_{600}) during the first 3 min. The activity is conveniently expressed in nmol cellobiose oxidized (dichlorophenolindophenol reduced) per min (molar absorption coefficient at pH 7.0: $\varepsilon_{600} = 1.61$ liters mol^{-1} cm^{-1}).

Endocellulase

Contaminating endocellulase activity can easily be determined with known standard methods or with a dyed carboxymethylcellulose (CMC-azure).[6] A 2% (w/v) CMC-azure solution in 0.1 M acetate buffer, pH 5.0 (0.5 ml), is mixed with enzyme (0.5 ml) and incubated at 40° for a variable length of time (up to 1 hr) depending on the enzyme concentration. The reaction is stopped with a CMC-precipitating solution[6] and the insoluble substrate is removed by centrifugation. The color of the supernatant is measured at 590 nm against a reagent blank. The response is linear with time and enzyme concentration up to an absorbance of 0.3. One unit of activity is arbitrarily defined as that amount of cellulase causing an absorbance increment of 0.01 in 1 min at pH 5.0 and 40°.

Culture Conditions

For maintenance and conidia production, *Sporotrichum thermophile* ATCC 42464 is grown for a week at 44° on a solid medium containing Difco Neopeptone (1%, w/v), Difco yeast extract (0.1%, w/v), maltose (2%, w/v), and agar (1.2%, w/v). Plates or slopes may be stored in a refrigerator (10°) for several weeks. Submerged cultures are prepared starting from a conidial suspension (one petri dish per liter of medium) in sterile saline to which a few drops of Tween 80 have been added. The spores are first brought to germination (10–11 hr incubation at 44° with stirring) in a complex medium containing KH_2PO_4 (0.1%, w/v), Difco casamino acids (0.1%, w/v), Difco yeast extract (0.1%, w/v), KCl (0.05%, w/v), $MgSO_4 \cdot 7H_2O$ (0.02%, w/v), and trace elements (pH 5.8 before

[6] B. V. McCleary, *Carbohydr. Res.* **86,** 97 (1980).

steam sterilization). The young mycelium is then filtered aseptically through Miracloth and used to inoculate a medium containing the following: KH_2PO_4 (0.1%, w/v), NH_4Cl (0.04%, w/v), ammonium tartrate (0.1%, w/v), Difco yeast extract (0.02%, w/v), KCl (0.05%, w/v) $MgSO_4 \cdot 7H_2O$ (0.02%, w/v), $FeCl_3$ (10 mg/liter), $CaCl_2 \cdot 2H_2O$ (100 mg/ liter), trace elements, and cellulose (0.5%, w/v). Different kinds of crystalline cellulose may be used, e.g., Avicel, Solka-Floc, filter paper, Whatman CC 31. The pH of the medium is adjusted to 5.7–5.8 before sterilization. The fermentation is carried out in a fermenter equipped with an automatic pH control unit under agitation (magnetic or mechanical) and sterile air inflation. Five liter cultures are easily run in a 7-liter fermenter (Chemap AG, Männedorf, CH) with a blade stirrer at 400 rpm and an air flow of 1 liter/min. Toward the end of the culture, i.e., after 6–7 hr incubation, foam formation has to be controlled by an antifoam additive. During the first 2 hr the pH of the medium is left to increase without back acid titration (from 5.8 to 5.3–6.4). In a second period, corresponding to the catabolic utilization of cellulose, the pH of the medium decreases: the acid production is then titrated to pH 6.7–6.8 by addition of 1 N ammonia. In the absence of an automatic pH control unit the culture may be grown in the same complex medium as described above for the germination of the spores but supplemented with cellulose to induce the enzyme. In order to harvest the culture at the highest enzyme yield (this goes through a maximum during the course of the culture and then declines) the cellobiose dehydrogenase in the culture filtrate is checked (see below) at regular intervals (e.g., 30 min). Maximum enzyme production, which may vary from 40 to 70 units (nmol/ml), is modulated by different factors such as the kind of cellulose in the medium, cellulose concentration, medium composition, and pH; it usually occurs after 9–12 hr of cultivation. The culture filtrate is conveniently collected by filtration through glass fiber filters and has a faint yellow color.

Enzyme Enrichment

Hollow Fiber Concentration and Ammonium Sulfate Precipitation

The culture fluid (e.g., from a 5-liter culture) is concentrated and dialyzed by hollow fiber ultrafiltration (Type H1P5, Amicon Corp., Danvers, MA) then salted out at 0° by addition of solid ammonium sulfate (51.6 g/ 100 ml) to 80% saturation. The precipitate is dissolved in a minimum of water and then desalted with BioGel P-6DG (Bio-Rad Laboratories, Richmond, CA). No serious loss of activity occurs during hollow fiber or BioGel treatment. After ammonium sulfate precipitation, however, a loss

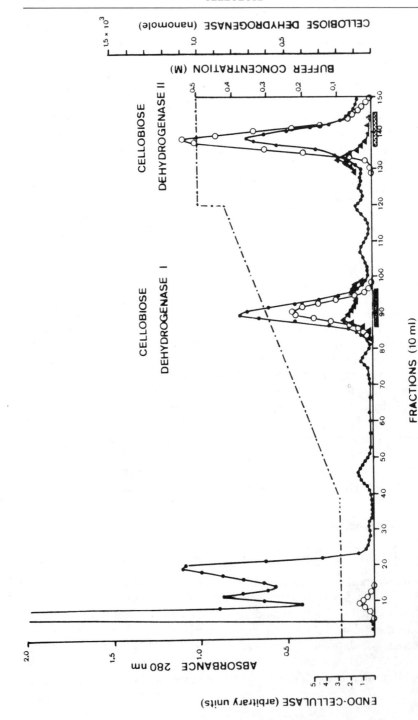

FIG. 1. DEAE-Trisacryl M chromatographic separation and enrichment of cellobiose dehydrogenase I and II. (●) Absorbance values at 280 nm; (○) cellobiose dehydrogenase (nmol min⁻¹ ml⁻¹: see text); (▲) endocellulase (arbitrary units min⁻¹ ml⁻¹); dotted lines: pooled fractions.

of activity may be occasionally observed. The original activity may be then restored by incubating the enzyme solution at 55° in a water bath for 1 hr. The crude enzyme may be stored freeze-dried for several months without loss of activity. Under the conditions thus described a 5-liter culture will yield between 400 and 800 mg crude enzyme powder.

DEAE-Trisacryl M Chromatography

A column (40 × 2.6 cm) of DEAE-Trisacryl M (Réactifs IBF, Ville-neuve-La-Garenne, France) equilibrated with 0.1 M ammonium acetate buffer, pH 5.5, is loaded with the crude enzyme (up to 1 g) and eluted with the same buffer (~500 ml) at a flow rate of 40 ml/hr, then with a linear gradient (made by mixing 500 ml of 0.1 M buffer with 500 ml of 0.5 M buffer) up to a 0.5 M buffer concentration followed by isocratic elution with 500 ml of 0.5 M buffer (see Fig. 1).

TABLE I
PURIFICATION OF CELLOBIOSE DEHYDROGENASE

Step	Volume (ml)	Total protein (Lowry) (mg)	Cellobiose dehydrogenase activity		Recovery (%)
			Units min^{-1} ml^{-1}	Units min^{-1} mg^{-1}	
Culture filtrate[a]	4900	882[b]	48	268	100
Hollow fiber concentration (Amicon, type HIP5)	105	420	2124	531	95
Ammonium sulfate (80% saturated) precipitation	22	409	9317	501	87
BioGel P-6DG desalting membrane ultrafiltration (Nucleopore, Type C 5000)	13.5	405	15404	513	88
DEAE-Trisacryl chromatography Cellobiose dehydrogenase I					
Active fractions	180	—	—	—	25[c]
Pooled fractions (membrane ultrafiltration)	20	25	2158	1726	18
Cellobiose dehydrogenase II					
Active fractions	190	—	—	—	54[c]
Pooled fractions (membrane ultrafiltration)	40	19.2	2282	4755	39

[a] Culture on 0.5% (w/v) Solka-Floc SW-40.
[b] Due to interference by phenolic compounds in the culture filtrate, the protein content (Lowry) is overestimated. The Bio-Rad procedure gives a lower value, i.e., 637 mg.
[c] Activity profile as in Fig. 1.

Cellobiose dehydrogenase, which is easily recognized by its yellow color, is recovered in three fractions: a very minor fraction eluted in the void volume and two main fractions (whose respective proportions are modulated by culture conditions), one (cellobiose dehydrogenase I) eluted approximately with 0.3 M buffer and the other (cellobiose dehydrogenase II) eluted later in a sharp peak with 0.5 M buffer. With this simple procedure the bulk of the cellulase (comprising endo- and exoglucanases) is eluted during the first isocratic step with 0.1 M buffer and is quantitatively separated from the cellobiose dehydrogenase-containing fractions. Nevertheless, some minor cellulase components comigrate with both cellobiose dehydrogenases as shown in Fig. 1. It is therefore advisable, before pooling the active fractions, to test them for their cellulase activity in order to keep the contaminating cellulase activity in the final enzyme preparation to a minimum. The pooled fractions are dialyzed and concentrated to the desired activity (between 1500 and 2500 units $min^{-1} ml^{-1}$) by ultrafiltration (Nucleopore membrane Type C 5000, Nucleopore Corp., Pleasanton, CA) and then stored frozen in small portions until needed. The cellobiose dehydrogenase II can easily be obtained free of contaminating cellulase and with a high specific activity. The cellobiose dehydrogenase I, however, has a lower specific activity but in spite of this its purity is fully compatible with the cellulase test described in chapter [10] of this volume. Table I gives the result of a typical enzyme enrichment obtained with the protocol just given.

[53] Cellobiose Dehydrogenase from *Sclerotium rolfsii*

By JAI C. SADANA and RAJKUMAR V. PATIL

Cellobiose + acceptor → cellobiono-δ-lactone + reduced acceptor

Assay Method

Principle. The enzyme-catalyzed reduction of 2,6-dichlorophenolindophenol (DCPIP) with D-cellobiose as substrate is measured by the decrease in absorbance at 600 nm.[1] A reference cell containing no cellobiose is used to compensate for any nonenzymatic reduction of the dye. For each molecule of cellobiose utilized, one molecule of DCPIP is reduced.

---

[1] J. C. Sadana and R. V. Patil, *J. Gen. Microbiol.* **131,** 1917 (1985).

Reagents

D-Cellobiose, 2.5 mM in 10 mM phosphate buffer, pH 6.3
2,6-Dichlorophenolindophenol (DCPIP), 2 mM
Phosphate buffer, 100 mM, pH 6.3
Cellobiose solution should be stored frozen
Phosphate buffer: K_2HPO_4–KH_2PO_4

Procedure. The enzyme is diluted just before assay in 100 mM phosphate buffer, pH 6.3, to obtain a concentration of 0.02–0.10 unit of enzyme per milliliter (see definition below). The reaction is initiated by adding enzyme. Two 1-ml cuvettes having a 1-cm path length are placed in the sample and reference positions of a recording spectrophotometer, equipped with a thermostatically controlled cuvette holder, maintained at 37°. Then 0.05 ml of dye solution is added to each cell, 0.9 ml of cellobiose solution to the cuvette in the sample position and 0.9 ml of phosphate buffer to the cuvette in the reference position. The enzyme solution (0.05 ml) is added to each cuvette to give a final volume of 1.0 ml, and they are mixed simultaneously. The decrease in absorbance at 600 nm is recorded at 2 min intervals. A control is run with boiled enzyme. The reduction of DCPIP causes a slight decrease in the pH of the reaction mixture. The maximal (initial) rate during the first 6 min is used for calculations. For purified enzyme preparations the cuvette without cellobiose can be omitted.

Application of the Assay to Crude Filtrate. The crude culture filtrate from *S. rolfsii* shows high β-D-glucosidase activity which hydrolyzes cellobiose to glucose. To overcome the interference by β-D-glucosidase, glucono-1,5-lactone at a final concentration of 2 mM is also included in the reaction mixture which inhibits β-D-glucosidase but not the cellobiose dehydrogenase.

Definition of Unit and Specific Activity. One unit of cellobiose dehydrogenase activity is defined as the amount of enzyme that catalyzes the reduction of 1 μmol of DCPIP (or oxidation of 1 μmol of cellobiose) per minute at 37°. Thus, 0.1 ml of a solution of enzyme containing 1 unit/ml would cause a rate of 1.85 A per minute, assuming a molar absorbance for the dye of 18,500.[2] The specific activity is defined as the number of units per milligram of protein. Protein is determined by the procedure of Lowry *et al.*[3]

---

[2] A. H. Phillips and R. G. Langdon, *J. Biol. Chem.* **237,** 2652 (1962).
[3] O. H. Lowry, N. J. Rosebrough, A. L. Farr, and R. J. Randall, *J. Biol. Chem.* **193,** 265 (1951).

Fungal Strain and Growth Conditions. Sclerotium rolfsii CPC 142[4] is grown for 14 days with cellulose-123 as the sole carbon source. It is maintained in a stock culture as described by Shewale and Sadana.[5] Stock cultures are stored at 28–30° on potato-dextrose agar (PDA) slants and subcultured once every 4 weeks.

The fungus is grown in a medium containing (in grams per liter) KH_2PO_4, 2.0; $(NH_4)_2SO_4$, 1.4; urea, 0.3; $MgSO_4 \cdot 7H_2O$, 0.3; $CaCl_2$, 0.3; cellulose-123, 10; Bacto-proteose peptone, 0.25; Difco yeast extract, 0.1; and 1 ml/liter of the trace metals[6] (in milligram per liter: $FeSO_4 \cdot 7H_2O$, 5.0; $MnSO_4 \cdot 7H_2O$, 1.56; $ZnSO_4 \cdot 7H_2O$, 3.34; $CoCl_2$, 2.0, and Tween 80, 0.33%) is also added. Large quantities are grown in a number of 1-liter Erlenmeyer flasks with 200 ml medium in each flask. The cultures are incubated at 29–30° for 14 days on a rotary shaker (150 rpm). Prior to autoclaving of the medium at 121° for 20 min the pH is adjusted to 6.5 with phosphoric acid. The media are inoculated with mycelia directly from 7-day PDA slants. The culture is harvested on day 14 by filtration through glass wool and centrifuging at 4000 *g* for 20 min. The clear supernatant fluid can be stored for long periods at 2–4°, or frozen at −15°, in the presence of 0.005% Merthiolate or 0.01% azide, without any loss of activity. The culture retains its enzyme activities (cellobiohydrolase, endo-β-glucanase, β-D-glucosidase, and cellobiose dehydrogenase) for over 6 years with frequent subculturing on PDA or when stored in a lyophilized state.

Purification Procedure

Step 1. Ammonium Sulfate Precipitation. The culture filtrate is concentrated by precipitating the proteins with solid ammonium sulfate at 90% saturation. The precipitate is suspended in a small volume of 50 m*M* citrate buffer, pH 4.8. Recovery of cellobiose dehydrogenase is around 80–90%. The precipitate can be stored for several months without any significant loss of activity.

Step 2. Fractionation by Gel Chromatography. The ammonium sulfate-precipitated enzymes are gel filtered on a Sephadex G-75 column (1.8 × 90 cm) for desalting and fractionation. The cellobiose dehydrogenase fraction (Fraction A, 110–165 ml) is eluted after the void volume along with β-D-glucosidase and endo-β-glucanase, and contains 75–80% of the cellobiose dehydrogenase activity. Fraction A is brown colored.

[4] *Sclerotium rolfsii* CPC 142 culture (NCIM No. 1084) is available from National Culture Industrial Microorganisms, National Chemical Laboratory, Poona 411 008, India.
[5] J. G. Shewale and J. C. Sadana, *Can. J. Microbiol.* **24**, 1204 (1978).
[6] M. Mandels, F. W. Parrish, and E. T. Reese, *J. Bacteriol.* **83**, 400 (1962).

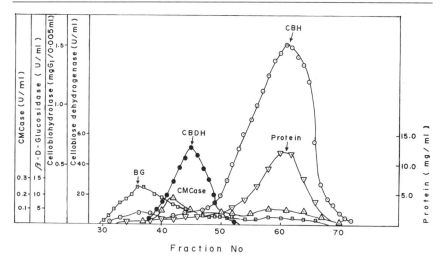

FIG. 1. Purification of cellobiose dehydrogenase on DEAE-Sephadex A-50 column. The concentrated enzyme solution from step 3 (ultrafiltration) is applied to the column (1.8 × 100 cm), previously equilibrated with 50 mM phosphate buffer, pH 7.3, and washed with the same buffer until no protein is detected. The column is eluted by 100 mM citrate buffer, pH 4.5. Fractions (2.5 ml) are collected at a flow rate 6–8 ml/hr. Fractions 40–50 show the highest cellobiose dehydrogenase activity. (●) Cellobiose dehydrogenase (CBDH); (○) cellobiohydrolase (CBH); (□) β-D-glucosidase; (▽) protein; and (△) endo-β-glucanase (CMCase).

Step 3. Ultrafiltration of Fraction A. Fraction A is concentrated with a Diaflo membrane XM50 ultrafiltration system (Amicon Corporation).

Step 4. DEAE-Sephadex A-50 Chromatography. The concentrated top enzyme solution from Step 3 is dialyzed in a collodion bag for 3–4 hr against 50 mM phosphate buffer, pH 7.3, and chromatographed on DEAE-Sephadex A-50 column (1.8 × 100 cm), previously equilibrated with 50 mM phosphate buffer, pH 7.3, and washed with the same buffer until no protein is detected. Cellobiose dehydrogenase is adsorbed on the column whereas endo-β-glucanase and β-D-glucosidase are not. Fractions (2.5 ml) are collected at a flow rate of 6–8 ml/hr. The column is eluted by 100 mM citrate buffer, pH 4.5. Fractions 40–50 show the highest cellobiose dehydrogenase activity and are collected (Fig. 1).

Step 5. Final Purification in Preparative Polyacrylamide Gel Electrophoresis. For preparative polyacrylamide gel electrophoresis, the procedure of Kerckaert[7] is used at pH 8.9. The combined fractions (fractions 40–50) are concentrated by lyophilization and dialyzed against 5 mM Tris–glycine buffer, pH 8.5. About 10–12 mg proteins with indicator dye

---

[7] J. P. Kerckaert, *Anal. Biochem.* **84,** 354 (1978).

TABLE I
PURIFICATION OF CELLOBIOSE DEHYDROGENASE FROM *Sclerotium rolfsii*

Purification step	Total protein (mg)	Total activity (U)	Specific activity [U (mg protein)$^{-1}$]	Purification (fold)	Yield (%)
Crude extract	17,202	33,540	1.95	1.00	100
Ammonium sulfate, 0–90% saturation	13,937	31,820	2.30	1.18	95
Sephadex G-75 chromatography Fraction A	7,815	26,570	3.40	1.74	79
Ultrafiltration of Fraction A (Amicon PM10)	6,051	211,180	3.50	1.79	63
DEAE-Sephadex A-50 chromatography	50	1,060	21.20	10.87	3
Preparative poly-acrylamide gel electrophoresis	8.5	229	26.90	13.79	0.7

(bromphenol blue) are loaded on slab gel (75 × 75 × 3 mm) GE-4 electrophoresis apparatus (Pharmacia, Sweden) containing 7.5% acrylamide. A current of 10–12 mA is applied and electrophoresis is carried out for 3–4 hr until the dye reaches the bottom of the gel. The gel is then removed and placed horizontally on the recovery cell assembly of an elution unit designed by Bodhe *et al.*[8] Vertical downward electrophoretic elution of the proteins across the thickness of the gel into foams is then carried out, applying 60–80 mA current for 1 hr. The proteins which collect into the foams are squeezed out and tested for activity after adjusting the pH to 6.3. The results of a typical purification of cellobiose dehydrogenase from *S. rolfsii* are presented in Table I.

Properties of the Purified Enzyme

Purity. The purified enzyme shows one protein band on polyacrylamide gel electrophoresis at pH 2.9 and 8.9, with or without SDS treatment, and isoelectric focusing in 7.5% polyacrylamide gel over the pH range 3.5–10.0.

Stability. The enzyme can be stored in 50 m*M* citrate buffer, pH 4.5 at −15° for several months with no or only a small loss of activity. The

[8] A. M. Bodhe, V. V. Deshpande, B. C. Lakshmikantham, and H. G. Vartak, *Anal. Biochem.* **123**, 133 (1982).

TABLE II
KINETIC CONSTANTS FOR CELLOBIOSE
DEHYDROGENASE FROM *Sclerotium rolfsii*[a]

Substrate	K_m (μM)	V_{max} [μmol min^{-1} (mg enzyme)$^{-1}$]
Electron donor		
Cellobiose	41.6	36.0
Cellotriose	62.5	16.6
Cellotetraose	66.6	15.4
Cellopentaose	83.3	13.3
Cellohexaose	100.0	12.5
Electron acceptor		
DCPIP	220	36.0
Cytochrome c[b]	384	0.017

[a] K_m and V_{max} values are calculated from Lineweaver–Burk plots, which are linear in all cases.
[b] The rate of reduction of cytochrome c is calculated from the molar absorbance at 550 nm (Δ, reduced minus oxidized = 18.7 × 10^3 mol^{-1} cm^{-1}).[2]

enzyme is most stable at pH 4.5–5.0. The enzyme is stable to repeated freezing and thawing.

Physical Properties. The relative molecular masses, estimated by gel filtration on BioGel 150, by electrophoresis using the slope method,[9] and by migration in SDS–polyacrylamide gels[10] are 63,000, 62,500, and 64,500, respectively. The enzyme is composed of a single polypeptide chain as carboxyamidomethylation of the reduced form of the enzyme on SDS–gel electrophoresis gives one protein band with a molecular weight corresponding to the native protein. The enzyme does not contain any flavin or heme component as no peak in the regions 445–455 and 405–415 nm is seen in the absorption spectrum of the enzyme. The pI of pure *S. rolfsii* enzyme is 5.18.

Chemical Properties. The enzyme is a glycoprotein containing 8.9% total carbohydrate; it contains 5.6 residues of glucosamine per molecule of enzyme but no galactosamine. The enzyme is high in acidic and low in basic amino acids, and contains no cystine or half-cystine.

Enzymatic Properties. The pH and temperature optima for the enzyme are 6.3–6.4 and 37°. The activation energy, calculated from an Arrhenius

[9] J. L. Hedrick and A. J. Smith, *Arch. Biochem. Biophys.* **126,** 155 (1968).
[10] K. Weber and M. Osborn, *J. Biol. Chem.* **244,** 4406 (1969).

plot, is 101.9 J mol^{-1}. Glucono-1,5-lactone (100 mM) does not inhibit the activity of the enzyme. One mole of DCPIP is reduced per mole cellobiose oxidized.

Substrate Specificity and Kinetics

Oxidizable Substrates. The enzyme oxidizes cellobiose and other cellodextrins (cellotriose to cellohexaose) and lactose. The relative rate of lactose oxidation compared to that of cellobiose under similar assay conditions is 40%. The turnover rate of cellobiose dehydrogenase, as represented by V_{max}, for cellobiose, cellotriose, cellotetraose, cellopentaose, and cellohexaose decreases, whereas the empirical K_m value increases with the chain length of the substrate (Table II).

Electron Acceptors. DCPIP is the most effective electron acceptor. Cytochrome c (*Candida krusei*) and ferricyanide are also reduced. With cellobiose as electron donor (2.5 mM) and at equivalent concentrations of electron acceptors, the rates of cytochrome c and ferricyanide reduction are 30 and 5% of that of DCPIP. NAD$^+$, NADP$^+$, FMN, FAD, p-benzoquinone, methylene blue, phenazine methosulfate, and molecular oxygen failed to accept electrons arising from the enzymatic oxidation of cellobiose.

[54] Cellobiose Dehydrogenase Produced by *Monilia* sp.

By R. F. H. DEKKER

Introduction

The hydrolysis of cellulose by cellulolytic microorganisms occurs through the concerted action of several hydrolytic enzymes[1] [e.g., exo- and endocellulases (1,4-β-D-glucanases) and β-D-glucosidases] which may be secreted extracellularly, located intracellularly, or bound to the cell wall. Nonhydrolytic enzymes (e.g., oxidative[2] and phosphorolytic[3])

[1] K.-E. Eriksson and T. M. Wood, *in* "Biosynthesis and Biodegradation of Wood Components" (T. Higuchi, ed.), pp. 469–503. Academic Press, New York, 1985.
[2] M. P. Vaheri, *J. Appl. Biochem.* **4,** 356 (1982).
[3] J. K. Alexander, *J. Biol. Chem.* **243,** 2899 (1968).

have also been reported to be involved in cellulose degradation, but little attention has been given to the role of these enzymes. The enzymes indirectly involved in the oxidation of cellulose include cellobiose oxidase,[4] cellobiose : quinone 1-oxidoreductase[5] [EC 1.1.5.1, cellobiose dehydrogenase (quinone)], and cellobiose dehydrogenase (acceptor)[6] (EC 1.1.99.18). These enzymes catalyze the oxidation of cellulose degradation products (e.g., cellobiose and cellooligosaccharides) using molecular oxygen or suitable electron acceptors, which, in turn, are reduced. Cellobiose oxidoreductase is implicated as being involved in both cellulose and lignin breakdown.[5] In this respect, the electron acceptors are certain quinones and phenoxy radicals produced by phenol oxidase that oxidize phenols arising from the enzymatic degradation of lignin.

Enzymes capable of oxidizing cellobiose to cellobiono-δ-lactone which is further hydrolyzed, spontaneously or in the presence of a lactonase, to cellobionic acid, have also recently been found whose natural electron, or hydrogen, acceptor has not yet been identified.[6–8] These enzymes, produced by cellulolytic, but ostensibly nonligninolytic fungi, are able to utilize dichlorophenolindophenol as an electron acceptor in the process of oxidizing cellobiose. Examples of fungi producing cellobiose dehydrogenase include a *Monilia* sp.,[6] *Sclerotium rolfsii,*[8] *Sporotrichum thermophile,*[7] and *Trichoderma reesei* QM 9414 (unpublished results).

A metabolic role for cellobiose dehydrogenase in cellulose degradation may be ascribed to alleviation of the inhibitory effect of cellobiose on cellobiohydrolase (an exocellulase) action, or in the regulation of synthesis of enzymes of the cellulase complex, or in the metabolism of cellobiose itself. Although the identity of the cellobiose-oxidizing agent *in vivo* is still unknown, and since cellobiose dehydrogenase is capable of reducing cytochrome *c* in the presence of cellobiose, the enzyme, which is cell or membrane bound (e.g., as in *Monilia* sp.), could function in an electron transport chain (membrane-associated), and in this way would be operative when the organism is grown on cellulose alone.

The following procedure describes the production, partial purification, and characterization of a cellobiose dehydrogenase produced by a species of *Monilia*. A zymogram method for the detection of cellobiose dehydrogenase activity is also described.

[4] A. R. Ayers, S. B. Ayers, and K.-E. Eriksson, *Eur. J. Biochem.* **90,** 171 (1978).
[5] U. Westermark and K.-E. Eriksson, *Acta Chem. Scand., Ser. B* **28,** 209 (1974).
[6] R. F. H. Dekker, *J. Gen. Microbiol.* **120,** 309 (1980).
[7] M. R. Coudray, G. Canevascini, and H. Meier, *Biochem. J.* **203,** 277 (1982).
[8] J. C. Sadana and R. V. Patil, *J. Gen. Microbiol.* **131,** 1917 (1985).

Experimental Procedures

Monilia sp.: Maintenance and Cultivation

The fungus used in this investigation was tentatively identified as *Monilia sitophila* (Mont.) Sacc., a highly cellulolytic fungus. Since this identification has not been confirmed the organism is referred to as a *Monilia* sp. The parent culture was isolated from decomposing sugarcane bagasse on solid agar Czapek Dox medium containing Whatman No. 1 filter paper as described elsewhere.[6]

Cultures were maintained on potato–dextrose–agar at 4°. Liquid cultures were grown in a composite synthetic nutrient medium containing 1% (w/v) microcrystalline cellulose (Avicel Type PH 101, FMC Corp., Philadelphia, PA) as described by Mandels and Reese[9] and Reese and Maguire.[10] Cultures were grown in shake flasks (250-ml Erlenmeyer flasks containing 100 ml nutrient medium) at 28° and shaken at 120 rpm. The flasks were inoculated with 1.5×10^6 spores. Conidia were separated from 2-week-old mycelium (grown on potato–dextrose–agar) by gently washing the mycelial mats with water containing 0.2% (w/v) Tween 80 and removing hyphae and mycelial fragments by filtration through a glass sinter (porosity G-3) filter. Spore counts were determined using a hemocytometer. Cultures were grown until the disappearance of particulate cellulose from the medium (8–10 days). Mycelium was removed by centrifugation (20,000 g/0.5 hr) and the supernatant (extracellular culture fluid, ECF) stored at 4° in the presence of 0.02% (w/v) sodium azide. The ECF served as the crude cellobiose dehydrogenase preparation.

Preparation of Intracellular and Mycelial Enzymes

Mycelium from liquid grown cultures was harvested by filtration on a Whatman GF/A glass fiber filter and washed with 3 volumes (50 ml each) of 100 mM citrate–phosphate (pH 7.0) buffer containing 0.02% (w/v) sodium azide and isotonic saline. The washings were discarded. The mycelium (a moist pad) was removed from the filter, suspended in 10–20 ml of the same buffer and homogenized using a Waring blender (1 min). The resulting homogenate was filtered, and the filtrate (i.e., the cytosol) collected; it served as the source of the intracellular enzyme preparation. The disrupted mycelial retentate was washed three times with the same buffer (50-ml aliquots), reduced to a moist pad on the filter by gentle suction, removed from the filter, and suspended in 10 ml buffer. This suspension served as the mycelial enzyme preparation.

[9] M. Mandels and E. T. Reese, *J. Bacteriol.* **73**, 269 (1957).
[10] E. T. Reese and A. Maguire, *Dev. Ind. Microbiol.* **12**, 212 (1971).

Assay Procedures

Soluble protein content was assayed by the Lowry method[11] as described by Hartree[12] using bovine serum albumin (BSA) as the standard. Mycelial protein content was determined by a modified Lowry method incorporating sodium dodecyl sulfate (SDS).[13] Mycelial protein was solubilized by heating a mycelial suspension (0.25 ml, in 50 mM phosphate buffer, pH 7.0) with 5% (w/v) SDS (0.25 ml) for 20 min at 100°. An aliquot of this was used to determine protein concentration. A separate protein standard curve using BSA as the standard was calibrated for this procedure.

Cellobiose dehydrogenase activity was assayed by measuring the decrease in absorbance of the electron acceptor, 2,6-dichlorophenolindophenol (DCPIP), spectrophotometrically at 600 nm. The reaction mixture consisted of DCPIP (0.05 ml, 1.25 mM in 10 mM citrate–phosphate buffer, pH 6.6), cellobiose (1.0 ml, 2.5 mM in the same buffer), and 0.05 ml of suitably diluted enzyme solution in a glass microcuvette. The decrease in absorbance at 600 nm was determined over a 4-min interval at 37° in a recording spectrophotometer equipped with a thermostatically controlled cuvette holder. When the crude enzyme preparation also contained enzymes which hydrolyze cellobiose, e.g., β-D-glucosidase, then it may be necessary to inhibit this contaminating enzyme activity by using D-glucono-1,5-lactone; a potent inhibitor of β-D-glucosidases.[14] However, because the half-life ($t_{1/2}$) of glucono-1,5-lactone at pH 6.6 is of the order of less than 5 min (glucono-1,5-lactone is spontaneously hydrolyzed to gluconic acid), it is necessary to perform the cellobiose dehydrogenase assay at a lower pH (usually pH optimum of the enzyme) where the $t_{1/2}$ is longer (e.g., at pH 4–5, the $t_{1/2}$ of glucono-1,5-lactone is of the order of 55–60 min).[14] Freshly prepared glucono-1,5-lactone solution was therefore added to the substrate reactants at a final concentration of 1 mM which totally inhibited[15] the strong β-D-glucosidase activity in the monilial enzyme fractions. At the lower pH (i.e., pH <5) reduction of DCPIP was followed spectrophotometrically at 520 nm, whereas for pH >5, 600 nm was used. The unit of cellobiose dehydrogenase activity is expressed as μmol DCPIP reduced/min/ml of enzyme. The molar absorption coefficient (ε_{600}) used for DCPIP at pH 6.6 was 1.51 × 10⁴ liters/mol/cm, whereas the ε_{520} for DCPIP at pH 4.5 was 3.965 × 10³ liters/mol/cm. In

[11] O. H. Lowry, N. J. Rosebrough, A. L. Farr, and R. J. Randall, *J. Biol. Chem.* **193**, 265 (1951).

[12] E. F. Hartree, *Anal. Biochem.* **48**, 422 (1972).

[13] H. Sandermann and J. L. Strominger, *J. Biol. Chem.* **247**, 5123 (1972).

[14] E. T. Reese and M. Mandels, *Dev. Ind. Microbiol.* **1**, 171 (1960).

[15] R. F. H. Dekker, *J. Gen. Microbiol.* **127**, 177 (1981).

experiments using electron acceptors other than DCPIP, the assay procedure was similar (using the various electron acceptors at 1 mg/ml) and the change in absorbance was monitored at their respective absorption maxima.

In assaying for mycelial cellobiose dehydrogenase activity, a mycelial suspension (0.25 ml) was added to the reaction substrates and incubated at 37°. The reaction was terminated after 10 min by the addition of 5 μl of 100 mM mercuric chloride. The inactivated suspension was centrifuged (48,000 g/10 min), and the absorbance read at 600 nm. In this case, the unit of cellobiose dehydrogenase activity is expressed as μmol DCPIP/min/mg mycelial protein.

Gel Filtration

ECF (7 ml, concentrated 10-fold by ultrafiltration) was applied to a column of Sepharose 4B (2.6 × 60 cm) equilibrated in 100 mM citrate–phosphate (pH 6.0) buffer and eluted with the same buffer. Fractions of 3 ml were collected and assayed for protein (A_{280}) and enzyme activity. Fractions containing cellobiose dehydrogenase activity were pooled, concentrated by ultrafiltration, and applied to a MW-calibrated column of BioGel P-100 (2.6 × 94.5 cm) equilibrated in 50 mM phosphate buffer (pH 7.0) and fractions of 3 ml collected.

Polyacrylamide Gel Electrophoresis and Isoelectric Focusing

Slab polyacrylamide gel electrophoresis and analytical isoelectric focusing in polyacrylamide gels was performed using a Multiphor 2117 unit (LKB-Produkter AB., Bromma, Sweden) as described in LKB Application Notes 250 and 306. Polyacrylamide gel ampholine plates of pH 3.5–10.0 were from LKB. Discontinuous polyacrylamide gel electrophoresis was performed by the method of Davis.[16] Protein within the gels was detected by staining with Coomassie Brilliant Blue R-250.

Zymogram Method for the Detection of Cellobiose Dehydrogenase

Cellobiose dehydrogenase activity was detected within polyacrylamide gels by a zymogram technique. Following electrophoresis or isoelectric focusing, the gel containing the enzyme was incubated in a petri dish with a reaction mixture consisting of DCPIP [10 mM, 4.0 ml contained in 100 mM citrate–phosphate (pH 6.6) buffer], cellobiose (2.5 mM,

---

[16] B. J. Davis, *Ann. N.Y. Acad. Sci.* **121,** 404 (1964).

40.0 ml in the same buffer), and glucono-1,5-lactone (20 mM, 2.0 ml in the same buffer) at room temperature, or 37°. Within 10–15 min the gels absorbed DCPIP which made them appear blue except for a colorless band where DCPIP had been reduced as a result of cellobiose dehydrogenase activity. The zymogram technique worked equally well with disc gels, gel slabs, and polyacrylamide gels containing pH ampholytes. When electrophoresis was carried out at pH >5 the dye incorporated into the gel appeared blue, whereas at pH <5, or when acid pH ampholytes were present in the gel, the gels appeared red (DCPIP functions not only as a redox indicator but also as a pH indicator). This effect did not, however, mask decoloration of the gel where cellobiose dehydrogenase activity was present. Visual detection of the enzyme was greatly facilitated by adjusting the pH of the reaction mixture by the addition of solid K_2HPO_4 which kept the DCPIP in the oxidized (i.e., blue) form.

Detection of cellobiose dehydrogenase activity by the zymogram procedure can also be used in screening cellulolytic microorganisms for the production of cellobiose dehydrogenases. DCPIP (1 mM final concentration) was incorporated into Czapek–agar medium containing Avicel (1%, w/v) and cellobiose (20 mM final concentration). The blue-colored agar plates were inoculated with fungal spores or mycelium, and incubated at 25°. Those microorganisms producing cellobiose dehydrogenase resulted in colorless zones in the agar around the growing colonies. The colorless zones within the agar are a consequence of the DCPIP being reduced during oxidation of cellobiose by the presence of cellobiose dehydrogenase.

Results and Discussion

Cellobiose dehydrogenase produced by the *Monilia* sp. has previously been demonstrated[6] to be an inducible enzyme. Cellulosic substrates that were rather resistant to attack by exo- and endocellulases, e.g., cotton, Avicel, and filter paper, were the strongest inducers of the enzyme. Cellulose preparations such as carboxymethylcellulose, hemicellulosic substrates (e.g., arabinoglucuronoxylans and galactomannans), and cellobiose were, by contrast, rather poor substrates for cellobiose dehydrogenase production.

The monilial enzyme was found to be bound to the mycelium from which it was released into the extracellular medium.

The time course of cellobiose dehydrogenase production by *Monilia* sp. when grown on 1% (w/v) Avicel is shown in Fig. 1. Maximum activity of both released and mycelial-bound enzyme occurred after 8–10 days growth. Enzyme activity was measured at pH 4.5 (optimal pH), and the

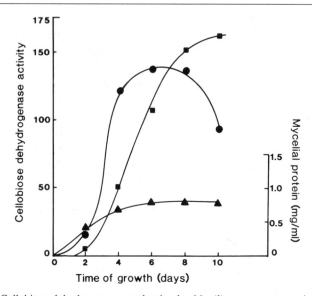

FIG. 1. Cellobiose dehydrogenase production by *Monilia* sp. grown on microcrystalline cellulose (Avicel). (■) Extracellular-released enzyme; (●) mycelial-bound enzyme; (▲) mycelial protein. Cellobiose dehydrogenase activity is expressed as μmol DCPIP reduced/min/ml enzyme, and as units/mg mycelial protein for the mycelial-bound enzyme.

absorbance read at 520 nm. Mycelial-bound enzyme activity is expressed as units of enzyme/mg mycelial protein.

ECF cellobiose dehydrogenase could be fractionated by gel permeation chromatography on Sepharose 4B, and contaminating β-glucosidase activity could be further separated from the enzyme by chromatography on BioGel P-100. Cellobiose dehydrogenase eluted as a single symmetrical peak on BioGel P-100 which corresponded to a molecular weight of 48,000. The enzyme could be further purified by isoelectric focusing using pH ampholytes 3.5–10.0 whereupon the enzyme migrated as a single protein band isoelectric at pH 5.3–5.5. The zymogram method was employed to detect cellobiose dehydrogenase activity in polyacrylamide gels following electrophoresis and isoelectric focusing, and confirmed the absence of isoenzymes.

The relative rates of oxidation of cellobiose, cellooligosaccharides, and other sugars are shown in Table I. Glucono-1,5-lactone, which was added to the reaction mixture to inhibit the strong β-D-glucosidase activity present,[15] did not itself affect the activity of the enzyme or inhibit the oxidation of cellobiose. The enzyme showed a high degree of specificity for cellobiose and cellooligosaccharides, but was also capable of oxidizing lactose and 4-glucosyl-β-D-mannose but to a lesser extent than cellobiose.

TABLE I

SUBSTRATES OXIDIZED BY CELLOBIOSE DEHYDROGENASE
FROM *Monilia* sp.[a]

Substrate[b]	Relative activity
Cellobiose	100
4-Methylumbelliferyl-β-D-cellobioside	70
Cellooligosaccharides (DP 3–6)[c]	67
Lactose	60
4-Glucosyl-β-D-mannose	47
4-Mannosyl-β-D-glucose	0
Sophorose	0
Maltose	0
Sucrose	0
Melibiose	0
Raffinose	0
Melizitose	0
Glucose	0
Mannose	0
Xylose	0
Glucono-1,5-lactone	0

[a] Seven-day-old culture grown on Avicel. Extracellular-released enzyme.
[b] Substrates were evaluated at a final concentration of 2 mM.
[c] DP denotes degree of polymerization.

The other carbohydrates listed in Table I could not serve as substrates for the enzyme.

Table II shows the specificity of cellobiose dehydrogenase toward several artificial electron acceptors that were capable of accepting electrons arising from the enzymatic oxidation of cellobiose. Of these, DCPIP appeared to be the most effective, but phenol blue (N,N-dimethylindoaniline), a related compound, and ferricyanide could also be reduced. Cytochrome c, a natural electron acceptor, was also capable of being reduced during cellobiose dehydrogenase oxidation of cellobiose. However, NAD$^+$, NADP$^+$, and none of the quinones investigated including p-benzoquinone were able to accept electrons arising from enzymatic oxidation of cellobiose. Previous work[6] reported that molecular oxygen was not consumed during oxidation of cellobiose by the *Monilia* sp. cellobiose dehydrogenase, nor was hydrogen peroxide produced. Thus the nature of the natural electron acceptor is still unknown.

Some of the physicochemical properties of the *Monilia* sp. extracellular-released cellobiose dehydrogenase are shown in Table III. These data

TABLE II

ELECTRON ACCEPTOR SPECIFICITY OF CELLOBIOSE DEHYDROGENASE FROM
Monilia sp.

Electron acceptor	Cellobiose dehydrogenase activity[a] (units/min/ml enzyme)
2,6-Dichlorophenolindophenol (DCPIP)	27.2
Phenol blue (*N,N*-dimethylindoaniline)	15.6
Cytochrome *c*	6.4[b]
Potassium ferricyanide	4.9
Methylene blue	0
NAD$^+$, NADP$^+$	0[c]
Coumarin (1,2-benzopyrone)	0
Vitamin E (*d*-α-tocopheryl acetate)	0
Quinones	
p-Benzoquinone	0
2,3-Dichloro-5,6-dicyano-p-benzoquinone	0
Ubiquinone (coenzyme Q$_{10}$)	0
Vitamin K$_3$ (menadione)	0
9,10-Anthraquinone	0
1,2,5,8-Tetrahydroxy-9,10-anthraquinone (quinalizarin)	0

[a] Seven-day-old culture grown on Avicel. Assayed at pH 6.6 using the extra-cellular-released enzyme, and absorbance measured at the respective λ$_{max}$.
[b] Rate of increase in absorbance at 530 nm.
[c] At 340 nm.

TABLE III

PHYSICOCHEMICAL PROPERTIES OF CELLOBIOSE
DEHYDROGENASE[a] FROM *Monilia* sp.

Properties	Unit
pH optimum	4.0–4.5
p*I*	5.3–5.5
Molecular weight	48,000
Apparent K_m[b]	
Cellobiose	$12.2 \times 10^{-6}\ M$
DCPIP	$8.0 \times 10^{-5}\ M$
Isoenzymes	1

[a] Extracellular-released enzyme.
[b] Determined at pH 6.6.

are not significantly different from that reported for cellobiose dehydrogenases produced by *Sclerotium rolfsii*[8] and *Sporotrichum thermophile.*[7]

As described above, the zymogram staining procedure was very effective in detecting cellobiose dehydrogenase activity in polyacrylamide gels following electrophoresis and isoelectric focusing. Furthermore, this experimental approach could be utilized in screening microorganisms for the production of cellobiose dehydrogenase when grown on agar plates containing cellulose, cellobiose, and DCPIP. Fungi that were successfully screened for the production of cellobiose dehydrogenase in this way included *Monilia* sp., *Trichoderma reesei* QM-9414, *Sclerotium rolfsii*, *Sporotrichum pulverulentum, S, thermophile,* and several species of *Stachybotrys, Cladosporium,* and *Chaetomium.*

[55] Cellobiose Dehydrogenase (Quinone)

By U. WESTERMARK and K.-E. ERIKSSON

Cellobiose dehydrogenase (quinone) (EC 1.1.5.1, cellobiose : quinone 1-oxidoreductase) has been found extracellularily in several wood-rotting and cellulolytic fungi. The enzyme was first discovered in the white-rot fungi *Sporotrichum pulverulentum* and *Polyporus versicolor*[1,2] and has subsequently been found in several other white-rot fungi[3,4] as well as in species of Ascomycetes[5,6] and the Fungi Imperfecti.[7]

The reaction mechanisms for cellobiose dehydrogenase are shown in Fig. 1. Cellobiose dehydrogenases from all the different fungi have so far turned out to have similar properties to the cellobiose dehydrogenase from *S. pulverulentum.* Therefore the purification procedure and enzyme properties described here will refer mainly to the *S. pulverulentum* dehydrogenase.[1,8]

[1] U. Westermark and K.-E. Eriksson, *Acta Chem. Scand., Ser. B* **28**, 204 (1974).
[2] U. Westermark and K.-E. Eriksson, *Acta Chem. Scand., Ser. B* **28**, 209 (1974).
[3] A. Hütterman and A. Noelke, *Holzforschung* **36**, 283 (1982).
[4] C. Sadana and R. V. J. Patil, *J. Gen. Microbiol.* **131**, 1917 (1985).
[5] R. F. H. Dekker, *J. Gen. Microbiol.* **120**, 309 (1980).
[6] P. Fähnrich and K. Irrgang, *Biotechnol. Lett.* **4**, 775 (1982).
[7] M.-R. Coudray, G. Canevascini, and H. Meier, *Biochem. J.* **203**, 277 (1982).
[8] U. Westermark and K.-E. Eriksson, *Acta Chem. Scand., Ser. B* **29**, 419 (1975).

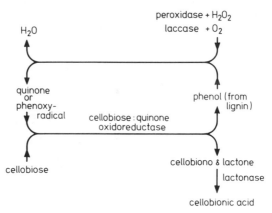

FIG. 1. Reaction mechanisms of the enzyme cellobiose dehydrogenase (quinone). From Westermark and Eriksson.[2]

Assay Method

Principle. The enzyme can be assayed spectrophotometrically by following the reduction of a suitable quinone.

Reaction Mixture. One micromole of the quinone 3-methoxy-5-*tert*-butylbenzoquinone-(1,2)(ε_{360} = 2180) and 2 μmol cellobiose are added to 300 μmol of acetate buffer (pH 4.5) and the enzyme solution for a total volume of 3 ml. The reaction is followed spectrophotometrically at 25° by measuring the decrease in absorbance at 360 nm. The rate of decrease is linear until 90% of the quinone is reduced to the corresponding phenol.

Alternative electron acceptors such as 2,6-diclorophenolindophenol (DCPIP) (λ_{max} 600 nm), 2-methoxybenzoquinone-(1,4) (λ_{max} 368 nm), and cerulignone (λ_{max} 472 nm) can also be used.

Definition of Enzyme Unit. One unit of cellobiose dehydrogenase activity is defined as the amount of enzyme that reduces 1 μmol of quinone per minute under the assay conditions.

Purification Procedure

Growth of the Organism

Sporotrichum pulverulentum (ATCC 32629, anamorph of *Phanerochaete chrysosporium*) is grown in a basal salt medium[9] with 3 g/liter powdered cellulose as a carbon source. For enzyme production 1-liter

[9] T. K. Kirk and A. Kelman, *Phytopathology* **55,** 739 (1965).

Erlenmeyer flasks containing 300 ml culture solution are used. Flasks are incubated at 25° for 8–9 days with reciprocal shaking. The mycelium is then removed by filtration through a glass filter.

Purification

The purification procedure[8] is summarized in Table I. Culture filtrate is concentrated about 30 times by ultrafiltration (membrane 4M10, Amicon, Cambridge, MA) and the concentrated solution centrifuged at 14,600 g to remove any precipitate which is formed. Solid ammonium sulfate is added to the supernatant fluid to 60% saturation and the resulting precipitate collected by centrifugation at 14,600 g for 30 min.

Desalting and Buffer Exchange

Gel filtration using a column of Sephadex G-25 (5 × 260 mm) is routinely used for desalting and change of buffer system.

Isoelectric Focusing

Column isoelectric focusing, carried out as described in the LKB manual (LKB, Bromma, Sweden), separates cellobiose dehydrogenase activity into three peaks.

Step 1. Ion-exchange chromatography on SP-Sephadex C-50 (Pharmacia) at pH 4.3 is used in the first purification step. The bulk of the proteins including about 70% of the cellobiose dehydrogenase activity is not re-

TABLE I
PURIFICATION PROCEDURE FOR CELLOBIOSE DEHYDROGENASE[a]

Purification step	Volume (ml)	Total activity (units, U)	Total A_{280}	Specific activity (U/A_{280})	Yield (%)	Purification (fold)
Culture filtrate	29,000	521	6,558	0.08	100	1
Ultrafiltrate	930	438	1,987	0.220	84	3
(NH₄)₂SO₄ precitated	17	379	802	0.47	73	6
Step 1						
SP-Sephadex, eluate	21	99	83	1.19	19	15
C-50 chromatography acetate, pH 4.3, void	360	256	656	0.39	49	5
Step 2						
Hydroxylapatite	9.8	31.4	2.45	12.8	6	160

[a] Reproduced from Westermark and Eriksson.[8]

tained on the column at this pH value. However, the one isoenzyme which is retained is subjected to further purification as follows. The column (20 × 340 nm) is equilibrated with 0.02 M sodium acetate buffer, pH 4.3, and the cellobiose dehydrogenase activity eluted (36 ml/hr) with a linear gradient of sodium chloride using a total volume of 400 ml of salt eluant with a final salt concentration of 0.5 M.

Step 2. Chromatography on a hydroxylapatite column is used in the second step of purification. After desalting, pooled active fractions from purification step 1 are passed through a BioGel HT (Bio-Rad, Richmond, CA) column (15 × 110 mm). Cellobiose dehydrogenase is readily adsorbed on the column leaving most of the contaminating proteins to pass unretarded. The column is initially equilibrated with 0.02 M phosphate buffer, pH 7.4, and cellobiose dehydrogenase eluted with 0.3 M NaCl at a flow rate of 5 ml/hr.

Purification of a cellobiose dehydrogenase from *Sclerotium rolfsii*[4] uses preparative polyacrylamide gel electrophoresis for the final purification instead of hydroxylapatite chromatography.

Purity. Purified enzyme gives one band in SDS–gel electrophoresis and is also homogeneous in analytical ultracentrifugation experiments.[8]

Properties[8]

Spectral Properties and Prosthetic Group

The purified enzyme is yellow and shows a typical flavin spectrum in the visible region with maxima at 375–380 and 457 nm. These maxima are reduced on addition of the substrate cellobiose. Boiling the enzyme for 20 min in the dark releases the flavin component which is analyzed by thin-layer chromatography using two different solvent systems: (1) 5% Na_2HPO_4 in water and (2) *tert*-butanol : water 6 : 4 (v/v). Detection by fluorescence under UV and comparison with authentic marker compounds reveals the flavin component to be flavin adenine dinucleotide (FAD).

Molecular Weight

Determination of the molecular weight of purified cellobiose dehydrogenase in an analytical ultracentrifuge using the sedimentation equilibrium method gives a value of 58,000 ± 2,000. The enzyme is dissolved in phosphate buffer (0.02 M, pH 7.4) and measurements made at 30,000 rpm for 24 hr at 20° using a Yphantis double-section cell. A partial specific volume of 0.72 is assumed.

Substrate Specificity

The number of carbohydrate substrates serving as electron donors is relatively small and the enzyme exhibits a high specificity for cellobiose and other cellodextrins (Table II). Lactose and 4β-glucosylmannose are also oxidised but not mannobiose and 4β-mannosylglucose indicating that the C-2-hydroxyl in the nonreducing part of the disaccharide must have the glucose conformation for activity. A hydroxymethyl group in one or both C-5 atoms in the disaccharide also appears to be a requirement for activity. The immediate product from the oxidation of cellobiose by the enzyme is cellobiono-δ-lactone. However, the fungal culture solution contains a lactonase that converts cellobiono-δ-lactone to cellobionic acid.

Specificity toward the quinone (or radical) substrate appears to be lower and several *o*- and *p*-quinones as well as oxidation products resulting from phenol oxidizing enzyme activity can serve as electron acceptors for cellobiose dehydrogenase.

TABLE II

OXIDATION OF DIFFERENT SUBSTRATES[a]

Substrate	Relative rate of oxidation
Cellobiose	100
Cellopentaose	100
Cellulose	0
Lactose (4-*O*-β-D-galactopyranosyl-D-glucose)	44
4β-Mannosylglucose (4-*O*-β-D-mannopyranosyl-D-glucose)	0
4β-Glucosylmannose (4-*O*-β-D-glucopyranosyl-D-mannose)	36
Mannobiose (4-*O*-β-D-mannopyranosyl-D-mannose)	0
Maltose	0
Sucrose	0
Xylobiose	0
Glucose	0
Galactose	0
Mannose	0
Gluconolactone	0
Arabinose	0
Xylose	0

[a] The reaction mixture contained purified cellobiose dehydrogenase, 300 μmol acetate buffer (pH 4.5), 1 μmol 3-methoxy-5-*tert*-butylbenzo-quinone-(1,2), 2 μmol of sugar in a total reaction volume of 3 ml. The cellulose sample had a degree of polymerization of 150 and about 8 μmol of reducing end groups. Reproduced from Westermark and Eriksson.[8]

Binding to Concanavalin

Cellobiose dehydrogenase binds readily to concanavalin A and can be eluted with α-methyl-D-mannoside indicating that the enzyme is a glycoprotein.

[56] Cellobiose Phosphorylase from *Cellvibrio gilvus*

By TAKASHI SASAKI

Cellobiose + $P_i \rightleftharpoons \alpha$-glucose 1-phosphate + glucose

Assay Method

Principle. The formation of the α-D-glucose 1-phosphate is monitored spectrophotometrically by coupling the phosphorolysis of cellobiose with pyridine nucleotide reduction in the presence of NADP, phosphoglucomutase, and glucose-6-phosphate dehydrogenase. Enzyme activity is directly related to the rate of change of absorbance at 340 nm due to NADPH production. Enzyme activity is also assayed by P_i formation.

Reagents

Tris–HCl buffer, 50 mM, pH 7.5
MgCl$_2$, 5 mM
P$_i$, 5 mM
Cellobiose, 5 mM
NADP, 0.2 mM
Dithiothreitol, 5 mM
Glcose-6-phosphate dehydrogenase (5000 IU/ml, Boehringer Mannheim Corporation)
Phosphoglucomutase (1000 IU/ml, Boehringer Mannheim Corporation)

Procedure. Enzyme activity is assayed by measuring spectrophotometrically the formation of glucose 1-phosphate from cellobiose. The reaction mixture, in a total volume of 1.0 ml, consisted of 50 mM Tris–HCl buffer, pH 7.5, 5 mM MgCl$_2$, 5 mM P$_i$, 5 mM cellobiose, 0.2 mM NADP, 5 mM dithiothreitol, 5 μl of glucose-6-phosphate dehydrogenase, 5 μl of phosphoglucomutase, and 10 μl of enzyme solution. The measurements are made in a recording spectrophotometer for 5 min at 20°. Enzyme activity is also assayed by P_i formation after precipitation of the protein with 0.1 M trichloroacetic acid.[1]

Definition of Unit and Specific Activity. One unit of activity is defined as 1 μmol of glucose 1-phosphate formed/min under the specified conditions. Specific activity is expressed as units per milligram of protein. Protein is determined by the method of Lowry *et al.*[2] with bovine serum albumin as standard.

Purification Procedure

Cell Growth. Cellvibrio gilvus (A.T.C.C. 13127) is grown at 30° in a modification of the medium described by Hulcher and King[3] containing the following (values in g/liter): KNO_3, 0.5; NaCl, 0.5; $MgCl_2$, 0.25; K_2HPO_4, 1; $(NH_4)_2SO_4$, 0.2; $FeSO_4$, 0.01; casamino acids, 5; yeast extract, 0.1; cellobiose, 2; and $CaCO_3$, 10. Cells are harvested in early stationary phase and washed with 50 mM Tris–HCl buffer, pH 7.0.

Step 1. Preparation of Crude Extract. A 10 g sample of *C. gilvus* resuspended in 50 ml of a buffer consisting of 50 mM Tris–HCl, pH 7.6, 1 mM EDTA, and 50 mM 2-mercaptoethanol and disrupted in a Bronson Sonifier (Danbury, CT) at 4°. Extracts are centrifuged for 30 min at 15,000 g and the supernatants are pooled.

Step 2. Protamine Treatment. Fifty microliter portion of 3% (w/v) protamine sulfate solution is added per ml of crude extract and the precipitate is removed by centrifugation. The enzyme is recovered by the addition of solid $(NH_4)_2SO_4$ at 4° until the solution is 2.7 M. The precipitate is collected by centrifugation, dissolved in 5 ml of 50 mM Tris–HCl buffer, pH 7.6, containing 10 mM 2-mercaptoethanol, and dialyzed overnight against the same buffer.

Step 3. Hydroxyapatite Chromatography. The dialyzed fraction is applied to a hydroxyapatite (Type II, Sigma Chemical Co., St. Louis, MO) column (2.5 × 30 cm) that is previously equilibrated with 10 mM KH_2PO_4–K_2HPO_4 buffer, pH 7.5, containing 10 mM 2-mercaptoethanol. The column is washed with about 300 ml of the same buffer and cellobiose phosphorylase is eluted with a linear gradient of KH_2PO_4–K_2HPO_4 buffer from 10 to 50 mM at pH 7.5, containing 10 mM 2-mercaptoethanol. Fractions containing enzyme activity are recovered by the addition of solid $(NH_4)_2SO_4$ to 3.3 M. The precipitate is dissolved in 50 mM Tris–HCl

[1] C. H. Fiske and Y. Subba Row, *J. Biol. Chem.* **66,** 375 (1925).
[2] O. H. Lowry, N. J. Rosebrough, A. L. Farr, and R. J. Randall, *J. Biol. Chem.* **193,** 265 (1951).
[3] F. H. Hulcher and K. W. King, *J. Bacteriol.* **76,** 565 (1958).

TABLE I

PURIFICATION OF CELLOBIOSE PHOSPHORYLASE FROM *Cellvibrio gilvus*[a]

Fraction	Protein (mg)	Activity (units)	Specific activity (units/mg of protein)	Purification (fold)	Yield (%)
Crude extract	136.5	19.8	0.1	1	100
Protamine sulfate	30.5	17.5	0.6	4.1	88.7
Hydroxyapatite	11.8	15.4	1.3	9.3	78.7
DEAE-Sephadex A-50	0.08	2.2	27.4	195.7	11.2

[a] Cellobiose phosphorylase activity is expressed as μmol of glucose 1-phosphate formed/min.

buffer, pH 7.5, containing 10 mM 2-mercaptoethanol and is dialyzed against the same buffer.

Step 4. DEAE-Cellulose Chromatography. The enzyme fraction from the hydroxyapatite column is then applied to a DEAE-Sephadex A-50 column (2 × 50 cm) previously equilibrated with 50 mM Tris–HCl buffer, pH 7.5, containing 10 mM 2-mercaptoethanol. Cellobiose phosphorylase is eluted with a linear gradient of NaCl from 100 mM (75 ml) to 500 mM (75 ml) in the same buffer. Fractions showing activity are pooled, concentrated, and stored at −20°.

The results of the purification procedure are summarized in Table I. An overall yield of about 11% of the initial activity is obtained with an increase in specific activity of 196-fold.

Properties

Physical Characteristics.[4] The purity of enzyme is monitored by polyacrylamide gel electrophoresis. Electrophoresis in the presence of 2-mercaptoethanol shows a single band after staining with Coomassie Blue, which is coincidental with the band of enzyme activity. Cellobiose phosphorylase activity is detected by the formation of a calcium phosphate precipitate.

The relative molecular weight (M_r) of cellobiose phosphorylase estimates by chromatography on Sephadex G-200 is ~280,000. SDS–polyacrylamide gel electrophoresis reveals the presence of a single band of protein and a molecular weight of 72,000, indicating that the enzyme consists of four subunits. The absorption spectrum of cellobiose phos-

[4] T. Sasaki, T. Tanaka, S. Nakagawa, and K. Kainuma, *Biochem. J.* **209**, 803 (1983).

phorylase shows a protein absorption maximum at 280 nm, but no sign of the presence of pyridoxal 5′-phosphate, which has an absorption in a region from 300 to 450 nm. Cellobiose phosphorylase does not contain pyridoxal 5′-phosphate as a cofactor. It is noteworthy that sucrose phosphorylase[5] and maltose phosphorylase[6] also give negative results, although all the α-glucan phosphorylases so far isolated from animals, plants, and microorganisms contain firmly bound pyridoxal 5′-phosphate and are totally inactive without it.

Specificity. Cellobiose, P_i and Mg^{2+} are required for the phosphorylation. Cellobiose phosphorylase of *C. gilvus* has a high specificity for cellobiose. The K_m values for cellobiose and P_i are 1.25 and 0.77 mM, respectively. Cellodextrin is not a substrate for cellobiose phosphorylase, neither is there any activity in the crude extract toward this oligosaccharide, indicating that *C. gilvus* does not possess a cellodextrin phosphorylase. No activity is observed with the following disaccharides: gentiobiose (β-1,6), laminaribiose (β-1,3), lactose (β-1,4), maltose (α-1,4), kojibiose (α-1,2), and sucrose (α-1,4).

This enzyme catalyzes the reversible phosphorolysis of cellobiose and has a following properties.[7] It has specificity for α-D-glucose 1-phosphate, but not the β-form, and it is active toward 10 different glucosyl acceptors, namely, D-glucose, 2-deoxyglucose, 6-deoxyglucose, D-glucosamine, D-mannose, D-altrose, L-galactose, L-fucose, D-arabinose, and D-xylose. The resulting disaccharides all contain β-1,4-glucosidic linkages. No activity is observed when the following compounds were examined as acceptors; L-arabinose, D-lyxose, D-ribose, 2-deoxyribose, L-xylose, D-allose, D-fructose, D-glucitol, *myo*-inositol, D-gluconate, 3-O-methyl-D-glucose, α-methyl-D-glucose, β-methyl-D-glucose, or N-acetyl-D-glucosamine.

Inhibitors. The enzyme activity is inhibited by nojirimycin (D-glucopiperidinose). Nojirimycin is known as an inhibitor of β-glucosidase which is produced by *Streptomyces reseochromogenes*.[8] It inhibits ~80% of enzyme activity at a concentration of 1.0 mM. The inhibition is competitive with cellobiose as determined from a Dixon plot,[9] and the K_i value is 45 μm. α-Oxogluconate, glucono-δ-lactone, methyl-α-D-glucoside, 6-phosphogluconate, and *p*-nitrophenyl-β-glucopyranoside are weakly inhibitory. The enzyme activity with N-ethylmaleimide at 500 μM results in

[5] D. J. Graves and J. H. Wang, "The Enzymes" (P. D. Boyer, ed.), 3rd Ed., p. 435. Academic Press, New York, 1972.

[6] A. Kamogawa, K. Kobayashi, and T. Fukui, *Agric. Biol. Chem.* **37**, 2813 (1973).

[7] J. K. Alexander, this series, Vol. 28, p. 944.

[8] S. Inouye, T. Tsuruoka, I. Ito, and T. Niida, *Tetrahedoron* **23**, 2125 (1968).

[9] M. Dixon, *Biochem. J.* **55**, 170 (1953).

a 56% inactivation and the enzyme is also inactivated completely by 20 μM p-chloromercuribenzoate.

Stability, Optimum. The enzyme shows low thermostability and is inactivated above 40°. Complete inactivation occurs after a 10 min inactivation at 60°. This enzyme is found to be quite labile during the latter stages of purification. The use of thiol reagent like 2-mercaptoethanol or dithiothreitol during purification is necessary to obtain acceptable recoveries in these steps. The storage of the purified enzyme at −20° without thiol reagent causes an appreciable loss of activity within 2 weeks. Optimum activity is found at pH 7.6, with a rapid decrease in activity both below and above this range.

[57] Cellulosomes from *Clostridium thermocellum*

By RAPHAEL LAMED and EDWARD A. BAYER

Biomass in the form of cellulose is the major constituent of plant matter, thereby comprising the earth's most abundant natural organic resource. Although the repeating unit of this biopolymer is a simple disaccharide (cellobiose), cellulose fibrils are organized into a complicated paracrystalline state. Due to the structural complexity thus formed, a single enzyme cannot degrade the substrate, and microorganisms which successfully grow on cellulose do so by producing a collection of different cellulases (endo- and exoglucanases) which act synergistically.

One of the most effective microbial cellulolytic systems is produced by the anaerobic thermophilic bacterium *Clostridium thermocellum*. The industrial potential of this organism has long been recognized; of note is its capacity to excrete large amounts of cellulolytic enzymes of particularly high specific activity on α-crystalline cellulose.[1,2]

Despite this fact, preparation of purified enzymes has proved exceptionally difficult. Most of the early work has therefore concentrated on characterizing the properties of crude enzyme preparations. In some cases, various endo-β-glucanases have been purified,[3–6] although their

[1] T. K. Ng, P. J. Weimer, and J. G. Zeikus, *Arch. Microbiol.* **114,** 1 (1977).
[2] E. A. Johnson, M. Sukojoh, G. Halliwell, A. Madia, and A. L. Demain, *Appl. Environ. Microbiol.* **43,** 1125 (1982).
[3] N. Ait, N. Creuzet, and P. Forget, *J. Gen. Microbiol.* **113,** 399 (1979).
[4] T. K. Ng and J. G. Zeikus, *Biochem. J.* **199,** 341 (1981).
[5] J. Petre, R. Longin, and J. Millet, *Biochimie* **63,** 629 (1981).

respective contribution to the total cellulolytic system has not been defined. Recently, various cellulase genes from *C. thermocellum* have been cloned into *Escherichia coli* and their products have been characterized.[7,8] But again their basic role in the cellulase system of the bacterium has yet to be elucidated.

In recent work, we have shown that the major enzymes responsible for cellulose degradation in *C. thermocellum* are arranged into a distinct multisubunit complex which we have called the cellulosome. The cellulosome appears both in an extracellular form and in a cell-associated form. The latter is considered to comprise a discrete cell surface organelle, which is also responsible for the adherence of the bacterium to its insoluble substrate.[9,10]

Assay Methods

Due to the complicated nature of both the substrate and the enzyme complex, various assays are commonly used to characterize the system; a single assay system would not provide a complete picture as to the nature of the enzyme. The fact that the cellulosome complex comprises an association of different cellulase types further complicates the situation.

Of the countless cellulase assays available which employ either cellulosic substrates of varying degrees of crystallinity (i.e., filter paper, microcrystalline cellulose, amorphous cellulose, etc.) or derivatized forms of cellulose (both soluble and insoluble), we have chosen two. One, true cellulase activity or the capacity of an enzyme preparation to completely (80–90%) hydrolyze the substrate is representative of the efficiency of a cellulase preparation to degrade relatively crystalline forms of cellulose, and the second gives the general level of endoglucanase activity.

Two nonenzymatic measurements are also presented which demonstrate other properties specifically characteristic of the cellulosome in *C. thermocellum*. The adsorption of the cellulosome to cellulose relates to the fundamental role of the cell-bound form in the adherence of the bacterium to its substrate. The availability of a cellulosome-specific antibody enables the quantification of the cellulosome in solution.

In addition to the assays mentioned above, other cellulase-related

---

[6] L. G. Ljungdahl, B. Pettersson, K.-E. Eriksson, and J. Wiegel, *J. Curr. Microbiol.* **9,** 195 (1983).

[7] P. Beguin, P. Cornet, and J. Millet, *Biochimie* **65,** 495 (1983).

[8] J. Millet, D. Pétré, P. Béguin, O. Raynaud, and J.-P. Aubert, *FEMS Microbiol. Lett.* **29,** 145 (1985).

[9] R. Lamed and E. A. Bayer, *Experientia* **42,** 72 (1986).

[10] R. Lamed and E. A. Bayer, *Adv. Appl. Microbiol.* **33,** 1 (1988).

activities have been employed to further characterize cellulolytic enzymes. In particular, chromogenic or fluorogenic (e.g., nitrophenyl or umbelliferyl, respectively) derivatives of glucose or cellobiose can be used to detect the corresponding β-glucosidase or cellobiohydrolase activity. The latter activity has been demonstrated in the cellulosome, while detectable levels of β-glucosidase activity have not been found in the purified cellulosome.

Endoglucanase Assay

Principle. Endoglucanase activity can be determined in the presence or absence of exoglucanase using a variety of water-soluble derivatized cellulose polymers. Derivatization interferes with or prevents exoglucanase activity. Most commonly, carboxymethylcellulose is used as a substrate and the resultant terminal reducing sugar moieties are determined by the classic dinitrosalicylic acid (DNS) method.[11]

Reagents

Carboxymethylcellulose (low viscosity), 2% solution[12]
Dinitrosalicylic acid reagent[13]

Procedure. A buffered solution (1 ml) containing carboxymethylcellulose is mixed with an enzyme sample (1 ml) in a 13-mm test tube. The reaction is initiated by immersing the test tubes into a water bath (60°). The suspension is incubated (usually for 30 min) and the reaction is terminated by the addition of 3 ml DNS reagent. The tubes are vortexed and heated for 10 min at 100°. After cooling in tap water for 5 min, the tubes are read directly in a Spectronic 20 or 21 spectrophotometer at 640 nm.

Definition of Unit. A unit of endoglucanase activity is defined (within the linear range) as the amount of enzyme which releases 1 μmol of reducing sugar (using glucose as a standard) per ml sample per min under the conditions indicated.

"True" Cellulase Assay

Principle. "True" cellulase activity is followed by the decrease in turbidity of a defined particulate (20 μm) preparation of microcrystalline

---

[11] G. L. Miller, R. Blum, W. E. Glenon, and A. L. Burton, *Anal. Biochem.* **2,** 127 (1960).
[12] Carboxymethylcellulose (20 g) is dissolved in 0.2 M sodium acetate buffer (pH 5) by overnight stirring at 25°.
[13] Dinitrosalicylic acid (10 g), phenol (2 g), sodium sulfite (0.5 g), sodium potassium tartarate (200 g), and NaOH (10 g) are brought to 1 liter with H$_2$O. The solution is stored under nitrogen at 4° in dark bottles.

cellulose. A relatively low level of cellulose (0.6 mg/ml) is used such that cellulolysis proceeds to completion within a relatively short time period with minimal interference from degradation products. The following is a modification of the original procedure of Johnson *et al.*[2]

Reagents

Avicel type PH 105, 20 μm particles (FMC Corp., Marcus Hook, PA), 3% suspension in H_2O
EDTA, 200 mM
$CaCl_2$, 1 M
Cysteine–HCl, 30 mg/ml[14]
Sodium acetate buffer, 0.5 M, pH 5.0

Procedure. The assay mixture contains 30 μl EDTA solution,[15] 30 μl $CaCl_2$ solution, 90 μl cysteine–HCl, 60 μl Avicel suspension,[16] 300 μl acetate buffer, 100 μl enzyme preparation,[17] and 2.4 ml H_2O. The suspension is incubated with intermittent shaking at 60°. After a 20-hr incubation period, the prestirred reaction mixture is measured spectrophotometrically at 660 nm (using 13-mm test tubes and a Spectronic 20 or 21 spectrophotometer). The results are plotted as the residual turbidity (percentage of original) as a function of the amount of protein added.

Definition of Unit. Extrapolation of the linear portion of the curve to 0% residual turbidity gives the amount of protein which exhibits 1 unit of "true" cellulase activity.

Antigenic Activity

Principle. Antigenic activity is determined by rocket immunoelectrophoresis of samples into agarose gel which contains anticellulosome antibody.[18,19] The height of the "rocket" is roughly proportional to the amount of cellulosome present in the sample. Qualitative differences in the cellulosome preparation can either be determined by crossed immunoelectrophoresis[20] or by rocket electrophoresis of fractions obtained by

---

[14] The solution is freshly prepared by dissolving 0.3 g cysteine hydrochloride in 8 ml double distilled water. The solution is brought to pH 9.5 with 5 N NaOH, and the volume is adjusted to 10 ml.

[15] EDTA is added to eliminate interference of the assay by heavy metals which bind the metal chelator more tenaciously than Ca^{2+}.

[16] The 3% microcrystalline cellulose suspension should be stirred vigorously while pipetting into the reaction mixture.

[17] A series of dilutions should be examined in order to avoid saturating amounts of enzyme.

[18] E. A. Bayer, R. Kenig, and R. Lamed, *J. Bacteriol.* **156**, 818 (1983).

[19] R. Lamed, E. Setter, and E. A. Bayer, *J. Bacteriol.* **156**, 828 (1983).

[20] B. Weeke, *Scand. J. Immunol.* **2** (Suppl. 1), 47 (1973).

gel filtration.[21] These assay methods will eventually be replaced by enzyme-linked immunoassay combined with monoclonal antibodies specific for individual components of the cellulosome.

Reagents

Barbital buffer, 0.02 M, pH 8.6,[22] used as diluent and running buffer
Agarose A (Pharmacia), 1% dissolved in barbital buffer
Reference antibody[23]
Anticellulosome antibody[24]

Procedure. Rocket immunoelectrophoresis is performed on rectangular glass plates (76 × 50 mm, Chance Bros. Ltd, Smethwick, UK). The plates are divided into three portions perpendicular to the 76 mm axis: a 40-mm upper gel containing reference antibody, a 20-mm intermediate gel containing anticellulosome antibody, and a 16-mm lower gel which contains the sample wells. Antibodies are present in the gel at a concentration of 100 μg/ml agarose, and about 2.0 ml agarose (kept at 45° before pouring onto the plates) is applied per 100 mm² area of the plate. The upper gel which contains reference antibody can be omitted, and samples can be electrophoresed directly into agarose gel which contains the cellulosome-specific antiserum. Samples of 8 μl are introduced into the sample wells, and electrophoresis is carried out at 200 V for 3 hr. Standard curves of serially diluted samples serve to determine the concentration of the cellulosome.

Definition of Unit. A unit of cellulosome-associated antigenic activity is defined as the amount of antigenic material necessary to form a 10-mm rocket in the intermediate gel. Alternatively, the height of the rocket formed by an unknown sample is compared to that formed by a known amount of purified cellulosome.

---

[21] B. Weeke, *Scand. J. Immunol.* **2** (Suppl. 1), 37 (1973).

[22] Stock solution contains 2.06 g sodium barbital, 0.4 g barbituric acid, and 0.1 g sodium azide per liter distilled water. The stock buffer is diluted with four volumes of distilled water before use.

[23] Reference antibodies are elicited by periodic intravenous injection of rabbits with mid-exponential phase cellobiose-grown cells of *C. thermocellum* YS. The immunoglobulin fraction is prepared by ammonium sulfate precipitation of the antiserum, and the resultant precipitate is adjusted to 10 mg/ml protein. The preparation is stored at −20°.

[24] Cellulosome-specific antibodies are prepared by exhaustive adsorption of the reference antibodies onto cells of an adherence-defective mutant (AD2).[18] Using this method, antibodies against cell surface antigens common to both wild-type and mutant cells are removed from the reference antibody preparation. The remaining immunoglobulin species are specific to wild-type YS. These have been found to be specific for the cellulosome, and in particular for its S1 subunit[19] (see also Fig. 2). The ammonium sulfate fraction is adjusted to 10 mg/ml and stored at −20°.

Purification Procedure

The steps designed for isolating the cellulose-binding factor from the extracellular medium are summarized in Fig. 1.[19] *C. thermocellum* is grown on cellulose under conditions whereby the latter is completely hydrolyzed. The cells are separated from the culture medium by centrifugation. Specific adsorption of the extracellular cellulosome is achieved by affinity chromatography of the cell-free supernatant fluids on insoluble microcrystalline cellulose. The eluent is further purified by gel filtration on Sepharose 4B.

Step 1. Growth of Cells and Preparation of Extracellular Material. A 50 liter culture of *C. thermocellum* strain YS is grown in a suitable fermentor (e.g., Biotec, Sweden). Other strains; i.e., ATCC 27405, NCIB 10682, and ATCC 31499 appear to behave in a similar manner. The culture medium contains the following additives per 50 liter deionized water: 25 g $MgCl_2$, 65 g NH_4SO_4, 1.75 liter 1 M potassium phosphate buffer, pH 7.4, 250 g yeast extract, 100 mg resazurin, and 250 g microcrystalline cellulose

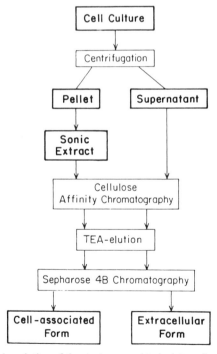

FIG. 1. Schematic description of the strategy used to isolate cell-associated and cell-free forms of the cellulosome.

(Merck). A solution (100 ml) of cysteine–HCl (15% w/v), pH 8.5 (autoclaved separately), is added immediately prior to inoculation. After flushing with a stream of nitrogen gas (continued at a rate of 100 ml/min throughout the fermentation), a 2-liter inoculum is introduced. Cells are grown for 30 hr at 60° under constant stirring at 100 rpm. The cells are centrifuged in a Sharples supercentrifuge (Sharples Centrifuges Ltd., Camberley, Surrey). The supernatant fluids are brought to a final concentration of 0.02% with NaN_3 and stored at 4° until processed further.

Step 2. Affinity Chromatography. A portion (23 liters) of the supernatant fluids is brought to pH 7.7 with 1 N NaOH, and Whatman cellulose CC 31 (276 g)[25] is added in bulk. The suspension is stirred mechanically for 1 hr at room temperature, and the cellulose is allowed to settle by gravity for a period of about 2 hr. The supernatant fluids are removed and saved for subsequent studies. The cellulose is washed on a Büchner funnel with 3 liters of 50 mM Tris–HCl buffer, pH 7.7 (Tris buffer). The cellulosome is eluted from the cellulosic matrix with 1.5 liters of a 1% solution of triethylamine,[26] and the eluent is immediately neutralized with 10% acetic acid.[27] In order to concentrate the protein, acetone (2.5 liters) is added at room temperature and the precipitate is dissolved in 150 ml Tris buffer.

Step 3. Sepharose 4B Chromatography. Preparative gel chromatography is carried out at room temperature on a Sepharose 4B column. The column dimensions are 2.5 × 80 cm, and the column is equilibrated and eluted with Tris buffer containing 0.05% NaN_3. The flow rate is 40 ml/hr, and fractions of 6.5 ml are collected. A sample (7 ml, 100 mg protein), obtained from affinity chromatography on cellulose, is applied to the column, and the major peak appears shortly after the position of the void volume. Closely associated with this peak is the bulk of both the cellulolytic and antigenic activities applied to the column. The major peak fractions (45 ml) are pooled and brought to 60% acetone. The precipitate is redissolved in 25 ml Tris buffer to a final protein concentration of 2.2 mg/ml.

---

[25] Other cellulose powders (e.g., Avicel and MM300) are also acceptable as affinity matrices.
[26] Double distilled water may also be used to elute the cellulose-adsorbed cellulosome. In general, the elution yields are usually higher using water. Concerning cellobiose-grown cells, this is of particular significance since triethylamine is not very effective in eluting the cellulosome from the column. However, we have experienced difficulties in maintaining adequate flow rates using water as an eluent. The cellulose column often becomes clogged, the exact reason(s) for which are not yet understood.
[27] For water-eluted columns, the solution is brought to 50 mM Tris–HCl, pH 7.7.

Properties of the Purified Cellulosome

Molecular Dimensions. The major form of the cellulosome comprises an apparently discrete multisubunit polypeptide complex of $M_r = 2.1 \times 10^6$. In addition, a vesicular form (which fractionates at the void volume of Sepharose 2B or 4B columns) and a lower molecular weight form ($M_r \simeq 1 \times 10^6$) can usually be separated from the major form by gel filtration. All of these forms display similar polypeptide patterns and also cross-react similarly with the anticellulosome antibody preparation.[24] The major cellulosome form is characterized by a distinct sharp 20 S peak (at 1 mg/ml protein) in the ultracentrifuge. In the electron microscope, molecular forms comprising particulate structures of relatively uniform size (18 nm) are revealed by negative staining.[19,28] The cellulosome appears to be composed of several different subunit types. In some cases, the cellulosome appears to be irregular in shape, and the reasons for the observed multiplicity in form may be explained by an innate or procedure-induced flexibility of the cellulosome structure. The molecule may thus consist of an organized conglomerate of different subunits.

Polypeptide Composition and Activities. The intact cellulosome exhibits specific endoglucanase activity of about 20 U/mg and true cellulase activity of approximately 100 U/mg. True cellulase activity is enhanced by calcium ions and thiols,[29] similar to the observed activation of the complete (crude) extracellular cellulase system in this organism.[2] Specific antigenic activity of the cellulosome corresponds to about 5 U/mg.

The cellulosome contains at least 14 polypeptides of molecular weight ranging from 48,000 to 210,000 (Fig. 2). The major bands in the cellulosome obtained from cellulose-grown cells are 210,000, 170,000, 150,000, 98,000, and 75,000. The 75,000 band is the dominant band in the YS strain. Zymograms using SDS electrophoretograms clearly indicate that about 8 of the 14 bands display regenerated endoglucanase activity.

Participation of a cellobiohydrolase has been implicated upon studying the product distribution generated by subcomplexes or subfractions obtained by partial dissociation with SDS as described in the next section.[28] The most antigenic polypeptide in the cellulosome is the largest S1 subunit (210,00 in the YS strain). This subunit is a glycoprotein containing about 50% carbohydrate, the majority of which is galactose. The S1 subunit does not exhibit measurable cellulolytic activity but may have an

[28] R. Lamed, E. Setter, R. Kenig, and E. A. Bayer, *Biotechnol. Bioeng. Symp.* **13**, 163 (1983).
[29] R. Lamed, R. Kenig, E. Setter, and E. A. Bayer, *Enzyme Microb. Technol.* **7**, 37 (1985).

Sugars Antigen CMCase Protein Subunit M$_r$(K)

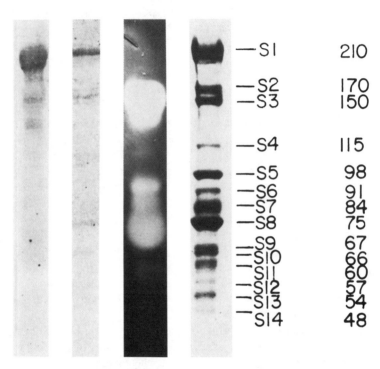

	M$_r$(K)
—S1	210
—S2	170
—S3	150
—S4	115
—S5	98
—S6	91
—S7	84
—S8	75
—S9	67
═S10	66
—S11	60
—S12	57
—S13	54
—S14	48

FIG. 2. Subunit composition and biochemical characterization of the cellulosome. Reproduced from Lamed and Bayer.[10] Cellulosome samples were subjected to SDS–polyacrylamide gel electrophoresis and stained for various activities. Protein content was demonstrated by Coomassie brilliant blue staining, and endoglucanase activity was determined by a carboxymethylcellulose overlay technique.[19] For antigenic activity, blot transfers of the gel were labeled successively with biotinylated cellulosome-specific antibodies and avidin–biotinyl peroxidase complexes. Sugar content was determined by oxidation of blots with periodate and labeling of the resultant aldehydes with enzyme hydrazide [J. M. Gershoni, E. A. Bayer, and M. Wilchek, *Anal. Biochem.* **146,** 59 (1985)]. Note that the S1 subunit appears to be a glycoprotein which contains most of the antigenic activity. In contrast, no endoglucanase activity is associated with this band.

essential structural role in the organization or assembly of the complex. A somewhat different cellulosome is produced by cellobiose-grown cells.[30] In this instance, the 75,000 band is no longer the major band; instead, the 66,000 and/or 67,000 band(s) dominate.

[30] E. A. Bayer, E. Setter, and R. Lamed, *J. Bacteriol.* **163,** 552 (1985).

Structural Stability. The complexed nature of the cellulosome is remarkably stable. Of various conditions which we have tested in attempts to interfere with interpeptide bonding (including urea, guanidine hydrochloride, and various detergents),[10,19] only treatment with boiling SDS effects significant dissociation of the complex into its subunits. By incubating the cellulosome with SDS at room temperature rather than boiling, an altered but reproducible SDS–polyacrylamide gel electrophoresis pattern is achieved.[28] In this case, some of the resultant bands represent subcomplexes which contain several but not all of the cellulosome subunits. Many of the subcomplexes are capable of exhibiting cellulase activity.

Comments

Various lines of evidence led to the recognition that the "cellulose-binding factor," which was initially isolated either from the culture medium or from whole cell extracts, is more than just an adherence factor. Notably, the tight association of the specific antigen with both endoglucanase and true cellulase activities led us to select the descriptive term "cellulosome" as an indication of its central role in the general process of cellulose degradation.[28]

The growth conditions—specifically, whether cells are grown on either cellulose or on cellobiose as insoluble or soluble carbon sources, respectively—are critical to the distribution of the cellulosome in the adherence-defective mutant AD2. Cellobiose-grown mutant cells (in contrast to the wild-type YS) lack the cellulosome on the surface and produce only minor quantities of the extracellular cellulosome. In the wild-type cell, the polypeptide composition (content and relative amount of each subunit) is also dependent on the carbon source.

The cellulosome exists in both cell-bound and cell-free forms. The cell-associated form is isolated from whole-cell sonicates[28] in a procedure analogous to that described above for the extracellular form (Fig. 1). The two forms are very similar regarding their respective activities, polypeptide composition, reaction of the S1 subunit with anticellulosome antibodies, and its sugar content (Fig. 2). A major difference is that the isolated cell-associated form contains larger quantities of the very-high-molecular-weight particulate fraction which appears at the void volume during gel filtration.

The mode of regulation and assembly of the multisubunit complex has yet to be elucidated. However, electron microscopic evidence has delineated the precise position of the cellulosome on the surface of *C. thermocellum*. When grown on soluble substrates, such as cellobiose, the cellulosome is centralized on large protuberant structures (50–200 nm) which decorate the cell surface at almost periodic intervals.[30] Growth of the bacterium directly on cellulose as a substrate confers a dramatic change in the constitution of the exocellular protuberances, which protract to yield an amorphous or fibrous network. These fibrous "contact corridors" mediate between the cellulosome (which is intimately attached to the cellulose matrix) and the bacterial cell surface.[9] Cellulolysis is performed by the cellulosome on the cellulose surface and the products are conveyed subsequently to the cell surface via the exocellular contact zones. These products, notably cellobiose, would then be taken up by the cell by an appropriate transport system.[31]

The cellulosome is thus considered to be a large discrete, multisubunit complex which exhibits a variety of activities, including cellulolytic, cellulose-binding, and antigenic activities. The antigenic and cellulolytic activities have been demonstrated to reside in separate components on the cellulosome. The molecular nature of the observed cellulose-binding activity, however, has yet to be determined. The complex comprises various different forms of cellulases, each of which may bear separate specificities toward different quaternary structures on the intricate cellulose substrate. The major organizational role of the cellulosomal structure may afford effective delivery to the substrate as well as enabling the proper orientation of the various complementary enzymes (e.g., exo- and endocellulases) which act synergistically on the cellulose. In addition, the complex may be structured such that various product intermediates are protected and their transfer to other cellulase components is facilitated for further hydrolysis. The cellulosome subunits may thus be viewed as an arrangement of cellulases and related activities within a defined supramolecular structure designed for highly efficient cellulose degradation.

In addition to *C. thermocellum* YS, work with several other strains has shown that the cellulosome is an integral part of this organism. There may be various strain-specific variations related to size, disposition of subunits within the complex, and arrangement of the cellulosome on the cell surface. There are also increasing reports in the literature which indicate that high-molecular-weight cellulolytic complexes associated in some manner with exocellular material (such as capsular material) may be a more general phenomenon. This appears to be especially true for many other cellulolytic anaerobic bacteria.[10]

[31] T. K. Ng and J. G. Zeikus, *J. Bacteriol.* **150,** 1391 (1982).

[58] Macrocellulase Complexes and Yellow Affinity Substance from *Clostridium thermocellum*

By LARS G. LJUNGDAHL, MICHAEL P. COUGHLAN, FRANK MAYER,
YUTAKA MORI, HIROMI HON-NAMI, and KOYU HON-NAMI

Introduction

Clostridium thermocellum is an anaerobic thermophilic spore-forming bacterium. It was described in 1923 by Viljoen *et al.*,[1] but a pure culture giving reproducible results was first obtained in 1948 by McBee.[2] A number of isolates of *C. thermocellum* have been described and they are summarized by Duong *et al.*[3]

The preferred substrates for *C. thermocellum* are cellulose and cellobiose, which are fermented to ethanol, acetate, lactate, CO_2, and H_2. The ability to ferment cellulose depends on an extracellular cellulolytic enzyme system, the formation of which is stimulated when the bacterium is grown on medium containing cellulose. The cellulolytic enzyme system consists of at least 14 different polypeptides, that form complexes.[4,5] The principle or basic unit size seems to be a complex from 2.6 to 4.2 million Da. Lamed *et al.*[4] coined the term "cellulosome" for this complex. However, much larger complexes, up to 100 million Da, have been isolated.[5] These larger complexes appear to be polycellulosomes and are found associated with or bound to the bacterial cell surface[6,7] and to the substrate, the cellulose fiber.[7]

In addition to producing the cellulolytic enzyme system *C. thermocellum* during growth on cellulose synthesizes a yellow affinity substance (YAS) that attaches to the cellulose fibers to form YAS-cellulose. The yellow coloring of cellulose by *C. thermocellum* has been noticed by

[1] J. A. Viljoen, E. B. Fred, and W. H. Peterson, *J. Agric. Sci.* **16,** 1 (1926).

[2] R. H. McBee, *J. Bacteriol.* **56,** 653 (1948).

[3] T.-V. C. Duong, E. A. Johnson, and A. L. Demain, *Top. Enzyme Ferment. Biotechnol.* **7,** 155 (1983).

[4] R. Lamed, E. Setter, R. Kenig, and E. A. Bayer, *Biotechnol. Bioeng. Symp.* **13,** 163 (1983).

[5] M. P. Coughlan, K. Hon-nami, H. Hon-nami, L. G. Ljungdahl, J. J. Paulin, and W. E. Rigsby, *Biochem. Biophys. Res. Commun.* **130,** 904 (1985).

[6] E. A. Bayer, E. Setter, and R. Lamed, *J. Bacteriol.* **163,** 552 (1985).

[7] F. Mayer, M. P. Coughlan, and L. G. Ljungdahl, *Annu. Meet. Am. Soc. Microbiol.* Abstr. K-112 (1986). (See also Ref. 25).

many investigators.[1,2,8,9] YAS seems to be intimately involved in the hydrolysis of cellulose. It facilitates the binding of the cellulolytic complexes to cellulose[10,11] and is produced prior to them.[12] In this chapter methods for the isolation of the cellulolytic complexes and YAS are given as well as some of their properties.

Culturing Methods for *Clostridium thermocellum*

Two strains of *C. thermocellum* are investigated in our laboratory. Strain JW20 (ATCC 31449) was isolated from a Louisiana cotton bale,[13,13a] and strain YM4 was obtained from soil collected in the volcanic area on Izu Peninsula, Japan.[14] Throughout the culturing the anaerobic technique originally described by Hungate is used.[15]

Strain JW20. This strain is grown in an atmosphere of N_2 at 60° in a medium containing per liter KH_2PO_4, 1.5 g; Na_2HPO_4, 2.5 g; NH_4Cl, 0.5 g; $(NH_4)_2SO_4$, 0.5 g; $MgCl_2 \cdot 5H_2O$, 0.09 g; $NaHCO_3$, 0.5 g; yeast extract (Difco), 3 g; vitamin solution, 0.5 ml; trace mineral solution, 5 ml; resazurin (0.1% w/v), 1 ml; reducing solution, 10 ml; and cellulose, 10 g. The vitamin solution contains in mg per 500 ml biotin, 20; *p*-aminobenzoic acid, 50; folic acid, 20; pantothenic acid, 50; nicotinic acid, 50; cyanocobalamin, 1; thiamin–HCl, 50; pyridoxine–HCl, 100; thioctic acid, 50; and riboflavin, 5. The composition of the reducing solution is as follows: NaOH (0.2 N), 200 ml; $Na_2S \cdot 9H_2O$, 2.5 g; cysteine–HCl $\cdot H_2O$, 2.5 g. The NaOH solution is boiled and then cooled during bubbling with N_2. Na_2S and cysteine are added, and the solution is autoclaved at 120° for 20 min. The mineral solution contains (mg/liter) nitrilotriacetic acid, 1500; $MgSO_4 \cdot 7H_2O$, 3000; $MnSO_4 \cdot H_2O$, 500; NaCl, 1000; $FeSO_4 \cdot 7H_2O$, 100; $Co(NO_3)_2 \cdot 6H_2O$, 100; $CaCl_2$ (anhydrous), 100; $ZnSO_4 \cdot 7H_2O$, 100;

[8] T. K. Ng, P. J. Weimer, and J. G. Zeikus, *Arch. Microbiol.* **114**, 1 (1977).

[9] J. Wiegel and M. Dykstra, *Appl. Microbiol. Biotechnol.* **20**, 59 (1984).

[10] L. G. Ljungdahl, B. Pettersson, K.-E. Eriksson, and J. Wiegel, *Curr. Microbiol.* **9**, 195 (1983).

[11] R. Lamed, R. Kenig, E. Setter, and E. A. Bayer, *Enzyme Microb. Technol.* **7**, 37 (1985).

[12] K. Hon-nami, M. P. Coughlan, H. Hon-nami, L. H. Carreira, and L. G. Ljungdahl, *Biotechnol. Bioeng. Symp.* **15**, 191 (1985).

[13] J. Wiegel and L. G. Ljungdahl, *in* "Anaerobe Fermentation-Klassische Prozesse mit neuen Aussichten" (H. Dellweg, ed.), p. 117. Verlag Versuchs- und Lehranstalt für Spiritusfabrikation und Fermentations technologie im Institut für Gärungsgewerbe und Biotechnologie, Berlin, 1979.

[13a] D. Freier, C. P. Mothershed, and J. Wiegel, *Appl. Environ. Microbiol.* **54**, 204 (1988).

[14] Y. Mori and K. Kiuchi, *Annu. Meet. Agric. Chem. Soc. Jpn.* p. 338 (Abstr.) (1984).

[15] L. G. Ljungdahl and J. Wiegel, *in* "Manual of Industrial Microbiology and Biotechnology" (A. L. Demain and N. A. Solomon, eds.), p. 84. Am. Soc. Microbiol., Washington, D.C., 1986.

$CuSO_4 \cdot 5H_2O$, 10; $AlK_2(SO_4)_3$ (anhydrous), 10; H_3BO_3, 10; Na_2-$MoO_4 \cdot 2H_2O$, 10; Na_2SeO_3 (anhydrous), 1; $NiCl_2 \cdot 6H_2O$, 50. The nitrilotriacetic acid is suspended in 500 ml H_2O and dissolved by titrating with 2–3 N KOH until the pH is stabilized at 6.5. The rest of the ingredients are added in the order given and the volume is adjusted to 1 liter. The cellulose is Solka-Floc (SW40), James River Corp., Berlin, NH, Munktell 400 powder, Grycksbo Pappersbruk, Sweden, or Avicel (microcrystalline cellulose, Type pH-105, average particle size 20 μm) FMC Corp., Philadelphia, PA. Resazurin is not included in the medium when isolation of YAS is intended.

Conveniently, culturing is done in 125-ml vials (Pierce Chemical Co., Rockford, IL) sealed with butyl rubber stoppers (Bellco Glass Inc., Vineland, NJ) and aluminum crimps or in 10-liter carboys with rubber stoppers and wire seals. Good growth is indicated by an increase in turbidity of the medium and by yellow coloring of the cellulose. The highest total cellulolytic activity (endo-β-glucanase) is obtained after 4–5 days. It stays at this level for at least 30 days when the culture is incubated at 60°. After the first 5-day period most cells sporulate and while further cell growth is not observed, hydrolysis of residual cellulose continues. Cultures grown at 60° can be stored in unopened (i.e., still anaerobic) vessels at 4° for several months without loss of cellulolytic activity. Such cultures are viable and are used to inoculate new cultures.

YAS, formed within 2–4 days after inoculation, colors the cellulose bright yellow. The YAS-cellulose, which is used to prepare affinity columns for purification of the cellulolytic complexes, is conveniently harvested at this stage.

Strain YM4. This strain digests cellulose more rapidly than does strain JW20 and produces 15-fold more endo-β-glucanase. It is grown in an atmosphere of CO_2 at 60° in a medium containing per liter K_2HPO_4, 0.45 g; KH_2PO_4, 0.45; $(NH_4)_2SO_4$, 0.9 g; NaCl, 0.9 g; $MgSO_4 \cdot 7H_2O$, 0.18 g; $CaCl_2 \cdot 2H_2O$, 0.12 g; yeast extract, 5 g; 0.1% hemin, 1 ml; 0.1% resazurin, 1 ml; vitamin solution, 2 ml; trace mineral solution, 10 ml; and cellulose (Avicel or Solka-Floc), 10 or 20 g. The vitamin solution contains in mg per liter: inositol, 1000; calcium pantothenate, 200; niacin, 200; pyridoxine–HCl, 200; thiamin–HCl, 200; *p*-aminobenzoate, 100; riboflavin, 100; cyanocobalamin 10; biotin, 5; and folic acid, 5. The trace mineral solution contains in mg per liter: $FeSO_4 \cdot 7H_2O$, 150; H_3BO_3, 100; $MnSO_4 \cdot H_2O$, 80; $ZnSO_4 \cdot 7H_2O$, 80; $Na_2MoO_4 \cdot 2H_2O$, 40; $Na_2WO_4 \cdot 2H_2O$, 40; KI, 20; $NiCl_2$, 10; $CoCl_2$, 10; Na_2SeO_3, 3; $CuSO_4 \cdot 5H_2O$, 4; and $AsCl_3$, 3. The pH of the medium is adjusted to 7.2 with NaOH, boiled, and bubbled with CO_2. Fifty milliliters of 8% Na_2CO_3 and 30 ml of freshly prepared 1% cysteine–HCl are then added. The medium is dispensed in 50 ml portions into 125-ml vials that are sealed with butyl rubber stoppers, crimped, and autoclaved. When preparing medium for large volume cultures the Na_2CO_3 and cysteine–HCl solutions are autoclaved separately and then added. Most cellulose is digested within 1 day after inoculation, when the culture is slowly agitated. Replacing CO_2 by N_2 in the gas phase considerably slows the fermentation rate.

The media described for strains JW 20 and YM4 are rich and can with certainty be simplified. It should be noted that Johnson et al.[16] with C. thermocellum, ATCC 27405, developed a defined minimal medium, which contains pyridoxamine, biotin, p-aminobenzoate, and vitamin B_{12} but no yeast extract. Morpholinopropane sulfonic acid is added as a buffer substance to control the pH at 7.4. As a nitrogen source, they use urea. A simple medium containing yeast extract, and morpholinopropane sulfonic acid is used by Bayer et al.[17] to grow C. thermocellum, strain YS.

Enzyme Assays

Carboxymethylcellulase (CMCase or endo-β-1,4-glucanase; EC 3.2.1.4) activity is conveniently assayed by a modification of the viscometric method of Almin et al.[18] Reaction mixtures (5 ml) containing 0.73% (w/v) CM-cellulose 7H3 SXF (Hercules, Inc., Wilmington, DE) in 91 mM triethanolamine–maleate buffer, pH 6, are preincubated for 20 min at 40° before addition of 50 μl of enzyme. Incubation is continued for 3 min and the viscosity at 40° is determined. Specific activity is expressed in arbitrary units of CMCase mg^{-1} protein. Ten of these units approximately correspond to the production of 0.4 μmol of reducing sugar groups per minute. This was established by assaying reducing sugars produced from the CM-cellulose using the dinitrosalicylic acid reagent according to Miller et al.[19]

The hydrolysis of crystalline cellulose such as Avicel, Munktell powder, or Solka-Floc is determined by assaying the liberation of reducing sugars[20] or by monitoring at 660 nm the decrease in turbidity during incubation of an aliquot of enzyme with 3 mg of substrate in 5 ml 50 mM triethanolamine–maleate buffer, pH 6, containing 2 mM disodium EDTA,

[16] E. A. Johnson, A. Madia, and A. L. Demain, Appl. Environ. Microbiol. 41, 1060 (1981).
[17] E. A. Bayer, R. Kenig, and R. Lamed, J. Bacteriol. 156, 818 (1983).
[18] K. E. Almin, K.-E. Eriksson, and C. Jansson, Biochim. Biophys. Acta 139, 248 (1967).
[19] G. L. Miller, R. Blum, W. E. Glennon, and A. L. Burton, Anal. Biochem. 2, 127 (1960).
[20] M. Dubois, K. A. Gilles, J. K. Hamilton, P. A. Rebers, and F. Smith, Anal. Chem. 28, 350 (1956).

TABLE I
PORTIONS OF ENDO-β-GLUCANASE (CMCase) ACTIVITY FREE IN THE CULTURE FLUID
AND BOUND TO YAS-CELLULOSE[a]

Days of fermentation	3	4	6	7	14	28
CMCase activity (% of total)[b]						
Free in culture fluid	27	35	44	49	85	91
Bound to YAS-cellulose	73	65	56	51	15	9

[a] During a fermentation from 3 to 28 days using *C. thermocellum*, JW20.

[b] Total CMCase activity was 540 units in the 4-day culture and 576 units in the 28-day culture, which indicates that very little cellulolytic enzyme was produced after day 4. The bound activity was released to the culture fluid as the cellulose was hydrolyzed. There is no evidence for proteinases in the culture fluid of *C. thermocellum*. The specific activity of the CMCase free in the culture fluid was 1.5 in the 3-day culture and 5.8 in the 28-day culture, and that of the bound enzyme was about 40 throughout the fermentation.

10 mM dithiothreitol, and 0.1% (w/v) $CaCl_2$.[21] Protein is determined by absorption at 280 nm or according to Bradford.[22]

Separation of Cellulolytic Active Material into Complexes (Cellulosomes) and Polypeptides

The fractionation of the cellulolytic enzyme (cellulase), of *C. thermocellum*, strain JW20, is based on the facts that the complexes (cellulosomes and polycellulosomes) bind tightly to YAS-cellulose in the presence of 5 mM or more concentrated buffer or salt solutions, but are easily extracted from YAS-cellulose with distilled water.[10,23] Before fractionation consideration must be given to the stage of the fermentation. As shown in Table I most of the cellulase is bound to YAS-cellulose during the early stage, but is released to the medium as the fermentation proceeds.[10,24] The following fractionation is based on previously described procedures,[10,12,23] and is outlined in Fig. 1. The fractionation of the cellulase is followed by assaying CMCase.

Separation of Bound and Free Cellulolytic Active Material from Strain JW20. The culture is chilled to below 25° (the cellulase of *C. thermocellum* has very little activity below 25°) and the YAS-cellulose con-

[21] E. A. Johnson and A. L. Demain, *Arch. Microbiol.* **137**, 135 (1984).

[22] M. M. Bradford, *Anal. Biochem.* **72**, 248 (1976).

[23] K. Hon-nami, M. P. Coughlan, H. Hon-nami, and L. G. Ljungdahl, *Arch. Microbiol.* **145**, 13 (1986).

[24] N. Ait, N. Creuzet, and P. Forget, *J. Gen. Microbiol.* **113**, 399 (1979).

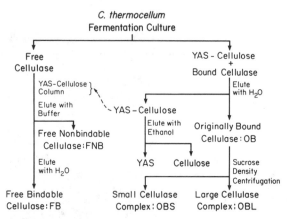

FIG. 1. Fractionation of cellulolytic enzyme complexes of *Clostridium thermocellum*. (From Hon-nami *et al.*[23] with permission of the publisher.)

taining bound cellulase is allowed to settle. The supernatant culture fluid containing bacterial cells or spores, fine cellulose particles, and free enzyme is carefully decanted. The decanted fluid is then centrifuged at 16,000 *g* for 15 min at 4° to remove the bacteria and fine cellulose particles. NaN$_3$ (final concentration 0.05%) is added to the supernatant which is then passed through a Pyrex glass (ASTM 40–60) filter to remove any remaining solids. The enzyme activity in the filtrate constitutes the total free enzyme activity. The settled YAS-cellulose containing the originally bound (OB) enzyme is poured into a column and is washed with 100 m*M* triethanolamine–maleate buffer, pH 6.85, until the wash is free of protein. The YAS-cellulose-bound cellulolytic enzyme is then eluted as a single peak with distilled water. The YAS-cellulose column when freed of enzyme is used as an affinity column for fractionation of the free cellulolytic activity of the culture fluid. The free enzyme has a specific activity from 1.5 to 5.8 (see Table I), whereas the OB enzyme has an activity of about 40.

Affinity Chromatography on YAS-cellulose of Free Enzyme from Strain JW20. The free enzyme in the culture fluid is resolved into two fractions by affinity chromatography on the YAS-cellulose column obtained in previous step after it is first equilibrated with 100 m*M* triethanolamine–maleate buffer, pH 6.85. One fraction termed free nonbindable cellulase (FNB) does not bind to the YAS-cellulose and passes through the column with the void volume. The second fraction of free enzyme binds to YAS-cellulose in the presence of buffer and so is termed free-bindable cellulase (FB). This fraction is eluted as a single peak from the column with distilled water. The CMCase specific activity of the FNB fraction varies from 0.8 to about 4 depending on the age of the culture from which it is obtained. The specific activity of the FB fraction is 40 and similar to that of the OB fraction.

Resolution of the Originally Bound (OB) Enzyme of C. thermocellum, JW20, into Originally Bound Large (OBL) and Originally Bound Small (OBS) Complexes. The originally bound cellulolytic enzyme eluted with distilled water from YAS-cellulose consists of a mixture of cellulolytic complexes of different sizes, but with the same or very similar enzymatic activity. The complexes can be separated either by using gel filtration or sucrose density centrifugation.[5,23]

Gel filtration is performed on a column (1.6 × 86 cm) consisting of Sepharose 4B (Pharmacia Fine Chemicals, Piscataway, NJ) equilibrated with 50 mM Tris–HCl buffer, pH 7.5, containing 0.1 M NaCl and 0.05% (w/v) NaN$_3$. The sample size loaded on the column conveniently is 3 ml containing about 20 mg of protein. Fractions of 1.9 ml are collected and assayed for protein and CMCase activity. The applied protein and CMCase activity are recovered in fractions 26–80, each fraction having the same specific activity of about 40. However, the CMCase as well as the protein emerge in several peaks of which two are major. The first of these elutes with the void volume (M_r at or higher than 20×10^6) and contains originally bound large enzyme complexes (OBL). The second major peak elutes around fraction 50 corresponding to an $M_r = 5 \times 10^6$. It is designated originally bound small enzyme complexes (OBS). It is apparent that OBS corresponds to the cellulosome[4] and that OBL complexes are polycellulosomes.[5,6] It should be noted that complexes of intermediate sizes are also found.

OBL and OBS complexes are also separated using sucrose density gradient centrifugation. An aliquot (0.2 ml) of enzyme is layered on top of 5 ml linear gradient (5–30%, w/v) of sucrose in 50 mM Tris–HCl buffer, pH 7.5, and centrifuged at 4° for 5 hr at 40,000 rpm in a Beckman ultracentrifuge (Model L2-65B) using an SW65 rotor. Fractions (0.25 ml) are collected and assayed for protein and CMCase activity. OBL complexes are found in fractions 8–11 and OBS complexes in fractions 15–18.

The OBS complexes and the free-bindable (FB) cellulase obtained by affinity chromatography on YAS-cellulase behave identically when subjected to gel filtration and sucrose density gradient centrifugation. FB like OBL and OBS complexes hydrolyze crystalline cellulose (Avicel). It is apparent that the FB fraction is comprised of complexes similar to those of the OBS fraction. By contrast the free nonbindable fraction (FNB) does not contain large complexes. It is a mixture of polypeptides including some with endo-β-glucanase activity, but it does not efficiently hydrolyze crystalline cellulose (Avicel).

Isolation of Originally Bound and Free-Bindable Cellulolytic Complexes from C. thermocellum, Strain YM4.[25] *C. thermocellum,* strain YM4, almost completely ferments 1% of cellulose in media within 1 day. Like strain JW20 it produces YAS and forms YAS-cellulose, however, it produces from 10 to 15 times more cellulolytic enzyme than does strain JW20. This enzyme is not of higher specific activity than that from strain JW20, there is just more of it. The reason for the high production of cellulase by strain YM4 is not known. However, it should be noted that by growing strain JW20 in the presence of bicarbonate and with CO_2 as gas phase instead of N_2 the CMCase activity more than doubled.[26]

One very noticeable difference between the cellulases of the two strains is that originally bound large complexes have not been found in cultures of YM4. However, OBS and FB complexes can be isolated from cultures of YM4 by methods identical with those described for isolation of the complexes obtained from strain JW20.

To isolate OBS complexes from strain YM4 it is obvious that the fermentation cannot be allowed to go to completion; YAS-cellulose must remain in the medium. It is separated from the supernatant culture fluid as described above. The YAS-cellulose containing the bound enzyme is washed with 0.1 M triethanolamine–maleate buffer until protein is no longer released. The OBS complexes are then eluted with distilled water.

The YAS-cellulose obtained from cultures of stain YM4 is generally too hydrolyzed to be used as an affinity column. Therefore, to isolate FB complexes from strain YM4 the affinity column prepared with YAS-cellulose from strain JW20 is used. The culture fluid is applied to the YAS-cellulose column previously equilibrated with 0.1 M buffer. The column is washed with the buffer and then the FB complexes are eluted with distilled water exactly as described above.

Properties of the Cellulase Complexes from *C. thermocellum,* Strains JW20 and YM4

Enzymatic Properties. The cellulolytic complexes OBL, OBS, and FB from *C. thermocellum* strain JW20, as well as OBS and FB from strain YM4 hydrolyze crystalline cellulose such as Avicel, Solka-Floc, and Munktell cellulose powder. The hydrolysis is dependent on Ca^{2+} and

[25] F. Mayer, Y. Mori, M. P. Coughlan, and L. G. Ljungdahl, *Appl. Environ. Microbiol.* **53**, 2785 (1987).
[26] H. Hon-nami, M. P. Coughlan, K. Hon-nami, and L. G. Ljungdahl, *Proc. R. Ir. Acad.* **87B**, 83 (1987).

dithiothreitol as was established by Johnson *et al.*[21,27] The products of the hydrolysis are cellobiose (90%) and glucose (10%). The complexes also effectively hydrolyze CM-cellulose, phosphoric acid-swollen Avicel, and 4-methylumbelliferyl-β-D-cellobioside in reactions that are not dependent on Ca^{2+} or dithiothreitol. However, the CMCase and 4-methylumbelliferyl-β-D-cellobioside activities are slightly stimulated by these agents.

The unfractionated originally bound complexes from strain JW20 are active between pH 4.5 and 7.5 with an optimum around pH 6. The temperature optimum for the same complexes is at 65° and there is very little activity below 30° and over 90°. The complexes are stable for 2 hr at 70° in 50 mM triethanolamine–maleate, pH 7.6. When incubated at 75 and 80° under the same conditions they lose 60 and 90%, respectively, of their activity within 1 hr.

Cellobiose effectively inhibits the hydrolysis of crystalline cellulose.[11,23,28] The extent of inhibition using the OB complexes of strain JW20 is 78 and 100% in the presence of 0.5 and 1% (w/v) of cellobiose, respectively. Xylose and glucose are much less inhibitory. The presence of 5% (w/v) of xylose or glucose inhibits the hydrolysis 33%, and 25%, respectively. Ethanol, one of the products produced by *C. thermocellum,* inhibits at a concentration of 5% the rate of Avicel hydrolysis to the extent of 30%. Sulfhydryl reagents, such as *o*-iodobenzoate, 5,5'-dithiobis-2-nitrobenzoic acid, *N*-ethylmaleimide, iodoacetic acid, and *p*-chloromercuribenzoic acid, inhibit the hydrolysis of Avicel but not of CM-cellulose by crude cellulase preparations of *C. thermocellum,* strain ATCC 27405.[21] This inhibition is prevented by dithiothreitol and other thiols, indicating an active role for sulfhydryl groups in the hydrolysis of crystalline cellulose but not of CM-cellulose. Other chemicals found to inhibit the *C. thermocellum* cellulase include $CuSO_4$ and *o*-phenanthroline.[21]

The cellulolytic complexes are remarkably stable. As mentioned cultures grown at 60° when stored anaerobically in sealed vials at 4° and also at room temperature maintain their viability and the integrity of the enzyme complexes for several months. The complexes are stable in 50% (w/v) ethanol at room temperature for at least 2 months. This solvent can substitute for distilled water for the elution of the OB complexes from YAS-cellulose. However, YAS is also extracted by the alcohol solution. It should be noted that 1% triethylamine is used for the elution of cellulosomes from cellulose and 66% (v/v) of acetone to concentrate them.[11,29]

[27] E. A. Johnson, M. Sakajoh, G. Halliwell, A. Madia, and A. L. Demain, *Appl. Environ. Microbiol.* **43,** 1125 (1982).
[28] E. A. Johnson, E. T. Reese, and A. L. Demain, *J. Appl. Biochem.* **4,** 64 (1982).
[29] R. Lamed, E. Setter, and E. A. Bayer, *J. Bacteriol.* **156,** 828 (1983).

Clearly the complexes are stable also in these solvents. It should be noted that the presence of 8 M urea during gel filtration of cellulosomes from strain YS on Sephacryl S-300 does not effect the elution pattern.[18] This indicates that urea is not an effective agent for the dissociation of the cellulosome into its polypeptides.

The inclusion of sodium dodecyl sulfate (SDS) in assay mixtures markedly decreases the ability of OB complexes to hydrolyze Avicel, but has little effect on CMCase activity.[23] Pretreatment of the complexes with SDS results in an irreversible loss of the activity against Avicel whereas as much as 86% of the original CMCase activity remains even after 10 days incubation at room temperature with 2% SDS. This indicates that maintenance of the structural integrity of the cellulase complexes is essential for hydrolysis of crystalline substrates whereas uncomplexed endoglucanases hydrolyze CM-cellulose even after SDS treatment.

Physical Properties. The originally bound large, originally bound small, and free-bindable cellulolytic complexes isolated from *C. thermocellum* strains JW20 and YM 4,[5,12,25] as well as the cellulosome and polycellulosome found in strain YS,[4,6] should be judged as "pure" enzyme complexes containing the activities required for the hydrolysis of crystalline cellulose to cellobiose as the major product. The complexes behave as single entities during gel filtration and sucrose gradient centrifugation and further "purification" has not increased their specific activities nor revealed the presence of any unrelated protein or material.

Treatment of the complexes by boiling in the presence of SDS and subsequent SDS–polyacrylamide electrophoresis either in tubes[30] or on slabs[31] reveal that the complexes consist of 14 or more different types of polypeptides, with M_r values from about 45,000 to 210,000.[4,6,11,23] (Fig. 2). There is a marked similarity between the polypeptide compositions of all the cellulolytic complexes, which indicates that the complexes are related. However, while the same types of polypeptides are found in all complexes, the relative amounts of these polypeptides differ from one complex to another. We noticed[23] that in the OBL complex from JW20 a 210,000-Da polypeptide is a major component, whereas in the OBS complex it is a relatively minor component. On the other hand polypeptides with M_r values from 45,000 to 60,000 are found at a greater concentration in the OBS complex than in the OBL complex. The different polypeptides in the complex may form subclusters. This was revealed by partially denaturing the cellulosome and subsequent SDS–polyacrylamide electrophoresis analysis of the separate subclusters.[4] Such subclusters have been

---

[30] K. Weber and M. J. Osborn, *J. Biol. Chem.* **244**, 4406 (1969).
[31] U. K. Laemmli, *Nature (London)* **227**, 680 (1970).

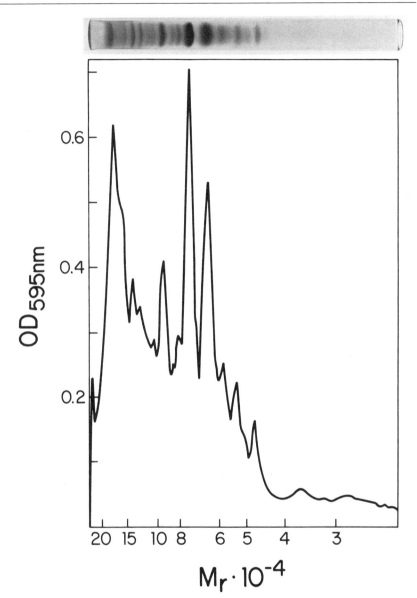

FIG. 2. SDS–polyacrylamide tube gel electrophoresis of originally bound cellulolytic complex from *C. thermocellum*, YM4. Staining was with Coomassie Brilliant Blue R250. The densitometric scan was performed at 595 nm.

observed in OB preparations from YM4 and JW20 using electron micros-copy.[25] Elution of the individual polypeptides from tube gels after SDS–polyacrylamide electrophoresis and assaying the eluates for CMCase ac-tivity revealed that at least four of the polypeptides possess this activity.[23] Using a CMC overlay technique for SDS gels Lamed et al.[29] demon-strated that not less than 10 of the polypeptides have CMCase activity. These observations fit well with results from genetic studies by Millet et al.,[32] who have cloned 10 distinct DNA fragments of C. thermocellum that code for seven endo-β-glucanases and three methylumbelliferyl-β-D-cel-lobiosidases.

The cellulolytic complexes in addition to being composed of polypep-tides also contains from 6 to 13% carbohydrate.[33] Sugars found in the complexes from strain JW20 and YM4 are xylose, mannose, galactose, glucose, and an amino sugar, probably N-acetylglucosamine. From analy-ses of different preparations and by extended dialysis of the complexes it is clear that the amounts of xylose, mannose, and glucose vary consider-ably in the complexes. However, galactose (from 62 to 92 μg · mg^{-1} of protein) and the amino sugar (from 5 to 10 μg · mg^{-1} of protein) are always found in relatively high amounts and apparently are integral parts of the complexes.

Electron Microscopic Observations. Lamed et al.[29] estimated the M_r for the cellulosome to be about 2 million using gel filtration. These cellulo-somes, when negatively stained and viewed with a transmission electron microscope, showed relatively uniform images with a size of 18 nm. As-suming that the images represent spherical particles this size corresponds to an $M_r = 2.6 \times 10^6$. Bayer et al.[6] in addition observed polycellulosomal structures bound to the cell surfaces of strain YS. Purified OBS and OBL complexes from strain JW20 have diameters of 21 and 61 nm, respec-tively, which corresponds to M_r values of 4×10^6 and 100×10^6 assuming that the complexes are spherical.[5] Intermediate size particles are also observed. More extensive investigations[7,25] using transmission electron microscopy reveal that OBL complexes and clusters of OBS are formed on the bacterial cell surface (Fig. 3) and that these complexes attach to and remain on the cellulose fibers even when the bacteria are no longer present (Fig. 4). These studies also indicate that OBS clusters are of two general types; one, TOBS, in which the polypeptides appear tightly packed and a second, LOBS, in which the polypeptides appear loosely packed. Furthermore, polypeptide complexes smaller than OBS com-

[32] J. Millet, D. Pétré, P. Béguin, O. Raynaud, and J.-P. Aubert, *FEMS Microbiol. Lett.* **29,** 145 (1985).

[33] Y. Mori and L. G. Ljungdahl, unpublished results.

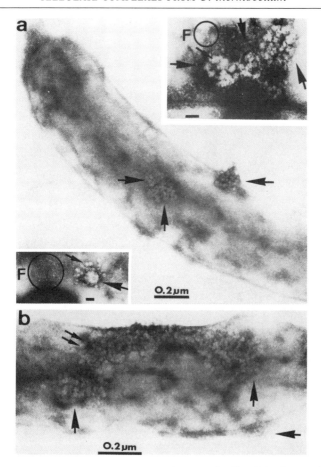

FIG. 3. Transmission electron microscopy images of cells of *C. thermocellum*, strain JW20. (a) Bacterium with large clusters of OBS complexes attached to the cell surface. The large inset shows three clusters of OBS particles attached to a cell surface; the clusters exhibit spherical shapes and are covered with fine fibers (F-circle). The small inset shows the pole of a bacterial cell (lower left side), and a nearby OBL complex surrounded by OBS complexes and by fibers. (b) Bacterial cell surface partially covered with clusters of tightly packed OBS complexes. (From Mayer *et al.*[25] with permission of the publisher.)

plexes exist. These appear to be dissociation products of LOBS particle (Fig. 5).

Isolation of the Yellow Affinity Substance (YAS)

YAS is conveniently isolated from 100 g of YAS-cellulose produced by *C. thermocellum* strain JW20. For this purpose the bacterium is grown

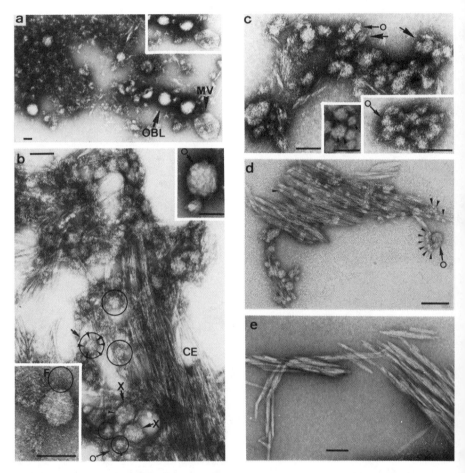

FIG. 4. Sample of YAS-cellulose with bound cellulase obtained from a culture of *C. thermocellum*, strain JW20, as revealed with transmission electron microscopy. (a) The bound cellulase is organized in prominent OBL complexes or as smaller entities not clearly visible. The insert shows two of the OBL complexes when tilted at an angle of 45°. The OBL complexes are obviously not perfectly spherical as previously was postulated,[5] but rather ellipsoidal. The size of the OBL complexes correspond to an M_r of about 80×10^6. MV is a membrane vesicle derived from an autolyzed bacterium. (b) Bundles of YAS-cellulose fibers with attached "tight" (O) and extra large "loose" (X) OBS complexes. The arrowheads show individual polypeptides in the complexes. CE designates YAS-cellulose fibers. Small inset: OBL complex consisting of "tight" OBS complexes. The OBL complex appears to be covered by a faint "skin" of unknown composition. Large inset: OBL complex and fibers with diameters from 1.5 to 2 nm, probably cellulose degradation products. (c) Small bundles of cellulose fibers with "tight" OBS complexes. Large inset: cluster of "tight" OBS complexes presumed formed from an OBL complex. Small inset: individual polypeptides visible inside "tight" OBS complexes. (d) Bundles of YAS-cellulose fibers after the first wash with distilled water. "Tight" OBS complexes are visible and also free polypeptides. (e) Bundles of YAS-cellulose after extensive washing with water. The cellulolytic complexes and polypeptides have been removed. (From Mayer *et al.*[25] with permission of the publisher.)

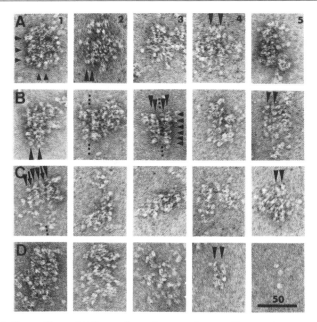

FIG. 5. Gallery of loose OBS complexes from *C. thermocellum* strain YM4. Row A: Full size structurally well-preserved loose OBS complexes. Arrowheads point toward ordered arrangements of individual polypeptides. Rows B and C: Complexes with clearly ordered arrangements of individual polypeptides packed in paracrystalline arrays. Row D: From left to right, remnants of increasingly degraded OBS complexes. (From Mayer *et al.*[25] with permission of the publisher.)

as described above in carboys with 10–15 liters of medium containing 1% (w/v) of cellulose (Solka-Floc or Avicel). Harvesting is done as soon as the cellulose is completely colored yellow (3–4 days). The procedure of purification of YAS is outlined in Fig. 1.

Step 1. Preparation of YAS-Cellulose. The culture to be harvested is chilled to room temperature and the YAS-cellulose is allowed to settle. The supernatant is decanted. The YAS-cellulose is placed on a Büchner filter and washed five times with about 100 ml of 50 mM triethanolamine–maleate, pH 6.85. This removes the free cellulolytic enzyme fractions designated FB and FNB. The YAS-cellulose still on the Büchner filter is now washed with distilled water until the wash is free of protein (about 500 ml). This removes the bound cellulolytic complexes. Remaining on the filter is YAS-cellulose that can be used for an affinity chromatography column to purify cellulolytic complexes of *C. thermocellum* or it can be used for the isolation of YAS.

Step 2. Ethanol Extraction of YAS. This and the following steps should be carried out with the exclusion of light and air as much as possible. YAS is extracted from the YAS-cellulose (still on the Büchner filter) with 100% ethanol (specpure grade). The extraction is continued until the extract is free of yellow colored material. This crude alcoholic extract containing YAS is heavily contaminated with the plasticizer dioctyl phthalate. Dioctyl phthalate is known to be a ubiquitous contaminant of materials including organic solvents, distilled water, food, and blood that are stored in plastic containers. In fact, all samples of cellulose from various suppliers we tested contain significant amounts of dioctyl phthalate. The extent of contamination may be definitely established by GC/mass spectroscopy. However, for routine purpose the 220/440 nm and 280/440 nm absorbance ratios may be used (see YAS properties below), since dioctyl phthalate absorbs only in the UV region, whereas YAS absorbs both in the UV and the visible regions.

Step 3. Separation of YAS from Dioctylphthalate by Hydrophobic Chromatography. The crude YAS in 100% ethanol is diluted with 3 volumes of anaerobic distilled water (previously boiled and bubbled with O_2-free nitrogen gas). Octyl-Sepharose CL-4B (Pharmacia Fine Chemicals, Piscataway, NJ) preequilibrated with anaerobic 20% (v/v) ethanol is

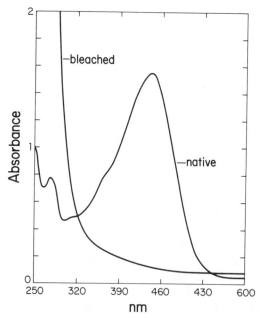

FIG. 6. The absorption spectrum of YAS in 50% (v/v) ethanol under anaerobic conditions and of the bleached (oxidized) material following addition of 15 μl of 30% (v/v) H_2O_2 per ml and exposure to sunlight. (From Hon-nami et al.[23] with permission of the publisher.)

added to the YAS extract in a quantity sufficient to absorb the bulk of the YAS. The suspension is mixed thoroughly while being bubbled vigorously with N_2. The Sepharose containing YAS is then allowed to settle. The supernatant still slightly yellowish is siphoned off. The Sephadex slurry is then poured into a column, which is washed with 2–3 volumes of anaerobic 20% ethanol. This wash, which is discarded, will contain resazurin if it was used as a redox indicator in the growth medium.

YAS is now eluted from the column using 40% (v/v) anaerobic ethanol. The elution continues until most of the YAS has been recovered or until the 220/440 nm or 280/440 nm ratios increase over 0.4 or 0.22, respectively, which will indicate the presence of dioctyl phthalate. The 40% ethanol fraction is now freed of ethanol by being dried at 55° under a stream of N_2 and subsequently freeze-dried to remove the water. This preparation has been used in further studies of the properties of YAS. Additional YAS can be obtained from the octyl-Sepharose column by elution with 100% ethanol. This fraction is heavily contaminated with dioctyl phthalate.

Properties of YAS

YAS is insoluble in water, *n*-hexane, and petroleum ether, slightly soluble in sodium bicarbonate solutions, ether, and chloroform, and very soluble in acetone, ethanol, methanol, and according to Lamed *et al.*[11] in methylethylketone.

A solution of freshly prepared (still anaerobic) YAS in ethanol is yellow/orange in color. The UV/visible spectrum (Fig. 6) shows a prominent peak at 440–445 nm, which disappears on oxidation and exposure to sunlight.

YAS does not contain metals as determined by plasma emission analysis, nor does it contain carbohydrates or amino acids, and it is not a flavin or flavin glycoside. Gel filtration of ethanolic solutions on Sephadex G-25 or BioGel P-2, P-4, or 6PDG all indicated an M_r value of about 1300. The elemental composition apparently is $C_{52}H_{94}O_{19}N$ (the oxygen content is a calculated figure), which corresponds to an M_r of 1036. Preliminary findings[34] using Raman, IR, and ^{13}C NMR spectroscopic techniques indicate that YAS is comprised of a ring structure with a long hydrocarbon tail that may contain carbonyl groups and possibly an NH moiety. YAS might be carotenoidlike.

[34] M. P. Coughlan, H. Hon-nami, J. DeHaseth, L. Carreira, and L. G. Ljungdahl, unpublished results.

Speculations

Results obtained in several laboratories demonstrate that *C. thermocellum* produces cellulolytic complexes of various sizes of which the basic unit appears to be the "cellulosome" or the OBS complex with an M_r from 2 to 4.5 million.[4,5] Cellulosomes form polycellulosomes or OBL complexes of M_r close to 100 million.[5]

According to our present concept of the cellulolytic enzyme system of *C. thermocellum* the bacterium forms clusters of OBS complexes, which assemble on the bacterial surface and form OBL complexes. These bind both to the cellulose and to the bacterium. The binding of the cellulosome to the bacterium has been shown by Lamed *et al.*[6,17] to involve a 210,000-Da polypeptide, that is part of the cellulolytic complexes. As the fermentation proceeds and the bacterium sporulates,[9] the OBL complexes disaggregate to form OBS complexes. These complexes in turn may desorb from the substrate and may then either rebind to the cellulose and again participate in the hydrolysis or they may further disaggregate to form polypeptides of which some maintain endo-β-glucanase activity as found in the free culture fluid.

YAS is present on the bacterium when it is growing on cellobiose, but it is produced in quantity only with cellulose as the substrate. The production of YAS precedes the production of cellulase. There is no doubt that YAS adheres to cellulose and that YAS-cellulose has a higher affinity for the cellulolytic complexes than cellulose itself.[10,11] However, there is no evidence for the involvement of YAS in the catalytic process. A possible function of YAS may be that of a signal substance. When YAS present in vegetative cells reacts with cellulose it may tell the cells that cellulose is present and that it is time to produce cellulase.

[59] Acid Proteases from *Sporotrichum pulverulentum*

By K.-E. ERIKSSON and B. PETTERSSON

Extracellular proteases are commonly produced among fungi[1] including several examples of wood-degrading Basidiomycetes.[2,3] In our studies

[1] K. Aunstrup, *Annu. Rep. Ferment. Processes* p. 125 (1978).
[2] W. H. E. Roschlan, *in* "Fibrinolytics and Antifibrinolytics" (F. Markwardt, ed.), pp. 337–450. Springer-Verlag, Berlin, 1978.
[3] H. Kumagai, M. Matsue, E. Majima, K. Tomuda, and E. Ishishima, *Agric. Biol. Chem.* **45**, 981 (1981).

of enzymes involved in cellulose degradation, it was frequently observed that high endo-1,4-β-glucanase activity was always accompanied by a high production of proteolytic enzymes when *Sporotrichum pulverulentum* was cultivated on cellulose. The two acidic proteases described here have been shown to play a role in the activation of the endo-1,4-β-glucanases in culture solutions of *S. pulverulentum* grown on cellulose.

Assay Method for the Proteases

Principle. Protease activity is assayed by measuring the increase in absorbance at 510 nm which occurs when the enzyme releases azodye from Azocol (Calbiochem, San Diego, CA), a collagen preparation to which is bound an azodye.

Reagents

Sodium acetate buffer, 50 mmol, pH 5.0
Azocol, 10 mg/ml of buffer

Procedure. A 2 ml Azocol–buffer suspension is incubated with an appropriate amount of enzyme in 1 ml solution at 30° for 1 hr and shaken every 15 min. The Azocol remaining after incubation is removed by centrifugation at 5800 rpm for 5 min and the absorbance of the supernatant at 520 nm is measured.

Purification Procedure

Growth of the Fungus. Sporotrichum pulverulentum (ATCC 32629, anamorph of *Phanerochaete chrysosporium*) is grown in shake cultures at 28° in 1-liter conical flasks containing 300 ml of a modified Norkrans' medium as described in chapter [41] in this volume but with the following amendments: the monobasic and dibasic potassium phosphate contents are reduced to 0.6 and 0.4 g/liter, respectively, ammonium phosphate content is 6.5 g/liter, urea content is 0.43 g/liter, cellulose powder content is 10 g/liter, and yeast extract 0.1 g/liter.

The flasks are inoculated with approximately 2.6 × 10⁶ spores/ml and cultivated for 10 days.

Step 1. Concentration of Culture Filtrates. After 10 days cultivation, the mycelium is filtered off through a wire sieve (Mesh DIN 20) supported on a Büchner funnel and residual cellulose is removed by filtration through a sintered glass funnel (L1) and a glass fiber filter (Sartorius 13400). Culture filtrate, normally 10 liters/batch, is cooled to 5° and concentrated to 200 ml using the Amicon TC ultrafiltration system, model 1.B (Amicon Corp., Lexington, MA) fitted with a Diaflc membrane UM10.

TABLE I
PURIFICATION OF PROTEASE I FROM CULTURE SOLUTION OF *S. pulverulentum*

Purification step	Volume (ml)	Total protein (A_{280} units)	Total protease activity (A_{520} units)	Specific activity (A_{520}/A_{280})	Purification factor	Yield (%)
Protease I DEAE-Sephadex A-50	260	851	691	0.81	3.9	26.3
Concentration (Millipore)	44	466	402	0.86	4.1	15.3
AcA 22 chromatography	198	72	345	4.80	28.6	13.2
Concentration + dialysis	164	64	320	5.00	30.0	12.2
Preparative flat-bed electro-focusing, pH 3–10	20	19	210	11.06	52.7	8.0
Column isoelectric focusing, pH 5–7	2.0	2.1	46	31.9[a]	151.9	1.8

[a] The specific activity as calculated from Fig. 4A of ref. 8 is considerably lower, about 1.0, due to the inhibitory effect of the ampholytes present in the protease assay solution; the specific activity of 31.9 is obtained when the ampholytes are removed.

The concentrate is dialyzed overnight against distilled water in collodion tubes (Membranfiltergesellschaft, Göttingen, FRG).

Concentrations between Purification Steps. Protease fractions obtained from each of the purification steps are pooled and concentrated at 0° to the desired volumes (cf. Tables I and II) by ultrafiltration using a Millipore immersible separator kit (Millipore Corp., Bedford, MA).

TABLE II
PURIFICATION OF PROTEASE II FROM CULTURE SOLUTION OF *S. pulverulentum*

Purification step	Volume (ml)	Total protein A_{280}	Total protease activity (A_{520})	Specific activity (A_{520}/A_{280})	Purification factor	Yield (%)
Protease II DEAE-Sephadex A-50	410	587	328	0.56	2.70	12.5
Concentration (Millipore)	50	402	300	0.75	3.60	11.4
AcA 22 chromatography	81	260	260	1.00	4.76	9.9
Concentration + dialysis	10	226	220	0.97	4.62	8.4
Preparative flat-bed electro-focusing, pH 3–10	20	12	130	10.80	51.40	2.0
Column isoelectric focusing, pH 5–7	2.0	1.2	32	26.6[a]	126.6	1.2

[a] The specific activity as calculated from Fig. 4B of ref. 8 is considerably lower, about 1.0, due to the inhibitory effect of the ampholytes present in the protease assay solution; the specific activity of 26.6 is obtained when the ampholytes are removed.

Step 2. Fractionation on DEAE-Sephadex. Initial fractionation is performed on a DEAE-Sephadex A-50 Column, 45 × 400 mm (Pharmacia Fine Chemicals, Uppsala, Sweden). The buffer system is 0.02–0.4 M ammonium acetate buffer, pH 5.0, and fractionation is achieved by a stepwise increase in the ionic strength as follows: fractions 0 ~ 150, 0.02 M buffer; 150 ~ 425, 0.1 M buffer; 425–end of protein absorbance at 280 nm 0.1 M buffer + 0.4 M NaCl. A peristaltic pump is used to maintain an even flow of 25–30 ml/hr and 12–15 ml fractions are collected. Two enzyme activities, protease I and protease II, are separated with the DEAE-Sephadex column and appropriate fractions are pooled according to Fig. 1.

Step 3. Fractionation on AcA 22. Each of the two protease solutions is concentrated to 10 ml by ultrafiltration as described above and the concentrated enzyme preparations run individually on an AcA 22 gel filtration column, 26 × 900 mm (LKB Ultragel AcA 22, LKB-Production AB, Stockholm, Sweden). The column is equilibrated and eluted with 0.05 M sodium acetate buffer, pH 5.0, using an upward flow rate of 3.8 ml hr^{-1} cm^{-2} to collect 10-ml fractions.

Step 4. Fractionation on Preparative Flat-Bed Electrofocusing. The chemicals, apparatus, preparation of gel, and the technique for preparative isoelectric focusing are the same as described by LKB (Application Note, 198). A 5% solution (100 ml) of Ultrodex (LKB-Production AB, Stockholm, Sweden) in 2% ampholine R, pH range 3.5–10, is poured on

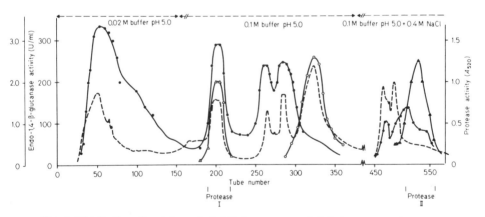

FIG. 1. Distribution of protein, endo-1,4-β-glucanase, exo-1,4-β-glucanase, and protease I and II activity after chromatography of concentrated culture solution from *S. pulverulentum* on DEAE-Sephadex A-50. The buffer was changed from 0.02 to 0.1 M ammonium acetate, pH 5.0, and to that containing 0.4 M NaCl at the points indicated. The fraction volume was 12–15 ml. (--) Absorbance 280 nm; (○) endo-1,4-β-glucanase; (●) exo-1,4-β-glucanase; (△) protease I; (▲) protease II.

the plate. This slurry is dried with the aid of a small fan until 36% of the water has evaporated. An electrode strip, soaked in 1 M H_3PO_4, is placed at the anodic side and another strip, soaked in 1 M NaOH, is placed at the cathodic side. The enzyme sample is mixed with the gel solution and applied to the flat-bed using a sample applicator (LKB-Production AB, Stockholm).

Electrofocusing is performed over a wide pH range, pH 3.5–10, with a constant power of 8 W and for a period of 16–18 hr. When the run is complete, a dry filter paper is placed on top of the gel bed. After 3 min the filter paper is removed and the number of protein bands estimated by staining the paper with 0.2% Coomassie Brilliant Blue R-250 dissolved in methanol/water/acetic acid (5 : 5 : 1). The same solvent mixture but without Coomassie Blue is used for destaining the paper. When the paper print of the protein distribution is ready, a fractionating grid is pressed through the gel bed. The pH values of the various gel fractions are measured with a surface pH electrode after which the gel sections are transferred with a spatula to small disposable syringes and diluted with 4 ml of cold distilled water. Protease activity and the absorbance at 280 nm are determined for all the fractions.

Step 5. Fractionation by Isoelectric Focusing in Econo Columns. Isoelectric focusing is performed using the method of Katsumata and Goldman[4] adapted to use Econo columns 0.7 × 15 cm (Bio-Rad Laboratories). The fractionation procedure is described in detail as follows.

Two Econo columns are connected with two 3-way nylon valves (Bio-Rad Laboratories). The upper column closures are fitted with platinum electrodes and cable connectors.[5] The stabilizing sucrose density gradient and balancing column are poured using the following procedure. Five milliliters of heavy cathode solution (10 ml of 70% sucrose, 0.1 ml concentrated H_2SO_4) is added to the balance column.[5] After the solution is passed several times in both directions through the connectors joining the two columns, by controlling the air pressure applied to the sample column,[5] the levels of liquid in the two columns are adjusted to 1 cm above the porous polyethylene bottom plates with excess material drained from the right valve. The left valve is closed to prevent flow between the two columns and the balance column is filled with medium cathode solution (15 ml 70% sucrose, 15 ml H_2O, 0.3 ml concentrated H_2SO_4) to the upper extent of the glass portion of the column. The reservoir of the balance column is filled to within 1 cm of the top with light cathode solution (30 ml H_2O, 0.1 ml concentrated H_2SO_4).

[4] M. Katsumata and A. S. Goldman, *Biochim. Biophys. Acta* **359,** 122 (1974).
[5] A. R. Ayers, S. B. Ayers, and K.-E. Eriksson, *Eur. J. Biochem.* **90,** 171 (1978).

The sample is prepared by applying 10% (1 ml) of the concentrated enzyme from the phenyl-Sepharose column on a PD10 column (Pharmacia Fine Chemicals) equilibrated with 0.8% ampholine ampholytes, pH 3.5–5.0 (LKB Produkter, Stockholm, Sweden). The sample is collected in a total of 2 ml of ampholytes. The sample column is prepared for the sample application by addition of 1 ml of heavy protection solution (1.5 ml 70% sucrose, 0.1 ml H_2O, 0.06 ml ampholine ampholytes, pH 3.5–5.0). The sample is applied in a step-gradient of sucrose prepared by mixing portions of the sample with heavy ampholyte solution (10 ml 70% sucrose, 0.4 ml ampholine ampholyte, pH 3.5–5.0) according to the following recipe: (1) 0.10-ml sample, 0.40 ml heavy ampholyte solution; (2) 0.15-ml sample, 0.35 ml heavy ampholyte solution; (3) 0.20-ml sample, 0.30 ml heavy ampholyte solution; (4) 0.25-ml sample, 0.25 ml heavy ampholyte solution; (5) 0.30-ml sample, 0.20 ml heavy ampholyte solution; (6) 0.35-ml sample, 0.15 ml heavy ampholyte solution; (7) 0.40-ml sample, 0.10 ml heavy ampholyte solution. Each step of progressively lower density in the sample gradient is gently layered upon the previous with a long-tipped Pasteur pipet. The upper portion of the sample gradient is protected from the strong basic conditions near the anode by filling the sample column to the upper extent of the glass portion of the column with light protection solution (1.3 ml H_2O, 0.2 ml 70% sucrose, 0.02 ml ampholine ampholytes, pH 3.5–5.0). The sample column reservoir is filled with light anode solution (20 ml H_2O, 0.08 g NaOH) to within 1 cm of the top.

The upper closures containing the electrodes are fitted to the tops of the columns and the passage between the columns is opened. The columns are suspended in an ice bath, the power supply is connected (anode to the left, low pH column, and cathode to the right, high pH, sample column), and power is applied according to the following regimen: 8 hr at 400 V (~0.7 mA), 16 hr at 600 V (~0.8 mA), and 24 hr at 1000 V (~0.9 mA). The sample column is fractionated by closing the connection between the two columns using the left-hand valve and collecting 3-drop fractions from the right-hand valve onto 1-ml aliquots of ice. Measurements of pH are made by equilibrating the fractions at 20° and measuring the pH with a combination glass electrode equilibrated and calibrated at the same temperature. After the addition of 1 ml of 50 mM sodium acetate buffer, pH 5.0, to each fraction, the absorbance at 280 nm and enzyme activity are measured. The peak fractions representing the peak absorbance at 280 nm coincide with the peaks of activity. These peaks are pooled and concentrated to 2 ml using immersible molecular separators (Millipore Corp.). The purified enzymes are placed in 50 mM sodium acetate buffer, pH 5.0, by passage through a PD10 column equilibrated with that buffer.

Data from a typical preparation of protease I and protease II, resulting in approximately 152- and 127-fold purifications respectively, and in yields of 1.8 and 1.2%, respectively, are summarized in Tables I and II.

Properties

General. Based on protein molecular weight standards, the molecular weights of protease I and protease II are estimated to be 28,000 and 26,000, respectively. The isoelectric points of protease I and protease II are pH 4.7 and 4.2, respectively.

The pH optima for protease I and protease II are 5.0 and 5.2, respectively. However, the enzymes are active over the pH range 3.5–6.5.

Substrate Specificity of Protease I and Protese II

Principle. The specificities of protease I and protease II are determined by analyzing the degradation products from human fibrinopeptide A. Human fibrinopeptide A is isolated under clotting conditions for highly purified human fibrinogen,[6] and contains 16 amino acid residues and has a molecular weight of 1536.

Reagents

Sodium acetate buffer, 50 mM, pH 5.0
Highly purified protease I giving rise to an absorbance of 0.25 at 520 nm after incubation of Azocol
Protease II giving rise to an absorbance of 0.275 at 520 nm after incubation of Azocol
Fibrinopeptide A, 5 mg/ml

Procedure. A solution containing 10 μl enzyme solution and 50 μl fibrinopeptide A solution is incubated overnight (18 hr) at 30°. Toluene is added as a preservative. References contain only the enzymes and 50 mM sodium acetate buffer at pH 5. After digestion of the fibrinopeptide A with the proteases, the respective hydrolyzates are freeze-dried. Prior to analysis, the freeze-dried hydrolyzates from the protease I and protease II incubations are dissolved in 5 and 50 μl, respectively, of distilled water and the samples subjected to both analytical and preparative thin-layer electrophoresis.[7] Identification of the fibrinopeptide A fragments obtained

---

[6] B. Blombäck, M. Blombäck, P. Edman, and B. Hessel, *Biochim. Biophys. Acta* **115,** 371 (1966).
[7] B. Blombäck, B. Hessel, J. Sadaaki, J. Reuterby, and M. Blombäck, *J. Biol. Chem.* **247,** 1496 (1972).

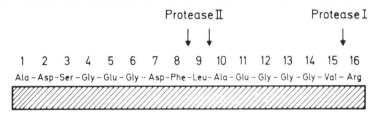

FIG. 2. Points of attack on human fibrinopeptide A by protease I and protease II.

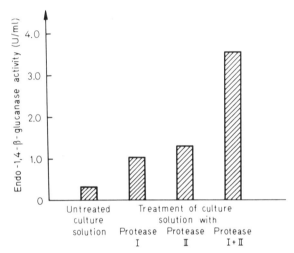

FIG. 3. Activation of endo-1,4-β-glucanases from *S. pulverulentum* after treatment with protease I and protease II.

after proteolysis with each individual protease is carried out according to the procedure of Blombäck *et al.*[7]

Two fragments are obtained with both protease I and protease II using thin-layer electrophoresis. The technique for identifying which amino acids are located at the peptide bonds susceptible to cleavage by the proteases is described in Ref. 8. It is clear from the results obtained that protease I splits off arginine in the C-terminal position, while protease II cleaves fibrinopeptide A on both sides of the leucine residue (Fig. 2).

Activation of Endo-1,4-β-glucanase with Protease I and Protease II

Cell-free culture solutions, obtained after 4 days cultivation of *S. pulverulentum* on 1% cellulose as the sole carbon source (cultivation

[8] K.-E. Eriksson and B. Pettersson, *Eur. J. Biochem.* **124,** 635 (1982).

technique, see above), are treated with protease I and protease II individually as well as with a mixture of both enzymes. Protease I increases endo-1,4-β-glucanase activity by a factor of 2.9, protease II by a factor of 3.9, and the mixture of protease I and protease II by a factor of 10.6 (Fig. 3). High reproducibility is obtained when the same culture solution is used. However, considerable variations in the increase of endoglucanase activity following protease treatment are seen between different culture solutions. The original cultivation time appears to be important in this context. Thus, the endoglucanases in the culture solution obtained after 4-day cultivation are activated to a much higher degree by protease treatment than the corresponding enzymes in a 10-day-old culture solution.

Section II

Hemicellulose

A. Preparation of Substrates for Hemicellulases
Articles 60 through 66

B. Analysis of β-Glucan and Enzyme Assays
Articles 67 through 69

C. Purification of Hemicellulose-Degrading Enzymes
Articles 70 through 93

[60] Purification of (1 → 3),(1 → 4)-β-D-Glucan from Barley Flour

By BARRY V. MCCLEARY

The major endosperm cell wall polysaccharide in barley and oats is a linear (1 → 3),(1 → 4)-β-D-glucan.[1,2] In barley it represents approximately 75% of the carbohydrate in endosperm cell walls. It is generally considered that the majority of the polysaccharide consists of two or three 1,4-β-linked D-glucosyl residues, joined by single 1,3-β-linkages.[2,3] The presence of lesser proportions of longer stretches of 1,4-β-linked D-glucosyl residues has been demonstrated[4,5] but the possible presence of sequences of 1,3-β-linked D-glucosyl residues[6-8] remains to be confirmed.

Although this mixed-linkage β-glucan represents only from 2.8 to 5.5% of the total weight of the barley grain, by virtue of its ability to produce solutions of high viscosity and gelatinous precipitates it is functionally significant in the malting and brewing processes.[9] Barley flour contains mixed-linkage β-glucan fractions which vary in their ease of extraction. The reasons for this have not been clearly defined but some authors suggest that it might be due to binding of a proportion of the glucan to protein,[10] or to peptide cross-linking.[11] Variation in the ease of extraction of mixed-linkage β-glucan may also be due to structural variation in the glucan itself, i.e., β-glucan subfractions with a high proportion of regions of sequential (1 → 4)-β-linked D-glucosyl residues could be expected to have a more celluloselike nature and thus to be less soluble.

Methods employed for the extraction and purification of barley β-glucan are generally modifications of the procedure described by Preece and Mackenzie.[12] In this procedure, after inactivation of enzymes by

[1] S. Peat, W. J. Whelan, and J. G. Roberts, *J. Chem. Soc.* p. 3916 (1957).

[2] F. W. Parrish, A. S. Perlin, and E. T. Reese, *Can. J. Chem.* **38**, 2094 (1960).

[3] P. Dais and A. S. Perlin, *Carbohydr. Res.* **100**, 103 (1982).

[4] W. W. Luchsinger, S.-C. Chen, and A. W. Richards, *Arch. Biochem. Biophys.* **112**, 531 (1965).

[5] J. R. Woodward, G. B. Fincher, and B. A. Stone, *Carbohydr. Polym.* **3**, 207 (1983).

[6] M. Fleming and D. J. Manners, *Biochem. J.* **100**, 4P (1966).

[7] M. Fleming and K. Kawakami, *Carbohydr. Res.* **57**, 15 (1977).

[8] G. N. Bathgate, G. H. Palmer, and G. Wilson, *J. Inst. Brew.* **80**, 278 (1974).

[9] C. W. Bamforth, *Brew. Dig.* **57**, 22 (1982).

[10] I. S. Forrest and T. Wainwright, *J. Inst. Brew.* **83**, 279 (1977).

[11] C. W. Bamforth, H. L. Martin, and T. Wainwright, *J. Inst. Brew.* **85**, 334 (1979).

[12] I. A. Preece and K. G. Mackenzie, *J. Inst. Brew.* **58**, 353 (1952).

refluxing the flour in boiling 80% (v/v) ethanol, the flour is extracted with water at 40°. Such a procedure gives $(1 \rightarrow 3),(1 \rightarrow 4)$-$\beta$-glucan containing minimal quantities of starch, but the amount of β-glucan extracted may be as little as 20% of that present in the flour. Extraction efficiency may be related to the time and conditions of storage of the grain or flour, or to conditions of pretreatment of the grain before extraction. Exhaustive extraction procedures have, however, been applied to isolated endosperm cell wall preparations[13,14] which are essentially devoid of starch and protein.

In the current chapter, a modification of the method of Preece and Mackenzie[12] which allows the large scale, essentially quantitative extraction of mixed-linkage β-glucan from barley flour is described. The same format can also be applied to the extraction of β-glucan from oat flour.

Extraction and Purification of $(1 \rightarrow 3),(1 \rightarrow 4)$-$\beta$-D-Glucan
   (Barley β-Glucan)

Step 1. Pretreatment and Inactivation of Endogenous Enzymes. Barley grain (var. Parwan) is milled to pass a 0.5-mm screen. Two hundred grams of this flour is suspended in 1 liter of 80% (v/v) aqueous ethanol and the slurry incubated in a steam bath for 20 min. The slurry is filtered and washed with 1 liter of 80% (v/v) aqueous ethanol and then with ethanol [95% (v/v)].

Step 2. Extraction at 40°. The alcohol-washed flour is suspended in 1 liter of water and incubated and stirred at 40° for 1 hr. The slurry is homogenized at minimum setting in a Waring blender for 1 min, centrifuged at 3000 g for 15 min, and the pellet suspended in 500 ml of water, homogenized again for 1 min, and centrifuged. The combined supernatants are treated with 1 ml of heat-treated Hitempase (a temperature stable α-amylase, available from Biocon Biochemicals Ltd., Carrigaline, Co. Cork, Ireland) at 80° for 30 min. Before use, Hitempase is incubated at 85° for 80 min to inactivate trace quantities of endo-β-glucanase activity. This treatment gives no significant loss in α-amylase activity.

Step 3. Purification of the 40° Extract. After incubation with Hitempase the solution is cooled to room temperature and treated with ammonium sulfate (30 g/100 ml) and stored at 4° for 20 hr. The precipitate which forms is collected by centrifugation at 3000 g for 10 min, suspended in 500

[13] G. M. Ballance and D. J. Manners, *Carbohydr. Res.* **61**, 107 (1978).
[14] B. Ahluwalia and E. E. Ellis, *in* "New Approaches to Research on Cereal Carbohydrates" (R. D. Hill and L. Munck, eds.), pp. 285–290. Elsevier, Amsterdam, 1985.

ml of 20% (v/v) aqueous ethanol, and blended using an Ultraturrax homogenizer. The precipitate is recovered by centrifugation at 3000 g for 10 min and washed again with 20% (v/v) aqueous ethanol and then with ethanol. The material recovered on centrifugation at 3000 g for 10 min is dissolved in water at 80° and filtered under vacuum through a bed of celite to remove a turbidity. The filtrate is treated with an equal volume of ethanol and the recovered precipitate is washed with ethanol and acetone and dried *in vacuo*.

Step 4. Extraction at 65°. The residue remaining after aqueous extraction at 40° is suspended in 1 liter of water at 65° and stirred gently at this temperature for 1 hr. The slurry is then homogenized in a Waring blender at minimum setting for 1 min and centrifuged at 3000 g for 15 min. The pellet is reextracted in 1 liter of water at 65° using a Waring blender (minimum setting, 1 min) and centrifuged. The combined supernatants are incubated with heat-treated Hitempase (2 ml) at 80° for 30 min.

Step 4. Purification of the 65° Extract. This is purified by the same format as employed for the 40° extract.

Step 5. Extraction at 90–95°. The pellet remaining after aqueous extraction at 65° is suspended in 1 liter of hot water and treated with 5 ml of heat-treated Hitempase. The slurry is incubated in a boiling water bath for 30 min during which time the temperature increases to 90–95° and there is complete depolymerization of starch (as shown by the iodine test). The slurry is homogenized in a Waring blender at minimum setting for 1 min and centrifuged at 3000 g for 15 min. The pellet is then reextracted and the combined supernatants treated with ammonium sulphate at a rate of 30 g/ 100 ml of solution.

Step 6. Purification of the 90–95° Extract. This is purified by the same format as employed for the 40° extract.

Step 7. Extraction with Sodium Hydroxide. The residue remaining on aqueous extraction at 90–95° is suspended in 1 liter of 5% (w/v) NaOH plus 0.5% (w/v) sodium borohydride and stirred gently at 22° for 16 hr. The slurry is homogenized in a Waring blender at minimum setting for 1 min and centrifuged at 3000 g for 15 min. The pellet is reextracted with 1 liter of the same solution and on centrifugation the supernatants are combined and neutralized with 50% (v/v) acetic acid. Crushed ice is added to keep the temperature below 40°. The solution is treated with 2 ml of heat-treated Hitempase at 80° for 30 min.

Step 8. Purification of the Sodium Hydroxide Extract. The same format as used for the 40° extract is employed.

Step 9. Recovery of the Residue. The residue after NaOH extraction is suspended in water and dialyzed against flowing tap water for 16 hr. It is then washed with ethanol and acetone, dried *in vacuo*, and samples analyzed for $(1 \rightarrow 3),(1 \rightarrow 4)$-$\beta$-glucan.[15]

Recovery and Purity

The relative amounts of total barley β-glucan extracted at 40, 65, and 90–95° and with NaOH from flour of the barley variety Parwan are 17, 60, 18, and 5%, respectively. Less than 1% of the β-glucan remains in the residue. However, these proportions vary for different barley varieties and even for the same variety depending on the milling procedure and possibly also on how the barley or flour is stored. Studies on a number of barley flour samples indicate that the proportion of total barley β-glucan in the 40° extract is 20–30%, in the 65° extract is 50–60%, in the 90–95° extract is 5–20%, and in the alkali extract is 5–20%. The amount remaining in the residue is 1–2%. The protein content of barley β-glucan extracted at 40, 65, and 90–95° is less than 0.6% (w/w) and contamination with arabinoxylan is less than 1% (w/w). The fractions are devoid of starch as shown by the absence of glucose on treating samples with amyloglucosidase and the lack of color formation on treatment with iodine/potassium iodide solution. Treatment of each of the fractions with highly purified cellulase [endo-$(1 \rightarrow 4)$-β-glucanase] from *Aspergillus niger* gives complete hydrolysis to oligosaccharides of DP < 5.

The purified β-glucan fractions differed only in their viscosity properties. In general, fractions extracted at higher temperatures had the same or higher intrinsic viscosities than those extracted at lower temperatures. Intrinsic viscosity values ranged between 2 and 7 dl/g but are generally in the range 4–6 dl/g. The patterns of amounts of oligosaccharides produced on hydrolysis by *Bacillus subtilis* lichenase of each of the water-soluble β-glucan fractions are indistinguishable as also are the patterns produced on hydrolysis by *Aspergillus niger* cellulase.

Mixed-linkage β-glucan extracted from the oat variety Cooba had properties very similar to that extracted from barley. Intrinsic viscosities of 7.3, 8.0, and 7.2 dl/g were obtained for the 40, 65, and 90–95° extracts, respectively. The patterns of amounts of oligosaccharides produced on hydrolysis by *Bacillus subtilis* lichenase or *Aspergillus niger* cellulase were very similar to those obtained on hydrolysis of the barley β-glucan fractions by these enzymes.

[15] B. V. McCleary and M. Glennie-Holmes, *J. Inst. Brew.* **91**, 285 (1985).

[61] Synthesis of β-D-Mannopyranosides for the Assay of β-D-Mannosidase and Exo-β-D-mannanase

By BARRY V. McCLEARY

Nitrophenyl-, naphthyl-, and methylumbelliferylglycosides are of value as convenient substrates for the assay of glycosidase and exoglycanase activity, and in enzyme-linked reactions they can be used to assay certain endoglycanases.[1] The preparation of such derivatives of β-D-mannopyranose, however, has posed considerable problems.

p-Nitrophenyl-β-D-mannopyranoside has been obtained from the reaction of penta-O-acetyl-α,β-D-mannopyranoside with p-nitrophenol under the conditions described by Helferich and Schmitz-Hillebrecht.[2] But, isolation of the β-anomer (7–8%) from the Helferich reaction mixture which contains mainly α-anomer (60–65%) is difficult. Rosenfeld and Lee[3] resolved this problem by chromatography of the deacetylated Helferich reaction products on the ion-exchange resin, Dowex 50. Alternatively, the anomeric mixture of p-nitrophenyl 2,3,4,6-tetra-O-acetyl-D-mannopyranosides can be separated by chromatography on a silica gel column.[4]

An alternative approach has involved the development of improved reaction procedures or new synthetic routes to increase the proportion of the β-anomer. Thus, Maley[5] produced p-nitrophenyl-β-D-mannopyranoside almost exclusively by reacting 4,6-di-O-acetyl-2,3-O-carbonyl-α-D-mannopyranosyl bromide with p-nitrophenol in the presence of silver oxide. A number of synthetic steps are involved and in the hands of the current author, yields at certain steps were quite low. The possibility of producing a variety of β-D-mannopyranosides by taking advantage of the effect of the particular leaving group at C-1 and the nonparticipating group at C-2 of D-mannopyranosyl derivatives to alter the rate and stereoselectivity of reactions at C-1 has been discussed.[6] However, application of this theory to the synthesis of p-nitrophenyl-β-D-mannopyranoside required a number of synthetic steps and chromatographic separation of intermediate reaction products.

[1] K. Wallenfels, B. Meltzer, G. Laule, and G. Janatsch, *Fresenius Z. Anal. Chem.* **301,** 169 (1980).
[2] B. Helferich and E. Schmitz-Hillebrecht, *Ber.* **66,** 378 (1933).
[3] L. Rosenfeld and Y.-C. Lee, *Carbohydr. Res.* **46,** 155 (1976).
[4] K. Kawaguchi and N. Kahimura, *Agric. Biol. Chem.* **40,** 241 (1976).
[5] F. Maley, *Carbohydr. Res.* **64,** 279 (1978).
[6] V. K. Srivastava and C. Schuerch, *Carbohydr. Res.* **79,** C13 (1980).

In contrast, the two-step procedure developed by Garegg et al.[7] gives high yields of p-nitrophenyl-β-D-mannopyranoside and the α-anomer is readily removed by recrystallization of the desired compound. A simple modification of this basic procedure to allow the synthesis of O-nitrophenyl, methylumbelliferyl, and naphthyl derivatives of β-D-mannopyranoside will now be described.[8] The methylumbelliferyl derivative is particularly useful for the measurement of trace quantities of β-D-mannosidase,[9] and the naphthyl derivative finds use in staining procedures specific for β-D-mannosidase.[8] In the procedure of Garegg et al.[7] D-mannose is first converted into 2,3 : 4,6-di-O-cyclohexylidene-α-D-mannopyranose, which upon treatment with triphenylphosphine, p-nitrophenol, and diethyl azodicarboxylate, followed by mild acid hydrolysis, yields crystalline p-nitrophenyl-β-D-mannopyranoside.

Synthesis of 1,1-Diethoxycyclohexane[10]

In a 2-liter round-bottomed flask, bearing a reflux condenser, is placed cyclohexanone (200 ml), triethyl orthoformate (416 ml), absolute ethanol (400 ml), and p-toluenesulfonic acid (250 mg). The flask is placed in a water bath and refluxed for 1 hr. Sodium ethoxide in ethanol is then added until the mixture is alkaline. The bulk of solvent is removed on a rotary evaporator below 35°. The resulting mixture is distilled through a Vigreux column to give diethyoxycyclohexane (251 g), boiling point 72–78° at 15 mm Hg.

1-Ethoxycyclohexene

In a 1-liter round-bottomed flask, bearing a Vigreux column, are placed diethoxycyclohexane (200 ml) and a catalytic amount of p-toluenesulfonic acid (25–50 mg). The flask is placed in an oil bath at 80° and evacuated to 15 mm Hg, and the resulting ethanol removed over 2–3 hr. The temperature of the oil bath is then raised and the 1-ethoxycyclohexene is distilled at a boiling point of approximately 65° at 15 mm Hg, with a yield of approximately 150 g.

[7] P. J. Garegg, T. Iversen, and T. Norberg, *Carbohydr. Res.* **73**, 313 (1979).
[8] B. V. McCleary, *Carbohydr. Res.* **101**, 75 (1982).
[9] P. J. Healy and B. V. McCleary, *Res. Vet. Sci.* **33**, 73 (1982).
[10] This was prepared exactly according to a recommendation by Dr. T. Iversen. Details given here were not presented in the original paper.[7]

2,3 : 4,6-Di-O-Cyclohexylidene-α-D-mannopyranose

This is prepared as described by Garegg et al.[7] 1-Ethoxycyclohexene (50.5 g) is added dropwise, over a period of 1 hr, to a stirred solution of D-mannose (18 g) and p-toluenesulfonic acid monohydrate (0.5 g) in dry N,N-dimethylformamide (250 ml) at 40°. The solution is kept at 13 mm Hg, and the ethanol formed is continuously removed. The solution is stirred at this pressure and temperature for a further 1 hr, diluted with diethyl ether, washed with saturated aqueous sodium hydrogen carbonate, and water, and then concentrated at reduced pressure. The crystalline residue is recrystallized from light petroleum (40–60°) dichloromethane to yield 2,3 : 4,6-di-O-cyclohexylidene-α-D-mannopyranose in a yield of approximately 17.0 g (49%), mp 174–175°, $[\alpha]_D^{20}$ − 33° (c = 1, chloroform).[7]

Nitrophenyl-, Naphthyl-, and Methylumbelliferyl-β-D-mannopyranosides

The procedure of Garegg et al.[7] for the synthesis of p-nitrophenyl-β-D-mannopyranoside is employed with the modification that p-nitrophenyl is replaced by equimolar quantities of O-nitrophenol, 1-naphthol, or methylumbelliferone. A solution of diethyl azodicarboxylate (6.5 g) in toluene (50 ml) is added dropwise over a period of 15 min to a stirred solution of 2,3 : 4,6-di-O-cyclohexylidene-α-D-mannopyranose (9 g), triphenylphosphine (10 g), and p-nitrophenol (5.2 g) [or O-nitrophenol (5.2 g), or finely ground 1-naphthol (5.4 g), or methylumbelliferone (6.6 g)] in toluene (200 ml) at room temperature. The solution is stirred for a further 1 hr at room temperature and then concentrated. A solution of the residue in acetic acid (100 ml) and water (25 ml) is heated at 100° for 2 hr and then concentrated at reduced pressure. The residue is dissolved in water and this solution washed with chloroform.

For p-nitrophenyl and O-nitrophenyl derivatives, concentration of the aqueous phase after chloroform washing gave either crystalline p-nitrophenyl-β-D-mannopyranoside or O-nitrophenyl-β-D-mannopyranoside. The latter was recovered in 25% yield and had a melting point of 161–163°. (Found: C, 46.44; H, 5.30; N, 4.47. $C_{12}H_{15}NO_8$ calc.: C, 47.84; H, 5.02; N, 4.65%.)

Naphthyl-β-D-mannopyranoside crystallizes in the aqueous phase while this is being extracted with chloroform. The aqueous phase is concentrated to dryness and the residue is twice recrystallized by dissolving (at 80°) in 50% aqueous acetone, increasing the acetone concentration to

90%, and storing the solution at 4° for 4 days. Yield, 30%; mp 237–239°. (Found: C, 62.61; H, 5.98. $C_{16}H_{19}O_6$ calc.: C, 62.76; H, 5.92%.)

Crude methylumbelliferyl-β-D-mannopyranoside crystallizes on concentration of the aqueous phase (after chloroform extraction). The product is dissolved in 50% aqueous acetone (0.5 g product/30 ml) at 80°. On cooling, the acetone concentration is increased to 80% and the solution stored at −20° for several weeks. Crystallization begins within 1 day; yield 26%; mp 240–242°. (Found: C, 56.25; H, 5.43. $C_{16}H_{19}O_8$ calc.: C, 56.80; H, 5.67%.)

Each of these substrates, after recrystallization, is devoid of the α-anomer as demonstrated by its resistance to hydrolysis by jack bean α-D-mannosidase (Sigma Chemical Co., Catalog No. M7275).

[62] Enzymatic Preparation of β-1,4-Mannooligosaccharides and β-1,4-Glucomannooligosaccharides

By Isao Kusakabe and Rihei Takahashi

Preparation of Mannooligosaccharides

Mannooligosaccharides [O-β-D-Man$p$-(1 → [4-O-β-D-Man$p$-1]$_n$ → 4)-D-Man$p$] have been prepared from partial acid and enzymatic hydrolyzates of plant mannans such as ivory nut, guaran, coffee bean, white spruce, and lucerne seeds.[1-6] This ordinary preparation method, however, does not yield a substantial quantity of oligosaccharides. We describe, in this section, the preparation method for mannooligosaccharides from copra mannan using the mannanase system from *Streptomyces* sp. No. 17.

Copra Mannan

The mannan was prepared by alkali extraction of coconut residual cake, which is a by-product of oil extraction from copra and was supplied by Blue Bar Inc. (Philippines). The residual cake contained about 60%

[1] G. O. Aspinall, R. B. Rashbrook, and G. Kessler, *J. Chem. Soc.* p. 215 (1975).
[2] R. L. Whistler and D. F. Durso, *J. Am. Chem. Soc.* **73**, 4189 (1951).
[3] A. Tyminski and T. E. Timell, *J. Am. Chem. Soc.* **82**, 2823 (1960).
[4] J. E. Courtois, F. Peter, and T. Kada, *Bull. Soc. Chem. Biol.* **40**, 2031 (1958).
[5] M. E. Henderson, L. Hough, and T. J. Painter, *J. Chem. Soc.* p. 3519 (1958).
[6] J. K. N. Jones and T. J. Painter, *J. Chem. Soc.* p. 669 (1957).

carbohydrate, 20% protein, and other components. The carbohydrate was composed of about 70% mannose (β-1,4-mannan), 20% glucose (cellulose), and small amounts of arabinose and galactose.

Twenty liters of 24% NaOH was added to 1 kg of the residual cake in a stainless-steel bucket, and the mixture was occasionally stirred to extract the mannan for 24 hr at room temperature. Then, the slurry was filtered through a cloth bag. The filtrate was neutralized with about 12 N H_2SO_4 at low temperature until the pH of the solution was about 5.5. The resultant precipitate (copra mannan), collected by centrifugation, and containing about 200 g of mannan, was dialyzed against tap water to remove salts.

Enzyme Source

A crude β-mannanase[7,8] (culture filtrate) from *Streptomyces* sp. No. 17 was used for a partial hydrolysis of the mannan.

Partial Hydrolysis of Copra Mannan

The mannan, 60 g (46.5 g as polymannose), was added to 2500 ml of the enzyme solution (total activity, 15,000 units) plus 4200 ml of water. The mixture was incubated at 40° and pH 6.8 for 3 hr with agitation at 120 rpm. After separation of nonliquefied substances by centrifugation, the supernatant solution containing 45.7 g of total sugar was concentrated to 1860 ml using a rotary evaporator at 45°. The sugar composition of the solution was 3.3% mannose, 42% β-1,4-mannobiose, 20% β-1,4-mannotriose, 13.3% β-1,4-mannotetraose, and 21.4% other oligosaccharides.[9]

Chromatographic Separation, Crystallization, and Properties
  of Mannooligosaccharides

The concentrate (1750 ml) obtained above was applied to a granular charcoal column (70 × 680 mm, 500 g of activated charcoal for chromatography, Wakô Pure Chemical Ind., Japan). After the column was washed well with water to remove mannose and salts, the oligosaccharides in the column were eluted with 5 liters each 2.5, 5, 7.5, 10, and 15% (v/v) ethanol. The effluent was collected in 500-ml fractions, and the sugar distribution of each fraction was examined by paper chromatography with the solvent system of *n*-butanol–pyridine–water (6:4:3).

[7] R. Takahashi, I. Kusakabe, A. Maekawa, T. Suzuki, and K. Murakami, *Jpn. J. Trop. Agric.* **27**, 140 (1983).
[8] I. Kusakabe and R. Takahashi, this volume [75].
[9] I. Kusakabe, R. Takahashi, K. Murakami, A. Maekawa, and T. Suzuki, *Agric. Biol. Chem.* **47**, 2391 (1983).

The mannobiose fractions (2.5 and 5% effluents) were combined and concentrated to a syrup to remove the ethanol. The mannobiose was further purified by chromatography on a granular-charcoal column (55 × 530 mm) with a linear gradient consisting of 8 liters of water and 8 liters of 7.5% ethanol. The effluent was collected in 275-ml fractions and the sugar composition of each fraction was examined by paper chromatography. Chromatographically identical fractions were combined and concentrated to syrup as above. Hot absolute ethanol was added to the syrup to bring the concentration to about 85% ethanol. Upon cooling, crystallization occurred, yielding 11.2 g of crystalline mannobiose. The mannobiose showed the following properties[9]: $[\alpha]_D = -8°$; mp $= 198°$.

The mannotriose fractions (7.5% ethanol effluent) were rechromatographed by the same method, with a linear gradient of 0 ~ 12% ethanol. The mannotriose fractions thus obtained were combined, concentrated, and crystallized by the method described above, and 5.2 g of crystalline mannotriose was obtained[9] ($[\alpha]_D = -22°$; mp $= 219°$).

The mannotetraose fractions (10% ethanol effluent) were rechromatographed on a charcoal column with a linear gradient of 0 ~ 18% ethanol, following the method previously described. The mannotetraose fractions were combined and concentrated, and 2.6 g of mannotetraose was crystallized. The sugar showed the following properties[9]: $[\alpha]_D = -29°$; mp $= 228°$.

Preparation of Glucomannooligosaccharides

The mannanase of *Streptomyces* has been observed to rapidly decrease the viscosity of viscous polysaccharides such as konjac glucomannan, guar gum, and locust bean galactoglucomannans. The enzyme is a typical endoenzyme. Therefore, the products arising from the enzymatic degradation of the substrate are heterooligosaccharides, that is, various glucomanno- or galactomannooligosaccharides. In this section, we describe the preparation[10] of glucomannooligosaccharides from konjac glucomannan.

Glucomannan

Konjac (*Amorphophalus konjac*) glucomannan was supplied by Tsuruta Shokuhin Kôgyô Co., Ltd. (Gunma-ken, Japan). The ratio of mannose to glucose of the mannan was 1 : 0.6.

---

[10] R. Takahashi, I. Kusakabe, S. Kusama, Y. Sakurai, K. Murakami, A. Maekawa, and T. Suzuki, *Agric. Biol. Chem.* **48**, 2943 (1984).

Enzyme Source

The culture filtrate (mannanase) of *Streptomyces* sp. No. 17 was used as the enzyme source. The preparation method was described previously.[7,8]

Partial Hydrolysis of Konjac Glucomannan

To 70 g of the mannan (58 g as anhydroglucomannose) was added about 1 liter of the mannanase solution (total units, 104,000) adjusted to pH 6.8, and the mixture was incubated at 45° with agitation, in a 1.5-liter glass vessel. Figure 1 shows a time course of the enzymatic reaction.

Chromatographic Separation, Crystallization, and Properties
  of Glucomannooligosaccharides

The enzymatic hydrolyzate after 24 hr reaction was heated to about 100° for 5 min to inactivate the enzyme. Then, the hydrolyzate was subjected to centrifugation to remove the insoluble materials. The resultant clear solution, containing 60 g of total sugar, was applied to a granular charcoal column (70 × 680 mm, 500 g of activated charcoal for chromatography). The column was then washed with 8 liters of water to remove mannose and salts. The oligosaccharides in the column were eluted by a linear gradient of 0 ~ 25% ethanol (total volume, 30 liters). The eluent was collected in 500-ml fraction tubes, and the sugar composition in each fraction tube was analyzed by paper chromatography with the solvent

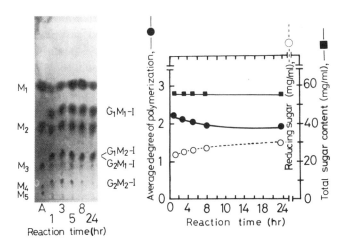

FIG. 1. Time course of hydrolysis of konjac glucomannan by a mannanase from *Streptomyces* No. 17.

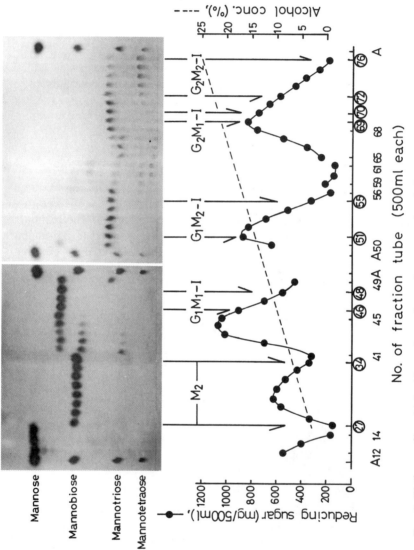

Fig. 2. Elution profiles obtained by carbon column chromatography of glucomannooligosaccharides resulting from the enzymatic hydrolysis of konjac glucomannan.

system of n-butanol–pyridine–water (6 : 4 : 3). Figure 2 shows the elution profiles of the glucomannan digest obtained by the column chromatography.

G_1M_1.[11] The contents of fractions 46 ~ 48 were combined and concentrated to a syrup. Hot absolute ethanol was added to the syrup to bring the concentration to about 85% ethanol. Upon cooling, crystallization occurred, and 2.4 g of crystalline G_1M_1 was obtained. The sugar showed the following properties: $[\alpha]_D = +5.4°$; mp = 136 ~ 138°.

G_1M_2. The contents of fractions 51 ~ 55 were combined, concentrated, and crystallized by the method described above; 3.5 g of crystalline G_1M_2 was obtained ($[\alpha]_D = -12.1°$; mp = 174 ~ 176°).

G_2M_1. As described above, 1.3 g of crystalline G_2M_1 was obtained from fractions 69 and 70 ($[\alpha]_D = -4.4°$; mp = 249 ~ 251°).

G_2M_2. The contents of fractions 72 ~ 76 were combined and concentrated, then applied to preparative paper chromatography on No. 527 filter paper sheets (Tôyô Roshi, Japan). Chromatographically identical sugar zones were extracted with water from the filter paper, and the resultant extract was decolorized with activated charcoal, followed by deionization with Amberlite IR-200c and IRA-68 resins. The purified sugar solution thus obtained was concentrated to a syrup, and 1.0 g of G_2M_2 was crystallized by the same method as described above ($[\alpha]_D = -9.8°$; mp = 188 ~ 191°).

[11] Abbreviations used: G_1M_1, O-β-D-glucopyranosyl-(1 → 4)-D-mannopyranose (epicellobiose); G_1M_2, O-β-D-glucopyranosyl-(1 → 4)-O-β-D-mannopyranosyl-(1 → 4)-D-mannopyranose; G_2M_1, O-β-D-glucopyranosyl-(1 → 4)-O-β-D-glucopyranosyl-(1 → 4)-D-mannopyranose; G_2M_2, O-β-D-glucopyranosyl-(1 → 4)-O-β-D-glucopyranosyl-(1 → 4)-O-β-D-mannopyranosyl-(1 → 4)-D-mannopyranose.

[63] Carob and Guar Galactomannans

By Barry V. McCleary

Galactomannans are reserve carbohydrates found in the endosperm cell walls of some legume seeds. They consist of a (1 → 4)-linked β-D-mannan backbone to which is attached single α-D-galactosyl residues at O-6 of some of the D-mannosyl residues.[1] The extent of D-galactosyl substitution of the D-mannan backbone varies from 0 to 10% in the "man-

[1] F. Smith and R. Montgomery, "The Chemistry of Plant Gums and Mucilages," p. 324. Reinhold, New York, 1959.

METHODS IN ENZYMOLOGY, VOL. 160

nans'' from palm seeds to almost complete substitution in the polymers from seed of some of the Trifolieae.[2,3] The average ratio of D-galactose to D-mannose and the pattern of distribution of the D-galactosyl groups along the mannan backbone appear to be species specific. These parameters also dictate the ease of dissolution of the polysaccharide and the degree of self-association and the extent of synergestic interaction with agarose, xanthan, and κ-carrageenan.

In this chapter, procedures for the quantitative extraction of hot and cold water-soluble fractions of carob galactomannan and of guar galactomannan will be described. Information on the fine structure and the solution properties will also be presented.

Carob Galactomannan

Extraction and Purification

Step 1. Inactivation of Endogenous Enzymes. Carob (*Ceratonia siliqua*) seed is milled to pass a 0.7-mm mesh screen. Ten grams of milled whole seed or of commercial flour (gum, locust bean, Sigma G0753) is suspended in 200 ml of 80% (v/v) aqueous ethanol and incubated in a boiling water bath for 10 min. The slurry is filtered and the residue washed with ethanol and acetone.

Step 2. Extraction and Purification of Cold Water-Soluble Carob Galactomannan. The solvent extracted flour is treated with 2 ml of ethanol, dispersed in 200 ml of cold water, and stored at 4° for 20 hr. A further 300 ml of water is added, the temperature adjusted to about 22° (room temperature), and the slurry is homogenized using a Waring blender at maximum speed for 2 min. On centrifugation at 4000 g for 30 min the supernatant is added to 2 volumes of ethanol and the pellet is reextracted (usually twice) in 400 ml of water at 22° until no further polysaccharide precipitates from the supernatant solution on addition to 2 volumes of ethanol. The combined precipitates are washed with ethanol and redissolved in water to a concentration of approximately 0.2% (w/v). On centrifugation at 15,000 g for 30 min the polysaccharide in the supernatant solution is precipitated with ethanol (2 volumes), washed with ethanol, acetone, and hexane, and dried *in vacuo*.

Step 3. Extraction and Purification of Hot Water-Soluble Carob Galactomannan. The pellet remaining after extraction of carob flour with cold water is suspended in hot water (500 ml) and incubated in a boiling

[2] I. C. M. Dea and A. Morrison, *Adv. Carbohydr. Chem. Biochem.* **31,** 241 (1975).
[3] P. M. Dey, *Adv. Carbohydr. Chem. Biochem.* **35,** 341 (1978).

water bath for 30 min. The slurry is homogenized by using a Waring blender at maximum setting for 2 min. The homogenate is centrifuged at 4000 g for 30 min and the pellet reextracted with 400 ml of hot water (usually twice) until no further polysaccharide precipitates from the supernatant solution on addition of 2 volumes of ethanol. The precipitated polysaccharide (hot water-soluble galactomannan) is purified as for the cold water-soluble galactomannan.

Recovery and Purity

The total yield of galactomannan from commercial carob flours ranges from 55 to 70% of flour weight, and the yield from whole-seed flour is 18–30% (w/w). The ratio of hot water-soluble to sold water-soluble galactomannan varies from 84 : 16 to 30 : 70. Acid hydrolysis and gas–liquid chromatography of the alditol acetates show that galactose and mannose together represent more than 97% of the sugars present in the polysaccharide samples. The presence of arabinose and xylose in the hydrolyzate (<3% of total monosaccharides) indicates that the major contaminating polysaccharide is arabinoxylan. This polysaccharide is not removed by precipitation of galactomannan from the copper complex[4] as the arabinoxylan also precipitates. However, if desired, it can be removed by treatment of a solution of galactomannan with highly purified endo-1,4-β-D-xylanase.[5] Galactomannans prepared by the described procedure are usually contaminated to an extent of <2% with protein and are suitable for use as substrates for the assays of β-mannanase.

Properties

In general, there appears to be insignificant differences between the hot water-soluble carob galactomannan fractions extracted from different seed varieties, different commercial flours, or from seed originating from a number of different countries.[6] The content of D-galactose is 18 ± 2%, the limiting viscosity is 12 ± 2 dl/g, and the degree of hydrolysis by *Aspergillus niger* β-mannanase is 26 ± 1%. The cold water-soluble fraction has properties which are quite distinct from those of the hot water-soluble fraction, but the variation in properties within this fraction are minimal: the D-galactose percentage is 25 ± 1%, the limiting viscosity is 10 ± 1 dl/g, and the degree of hydrolysis by *A. niger* β-mannanase is 20 ± 2%.

[4] P. Andrews, L. Hough, and J. K. N. Jones, *J. Am. Chem. Soc.* **74,** 4029 (1952).
[5] T. S. Gibson and B. V. McCleary, *Carbohyd. Polym.* **7,** 225 (1987).
[6] B. V. McCleary, A. H. Clark, I. C. M. Dea, and D. A. Rees, *Carbohydr. Res.* **139,** 237 (1985).

Treatment of each of the polysaccharide fractions with guar seed α-galactosidase (20 U/10 mg polysaccharide) plus *A. niger* β-mannanase (20 U/10 mg polysaccharide) at 40° for 8 hr gives complete hydrolysis to D-galactose, (1 → 4)-β-D-mannobiose and (1 → 4)-β-D-mannotriose with trace quantities of D-mannose, consistent with the accepted structure for this polysaccharide.[1,7]

The distribution of D-galactosyl residues along the D-mannan backbone (fine structure) of the carob galactomman fractions has been studied by a detailed computer analysis[6] of the amounts and structures of oligosaccharides released on hydrolysis of the polymers with highly purified β-mannanases isolated from germinating guar seed and from *Aspergillus niger* preparations.[8] Oligosaccharides released on hydrolysis of the carob galactomannan fractions with either of the β-mannanases are quantitatively separated by gel permeation chromatography and characterized using a combination of enzymatic, NMR and chemical procedures.[9,10] From these studies a pattern of binding between the β-mannan chain and the enzyme can be deduced[8] allowing the development of computer programs which account for the specific subsite-binding requirements of the β-mannanases.[6] Using these programs, and others which simulate the synthesis of galactomannan, we have shown that the D-galactose in carob galactomannan and in the hot and cold water-soluble fractions of carob galactomannan is distributed in a nonregular pattern with a high proportion of substituted couplets, lesser amounts of triplets, and an absence of blocks of substitution. The probability of sequences in which alternate D-mannosyl residues are substituted is low. The probability distribution of block sizes of unsubstituted D-mannosyl residues indicates that there is a higher proportion of blocks of intermediate size than would be present in a galactomannan with a statistically random D-galactose distribution.

Guar Galactomannan

Extraction and Purification

Commercial guar flour (Sigma G4129) or guar seed (milled to pass a 0.7-mm screen) is treated with boiling aqueous ethanol and extracted exactly as for carob flour samples. However, essentially all the galactomannan can be extracted by homogenizing the hydrated flour in water at

[7] T. J. Painter, *Lebensm.-Wiss. Technol.* **15**, 57 (1982).
[8] B. V. McCleary and N. K. Matheson, *Carbohydr. Res.* **119**, 191 (1983).
[9] B. V. McCleary, E. Nurthen, F. R. Taravel, and J.-P. Joseleau, *Carbohydr. Res.* **118**, 91 (1983).
[10] B. V. McCleary, F. R. Taravel, and N. W. H. Cheetham, *Carbohydr. Res.* **104**, 285 (1982).

25° using either a Waring blender at maximum setting, or an Ultraturrax homogenizer. Precipitation as the copper complex gives no purification beyond that obtained using the procedure described for the preparation of cold water-soluble carob galactomannan. Some guar flour samples contain a small proportion of galactomannan which does not extract in water at 25°. This material can not be extracted even at 90–95°, but can be solubilized by treatment of the homogenate with *A. niger* β-mannanase (~10 U/g original flour weight) at 40° for 30 min. The released, low DP material has the same ratio of D-galactose : D-mannose as galactomannan extracted at 25°, indicating that insolubility is not due to a low proportion of D-galactose in the polymer, but rather must be due to some form of chemical cross-linking.

Recovery and Purity

An analysis of whole-seed flour from 27 different guar seed varieties[6,11] indicates that the galactomannan content ranges from 23 to 31% (w/w). The galactomannan content of commercial guar flour is in the range of 70–80% (w/w). The purified polysaccharide samples contain galactose and mannose as the major monosaccharide components (>96%) with lesser quantities of arabinose plus xylose (<2%). Most samples are contaminated with protein to an extent of 1–2% (w/w).

Properties

Galactomannan extracted from 27 different guar seed varieties and from two commercial guar flours shows very similar properties. For each sample, the D-galactose content is 38 ± 1%. Hydrolysis of each galactomannan sample with *A. niger* β-mannanase and chromatography of the hydrolysis products on BioGel P-2 give patterns of amounts of oligosaccharides which are indistinguishable, indicating that the fine structures are the same. Further, computer analysis of the patterns of amounts and structures of the oligosaccharides produced shows that the pattern of distribution of D-galactosyl residues along the D-mannan backbone of guar galactomannan is not regular, blocklike, or statistically random. Rather, the distribution appears to be nonregular with a high proportion of substituted couplets. Treatment of the galactomannan samples with a mixture of guar seed α-galactosidase II and *A. niger* β-mannanase gives complete hydrolysis to galactose, mannobiose, and mannotriose with traces of mannose, consistent with the accepted structure for this galactomannan as a $(1 \rightarrow 4)$-β-D-mannan backbone substituted with $(1 \rightarrow 6)$-α-linked D-galactosyl residues.[1]

[11] B. V. McCleary, *Lebensm.-Wiss. Technol.* **14,** 188 (1981).

[64] Xylobiose and Xylooligomers

By JÜRGEN PULS, ANNEGRET BORCHMANN,
DIETER GOTTSCHALK, and JÜRGEN WIEGEL

Introduction

Xylobiose and xylooligosaccharides are derived from xylan, which after cellulose is the next most abundant renewable resource. Research on enzymatic hydrolysis of xylans needs pure xylooligomers as model compounds for optimization of hydrolysis processes. Recent results on fermentative utilization of xylans from different sources revealed that not only endo-1,4-β-xylanases (EC 3.2.1.8) and β-xylosidases (EC 3.2.1.37) are engaged in total xylan breakdown. The presence of enzymes responsible for cleavage of side groups, e.g., α-L-arabinofuranosidases (EC 3.2.1.55),[1] acetyl esterases,[2] and α-glucuronidases[3] significantly enhances the hydrolytic action of glycanases. Lack of one key enzyme may result in an accumulation of oligomeric degradation products, which cannot be hydrolyzed further.[4] Therefore mixed oligosaccharides carrying arabinofuranosyl, O-acetyl, or 4-O-methylglucuronosyl substituents are of increasing interest in addition to unsubstituted xylooligomers.

Xylan Structure

Xylans from land plants are composed of chains of 1,4-linked β-D-xylopyranose residues. The simplest form of xylan was reported to be esparto xylan consisting of β-1,4-linked xylopyranosyl units without any substituents but with some branching points.[5] According to their origin xylans of hardwoods, softwoods, or grasses have different substituents. The major substituents of xylans from straw and grasses are single arabinofuranosyl units attached to some C-3 positions of the main xylan chain.[6] These xylans carry smaller portions of D-glucopyranosyluronic

[1] L. C. Greve, J. M. Labavitch, and R. E. Hungate, *Appl. Environ. Microbiol.* **47,** 1135 (1984).

[2] K. Poutanen, J. Puls, and M. Linko, *Appl. Microbiol. Biotechnol.* **23,** 487 (1986).

[3] J. Puls, O. Schmidt, and C. Granzow, *Enzyme Microb. Technol.* **9,** 83 (1987).

[4] J. Wiegel, C. P. Mothershed, and J. Puls, *Appl. Environ. Microbiol.* **49,** 656 (1985).

[5] S. K. Chanda, E. L. Hirst, J. K. N. Jones, and E. G. V. Percival, *J. Chem. Soc.* p. 1289 (1950).

[6] K. C. B. Wilkie, *Adv. Carbohydr. Chem. Biochem.* **36,** 215 (1979).

METHODS IN ENZYMOLOGY, VOL. 160

acid units or of their 4-methyl esters, or both, attached to C-2 positions. In contrast the 4-O-methylglucuronic acid side group is the main substituent in wood xylans. Additionally softwood xylans carry smaller portions of single arabinofuranosyl units whereas hardwood xylans are acetylated.[7] For these reasons, the origin of a xylan has a major influence on the nature of the resulting xylan fragments after partial hydrolysis.

Isolation of Xylans

Xylans from different plant species are commercially available. For several reasons, however, it might be useful to start the preparation of xylooligomers from the very beginning, the lignified plant material. Only xylans from less lignified plant material, e.g., grasses, can be obtained in high yields by direct alkaline extraction. This procedure has the disadvantage of dissolving not only xylans, but also part of the lignin and other extractives. For xylan isolation most workers prefer to use a highly delignified pulp, the so-called holocellulose.[8-10] Holocellulose is defined as the water-insoluble carbohydrate portion including cellulose and hemicelluloses. It is usually prepared by removal of lignin with sodium chlorite,[11] or by several treatments with chlorine in ice water followed by exhaustive extraction of the chlorinated lignin with alcoholic ethanolamine.[12] Arabinoxylans from grasses and 4-O-methylglucuronoxylans from hardwoods may be directly extracted from holocellulose with 4.5%[13] or higher[14] NaOH concentrations. Isolation of arabino-4-O-methylglucuronoxylan from softwood demands a more complicated procedure after extraction of all hemicelluloses with 24% Ba(OH)$_2$. Separation of galactoglucomannan from xylan needs repeated precipitation of the insoluble galactoglucomannan Ba(OH)$_2$ complexes.[15] Should (O-acetyl-4-O-methylglucurono) xylans be required, this compound has to be extracted from hardwood holocellulose by dimethyl sulfoxide treatment,[16] because alkali solutions bring about complete deacetylation.

[7] T. E. Timell, *Wood Sci. Technol.* **1**, 45 (1967).

[8] G. J. Ritter and E. F. Kurth, *Ind. Eng. Chem.* **25**, 1250 (1933).

[9] T. E. Timell, *Methods Carbohydr. Chem.* **5**, 135 (1965).

[10] A. Ebringerová, A. Kramár, F. Rendos, and R. Domansky, *Holzforschung* **21**, 74 (1967).

[11] H. H. Dietrichs and K. I. Zschirnt, *Holz Roh-Werkst.* **30**, 66 (1972).

[12] T. E. Timell and E. C. Jahn, *Sven. Papperstidn.* **54**, 831 (1951).

[13] A. Ebringerova, A. Kramár, and R. Domanski, *Holzforschung* **23**, 89 (1969).

[14] T. E. Timell, *Sven. Papperstidn.* **65**, 435 (1962).

[15] T. E. Timell, *Tappi* **44**, 88 (1961).

[16] T. E. Timell, *J. Am. Chem. Soc.* **82**, 5211 (1960).

Partial Hydrolysis of Xylans

Fragmentation of xylans into oligomers is accomplished by acid or enzymatic degradation. The desired end product influences the choice of catalyst and reaction conditions (Table I). Uronic acid-substituted xylooligomers have generally been prepared by partial acid hydroly-

TABLE I
PREPARATION OF XYLOBIOSE AND XYLOOLIGOSACCHARIDES

Main products	Source of xylan	Catalyst	Fractionation	Reference
X_2-X_6	Hardwood (commercial)	Xylanase from *Streptomyces* sp.	Charcoal chromatography	20
X2	Hardwood (commercial)	Xylanase from *Streptomyces* sp.	Only active carbon treatment and deionization necessary	23
X, X_2-X_5 (4-O-Me-GlcA)X_3 to (4-O-Me-GlcA)X_6	White birch holocellulose after 24% KOH extraction	Commercial pectinase preparation	Charcoal chromatography	14
Mixed arabinose–xylose oligosaccharides from X_2A-X_6A	Wheat-straw xylan	Purified enzyme from *Myrothecium verrucaria*	Charcoal chromatography	19
(4-O-Me-GlcA)X_2	Aspen holocellulose after 24% KOH extraction	90% formic acid, 30 min, room temperature 45% formic acid 255 min, 100°	Separation of neutral and acidic sugars by Dowex 1-X4 (OAc⁻) ion-exchange resin, charcoal chromatography	17
Xylooligomers with one or two 4-O-Me-GlcA substituents	Larch wood holocellulose after successive KOH extraction	0.125 M sulfuric acid, 15 hr, 90°	Separation of neutral and acidic sugars by Dowex 1-X8 (OAc⁻) ion-exchange resin, anion-exchange chromatography in acetate medium	18
X_2-X_6 X_7	Corncob xylan	Fuming hydrochloric acid, 0°	Charcoal chromatography	21 22

sis.[17,18] The α-1,2-glucosidic bond of the 4-O-methylglucuronic acid side group turned out to be very stable under the action of acids and most enzyme preparations. Moreover the 4-O-methylglucuronic acid residue showed a stabilizing effect for the neighboring xylosidic bonds of the xylan main chain during hydrolysis with 45% formic acid[17] or enzymes.[14] The fact that arabinose units are furanosidically linked makes them particularly sensitive toward acid hydrolysis. This linkage is more readily hydrolyzed than xylopyranoside linkages. Therefore mixed arabinose–xylose oligosaccharides have mainly been prepared under the action of xylanases.[19] Unsubstituted xylooligomers have been prepared by xylanase catalysis[20] or with fuming hydrochloric acid under mild conditions.[21,22] Unless pure xylanases are used, xylose was one of the main end products eliminated by chromatographic fractionation or by yeasts.[23] Mild acid and enzymatic hydrolysis of hardwood xylan will result in a series of neutral and uronic acid-substituted oligomers. If only xylobiose and neutral xylooligomers are desired the following elegant procedure may be useful. Beechwood xylan was hydrolyzed by a 0.1% solution of a commercial enzyme preparation (Pectinesterase 7020, Röhm Darmstadt, FRG) yielding xylobiose and the pentaoligouronic acid (4-O-Me-GlcA)X_4 as the main products of degradation (Fig. 1). In step two a crude α-glucuronidase from *Agaricus bisporus* ATCC 44736[3] was added, which cleaved off the 4-O-methylglucuronic acid side group, thus increasing the yield of neutral sugars. However, poor β-xylosidase activity in this preparation reduced the advantage of the first step (Fig. 2) since some xylose was formed. O-Acetyl-substituted xylooligomers may be obtained by enzymatic hydrolysis of (O-acetyl-4-O-methylglucurono)xylan in the absence of esterase activity.[2,24] So far, these oligomers have not been fractionated and chemically characterized.

Fractionation of Xylobiose and Xylooligosaccharides

Early work on isolation of xylooligosaccharides used preparative paper chromatography or successive chromatographic fractionation on charcoal columns.[25] Today these techniques are believed to be less satis-

[17] N. Roy and T. E. Timell, *Carbohydr. Res.* **6**, 482 (1968).

[18] K. Shimizu, M. Hashi, and K. Sakurai, *Carbohydr. Res.* **62**, 117 (1978).

[19] C. T. Bishop and D. R. Whitaker, *Chem. Ind.* p. 119 (1955).

[20] I. Kusakabe, T. Yasui, and T. Kobayashi, *Nippon Nogei Kaguku Kaishi* **49**, 338 (1975).

[21] R. L. Whistler and C. C. Tu, *J. Am. Chem. Soc.* **74**, 3609 (1952).

[22] R. L. Whistler and C. C. Tu, *J. Am. Chem. Soc.* **75**, 645 (1953).

[23] I. Kusakabe, T. Yasui, and T. Kobayashi, *Agric. Biol. Chem.* **39**, 1355 (1975).

[24] P. Biely, J. Puls, and H. Schneider, *FEBS Lett.* **186**, 80 (1985).

[25] R. L. Whistler and D. J. Durso, *J. Am. Chem. Soc.* **72**, 677 (1950).

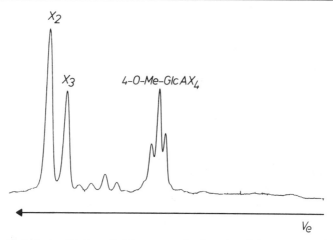

Fig. 1. Hydrolysis of 4-O-methylglucuronoxylan from beech by a commercial enzyme preparation (Pectinesterase 7020, Röhm Darmstadt, F.R.G.). Gel chromatographic separation on BioGel P-4 (-400 mesh), 100 × 2.5 cm, and TSK HW50S (E. Merck, Dormstadt, F.R.G.; EM Science, Gibbstown, NJ), 100 × 2.5 cm coupled in line. Mobile phase: 0.05 M Tris–HCl buffer, pH 7.8, 48 ml/hr. Detection: 0.1% 3,5-dihydroxytoluene in 65% sulfuric acid in a system shown in Fig. 3.

factory as far as separation efficiency and time are concerned. Gel permeation chromatography on dextran-based gels, polyacrylamide, and vinyl polymers has been successfully introduced.[26–28] A preparative system for oligomeric sugars is outlined in Fig. 3. The column was operated with 0.1 M acetic acid at a flow rate of 2 ml/min. Only a small part of the column eluate was used for specific detection of carbohydrates by postcolumn derivatization with 0.1% (w/v) 3,5-dihydroxytoluene in 65% sulfuric acid in an autoanalyzer system. The main stream was recovered in a fraction collector. Xylooligomers were separated according to their molecular weight. In order to avoid overlapping of neutral and uronic acid-substituted xylan fragments it was useful to separate neutral and acidic oligomers with Dowex 1-X4 exchange resins in acetate form[14] before the chromatographic run. Neutral sugars were removed with water, after which the acidic sugars were recovered with 30% aqueous acetic acid.

Analysis of Xylobiose and Xylooligosaccharides

Generally, methods of fractionation can also be applied for analysis of oligomeric sugars. This is especially valid for gel permeation chromatog-

[26] J. Havlicek and O. Samuelson, *Carbohydr. Res.* **22,** 307 (1972).
[27] M. John, J. Schmidt, C. Wandrey, and H. Sahm, *J. Chromatogr.* **247,** 281 (1982).
[28] H.-U. Körner, D. Gottschalk, J. Wiegel, and J. Puls, *Anal. Chim. Acta* **163,** 55 (1984).

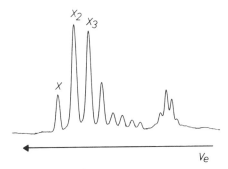

Fig. 2. Hydrolysis of xylooligomers shown in Fig. 1 by a culture filtrate of *A. bisporus*. Separation and detection conditions as in Fig. 1.

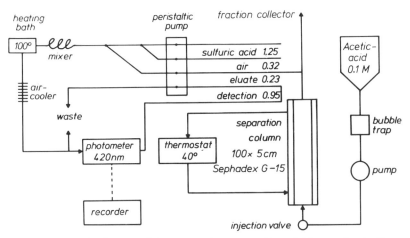

Fig. 3. Schematic diagram of the preparative system for xylobiose and xylooligosaccharides. For details, see text.

raphy where columns of smaller diameter and gels of smaller and more uniform particle size may increase the resolution.[28] A special effect is obtained with a polyacrylamide gel (BioGel P-4) and 0.05 M Tris–HCl buffer, pH 7.8; the neutral series of xylooligosaccharides are eluted according to size-exclusion principles, whereas the acidic oligomers are separated by partition principles. Only xylooctaose would be superimposed by 4-O-methylglucuronic acid (Fig. 4). Modern analysis techniques include partition chromatography on cation-exchange resins for separation of oligomers and arabinoxylooligomers[29] and anion-exchange chro-

[29] J. Schmidt, M. John, and C. Wandrey, *J. Chromatogr.* **213**, 151 (1981).

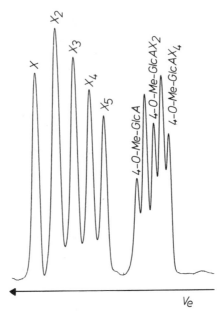

FIG. 4. Separation of an artificial mixture of xylose to xylopentaose and 4-O-methylglucuronic acid to (4-O-methylglucuronosyl)xylotetraose (0.25 mg each). Separation and detection system as in Fig. 1.

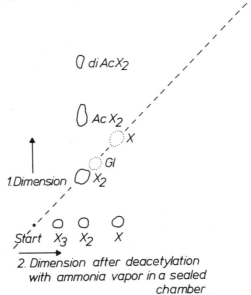

FIG. 5. Thin-layer chromatography of neutral products of xylanase hydrolysis of (O-acetyl-4-O-methylglucurono) xylan. TLC ready-plate 20 × 20 cm Cellulose G 1440 (Schleicher & Schüll, Dassel, F.R.G.), using ethylacetate/acetic acid/water (18:7:8). Staining of sugars with aniline phthalate.

TABLE II

CHARACTERISTICS OF XYLOBIOSE AND XYLOOLIGOSACCHARIDES

Sample	Specific rotation $[\alpha]_D$	Melting point (°C)	Reference
X_2	−25.5	185–186	22
	−25.6	184.5–185	31
	−24.8		19
X_3	−47.0	205–206	22
	−51		20
	−48.1	215–216	31
X_4	−60.0	219–220	22
	−63		20
	−61.9	224–226	31
X_5		231–232	22
	−73		20
	−71.7	240–242	31
X_6		236–237	22
	−78		20
	−78.5	237–242	31
X_7	−84		20
X_2A	−19.3		19
X_3A	− 4.0		19
X_3A_2	−51.0		19
X_5A	−33.8		19
X_6A	− 3.26		19
(4-O-Me-GlcA)X	+70−+110		31, 32
(4-O-Me-GlcA)X_2	+59	183	33, 25
(4-O-Me-GlcA)X_3	+23.4		31
(4-O-Me-GlcA)X_4	+ 0.6		31
(4-O-Me-GlcA)X_5	−11.8		31
(4-O-Me-GlcA)X_6	−20.8		31
(4-O-Me-GlcA)X_7	−25.7		31

matography in acetate media[3] with improved resins[28,30] for the quantitative analysis of mixtures containing xylooligouronic acids. O-Acetyl-substituted xylooligomers have been successfully identified by two-dimensional thin-layer chromatography.[24] Figure 5 shows a thin-layer chromatogram of neutral products of xylanase hydrolysis of (O-acetyl-4-O-methylglucurono)xylan. Thin-layer plate initially runs in one direction. Spraying with aniline phthalate reveals the presence of xylose, xylobiose, as well as of substances running faster than these compounds. These substances carry O-acetyl substituents, which can be demonstrated by

[30] S. Johnson and O. Samuelson, *Anal. Chim. Acta* **36**, 1 (1966).

changes in their mobilities during the run in the second direction, after deacetylation with ammonia vapor in a sealed chamber. A standard sugar mixture was added at the starting point for development only in the second dimension.

Characteristics of Xylobiose and Xylooligosaccharides

Specific rotations and melting points of these compounds are listed in Table II.[31-33] Reported values are sometimes divergent, therefore, in some cases more than one value is listed. Rates of movement on the paper chromatogram relative to D-xylose (R_x values) might be useful for identification of unknown compounds. These values have been reported by different workers.[14,19] However, different solvent systems have sometimes been used complicating any comparison, therefore R_x values have not been added to Table II.

[31] R. H. Marchessault and T. E. Timell, *J. Polym. Sci, Part C* **2**, 49 (1963).
[32] J. K. Hamilton and N. S. Thompson, *J. Am. Chem. Soc.* **79**, 6464 (1957).
[33] J. Puls, unpublished results.

[65] Remazol Brilliant Blue–Xylan: A Soluble Chromogenic Substrate for Xylanases

By PETER BIELY, DANA MISLOVIČOVÁ, and RUDOLF TOMAN

The substrate is derived from a water-soluble beechwood 4-*O*-methyl-D-glucurono-D-xylan (xylan) which does not require any modifications to preserve the solubility and good substrate properties after covalent coupling to an appropriate amount of the dye Remazol Brilliant Blue R. The conjugate of the polysaccharide with the dye, abbreviated as RBB–xylan, represents a convenient substrate for determination of activity and detection of xylanases.[1-3]

The assay is based on photometric measurements of the enzyme-released dyed fragments soluble in the presence of 2 volumes of ethanol, which precipitates the original substrate and its high-molecular-weight fragments. This principle offers to measure xylanase activity in cell-free extracts and media containing larger amount of reducing compounds

[1] P. Biely, D. Mislovičová, and R. Toman, *Anal. Biochem.* **144**, 142 (1985).
[2] P. Biely, O. Markovič, and D. Mislovičová, *Anal. Biochem.* **144**, 147 (1985).
[3] V. Farkaš, M. Lišková, and P. Biely, *FEMS Microbiol. Lett.* **28**, 137 (1985).

which would interfere with classical xylanase activity determination. Xylanase activity on the cell surface and on isolated membranes and organelles can also be followed in the presence of viable cells consuming xylose and xylooligosaccharides.

The detection of xylanase activity in gels employs transparent agar replicas containing RBB–xylan. Diffusion of dyed fragments released in the place of the enzyme into the separation gel, or their selective removal from the agar replicas by the solvent system used for the precipitation of unhydrolyzed RBB–xylan in a solution, represents the basis for enzyme detection. A great advantage of the technique is that the substrate present in a 2% agar gel does not precipitate as it does in a solution, so that the agar replicas remain transparent during and after the destaining of the zones of the enzyme-depolymerized substrate.

Preparation of RBB–Xylan

Isolation of Beechwood Xylan.[4] Air-dried, beechwood sawdust (1000 g, particle size 0.5–2 mm) stirred in 10 liters of water is heated to 50°. Acetic acid (99%, 140 ml) and sodium chlorite (700 g) are added gradually in three equal portions and the mixture is heated to 70° and maintained at this temperature under cooling for 1 hr. The delignification procedure is effected in a well-ventilated fume hood. The mixture is then filtered through cloth on a Büchner funnel and washed with water (4 × 2 liters). Wet sawdust is transferred to a beaker and extracted under stirring with a 1% aqueous solution of ammonium hydroxide (7 liters) at room temperature for 2 hr. The extract is filtered off and sawdust is stirred in 5% aqueous solution of sodium hydroxide (7 liters) at room temperature for 2 hr. The sodium hydroxide extract is collected by filtration through a Büchner funnel and the latter extraction is repeated. The combined extracts are added to 3 volumes of 96% ethanol, the precipitate is allowed to settle, and, after decantation of as much of the supernatant as possible, collected by filtration. The precipitate is washed with 75–80% ethanol containing 2% of acetic acid to neutralize the residual alkali and with 80% ethanol until the ash content in the filtrate, is constant. The wet polysaccharide is suspended in water and lyophilized; yield 195 g. Isolated 4-*O*-methyl-D-glucurono-D-xylan had $[\alpha]_D^{22} - 70°$ (c 0.53; water), a number-average molecular weight ($\bar{M}_n$) 18,600, and a D-xylose content 79.1%. Uronic acid and methoxyl group contents were 19.3 and 2.93%, respectively, which corresponds to a D-xylose to 4-*O*-methyl-D-glucuronic acid ratio of 7 : 1.

---

[4] A. Ebringerová, A. Kramár, F. Rendoš, and R. Domanský, *Holzforschung* **21**, 74 (1967).

Coupling of Remazol Brilliant Blue R to Xylan.[1] The dye (Hoechst AG, FRG) (100 g) is dissolved under stirring in a solution of beechwood xylan (100 g) in 2.5 liters of water prepared under heating (40–60°). The dark solution is mixed first with 0.5 liter of aqueous solution of sodium acetate (27 g) and then with 1 liter of 6% (w/v) NaOH. The mixture is stirred at room temperature for 1 hr. The product is precipitated by 2 volumes of ethanol (8 liters), filtered off, and washed with a mixture ethanol–0.05 M sodium acetate, 2:1 (v/v) until the filtrate is practically colorless. The product is reprecipitated from 4 liters of 0.05 M sodium acetate solution with 8 liters of ethanol, washed again as above, then desalted by washing with ethanol–water, 4:1 (v/v), dissolved in water, and lyophilized. The yield of RBB–xylan is approximately 110 g and the dye content in the product 13.2% (determined photometrically using $\varepsilon_{595} = 9.25 \times 10^3 \ M^{-1} \ cm^{-1}$) which corresponds to a ratio of the dye to sugar residue of 1:20.

By changing the weight ratio of the dye to xylan in the reaction mixture from 0.15 to 2.0, products containing 2.5–25.0% of the dye can be obtained.

Xylanase Assay[1]

Reagents

RBB–xylan, solution in 0.05 M acetate buffer, pH 5.4, 5.75 mg/ml (the concentration corresponds to 5 mg/ml of undyed xylan) or RBB–xylan solution in 0.1 M acetate buffer, pH 5.4, 11.5 mg/ml
Ethanol, 96% or absolute

Procedure. Enzyme solution or suspension (10–50 µl) is mixed with 0.5 ml of preheated (30°) solution of RBB–xylan (5.75 mg/ml), or 0.25 ml of enzyme solution or suspension is mixed with 0.25 ml of RBB–xylan solution in 0.1 M acetate buffer, and incubated at 30°. The reaction is terminated by addition of 1 ml of ethanol to precipitate unhydrolyzed RBB–xylan and its high-molecular-weight fragments. After standing for 30 min at room temperature (thermal equilibration of samples) the precipitated substrate is removed by centrifugation at 2000 g for 1.5 min and the absorbance of supernatants is measured at 595 nm against the respective substrate blank. In the case of colored enzyme solutions, enzyme blanks have to be measured as well.

With the exception of early stages of the reaction at low enzyme–substrate ratios, the liberation of dyed fragments soluble in 63% ethanol proceeds linearly with the time of incubation until about one-third of the total dye is solubilized (Fig. 1). A different plotting of the data in Fig. 1 would show that the amount of dyed fragment released during shorter

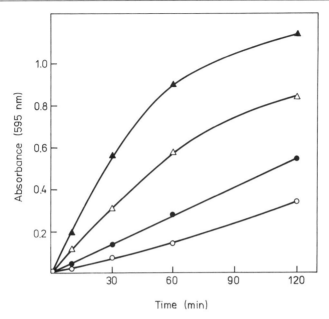

FIG. 1. Release of dyed fragments soluble in 63% ethanol from RBB–xylan during incubation with *Aspergillus niger* xylanases at concentrations 1.25 mU/ml (○), 2.5 mU/ml (●), 6.25 mU/ml (△), and 12.5 mU/ml (▲). Substrate concentration in the mixtures prior to precipitation with ethanol, 5.75 mg/ml. Reproduced with permission from Biely *et al.*[1]

incubations is also proportional to the enzyme concentration. The standard deviation of the assay was calculated to be ±13%. This is, however, true only under conditions of constant ionic strength and temperature of the final precipitation mixture. These two parameters influence considerably the solubility of dyed polysaccharide fragments. Addition of salts to the precipitation mixture may partially eliminate the salt effect,[5] however, the sensitivity of the assay then decreases.

In terms of absorbance values, the assay is approximately six times less sensitive than the xylanase assay in which reducing sugars are determined by the Somogyi–Nelson procedure.[1]

Detection of Xylanase in Gels[2]

Materials

RBB–xylan, aqueous solution, 150 mg/10 ml (heating to 60° and grinding with a glass rod accelerates dissolution)

[5] B. V. McCleary, *Carbohydr. Res.* **86**, 97 (1980).

Acetate buffer, 0.2 M, pH 4.0–5.4
Agar, ethanol
Glass plates, 30 × 15 cm
Plastic sheets, 30 × 15 cm
Plastic bars (spacers) 30 × 1 × 0.075 cm

Preparation of Agar Replicas. The aqueous solution of RBB–xylan (10 ml) is mixed with 20 ml of hot 3% agar solution in 0.2 M acetate buffer and, without degassing, poured between two plastic sheets mounted by means of water on glass plates and separated by plastic space bars (0.75 mm thickness). The agar–substrate gel can also be prepared on glass plates or gel-bond sheets simply by pouring and spreading the hot solution.

Procedure. The gel with separated proteins (any flat electrophoresis or isoelectric focusing agarose or polyacrylamide gel) is layered over an agar–RBB–xylan gel preheated to about 35–40° over a hot plate, and incubated at room temperature until the first most active enzyme zones become clearly visible against a white light. The presence of xylanase leads to a change in the shade of the blue background due to a more rapid diffusion of enzyme-released dyed fragments compared to the unhydrolyzed polysaccharide. The agar layer is then separated from the enzyme-containing separation gel and immediately or after further incubation in a wet chamber without further contact with the separation gel, dipped into the solvent ethanol–0.05 M acetate buffer (pH 5.4) 2 : 1 (v/v). The enzyme-degraded substrate zones are continuously destained as a result of the solubilization of the dyed fragments. The duration of the destaining depends on the extent of substrate degradation and may take 2–20 hr. An example of the detection of xylanases of the fungus *Schizophyllum commune* is shown in Fig. 2.[6]

The enzyme-released dyed polysaccharide fragments diffuse from the detection gel into the separation gel where they stain the enzyme zones, however, with a much lower contrast and sharpness in comparison with the zones in the destained agar replicas. The staining of the enzyme zones in the separation gel is important when enzyme forms have to be isolated.

Specificity of Detection. RBB–xylan is a substrate specific for endo-1,4-β-xylanases. Enzymes attacking xylan in an exo-fashion are not detected. RBB–xylan is, however, hydrolyzed by some nonspecific endo-1,4-β-glucanases (cellulases).

Documentation. Zymograms are dried on transparent plastic sheets after soaking in the destaining solution supplied with 5% glycerol. Photog-

[6] P. Biely, C. R. MacKenzie, and H. Schneider, this volume [90].

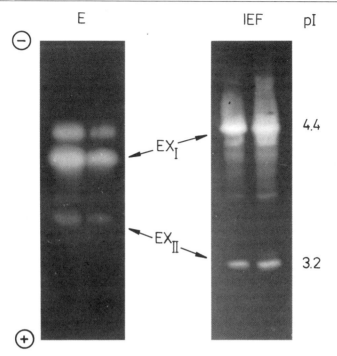

Fig. 2. Xylanases produced by *Schizophyllum commune* during growth on cellulose, as resolved by thin-layer agarose electrophoresis at pH 8.6 (E) and isoelectric focusing on polyacrylamide gel in the pH range 2.5–5 (IEF). The enzymes were detected using agar–RBB–xylan replicas. EX_I and EX_{II} are the two xylanases which were isolated by ion-exchange HPLC of the crude cellulolytic system.[6]

raphy of wet or dried agar replicas against a diffuse white light gives satisfactory results. The dried replicas are most stable when the blotted enzymes are denatured in the destaining solution containing 5% trichloroacetic acid.

Detection of Microbial Producers of Xylanase.[3] Solid media supplied with 0.2% RBB–xylan (can be autoclaved) are suitable for the detection of microbial producers of extracellular xylanase. Around stiches or cell colonies of xylanase producing microorganisms pale blue zones are formed as a result of radial diffusion of the secreted enzyme and of enzyme-released dyed fragments of RBB–xylan.

[66] Preparation of L-Arabinan and 1,5-L-Arabinan

By Kiyoshi Tagawa and Akira Kaji

L-Arabinan is a polymer of L-arabinose, which is found in nature wherever pectic substances occur. The molecule has a structure consisting of a chain of α-1,5-L-linked arabinofuranose units to which L-arabinofuranose units are attached mainly at position 3 in the α-configuration to form one unit side chains.[1-3]

1,5-L-Arabinan is a straight chain polymer of L-arabinofuranose units derived from L-arabinan by the action of an α-L-arabinofuranosidase which preferentially splits side chains from the L-arabinan molecule.[2,4,5]

Preparation of L-Arabinan

Beet pulp (1 kg) is mixed with 8 liters of water containing 200 g of calcium hydroxide, and the mixture is heated at 100° for 12 hr. The slurry is filtered through cloth and the filtrate is centrifuged. The clear solution is then acidified with acetic acid to pH 4.0, and the resulting precipitate is centrifuged to remove degraded pectic acid. The solution is concentrated under reduced pressure to 1 liter and the insoluble material is centrifuged off. On addition of ethanol (10 liter), crude L-arabinan, contaminated with calcium acetate, is precipitated. The precipitate is dissolved in water (300 ml) and reprecipitated with ethanol. On vacuum dehydration, the crude L-arabinan is obtained as a yellow glass (13 g, $[\alpha]_D^{25} - 87°$, L-arabinose content 73.7%, D-galactose content 22.0%).

The crude L-arabinan (10 g) is dissolved in water (200 ml) and applied to a column (3.8 × 53 cm) of DEAE-cellulose (OH⁻ form), and eluted with water. The fraction rich in polysaccharide is collected and concentrated to 200 ml under reduced pressure. The concentrated solution is passed through a column (3.8 × 60 cm) of Sephadex G-50 under gravity flow and eluted with water. As shown in Fig. 1, polysaccharides having a high content of L-arabinose are eluted faster than those containing D-galactose

---

[1] E. L. Hirst and J. K. N. Jones, *J. Chem. Soc.* p. 496 (1938), p. 1221 (1947), and p. 2311 (1948).
[2] K. Tagawa and A. Kaji, *Carbohydr. Res.* **11**, 293 (1969).
[3] J. C. Villettaz, R. Amado, and H. Neukom, *Carbohydr. Polym.* **1**, 101 (1981).
[4] A. Kaji, K. Tagawa, and K. Matsubara, *Agric. Biol. Chem.* **31**, 1023 (1967).
[5] K. Tagawa, *J. Ferment. Technol.* **48**, 730 (1970).

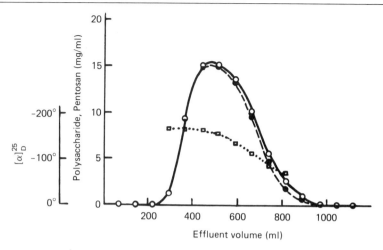

FIG. 1. Fractionation of the partially purified L-arabinan on Sephadex G-50. Polysaccharide content (○); pentosan content (●). Optical rotation $[\alpha]_D^{25}$ (□).

residues. Therefore, the fraction eluted at 320–560 ml (in Fig. 1) is collected, concentrated, and the polysaccharides are precipitated with ethanol. On vacuum dehydration, the purified L-arabinan is obtained as a hygroscopic material (2.3 g, $[\alpha]_D^{25}$ − 161°, L-arabinose content 97.8%, D-galactose content 1.1%).

Preparation of 1,5-L-Arabinan

The purified L-arabinan (10 g) is dissolved in 0.02 M citrate–phosphate buffer, pH 4.0 (400 ml), and purified α-L-arabinofuranosidase from *Aspergillus niger*[4,6] (500 units) is added. The reaction takes place at 40° and is monitored by the increase in reducing value. After 30% hydrolysis is attained (about 6 hr of incubation) the mixture is cooled to 20° and incubation is carried out at this temperature for an additional 12 hr. These treatments lead to formation of an amorphous precipitate and the extent of hydrolysis reaches 38%. The reaction mixture is then heated for 5 min at 100° to inactivate the enzyme, and ethanol is added to 80% concentration. The resulting precipitate is collected, dissolved in hot water (200 ml), and reprecipitated by keeping the solution for 24 hr at 2°. On vacuum dehydration, 1,5-L-arabinan is obtained as a white powder (5.6 g, $[\alpha]_D^{25}$ − 148°, L-arabinose content 98.3%, D-galactose content traces).

[6] A. Kaji and K. Tagawa, *Biochim. Biophys. Acta* **209,** 456 (1970).

TABLE I
PROPERTIES OF L-ARABINAN AND 1,5-L-ARABINAN

Property	L-Arabinan	1,5-L-Arabinan
Sugar components	L-Arabinose (97.8%)	L-Arabinose (98.3%)
	D-Galactose (1.1%)	
Solubility in water	Very soluble, 0.2 g/ml	3.5×10^{-3} g/ml at 25°, soluble after heating at above 90°
$[\alpha]_D^{25}$ (water, c, 1.0)	−161°	−148°
Intrinsic viscosity $[\eta]$	19.5 g cm^{-1}	23.7 g cm^{-1}
Density	1.593 g cm^{-3}	1.524 g cm^{-3}
Periodate consumption (per sugar residues, in 48 hr)	0.697 mol	1.067 mol

Properties

The L-arabinan preparation is very sticky and soluble in water but insoluble in alcohols, ethers, and acetone. The 1,5-L-arabinan preparation is a white powder and insoluble in cold water but it is solubilized by heating the solution above 90°. Both the L-arabinans have a strong levorotation in aqueous solution and are easily hydrolyzed with weak acid and by α-L-arabinofuranosidases[2,4,5,7–11] into L-arabinose, suggesting that all of the L-arabinose residues are in the furanose form and are connected by α-glycosidic linkages.

Structural studies on L-arabinan[1–3,12–17] and 1,5-L-arabinan[2,18] by methylation analysis, periodate oxidation, and ^{13}C NMR spectroscopy reveal that L-arabinan is a highly branched polymer of L-arabinofuranose units consisting of a α-1,5-linked main chain to which single unit side chains are joined through α-L-glycosidic linkages. These methods show that 1,5-L-

[7] A. Kaji, K. Tagawa, and T. Ichimi, *Biochim. Biophys. Acta* **171**, 186 (1969).
[8] A. Kaji and O. Yoshihara, *Biochim. Biophys. Acta* **250**, 367 (1971).
[9] F. Laborda, A. H. Fielding, and R. J. W. Byrde, *J. Gen. Microbiol.* **79**, 321 (1973).
[10] K. Schwabe, A. Grossmann, B. Fehrmann, and B. Tschiersch, *Carbohydr. Res.* **67**, 541 (1978).
[11] M. Tanaka, A. Abe, and T. Uchida, *Biochim. Biophys. Acta* **658**, 377 (1981).
[12] L. Hough and D. B. Powell, *J. Chem. Soc.* p. 16 (1960).
[13] P. S. Ócolla, *Methods Carbohydr. Chem.* **5**, 389 (1965).
[14] J. P. Joseleau, G. Chambat, M. Vignon, and F. Banoud, *Carbohydr. Res.* **58**, 165 (1977).
[15] J. P. Joseleau, G. Chambat, and M. Lanvers, *Carbohydr. Res.* **122**, 107 (1983).
[16] M. McNeil, A. G. Darvill, and P. Albersheim, *Prog. Chem. Org. Naturstoffe* **37**, 191 (1979).
[17] P. Capek, R. Toman, A. Kardošová, and J. Rosik, *Carbohydr. Res.* **117**, 133 (1983).
[18] S. C. Churms, E. H. Merrifield, A. M. Stephen, D. R. Walwyn, A. Polson, K. J. Merwe, H. S. C. Spies, and N. Costa, *Carbohydr. Res.* **113**, 339 (1983).

arabinan is a linear polymer of L-arabinofuranose units with α-1,5-glycosidic linkages. The properties of L-arabinan and 1,5-L-arabinan preparations obtained from beet pulp are summarized in Table I.

Comment

Although the amount of D-galactose, if present, has not been clearly established, all L-arabinans obtained from various plant tissues seem to have the same properties and structural features as described above, the differences in detailed structures reside in the number of side chains and/or their linkage positions. On the other hand, an attempt has been made by Kochetkov and co-workers[19,20] to prepare L-arabinan and 1,5-L-arabinan by chemical synthesis. They have synthesized a highly branched polymer of L-arabinofuranose by polymerization of tricyclic ortho ester β-L-arabinofuranose 1,2,5-orthobenzoate using mercuric bromide as a catalyst in nitromethane and thereafter deacylation with dilute sulfuric acid. They have also synthesized a linear polymer of L-arabinofuranose from 3-O-benzoyl-1,2-O-[(1-endo-cyano)ethylidene]-5-O-trityl-β-L-arabinofuranose by polymerization in dichloromethane in the presence of triphenylmethylium perchlorate and by deacylation with sodium methoxide. The properties of these synthesized L-arabinan and 1,5-L-arabinan seem to be close to each arabinan described above.

[19] N. K. Kochetkov, A. F. Bochkov, and I. G. Yazlovetsky, *Carbohydr. Res.* **9**, 49 (1969).
[20] L. V. Backinowsky, S. A. Nepogod'ev, and N. K. Kochetkov, *Carbohydr. Res.* **137**, C1 (1985).

[67] Measurement of (1 → 3),(1 → 4)-β-D-Glucan

By BARRY V. MCCLEARY, I. SHAMEER, and M. GLENNIE-HOLMES

The major carbohydrate component of the endosperm cell walls of barley and oat grain is a mixed-linkage (1 → 3),(1 → 4)-β-D-glucan commonly termed barley β-glucan.[1,2] This glucan represents from 2.5 to 5.5% (w/w) of the whole seed weight and about 75% of the carbohydrate present in the endosperm cell walls.[3,4] Barley β-glucan forms highly vis-

[1] I. A. Preece and K. G. Mackenzie, *J. Inst. Brew.* **58**, 353 (1952).
[2] C. W. Bamforth, *Brew. Dig.* **57**, 22 (1982).
[3] G. B. Fincher, *J. Inst. Brew.* **81**, 116 (1975).
[4] I. S. Forest and T. Wainwright, *J. Inst. Brew.* **83**, 279 (1977).

cous aqueous solutions and gelatinous suspensions. In the brewing industry it can lead to diminished rates of wort and beer filtration and to the formation of hazes, precipitates, and gels in stored beer.[5-8] In chicken diets these polymers display antinutritional properties through their "gumminess" and indigestibility which severely affect food intake.[9,10]

In an attempt to alleviate the problems caused by barley β-glucan in the brewing and animal feed industries, various approaches have been adopted including the breeding of barley varieties low in this component, the use of only well-modified malts in brewing, and the addition of enzymes active on barley β-glucan. Whichever approach is adopted there is a need for a simple and reliable assay for this component. Various methods have been developed,[11-15] but as yet none has been adopted as a standard procedure. Reasons for this include the lack of specificity or reliability of the assay or the tedious nature of the assay format which limits the number of samples which can be processed in a given time. Other assays appear very useful but employ crude enzyme preparations.[16,17] An assay procedure which overcomes these limitations has been recently described by the authors. In this assay, highly purified endo-1,3(4)-β-glucanase (lichenase) and β-glucosidase[18,19] are employed. The glucan is depolymerized by lichenase to oligosaccharides, these oligosaccharides are quantitatively hydrolyzed by β-glucosidase to glucose, and this is specifically measured using glucose oxidase/peroxidase reagent.

Purification of Enzymes

Lichenase. This enzyme is purified from bacterial amylase Novo 240 L (Novo Industries, Denmark) by pH treatment and by chromatography on DEAE-cellulose and on CM-Sepharose CL-6B.[18,20] The purified enzyme

[5] P. Gjertsen, *Proc. Am. Soc. Brew. Chem.* p. 113 (1966).
[6] W. W. Luchsinger, *Brew. Dig.* **42**, 56 (1967).
[7] K. Schuster, L. Narziss, and J. Kumada, *Brauwissenschaft* **20**, 185 (1967).
[8] R. W. Scott, *J. Inst. Brew.* **78**, 179 (1972).
[9] G. S. Burnett, *Br. Poult. Sci.* **7**, 55 (1966).
[10] B. Gohl, S. Alden, K. Elwinger, and S. Thomke, *Br. Poult. Sci.* **19**, 41 (1978).
[11] M. A. Anderson, J. A. Cook, and B. A. Stone, *J. Inst. Brew.* **84**, 233 (1978).
[12] H. L. Martin and C. W. Bamforth, *J. Inst. Brew.* **87**, 88 (1981).
[13] D. T. Bourne, T. Powlesland, D. Smith, and A. Morgan, *J. Inst. Brew.* **88**, 135 (1982).
[14] M. Fleming, D. J. Manners, R. M. Jackson, and S. C. Cooke, *J. Inst. Brew.* **80**, 399 (1974).
[15] R. J. Henry, *J. Inst. Brew.* **90**, 178 (1984).
[16] P. Aman and K. Hesselman, *J. Cereal Sci.* **3**, 231 (1985).
[17] B. Ahluwalia and E. E. Ellis, *J. Inst. Brew.* **90**, 254 (1984).
[18] B. V. McCleary and M. Glennie-Holmes, *J. Inst. Brew.* **91**, 285 (1985).
[19] B. V. McCleary and E. Nurthen, *J. Inst. Brew.* **92**, 168 (1986).
[20] B. V. McCleary, this volume [70].

is adjusted to a concentration of 50 U/ml and stored at −20°. This preparation is devoid of enzymes active on starch, maltosaccharides, cellulose, and sucrose.

β-Glucosidase. Purification involves chromatography on Ultrogel AcA 44, DEAE-Sepharose CL-6B, and Sulfopropyl-Trisacrylamide.[21] The purified enzyme is adjusted to a concentration of 2 U/ml and stored at −20°. This preparation is devoid of enzymes active on starch, cellulose, and maltosaccharides and activity on sucrose is so low as not to interfere with the assay.

Glucose Oxidase/Peroxidase Reagent[22]

Solution A

Disodium hydrogen orthophosphate dodecahydrate, 24.8 g
Sodium dihydrogen orthophosphate dihydrate, 12.4 g
Benzoic acid, 4.0 g
p-Hydroxybenzoic acid, 3.0 g
These chemicals are dissolved in 1800 ml of distilled water by stirring at room temperature and the volume is adjusted to 2 liters. The solution is stored at 4° and is stable for at least 12 months.

Solution B. One hundred milligrams of glucose oxidase (Boehringer Cat. No. 105147, 250 U/mg) is dissolved in 4 ml of distilled water and then stabilized by the addition of 2 g of finely ground ammonium sulfate. The enzyme is stable at 4°.

Solution C. Peroxidase[23] (Boehringer Cat. No. 108073, 250 U/mg).

Solution D. Two hundred milligrams of 4-aminoantipyrine (Sigma Chemical Co., Cat. No. A4382) is dissolved in 10 ml of distilled water. This solution is made up just before preparation of the working solution.

Working Solution

Solution A, 200 ml
Solution B, 0.2 ml
Solution C, 250 U
Solution D, 1.0 ml

---

[21] B. V. McCleary and Joan Harrington, this volume [71].

[22] A. B. Blakeney and N. K. Matheson, *Starke* **36**, 265 (1984).

[23] It is essential that high-purity glucose oxidase and peroxidase are used. Alternatively, glucose can be measured using the hexokinase/glucose-6-phosphate dehydrogenase method of Boehringer Mannheim GmbH or using a highly specific glucose oxidase/peroxidase-based kit from Biocon (Australia) Pty. Ltd., 31 Wadhurst Drive, Boronia, Victoria 3155, Australia.

The four solutions are mixed and stored in the dark at 4°. This glucose oxidase/peroxidase reagent is stable and gives similar standard curves for about 3 months.

Controls and Precautions

1. With each set of determinations, reagent blanks and glucose standards (50 and 100 μg) are included, in duplicate. Glucose standard solutions (1 mg/ml) are prepared in 0.2% benzoic acid and are stable at room temperature.

2. With each set of determinations at least one standard cereal flour, malt flour, wort, or beer is included.

3. With each new batch of glucose oxidase/peroxidase reagent the time for maximum color formation with 100 μg glucose is checked. This is usually 20–25 min.

4. It is imperative that lichenase is not cross-contaminated with β-glucosidase (the reverse is not a problem).

Assay Procedure for Barley and Oats

1. Barley or oat grain is milled to pass a 0.5-mm screen.

2. Samples of flour of known moisture content (~0.5 g) are accurately weighed into polypropylene tubes (35 ml capacity, 28 × 87 mm, Kayline Plastics, South Australia).

3. To each tube is added an aliquot (1 ml) of 50% v/v aqueous ethanol to aid in the subsequent dispersion of samples.

4. Five milliliters of 20 mM sodium phosphate buffer (pH 6.5) is added and the tubes are stirred on a vortex mixer.

5. Tubes are incubated in a boiling water bath for 2 min, removed, and vigorously stirred on a vortex mixer and then heated for a further 3 min. Mixing after 2 min prevents the formation of lumps of gelatinous material.

6. The tubes are cooled to 40° and 0.2 ml of lichenase (10 U) is added to each tube, the tubes capped, stirred, and incubated at 40° for 1 hr.

7. The volume in each tube is adjusted to 30.0 ml by the addition of distilled water.

8. The contents of the tubes are mixed thoroughly and an aliquot from each tube is filtered through a Whatman No. 41 filter circle or centrifuged at 3000 g for 5 min.

9. From each filtrate, 0.1-ml aliquots are carefully and accurately transferred to the bottom of three test tubes.

10. To one of these (the blank) is added 0.1 ml of 50 mM sodium acetate buffer (pH 4), while to the other two, 0.1 ml of β-glucosidase (0.2

U) in 50 mM sodium acetate buffer (pH 4) is added and the tubes are incubated at 40° for 15 min.

11. Three milliliters of glucose oxidase/peroxidase reagent is then added to each tube at 30-sec time intervals and each tube is incubated at 40° for exactly 20 min.

12. The absorbance at 510 nm for each sample is measured at 30-sec time intervals in the same sequence as step 11.

Assay Procedure for Malt

1. To approximately 1.0 g (accurately weighed) of malt flour (milled to pass a 0.5-mm screen) or lyophilized barley samples removed during the malting process, add 5.0 ml of 50% (v/v) aqueous ethanol.

2. Incubate in a boiling water bath for 5 min. Mix the contents on a vortex stirrer and add a further 5.0 ml of 50% (v/v) aqueous ethanol.

3. Centrifuge for 10 min at 2000 g and discard the supernatant.

4. Resuspend the pellet in 10 ml of 50% (v/v) aqueous ethanol, centrifuge, and discard the supernatant (as in step 3).

5. Suspend the pellet in 5 ml of 20 mM sodium phosphate buffer (pH 6.5).

6. Assay for β-glucan as per the procedure for barley and oat-grain flour from step 5.

Assay Procedure for Wort and Beer

1. Degas beer by heating an aliquot to ~80° in a boiling water bath. Allow to cool.

2. To 5 ml of wort or degassed beer in a preweighed centrifuge tube add 2.5 g of powdered ammonium sulfate.

3. Dissolve the ammonium sulfate by careful inversion; avoid frothing.

4. Allow to stand for about 20 hr at 4°, centrifuge at 1000 g for 10 min in a rotor with swing-out buckets. Discard the supernatant.

5. Resuspend the pellet by vigorously vortexing with 1 ml of 50% (v/v) aqueous ethanol; add a further 10 ml of 50% (v/v) aqueous ethanol, mix well.

6. Centrifuge for 5 min at 1000 g in a rotor with swing-out buckets. Discard the supernatant.

7. Repeat the ethanol washing procedure by resuspending the pellet and centrifuging as in steps 5 and 6. Discard the supernatant.

8. Resuspend the pellet in 20 mM sodium phosphate buffer (pH 6.5) to give a final sample weight of 4.8 g (~4.8 ml volume).

9. Add 0.2 ml of lichenase (10 U), incubate at 40° for 1 hr, and proceed as per the procedure for barley and oat flour starting from step 8.

Calculations

Barley, Oats, and Malt

$$\beta\text{-Glucan, }\% \text{ (w/w)} = \Delta_E \times F \times 300 \times \frac{1}{1000} \times \frac{100}{w} \times \frac{162}{180}$$

$$= \Delta_E \times (F/w) \times 27$$

Wort and Beer

$$\beta\text{-Glucan, (mg/l)} = \Delta_E \times F \times 10{,}000 \times \frac{1}{1000} \times \frac{5}{5} \times \frac{162}{180}$$

$$= \Delta_E \times F \times 9$$

Δ_E, absorbance of reaction minus absorbance of blank; 1/1000, conversion from micrograms to milligrams; 162/180, adjustment from free glucose to anhydro glucose; 300, volume correction (i.e., 0.1 ml taken from 30 ml); 100/w, factor to express β-glucan content as a percentage of dry flour weight; w, the calculated dry weight (mg) of the sample analyzed.

TABLE I
β-GLUCAN CONTENT OF OAT VARIETIES

Oat variety	β-Glucan content (%) (w/w)
Lort	4.3 ± 0.08
Bulban	3.1 ± 0.03
Cooba	4.9 ± 0.09
Dolphin	4.2 ± 0.06
Hill	4.0 ± 0.12
Carbeen	3.9 ± 0.09
Stout	5.4 ± 0.02
Trisperria	5.5 ± 0.06
Mortlock	4.3 ± 0.04
Cassia	4.3 ± 0.02
Coolabah	4.1 ± 0.11
Moore	3.5 ± 0.06
Barmah	3.8 ± 0.06
Swan	4.3 ± 0.02
West	3.9 ± 0.11
Blackbutt	4.5 ± 0.02

Moisture content can be determined by near infrared reflectance, using a moisture meter or by drying samples at 80° for 20 hr; 5/5, volume correction factor. Five milliliter samples were treated with ammonium sulfate and the volume was readjusted to 5 ml before removal of aliquots for analysis; 10,000, volume adjustment factor, 0.1-ml aliquots are analyzed but the results are expressed as mg/liter of sample.

Accuracy and Reliability

The described method is suitable for the routine analysis of mixed-linkage β-glucan in cereal flours, malt, wort, and beer. Up to 50 samples can be analyzed by a single operator in a day. Evaluation of the technique over a period of days indicates a mean standard error of 0.1 for cereal flour samples containing 3.8 and 4.6% (w/w) β-glucan content.

Typical results for oat flour samples are shown in Table I.

[68] Measurement of Acetylxylan Esterase in *Streptomyces*

By K. G. JOHNSON, J. D. FONTANA, and C. R. MACKENZIE*

$$O\text{-Acetyl-1,4-}\beta\text{-xylooligosaccharides} \xrightarrow{\text{H}_2\text{O}} 1,4\text{-}\beta\text{-xylooligosaccharides} + \text{acetic acid}$$

Incomplete hydrolysis of hemicellulosic materials by xylanases and xylobiases may result from substitution of the xylan backbone by a variety of moieties including arabinose, uronic acids, cinnamate-based esters, and O-acetyl groups.[1-3] Acetylxylan esterase occurs in a number of microorganisms and is known to act cooperatively with xylanases in the depolymerization of xylan.[4] This particular enzyme activity has been determined either by release of acetic acid from partially purified preparations of acetylxylan from hardwoods or by measuring the release of 4-nitrophenol from 4-nitrophenyl acetate.[5] The latter activity has been referred to as acetylesterase (EC 3.1.1.6),[1] but may or may not be identi-

* All three authors are affiliated with National Research Council of Canada.

[1] P. Biely, *Trends Biotechnol.* **3**, 286 (1985).
[2] A. G. Williams and S. E. Withers, *J. Appl. Bacteriol.* **51**, 375 (1981).
[3] P. J. van Soest, *Agric. Environ.* **6**, 135 (1981).
[4] P. Biely, C. R. MacKenzie, and H. Schneider, *Bio/Technology* **4**, 731 (1986).
[5] P. Biely, J. Puls, and H. Schneider, *FEBS Lett.* **186**, 80 (1985).

cal with acetylxylan esterase activity determined with acetylxylan as substrate. Since the lack of readily available sources of more natural substrates makes difficult the estimation of acetylxylan esterase activities, a method has been developed to prepare chemically acetylated xylan from various commercial xylans. These materials, along with 4-nitrophenyl acetate, have been used for the measurement of acetylxylan esterase produced by a variety of *Streptomyces* species.

Preparation of Acetylxylan

Larchwood and oat spelts xylan are obtained from Sigma Chemical Company while aspen xylan was prepared by the method of Jones *et al.*[6] Ten grams of dry xylan is slowly added to 250 ml of dimethyl sulfoxide with gentle stirring at ambient temperature. After the xylan is evenly suspended, the preparation is heated to 55° over a 20-min period until solubilization appears complete. Solid potassium borate is then added to a final concentration of 0.8% (w/v) after which the preparation is stirred for an additional 10 min at 55°. Two hundred milliliters of acetic anhydride, preheated to a temperature of 60°, is then added with stirring over a 5-min period to the dissolved xylan preparation. After all the acetic anhydride has been added, the reaction mixture is placed in dialysis tubing, and is dialyzed against running tap water at 4° until no odor of either dimethyl sulfoxide or acetic anhydride remains, usually 5 days. Following 24 hr of dialysis against distilled water, the material is lyophilized.

Notes. Pyridine, a solvent frequently encountered in acetylation procedures, is not recommended in the above procedure since its presence leads to darkening and insolubilization of the preparations. No special precautions such as distillation of acetic anhydride are required to maintain optimum reactivity, although the use of freshly opened anhydride is preferrable.

Estimation of Acetic Acid by High-Performance Liquid Chromatography (HPLC)

Acetic acid is quantified by HPLC using a 7.8×300 mm HPX-87H column (Bio-Rad, Richmond, CA) maintained at 25°. Samples (50 μl) are eluted with 0.01 N H_2SO_4 at a flow rate of 0.5 ml/min and the resulting peaks are detected with a Waters R401 refractometer. The retention time for acetic acid is 13.80 min in this system.

[6] J. K. N. Jones, C. B. Purves, and T. E. Timell, *Can. J. Chem.* **39**, 1059 (1961).

Enzyme Assay Methods

Arylesterase (Acetylesterase) Using 4-Nitrophenylacetate as Substrate

Reagents

10 μmol/ml 4-nitrophenyl acetate (Sigma Chemical Co.) dissolved in
dimethyl sulfoxide 0.5 M potassium phosphate buffer, pH 6.0
0.5 M potassium phosphate buffer, pH 6.0

Procedure. To test tubes are added 100 μl buffer, appropriate enzyme
dilutions, and distilled water in a total volume of 900 μl. To initiate the
reaction, 100 μl of 4-nitrophenyl acetate is added. After exactly 5 min
incubation at 50°, the absorbance at 420 nm is determined. One enzyme
unit is the amount of enzyme releasing 1 μmol of 4-nitrophenol per minute
at pH 6.0 and 50° under the above assay conditions.

Notes. A reagent blank is prepared and incubated along with test
samples, and the absorbance at 420 nm is determined against a water
blank. When dealing with crude or highly colored enzymes such as pig-
mented culture filtrates, it is necessary to prepare reaction mixtures lacking substrate. For calculation of results, both reagent blank and enzyme
control readings must be subtracted from test readings. Concentration of
4-nitrophenol is obtained from standard curves prepared under assay conditions and is linear up to 0.6 μmol of 4-nitrophenol.

Acetylxylan Esterase Using Acetylated Xylan as Substrate

Reagents

10% (w/v) acetylated xylan suspension in distilled water
0.5 M potassium phosphate buffer, pH 6.0
1 N sulfuric acid

Procedure. To test tubes are added 100 μl of buffer, appropriate enzyme dilutions, and distilled water in a total volume of 300 μl. After
initiation of reaction by the addition of 200 μl of substrate, samples are
incubated at 50° for a designated time interval, usually 20 min. Reactions
are terminated by the addition of 10 μl of sulfuric acid, are transferred to
microcentrifuge tubes, and are centrifuged at 8000 g for 2–3 min to sediment residual acetylxylan. Clarified supernatant fractions are removed for
analysis by HPLC. If not immediately analyzed, samples are stored at
$-20°$. One enzyme unit is the amount of enzyme releasing 1 μmol acetic
acid per minute at pH 6.0 and 50° under the above assay conditions.

Notes. To obtain an accurate measurement of acetylxylan esterase
activity, it is necessary to use several enzyme dilutions to ensure that a

linear relationship between enzyme concentration and acetic acid release exists. This linear range varies for each of the acetylxylan esterases from the streptomycetes studied here. Since substrate inhibition occurs in some of the streptomycete enzyme systems (see below), it is particularly important to ascertain the linear range of assay.

Xylanase Activity

Xylanase activity is measured as previously described[7] using xylan prepared for aspen wood by the method of Jones *et al.*[6] One enzyme unit is the amount of enzyme releasing 1 μmol of reducing sugar at pH 6.0 and 50°.

Analytical Techniques

Protein is measured by the method of Bradford[8] using γ-globulin as standard. Reducing sugars are measured colorimetrically[9] using xylose as standard.

Growth of *Streptomyces*

Streptomyces flavogriseus 45-CD (ATCC 33331), *Streptomyces olivochromogenes* [National Research Culture Collection (NRCC) 2258], *Streptomyces* strain R39 (NRCC 2811), *Streptomyces* strain C248 (NRCC 3020), and *Streptomyces* strain C254 (NRCC 3021) are grown in a proteose peptone–yeast extract–mineral salts medium (1 liter/4 liters baffled flask) designated IAF.[10] Enzyme production cultures are grown at 37° for 72–96 hr in a gyratory shaker operated at 250 cycles per minute. Inocula (5% v/v) for such cultures are obtained from 48-hr cultures grown in Trypticase Soy Broth (BBL Microbiology Systems). When used, xylan, sugarcane bagasse, or arabinogalactan are added to the basal IAF medium at a 1% (w/v) final concentration.

Evaluation of the Acetylation Procedure

Nuclear magnetic resonance (NMR) by ^{13}C NMR using a 50 MHz spectrophotometer and infrared (IR) spectrophotometry analyses using a Digilab FTS-15 Fourier transform spectrophotometer are used initially to

[7] C. R. MacKenzie, D. Bilous, and K. G. Johnson, *Can. J. Microbiol.* **30,** 1171 (1984).
[8] M. Bradford, *Anal. Biochem.* **72,** 248 (1976).
[9] M. Somogyi, *J. Biol. Chem.* **195,** 19 (1952).
[10] M. Ishaque and D. Kleupfel, *Can. J. Microbiol.* **26,** 183 (1980).

confirm acetylation of the xylan preparations. O-acetylated larchwood xylan presents δ values of 174.4/173.8 and 21.4/21.2 ppm for carbonyl and methylene carbons; δ is 31.1 ppm for CD_3COCD_3 as internal standard. The IR spectra obtained in KBr disks reveal a strong band at 1735 cm^{-1} corresponding to carbonyl stretching of an ester group.

Yields of acetylated xylan vary from 12 to 85% by weight of the original xylan (Table I). Larchwood xylan clearly delivers the highest yields.

The amount of acetylation achieved is determined by measurement of enzymically released acetic acid from the acetylxylan preparations. For this estimation, filtrate from a 72-hr culture of *S. olivochromogenes* concentrated by a factor of 20 using an Amicon ultrafiltration system equipped with a YM-10 membrane serves as the enzyme source. Reaction mixtures containing 100 μl crude enzyme (approximately 8 mg/ml protein), 100 μl 0.5 M potassium phosphate buffer, pH 6.0, 200 μl of 10% (w/v) test acetylxylan, and distilled water in a total volume of 0.5 ml are incubated at 50° for 1–4 hr. Following incubation, the reactions are terminated and the amount of acetic acid is determined by HPLC as described above using duplicate 50-μl aliquots of the clarified supernates. Reaction mixtures lacking enzyme serve as controls for background acetic acid. Estimation of the amount of acetylation in the larchwood, oat spelts, and aspen xylan preparations appears in Table I. That all available acetic acid is enzymatically released from these substrates is evidenced by the fact that acetic acid content of the 1, 2, 3, and 4 hr reaction mixtures does not vary significantly.

Production of Acetylxylan Esterase by *Streptomyces* Species under Different Growth Conditions

Levels of extracellular enzyme activities hydrolyzing acetylxylan and 4-nitrophenyl acetate are just detectable at 24 hr growth but rise quickly thereafter to a maximum at 72 hr and decline slightly at 96 hr. Enzyme activities observed in 72-hr cultures are presented in Table II. Several conclusions can be drawn from these data. First, the low constitutive levels of extracellular enzyme activity can be enhanced greatly by supplementation of the growth media with plant cell wall materials. Oat spelts xylan is the most efficacious supplement for the production of high levels of acetylxylan esterase and arylesterase. Little relationship appears to exist between the enzyme activities measured with the acetylxylan and the 4-nitrophenyl acetate substrates. If one were measuring the same enzyme activity with these two substrates, a constant ratio between the specific enzyme activities under varying growth conditions would be ex-

TABLE I
EVALUATION OF THE ACETYLATION PROCEDURE

Acetylxylan preparation	Yield[a] (%)	Acetic acid released (μg/assay)	Acetylation[b] (%)
Larchwood xylan	85.0	3804	19.0
Oat spelts xylan	12.0	515	2.57
Aspen xylan	53.0	2464	12.32

[a] Results are expressed as percentage of weight of product acetylxylan from input xylan.
[b] Expressed as percentage by weight.

pected. This, for at least the streptomycete systems studied here, is not the case. The large differences in enzyme activity as measured with 4-nitrophenyl acetate and acetylxylan substrates observed in *Streptomyces* strains C248 and C254 strongly demonstrate that acetylxylan is the preferred substrate for measurement of acetylxylan esterase activity.

Fractionation of S. olivochromogenes Enzymes

Step 1. S. olivochromogenes (5 liters) is grown for 72 hr in IAF medium containing 1% (w/v) oat spelts xylan. The filtrate is concentrated approximately 25-fold using a Pellicon ultrafiltration unit equipped with 10 NMWL membranes. Following concentration, the preparation is subjected to two cycles of dialysis *in situ* in the ultrafiltration unit using 1 liter of distilled water for each cycle. The dialyzed material is then further concentrated to 41 ml in an Amicon ultrafiltration cell using a PM10 membrane.

Step 2. Concentrated material (20 ml; 25 mg/ml protein) is applied to a 2.6 × 25 cm column of DEAE-BioGel A equilibrated with 10 mM potassium phosphate buffer, pH 7.0, and is eluted with the same buffer at 4°. Acetylxylan esterase does not adsorb to this ion-exchange material but is largely separated from an undesirable black pigment which is strongly bound to the gel.

Step 3. Two milliliters (0.45 mg/ml protein) of the above eluate is applied to a 7.5 × 150 mm CM-3SW HPLC column (LKB) equilibrated with 10 mM potassium phosphate buffer, pH 6.0. The sample is eluted with 15 ml of equilibration buffer, followed by a 0–0.3 M linear gradient of NaCl (30 ml) in the same buffer. The eluant is monitored continuously for protein at 280 nm and 1-ml fractions are collected. Suitable aliquots of each fraction are assayed for their content of xylanase, acetylxylan ester-

TABLE II

PRODUCTION OF ACETYLXYLAN ESTERASE AND ARYL ESTERASE BY *Streptomyces* SPECIES[a]

Organism	Growth supplement	Acetylxylan esterase		Aryl esterase		A/B
		Units/ml	Specific activity (A)[b]	Units/ml	Specific activity (B)[b]	
S. *flavogriseus*	None	0.127	0.35	0.01	0.053	6.66
45-CD	Oat spelts xylan	4.42	12.27	0.04	0.111	110.5
	Sugarcane bagasse	3.11	8.64	0.08	0.229	37.7
S. *olivochromogenes*	None	0.05	0.09	0.063	0.112	0.83
	Oat spelts xylan	5.86	17.23	1.08	3.19	2.63
	Sugarcane bagasse	3.85	8.38	0.22	0.48	17.46
Strain R39	Oat spelts xylan	[c]	—	0.256	2.04	—
	Arabinogalactan	[c]	—	0.158	0.83	—
Strain C248	Oat spelts xylan	34.91	129.3	0.018	0.069	1874.0
Strain C254	Oat spelts xylan	38.80	110.9	0.194	0.41	94.6

[a] All cultures were grown in 1AF media with indicated supplements (1% w/v) for 72 hr.
[b] Specific activity is expressed as units/mg protein.
[c] Below levels of detection.

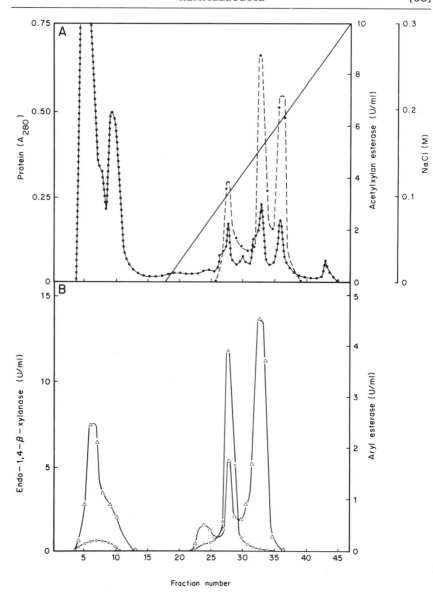

FIG. 1. Fractionation by cation exchange HPLC on CM-3SW of *S. olivochromogenes* xylan hydrolyzing enzymes. (A) Protein (●——●); acetylxylan esterase (●---●); solid line, NaCl concentration. (B) Xylanase (△); aryl esterase (○).

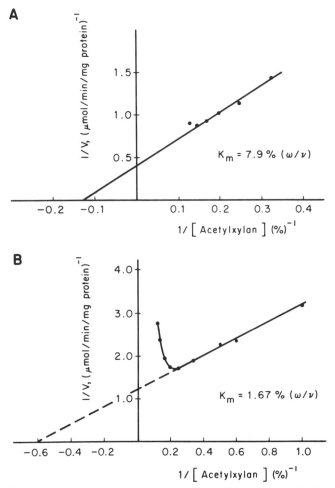

FIG. 2. Determination of acetylxylan esterase K_m for *S. flavogriseus* 45-CD (A) and *S. olivochromogenes* (B).

ase, and arylesterase activities (Fig. 1). Three distinct cationic fractions acting on acetylxylan elute at 0.1, 0.155, and 0.19 M NaCl, respectively (Fig. 1A). The first two of these acetylxylan esterases coelute with endoxylanase (Fig. 1B). A third major endoxylanase activity appears in the wash fraction of the column along with small quantities of aryl esterase. A single aryl esterase component coelutes with endoxylanase and acetylxylan esterase at 0.1 M NaCl. This fractionation, therefore, conclusively demonstrates that enzyme activities hydrolyzing acetylxylan and 4-nitrophenyl acetate are distinct in some systems.

Determination of K_m

Acetylxylan prepared from larchwood xylan is used to determine the K_m of crude enzyme preparations. K_m values are obtained from Lineweaver–Burk plots of $1/V$ versus $1/S$ using substrate concentrations from 2.5 to 10% (w/v) for *S. flavogriseus* 45-CD acetylxylan esterase, and from 1 to 5% acetylxylan for *S. olivochromogenes* acetylxylan esterase. *S. flavogriseus* 45-CD enzyme has a very high K_m of 7.9% (w/v) acetylxylan and a V_{max} of 26 μmol acetic acid released per minute per mg protein (Fig. 2A). K_m and V_{max} values for the *S. olivochromogenes* enzyme (Fig. 2B) are much lower (1.67% acetylxylan and 0.82 μmol acetic acid released per minute per mg protein). In addition, the acetylxylan esterase from the latter system presents kinetic patterns consistent with substrate inhibition. As such, it would be prudent to establish the level at which substrate inhibition occurs in systems from other organisms before selecting the concentration of acetylxylan to be used in standard assays.

Acknowledgments

The authors gratefully acknowledge the technical assistance of D. Bilous and S. De Souza, and thank H. Casal for performing IR analyses.

[69] α-4-*O*-Methyl-D-glucuronidase Component of Xylanolytic Complexes[1]

By J. D. Fontana,* M. Gebara, M. Blumel, H. Schneider,* C. R. MacKenzie,* and K. G. Johnson*

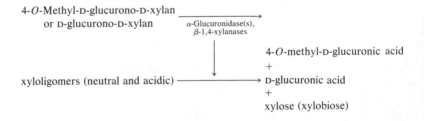

Xylans commonly bear substituents on the main xylan chain. These include 4-*O*-methyl-D-glucuronic acid, D-glucuronic acid, arabinose, and

[1] Issued as National Research Council of Canada Publication No. 27893.
* National Research Council of Canada.

acetic acid. Analysis of such xylans in general, and of D-glucuronoxylans in particular, is complicated by the resistance to acid hydrolysis of 2-O-α-(4-O-methyl-D-glucopyranosyluronic)-D-xylose, the aldobiuronic building block of heteroxylans. As a result of this resistance, acidic hydrolysis conditions that result in complete recovery of free xylose are difficult to achieve, as evidenced by the hydrolysis of grasses by aqueous phosphoric acid.[2] Enzymatic hydrolysis represents an alternate and more efficient method of hydrolysis. However, while the α-D-glucuronide moiety of synthetic glycosides is susceptible to enzymatic hydrolysis by *Patella vulgata* visceral hump enzymes,[3] α-(4-O-methyl)-D-glucuronidases from organisms of biotechnological interest have yet to be fully characterized. Some free 4-O-methyl-D-glucuronic acid arises along with xylooligosaccharides after digestion of hemicellulosic materials with *Trichoderma*,[4] other fungal and snail hydrolases,[5,6] and *Thermoanaerobacter*.[7] The specificity of enzymes that remove uronyl substituents from heteroxylans with respect to activity against 4-O-methylated and unmethylated glucuronic acid is unknown.

The present chapter describes chromatographic and electrophoretic techniques for the detection and measurement of α-(4-O-methyl)-D-glucuronidase and D-glucuronidase activities with specific reference to heteroxylans.[8]

Enzyme Assay

Preparation of Substrates and Reference Compounds

Larchwood xylan (Sigma Chemical Co.) is a readily available xylan containing 4-O-methyl-D-glucuronyl residues. A 3% w/v solution of the commercial material is precipitated twice with 3 volumes of methanol to

[2] J. D. Fontana, J. B. C. Correa, J. H. Duarte, A. M. Barbosa, and M. Blumel, *Biotechnol. Bioeng. Symp.* **14,** 175 (1984).
[3] G. A. Levy and C. A. Marsh, *Adv. Carbohydr. Chem.* **14,** 425 (1959).
[4] R. F. H. Dekker, *Biotechnol. Bioeng.* **25,** 1127 (1983).
[5] J. D. Fontana, M. A. L. Feijao, and J. H. Duarte, *Cien. Cult. (Sao Paulo) Suppl.* **28a,** 247 (1976).
[6] J. D. Fontana, A. M. Barbosa, M. Blumel, and M. Gebara, *Arq. Biol. Technol.* **29,** 108 (1986).
[7] P. J. Weimer, *Arch. Microbiol.* **143,** 130 (1985).
[8] Abbreviations used: ara, L-arabinose; xyl, D-xylose; X_2, 1,4-D-xylobiose; X_3, xylotriose; X_4, xylotetraose; GlcUA, D-glucuronic acid; mGlcUA, 4-O-methyl-D-glucuronic acid; mABU, 4-O-methylaldobiouronic acid; mATU, 4-O-methylaldotriouronic acid; mATtU, 4-O-methylaldotetrouronic acid; L, glucuronolactone; Glc, D-glucose; CB, cellobiose; p-β-Galp, phenyl-β-D-galactopyranoside.

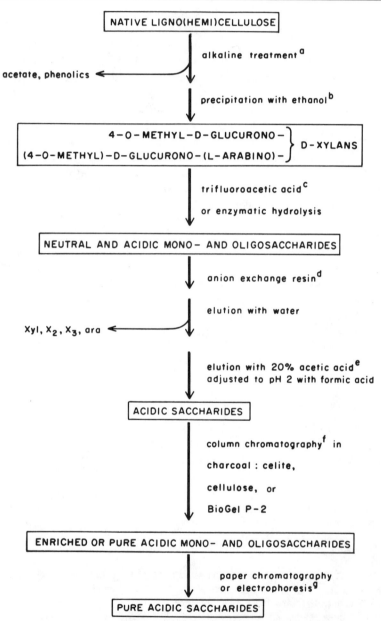

FIG. 1. Sequence of steps in preparation of hetero-D-xylans and their acidic mono- and oligosaccharides from angiosperm biomass. (a) Extraction of hemicelluloses using 10 volumes of 8% (w/v) KOH under N_2 at room temperature with occasional mixing. (b) Extracted hemicelluloses are precipitated with 4 volumes of ethanol after neutralization with acetic

remove a yellow pigment, and is adjusted to 2% w/v with water. The solution is then heated to 80°, and, while hot, is clarified by filtration through a Millex-GS 0.22-μm membrane (Millipore Co., Bedford, MA).

An alternative substrate, and one which contains appreciable amounts of 4-O-methyl-D-glucuronyl as well as D-glucuronyl residues, is the total hemicellulose fraction from wheat bran. The bran is first treated to reduce the content of starch and pectin by suspending 2 g in 100 ml of 1 M potassium acetate, autoclaving at 120° for 15 min, and then collecting and washing the insoluble residue with water on a filter. The hemicelluloses are subsequently extracted using alkali, as outlined in Fig. 1.[9-12]

Standards of 4-O-methyl-D-glucuronic acid, its albiouronic acid, and higher oligomers up to a degree of polymerization of 4 are prepared by hydrolysis of 2% w/v larchwood xylan in 1.2 M trifluoroacetic acid at 85–90° for 2 hr. The hydrolyzate is freed of acid by evaporation at low pressure and is then fractionated using an anion-exchange column (~200 mg of total sugar/ml of wet resin; see Fig. 1). The purpose of the fractionation is to separate the acidic (uronyl components) from neutral material, in order to simplify interpretation of the results of subsequent electrophoretic and chromatographic steps. The acidic fraction is subsequently filtered through a BioGel P-2 column (90 × 1.0 cm; ~20 mg of acidic sugar) using water as eluant. Pooled fractions of each peak are then subjected to preparative electrophoresis (conditions given in legend of Fig. 1 and Fig. 3). Preparations enriched in aldobiouronic acid and its homologs of higher molecular weight can be obtained as well from raw sugarcane bagasse after hydrolysis with 0.1–0.2% H$_3$PO$_4$ at 10 atm (185°) for 5 min.[2] In this case, fractionation by adsorption on charcoal-Celite (Fig. 1) instead of BioGel is preferable, since the initial elution with water eliminates phos-

---

[9] K. C. B. Wilkie, in "Biochemistry of Plant Cell Walls" (C. T. Brett and J. R. Hilman, eds.), pp. 1–38. Cambridge University Press, London, 1985.
[10] T. E. Timmel, *Methods Carbohydr. Chem.* **1**, 301 (1962).
[11] R. L. Whistler and J. N. BeMiller, *Methods Carbohydr. Chem.* **1**, 47 (1962).
[12] J. D. Fontana, J. D. Duarte, C. B. Gallo, M. Iacomina, and P. A. J. Gorin, *Carbohydr. Res.* **143**, 175 (1985).

---

acid. Air drying of the preparation is avoided to obtain better resolubilization from the wet precipitate or lyophilized material.[9] (c) Fifty volumes of 1.2 M TFA at 85–90° for 2 hr. (d) Dowex 2-X8 resin in acetate form. (e) Eluates are freed from volatile acids in a vacuum centrifuge, by lyophilization, or by evaporation under a N$_2$ stream. (f) When used, the charcoal:celite column is gradient eluted with aqueous ethanol.[10] The cellulose column is packed in acetone[11] and eluted with acetone containing increasing amounts of water.[12] BioGel P-2 gel filtration is performed according to the supplier's instructions (Bio-Rad Laboratories, Richmond, CA). (g) Whatman 3MM paper with solvent A or buffer E. This last step is optional, depending upon the degree of purity required.

phoric acid and its salts. The presence of 4-O-methyl-D-glucuronic acid in fractions of the acid hydrolyzate can be confirmed by reductive conversion to the 4-O-methyl-D-glucose derivative.[13]

Enzyme Sources. *Megalobulimus paranaguensis* is a terrestrial snail found on the coast of the State of Parana in Brazil. Its gastric juice is a rich source of several hydrolases[5] suitable for the degradation of plant cell walls and for the preparation of yeast and mold protoplasts. Each adult specimen (80–120 g) provides 4–6 ml of gastric juice with average solids content of 15%, most of which is protein. Snail juice obtained from animals which have fasted for 3 days is centrifuged at 10^4 g for 15 min, and the supernate is retained. Subsequent dialysis is not used since the cellulase activity present digests dialysis membranes.

Crude hemicellulase from *Dactylium dendroides*,[14] a galactose-oxidase producing mold (formerly misnamed *Polyporus circinatus*[15]), is an extracellular enzyme preparation obtained as the filtrate of culture media. The medium used for growth consists of Vogel's salts[16] and 1% (w/v) hemicellulose A from the stem of *Mimosa scabrella*,[17] a leguminous tree which is widespread in the south of Brazil and neighboring countries. One percent (w/v) larchwood xylan supplemented with 0.2% w/v yeast nitrogen base (Difco Co., Detroit, MI) can also serve as the growth medium.

Streptomyces olivochromogenes enzymes are obtained as ultrafiltration concentrates of 72-hr culture supernates as described in chapter [68] in this volume.

Reagents

Larchwood xylan or total hemicelluloses of wheat bran, 2% (w/v) in water
Sodium acetate buffer, 0.2 M, pH 4.8
Anhydrous methanol
Dowex 2-X8 resin, acetate form
Hydroxylamine, 2.5% (w/v) in dry pyridine containing phenyl-β-D-galactopyranoside, 0.6% (w/v) as internal standard
N,O-Bis(trimethylsilyl)trifluoroacetamide (BSTFA)

---

[13] F. Smith, *J. Chem. Soc.* p. 2646 (1951).
[14] A. M. Barbosa, M.S. thesis. Universidade Federal do Parana, Curitiba, Parana, Brazil (1984).
[15] D. Amaral, F. Kelly-Falcoz, and B. L. Horecker, this series, Vol. 9, p. 87.
[16] H. S. Vogel, *Microb. Genet. Bull.* **13**, 42 (1956).
[17] J. D. Fontana, J. B. C. Correa, and M. Gebara, *Cien. Cult. (Sao Paulo) Suppl.* **28a**, 492 (1976).

Procedure

Enzyme Hydrolysis and Fractionation. To a 1.5-ml polypropylene conical test tube are added 0.25 ml of substrate, 0.125 ml of buffer, appropriate levels of enzyme activity (0.2–1.0 mg of protein), and distilled water in a final volume of 0.5 ml. After 1 and 12 hr, at 35° for snail or fungal enzymes or at 50° for streptomycete enzyme, the reactions are terminated by the addition of 1 ml of chilled methanol, and the reaction mixture is centrifuged to remove material precipitated by methanol. The supernatant fraction is then applied to a small Dowex-2 (acetate) column (plastic tip of a 5-ml Eppendorf pipet plugged with glass wool and partially filled with resin) and is sequentially eluted with water and 20% acetic acid adjusted to pH 2 with formic acid. A fraction of 2.0–2.5 ml (~2.5 column volumes) is collected for each elution solvent (Fig. 1).

Gas-Liquid Chromatography (GLC). The acidic eluate, or the whole methanolic supernate in instances where the concentration of neutral components is relatively low, is brought to dryness in a vacuum centrifuge (Savant Instruments Inc., Hicksville, NY) operated at ambient temperatures. Heating of the sample is avoided to reduce the potential for lactone formation. The reducing sugars are then converted to their silylated oxime derivatives by the addition of 0.1–0.2 ml of hydroxylamine reagent, heating to 65° for 30 min with occasional mixing on a vortex mixer, followed by the addition an equal volume of BSTFA reagent under the same heating and stirring conditions as for the addition of hydroxylamine. If samples are not to be analyzed immediately, derivatization is carried out in 1-ml glass vials sealed with Teflon-coated rubber caps, in order to avoid penetration of moisture. Chromatography is carried out using a 1 μl injection on a glass SE-30 column (2 m long × 2 mm i.d.) with a He carrier gas flow of 45 ml/min and a temperature program consisting of 4 min at 180° followed by heating at 32°/min to 252°. Alternatively, a fused silica capillary OV-17 column is used (25 m long × 0.32 mm i.d., 0.25 μm film thickness) with a N_2 carrier flow of 1.0 ml/min and a temperature program consisting of 2 min at 180° followed by heating at 4°/min to 240°. Detection is by flame ionization.

The chromatographic properties of neutral and acidic monosaccharides and disaccharides are summarized in Table I. An indication that 4-O-methyl-D-glucuronic acid is the product of both snail and streptomycete enzyme action is provided, using both chromatographic columns, by the linear increase with incubation times of 30–120 min of the area of peaks with retention times for the 4-O-methylglucuronic acid derivative. Confirmation of the identity of the 6.5 and 2.6 min peaks as the appropri-

TABLE I

GAS LIQUID CHROMATOGRAPHIC ANALYSIS OF NEUTRAL AND ACIDIC
MONO- AND DISACCHARIDES FROM HETEREO-D-XYLAN
AS OXIME-SILYL DERIVATIVES

	Retention time of standards (min)					
Column	p-β-Galp^a	Xyl	mGlcUA	GlcUA	X_2	mABU
OV-17	13.9	3.7	6.5	7.2	15.0	18.2
SE-30	6.5	1.7	2.6	3.5	9.2	11.6

[a] Internal standard.

ate silyl derivative is made by GC–mass spectrometry (Hewlett-Packard 5985 system) employing methane chemical ionization and electron impact.

Note. It has been reported that alkaline extraction of 4-O-methyl-D-glucurono-D-xylan may result in β-elimination of some uronyl moieties,[9] generating an unsaturated hexuronate. However, no differences are observed between the acid or enzymatically released products with the xylans used.

The syn and anti isomers of the oxime of the silyl derivatives of the reducing sugars[18] in the hydrolysates are not separated under the conditions used. However, a small companion peak appears along with that for the 4-O-methyl-D-glucuronic derivatives when using larger injection volumes.

Thin-Layer Chromatography. The plates used are DC-Alufolien silica gel 60 chromatoplates (Merck, Darmstadt). Elution is with the following solvent systems: (A) ethyl acetate:acetic acid:formic acid:water (9:3:1:4); (B) ethyl acetate:acetic acid:2-propanol:water (18:7:5:10); (C) 2-propanol:nitromethane:ethyl acetate:water (6:1:1:2); and (D) ethyl acetate:acetic acid:2-propanol:formic acid:water (25:10:5:1:15). All formulations are expressed in volumes. Visualization is accomplished by spraying with 0.2% (w/v) orcinol in sulfuric acid:methanol (10:90 ml), using ~15 ml of solution per plate, followed by heating at 105° for 3–5 min. Pentoses, hexoses, and uronic acids can be differentiated by their violet, red-violet, and blue-violet to gray colors, respectively. Naphthoresorcinol or carbazole can substitute for orcinol when a more differentiated staining is desired for uronic acids (deep blue). To monitor uniformity of migration of the solvent front, a

[18] D. R. Knapp, "Handbook of Analytical Derivatization Reactions," p. 572. Wiley, New York, 1979.

TABLE II
THIN-LAYER CHROMATOGRAPHIC RESOLUTION OF NEUTRAL AND ACIDIC MONO- AND
OLIGOSACCHARIDES FROM HETERO-D-XYLAN

Solvent system	Mobility of standards relative to D-xylose										
	Ara	X_2	X_3	X_4	GlcUA	mGlcUA	mABU	mATU	L	Glc	CB
A	0.84	0.76	0.58	0.43	0.72	0.93	0.79	0.63	1.23	0.81	0.48
B	0.84	0.74	0.57	0.42	0.50	0.65	0.54	0.35	1.28	—	0.55
C	0.83	0.87	0.78	0.63	0.20	0.48	0.37	—	1.20	—	0.71
D	0.87	0.76	0.61	0.50	0.67	0.83	0.69	0.57	1.29	—	0.80

0.1% ethanolic solution of thymol blue is spotted on marginal and central positions of the sample application line. The thymol blue spot is also useful in judging the amount of the acid spray needed, as the color changes to deep pink when sufficient acid has been applied.

An illustrative run showing the resolution which can be achieved is presented in Fig. 2. Mobility data for xylose, xylobiose, GlcUA, mGlcUA, and mABU are in Table II, which also includes their migration properties for solvents other than that used in Fig. 2.[19] Because of the dependence of mobility on solvent, a change in solvent can improve separation, depending on the composition of the hydrolyzate. The example in Fig. 2 also shows the enhancement resulting from prior separation of the neutral and acid components before analysis by comparing total products of larchwood xylan depolymerization on longer incubation (12 hr) with snail (lane 2) and streptomycete (lane 15) enzymes and their respective neutral (lanes 3 and 16) and acidic counterparts (lanes 4 and 17) obtained from resin fractionation.

The results seen in Fig. 2 illustrate that the two enzyme preparations are able to deglucuronylate heteroxylan. However, their hydrolytic capability differs in that the final neutral products formed are monomers using snail enzymes and dimers using the streptomycete enzymes. The persistence of acidic homologs with degrees of polymerization varying from 5 (lane 4) to 2 (lane 16) indicates that removal of 4-O-methyl-D-glucuronyl substituents is a limiting step for the complete enzymatic biodegradation of heteroxylans, as is the case with acid hydrolysis. Accordingly, enzyme systems for complete degradation of glucurono-D-xylans should contain sufficient activities hydrolyzing xylose–xylose linkages in addition to deglucuronidating activities.

Paper Electrophoresis. Paper electrophoresis is carried out on 25 cm

[19] E. Chornet, C. Vanasse, and R. P. Overend, *Entropie* **130/131,** 89 (1986).

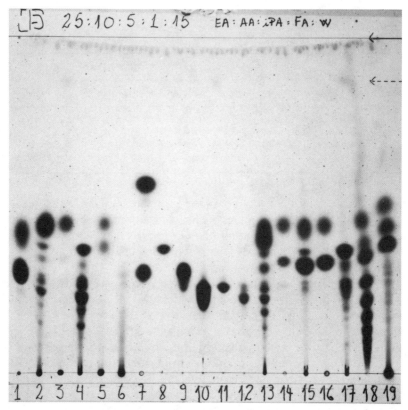

FIG. 2. Thin-layer chromatography of neutral and acidic products of hetero-D-xylan hydrolysis. Solvent D employed with detection by the orcinol reagent. Standards (listed in order of decreasing migration when mixtures applied) are 1, ara + CB; 7, L + GlcUA; 8, mGlcUA; 9 and 10, mABU and mATU from larchwood xylan; 11 and 12, acidic oligosaccharides from sugarcane bagasse; 14, xyl + X_2; 18, steam-treated aspen prepared using the University of Sherbrooke process. Aspen wood suspensions containing 12–20% by weight of solids are prepared from ground wood and water. These are subjected to repeated passage through a homogenizing valve to produce a homogeneous suspension which is then rapidly heated with saturated steam, pumped through a homogenizing value with a pressure differential of 4800 psi, and subsequently held in a plug flow reactor for 40 sec using the technique described.[19] Samples: 2, 3, and 4, whole digest, neutral, and acidic fractions from snail enzyme digestion of larchwood xylan, respectively; 4 and 5, neutral and acidic fraction of wheat bran hemicelluloses, respectively, incubated with the same enzyme; 13, duplicate of sample 3, enzyme frozen and thawed; 15, 16 and 19, whole digest, neutral, and acidic fraction of larchwood xylan digested with streptomycete enzyme, respectively; 19, wheat bran hemicelluloses degraded by the same enzyme. Upper arrow, solvent front. Dashed arrow, thymol blue.

TABLE III
PAPER ELECTROPHORETIC RESOLUTION OF ACIDIC MONO- AND
OLIGOSACCHARIDES FROM HETERO-D-XYLAN

Buffer system	Mobility relative to D-glucuronic acid				
	Xylose	mGlcUA	mABU	mATU	mATtU
Pyridine–acetate	—	0.94	0.73	0.60	0.51
Tetraborate	0.82	0.72	0.63	0.50	0.41

wide Whatman No. 1 paper under the following conditions. System E: 10% pyridine–0.4% acetic acid in water (v/v/v), pH 6.5, at 1.8 kV and 20 mA for 50–70 min. System F: 50 mM sodium tetraborate, pH adjusted to 10.0 with NaOH 10 M, at 0.9 kV and 35 mA for 40–60 min. Bromphenol blue is used as a control for uniformity of electrophoretic migration and to monitor movement of the front. Fifteen centimeters of migration is sufficient. Visualization of separated compounds in system E is accomplished by immersion of the dried electrophoretogram sequentially in (1) saturated silver nitrate : acetone (0.5 : 99 ml), (2) 10 M NaOH : ethanol (1 : 99 ml), and (3) 5% (w/v) sodium thiosulfate, followed by thorough washing with distilled water. Double immersion in the silver nitrate reagent and a 2-fold concentrated ethanolic sodium hydroxide reagent is employed for system F.

The separation achieved in the two buffer systems for glucuronic acid and its 4-methyl homolog is shown in Fig. 3 (see also Table III). The system employing borate buffer provides sharper resolution (part B, lane 7), because the C-4 O-methyl substituent hinders complexation with borate. Nevertheless, the two acids can be differentiated in the pyridine–acetate buffer (part A, lane 7). Differentiation in the latter system is a result of the weakening effect of the methyl group on the dissociation of the C-6 carboxyl group, which results in different electrophoretic mobility for the acids. An advantage of pH 6.5 buffer is the convenient profile generated for the complete series of acidic components. Neutral saccharides fail to move in this buffer, except for the small displacement resulting from capillary action and electroendosmotic effects.

M. paranaguensis α-4-O-methyl-β-D-glucuronidase activity undergoes inactivation on repeated freezing and thawing (lane 3 in both parts of Fig. 3) and to a lesser extent after lyophilization. Such inactivation may account for the lack of demonstrable 4-O-methylglucuronidase activity reported in other snail preparations.[3] However, fresh preparations from M. paranaguensis show an additional activity hydrolyzing small amounts

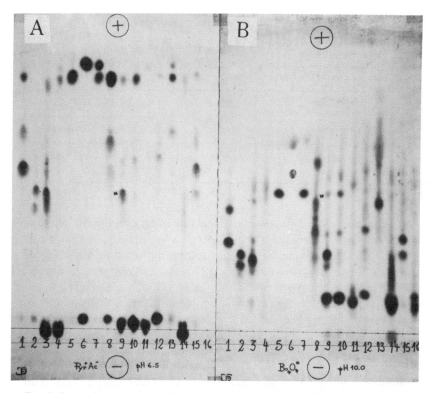

FIG. 3. Paper electrophoresis of neutral and acidic products of hetero-D-xylan hydrolysis. Buffer systems used are E (pyridine acetate; a) or F (tetraborate, b). Sugars are detected
with alkaline silver nitrate. Central mark (*) denotes bromphenol blue. In order of decreasing mobility, standards are 1, mGlcUA, aldobi- and aldotriouronic acids; 5, mGlcA; 6,
GlcUA and xylose; 7, mixture of GlcUA and mGlcUA; 8, crude acid fraction of *M. scabrella*
hemicellulose hydrolyzed with TFA; 13, acidic fraction from TFA hydrolysis of wheat bran
total hemicellulose; 15, acidic fraction of aqueous phosphoric acid hydrolysis of cane bagasse (pH 2.0, 10 atm, 5 min). Samples 2 (1-hr incubation), 3, and 4 (12-hr incubation) were
obtained from snail enzyme digests using larchwood xylan as substrate for the first two
samples, and using total hemicellulose from wheat bran for sample 4. (Sample 2 was cleared
of neutral sugars by anion-exchange chromatography prior to electrophoresis.) Samples 9 (1-
hr incubation), 10, and 11 (12-hr incubation) were obtained using streptomycete enzymes
derived from cultures grown in 1% (w/v) sugarcane bagasse. The first two lanes are hydrolyzates of larchwood xylan and the third of wheat bran hemicelluloses. Sample 12 is also a
hydrolyzate of larchwood xylan, but the streptomycete enzymes used were obtained from
cultures grown in 1% (w/v) oat spelts xylan. Sample 14 is the whole hydrolyzate of *M.
scabrella* hemicellulose A with fungal enzymes. Sample 16 (A) is a pool of the control
incubations for all enzymes and substrates. Sample 16 (B) is steam treated birchwood.

Note: Alkaline "peeling" from the nonreducing end of hemicelluloses can be prevented
by the inclusion of sodium borohydride in the extractant. However, the reducing agent can
cause conversion of some of the 4-*O*-methylated-D-uronyl substituents to 4-*O*-methyl-D-
glucose, if they are esterified at the C-6 carboxyl group. The latter methylated sugar is
susceptible to alkali action, including β-elimination to hexenuronate.[9] Controlled chlorina-
tion to promote lignin oxidation and extractibility, followed by dimethyl sulfoxide extraction, is an alternative approach to obtain the so-called delignified "native" hemicelluloses.

of GlcUA (Fig. 3, part A, lane 4) from hemicelluloses bearing both kinds of uronic acid such as that derived from wheat bran.[20] TFA-released products from these pentosans appear on lane 13 of the same figure). No enzymatic 4-*O*-demethylation would be expected when wheat bran hemicelluloses are digested with snail hydrolases since GlcUA is never seen in enzymatic hydrolyzates of larchwood polymer. The fungal enzyme (lane 14 of both parts of Fig. 3) shows less α-4-*O*-methyl-D-glucuronidase activity compared with snail and streptomycete enzymes. This is believed to result from an overestimation of the protein content of the preparation caused by interference from medium components.

Note. In the present study, *S. olivochromogenes* enzymes display no significant activity against synthetic *p*-nitrophenyluronides.[21] While fungal α-glucuronidase activity is barely detectable, the snail enzyme is very active toward both configomers, particularly the *p*-nitrophenyl-β-D-glucuronide. The α-4-*O*-methyl-D-glucuronidase component of both snail and streptomycete enzyme preparations is active over a wide range of pH values from 3.5 to 6.5, activity being dramatically reduced in the alkaline pH range. In comparative enzyme reactions with all substrates used at ~4 mg/ml total carbohydrate (therefore mABU > mATU ≫ larchwood xylan, in terms of effective mGlcUA concentration), the simpler acidic di- and trisaccharide building blocks do not undergo extensive hydrolysis. Whether this is a result of there being insufficient xylanase content in the crude preparations to produce lower molecular weight substrates for glucuronidase action, or whether glucuronidase action is subject to product inhibition is unclear at present. However, preliminary studies using *S. olivochromogenes* glucuronidases freed of contaminating xylanase activity indicate that little or no detectable hydrolysis of high-molecular-weight glucurono-D-xylans occurs.

Acknowledgments

J.D.F. dedicates this work to his Ph.D. advisor, Dr. Luis F. Leloir (IIB-FC/FCEyN-UBA, Argentina), on the occasion of his 80th birthday. We thank Dr. R. P. Overend for providing the steam-treated aspen. The technical assistance of Sue Levy-Rick, J. Labelle, H. Turner, D. Bilous, F. Cooper, Celia MacKenzie, J. Pie, and R. P. Rocha (UFPR) is acknowledged. Financial support was provided by CNPq/FINEP/PADCT-SBIO/SPIN, FDCTC-UFPR (Brazil) and the National Research Council of Canada.

[20] S. G. Ring and R. R. Selvendran, *Phytochemistry* **19,** 1723 (1980).
[21] C. A. Marsh, *in* "Glucuronic Acid: Free and Combined" (G. J. Dutton, ed.), p. 60. Academic Press, New York, 1966.

[70] Lichenase from *Bacillus subtilis*

By BARRY V. MCCLEARY

Lichenase (EC 3.2.1.73, endo-1,3(4)-β-glucanase) is a highly specific endo-β-glucanase, restricted in its substrate range to the mixed-linkage β-D-glucans from Icelandic moss (lichenan), barley, or oat-endosperm cell walls (barley β-glucan), and to reduced pneumococcal polysaccharide (RSIII polysaccharide).[1-4] The enzyme from *Bacillus* culture filtrates is termed lichenase based on its demonstrated action on lichenan, whereas that from germinated barley is termed malt β-glucanase. Both enzymes have similar action patterns. Lichenase has no action on 1,4-β-D-glucan (CM-cellulose) or 1,3-β-D-glucans (laminaran).[1-4] Hydrolysis of barley β-D-glucan by malt β-glucanase[4,5] or *Bacillus pumilus* lichenase[6] yields 3-β-cellobiosyl-D-glucose and 3-β-cellotriosyl-D-glucose in a molar ratio of about 2:1 and these make up 80–90% of the oligomers produced. The remainder includes an insoluble fraction and oligomers of higher degree of polymerization with 1,4-linkages and a 3-linked glucose at the reducing end.

The specificity of lichenase makes it particularly useful in the enzymatic quantification of mixed linkage β-glucan in barley and oats.[7-9] Barley β-glucan gives solutions of high viscosity which can lead to problems in the brewing industry such as reduced rate of wort and beer filtration. It can also lead to haze, precipitate, and gel formation in stored beer. These polymers also have antinutritional properties, particularly in chicken diets where their "gumminess" and indigestibility severely affect food intake.

Lichenase has been purified from crude *Bacillus* enzyme preparations[7,8-11] and from crystalline α-amylase preparations[11] (derived from

[1] E. A. Moscatelli, E. A. Ham, and E. L. Rickes, *J. Biol. Chem.* **236**, 2858 (1961).

[2] E. T. Reese and A. S. Perlin, *Biochem. Biophys. Res. Commun.* **12**, 194 (1963).

[3] M. A. Anderson and B. A. Stone, *FEBS Lett.* **52**, 202 (1975).

[4] W. W. Luchsinger, S.-C. Chen, and A. W. Richards, *Arch. Biochem. Biophys.* **112**, 524 (1965).

[5] J. R. Woodward, G. B. Fincher, and B. A. Stone, *Carbohydr. Polym.* **30**, 207 (1983).

[6] H. Suzuki and T. Kaneko, *Agric. Biol. Chem.* **40**, 577 (1976).

[7] M. A. Anderson, J. A. Cook, and B. A. Stone, *J. Inst. Brew.* **84**, 233 (1978).

[8] R. Henry, *J. Inst. Brew.* **90**, 178 (1984).

[9] B. V. McCleary and M. Glennie-Holmes, *J. Inst. Brew.* **91**, 285 (1985).

[10] E. L. Rikes, E. A. Ham, E. A. Moscatelli, and W. H. Ott, *Arch. Biochem. Biophys.* **69**, 371 (1962).

[11] D. J. Huber and D. J. Nevins, *Plant Physiol.* **60**, 300 (1977).

Bacillus preparations) by conventional chromatographic procedures, which, by their nature, are time consuming and gave a limited yield of enzyme. A procedure[9] which takes advantage of the affinity binding of lichenase to DEAE-cellulose at high pH values will be described here. With this procedure approximately 2 g (~200,000 U) of electrophoretically homogeneous lichenase can be prepared within 3 days using two simple chromatographic steps.

Assay Method

Lichenase is routinely assayed using carboxymethyl-barley β-glucan dyed with Remazol Brillant Blue R.[12-14] Pure enzyme preparations are standardized by the Somogyi–Nelson reducing sugar assay employing barley β-glucan as substrate. In this latter assay, suitably diluted enzyme preparation (0.05 ml) is incubated with 0.5 ml of barley β-glucan (5 mg/ml) in 0.1 M sodium phosphate buffer (pH 6.5) at 40° for 0–10 min. The reaction is terminated by the addition of 0.5 ml of Somogyi–Nelson copper reagent and released reducing sugar equivalents determined.[9,15]

Purification of Lichenase

Step 1. pH Treatment. One liter of bacterial amylase Novo 240 L (Novo Industrias, Denmark) is diluted with an equal volume of cold water and the pH adjusted to 3.5 by careful addition of 1 M HCl with vigorous stirring. The solution is stored at 4° for 2 hr and centrifuged at 3500 g for 10 min. The supernatant solution is treated with 100 ml of 1 M sodium phosphate buffer (pH 6.5) and adjusted to pH 8 by careful addition of 1 M NaOH. On storage at 4° for 2 hr, the solution is centrifuged at 3500 g for 30 min to remove a fine colloidal precipitate.

This initial purification step results in a more than 95% removal of α-amylase which at pH 3.5 is unstable and precipitates from solution. Lichenase is quite stable at this pH at 4°.

Step 2. DEAE-Cellulose Column Chromatography. The pH-treated and centrifuged solution is applied to a DEAE-cellulose column (4.2 × 22 cm) previously equilibrated with 50 mM sodium phosphate buffer (pH 8). The column is then washed with 500 ml of 0.5 M KCl in 50 mM sodium phosphate buffer (pH 8) and then with 2.5 liters of 50 mM sodium phos-

[12] B. V. McCleary and I. Shameer, *J. Inst. Brew.* **93**, 87 (1987).
[13] B. V. McCleary, *Carbohydr. Polym.* **6**, 307 (1986).
[14] B. V. McCleary, this volume [160].
[15] M. Somogyi, *J. Biol. Chem.* **195**, 19 (1952).

phate buffer (pH 8). Lichenase is eluted by washing the column with 0.1 M sodium acetate buffer (pH 4). The recovered enzyme is diluted with an equal volume of water and adjusted to pH 4.5.

At pH 8, lichenase displays a specific affinity binding to DEAE-cellulose and is not displaced even by high concentrations of KCl. Contaminating proteins are thus removed by washing the column with 0.5 M KCl in phosphate buffer followed by exhaustive washing with 50 mM phosphate buffer (pH 8). Under the same conditions lichenase does not bind to DEAE-Sepharose CL-6B. In fact, binding to this support material does not occur even in the presence of low concentrations (20 mM) of sodium phosphate buffer (pH 8).

Step 3. CM-Sepharose CL-6B Column Chromatography. The diluted enzyme is applied to a CM-Sepharose CL-6B column (2.5 × 10 cm) equilibrated with 50 mM sodium acetate buffer (pH 4.5) and the column eluted with 600 ml of a linear KCl gradient (0–0.5 M). Lichenase elutes as a single sharp peak at a KCl concentration of 0.3 M, paralleling the pattern of elution of protein. The increase in specific activity is insignificant, but this step serves to concentrate the enzyme and contamination with α-amylase is reduced by virtue to the instability of the enzyme under these chromatographic conditions. This enzyme is stored either at −20° or is treated with ammonium sulphate (50 g/100 ml) and stored at 4°.

Recovery and Purity

The purification procedure is summarized in Table I. The enzyme is purified 64-fold from the crude preparation with a final specific activity of 115 U/mg on barley β-glucan at pH 6.5 and 40°. The overall recovery is approximately 2 g (~200,000 U) of an electrophoretically homogeneous enzyme. The enzyme appears as a single protein band on SDS–gel electrophoresis (MW 24,000) but as two major (pI values 8.15 and 8.4) and several minor protein bands (pI values 6.5–7.4) on isoelectric focusing. The purified enzyme has no detectible activity (<1 ppm) on CM-pachyman, CM-cellulose 4M6F, p-nitrophenyl-β-D-glucopyranoside, or p-nitrophenyl-α-D-glucopyranoside, and the activities on soluble starch and maltose are less than 0.0005% the activity on barley β-glucan.

Properties

The purified enzyme has an optimal pH for activity of 6.5–7.0 with half maximal values at pH 5.2 and 8.0. In 0.2 M citrate/phosphate (pH 2.5–7.5) or phosphate/Tris–HCl (pH 5.5–9.0) buffers it is completely stable on storage at 4 or 20° for 20 hr. It is also stable to repeated freeze/thaw

TABLE I
PURIFICATION OF LICHENASE FROM *Bacillus subtilis*

Step	Total activity (units)	Total protein (mg)	Specific activity U/mg	Yield (%)	Purification (fold)
Original solution (1 liter)	454,000	248,000	1.8	—	1.0
pH treatment	440,750	34,850	12.6	97	7.0
DEAE-cellulose					
Fraction I[a]	57,600	568	101.4	12.7	56.3
Fraction II	144,000	1,333	108.0	31.7	60.0
CM-Sepharose CL-6B					
Fraction I	55,400	482	114.9	12.2	63.8
Fraction II	141,450	1,222	116.0	31.2	64.4

[a] On chromatography of the enzyme on DEAE-cellulose, some activity eluted while the column was being washed with 50 mM sodium phosphate buffer (pH 8). This fraction, (I) had the same physical, kinetic, and electrophoretic properties as the major fraction, (II), which eluted on washing the column with 100 mM sodium acetate buffer (pH 4).

cycles. Lichenase shows maximal activity at 60°, and on storage for 15 min at pH 6.5, is stable at temperatures up to 60°. K_m and V_{max} values with barley β-glucan as substrate are 0.3 mg/ml and 118 U/mg, respectively. Barley β-glucan is hydrolyzed to an extent of ~16% yielding mainly tri-saccharide (53%) and tetrasaccharide (25%) fractions, with lesser amounts (17%) of oligomers of higher DP (5–15) and an insoluble precipitate (3%). Small amounts of disaccharide are also present. The structures of the tri- and tetrasaccharides produced by *Bacillus pumilus* endo-1,3(4)-β-glucanase have been characterized by Suzuki and Kaneko.[6]

[71] Purification of β-D-Glucosidase from *Aspergillus niger*

By BARRY V. MCCLEARY and JOAN HARRINGTON

β-D-Glucosidases (EC 3.2.1.21) have been isolated and purified from a number of fungal culture filtrates.[1–8] These enzymes vary markedly in

[1] J. C. Sadana, J. G. Shewale, and R. V. Patil, *Carbohydr. Res.* **118**, 205 (1983).
[2] I. M. Tavobilov, N. A. Rodinova, and A. M. Bezborodov, *Biochemistry (Engl. Transl.)* **49**, 847 (1984).
[3] G. Halliwell and R. Vincent, *Proc. Bioconversion Symp.* p. 24 (1980).
[4] F. E. Cole and K. W. King, *Biochim. Biophys. Acta* **81**, 122 (1964).

their specificity. Thus, the four β-glucosidases from *Sclerotinium rolfsii*[1] have a strict requirement for a D-gluco configuration but can tolerate a wide range of aglycons, provided the D-glucosyl residue of the substrate has a β-configuration. The nature of the aglycon affects the rate of hydrolysis. All four enzymes hydrolyze positional isomers of cellobiose. In contrast, a number of β-glucosidases, including those from almond emulsin[9] and from *Aspergillus niger* 15,[2] do not have a strict requirement for a D-gluco configuration. The enzyme from almond emulsin catalyzes hydrolysis of both β-D-glucopyranosides and β-D-galactopyranosides and evidence has been obtained for the involvement of different enzyme active sites. An enzyme purified to homogeneity from culture filtrates of *Aspergillus niger* 15[2] has a very broad specificity with activity on β-D-glucosides, β-D-xylosides, β-D-galactosides, and β-L-arabinosides.

It is the opinion of the authors that β-glucosidase in combination with a specific endo-β-glucanase could find widespread application in the quantification of a range of β-D-glucans such as $(1 \rightarrow 4)$-β-D-glucan, $(1 \rightarrow 3)$-β-D-glucan, $(1 \rightarrow 3),(1 \rightarrow 4)$-$\beta$-D-glucan, and $(1 \rightarrow 3),1 \rightarrow 6)$-$\beta$-D-glucan. Together with endo-1,4-β-D-mannanase and β-D-mannosidase it may also prove useful in the measurement of β-D-glucomannans. A method for the assay of $(1 \rightarrow 3),(1 \rightarrow 4)$-$\beta$-D-glucan has already been developed by the authors[10] using a highly purified β-D-glucosidase from a commercially available *Aspergillus niger* enzyme preparation, Novozym 188. In this chapter we shall describe the purification of this enzyme and report on some of its properties.

Assay Method

Using p-Nitrophenyl-β-D-glucopyranoside. Suitably diluted enzyme preparation (0.1 ml) is incubated with 0.1 ml of 10 mM p-nitrophenyl-β-D-glucopyranoside in 0.1 M sodium acetate buffer (pH 4.0) at 40° for 1–5 min. The reaction is terminated and color developed by the addition of 3 ml of 2% (w/v) aqueous sodium carbonate. The absorbance at 410 nm is measured and the amount of released p-nitrophenol determined by refer-

[5] T. Hirayama, S. Horie, H. Nagayama, and K. Matsuda, *J. Biochem.* (*Tokyo*) **84**, 27 (1978).
[6] S. Murao and R. Sakamoto, *Agric. Biol. Chem.* **43**, 1791 (1979).
[7] L. H. Li, R. M. Flora, and K. W. King, *Arch. Biochem. Biophys.* **111**, 439 (1965).
[8] C. S. Gong, M. R. Ladisch, and G. T. Tsao, *Biotechnol. Bioeng.* **19**, 959 (1977).
[9] A. K. Grover, D. D. Macmurchie, and R. J. Cushley, *Biochim. Biophys. Acta* **482**, 98 (1977).
[10] B. V. McCleary and M. Glennie-Holmes, *J. Inst. Brew.* **91**, 285 (1985).

ence to a p-nitrophenol standard curve [prepared in 2% (w/v) aqueous sodium carbonate]. Activity on other nitrophenyl glycosides was assayed by the same procedure.

Using Oligosaccharides and Aromatic Glucosides. Suitably diluted enzyme preparation (0.05 ml) is incubated with 0.5 ml of 10 mM substrate in 0.1 M sodium acetate buffer (pH 4.0) at 40° for 5–15 min. The reaction is terminated by incubating the reaction tubes at 100° for 1 min and re- leased D-glucose is measured using the glucose oxidase/peroxidase proce- dure.[11] Substrates employed include cellobiose, -triose, -tetraose, and -pentaose; gentiobiose; $(1 \rightarrow 4)$-β-D-glucosyl-D-mannose and $(1 \rightarrow 4)$-β- D-glucosyl-D-mannobiose; 3-O-β-D-cellobiosyl-D-glucose, 3-O-β-D-cello- triosyl-D-glucose, and 3^2-β-D-glucosyl-D-cellobiose; the heptasaccharide released on hydrolysis of tamarind amyloid by *Trichoderma reesei* cellu- lase [i.e., 6^2, 6^3, 6^4-tri-O-α-D-xylosyl-$(1 \rightarrow 4)$-β-D-cellotetraose]; and a range of aromatic glucosides including arbutin, salicin, phloridzin, and esculin hydrate.

Purification Procedure

Step 1. Preparative Chromatography on Ultrogel AcA34. Novozym 188 (Novo Industrias, Denmark, 150 ml) is diluted 2-fold with distilled water, centrifuged at 15,000 g for 15 min, and the supernatant is applied (in six separate lots) to a column of Ultrogel AcA34 (3.5 × 40 cm) equili- brated with 10 mM sodium acetate buffer (pH 4.5). Fractions active in β- glucosidase and with a low maltase activity are combined and stored in the frozen state.

Step 2. SP-Tris Acrylamide Column Chromatography. The enzyme recovered from Ultrogel AcA 34 chromatography is treated with sodium citrate buffer (pH 3) to a final concentration of 10 mM and adjusted to pH 3 by careful addition of 1 M HCl. The solution is applied to an SP-Tris Acrylamide column (2.5 × 10 cm) equilibrated with 10 mM sodium citrate buffer (pH 3) at 4° and the column is subjected to linear gradient elution with 1 liter of this buffer containing 0–0.5 M KCl (Fig. 1). The fractions active in β-glucosidase are pooled and treated with ammonium sulfate (50 g/100 ml) and stored at 4° for 2 hr. The suspension is centrifuged at 15,000 g for 10 min and recovered pellet dissolved in a minimum volume of distilled water (~20 ml).

Step 3. Ultrogel AcA 44 Column Chromatography. β-Glucosidase re- covered from SP-Tris Acrylamide chromatography is applied to a column (2.5 × 90 cm) of Ultrogel AcA 44 and the column is eluted with 20 mM

[11] A. B. Blakeney and N. K. Matheson, *Starke* **36**, 265 (1984).

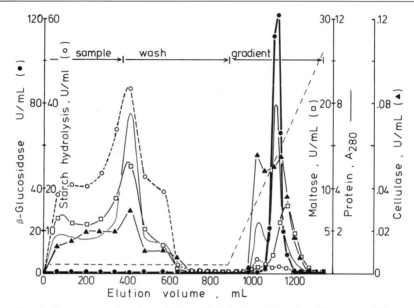

FIG. 1. Chromatography of partially purified β-glucosidase from Novozyme 188 on an SP-Tris Acrylamide column (2.5 × 10 cm). The column is eluted with 1 liter of a linear KCl gradient (0–0.5 M) in 10 mM sodium citrate buffer (pH 3).

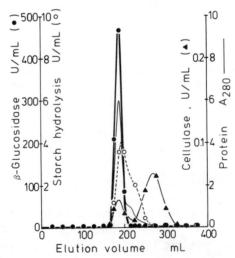

FIG. 2. Chromatography of β-glucosidase on an Ultrogel AcA 44 column (2.5 × 90 cm). The column is eluted with 20 mM sodium phosphate buffer (pH 6.5).

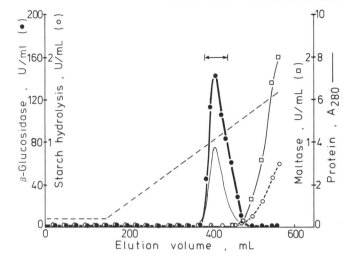

FIG. 3. Chromatography of β-glucosidase on a DEAE-Sepharose CL-6B column (2.5 × 10 cm). The column is eluted with a linear KCl gradient (0–0.3 *M*) in 20 m*M* sodium phosphate buffer (pH 6.5).

sodium phosphate buffer (pH 6.5) (Fig. 2). The active fractions are collected and pooled.

Step 4. DEAE-Sepharose Column Chromatography. The active fraction from the previous column run is applied directly to a column (2.5 × 10 cm) of DEAE-Sepharose CL-6B equilibrated with 20 m*M* sodium phosphate buffer (pH 6.5) at 4°, and the column eluted with 600 ml of a linear KCl gradient (0–0.3 *M*) (Fig. 3). This chromatographic step removes the last traces of maltase and amyloglucosidase activity from the β-glucosidase. Two fractions across the peak are usually collected. The first one (shown in Fig. 3) is completely devoid of maltase and amyloglucosidase; the second fraction still contains traces of these activities and consequently is usually rechromatographed. The recovered enzyme is treated with ammonium sulfate (50 g/100 ml) and stored at 4°. Alternatively, the enzyme is dialyzed against 20 m*M* sodium acetate buffer (pH 4) and stored in the frozen state in polypropylene containers.

Purity and Properties

Recovery and Purity. The described purification procedure, summarized in Table I, results in a 67-fold overall purification with about 60% recovery of activity. The recovered enzyme has a specific activity of 53 U/mg on 10 m*M* *p*-nitrophenyl-β-D-glucopyranoside at pH 4 and 40°, and

pH

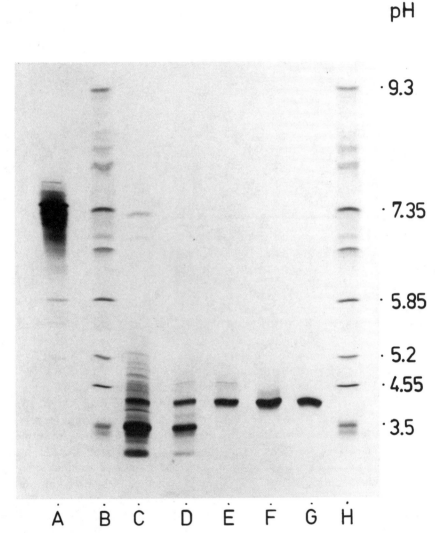

FIG. 4. Patterns obtained on isoelectric focusing of Novozym 188 β-glucosidase at various stages of purification. A, hemoglobin; B, isoelectric focusing standards; C, crude Novozym 188 preparation; D, β-glucosidase through Ultrogel AcA 34; E, enzyme off SP-Tris Acrylamide; F, enzyme through Ultrogel AcA 44; G, enzyme off DEAE-Sepharose CL-6B; and H, isoelectric focusing standards.

TABLE I
PURIFICATION OF β-GLUCOSIDASE

Step	Total activity (units)	Total protein (mg)	Specific activity (units/mg)	Yield (%)	Purification (fold)
Original solution	18,810	24,000	0.78	—	1
1. Ultrogel AcA 34 (bulk)	17,212	1,800	9.6	92	12.3
2. SP-Tris Acrylamide	14,596	423	34.5	78	44.2
3. Ultrogel AcA 44	13,212	267	49.5	70	63.5
4. DEAE-Sepharose CL-6B					
I	10,875	206	52.8	58	67.7
II	1,300	25	52.0	7	66.7

TABLE II
RELATIVE INITIAL RATES OF HYDROLYSIS OF GLUCOSIDES AND NITROPHENYL-
GLYCOSIDES BY A. *niger* β-GLUCOSIDASE

Substrate	Linkage of D-glycosyl group	Relative initial rate of hydrolysis (%)
Cellobiose	β-(1 → 4)	100
Cellotriose	β-(1 → 4)	109
Cellotetraose	β-(1 → 4)	89
Cellopentaose	β-(1 → 4)	87
Gentiobiose	β-(1 → 6)	53
Sophorose	β-(1 → 2)	43
Gentiotriose	β-(1 → 6)	64
$6^2,6^3,6^4$-Tri-O-α-D-xylosyl-(1 → 4)-β-D-cellotetraose	β-(1 → 4)	0
3-O-β-D-Cellobiosyl-D-glucose	β-(1 → 4),(1 → 3)	100
3-O-β-D-Cellotriosyl-D-glucose	β-(1 → 4),(1 → 3)	87
3^2-β-D-Glucosyl-cellobiose	β-(1 → 3),(1 → 4)	111
(1 → 4)-β-D-Glucosyl-D-mannose	β-(1 → 4)	35
(1 → 4)-β-D-Glucosyl-D-mannobiose	β-(1 → 4)	38
Arbutin (hydroquinone → Glc)	β	22
Salicin (salicyl → Glc)	β	26
Phloridzin (phloretin → Glc)	β	u.d.[a]
Esculin hydrate (6,7-dihydroxycoumarin-6-yl → Glc)	β	65
Methyl-β-D-glucopyranoside	β	15
p-Nitrophenyl-β-D-glucopyranoside	β	46
o-Nitrophenyl-β-D-xylopyranoside	β	1.2
o-Nitrophenyl-β-D-galactopyranoside	β	0.2
p-Nitrophenyl-β-D-mannopyranoside	β	0.1
p-Nitrophenyl-α-D-glucopyranoside	α	<0.1

[a] u.d., undetectible.

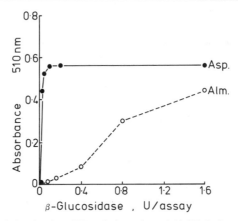

FIG. 5. Hydrolysis by *A. niger* (●) and almond emulsin (○) β-glucosidases of the oligo-saccharide mixture produced on treatment of (1 → 3),(1 → 4)-β-D-glucan by 1,3(4)-β-D-glucanase. Oligosaccharide mixture (35 μg in 0.1 ml) is incubated with β-glucosidase (0–1.6 U) at 40° for 15 min and the released D-glucose determined using glucose oxidase/peroxidase reagent mixture.

116 U/mg on 50 mM cellobiose. The extent of contamination with other activities is α-amylase, 0.001%; amyloglucosidase, <0.001%; maltase, <0.01%; endo-(1 → 4)-β-D-glucanase, <0.001%; and invertase, 0.3%. The enzyme appears as a single sharp band on isoelectric focusing (pI 4.0) (Fig. 4) and on SDS–gel electrophoresis (MW ~ 118,000).

Properties. *A. niger* β-glucosidase has a molecular weight of 118,000 ± 2000 and an isoelectric point of 4.0. The optimal pH for activity is 4.0. The enzyme showed no loss in activity on storage at pH 2.5–9.0 for 24 hr at 4°, but at 22° a loss of activity occurred above pH 7.5. At pH 9 and 22°, 10% of activity was lost in 24 hr. β-Glucosidase is stable at temperatures up to 65° (15 min at pH 4.5) and shows maximal activity at 70–75°. It is stable to repeated freezing and thawing.

Kinetic Properties and Specificity. *A. niger* β-glucosidase readily hydrolyzes a wide range of substrates. The enzyme can tolerate a wide range of aglycons but the rates of hydrolysis vary (Table II). Cellooligosaccharides of varying DP are hydrolyzed at similar rates, but as the DP increases from 2 to 3 the K_m decreases. K_m values obtained for cellobiose, triose, and tetraose are 1.89, 0.50, and 0.51 mM, respectively. The K_m for gentiobiose is 1.88 mM and that for *p*-nitrophenyl-β-D-glucopyranoside is 0.8 mM. *A. niger* β-glucosidase is far more effective in hydrolyzing (1 → 4)-β-D-glucosyl-D-mannose and mixed linkage β-glucooligosaccharides than is almond emulsin β-glucosidase.[12] The relative rates of hydrolysis

[12] B. V. McCleary and N. K. Matheson, *Carbohydr. Res.* **119,** 191 (1983).

by *A. niger* and highly purified almond emulsin β-glucosidases of the oligosaccharide mixture produced on treatment of barley β-glucan with endo-1,3(4)-β-D-glucanase (lichenase) is shown in Fig. 5. This purified *A. niger* β-glucosidase, unlike that reported by Tvaobilov *et al.*,[2] has little action on nitrophenyl glycosides other than *p*-nitrophenyl-β-D-glucopyranoside. Nitrophenyl-β-D-xyloside is hydrolyzed at only 2–3% the rate of the nitrophenyl-β-D-glucoside by the currently reported β-glucosidase, whereas that reported by Tavobilov *et al.*[2] from *A. niger*-15 hydrolyzes the β-xyloside at 4.5 times the rate of *p*-nitrophenyl-β-D-glucoside.

The high specificity of *A. niger* β-glucosidase for the D-gluco configuration and its ability to readily release β-linked D-glucose from a wide range of substrates and linkage types indicate that this enzyme could be of particular value in the structural characterization of D-glucose containing oligosaccharides and polysaccharides.

[72] Exo-1,4-β-mannanase from *Aeromonas hydrophila*

By Toshiyoshi Araki and Manabu Kitamikado

$$1,4\text{-}\beta\text{-Mannan} + H_2O \rightarrow \text{Man-}\beta(1 \rightarrow 4)\text{Man}$$

A novel exo-1,4-β-mannanase was isolated from the culture fluid of strain No. F-25 of *Aeromonas hydrophila*.[1] The enzyme removes mannobiose residues successively from the nonreducing end of the 1,4-β-mannan link in polysaccharides of three or more 1,4-β-linked D-mannose units. It has no action on mannobiose, mannosylmannitol, and *p*-nitrophenyl-β-D-mannopyranoside.[2] The organism also secretes an endo-1,4-β-mannanase in addition to the exo-1,4-β-mannanase in the culture fluid.[3] These enzymes can be separated by DEAE-Sephadex A-50 column chromatography.

Assay Methods

Principle. The exo-1,4-β-mannanase hydrolyzes 1,4-β-mannan and mannotriose, while the endo-1,4-β-mannanase cleaves 1,4-β-mannan but not mannotriose. Therefore, 1,4-β-mannan from *Codium fragile* and man-

[1] T. Araki and M. Kitamikado, *Bull. Jpn. Soc. Sci. Fish.* **47**, 753 (1981).
[2] T. Araki and M. Kitamikado, *J. Biochem. (Tokyo)* **91**, 1181 (1982).
[3] T. Araki, *J. Fac. Agric., Kyushu Univ.* **27**, 89 (1983).

notriose are used as substrates to distinguish between these two enzymes. The assay procedure is based on the measurement of reducing power of mannobiose or the other mannooligosaccharides released from mannotriose or 1,4-β-mannan, and the reducing sugar released is determined by the Somogyi–Nelson method.[4]

Reagents

Mannotriose solution, 1.0 mg/ml
1,4-β-Mannan suspension, 10 mg/ml
Sodium acetate buffer, 0.1 M, pH 6.0 or pH 5.5

Procedure. The standard assay mixture (1 ml) for determining mannotriose-hydrolyzing activity contains 0.25 ml of mannotriose solution, 0.25 ml of sodium acetate buffer, pH 6.0, enzyme solution, and water. The volume of 2.5 ml for determining 1,4-β-mannan-hydrolyzing activity contains 1 ml of 1,4-β-mannan suspension, 1 ml of sodium acetate buffer, pH 5.5, enzyme solution, and water. After incubation at 37° for 10 min, the reaction is stopped by placing the mixture in a boiling water bath for 5 min. The reducing sugar produced is then determined by the Somogyi–Nelson method,[4] and is expressed as mannose.

Definition of Unit and Specific Activity. One unit of exo- or endo-1,4-β-mannanase activity is defined as the amount of enzyme that produces a reducing capacity equivalent to 1 μmol of D-mannose, from the substrate per minute, under the above conditions. Specific activity is defined as units per milligram of protein.

Materials

Organism. The organism used is strain No. F-25 of *Aeromonas* sp. isolated by us in 1973 from intestinal contents of the rainbow trout, *Salmo gairdnerii,*[5] and belongs to the *Aeromonas hydrophila* subspecies *anaerogenes,* according to the 8th edition of *Bergey's Manual of Determinative Bacteriology.*[1,6] The strain is currently maintained in our laboratory as a slant culture or as lyophilized cells. The liquid medium used for culturing the organism is composed of 1.0% peptone, 0.1% yeast extract, 0.05% $MgSO_4 \cdot 7H_2O$, 0.2% K_2HPO_4, 0.05% KH_2PO_4, 0.5% NaCl, and 0.5% codium 1,4-β-mannan or 0.5% powder of *Amorphophallus konjac,* pH 7.0.

[4] M. Somogyi, *J. Biol. Chem.* **195,** 19 (1952).
[5] T. Araki and M. Kitamikado, *Bull. Jpn. Soc. Sci. Fish.* **44,** 1135 (1978).
[6] R. E. Buchanan and N. E. Gibbons, *in* "Bergey's Manual of Determinative Bacteriology" (R. H. W. Schubert, ed.), 8th Ed., p. 345. Williams & Wilkins, Baltimore, Maryland, 1974.

Substrates. Codium 1,4-β-mannan (Man : Glc, 95 : 5),[7] coffee 1,4-β-mannan (Man : Gal, 98 : 2),[8] konjac glucomannan (Man : Glc, 3 : 2),[9] and guar gum galactomannan (Man : Gal 2 : 1)[10] are prepared from *Codium fragile* (a species of green seaweed), coffee beans, *Amorphophallus konjac,* and guar gum, respectively. Mannooligosaccharides composed of 1,4-β-linked D-mannose units are prepared from hydrolysis products of codium 1,4-β-mannan with the endo-1,4-β-mannanase from strain No. F-96 of *Bacillus* sp.[1] These oligosaccharides are fractionated by chromatography on charcoal and Sephadex G-15 columns, and purified by preparative paper chromatography. Reduction of mannose and mannooligosaccharides is carried out as described by Cole and King.[11]

Purification Procedure

The exo-1,4-β-mannanase activity is measured, using mannotriose as the substrate. Unless otherwise indicated, all procedures are carried out at 4°.

Step 1. Preparation of Culture Fluid. The cells taken from the slant culture are inoculated into 20-ml aliquots of liquid medium in 50-ml flasks, and the preparation is then incubated at 25° for 1 day. Each culture is then transferred to 2000 ml of liquid medium in a 5000-ml flask, and incubated on a reciprocal shaker (70 cycles/min) with stirring at 25° for 3 days. The cultures are centrifuged at 15,000 *g* for 30 min. From 8000 ml of liquid culture, 7700 ml of clear culture fluid is obtained.

Step 2. Ammonium Sulfate Precipitation. The culture fluid is adjusted to 75% saturation by the addition of 51.6 g of solid ammonium sulfate per 100 ml, then left to stand overnight. The precipitate is collected by centrifugation, and dissolved in 200 ml of distilled water.

Step 3. Chromatography on DEAE-Sephadex A-50. The enzyme solution from the previous step is dialyzed against 10 mM sodium acetate buffer, pH 6.0, containing 2 mM calcium acetate for 2 days. The dialyzed enzyme solution is applied to a column (2.2 × 42 cm) of DEAE-Sephadex A-50 equilibrated with the same buffer. After being washed with 6 bed volumes of the same buffer, the column is eluted with 40 mM sodium acetate buffer, pH 6.0, containing 2 mM calcium acetate. The exo-1,4-β-mannanase, which hydrolyzes both codium 1,4-β-mannan and manno-

[7] J. Love and E. Percival, *J. Chem. Soc.* p. 3345 (1964).
[8] M. L. Wolfrom, M. L. Lavar, and D. L. Patin, *J. Org. Chem.* **26,** 4533 (1961).
[9] N. Sugiyama, H. Shimahara, and T. Andho, *Bull. Chem. Soc. Jpn.* **45,** 561 (1972).
[10] E. Heyne and R. Whistler, *J. Am. Chem. Soc.* **70,** 2249 (1948).
[11] F. E. Cole and K. W. King, *Biochim. Biophys. Acta* **81,** 122 (1964).

triose, appears in tubes 52–85. The column is finally eluted with a continuous linear gradient formed from 500 ml of 40 mM sodium acetate buffer, pH 6.0, containing 2 mM calcium acetate and 500 ml of 0.4 M NaCl in the same buffer. The endo-1,4-β-mannanase, which cleaves codium 1,4-β-mannan but not mannotriose, is eluted in tubes 116–140. A typical elution profile of these two enzymes is illustrated in Fig. 1. The exo-1,4-β-mannanase fractions are pooled and concentrated to 50 ml by the use of polyethylene glycol (MW 20,000), and then further purified in the following steps.

Step 4. Chromatography on CM-Sephadex C-50. All the buffers used in this step contain 1 mM mercaptoethanol. The exo-1,4-β-mannanase solution obtained above is dialyzed against 20 mM sodium acetate buffer, pH 5.5. The dialyzed enzyme is applied to a column (1.2 × 10 cm) of CM-Sephadex C-50 equilibrated with the same buffer, and then eluted with a continuous linear gradient generated from 250 ml each of 20 mM sodium acetate buffer, pH 5.5, and 0.5 M sodium acetate buffer, pH 6.2. After the

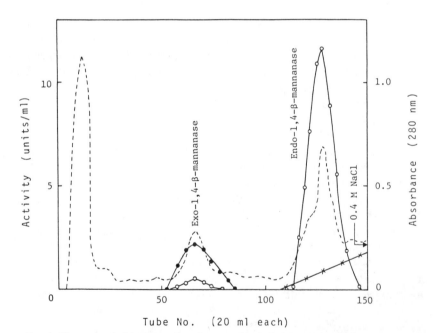

Fig. 1. Chromatographic separation of exo- and endo-1,4-β-mannanase from *Aeromonas hydrophila* on DEAE-Sephadex A-50. (●) Mannotriose-hydrolyzing activity; (○) codium 1,4-β-mannan-hydrolyzing activity; (×) concentration of NaCl; (--) absorbance at 280 nm; tubes 0–50, 10 mM sodium acetate buffer of pH 6.0; tubes 51–100, 40 mM sodium acetate buffer of pH 6.0; tubes 101–150, 0.4 M NaCl linear gradient.

active fractions are pooled, calcium acetate is added to a final concentration of 2 mM. The solution is then concentrated to 6 ml in an Amicon ultrafiltration apparatus with a UM30 membrane.

Step 5. Gel Filtration on Sephadex G-100. The above enzyme solution is applied to a column (2.2 × 100 cm) of Sephadex G-100 equilibrated with 20 mM sodium acetate buffer, pH 6.0, containing 2 mM calcium acetate and 0.1 M NaCl. Elution is performed with the same buffer. Active fractions are pooled and concentrated to 10 ml by ultrafiltration.

Step 6. Chromatography on DEAE-Sephadex A-50. The enzyme solution from the previous step is dialyzed against 20 mM Tris–HCl buffer, pH 7.5, containing 2 mM calcium acetate. The dialyzed enzyme is applied to a column (1.5 × 30 cm) of DEAE-Sephadex A-50 equilibrated with the same buffer. The column is eluted with a linear NaCl gradient between 200 ml of 20 mM Tris–HCl buffer, pH 7.5, containing 2 mM calcium acetate and 200 ml of 0.4 M NaCl in the same buffer. Fractions containing the activity are pooled and dialyzed against 50 mM sodium acetate buffer, pH 6.0. The dialyzed solution is used as the purified exo-1,4-β-mannanase.

Properties

Purity. The purification and yield of the exo-1,4-β-mannanase are summarized in Table I. The purified preparation yields a single band on the polyacrylamide disc gel electrophoresis.

TABLE I
PURIFICATION OF EXO-1,4-β-MANNANASE FROM *Aeromonas hydrophila*

Fraction	Total protein (mg)	Total activity[a] (units)	Specific activity (units/mg)	Yield (%)
Culture fluid (7700 ml)	38,300	836	0.0218	100
Ammonium sulfate	374	950	2.54	114
DEAE-Sephadex A-50	49.8	1,180	23.7	141
CM-Sephadex C-50	12.9	618	47.9	73.9
Sephadex G-100	2.99	233	77.9	27.9
DEAE-Sephadex A-50	1.71	148	86.5	17.7

[a] Mannotriose is used as the substrate for assay of the enzyme.

Stability. The enzyme at a concentration of at least 0.5 mg of protein per milliliter is stable at $-20°$ for over 1 year.

Isoelectric Point. The isoelectric point is pH 5.9.

Molecular Weight. The apparent molecular weight is estimated to be 64,000, by Sephadex G-100 gel filtration.

Effect of pH. The optimum pH and the pH stability curve are determined by using 50 mM sodium acetate (pH 3.2–6.2), phosphate (pH 5.5–8.0), Tris–HCl (pH 7.0–9.0), and glycine–NaOH (pH 8.0–11.0) buffers in the assay system. The maximal activity is observed at approximately pH 6.0. The enzyme is stable between pH 5.0 and 8.5.

Effect of Temperature. The enzyme, kept in 50 mM sodium acetate buffer, pH 6.0, for 15 min, retains full activity at temperatures up to 45°, but loses all activity at temperatures above 55°.

Inhibitors. The enzyme activity is almost completely inhibited by 1 mM Ag^+, Hg^{2+}, Cu^{2+}, or Pb^{2+}, while Zn^{2+}, Fe^{3+}, EDTA, and PCMB give 30–50% inhibition, at the same concentration. Na^+, Ca^{2+}, Mg^{2+}, and Mn^{2+} have no significant effects on the enzyme activity, at the same concentration.

Kinetic Parameters. The apparent K_m values for mannotriose, mannotetraose, and mannopentaose are 5.1×10^{-4}, 2.4×10^{-4}, and 1.3×10^{-4} M, respectively. When the relative rates of action of the exo-1,4-β-mannanase on mannooligosaccharides and reduced oligosaccharides are compared by setting arbitrarily the rate for mannopentaose at 100, those for mannotetraose, mannotriose, mannotriosylmannitol, and mannobiosylmannitol are 78, 41, 20, and 0.08, respectively.

Action Pattern. The exo-1,4-β-mannanase hydrolyzes codium and coffee 1,4-β-mannans to give only mannobiose, but does not cleave konjac glucomannan or guar gum galactomannan. The extents of hydrolysis of codium and coffee 1,4-β-mannans by the enzyme (1.0 unit) are 13.8 and 3.6%, respectively, after 48 hr of incubation at 37°. The enzyme cleaves mannotriose and mannopentaose to give mannobiose and mannose, and forms mannobiose from mannotetraose. It also hydrolyzes mannobiosylmannitol and mannotetraosylmannitol to produce mannobiose and mannitol, and forms mannobiose and mannosylmannitol from mannotriosylmannitol, as shown in Fig. 2. The exo-1,4-β-mannanase possesses the ability to carry out transglycosylation, in that the enzyme attacks mannotriose to give a small amount of mannopentaose in addition to mannobiose and mannose, and acts on mannotetraose to form a small amount of mannohexaose with mannobiose, in the early stages of the reaction. The enzyme cannot cleave mannobiose, mannosylmannitol, and p-nitrophenyl-β-D-mannopyranoside.

In conclusion, the exo-1,4-β-mannanase from strain No. F-25 of *Aero-*

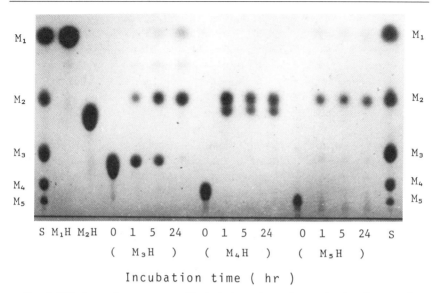

S M₁H M₂H 0 1 5 24   0 1 5 24   0 1 5 24 S
           (M₃H   )   (M₄H   )   (M₅H   )

Incubation time (hr)

FIG. 2. Thin-layer chromatogram of hydrolysis products of several reduced mannooligo-saccharides with exo-1,4-β-mannanase. S, standard; M_1–M_5, mannose–mannopentaose; M_1H–M_5H, mannitol–mannotetraosylmannitol. The reaction mixture consists of 0.5 ml of 10 mM sodium acetate buffer, pH 6.0, containing 1% reduced oligosaccharide and 0.5 ml of the enzyme solution (1 unit), and it is incubated at 37°. The plate is developed with *n*-butanol : 2-propanol : water (50 : 25 : 20, v/v), and sprayed with an anisaldehyde reagent.

monas hydrophila is a novel glycosidase which successively removes the mannobiose residue from the nonreducing end of the 1,4-β-mannan molecule. The enzyme is, therefore, classified as 1,4-β-D-mannan mannobiohydrolase.

[73] Exo-β-D-mannanase from *Cyamopsis tetragonolobus* Guar Seed

By BARRY V. MCCLEARY

Exo-β-mannanase (β-D-mannoside mannohydrolase, EC 3.2.1.25) catalyzes the hydrolytic cleavage of single, nonreducing terminal, D-mannosyl residues from 1,4-β-D-mannooligosaccharides and from synthetic β-linked D-mannosyl glycosides.[1] In several respects the action pattern of

[1] B. V. McCleary, *Carbohydr. Res.* **101,** 75 (1982).

this enzyme is very similar to that of β-mannosidase[2] (also classified as β-D-mannoside mannohydrolase), but there are differences in the nature of the preferred substrate and the anomeric configuration of the released D-mannose. The only detailed report of an exo-β-mannanase is that of the enzyme from guar seeds[1] (*Cyamopsis tetragonolobus*) although other studies[1,3] indicate that the β-D-mannosylmannohydrolase enzymes from a range of seeds may in fact have a similar action pattern to the guar seed enzyme.

Early research[4,5] on the enzymatic degradation of galactomannan clearly demonstrated the need for three enzymes, α-galactosidase, endo-β-mannanase, and β-D-mannoside mannohydrolase (β-mannosidase or exo-β-mannanase). However, attempts to demonstrate the presence of the latter enzyme in extracts of seeds known to contain galactomannan as a reserve polysaccharide have met with limited success. The currently described research shows that there were several reasons for this. Many of the problems appear to be due to a relatively specific interaction of these enzymes with other relatively insoluble protein material present in the cotyledons of these seeds. Further, the exo-β-mannanase present in seed endosperms is distributed throughout the endosperm tissue (which is mainly galactomannan),[1,6] and effective extraction requires prior dissolution of this material either mechanically or using endo-β-mannanase. Demonstration of the presence of this enzyme was also complicated by the fact that the enzyme is unstable below pH 5 and has no action on reduced mannotriose, which has been employed as substrate in some studies.[7]

Assay Method

Principle. exo-β-Mannanase activity can be assayed using β-D-mannotetraitol or β-D-mannopentaitol as substrate and measuring the release of D-mannose by a suitable reducing sugar procedure. A more convenient assay employs *p*-nitrophenyl β-D-mannoside and measures the rate of cleavage of glycosidic bonds by the rate of release of *p*-nitrophenol.

Reagents and Procedure. These are the same as employed for the assay of β-mannosidase[8] except that the substrate is prepared in 0.1 *M* sodium acetate buffer at pH 5.5.

[2] B. V. McCleary, *Carbohydr. Res.* **111**, 297 (1983).
[3] F. B. F. Ouellette, M.S. thesis. University of Calgary (1984).
[4] E. T. Reese and Y. Shibata, *Can. J. Microbiol.* **11**, 167 (1965).
[5] J. S. G. Reid and H. Meier, *Planta* **112**, 301 (1973).
[6] B. V. McCleary, *Phytochemistry* **22**, 649 (1983).
[7] R. Somme, *Acta Chem. Scand.* **25**, 759 (1971).
[8] B. V. McCleary, this volume [76].

Purification Procedure

Germination of Guar Seed. Five hundred grams of guar seed obtained from the previous year's crop and stored over desiccant at 4° is used. The seed is surface sterilized by soaking in 4 liters of 0.5% (v/v) sodium hypochlorite solution for 15 min. The seed is then collected on nylon mesh, rinsed thoroughly with demineralized water, and soaked in demineralized water for 6 hr. Over this period the water is changed several times and finally the seed is rinsed thoroughly with demineralized water, collected on a screen, and excess water allowed to drain. This process removes compounds which inhibit seed germination. The seed is spread evenly over sheets of paper towel in large metal trays (30 × 80 cm), overlain with more towel followed by another layer of seeds and towel. The paper towel is moistened with demineralized water from a wash bottle and the trays covered with aluminum foil and stored at 25° for 2.5 days. By this time the bulk of the reserve galactomannan in seed endosperms is mobilized, facilitating enzyme extraction.

Extraction of Enzymes. The germinated seed is homogenized in 5 liters of distilled water in a Waring blender at maximum setting and the homogenate stored at 30° for 60 min to allow degradation of residual galactomannan. The slurry is then rehomogenized and squeezed through fine nylon mesh. The filter-cake is stored at 4° and the filtrate is centrifuged at 3500 *g* for 30 min. The supernatant which contains most of the endo-β-mannanase and α-galactosidase activity is either discarded or stored at −20° awaiting purification of these enzymes. The pellet obtained on centrifugation is resuspended in 500 ml of a solution of 0.2 *M* NaCl in 50 m*M* Tris–HCl buffer (pH 8) and combined with the filter-cake. A further 2 liters of 0.2 *M* NaCl in 50 m*M* Tris–HCl buffer (pH 8) is added and the slurry is homogenized in a Waring blender at maximum setting for 60 sec and filtered through nylon mesh. The filtrate, which contains most of the exo-β-mannanase activity, is centrifuged at 3500 *g* for 20 min and the supernatant dialyzed against 10 liters of 20 m*M* Tris–HCl buffer (pH 8). The external solution is changed three times over 48 hr.

The conditions of extraction of guar seed exo-β-mannanase have a significant effect on the subsequent ease of purification. Effective extraction is achieved only at salt concentrations in excess of 0.2 *M*. However, if the seed material is extracted at this salt concentration in a solution buffered at pH 5.0–5.5, essentially all of the enzyme precipitates from solution on dialysis. In contrast, if the enzyme is extracted with 0.2 *M* sodium chloride buffered at pH 8, 60–80% of the exo-β-mannanase remains soluble on dialysis. The insolubility of the enzyme after extraction and dialysis at pH 5.0–5.5 is considered to be due to interaction with other protein at these pH values. In a purified form, guar exo-β-mannanase is quite soluble in solutions of low pH and low salt concentration. The extraction procedure described here takes advantage of the limited ability of water to extract the enzyme from crude seed homogenates. Preliminary extraction with water removes most of the α-galactosidase and endo-β-mannanase but only 20–30% of the exo-β-mannanase. The remaining exo-β-mannanase is extracted with 0.2 M NaCl in 20 mM Tris–HCl (pH 8).

Purification Procedure

Step 1. DEAE-Cellulose Treatment. The dialyzed enzyme solution is centrifuged at 3500 g for 20 min and the supernatant solution is treated with approximately 500 g of wet-cake DEAE-cellulose which has been preequilibrated with 20 mM Tris–HCl buffer (pH 8). The DEAE-cellulose is added with gentle stirring over 10 min and the slurry is poured in two separate lots onto a sintered-glass funnel (10 × 7 cm) and washed with 4 liters of 20 mM Tris–HCl buffer (pH 8) under a slight vacuum. Exo-β-mannanase is eluted by washing with 0.2 M NaCl plus 20 mM Tris–HCl buffer (pH 8). Fractions (approximately 100 ml) of the eluate are collected and assayed for activity. The active fractions are pooled and dialyzed against 10 liters of 20 mM Tris–HCl buffer (pH 8) at 4° for 20 hr.

This purification step removes the protein material which interacts with exo-β-mannanase and renders it insoluble at low pH values.

Step 2. DEAE-Cellulose Column Chromatography. The dialyzed solution is applied to a DEAE-cellulose column (3.7 × 17 cm) previously equilibrated with 20 mM Tris–HCl buffer (pH 8). The column is washed with 500 ml of the same buffer and is then subjected to linear gradient elution with this buffer containing 0–0.4 M NaCl (total volume, 2 liters). Exo-β-mannanase elutes as a single peak at a salt concentration of 0.15 M. Endo-β-mannanase elutes at the same salt concentration whereas α-galactosidase I elutes at 0.18 M NaCl and α-galactosidase II elutes at 0.25 M NaCl. The fractions active in exo-β-mannanase are combined and treated with solid ammonium sulfate at a rate of 50 g/100 ml (80% saturation) with gentle stirring at 4° for 4 hr. Precipitated protein is collected by centrifugation at 12,000 g for 10 min, redissolved in a minimum volume (approximately 15 ml) of 0.2 M NaCl plus 20 mM Tris–HCl buffer (pH 8), and dialyzed against 1 liter of the same solution at 4° for 4 hr.

Step 3. Ultrogel AcA 44 Column Chromatography. The dialyzed enzyme preparation is centrifuged at 12,000 g for 10 min and applied to an Ultrogel AcA 44 (LKB Productur) column (2.5 × 88 cm) and eluted with 0.2 M NaCl in 20 mM Tris–HCl buffer (pH 8). The peak of enzyme activity elutes at a volume of 310 ml (the column total volume is 430 ml). The active fractions are combined and concentrated by dialysis against polyethylene glycol 4000 and then dialyzed against 10 mM Tris–HCl buffer (pH 8) at 4° for 16 hr.

Step 4. Polybuffer Exchanger PBE 118 Column Chromatography. The dialyzed enzyme is divided into three equal aliquots. One of these is treated with diethylamine–HCl buffer (pH 11.0) to a final concentration of 25 mM and immediately applied to a Polybuffer Exchanger PBE 118 (Pharmacia Inc.) column equilibrated with 25 mM diethylamine–HCl (pH 11) and eluted with 160 ml of Pharmalyte pH 8–10.5 HCl (Pharmacia Inc.) (diluted 1 : 45 and adjusted to pH 8). Exo-β-mannanase elutes at pH 9.4. The column is washed with 1 M NaCl to remove remaining proteins, including α-galactosidase and exo-β-mannanase. The recovered exo-β-mannanase is concentrated by dialysis against polyethylene glycol 4000 and chromatographed on a column of Sephadex G-25 in 5 mM Tris–HCl (pH 8) to remove most of the Pharmalyte pH 8–10.5 HCl. The enzyme is stored at −20° in polypropylene containers.

Purity and Properties

The purification procedure, summarized in Table I, results in a 104-fold overall purification with about 7% recovery of activity. The final specific activity is 5.7 U/mg on *p*-nitrophenyl-β-D-mannopyranoside and 57.9 U/mg on mannohexaitol at 40° and pH 5.5.

TABLE I
PURIFICATION OF EXO-β-MANNANASE FROM GERMINATED GUAR SEEDS

Step	Total activity (units)	Total protein (mg)	Specific activity (units/mg)	Yield (%)	Purification (fold)
Crude extract	2275	15,400	0.15	100	1
Dialysis (pH 8)	1982	12,350	0.16	87.1	1.1
Centrifugation (3500 *g*)	1251	6,800	0.19	54.9	1.3
1. DEAE-Cellulose (bulk)	832	3,180	0.26	36.6	1.7
2. DEAE-Cellulose (column)	575	327	1.76	25.3	11.7
3. Ultrogel AcA 44	281	39	7.21	12.4	48.1
4. Polybuffer Exchanger PBE 118[a]	169	10.8	15.65	7.4	104.0

[a] Enzyme chromatographed in three separate lots.

Guar seed contains exo-β-mannanase activity in both the cotyledons and in endosperm tissue. The enzymes, which can be prepared separately by seed dissection before enzyme extraction, are chromatographically indistinguishable and appear to have the same action patterns. The exo-β-mannanase purified by the described procedure appears as a single sharp band on SDS–gel slab electrophoresis using a 10% cross-linked gel and an imidazole–phosphate buffer system.[9] The molecular weight by SDS–gel electrophoresis and by gel filtration on Ultrogel AcA 44 is estimated to be 59,000 ± 2,000. Exo-β-mannanase is unstable to isoelectric focusing.[1] Sample applied near the cathode side of the gel migrates toward the cathode and is partially focused. However, some of the sample protein precipitates from solution at the end of the paper application-tab. Sample applied at the anode side of the gel precipitates from solution as it migrates toward the cathode. The precipitated protein still contains enzyme activity, as shown by its ability to hydrolyze naphthyl-β-D-mannopyranoside. The purified enzyme was devoid (<1 ppm) of α-galactosidase and endo-β-mannanase activities.

Properties. Guar seed exo-β-mannanase has a molecular weight of 59,000 ± 2,000, an isoelectric point (pI) of 9.4, and contains 7% of carbohydrate. It shows optimal activity between pH 5 and 6 and at 52°. It is stable at temperatures up to 45° on incubation at pH 5.5 for 15 min, and is stable in the pH range 5–8 on storage at 40° for 30 min. On hydrolysis of mannooligosaccharides the released D-mannose is in the α-anomeric configuration. The ultraviolet absorption spectrum in 5 mM Tris–HCl buffer (pH 8) shows a maximum absorption at 280 nm. The $E^{1\%}_{1cm}$ at 280 nm value is 2.05 where the protein concentration is determined by the Folin/Lowry method.[10] The extent of inhibition by 1 mM metal ions is Hg^{2+}, 100%; Ag^{+}, 100%; Cu^{2+}, 40%; and Zn^{2+}, 18%. β-D-Mannotriitol at a concentration of 10 mM does not inhibit, whereas D-mannose does (K_i 23.8 mM).

Kinetic Properties and Specificity. The preferred natural substrates for guar seed exo-β-mannanase are β-D-mannooligosaccharides of degree of polymerization (DP) 5 or 6. The relative initial rates of hydrolysis of β-D-mannooligosaccharides of degree of polymerization (DP) 2, 3, 4, and 5 at a concentration of 6.7 mg/ml are 12, 20, 52, and 100%, respectively. Reduction of the terminal β-D-mannosyl residue with borohydride greatly affects the susceptibility of the oligosaccharides to hydrolysis. β-D-Mannobiitol is totally resistant to hydrolysis and β-D-mannotriitoil is hydro-

[9] H. Fehrnstrom, LKB Application Note 306, SDS and Conventional Polyacrylamide Gel Electrophoresis with LKB 2117 Multiphor.

[10] O. H. Lowry, N. J. Rosebrough, A. L. Farr, and R. J. Randall, *J. Biol. Chem.* **193**, 265 (1951).

TABLE II
KINETIC CONSTANTS OF GUAR SEED EXO-β-MANNANASE

Substrate	K_m (mM)	V_{max} (U/mg)	V_{max}/K_m
β-D-Mannobiitol	—	0	—
β-D-Mannotriitol	80.0	2.1	0.026
β-D-Mannotetraitol	12.0	31.8	2.65
β-D-Mannopentaitol	2.8	57.9	20.68
β-D-Mannohexaitol	2.8	57.9	20.68
p-Nitrophenyl-β-D-mannopyranoside	0.5	15.6	31.20
o-Nitrophenyl-β-D-mannopyranoside	3.0	60.2	20.07
Methylumbelliferyl-β-D-mannopyranoside	0.5	68.8	137.60

lyzed at only one twenty-fifth the rate for β-D-mannotriose. β-D-Mannotetraitol is cleaved at 40% of the rate of β-D-mannotetraose whereas the rate of hydrolysis of β-D-mannopentaitol and β-D-mannopentaose is the same. Rapid cleavage by this enzyme thus requires binding across approximately 5 D-mannosyl residues.

The kinetic constants for synthetic substrates and reduced β-D-mannooligosaccharides are given in Table II. As the DP increases from 2 to 5 the K_m decreases and V_{max} increases. The V_{max}/K_m values for reduced oligosaccharides of DP 3, 4, 5, and 6 are 0.026, 2.65, 20.68, and 20.68, respectively.

The introduction of branch points into β-D-mannooligosaccharides greatly affects their susceptibility to hydrolysis by exo-β-mannanase. β-D-Mannobiose substituted on C-6 of the nonreducing D-mannosyl residue by α-linked D-galactose (i.e., 6²-α-D-galactosyl β-D-mannobiose) is not hydrolyzed. This oligosaccharide is also resistant to hydrolysis by β-mannosidase. However 6¹-α-D-galactosyl β-D-mannobiose and 6³-α-D-galactosyl β-D-mannotetraose, which are hydrolyzed by β-mannosidase at approximately 4% the rate of β-D-mannobiose, also are completely resistant to hydrolysis by exo-β-mannanase. 6¹-α-D-Galactosyl β-D-mannotriose is hydrolyzed by exo-β-mannanase at a rate similar to that for β-D-mannotriose.

Physiological studies[6] clearly demonstrate that exo-β-mannanase plays a key role in the hydrolysis of galactomannan in germinating guar seed. These studies also indicate that the three enzymes, exo-β-mannanase, endo-β-mannanase, and α-galactosidase are sufficient to account for galactomannan mobilization in the seeds.

[74] β-D-Mannanase

By BARRY V. MCCLEARY

β-D-Mannanase (EC 3.2.1.78, mannan endo-1,4-β-mannosidase) enzymes are endohydrolases which cleave randomly within the 1,4-β-D-mannan main chain of galactomannan, glucomannan, galactoglucomannan, and mannan.[1,2] Hydrolysis of these polysaccharides is affected by the degree and pattern of substitution of the main chain by α-D-galactosyl residues[3] (galactomannan and galactoglucomannan) and by the pattern of distribution of D-glucosyl residues within the main chain (glucomannan and galactoglucomannan). In glucomannan, the pattern of distribution of O-acetyl groups may also affect the susceptibility of the polysaccharide to hydrolysis. Insoluble, crystalline mannan is quite resistant to hydrolysis.

For the β-mannanases described here, the oligosaccharides of degree of polymerization (DP) 2–9 produced on hydrolysis of galactomannan and those of DP 2–4 released on cleavage of glucomannan have been quantitatively separated and structurally characterized.[4,5] The presence of certain structures and the absence of others has allowed us to develop models defining the specific subsite binding requirements for hydrolysis by each enzyme.[6] Thus, for effective hydrolysis at point X (Fig. 1) by a β-mannanase from guar seed (Cyamopsis tetragonolobus), binding across five D-mannosyl residues is required. At sugar residues B and D, binding occurs to the hydroxymethyl edge of the pyranose ring, while for sugar residues A, C, and E, binding is to the dihydroxyl edge. D-Galactosyl substitution on sugar residues B, C, or D prevents hydrolysis at point X. Replacement of D-mannosyl residue C or E by a D-glucosyl residue also prevents hydrolysis. In contrast, hydrolysis of galactomannan by a β-mannanase from Aspergillus niger culture filtrates requires binding across only four D-mannosyl residues with binding to the 2,3-hydroxyl edge of sugar residues C and E and to the hydroxymethyl edge of residues B and D. D-Galactosyl substitution on sugar residues B or D prevents cleavage

[1] N. K. Matheson and B. V. McCleary, "The Polysaccharides" (G. O. Aspinall, ed.), Vol. 3, p. 1. Academic Press, New York, 1985.

[2] B. V. McCleary and N. K. Matheson, Adv. Carbohydr. Chem. Biochem. 44, 147 (1986).

[3] B. V. McCleary, Carbohydr. Res. 71, 205 (1979).

[4] B. V. McCleary, F. R. Taravel, and N. W. H. Cheetham, Carbohydr. Res. 1045, 285 (1982).

[5] B. V. McCleary, E. Nurthen, F. R. Taravel, and J. P. Joseleau, Carbohydr. Res. 118, 91 (1983).

[6] B. V. McCleary and N. K. Matheson, Carbohydr. Res. 119, 191 (1983).

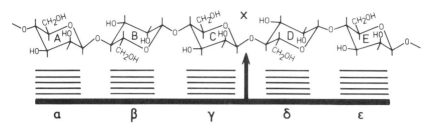

FIG. 1. Schematic representation of subsite binding between β-mannanase and the 1,4-β-D-mannan chain.

at X, but, unlike guar seed β-mannanase, substitution on sugar residue C has no effect on cleavage at point X. These results are consistent with theoretical predictions that the favored conformation of the 1,4-β-D-mannan chain is a ribbonlike structure with a 2-fold axis.[7] Such a conformation places neighboring D-galactosyl residues on the opposite side of the mannan main chain.

The β-mannanases from a range of other legume seeds,[6,8] from *Bacillus subtilis* and *Irpex lacteus* culture filtrates, and from the gut solution of *Helix pomatia* have action patterns on galactomannan which are intermediate between those of *A. niger* and guar seed β-mannanases, β-Mannanases with different action patterns have been used to characterize the "fine structures" (i.e., the distribution of D-galactosyl residues along the D-mannan backbone) of a range of galactomannans,[6,9] and these fine structural differences have been related to the different extents of interaction of these polysaccharides with agar, κ-carrageenan, and xanthan.[10]

In this chapter, methods for the purification of β-mannanase from a range of sources will be described. Since, in most cases, purification can be effected by affinity chromatography on glucomannan immobilized on aminohexane-Sepharose 4B,[8,11] preparation of this material will be described in detail.

Preparation of Glucomannan-AH-Sepharose

Dracena draco Glucomannan. Seed of *Dracena draco* (dragonwood tree) can be obtained from the Royal Botanic Gardens, Sydney or from

[7] P. R. Sundararajan and V. S. M. Rao, *Biopolymers* **9**, 1239 (1970).

[8] B. V. McCleary, *Phytochemistry* **18**, 757 (1979).

[9] B. V. McCleary, A. H. Clark, I. C. M. Dea, and D. A. Rees, *Carbohydr. Res.* **139**, 237 (1985).

[10] I. C. M. Dea, A. H. Clark, and B. V. McCleary, *Carbohydr. Res.* **147**, 275 (1986).

[11] B. V. McCleary, *Phytochemistry* **17**, 651 (1978).

Flamingo Enterprises Pty. Ltd., P.O. Box 1037, East Nowra, N.S.W., 2540, Australia.

Seed is milled to pass a 2-mm screen, the flour (100 g) is extracted with aqueous ethanol (80% v/v) under reflux and then suspended in 2.5 M sodium hydroxide (1 liter) plus sodium borohydride (1% w/v). The suspension is stirred at room temperature for 16 hr and then homogenized in a Waring blender. On centrifugation (3000 g, 15 min) the supernatant is collected and the pellet is reextracted twice with 500 ml of the same solvent. The combined supernatants are carefully neutralized by the addition of 5 N hydrochloric acid together with crushed ice to keep the temperature below 50°. On neutralization, the solution is centrifuged to remove a dark brown precipitate. Fehling's solution (~40 ml) is added to the supernatant solution while vigorously stirring. The copper–glucomannan complex which forms is collected by centrifugation (3000 g, 10 min) and the supernatant discarded. The pellet is resuspended in ice-cold water and redissolved by the careful addition of 1 N HCl together with vigorous blending. The complex which reforms on the addition of sufficient 1 N NaOH to make the solution alkaline is collected by centrifugation. This process is repeated until the supernatant solution is essentially colorless (usually two or three times). The copper–glucomannan complex is then resuspended in ice-cold water, dissolved by the addition of 1 N HCl, and precipitated from solution by the addition of ethanol (2 volumes). The precipitate is collected by centrifugation (3000 g, 10 min), washed twice with acidic ethanol (1% HCl in ethanol) and then with ethanol and acetone, and dried *in vacuo*. Yield 30%.

Aminohexane Sepharose (AH-Sepharose).[12] All operations must be performed in a well-ventilated fume hood! Dissolve 5 g of 1,6-diaminohexane in 30 ml of distilled water and adjust the pH to 9.5. Wash Sepharose 4B (50 ml packed bed volume) on a sintered glass funnel with 500 ml of distilled water and then with 500 ml of 1 M sodium carbonate solution. The Sepharose 4B is then suspended in 150 ml of 1 M sodium carbonate solution. Dissolve 10 g of cyanogen bromide in 15 ml of methyl cyanide and add this to a vigorously stirred suspension of the Sepharose 4B. Continue stirring for 2 min and then pour the slurry onto a sintered glass funnel on a Büchner flask. The gel is washed, under vacuum with 10 volumes of ice-cold water followed by 10 volumes of ice-cold 0.2 M sodium hydrogen carbonate solution. The activated Sepharose 4B and the 1,6-diaminohexane solution are then slurried and gently stirred at 4° in a sealed Quickfit flask for 16 hr. The aminohexane-Sepharose 4B (AH-Sepharose) slurry is then washed with 10 volumes of water and then with 5 volumes of 1 M sodium hydrogen carbonate.

---

[12] S. C. March, I. Parikh, and P. Cuatrecasas, *Anal. Biochem.* **60**, 149 (1974).

Glucomannan-AH-Sepharose 4B. All operations must be performed in a well-ventilated fume hood! *Dracena draco* glucomannan (6 g) is dissolved in 50 ml of 10% NaOH. On dissolution, the solution is neutralized by the addition of 5 *N* HCl. If a precipitate forms, this is removed by centrifugation (15,000 g, 10 min). The glucomannan solution is added to 200 ml of 1 *M* sodium hydrogen carbonate solution in a 500-ml Quickfit flask.

While this solution is vigorously stirred, add 3 g of cyanogen bromide dissolved in 5 ml of methyl cyanide. Stir vigorously for 2 min and then add the washed AH-Sepharose. Seal the flask and gently stir the contents for 16 hr at 4°. The slurry is then poured onto a sintered glass funnel attached to a Büchner flask (in a well-ventilated fume hood) and washed with 10 volumes of distilled water followed by 5 volumes of 0.5 *M* KCl in 0.1 *M* sodium acetate buffer (pH 5).

The conjugate is suspended in a solution of 0.5 *M* KCl in 0.1 *M* sodium acetate buffer and poured into a column (1.7 × 15 cm) and the support washed with the same buffer/salt solution at 4°.

Assay of β-Mannanase

β-Mannanase in column chromatographic eluates is routinely assayed using Remazol Brilliant Blue (RBB) dyed carob galactomannan as substrate[13] (see chapter [63] in this volume). However, pure enzyme preparations are always standardized by the Somogyi–Nelson reducing sugar assay[14,15] employing carob galactomannan as substrate.

RBB–Carob Galactomannan Assay Procedure. Suitably diluted enzyme preparation (0.5 ml) is incubated with dyed carob galactomannan (1 ml, 1%) in 0.3 *M* sodium acetate buffer (pH 5) at 40° for 10 min. The reaction is stopped by the addition of 3 ml of ethanol and the mixture is stirred and equilibrated to room temperature (5 min) before centrifugation at 1000 g for 10 min. The enzyme reaction is monitored by increased absorbance (590 nm) of the supernatant solution.

Reducing Sugar Assay Procedure. The enzyme preparation (0.05 ml) is incubated with carob galactomannan solution (0.5 ml 0.2%) in 0.1 *M* sodium acetate buffer (pH 5) for 2–10 min at 40°. The reaction is stopped by the addition of Somogyi–Nelson copper reagent[12,13] (0.5 ml) and color developed by the Somogyi–Nelson reducing sugar assay. Blanks and mannose standards (25 and 50 μg) are developed concurrently and activity is expressed as μmol of mannose reducing sugar equivalents released per minute per ml of enzyme solution at pH 5 and 40°.

---

[13] B. V. McCleary, *Carbohydr. Res.* **67**, 213 (1978).
[14] M. Somogyi, *J. Biol. Chem.* **195**, 19 (1952).
[15] B. V. McCleary and M. Glennie-Holmes, *J. Inst. Brew.* **91**, 285 (1985).

Lucerne Seed β-Mannanase B

Purification

Seed Germination and Extraction. Lucerne (*Medicago sativa*) seed (600 g) is surface sterilized by addition to 2 liters of 0.5% (v/v) sodium hypochlorite solution. After 10 min the hypochlorite solution is poured off and the seed collected on a screen and washed thoroughly with demineralized water. The seed is suspended in 4 liters of demineralized water and allowed to soak for 2 hr. The water is poured off and this washing and soaking process is repeated several times over 7 hr. This removes compounds which inhibited seed germination. The seed is then washed thoroughly and spread thinly over sheets of moist paper towel in large metal trays (30 × 80 cm). The seed is overlain with paper towel followed by another layer of seeds and more paper. The paper is moistened with demineralized water from a wash bottle, taking care not to add excess. The trays are covered with aluminum foil and stored at room temperature (~23°) for 3 days.

The germinated seed is homogenized in 5 liters of 0.1 M sodium acetate buffer (pH 4.5) using a Waring blender. The homogenate is incubated at 35° for 1 hr to allow depolymerization of polysaccharides which increase the mash viscosity. The slurry is homogenized again and the slurry centrifuged (3000 g, 20 min). The clear supernatant is treated with 500 g of ammonium sulfate per liter of solution and stored at 4° for 2 hr. The pellet recovered on centrifugation (3000 g, 20 min) of this suspension is dissolved in distilled water and dialyzed against 10 liters of distilled water for 20 hr at 4°.

Fractionation Procedure. The clear supernatant solution obtained on centrifugation of the dialyzed enzyme preparation at 15,000 g for 10 min is treated with 2 M Tris–HCl buffer (pH 8.0) to give a final concentration of 20 mM. The solution is applied to a preequilibrated column (3.5 × 15 cm) of DEAE-Sepharose CL-6B and protein is eluted with a linear KCl gradient (0–0.5 M, total volume 3 liters) in 20 mM Tris–HCl buffer (pH 8.0). The fractions active in β-mannanase are pooled, treated with ammonium sulfate (50 g/100 ml), and stored at 4° for 2 hr. The precipitate recovered by centrifugation is dissolved in a minimum volume of distilled water (10–20 ml) and dialyzed for 20 hr against 0.5 M KCl in 0.1 M sodium acetate buffer (pH 4.5, 1 liter). The dialyzed enzyme is chilled and then applied directly to the column containing glucomannan-AH-Sepharose 4B (1.7 × 15 cm) and eluted with an ice-cold solution of 0.5 M KCl in 0.1 M sodium acetate buffer (pH 4.5). The active fraction, which elutes well behind the major peak of protein, is collected, dialyzed against 5 mM sodium acetate buffer (pH 4.5), and stored in the frozen state. This enzyme is concentrated by application to a small bed (1.5 × 3.0 cm) of DEAE-Sepharose CL-6B equilibrated against 20 mM Tris–HCl buffer (pH 8) followed by elution with 0.5 M KCl.

Recovery and Purity. Two peaks of β-mannanase activity (A and B) are separated on chromatography of lucerne seed extracts on DEAE-Sepharose CL-6B.[13] One of these (peak A) represents only 10–20% of the total activity and can be further fractionated by chromatography on CM-cellulose and Sephadex G-100.[16,17] Peak B (80–90% of total activity) elutes from DEAE-Sepharose CL-6B at a salt concentration of 0.22 M. Further chromatography of this fraction on glucomannan-AH-Sepharose 4B yields a homogeneous enzyme with an activity of 129.3 U/mg protein when assayed on carob galactomannan substrate (2 mg/ml) in 0.1 M sodium acetate buffer (pH 4.5) at 40°. The overall recovery of β-mannanase B was 77% with a yield of 5 mg of enzyme protein from 600 g original weight of lucerne seed. Purified lucerne seed β-mannanase B appears as a single protein band on isoelectric focusing (pI 4.5) and on SDS–gel electrophoresis (MW 41,000).[8,11]

Properties

Optimal pH for activity of β-mannanase B is pH 4.5 and the enzyme is stable in the pH range 4–8 and below 50°. With carob galactomannan as substrate the K_m value is 0.9 mg/ml. The enzyme gives a rapid decrease in viscosity of carob galactomannan solutions but has limited ability to depolymerize highly substituted galactomannans.

Guar Seed β-Mannanase[16,18]

Purification

Seed Germination and Extraction. Guar (*Cyamopsis tetragonolobus*) seed (500 g) was surface sterilized, germinated, and the enzymes extracted as for lucerne seed.

Fractionation Procedure. The same format as used for lucerne seed β-mannanase B is employed. On chromatography on DEAE-Sepharose CL-6B a double peak of β-mannanase activity elutes at salt concentrations between 0.14 and 0.18 M. Fractions over this entire peak are pooled,

---

[16] B. V. McCleary and N. K. Matheson, *Phytochemistry* **14,** 1187 (1975).

[17] N. K. Matheson, D. M. Small, and L. Copeland, *Carbohydr. Res.* **82,** 325 (1980).

[18] B. V. McCleary, *Phytochemistry* **22,** 649 (1983).

concentrated, and chromatographed on glucomannan-AH-Sepharose as for lucerne seed β-mannanase B.

Recovery and Purity. Typically, recoveries of 15–20 mg of enzyme protein, representing approximately 50% of the activity in crude extracts, are obtained. The enzyme preparation recovered from substrate affinity chromatography appears as a single protein band on SDS–gel electrophoresis (MW 41,700), but on isoelectric focusing the preparation is fractionated into two major protein bands (pI 5.35 and 6.1) and three minor protein bands (pI 4.8, 5.8, and 6.2), each of which has enzyme activity. The two major bands of β-mannanase can be separated by chromatofocusing on Polybuffer Exchanger PBE 94 (Pharmacia Inc.). Enzymes are applied to the column (1.2 × 6 cm) in 25 mM imidazole buffer (pH 7.4) and eluted with Polybuffer 74-HCl (diluted 1.8 and adjusted to pH 4).

Properties

The two major protein bands and one of the minor components (pI 5.8) (recovered by slicing isoelectric focusing gels) have very similar properties. They show optimal activity at pH 4–5 and are stable at pH 4–8 and below 50°. The specific activity of the whole preparation is 121 U/mg protein at pH 4.5 and 40° with 0.2% carob galactomannan as substrate. The whole preparation and the separated components have K_m values between 0.3 and 0.5 mg/ml with carob galactomannan as substrate.

Guar seed β-mannanase catalyzes a rapid decrease in viscosity of carob galactomannan solutions, but of all the β-mannanases studied in this laboratory this enzyme is the least effective in depolymerizing galactomannans, particularly those with a high proportion of D-galactose. Guar seed β-mannanase plays a key role in the mobilization of galactomannan reserves in the germinating seed.

Helix pomatia β-Mannanase[8]

Purification

Fractionation Procedure. Crude *Helix pomatia* gut preparation as obtained from Sigma Chemical Co. (β-glucuronidase Type H-2, Crude solution GO876) or from Boehringer-Mannheim (β-glucuronidase/arylsulfatase; Cat. No. 127698) has an activity of ~200 U of β-mannanase per milliliter of crude solution. Ten milliliters of this solution is chilled to 4° and applied directly to a glucomannan-AH-Sepharose affinity column (1.7 × 15 cm) equilibrated against M KCl in 0.1 M sodium acetate buffer (pH 5), at 4° and eluted with the same solvent. The β-mannanase enzyme elutes from the column, without the addition of soluble substrate, well after all the other protein had passed through. A pure preparation of β-mannanase is obtained within 4 hr.

Recovery and Purity. A recovery of essentially 100% is obtained with overall purification of 15-fold. On SDS–gel electrophoresis there is a minor protein band (MW 40,000) and a single major protein band (MW 37,000). However, a number of protein bands are obtained on isoelectric focusing of the affinity purified preparation; the three major bands have p*I* values of 7.7, 7.4, and 7.0, whereas the minor bands have p*I* values between 5.0 and 7.2. The major protein bands obtained on isoelectric focusing have been separated by slicing the electrofocusing gel and extracting the enzyme after homogenizing the gel slices.

Properties

The affinity purified enzyme mixture showed optimal activity at pH 4.5–5.5 and was stable between pH 5 and 8 and below 50°. A K_m of 0.3 mg/ml on carob galactomannan substrate was obtained. However, the individual β-mannanase components within this mixture show considerable variation in their ability to hydrolyze highly substituted galactomannans, particularly at points of single unsubstituted D-mannosyl residues. The ratio of the initial rates of hydrolysis of *Leucaena leucocephala* compared to carob galactomannan by the different β-mannanase components range from 37 to 100%. *Leucaena leucocephala* galactomannan is highly substituted by D-galactose with a high frequency of regions of the repeating unit—[Man-Man(Gal)]$_n$—and a very low occurrence of totally unsubstituted regions.[3] In contrast, carob galactomannan has a more open, less substituted main chain.

Bacillus subtilis β-Mannanase[6,19]

Purification

Culturing and Fractionation Procedures. Bacillus subtilis TX1 is cultured as described by Reese and Shibata[20] with carob galactomannan (0.5% w/v) as the carbon source. The same type of β-mannanase has been found in four other *B. subtilis* strains. Six liters of culture solution, containing ~3000 units of β-mannanase activity, is centrifuged at 3000 *g* for 30 min and the supernatant dialyzed for 18 hr against flowing tap water.

[19] S. Emi, J. Fukumoto, and T. Yamamoto, *Agric. Biol. Chem.* **36**, 991 (1972).
[20] E. T. Reese and Y. Shibata, *Can. J. Microbiol.* **11**, 167 (1965).

The solution is adjusted to pH 8 by the addition of Tris–HCl buffer to a concentration of 20 mM, and passed through a bed of preequilibrated DEAE-Sepharose CL-6B (10 × 7 cm) in a sintered glass funnel. The enzyme is recovered by washing with 0.2 M KCl (800 ml) and concentrated by rotary evaporation (below 40°) to a volume of 100 ml. The solution is further concentrated (to 10 ml) by dialysis against polyethylene glycol 4000 and dialyzed against 1 M KCl in 0.1 M sodium acetate buffer (pH 5) at 4°. The enzyme preparation (10 ml, 2000 units) is applied to a glucomannan-AH-Sepharose affinity column (1.7 × 15 cm) which is eluted with 1 M KCl in 0.1 M sodium acetate buffer (pH 5) at 4°. The enzyme elutes without the addition of soluble substrate, well behind the initial protein peak.

Recovery and Purity. A recovery of enzyme activity of ~50% from the crude extract is usually obtained. The recovered enzyme (~4 mg) has a specific activity of 514 U/mg protein with carob galactomannan (0.2%) as substrate, at pH 5.0 and 40°. A single protein band (MW 37,000) is obtained on SDS–gel electrophoresis, whereas, on isoelectric focusing, a major protein band (pI 5.1) and two very minor bands (pI 5.0 and 5.5) are apparent. As far as can be determined, all protein bands have β-mannanase activity.

Properties

The enzyme shows optimal activity at pH 5–6 and is stable in the pH range 5–8 and below 50°. It has a specific activity of 514 U/mg and a K_m value of 1.1 mg/ml with carob galactomannan as substrate. It gives a rapid decrease in viscosity of carob galactomannan solutions but has limited ability to hydrolyze galactomannans highly substituted by galactose. The action of this enzyme on galactomannan and glucomannan, as shown by a characterization of the reaction products,[6,8] is similar to that of lucerne seed β-mannanase B.

Aspergillus niger β-Mannanase[21,22]

Purification

Fractionation Procedure. Hemicellulase preparation (100 g) from Miles Laboratories, Elkhart, IN is suspended in 500 ml of 20 mM Tris–HCl buffer (pH 8.5), stirred for 10 min, and centrifuged at 3000 g for 20

[21] Y. Tsujisaka, K. Hiyama, S. Takenishi, and J. Fukumoto, *Nippon Nogei Kagaku Kaishi* **46**, 155 (1972).

[22] K. E. Eriksson and M. Winell, *Acta Chem. Scand.* **22**, 1924 (1968).

min. The pH is adjusted to 8.5 by dropwise addition of 1 N sodium hydroxide and the solution applied to a column (4 × 15 cm) of DEAE-Sepharose CL-6B preequilibrated with 20 mM Tris–HCl (pH 8.5). The column is eluted with 3 liters of a linear KCl gradient (0–0.5 M) in 20 mM Tris–HCl (pH 8.5) (Fig. 2). The active fraction, which elutes at a KCl concentration of 0.3 M, is treated with ammonium sulfate at a rate of 60 g/100 ml and the solution stirred at 4° for 2 hr. The protein pellet recovered on centrifugation of the suspension at 15,000 g for 10 min is redissolved in a minimum volume of water (~40 ml), centrifuged (15,000 g, 10 min), applied to a column (4 × 92 cm) of Ultrogel AcA 54 (LKB Produktur), and eluted with 20 mM acetate buffer (pH 4.5) (Fig. 3). The fraction active in β-mannanase is treated with 1 M citrate buffer (pH 3.0) to give a final citrate concentration of 10 mM. The pH is adjusted to 3.0 by dropwise addition of 1 N HCl to an ice-cold solution of the enzyme. This solution is applied directly to a column (2.8 × 10 cm) of Whatman Cellulose Phosphate P1 (coarse, fibrous cation exchanger) preequilibrated with 10 mM citrate buffer (pH 3) at 4° and the column eluted with 800 ml of a linear KCl gradient (0–0.3 M) in 10 mM citrate buffer (pH 3) (Fig. 4).

β-Mannanase which elutes at a KCl concentration of 0.22 M, is treated with ammonium sulfate at a rate of 60 g/100 ml enzyme solution, at 4° for 2 hr. The suspension is centrifuged at 15,000 g for 10 min and the recovered pellet dissolved in a minimum volume of distilled water (~30 ml). This is applied to the column (4 × 92 cm) of Ultrogel AcA 54 and the column eluted with 20 mM acetate buffer (pH 4.5) (Fig. 5). The fractions active in

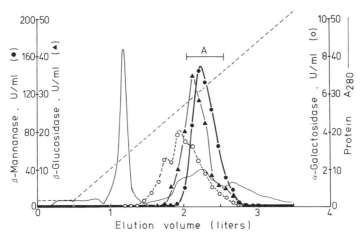

FIG. 2. Chromatography of Myles hemicellulase preparation on a DEAE-Sepharose CL-6B column (4 × 15 cm). The column is washed with a linear KCl gradient (0–0.5 M) in 20 mM Tris–HCl (pH 8.5).

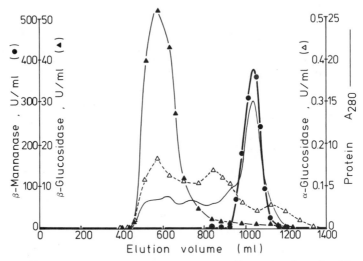

FIG. 3. Chromatography of β-mannanase on an Ultrogel AcA 54 column (4 × 92 cm). The column is eluted with 20 mM sodium acetate buffer (pH 4.5).

β-mannanase are pooled, treated with ammonium sulfate at a rate of 60 g/100 ml, and stirred gently at 4° for 2 hr. The suspension is centrifuged at 15,000 g for 10 min and the supernatant discarded. The pellet is redissolved in a minimum volume of distilled water and then treated with

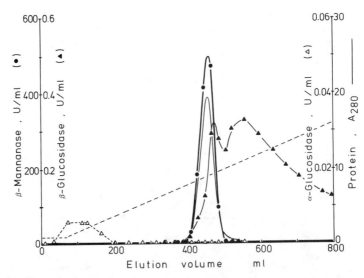

FIG. 4. Chromatography of β-mannanase on a cellulose phosphate column (2.8 × 10 cm). The column is eluted with a linear KCl gradient (0–0.3 M) in 10 mM citrate buffer (pH 3).

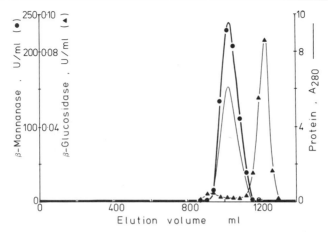

FIG. 5. Chromatography of β-mannanase on an Ultrogel AcA 54 column (4 × 92 cm). The column is eluted with 20 mM sodium acetate buffer (pH 4.5).

saturated ammonium sulfate by dropwise addition until a slight turbidity begins to form. The solution is centrifuged at 15,000 g for 10 min and the supernatant solution stored at 4°. After about 4 months crystals of *Aspergillus niger* β-mannanase begin to form. These crystals are recovered by centrifugation at low speed (1000 g, 5 min) and washed with 60% ammonium sulfate solution. The crystals are pencil shaped (Fig. 6).

Recovery and Purity. The described purification procedure summarized in Table I results in a 5.9-fold overall purification with 28% recovery of activity. The enzyme appears as a single but rather diffuse band (pI 4.0) on isoelectric focusing and as a single band on SDS–gel electrophoresis (MW 45,000). The extent of contamination with other activities is cellulase, 0.05%; α-amylase, 0.03%; α-glucosidase, β-glucosidase, β-mannosi-

TABLE I
PURIFICATION OF *Aspergillus niger* β-MANNANASE

Step	Total activity (units)	Total protein (mg)	Specific activity (U/mg)	Yield (%)	Purification (fold)
Crude extract[a]	71,500	11,000	6.5	100	1
1. DEAE-Sepharose CL-6B (pH 8)	62,560	3,841	16.3	87.5	2.5
2. Ultrogel AcA 54 (pH 4.5)	42,750	1,235	35.9	59.8	5.5
3. Phosphocellulose (pH 3)	27,540	721	38.2	38.5	5.9
4. Ultrogel AcA 54 (pH 4.5)	19,775	518	38.2	27.7	5.9
5. Crystallization	—	—	39.6	—	6.1

[a] From 100 g of Miles hemicellulase 100,000.

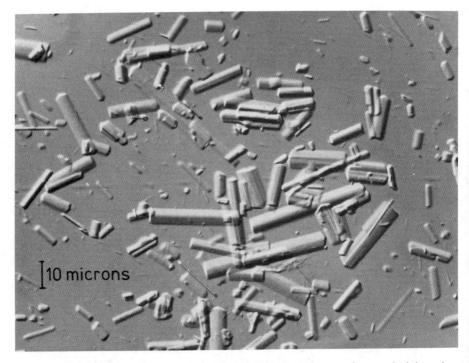

Fig. 6. Crystalline *Aspergillus niger* β-mannanase. Crystals were photographed through a Leitz Ortholux 2 microscope. The method used was Interference Contrast (160×) with Ilford Pan F film developed in Perceptol.

dase, α-galactosidase, β-xylosidase, and α-L-arabinofuranosidase, each less than 0.001%.

Aspergillus niger β-mannanase has been purified by substrate affinity chromatography on glucomannan-AH-Sepharose,[8] but binding to the affinity support is very weak and purification thus requires repeated chromatography and small amounts of enzyme can be purified.

Properties

The purified enzyme has a specific activity of 39 U/mg protein on 0.2% carob galactomannan at pH 4, and a K_m value of 0.1 mg/ml on this substrate. It is stable in the pH range 3–8 and below 70°. This enzyme is very effective in the depolymerization of even highly substituted galactomannans. Its action on galactomannans, glucomannans, and mannooligosaccharides has been studied in detail[5,6] and a computer model developed to simulate the subsite binding requirements and catalytic behavior.[9]

Irpex lacteus β-D-Mannanase

Purification

Ten grams of commercial Driselase preparation (Kyowa, Hakko Ko-gyo Co. Ltd., Japan) is suspended in 200 ml of 0.1 M Tris–HCl buffer (pH 8) and stirred for 10 min. The suspension is centrifuged at 15,000 g for 15 min and the supernatant solution chilled to 4° and dialyzed against 5 liters of 10 mM Tris–HCl buffer (pH 8) at 4° for 16 hr. Dialysis is performed under these conditions to prevent digestion of the dialysis sac by cellulase activity. This solution is chromatographed on a preequilibrated column (2.5 × 15 cm) of DEAE-Sepharose CL-6B and eluted with 1 liter of a linear KCl gradient (0–0.2 M) in 10 mM Tris–HCl (pH 8). β-Mannanase preparation (~120 U) is applied to a glucomannan-AH-Sepharose affinity column (1.7 × 15 cm) at 4° and the column eluted with ice-cold 0.5 M KCl in 0.1 M sodium acetate buffer (pH 5). The β-mannanase does not bind irreversibly to the affinity column, but rather is retarded to various de-grees, depending on both the amount of enzyme added and the prior use of the column. The recovered β-mannanase is concentrated by dialysis against polyethylene glycol 4000, dialyzed against distilled water, and stored in polypropylene containers at −20°. Samples are prepared for isoelectric focusing and electrophoresis by lyophilizing appropriate aliquots and redissolving in distilled water to a concentration of 10 mg/ml.

Recovery and Purity. The described purification method results in a 120-fold overall purification with 45% recovery of activity. The final spe-cific activity is 30.2 U/mg on a soluble β-D-mannan substrate and 59.1 U/mg on carob galactomannan (0.2% w/v). The enzyme appears as a single protein band on SDS–gel electrophoresis (MW 53,000), but on isoelectric focusing, two major (pI values 5.0 and 5.5) and several minor protein bands are apparent. The purified enzyme is completely devoid (<1 ppm) of α-galactosidase, β-mannosidase, β-glucosidase, and cellulase activ-ities.

Properties

The two major β-mannanase fractions separated by isoelectric focus-ing of affinity purified *Irpex lacteus* β-mannanase preparation have essen-tially identical properties and kinetic parameters. The optimal pH for activity is pH 3.0, with half maximal activities at pH 2.5 and 6.0. On storage for 18 hr at 4°, the enzymes show no loss in activity in the pH range 4–10, but lost 50% of its activity at pH 2.5. On storage for 18 hr at 40°, both enzymes are stable in the pH range 4–6, but show a major loss of

TABLE II
KINETIC PROPERTIES OF *I. lacteus* β-MANNANASE

Source of substrate	Gal/Man ratio	K_m (%, w/v)	Relative V_{max} (U/mg)
Mannan			
Livistona australis			
Soluble	0/100	0.005	29.9
Insoluble	0/100	0.11	29.9
Galactomannans			
Carob	23/77	0.03	59.1
Delonix regia	22/78	0.03	59.1
Cassia didymobotryia	22/78	0.03	59.1
Honey locust	27/73	0.03	59.1
Soybean	32/68	0.03	59.1
Hardenbergia violacea	34/66	0.06	54.0
Leucaena leucocephala	38/62	0.07	52.0
Guar	38/62	0.15	49.0
Lucerne galactomannan, pretreated	43/57	0.08	52.0
with α-galactosidase	37/63	0.06	55.0
	33/67	0.03	59.1
	21/79	0.02	59.1
	19/81	0.01	59.1
	5/95	0.005	59.1

activity at pH values above and below this range. The enzymes are stable at temperatures up to 60° on incubation at pH 5.0 for 30 min.

Irpex lacteus β-mannanase, in common with the enzyme from *Aspergillus niger,* has a greater ability to hydrolyze highly substituted galactomannans than do the enzymes from legume seeds or from *Bacillus subtilis.* The relative initial rate of hydrolysis of *Leucaena leucocophala* galactomannan is 82% that of carob galactomannan. The kinetic constants for a range of galactomannans with varying degrees of substitution of the 1,4-β-D-mannan backbone by α-linked D-galactose are shown in Table II. As the D-galactose content of the polysaccharide increases to 32%, no significant change in the K_m or relative V_{max} values is observed. However, as the D-galactose content approaches 34–38%, the K_m values double and the relative V_{max} values decrease by 10–20%.

Acknowledgment

I thank J. M. Stubbs, Unilever Research, Colworth Laboratory, Bedford, U.K., for photographs of crystalline *A. niger* β-mannanase.

[75] β-Mannanase of *Streptomyces*

By ISAO KUSAKABE and RIHEI TAKAHASHI

β-Mannanase refers to hydrolytic enzymes capable of hydrolyzing the $(1 \rightarrow 4)$-β-D-mannopyranosyl linkages of the $(1 \rightarrow 4)$-β-D-mannans, namely, mannan, galactomannan, glucomannan, and galactoglucomannan. Mannanases of this type have been assigned EC 3.2.1.78 [mannan endo-1,4-β-mannosidase, $(1 \rightarrow 4)$-β-D-mannan mannanohydrolases, endomannanases]. The enzymes have been reported to be produced by various microorganisms including bacteria,[1] fungi,[1] and *Streptomyces*,[2,3] and also to occur in animals[1] and plants.[1] In this chapter, we describe the isolation of mannanase-producing *Streptomyces*, a purification procedure, and some properties of the enzyme.

Assay Method

Copra mannan[2] was used as a substrate for the determination of mannanase activity. The mannan (galactose : mannose, 1 : 14) was insoluble in water and composed of the main chain of 1,4-linked β-D-mannosyl residues to which are attached single α-D-galactosyl branches at the O-6 of mannosyl residue. The mannanase activity was determined as follows,[2] a reaction mixture containing 129.4 mg the mannan (equivalent to 100 mg of polymannose), 4 ml of McIlvaine buffer solution (pH 6.8), and 5 ml of water was taken into an L-form tube. The tube was preincubated at 40° for 10 min on a Monod shaker with agitation at speed rate of 60 oscillations per min. One milliliter of enzyme solution was added to the mixture, then the mixture was incubated for 30 min at the same conditions. One milliliter of the reaction mixture after the incubation was taken into 5 ml of Somogyi's reagent[4] in a test tube. The test tube was heated in a boiling water bath for 20 min, and the reducing power produced by enzymatic reaction was determined as mannose by Somogyi's method.[4] Ten units of enzyme activity was defined[5] as the amount of enzyme that released 5.5

[1] R. F. H. Dekker and G. N. Richards, *Adv. Carbohydr. Chem. Biochem.* **32**, 299 (1976).
[2] R. Takahashi, I. Kusakabe, A. Maekawa, T. Suzuki, and K. Murakami, *Jpn. J. Trop. Agric.* **27**, 140 (1983).
[3] R. Takahashi, I. Kusakabe, H. Kobayashi, K. Murakami, A. Maekawa, and T. Suzuki, *Agric. Biol. Chem.* **48**, 2189 (1984).
[4] M. Somogyi, *J. Biol. Chem.* **160**, 61 (1945).
[5] R. Takahashi, Ph.D. dissertation. Tokyo University of Agriculture, Tokyo, Japan (1983).

TABLE I
Composition of Media[a]

Medium	Mannan	Peptone	Yeast ex.	KH_2PO_4	$(NH_4)_2HPO_4$	$MgSO_4 \cdot 7H_2O$	NH_4NO_3	C.S.L.	pH
First screening	1.0	0.1	0.05	0.1	0.1	0.05	—	—	5.5
Second screening	1.0	0.6	0.1	1.0	—	0.05	0.5	0.5	5.5
Enzyme production	2.0	0.9	0.1	1.0	—	0.05	—	0.5	5.5

[a] Data in percentage. C.S.L., corn steep liquor.

mg as mannose to 10 ml of the reaction mixture per 30 min under the above conditions.

Isolation of Mannanase-Producing Microorganisms

We are able to isolate the microorganisms in the following way.[2] Soil samples were collected from various districts, as isolation source for mannanase-producing microorganisms. The first screening was done on the basis that the mannanase-producing microorganisms formed clear zone around the colony when grown on the agar plate of the first screening medium (Table I). One drop of soil suspension diluted 100- to 1000-fold with sterilized water was spread on the agar plate, and was incubated at 35° for 5 or 6 days. The microorganisms forming clear zone on the plate were picked up, maintained on slants of Bennet agar[6] at room temperature, and subjected to a second screening.

One loop of the isolate was inoculated into 70 ml of second screening medium (Table I) in a 500-ml shaking flask (Sakaguchi flask) and incubated at 35° for 4 days on a reciprocal shaker (130 oscillations/min). The culture broth thus obtained was filtered through Tôyô-Roshi No. 2 filter paper. The mannanase activity of the filtrate was determined by the method described above.

Production of Mannanase

In this chapter, *Streptomyces* sp. No. 17 as a representative strain was used in the production of mannanase. Because the strain showed high stability for the enzyme productivity.

[6] B. K. Kondankai (ed.), "Biseibutsu-gaku Jikkenho," pp. 96 and 423. Kodansha Press, Japan, 1975.

Two liters of enzyme production medium (Table I) was placed into a 3-liter jar fermentor and sterilized at 120° for 15 min in an autoclave. After cooling, 100 ml of seed culture, which had been grown in the same medium in 500 ml shaking flask at 35° for about 3 days on the reciprocal shaker, was added to the fermentor. The cultivation was carried out at 35° with aeration of 700 ml/min and agitation of 600 rpm. The mannanase activity in the culture filtrate reached the maximum at 116 ~ 120 hr and showed an activity of 220 units/ml.[2] The culture broth after 120 hr cultivation was filtered with a Büchner funnel, and the resultant filtrate was used as an enzyme source.

Purification Procedure

The culture filtrate (mannanase) of *Streptomyces* sp. No. 17 obtained above was dialyzed against 1 M citric acid–NaOH buffer solution (pH 6.0). The mannanase (46,190 units) was applied to a column (25 × 300 mm) of ECTEOLA-cellulose (Bio-Rad Laboratories, Richmond, CA) equilibrated with the same buffer solution, and the column was washed with 600 ml of the same buffer solution. Elution was then carried out with a linear gradient of NaCl starting 0–0.5 M (total volume, 1000 ml) at the flow rate of 30 ml/hr. As shown in Fig. 1, the mannanase activity was separated into four activities: mannanase I, II, III, and IV. Mannanase IV, which accounted for 64.4% of total mannanase activity and was the

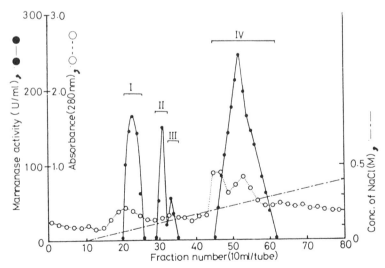

FIG. 1. Chromatography of the mannanase from *Streptomyces* No. 17 on ECTEOLA-cellulose column.

main enzyme of the mannanase system, was concentrated by ultrafiltration with a Toyo ultrafilter (UK-10) to 5 ml for further purification.

A column (40 × 900 mm) filled with BioGel P-100 (Bio-Rad Laboratories) was conditioned with 0.02 M citrate–NaOH buffer solution (pH 6.0) containing 0.10 M NaCl. The mannanase (14,000 units) was then applied to the column and eluted with the same buffer solution at the flow rate of 20 ml/hr.[3] Fractions of 10 ml were collected. Fraction Nos. 59 ~ 70 containing mannanase activity were combined. The purified mannanase IV gave a single band[3] in staining for protein on polyacrylamide disc gel electrophoresis at pH 8.

Properties of Mannanase IV

The purified mannanase IV from *Streptomyces* sp. No. 17 had the following properties.[3] Optimum pH and temperature for the activity of the enzyme were 6.8 and 57°, respectively. It was stable up to 45° when examined at pH 6.8 for 30 min, and lost only 15% of its activity at 70° for 30 min at pH 6.8. The isoelectric point and molecular weight of the enzyme were pH 3.65 and 42,900, respectively. The enzyme contained amino acids in the order of glycine > serine > aspartic acid and alanine > glutamic acid. The enzyme was completely inactivated by treatment with Al^{3+}, Cu^{2+}, Ag^+, Hg^+, Fe^{2+}, and Fe^{3+}, and was 50–90% inactivated by Cd^{2+}, Sn^{2+}, and Zn^{2+}; Ni^{2+}, Na^+, K^+, Mg^{2+}, and Pb^{2+} did not affect its activity. The enzyme did not hydrolyze β-1,4-mannobiose, but hydrolyzed slightly β-1,4-mannotriose to produce mannose and the mannobiose. β-1,4-Mannotetraose was hydrolyzed to produce either the mannobiose or mannose and the mannotriose. In addition, mannanase I, II, and IV hydrolyzed copra mannan to mainly produce mannose, mannobiose, and mannotriose at the final stage of enzyme reaction.

[76] β-D-Mannosidase from *Helix pomatia*

By BARRY V. McCLEARY

β-Mannosidase[1] (β-D-mannoside mannohydrolase, EC 3.2.1.25) is a glycosidase which cleaves single D-mannosyl residues from the nonreducing end of β-1,4-linked D-mannooligosaccharides and similarly linked

[1] P. M. Dey, *Adv. Carbohydr. Chem. Biochem.* **35,** 341 (1978).

sugar glycosides. This enzyme has found widespread use in structural studies of the core glycopeptides of glycoproteins[2-4] and the oligosaccharides accumulated or excreted in various storage disorders in animals.[5] It has also been used to characterize the structures of the D-galactose-containing 1,4-β-D-mannooligosaccharides produced on enzymatic or acid hydrolysis of galactomannans[6,7] and oligosaccharides produced on partial hydrolysis of glucomannans[8] and galactoglucomannans.

Although β-mannosidase has been reported to occur in a wide range of plant[2,9-12] and animal tissues,[3,13-16] and in culture filtrates or mycelia of several microorganisms,[4,17-20] only a few highly purified preparations of this enzyme have been reported.[4,18] This is due in part to the low levels of activity in the source materials and, in some cases, to the instability of the enzyme. Thus, with few exceptions, the β-mannosidases used in structural studies have been only partially purified and had quite low specific activities. Reports on the substrate specificity and kinetics of β-mannosidase are limited and restricted mainly to the action on core glycopeptides and derived oligosaccharides.

Commercially available, gut solution of *Helix pomatia* is a very useful source of β-mannosidase. The level of activity in the crude solution is high and the enzyme can be obtained readily in an almost homogeneous form by three simple chromatographic steps.[21]

[2] Y.-T. Li and Y. C. Lee, *J. Biol. Chem.* **247,** 3677 (1972).
[3] T. Sukeno, A. Tarentino, T. Plummer, and F. Maley, *Biochemistry* **11,** 1493 (1972).
[4] C. C. Wan, J. E. Muldrey, S.-C. Li, and Y.-T. Li, *J. Biol. Chem.* **251,** 4384 (1976).
[5] H. M. Flowers and N. Sharon, *Adv. Enzymol.* **48,** 29 (1979).
[6] B. V. McCleary, F. R. Taravel, and N. W. H. Cheetham, *Carbohydr. Res.* **104,** 285 (1982).
[7] B. V. McCleary, E. Nurthen, F. R. Taravel, and J.-P. Joseleau, *Carbohydr. Res.* **118,** 91 (1983).
[8] B. V. McCleary and N. K. Matheson, *Carbohydr. Res.* **119,** 191 (1983).
[9] J. Schwartz, J. Sloan, and Y. C. Lee, *Arch. Biochem. Biophys.* **137,** 122 (1970).
[10] J. S. G. Reid and H. Meier, *Planta* **112,** 301 (1973).
[11] G. Franz, *Planta Med.* **36,** 68 (1979).
[12] C. W. Houston, S. B. Latimer, and E. D. Mitchell, *Biochim. Biophys. Acta* **370,** 276 (1974).
[13] E. T. Reese and Y. Shibata, *Can. J. Microbiol.* **11,** 167 (1965).
[14] T. Muramatsu, *Arch. Biochem. Biophys.* **115,** 427 (1966).
[15] K. Sugahara, T. Okumura, and I. Yamashina, *Biochim. Biophys. Acta* **268,** 488 (1972).
[16] H. B. Bosmann, *Biochim. Biophys. Acta* **258,** 265 (1972).
[17] Y. Hashimoto and J. Fukumoto, *Nippon Nogei Kagaku Kaishi* **43,** 564 (1969).
[18] Y. Sone and A. Misaki, *J. Biochem. (Tokyo)* **83,** 1135 (1978).
[19] A. D. Elbein, S. Adya, and Y. C. Lee, *J. Biol. Chem.* **252,** 2026 (1977).
[20] S. Bouquelet, G. Spik, and J. Montreuil, *Biochim. Biophys. Acta* **522,** 521 (1978).
[21] B. V. McCleary, *Carbohydr. Res.* **111,** 297 (1983).

Assay Method

Principle. The most convenient assay for β-mannosidase activity employs *p*-nitrophenyl-β-D-mannopyranoside as substrate and measures the rate of cleavage of glycosidic bonds by the rate of release of *p*-nitrophenol. Action on reduced 1,4-β-D-mannooligosaccharides is measured by the increase in D-mannose reducing sugar equivalents.

Reagents

p-Nitrophenyl-β-D-mannopyranoside, 10 mM in 0.1 M sodium acetate buffer, pH 4.5
Sodium carbonate solution, 2% (w/v)

Procedure. Suitably diluted enzyme preparation (0.1 ml) is incubated with 0.1 ml of 10 mM *p*-nitrophenyl-β-D-mannopyranoside in 0.1 M sodium acetate buffer (pH 4.5) at 40° for 1–5 min. The reaction is terminated and color developed by the addition of 3 ml of 2% aqueous sodium carbonate. The absorbance at 410 nm is measured and the amount of released *p*-nitrophenol determined by reference to a *p*-nitrophenol standard curve [prepared in 2% (w/v) aqueous sodium carbonate].

Staining for β-Mannosidase in Electrophoresis Gels

Principle. α-Naphthol released on hydrolysis of naphthyl β-D-mannopyranoside by β-mannosidase reacts with Fast Blue BB dye to give an insoluble brown compound which deposits in the gel and locates the β-mannosidase activity.

Substrate. Naphthyl-β-D-mannopyranoside[22] (50 mg) is dissolved in a mixture of 2.5 ml of acetone and 2.5 ml of 0.05 M sodium acetate buffer (pH 5.5). This solution can be stored at 4° for more than 12 months. Substrate which crystallizes from solution redissolves on warming the solution.

Procedure. Two milliliters of 1 M sodium acetate buffer (pH 5.5) and 0.2 ml of substrate solution are added to 6 ml of water. To this solution, 30 mg of Fast Blue BB (salt) is added and dissolved by stirring. The solution is filtered and the filtrate poured onto the gel to be stained. Staining usually takes 10–30 min at 40°.

Purification Procedure

Step 1. Dialysis. Ten milliliters of crude, snail-gut solution (Sigma Chemical Co., G0876) is diluted with an equal volume of ice-cold, 200

---

[22] B. V. McCleary, *Carbohydr. Res.* **101,** 75 (1982); see also McCleary, this volume [61].

mM sodium phosphate buffer (pH 6.5) and the solution is dialyzed against ice-cold, 10 mM sodium phosphate buffer (pH 6.5) for 16 hr.

Step 2. DEAE-Cellulose Column Chromatography. The dialyzed solution is applied to a DEAE-cellulose column (3 × 22 cm) previously equilibrated with 10 mM sodium phosphate buffer (pH 6.5). The column is subjected to linear gradient elution with this buffer containing 0–0.2 M NaCl (total volume 2 liters). β-Mannosidase elutes at a salt concentration of 0.06 M whereas endo-β-mannanase does not bind to the column and α-galactosidase elutes at a salt concentration of 0.1 M. The fractions active in β-mannosidase are combined and concentrated by dialysis against polyethylene glycol 4000 for 4 hr at 4°.

Step 3. Ultrogel AcA 44 Column Chromatography. The concentrated enzyme (approximately 10 ml) is applied to an Ultrogel AcA 44 (LKB Produktur) column (2.5 × 88 cm) preequilibrated with 20 mM sodium acetate buffer (pH 5.0). The column is washed with the same buffer and β-mannosidase elutes just behind the column void volume. α-Galactosidase elutes before β-mannosidase and endo-β-mannanase elutes near the column total volume. The recovered β-mannosidase is devoid of endo-β-mannanase (<1 ppm) and contamination with α-galactosidase is approximately 0.05% of the β-mannosidase activity.

Step 4. CM-Cellulose Column Chromatography. Active fractions from the previous chromatographic step are combined, adjusted to pH 4, and applied directly to a column (1.2 × 12 cm) of CM-cellulose equilibrated with 20 mM sodium acetate buffer (pH 4). The column is washed with a linear gradient of NaCl (0–0.2 M) in 20 mM sodium acetate buffer (pH 4). β-Mannosidase elutes at a salt concentration of 0.05 M and α-galactosidase at 0.10 M. The active fractions are combined, concentrated in a Diaflo Ultrafiltration cell with a UM10 membrane, washed with 5 mM sodium acetate buffer (pH 5), and stored in the frozen state.

Purity and Properties

Recovery and Purity. The described purification procedure, summarized in Table I, results in 163-fold overall purification with 37% recovery of activity. The final specific activity is 101.4 U/mg on *p*-nitrophenyl-β-D-mannopyranoside at pH 4.5 and 40°. The enzyme appears as a single protein band (MW 94,000 ± 2,000) on SDS–gel electrophoresis, but as three protein bands with very similar isoelectric points (pI values ~4.7) on isoelectric focusing.[21] Staining of isoelectric focusing gels for β-mannosidase activity using the naphthyl-β-D-mannopyranoside/Fast Blue BB stain shows that each of these bands has enzyme activity. The degree of contamination of β-mannosidase with other activities is α-galactosidase,

TABLE I
PURIFICATION OF β-MANNOSIDASE FROM *Helix pomatia*

Step	Total activity (units)	Total protein (mg)	Specific activity (units/mg)	Yield (%)	Purification (fold)
Crude extract	498	800	0.62	100	1
1. Dialysis (pH 6.5)	494	569	0.87	99	1.4
2. DEAE-Cellulose (pH 6.5)	363	51.5	7.05	73	11.4
3. Ultrogel AcA 44 (pH 5)	290	4.9	59.20	58	95.5
4. CM-Cellulose (pH 4)	183	1.8	101.70	37	164.0

<0.01%; β-glucosidase, ~0.02%; β-galactosidase, β-xylosidase, and 2-acetamido-2-deoxy-D-glucosidase activities, <0.001%; and endo-β-mannanase, undetectable (<1 ppm).

Properties. The enzyme has a molecular weight of 94,000 ± 2,000, an isoelectric point of 4.7, and contains 3.1% carbohydrate. Optimum activity is shown at pH 4, but, on extended incubation, the enzyme slowly loses activity at this pH. At pH 4.5, the enzyme is stable to extended incubation at 45°, but there is a rapid loss of activity above 50°. Snail β-mannosidase is unstable above pH 7. Chromatography of the enzyme on DEAE-cellulose at pH 8 results in a considerable loss of activity. Of a range of metal ions tested, only Hg^{2+}, Ag^+, Ca^{2+} caused appreciable loss of activity; with mM metal ion at 40° for 10 min, the activity losses are 50, 70, and 27%, respectively. *N*-ε-Aminocaproyl-β-D-mannopyranosyl-amine (mM) caused no inhibition and is thus of no value for the preparation of affinity-chromatography materials.

Kinetic Properties and Specificity. The initial rates of hydrolysis of 1,4-β-D-mannooligosaccharides of degree of polymerization (DP) 2–5 and of reduced mannooligosaccharides of DP 3–5 are essentially the same. β-D-Mannobiitol is hydrolyzed at one-ninetieth the rate of β-D-manno-biose.

The kinetic constants for reduced β-D-mannooligosaccharides and some synthetic substrates are shown in Table II. The K_m and V_{max} values for reduced oligosaccharides of DP 3–6 are essentially the same. *p*-Nitrophenyl-, *O*-nitrophenyl-, methylumbelliferyl-, and naphthyl-β-D-mannopyranosides are readily hydrolyzed. The enzyme has a greater affinity for these substrates than for β-D-mannooligosaccharides.

TABLE II
KINETIC CONSTANTS OF *Helix pomatia* β-MANNOSIDASE

Substrate	K_m (mM)	V_{max} (U/mg)	V_{max}/K_m
p-Nitrophenyl-β-D-mannopyranoside	1.43	101.4	70.9
O-Nitrophenyl-β-D-mannopyranoside	2.33	151.9	65.2
Methylumbelliferyl-β-D-mannopyranoside	0.91	50.4	55.4
Naphthyl-β-D-mannopyranoside	3.22	76.8	23.9
β-D-Mannobiitol	n.d.[a]	n.d.	—
β-D-Mannotriitol	12.5	40.2	3.22
β-D-Mannotetraitol	12.5	38.4	3.07
β-D-Mannopentaitol	12.5	40.2	3.22
β-D-Mannohexaitol	12.5	39.6	3.17

[a] Hydrolysis of mannobiitol was so slow that accurate kinetic data could not be obtained.

Snail β-D-mannosidase is particularly useful in the characterization of the D-galactose containing mannooligosaccharides produced by acid or enzymic hydrolysis of galactomannans.[6–8] The enzyme sequentially removes single D-mannosyl residues from the nonreducing end of such oligosaccharides, up to, but not beyond, the D-mannosyl residue carrying the D-galactosyl substituent. It thus allows the location of D-galactosyl residues on D-mannooligosaccharides. β-D-Mannotriose substituted by D-galactose on the reducing end residue (6^1-α-D-galactosyl β-D-mannotriose) is rapidly hydrolyzed (at the same initial rate as β-D-mannobiose) to 6^1-α-D-galactosyl β-D-mannobiose but the removal of the D-mannosyl residue adjacent to that substituted with D-galactose proceeds at only 4% the rate of removal of the first residue. 6^3-α-D-Galactosyl β-D-mannotetraose is hydrolyzed (at 4% the rate of β-D-mannobiose) to 6^3-α-D-galactosyl β-D-mannotriose which is resistant to further hydrolysis. 6^1-α-D-Galactosyl β-D-mannotriose is hydrolyzed to α-D-galactosyl-D-mannose and D-mannose under conditions where 6^2-α-D-galactosyl β-D-mannobiose is completely resistant to hydrolysis, clearly demonstrating the inability of the enzyme to bypass a D-mannosyl residue substituted with D-galactose. This enzyme, together with α-galactosidase and endo-β-D-mannanase, can be used to characterize the structures of more complex oligosaccharides such as 6^1,6^3,6^4-tri-α-D-galactosyl β-D-mannopentaose.[7]

[77] α-Mannanase from *Rhodococcus erythropolis*

By I. Ya. Zacharova, V. Y. Tamm, and I. N. Pavlova

It is known that some microbial enzymes[1-5] are capable of digesting yeast α-mannans. One such enzyme has been found by Pavlova and co-workers[6] in soil microorganism that was classified as *Rhodococcus erythropolis*.[7] The production, purification, and characteristics of this enzyme possessing properties of α-mannanase are described below.

Assay Method

Principle. α-Mannanase is assayed by measuring the liberation of mannose from bakers' yeast mannan. The mannose released is determined by reducing sugar according to the method of Somogyi.[8]

Reagents

Acetic buffer, 0.05 M, pH 5.8
$CaCl_2$, 20 mM
Bakers' yeast mannan, 5 mg/ml
Alkaline copper reagent[8] (for 1 liter)
  4 g of $CuSO_4 \cdot 5H_2O$
  24 g of anhydrous Na_2CO_3
  16 g of $NaHCO_3$
  12 g of Rochelle salt
  180 g of anhydrous Na_2SO_4
Nelson's reagent[9] (for 0.5 liter)
  25 g of $(NH_4)_2MoO_4$

[1] G. H. Jones and C. E. Ballou, *J. Biol. Chem.* **243,** 2442 (1968).
[2] S. Yamamoto and S. Nagasaki, *Agric. Biol. Chem.* **39,** 1981 (1975).
[3] T. Nakajima, S. K. Maitra, and C. E. Ballou, *J. Biol. Chem.* **251,** 174 (1976).
[4] Z. Zouchova and J. Kocourek, *Folia Microbiol.* **22,** 98 (1977).
[5] E. Ichishima, M. Arai, Y. Shigimatsu, H. Kumagai, and R. Sumida-Tanaka, *Biochim. Biophys. Acta* **658,** 45 (1981).
[6] I. N. Pavlova, I. Y. Zaccharova, and N. Z. Tinjanova, *Autor. Svidet.* 958498, B.I. 34 (1982).
[7] I. N. Pavlova, I. Y. Zaccharova, and N. Z. Tinjanova, *Microbiology* (*Engl. Transl.*) **52,** 735 (1983).
[8] M. Somogyi, *J. Biol. Chem.* **195,** 19 (1952).
[9] Z. Dische, *in* "Methods in Carbohydrate Chemistry" (R. L. Whistler and M. L. Wolfrom, eds.) (in Russian), p. 53. Mir, Moscow, 1967.

21 ml of conc. H_2SO_4
3 g of $Na_2HAsO_4 \cdot 7H_2O$

Procedure. The standard assay mixture for determining α-mannanase activity contains the following (2 ml): 500 μg (0.1 ml) of mannan, 2 μmol of $CaCl_2$ (0.1 ml), 0.1 ml of enzyme solution, and 0.05 *M* acetic buffer, pH 5.8, to volume. After incubation for 10 min (with purified enzyme) or 1 hr (with crude extract) at 37°, the reaction is stopped by addition of 2 ml of the alkaline copper reagent. The tubes are shaken vigorously and heated in a boiling water bath for 25 min. After the tubes cooled, 1 ml of Nelson's reagent is added, the tubes are stirred vigorously, and the contents diluted to 10 ml. After 10 min the optical density is read at 560 nm. The amount of reducing sugar released by the enzyme is determined from the curve with mannose as standard.

Definition of Unit and Specific Activity. One unit of activity is defined as the amount of the enzyme required for the release of 1.0 μmol of mannose per minute at 37°. Specific activity is defined as units per milligram of protein.

Purification Procedure[10]

Step 1. Crude Extract. Rhodococcus erythropolis[11] is grown in liquid medium, 100 ml per 500-ml shaking flask, and incubated at 28° for 40 hr on a reciprocal shaker. The culture medium has the following composition (g/liter): 10 of dried cells of bakers' yeast, 0.3 of $(NH_4)_2SO_4$, 0.2 of $MgSO_4$, 0.08 of $CaCl_2 \cdot 2H_2O$, 7.54 of K_2HPO_4, 2.32 of KH_2PO_4. After 40 hr the cells are removed by centrifugation and the supernatant is fractionated in the following steps. All the following procedures are performed at 4°.

Step 2. Ammonium Sulfate Precipitation. The culture supernatant is brought to 80% saturation by adding solid ammonium sulfate (561 g/liter) with stirring, and after several hours the resulting precipitate is collected by centrifugation at 10,000 *g* for 20 min. The precipitate is dissolved in a small volume of water and dialyzed first against distilled water and then against 0.05 *M* acetic buffer, pH 5.8, with 0.1 m*M* $CaCl_2$ (buffer A).

[10] V. E. Tamm and I. N. Pavlova, *Microbiol. J.* (*Kiev*) **47**, 72 (1985).
[11] This strain was isolated from forest soil and identified as *Rhodococcus* (*Nocardia*) *erythropolis* by us. The microorganism differs from typical cultures of species *R. erythropolis* in that it is capable of growth on media containing yeast mannan as a sole carbon source. The strain with mannanase activity is characterized in detail in *Microbiology* (*Moscow*) **52**, 735 (1983).[7] At present the culture is deposited at the Central Museum of Industrial Microorganisms of the Institute of Genetics and Selection of Industrial Microorganisms, Moscow. Its collection number is CMPM S-555. The strain is covered by Patent N 958498, registered May 14, 1982.

Step 3. Chromatography on DEAE-Cellulose. The enzyme solution (100–150 mg of protein) is concentrated by ultrafiltration with nitrogen in a FM-02-200 apparatus using a UEM-150 membrane "Vladipor" (PM10, Amicon) and dialyzed against 0.05 M phosphate buffer, pH 7.0, containing 0.1 mM CaCl$_2$ (buffer B). The solution of enzyme is applied to a column (2 × 20 cm) of DEAE-cellulose (Serva) equilibrated with buffer B and is eluted with the starting buffer until there is no protein in the effluent. The elution is then carried out at a flow rate 0.5 ml/min sequentially with 200-ml batches of buffer B containing 0.3 and 0.8 M NaCl. The mannanase activity elutes in the buffer containing 0.8 M sodium chloride. The active fractions obtained from the DEAE-cellulose column and purified 3- to 4-fold from the previous step are pooled, concentrated to approximately 5 ml by pressure over a membrane, and chromatographed on Sepharose 6B (Pharmacia, Uppsala, Sweden).

Step 4. Gel Filtration on Sepharose 6B. The enzyme solution from step 3 is dialyzed against buffer A and applied to a column of Sepharose 6B (4.5 × 40 cm) equilibrated with the same buffer. The elution is carried out with buffer A at a flow rate 0.75 ml/min. The mannanase activity is eluted just after the void volume. The active fractions located in the first of the two protein peaks are pooled, concentrated by ultrafiltration, and subjected to isoelectrofocusing.

Step 5. Isoelectric Focusing in Borax-Polyol System. Isoelectrofocus-

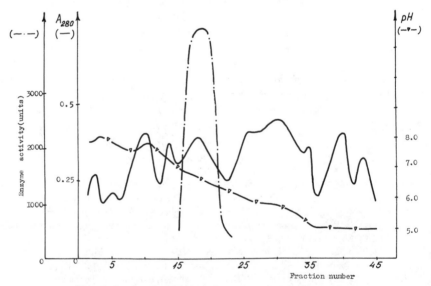

FIG. 1. Isoelectric focusing of mannanase *R. erythropolis.*

TABLE I
PURIFICATION OF α-MANNANASE FROM Rhodococcus erythropolis

Fraction	Total units	Total protein[a] (mg)	Specific activity (units/mg)	Recovery (%)	Purification (fold)
Culture fluid	56,440	802	70.0	100	—
$(NH_4)_2SO_4$	52,500	238	220.0	93	3.1
DEAE-cellulose[b]	42,430	58	729.8	75	10.4
Sepharose 6B	36,800	8.2	4,487.7	65	64.1
Electrofocusing	18,010	1.3	13,892.0	32	198.4

[a] Protein was determined by the method of O. H. Lowry, N. J. Rosebrough, A. L. Farr, and R. J. Randall, J. Biol. Chem. 193, 265 (1951).

[b] The enzyme solution from step 2 is applied to the DEAE-cellulose column twice and active fractions from both procedures are combined at the fourth step.

ing is carried out as described by Agitsky et al.[12] The enzyme solution from step 4 (8 mg of protein) is dialyzed against distilled water, applied to 50-ml column with a sucrose density gradient (0–40%) and mannitol–boric acid–borax buffer giving a gradient of pH range from 4.9 to 7.5, and subjected to isoelectrofocusing with current of 2.7 mA and electric potential of 20 V/cm for 20 hr. The protein material resolving into 10 peaks focusing with different pI values is collected in 1-ml fractions. The fractions demonstrating mannanase activity (fractions 15–22) are focused at pH 6.5–6.8 (Fig. 1). The maximum activity is revealed at pH 6.6. The major active fractions (fractions 16–20) are pooled, dialyzed against buffer A, and stored in frozen state (at −20°) without loss of activity for at least 1 year.

The procedures for preparing the purified mannanase Rhodococcus erythropolis are summarized in Table I.

Properties[13–15]

Purity. The mannanase from step 5 reveals a single band using electrophoresis on Tris–glycine and sodium dodecyl sulfate (SDS)–polyacrylamide gels.[16,17] This band stains as protein and carbohydrate (it is stained

[12] G. Y. Agitsky, V. F. Petrenko, G. V. Troitsky, and L. S. Gigys, in "Electrophoretic Methods of Protein Analysis," p. 44. Nauka, Novosibirsk, USSR, 1981.

[13] V. E. Tamm, I. N. Pavlova, and I. Y. Zaccharova, Microbiol. J.(Kiev) 47, 7 (1985).

[14] I. N. Pavlova and V. E. Tamm, Microbiol. J. (Kiev) 49, 11 (1987).

[15] E. A. Kovalenko, Microbiol. J. (Kiev) 46, 65 (1984).

[16] B. J. Davis, Ann. N.Y. Acad. Sci. 121, 404 (1964).

[17] K. Weber and M. Osborn, J. Biol. Chem. 244, 4406 (1969).

positively with the periodic acid–Schiff procedure[18]) and demonstrates mannanase activity.

Physical and Chemical Properties. SDS–polyacrylamide gel electrophoresis indicates that the enzyme is a single polypeptide chain with molecular weight 30,000. Exactly such a value is estimated by sedimentation equilibrium centrifugation. Based upon gel filtration on Sephadex G-150 in the presence of detergent, the molecular weight of mannanase is 32,000. The enzyme associates into the aggregates with M_r 1,000,000 (as determined by gel filtration on Sepharose 6B) if the enzyme concentration is not less than 0.1 mg/ml.

The homogeneous mannanase is apparently a glycoprotein since its single protein band on polyacrylamide gel electrophoresis is stained positively with the periodic acid–Schiff procedure. The carbohydrate content of the enzyme is 15%; mannose, glucose, and galactose are the major sugar constituents in a ratio of 6 : 5 : 2, as determined by gas–liquid chromatography.

The apparent isoelectric point determined by isoelectrofocusing is 6.6.

Stability. The mannanase from *Rhodococcus erythropolis* is relatively stable exhibiting no loss of activity on repeated freezing and thawing. The enzyme is stable over a wide pH range (pH 4.8–8.0, acetic or phosphate buffer) for 5 hr at 20°. In phosphate buffer pH 7.0, it loses 30% of activity in 1 week at 4°. At 60° mannanase is fully inactivated in 5 min.

The mannanase thermostability increases remarkably in the presence of 0.1–1.0 mM CaCl$_2$: at 50° its half inactivation times are 45 and 4 min with and without calcium ions, respectively. The homogeneous enzyme (0.2 mg protein per milliliter) has been kept at $-20°$ for at least 1 year with no loss of activity.

Optimum pH and Temperature. The pH range for mannanase extends from 4.3 to 9.0. The enzyme has a pH optimum at 5.8 and exhibits 50% of its maximal activity at pH 4.9 and 8.2, utilizing mannan of bakers' yeast as substrate. Maximum activity is obtained in acetate buffer.

When the standard assay is run at varying temperatures, the enzyme exhibits maximal activity at 45°. Mannanase activity increases almost in direct proportion to temperature up to 45°; at temperature higher than 50° the enzyme is rapidly inactivated.

The mannanase exhibits an activation energy of 6640 cal/mol.

Activators and Inhibitors. In an extensive survey of metal ions it was found that at 1 mM concentration of effectors, Ca^{2+} and Mg^{2+} are moderate activators (30–85% activation), Zn^{2+}, Pb^{2+}, Ni^{2+}, Ag^+, and Mn^{2+} are moderate inhibitors (50–80% inactivation), Fe^{2+} and Cu^{2+} inhibit the en-

---

[18] G. Fairbanks, F. L. Steck, and D. F. H. Wallach, *Biochemistry* **10**, 2606 (1971).

zyme to 6–8% of control, and Hg^{2+} inactivates completely. The enzyme is strongly inhibited by EDTA and *p*-chloromercuribenzoate. The exhibit K_i values of 5.59×10^{-5} and 3×10^{-4} M, respectively. Activity in the standard assay is inhibited 80% by 0.1 mM EDTA and fully by 1 mM *p*-chloromercuribenzoate.

Kinetics. The rate of hydrolysis of yeast mannan is directly proportional to the enzyme concentration up to 2 μg under standard assay conditions. Mannose release is linear for 20 min at 37°. The hydrolysis appears to obey normal Michaelis–Menten kinetics. From initial velocity measurements at pH 5.8 an apparent K_m for bakers' yeast mannan is 0.27 mg/ml; V_{max} value is 0.69 μmol of mannose per milligram of protein per minute.

Specificity. Numbers of yeast and plant mannans containing different α- and β-mannosidic linkages including highly branched and linear polymers as well as plant heteropolymers terminating in β-D-mannopyranosyl residues were tested as potential substrates (Table II).

Substrate specificity studies indicate that the enzyme can hydrolyze the linear and branched yeast mannans containing terminal α-1,2- and α-1,3-linked mannopyranosyl residues. The reaction occurs with a rapid increase in reducing sugar in the reaction mixture and an insignificant decrease in viscosity. The main product of the reaction is mannose as determined by paper chromatography in the solvent mixture butanol–

TABLE II

SUBSTRATE SPECIFICITY OF α-MANNANASE FROM *Rhodococcus erythropolis*

Substrate	Kind of mannosidic linkage	Producing of mannose
Mannan from *Saccharomyces cerevisiae*	α-1,6; α-1,2; α-1,3[a]	+
Mannan from *Candida pseudotropicalis*	α-1,6; α-1,2[b]	+
Mannan from *Cryptococcus laurentii*	α-1,3[c]	+
Mannan from *Rhodotorula rubra* (extracellular)	β-1,3; β-1,4[d]	−
Mannan from red algae	α-1,3[e]	−
Coffee mannan	β-1,4	−
Galactomannan from iceland lichen	β-1,4	−
Glucomannan from water-lily	β-1,4	−

[a] C. Ballou, *Adv. Microbiol. Physiol.* **14**, 93 (1976).
[b] N. P. Elinov and G. A. Vitovskaja, *Biochemistry (Engl. Transl.)* **30**, 933 (1965).
[c] N. P. Elinov, G. A. Vitovskaja, V. G. Kaloshin, and T. M. Kolotinskaja, *Biochemistry (Engl. Transl.)* **39**, 787 (1974).
[d] N. P. Elinov and G. A. Vitovskaja, *Biochemistry (Engl. Transl.)* **36**, 1187 (1971).
[e] A. I. Usov, K. S. Adamjanz, S. V. Jarotsky, A. A. Anashina, and N. K. Kotshetkov, *J. Gen. Chem. USSR (Engl. Transl.)* **44**, 416 (1974).

acetic acid–water (4 : 1 : 5). Thus, mannanase splits off mannose residues from the nonreducing end(s) of the mannan chain(s). It therefore acts as an exoglycosidase that is capable of hydrolyzing the high-molecular polymers of mannose. Microbial mannanase liberates free mannose from various yeast α-mannans at different rates and cleaves the polymers containing the terminal α-1,2 mannosidic linkages more effectively than the mannan chains terminating in α-1,3-linked units. The relative rates of releasing of mannose from mannans *Candida pseudotropicalis* (there are no α-1,3 linkages), *Saccharomyces cerevisiae* (with α-1,2 and α-1,3 terminal linkages in side chains), and *Cryptococcus laurentii* (α-1,3-linked polymer, obtained from heteropolysaccharide) are 100, 35, and 14, respectively. The mannan from red algae (*Nemalion vermiculare*) consisting of α-1,3-linked mannopyranosyl units only is stable to mannanase action likely because of the presence of $NaSO_3^-$ groups in chain. The enzyme does not cleave β-mannans and plant heteropolymers containing β-1,4-linked terminal residues of mannose.

In the homogeneous enzyme using *p*-nitrophenylglycosides as substrates, only traces of α-D-mannosidase (0.001 units/mg), and no β-D-mannosidase and other exoglycosidases activities can be detected.

Lectin Properties. In the hemagglutination assay applied to the study of lectin-induced clot formation, mannanase demonstrates the properties of lectin due to its ability to agglutinate trypsin-treated or neuraminidase-treated rabbit erythrocytes at 4 and 20°. There is no correlation between the enzyme and lectin activities of mannanase: the agglutination is not inhibited by the inhibitors of enzyme activity (EDTA or *p*-CMB), and mannanase agglutinates rabbit red blood cells after heating for 1 hr at 60° which abolishes its catalytic activity completely.

Using different carbohydrates to inhibit hemagglutination, it was found that agglutination of rabbit erythrocytes by mannanase is inhibited specifically by mannose. The mannanase shows the ability to bind mannose under conditions precluding enzymatic activity. It is possible that mannanase contains two active centers which have the common site for recognizing and binding of carbohydrate substrate.

[78] α-D-Galactosidase from Lucerne and Guar Seed

By BARRY V. MCCLEARY

α-Galactosidase (α-D-galactoside galactohydrolase, EC 3.2.1.22) has been shown to occur in a wide range of plants and animals and to be synthesized by microorganisms.[1] This enzyme has been purified from several sources using conventional chromatographic procedures and a range of affinity supports.[1-4] Of the affinity procedures, that employing N-ε-aminocaproyl-α-D-galactopyranosylamine coupled to Sepharose 4B as described by Harpaz *et al.*[2,3] is effective and reliable and can be used to purify α-galactosidase from a wide range of biological materials.

α-Galactosidases are active on a wide range of substrates containing α-linked D-galactosyl residues such as the naturally occurring oligosaccharides melibiose, raffinose, stachyose, and verbascose; synthetic substrates such as p-nitrophenyl-α-D-galactopyranose and methyl α-D-galactopyranose; and polymeric material such as legume seed galactomannans[5] and the D-galactose-containing oligosaccharides attached to red blood cells.[3] However, the relative rates of cleavage of these substrates by different α-galactosidases vary markedly.[1] The only α-galactosidases which effectively remove D-galactose from legume seed galactomannans are derived from seed sources, and generally from the endosperm tissue of legume seeds which contain galactomannan reserves. These enzymes are involved in the removal of D-galactosyl residues from galactomannan during seed germination.[6,7]

The affinity technique described by Harpaz *et al.*[2,3] was employed by these authors to purify α-galactosidase from green coffee beans and from soybean seed. However, neither of these materials is a good source of this activity. In the current chapter I will describe the large-scale purification of α-galactosidases with high activity on galactomannan from germinated seeds of lucerne and guar employing the affinity matrix developed by these authors.

[1] P. M. Dey and J. B. Pridham, *Adv. Enzymol.* **36**, 91 (1972).

[2] N. Harpaz, H. M. Flowers, and N. Sharon, *Biochim. Biophys. Acta* **341**, 213 (1974).

[3] N. Harpaz and H. M. Flowers, this series, Vol. 34, p. 347.

[4] C. A. Mapes and C. C. Sweeley, *J. Biol. Chem.* **248**, 2461 (1973).

[5] P. M. Dey, *Adv. Carbohydr. Chem. Biochem.* **37**, 283 (1980).

[6] P. M. Dey, *Adv. Carbohydr. Chem. Biochem.* **35**, 341 (1978).

[7] H. Meier and J. S. G. Reid, *in* "Encyclopedia of Plant Physiology: Plant Carbohydrates I. Intracellular Carbohydrates" (F. A. Loewus and W. Tanner, eds.), Vol. 13A, p. 418. Springer-Verlag, New York, 1982.

METHODS IN ENZYMOLOGY, VOL. 160

Assay Method

Using p-Nitrophenyl-α-D-Galactopyranoside. Suitably diluted enzyme preparation (0.1 ml) is incubated with 0.1 ml of 10 mM p-nitrophenyl-α-D-galactopyranoside in 0.1 M sodium acetate buffer (pH 4.5) at 40° for 1–5 min. The reaction is terminated and color developed by the addition of 3 ml of 2% (w/v) aqueous sodium carbonate. The absorbance at 410 nm is measured and the amount of released p-nitrophenol determined by reference to a p-nitrophenol standard curve [prepared in 2% (w/v) aqueous sodium carbonate].

Using Galactomannan as Substrate. Suitably diluted enzyme preparation (0.1 ml) is incubated with 0.5 ml of 0.2% (w/v) guar or carob galactomannan in 0.1 M sodium acetate buffer (pH 4.5) for 5–10 min at 40°. The reaction is terminated by the addition of p-hydroxybenzohydrazide solution[8] (5 ml) and the color is developed by incubating the tubes at 100° for exactly 6 min.

Staining for α-Galactosidase in Electrophoresis Gels

Substrate and Procedure. Naphthyl-α-D-galactopyranoside is employed as substrate. Substrate preparation and staining procedure is the same as that described for β-mannosidase.[9]

Guar Seed α-Galactosidase

Extraction and Purification

Step 1. Seed Germination and Extraction. Guar (*Cyamopsis tetragonolobus*) seed (500 g) is surface sterilized as described for lucerne seed[10] and germinated at 25° for 3 days.[11] The seed is then homogenized in 3 liters of 0.1 M sodium acetate buffer (pH 4.5) using a Waring blender and incubated at 35° for 1 hr to allow depolymerization of extracted galactomannan. The slurry is homogenized again and filtered through fine nylon mesh and the filter-cake squeezed to remove essentially all the free liquid. The filtrate is centrifuged (3500 g, 30 min) and the clear supernatant is dialyzed against three changes of ice-cold 20 mM sodium acetate buffer (pH 4.5, 10 liters) during 48 hr. After dialysis, the solution is centrifuged (3500 g, 20 min) and treated with Tris–HCl buffer (pH 8) to a final

---

[8] M. Lever, *Biochem. Med.* **7,** 274 (1973).

[9] B. V. McCleary, this volume [76].

[10] B. V. McCleary, this volume [74].

[11] B. V. McCleary, *Phytochemistry* **22,** 649 (1983).

concentration of 30 mM, with pH adjustment to pH 8 using 1 M sodium hydroxide.

Step 2. DEAE Sepharose CL-6B Column Chromatography. The enzyme solution at pH 8 is applied to a DEAE-Sepharose CL-6B column (3.7 × 17 cm) equilibrated with 30 mM Tris–HCl buffer (pH 8) at 4° and the column is washed with 30 mM Tris–HCl buffer (500 ml, pH 8) followed by linear gradient elution with 3 liters of this buffer containing 0–0.4 M NaCl. The fractions active in α-galactosidase in the major peak eluting from the column (α-galactosidase II) are pooled and treated with ammonium sulfate (50 g/100 ml) and stored at 4° for 2 hr. The suspension is centrifuged at 15,000 g for 10 min and the recovered pellet is dissolved in a minimum volume of distilled water (~50 ml). This solution is dialyzed against 5 liters of distilled water for 16 hr at 4° and then against a solution of 0.5 M sodium chloride in 0.1 M sodium acetate (pH 4.5) for 4 hr, and then centrifuged (15,000 g, 10 min) to remove an insoluble precipitate.

Step 3. Affinity Chromatography. The enzyme solution is applied directly to a column (2.5 × 35 cm) of N-ε-aminocaproyl-α-D-galactopyranosylamine   (N-6-aminohexanoyl-α-D-galactopyranosylamine)-Sepharose 4B (prepared according to the method of Harpaz *et al.*[2,3]) at 4° and the column eluted with 0.5 M KCl plus 20% of ethylene glycol in 0.1 M sodium acetate (pH 4.5) at 4°.[12] α-Galactosidase is eluted without the addition of galactose, well behind the peak of nonretarded protein.

Step 4. Enzyme Concentration. The combined active fractions are dialyzed against 20 mM Tris–HCl buffer for 16 hr and applied to a DEAE-Sepharose CL-6B column (1.2 × 3 cm). The enzyme is eluted with a minimum volume of 0.5 M KCl in 20 mM Tris–HCl (pH 8) and dialyzed against distilled water at 4° for 16 hr to remove salt. The enzyme is stable to repeated freeze/thaw cycles and thus can be stored in the frozen state.

Purity and Properties

Recovery and Purity. Two peaks of α-galactosidase activity, termed I and II, are separated on chromatography of guar seed extract on DEAE-Sepharose CL-6B. They represent approximately 10% (I) and 90% (II) of the total α-galactosidase activity.[12] In the seed, α-galactosidase I occurs in the cotyledon embryo and α-galactosidase II occurs in the endosperm and is synthesized during seed germination. Affinity purified α-galactosidase II appears as a single protein band on isoelectric focusing (pI 3.7) and on SDS–gel electrophoresis (MW 40,500). It has a specific activity of 52.1 U/mg protein when assayed on p-nitrophenyl-α-D-galactopyranose substrate (10 mM) in 0.1 M sodium acetate buffer (pH 4.5) at 40°.

[12] B. V. McCleary, R. Amado, R. Waibel, and H. Neukom, *Carbohydr. Res.* **92**, 269 (1981).

Properties. Guar seed α-galactosidase II displays optimal activity at pH 4.5–5.0 and at 45°. On extended incubation, the enzyme is unstable at temperatures above 40°. This enzyme is very effective in the removal of D-galactose from galactomannans. On incubation of 40 units (on *p*-nitrophenyl-α-D-galactopyranoside) of this enzyme with a solution of guar seed galactomannan (20 ml, 0.1% w/v), the D-galactose content of the polysaccharide was reduced from 38 to 0.5% within 3 hr, and an insoluble mannan precipitate formed. In the initial stages of hydrolysis, guar seed α-galactosidase II preferentially removes D-galactose residues from alternate D-mannosyl residues, which, with the D-mannan backbone in the preferred 2-fold conformation, is consistent with the enzyme moving along one face of the galactomannan molecule.[13] In combination with β-mannanase, guar seed α-galactosidase II can be used to quantitatively estimate the galactomannan content of guar seed varieties or milling fractions.[14,15]

Lucerne Seed α-Galactosidase

Extraction and Purification

Step 1. Seed Germination and Extraction. Lucerne (*Medicago sativa*) seed (600 g) is surface sterilized, germinated, and extracted as described for the purification of lucerne seed β-mannanase.[10]

Step 2. Enzyme Purification. The clear supernatant solution obtained on centrifugation (15,000 *g*, 10 min) of the dialyzed enzyme preparation is treated with 1 *M* potassium phosphate buffer (pH 6.5) to give a final concentration of 10 m*M*. The solution is applied to a preequilibrated column (4 × 17 cm) of DEAE-Sepharose CL-6B and protein is eluted with a linear KCl gradient (0–0.3 *M*, total volume 3 liters) in 10 m*M* phosphate buffer (pH 6.5). Two peaks of α-galactosidase activity, I (formerly termed A)[14] and II (formerly termed C) elute. Active fractions associated with each peak are pooled separately and treated with ammonium sulfate (50 g/100 ml) and stored at 4° for 2 hr. The precipitate is recovered, dissolved, dialyzed, and chromatographed on a column (2.5 × 35 cm) of *N*-ε-aminocaproyl-α-D-galactopyranosylamine-Sepharose 4B[2,3] as described for guar seed α-galactosidase. The results of a typical purification are shown in Table I.

[13] B. V. McCleary, I. C. M. Dea, J. Windust, and D. Cooke, *Carbohydr. Polym.* **4,** 253 (1984).
[14] B. V. McCleary, *Carbohydr. Res.* **71,** 205 (1979).
[15] B. V. McCleary, *Lebensm.-Wiss. Technol.* **14,** 188 (1981).

TABLE I

SEPARATION AND PURIFICATION OF LUCERNE SEED α-GALACTOSIDASES I AND II

Step	Total activity (units)	Total protein (mg)	Specific activity (U/mg)	Yield (%)	Purification (fold)
Crude extract	3114	45,240	0.07	100	1
1. Ammonium sulfate treatment (0–80%)	3102	4,314	0.72	99	10
2. DEAE-Sepharose CL-6B (pH 6.5)					
α-Galactosidase I	1210	816	1.48	39	21
α-Galactosidase II	1437	251	5.73	46	82
3. Affinity column					
α-Galactosidase I	1060	13	81.54	33	1165
α-Galactosidase II	1234	16	77.13	40	1102

Purity and Properties

Recovery and Purity. Two peaks of α-galactosidase activity are separated on DEAE-Sepharose CL-6B chromatography of lucerne seed extract. As for guar seed extract, peak I is composed of α-galactosidases

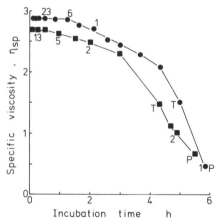

FIG. 1. Effect of galactose removal on the solution viscosity and solubility of galactomannans from guar (●) and carob (■). Galactomannan solution (17 ml, 0.1%) in 0.1 M sodium acetate buffer (pH 4.5) was incubated with lucerne seed α-galactosidase I (0.4 U on this substrate) in an Ubbelohde suspended level viscometer at 40°. Samples were removed for determination of released galactose by the p-hydroxybenzohydrazide reducing sugar method. Numbers represent the galactose content of the remaining polysaccharide. At point T, the solution was very turbid; at point P, a precipitate had formed.

from cotyledon embryo tissue, whereas peak II is endosperm-derived α-galactosidase. The combined recovery of α-galactosidases I plus II is greater than 70% of that present in the crude extract (Table I) and the extent of purification is in excess of 1000-fold. Both α-galactosidases I and II are devoid of other glycanase and glycosidase activities (<0.002%). Contamination of α-galactosidase I with endo-β-mannanase was less than one part in 10^6 but complete removal of this activity from α-galactosidase II required a second chromatographic purification on the affinity column support.

Properties. Affinity purified α-galactosidase II appears as a single band on isoelectric focusing (pI 4.6) and on SDS–gel electrophoresis (MW = 21,000). It shows optimal activity at pH 4.5–5.5 and is unstable at temperatures above 45°. In contrast, α-galactosidase I appears as a number of protein bands on isoelectric focusing (pI values 5.8–6.6) but as a single broad band on SDS–gel electrophoresis (MW = 33,000). The enzymes catalyze the hydrolysis of a wide range of substrates with nonreducing terminal D-galactose, including melibiose, raffinose, stachyose, galactomannan, and p-nitrophenyl-α-D-galactopyranoside, but the relative initial rates of hydrolysis vary significantly. Both enzymes are very effective in the removal of D-galactose from galactomannans (Fig. 1). Galactomannans with widely different D-galactose contents and varying patterns of distribution of D-galactosyl residues along the D-mannan backbone are hydrolyzed at essentially the same rate.[14]

[79] Xylanase of *Bacillus pumilus*

By Hirosuke Okada and Atsuhiko Shinmyo

Xylan, the primary component of pentosan, consists of a backbone chain of 1,4-linked β-D-xylopyranosyl residues and side chains of arabinose, glucuronic acid, or methylglucuronic acid. The backbone is hydrolyzed by endoxylanase or exoxylanase and the xylooligosaccharides formed are hydrolyzed to xylose by β-xylosidase. Xylanase production has been reported in many microorganisms including fungi, yeasts, and bacteria. Since natural xylan is a heterogeneous polysaccharide, the hydrolysis of side chains is not well understood.

Bacillus pumilus is a microbe producing potent xylan degrading enzymes. In our laboratory, endoxylanase (1,4-β-D-xylan xylanohydrolase, EC 3.2.1.8) and β-xylosidase (1,4-β-D-xylan xylohydrolase, EC 3.2.1.37, xylan 1,4-β-xylosidase) of *B. pumilus* IPO,[4] isolated from a rice field, were purified, and hydrolysis of xylan to xylose was found to be accomplished by sequential reaction of these two enzymes.[1] A chromosomal DNA fragment of *B. pumilus* IPO containing genes of both enzymes was cloned in *Escherichia coli*[2] and the complete DNA sequences of both genes were determined.[3,3a] Here, we describe the purification and properties of the xylanase of *B. pumilus* IPO.[4]

Assay Methods

Principle. Xylanase activity is measured from the reducing sugars liberated from larchwood xylan (Sigma, St. Louis, MO, approximate molecular weight 20,000). Reducing sugar is measured by the dinitrosalicylic acid method.[5]

Reagents

Potassium phosphate buffer, pH 6.5
Larchwood xylan (Sigma)
Xylose
3,5-Dinitrosalicylic acid reagent

Procedure. Xylan is suspended in 50 mM potassium phosphate buffer at 1% (w/v) and boiled for 10 min to dissolve it. The reaction mixture, consisting of 1 ml of xylan solution and 0.5 ml of enzyme solution, is incubated at 40° for 10 min. The reaction is stopped by adding 3 ml of 3,5-dinitrosalicylic acid reagent. Reducing sugar present without incubation at 40° is subtracted from that with incubation. One unit of xylanase is defined as the amount of enzyme that liberates 1 μmol of xylose equivalent in 1 min.

When a crude enzyme solution is used, liberation of reducing sugar without xylan must be checked. *B. pumilus* IPO excretes xylanase in the culture medium, but accumulates β-xylosidase in the cytoplasm. On the other hand, the *E. coli* clone which harbors the plasmid coding for the xylanase and β-xylosidase genes accumulates both enzymes in its cells. Xylanase activity determined in *E. coli* cell extracts is overestimated

[1] W. Panbangred, A. Shinmyo, S. Kinoshita, and H. Okada, *Agric. Biol. Chem.* **47,** 957 (1983).

[2] W. Panbangred, T. Kondo, S. Negoro, A. Shinmyo, and H. Okada, *Mol. Gen. Genet.* **192,** 335 (1983).

[3] E. Fukusaki, W. Panbangred, A. Shinmyo, and H. Okada, *FEBS Lett.* **171,** 197 (1984).

[3a] H. Moriyama, E. Fukusaki, J. Cabrera Crespo, A. Shinmyo, and H. Okada, *Eur. J. Biochem.* **166,** 539 (1987).

[4] *B. pumulis* IPO is deposited in the Fermentation Research Institute Yatabe-cho, Higashi-gun, Ibaraki, Japan, and is designated culture 4. *E. coli* plasmid pOXN391 in the same collection is designated Ferm P-6996.

[5] G. L. Miller, *Anal. Chem.* **31,** 426 (1959).

because of the presence of a β-xylosidase which produces reducing sugars from the xylooligosaccharides formed by the xylanase reaction. When purified β-xylosidase is added to xylanase solution at an activity ratio of 3 : 1, the apparent xylanase activity is about 1.2 times that without β-xylosidase.

Purification

Highly purified enzyme with a specific activity in the range of 1700–1900 units/mg protein can be obtained from the culture fluid by a few purification steps.

Culture Conditions. The enzyme production medium consists of 0.5% larchwood xylan (Sigma), 0.5% Bacto-yeast extract (Difco), 0.2% NH_4NO_3, 0.2% KH_2PO_4, and 0.02% $MgSO_4 \cdot 7H_2O$, pH 6.8. *B. pumilus* IPO cells are grown in a test tube containing 10 ml L-broth [10 g Bacto-tryptone (Difco), 5 g Bacto-yeast extract, 5 g NaCl, and 1 g glucose in 1 liter] overnight at 30° with shaking and the whole culture broth is transferred to 500 ml of production medium in a 3-liter Sakaguchi flask. The flask is incubated at 30° for 36 hr in a water bath with shaking. Then the culture fluid is obtained by centrifugation at 10,000 *g* for 15 min and used for xylanase purification. The following procedures are performed at <4°.

Step 1. The protein fraction in 2 liters of the culture fluid which precipitated in the ammonium sulfate concentration range between 0.2 and 0.6 saturation is collected as follows; solid ammonium sulfate (114 g/liter) is added to the culture fluid with stirring. After 3 hr, the supernatant is obtained by centrifugation at 15,000 *g* for 30 min, and further ammonium sulfate is added (262 g/liter). The precipitate is collected after 3 hr by centrifugation, dissolved in 200 ml of 50 m*M* potassium phosphate buffer, pH 6.5, and dialyzed for about 20 hr against several changes of 5 liters of the same buffer. The resultant insoluble materials, which contain residual xylan, are removed by centrifugation.

Step 2. The dialyzed enzyme solution is applied to a DEAE-Sephadex column (5.5 × 25 cm) equilibrated with 50 m*M* phosphate buffer, pH 6.5 (the same phosphate buffer pH and composition are used in all other steps). On elution with the same buffer at a flow rate of 200–250 ml/hr, xylanase activity appears in the void volume, whereas most of the other proteins and colored materials are adsorbed onto the gel. Fractions of 50 ml are collected. The fractions containing specific activity greater than 100 units/mg protein are pooled to yield a total volume of about 500 ml with an enzyme activity recovery of 60% in this step.

Step 3. The pooled active fraction in Step 2 is applied to a CM-Sephadex C-50 column (5.5 × 20 cm) which has been equilibrated with the buffer. The column is washed with 1 liter of the phosphate buffer and the enzyme is eluted from the column with a linear gradient established between 500 ml of the phosphate buffer and 500 ml of the buffer containing 0.6 M NaCl. Fractions of 20 ml are collected at the flow rate of 200 ml/hr. Xylanase activity is eluted at about 0.22 M NaCl concentration. The fractions containing specific activity greater than 800 units/mg protein are pooled to yield a total volume of 520 ml with 87% recovery of enzyme activity in this step. The solution is concentrated to a volume of 20 ml to 25 ml by ultrafiltration using a Toyo UM-10 membrane, and dialyzed against three changes of 2 liters of the phosphate buffer.

Step 4. The dialyzate in Step 3 is added ammonium sulfate to 40% of saturation by adjusting the pH at 6.5 with 1 N KOH, and applied to a column (1.5 × 40 cm) of TSK HW polyvinyl gel (Toyosoda, Tokyo) which has been equilibrated with the phosphate buffer containing 40% saturation of ammonium sulfate. The column is washed with 200 ml of the same equilibration buffer solution. Then the enzyme is eluted with a linear gradient of ammonium sulfate from 40% saturation to 0%, each in 200 ml of the phosphate buffer. The flow rate is 20 ml/hr and 5-ml fractions are collected. The enzyme activity is eluted at 13% saturation of ammonium sulfate with the specific activity of about 1800 units/mg protein. Recovery of the activity in this step is about 70%.

Approximately 90-fold purification is achieved with an overall yield of 22%. A summary of the purification procedures is presented in Table I.

Preparation of Xylanase from E. coli Cells Harboring the Plasmid. Five hundreds milliliters of L-broth containing 50 μg/ml of ampicillin in a 3-liter Sakaguchi flask is inoculated with 5 ml of culture broth of *E. coli* C600 harboring the hybrid plasmid pOXN391 which codes for the xylanase gene of *B. pumilus* IPO and the ampicillin-resistance gene,[3] grown at 37° overnight in the same medium, and incubated for 16–20 hr at 37° with shaking. *E. coli* (pOXN391) produces xylanase constitutively,

TABLE I
PURIFICATION OF XYLANASE

Step	Volume (ml)	Total protein (mg)	Total activity (units × 10³)	Specific activity (units/mg)	Purification (fold)	Yield (%)
1. Culture fluid	2080	8740	175	20	1	100
2. 20–60% (NH₄)₂SO₄	220	1830	110	60	3	63
3. DEAE-Sephadex A-50	600	234	67.2	287	14	38
4. CM-Sephadex C-50	20	47	56.6	1200	60	32
5. TSK HW-65	10	22	39.1	1780	89	22

whereas *B. pumilus* IPO xylanase is induced by xylan and also by xylose. Cells are harvested from 2 liters of culture by centrifugation at 10,000 *g* for 10 min and washed once with 200 ml of ice-cold 50 m*M* potassium phosphate buffer, pH 6.5. The following procedures are done at <4°. Washed cells are suspended in 25 ml of the same buffer solution and disintegrated by passing through a French pressure cell at 450–500 kg/cm². The cell lysate is centrifuged at 10,000 *g* for 30 min and supernatant is again centrifuged at 40,000 *g* for 1 hr to yield clear supernatant. Xylanase is purified from the clear supernatant by the above procedure.

Properties

Homogeneity and Molecular Weight. The purified enzyme preparation gave a single protein band on polyacrylamide disc gel electrophoresis with or without sodium dodecyl sulfate (SDS). The purity was estimated at more than 95%. No carbohydrate was detected by acid fuchsin staining of the gel. The molecular weight of the purified xylanase was 24,000 by SDS–polyacrylamide gel electrophoresis and 20,000 by equilibrium sedimentation, suggesting a single polypeptide chain. From the nucleotide sequence of the cloned xylanase gene in *E. coli* it was deduced that xylanase consists of 201 amino acid residues with a molecular weight of 22,384.

Stability and Activity. The purified xylanase had an activity and stability optimum at pH 6.5. The enzyme retained more than 85% of its activity after standing at 40° for 30 min at pH 8.5 or 5.0. The maximum xylanase activity was observed at temperatures between 45 and 50° over a period of 10 min. The enzyme lost half of its activity when kept at 50° for 30 min, and the decrease was in accord with monomolecular kinetics. All activity was lost at 60° in 15 min.

The xylanase activity was influenced by the salt concentration of the reaction mixture. In 10 m*M* phosphate buffer the activity was about 20% of the maximum value, which was obtained in 50–60 m*M* phosphate buffer. Sodium chloride added to 10 m*M* phosphate buffer affected the activity similarly.

Amino Acid Sequence. The amino acid sequence of prexylanase, deduced from the DNA sequence is shown in Fig. 1. This is the first example of the total amino acid sequence of a xylanase. The protein is probably synthesized in the form of a prexylanase with a signal sequence consisting of 27 amino acids, of which 3 are basic amino acid residues in the region near N-terminus and 18 are hydrophobic amino acid residues. The signal sequence might be processed between Ala^{-1} and Arg^{+}. The N-terminal amino acid sequence of the purified xylanase was NH$_2$-Arg-Thr-Ile-Thr

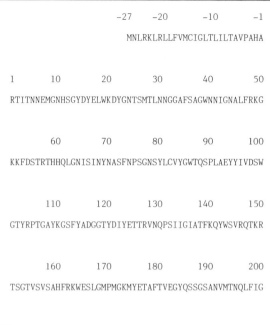

```
                -27      -20       -10       -1

                MNLRKLRLLFVMCIGLTLILTAVPAHA


   1        10        20        30        40        50

   RTITNNEMGNHSGYDYELWKDYGNTSMTLNNGGAFSAGWNNIGNALFRKG


            60        70        80        90        100

   KKFDSTRTHHQLGNISINYNASFNPSGNSYLCVYGWTQSPLAEYYIVDSW


           110       120       130       140       150

   GTYRPTGAYKGSFYADGGTYDIYETTRVNQPSIIGIATFKQYWSVRQTKR


           160       170       180       190       200

   TSGTVSVSAHFRKWESLGMPMGKMYETAFTVEGYQSSGSANVMTNQLFIG


        201

        N
```

FIG. 1. The complete amino acid sequence of the prexylanase. The 27 amino acid signal peptide indicated by minus numbers might be processed between Ala and Arg. Amino acids are indicated by one-letter symbols.

and the C-terminal sequence was Ile-Gly-Asn-COOH by sequential Edman degradation and carboxypeptidase digestion, respectively. The amino acid composition from the purified enzyme agreed well with that deduced from the DNA sequence.

Hydrolysis of Xylan and Xylooligosaccharides. The maximum degree of hydrolysis of xylosidic linkage of larchwood xylan (molecular weight 20,000) by the purified xylanase was about 25%. The end products were oligosaccharides, corresponding xylobiose (X_2), xylotriose (X_3), xylotetraose (X_4), and higher oligomers. At the early stages of the reaction, xylose oligomers higher than xylopentaose were the only products. Authentic xylobiose was not hydrolyzed. Incubation with X_3 yielded X_4 and X_2, while with X_4 as the substrate, X_2, X_3, X_4, and higher oligomers were detected, although the reaction rate was much slower than that toward xylan. The production of X_4 from X_3, and of the higher oligomers from X_4 indicates that xylanase has trans-xylosidation activity.

[80] Xylanase of *Cryptococcus albidus*

By Peter Biely and Mária Vršanská

$$1,4\text{-}\beta\text{-}D\text{-Xylan} \xrightarrow{H_2O} 1,4\text{-}\beta\text{-}D\text{-xylooligosaccharides} + D\text{-xylose}$$

Cryptococcus albidus is a noncellulolytic saprophytic yeast capable of growing on plant xylans as a sole carbon source.[1,2] For utilization of the polysaccharide the strain produces an inducible xylan-degrading enzyme system composed of three components differing in function and cellular localization.[3] The component responsible for hydrolysis of xylan in the medium is an extracellular endo-1,4-β-xylanase, purification and properties of which are described in the present chapter. It is the first xylanase with established subsite structure of the substrate binding site.

Assay[4]

Reagents

Acetate buffer, 0.1 and 0.05 M, pH 5.4

4-*O*-Methyl-D-glucurono-D-xylan from beechwood, hornbeamwood, or larchwood (fractions soluble in hot water) 0.4% solution in 0.1 M acetate buffer or 0.2% solution in 0.05 M acetate buffer

D-Xylose, 1 mM calibration solution in 0.05 M acetate buffer

Somogyi and Nelson reagents for determination of reducing sugars[5]

Procedure. Xylan solution (0.25 ml, 0.4%) is mixed with 0.25 ml of enzyme solution, or 0.5 ml of 0.2% xylan solution is mixed with 10–50 μl of enzyme solution and incubated at 30°. The reaction is terminated by addition of 0.5 ml of the Somogyi reagent. The mixture is heated for 10 min on a boiling water bath, then cooled under tap water and vigorously mixed with 0.5 ml of the Nelson reagent. After 15 min standing with occasional stirring, the samples are centrifuged at 2000 g to remove a fine precipitate. Absorbance at 560 nm of the supernatants is measured.

[1] P. Biely, Z. Krátký, A. Kocková-Kratochvílová, and Š. Bauer, *Folia Microbiol.* 23, 366 (1978).
[2] T. D. Leathers, C. P. Kurtzman, and R. W. Detroy, *Biotechnol. Bioeng. Symp.* 14, 225 (1984).
[3] P. Biely, *Trends Biotechnol.* 3, 286 (1985).
[4] P. Biely, M. Vršanská, and Z. Krátký, *Eur. J. Biochem.* 108, 313 (1980).
[5] L. G. Paleg, *Anal. Chem.* 31, 1902 (1959).

Blanks to correct the values for reducing power of the substrate and the enzyme solutions are run in parallel and the absorbance of samples corrected accordingly. Calibration is done in the range 0.05–0.5 μmol of xylose.

The release of reducing sugars from xylan during incubation with enzyme follows linearity with time until about 0.2 μmol of xylose equivalents had been released. The amounts of the enzyme and time of incubation are chosen to fill the above requirement.

Definition of Unit. One unit is defined as the amount of enzyme which liberates from xylan 1 μmol equivalents of xylose in 1 min under the given conditions.

Enzyme Purification

Enzyme from Xylan Growth Medium[4]

Growth of the Yeast. Strain *Cryptococcus albidus* CCY 17-4-1 is grown for 4 days on a shaker at 27° in a synthetic medium containing in g/1000 ml: yeast nitrogen base 6.7, L-asparagine 2.0, KH_2PO_4 5.0, and xylan (beechwood 4-*O*-methyl-D-glucurono-D-xylan[6]) 10.0. A 3-day culture grown in the same medium containing 1% glucose instead of xylan is used as inoculum.

Concentration of the Culture Fluid. The culture is centrifuged, and the cell-free supernatant concentrated by distillation *in vacuo* (30–35°) to one-fifth of the original volume. All further steps are done at 4°. After desalting by dialysis (4 days against two changes of distilled water containing 0.01% sodium azide), the solution is freeze-dried to give about 3.5 g of crude xylanase from 1000 ml of the culture. Most of this material is represented with residual xylan.

DEAE-Cellulose Chromatography. The crude xylanase preparation (1.5 g) is partially dissolved in 50 ml of 0.05 M phosphate buffer, pH 7, the insoluble residue is removed by centrifugation, and the clear, viscous supernatant applied on a column of DEAE-Cellulose (Whatman DE-1, 3 × 25 cm) equilibrated in 0.05 M phosphate buffer, pH 7. The column is eluted first with equilibrating buffer (240 ml) and then with a linear gradient of NaCl in the same buffer at a rate of 18 ml/hr. As can be seen in Fig. 1, the major portion of xylanase is not retained on the column and is considerably purified from material absorbing at 280 nm and from xylan remnants.

[6] A. Ebringerová, A. Kramár, F. Rendoš, and R. Domanský, *Holzforschung* **21**, 74 (1967).

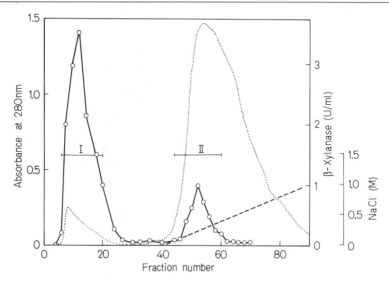

FIG. 1. Fractionation of crude xylanase on DEAE-Cellulose. Absorbance at 280 nm (dotted line), activity of xylanase (○), and concentration of NaCl in the effluent (dashed line). Reproduced with permission from Biely *et al.*[4]

CM-Sephadex Chromatography. The xylanase fraction I from the DEAE-Cellulose step (Fig. 1) is adjusted to pH 5.0 with 50% (v/v) acetic acid and poured on a CM-Sephadex C-50 column (1.8×20 cm) equilibrated with 0.05 M acetate buffer, pH 5.0. The column is eluted first with equilibrating buffer at a rate of 24 ml/hr. The enzyme is eluted as a sharp protein peak at 0.45 M NaCl (Fig. 2). The active fractions are pooled, desalted by dialysis, and used for further investigations.

The results of a typical purification are summarized in Table I. The low yields are caused mainly by interference of unused xylan present in

TABLE I

PURIFICATION OF XYLANASE SECRETED BY THE CELLS DURING GROWTH ON
BEECHWOOD XYLAN

Step	Volume (ml)	Total activity (U)	Total protein (mg)	Specific activity (U/mg)	Degree of purification	Yield (%)
Culture fluid	2600	2300	1800	1.28	1	100
DEAE-Cellulose (fraction I)	260	470	63	7.5	5.8	20.4
CM-Sephadex	24	204	3.7	55.1	43.0	8.9

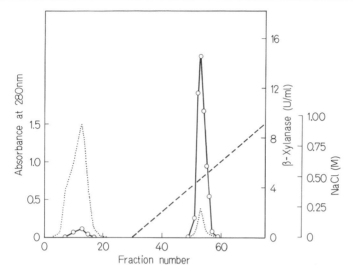

FIG. 2. CM-Sephadex chromatography of xylanase fraction I from the DEAE-cellulose step (Fig. 1). Absorbance at 280 (dotted line), activity of xylanase (○), and concentration of NaCl in the effluent (dashed line). Reproduced with permission from Biely *et al.*[4]

the culture fluid with the enzyme during concentration of the culture fluid and the first step of the purification.

Enzyme Induced by Methyl-β-D-Xylopyranoside

Growth of the Yeast. Strain *Cryptococcus albidus* CCY 17-4-1 is grown for 2 days on a shaker at 27° in a synthetic medium containing in g/1000 ml: yeast nitrogen base 6.7, L-asparagine 2.0, KH_2PO_4 5.0, and D-xylose 10.0.

Induction of the Enzyme. The cells harvested by centrifugation from 1000 ml of xylose medium are suspended in 500 ml of the above medium, however free of D-xylose and supplied with methyl-β-D-xylopyranoside at a concentration 0.5 mg/ml. The cell suspension is incubated on a shaker at 27° for 42 hr.

Concentration of the Induction Medium. The cells are removed by centrifugation and the supernatant is evaporated *in vacuo* to 80 ml and desalted by dialysis against two changes of distilled water containing 0.01% sodium azide at 4°.

CM-Sephadex Chromatography. The concentrated and desalted induction medium is poured on a CM-Sephadex C-50 column (1.8 × 20 cm) equilibrated with 0.05 *M* acetate buffer, pH 5.0. The column is eluted with 240 ml of a linear gradient of NaCl (1.0 *M* is the upper concentration

TABLE II

PURIFICATION OF XYLANASE INDUCED BY METHYL-β-D-XYLOPYRANOSIDE

Step	Volume (ml)	Total activity (U)	Total protein (mg)	Specific activity (U/mg)	Degree of purification	Yield (%)
Induction medium	480	816	120	6.8	1	100
Concentrated induction medium	80	760	104	7.3	1.07	93.1
CM-Sephadex	33	290	6.8	42.6	6.3	35.5

limit) and fractions of 4 ml are collected at 10 min intervals. The enzyme is eluted as a sharp peak at 0.45 M NaCl similar to the enzyme isolated from the xylan growth medium (Fig. 2). After desalting by dialysis the fraction is used as a pure xylanase preparation.

The summary of purification is shown in Table II. The enzyme induced under nongrowing conditions is obtained in a better yield than that produced during growth on xylan. Moreover, the purification is not complicated by the presence of residual xylan and extracellular polysaccharides produced by the cells during growth.

Properties

Purity. Both enzyme preparations appear homogeneous on gel-permeation chromatography and give a single diffuse protein and activity band on electrophoresis in polyacrylamide gel (pH 8.9). Identity and purity of both xylanase preparations are best demonstrated by isoelectric focusing in polyacrylamide gel in pH range 3–7 (Fig. 3).[7] One major protein band (pI ~ 5.0) and several minor protein bands (pI 3.5–4.5) exactly coincide with the enzyme activity. The reason for the multiplicity of forms remains to be solved. All other properties are reported for the enzyme isolated from the xylan-spent medium.

Effect of pH. The enzyme shows optimum activity at pH 5.4 with xylan or 4-nitrophenyl-β-D-xylopyranoside as substrate.

Stability. In solutions preserved against microbial contamination by 0.02% sodium azide, the enzyme is stable at 4° for several months. In a frozen state the enzyme is stable indefinitely. The half-life of the enzyme present in 0.05 M acetate buffer, pH 5.4, containing 0.1 M NaCl was 7 min at 50°.

---

[7] P. Biely, O. Markovič, and D. Mislovičová, *Anal. Biochem.* **144,** 147 (1985).

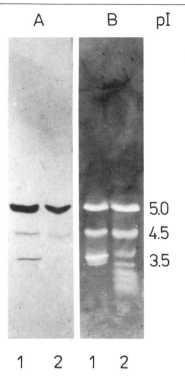

FIG. 3. Isoelectric focusing of purified xylanase on thin-layer polyacrylamide gel (pH range 3–7). (A) Detection for proteins with Commassie Brillant Blue R-250; (B) detection for enzyme activity with covalently dyed xylan[7] (more sensitive than the protein detection). Lane 1, enzyme induced with methyl-β-D-xylopyranoside; lane 2, enzyme produced during growth on xylan. pI values are indicated on the side.

Molecular Properties. Gel filtration on calibrated columns of BioGel A1.5m and Superose 6 (Pharmacia FPLC system) gave molecular weights of 26,000 and 28,000, respectively. The value obtained by SDS–polyacrylamide gel electrophoresis is 48,000, however.[8] The enzyme is glycosylated as shown by its interaction with concanavalin A.[9] Deglycosylation leads to a 27% decrease in the molecular weight.[8] The nature of the carbohydrate moiety has not been established yet.

Specificity. The enzyme hydrolyzes in endo fashion all basic types of plant xylans: 4-O-methyl-D-glucurono-D-xylan, L-arabino-D-xylan, and more linear grass 1,4-β-xylan or larger 1,4-β-linked xylooligosaccharides

[8] R. Morosoli, *Biochim. Biophys. Acta* **826,** 202 (1985).
[9] A. Peciarová and P. Biely, *Biochim. Biophys. Acta* **716,** 391 (1982).

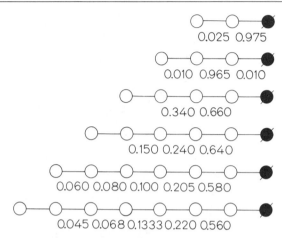

Fig. 4. Bond cleavage frequencies of 1,4-β-[1-³H]xylooligosaccharides with xylanase of *C. albidus* at 0.025 m*M* concentration of the substrates. (○) A nonreducing xylosyl residue; (●) xylose residue 1-³H-labeled at the reducing end. Reproduced with permission from Biely *et al.*[13]

affording a series of lower xylooligosaccharides and xylose. The enzyme also attacks rhodymenan (xylan containing 1,4-β and 1,3-β linkages) liberating, besides 1,4-β-linked products, isomeric xylooligosaccharides containing a 1,3-β linkage.[10] The enzyme is capable of hydrolyzing 1,2-β- and 1,3-β-xylosidic linkages which follow a 1,4-β-xylosidic bond in the direction toward the reducing end of an oligosaccharide.[11]

The enzyme catalyzes degradation of aryl-β-D-xylopyranosides, although it does not hydrolyze xylobiose and alkyl-β-D-xylopyranosides. The reaction proceeds well only at high substrate concentrations and its rate shows a sigmoidal dependence on substrate concentration.[12] The degradation of aryl-β-xylopyranosides does not proceed as a simple hydrolysis and involves a series of xylosyl transfer reactions.

Substrate Binding Site.[13] The basic input data for the calculation of subsite affinities leading to the image of the substrate binding site are the bond cleavage frequencies of linear 1,4-β-xylooligosaccharides and their V/K_m parameters referred to equal enzyme concentration.[14] The bond

[10] P. Biely and M. Vršanská, *Eur. J. Biochem.* **129,** 645 (1983).
[11] M. Vršanská, P. Biely, and J. Hirsch, *Proc. Bratislava Symp. Saccharides, 2nd* p. 57 (Abstr.) (1984).
[12] P. Biely, M. Vršanská, and Z. Krátký, *Eur. J. Biochem.* **112,** 375 (1980).
[13] P. Biely, Z. Krátký, and M. Vršanská, *Eur. J. Biochem.* **119,** 559 (1981).
[14] T. Suganuma, R. Matsuno, M. Ohnishi, and K. Hiromi, *J. Biochem.* (*Tokyo*) **84,** 293 (1978).

TABLE III
KINETIC PARAMETER V/K_m REFERRED TO A
UNIT CONCENTRATION OF *C. albidus* XYLANASE
(1 U/ml) FOR 1,4-β-XYLOOLIGOSACCHARIDES[a]

Oligosaccharide	V/K_m (min^{-1} U^{-1} ml)
Xylotriose	0.0124
Xylotetraose	1.84
Xylopentaose	2.74
Xylohexaose	3.07
Xyloheptaose	3.23
Xylooctaose	1.11

[a] Reproduced with permission from Biely *et al.*[13]

cleavage frequencies determined as initial product ratios of degradation of [1-^{3}H]xylooligosaccharides by xylanase at low substrate concentration (0.025 mM) are shown in Fig. 4. The enzyme shows a strong preference to cleave all tested substrates with the exception of xylotriose, at the second glycosidic linkage from the reducing end. Xylotriose appears to be hydrolyzed about 100 times more slowly than xylotetraose (Table III). For oligosaccharides higher than xylotetraose the values V/K_m increase only slightly in going to xyloheptaose.

The substrate binding site of the yeast xylanase appears to be composed of four subsites (Fig. 5) of which only two, the outer subsites II and $-$II, show a strong affinity to bind xylosyl residues of 1,4-β-xylooligosaccharides. The catalytic groups are localized in the middle which is in agreement with the fact that xylotetraose is the smallest substrate that is hydrolyzed rapidly and, at low concentrations, almost exclusively to xylobiose. The negative value for the sum of affinities of subsites adjacent to the catalytic groups accounts for the negligible susceptibility of xylobiose to direct hydrolysis.[15]

Mechanism of Substrate Degradation.[15] Bond cleavage frequencies of xylotriose, xylotetraose, and xylopentaose as shown in Fig. 4 are valid only for low substrate concentrations. With increasing substrate concentration the bond cleavage frequencies change dramatically and reactions other than a simple hydrolysis take place. The substrates interact with the enzyme binding site at different, shifted positions, and at such a shifted binding, xylosyl, xylobiosyl, and xylotriosyl transfer reactions occur and

[15] P. Biely, M. Vršanská, and Z. Krátký, *Eur. J. Biochem.* **119**, 565 (1981).

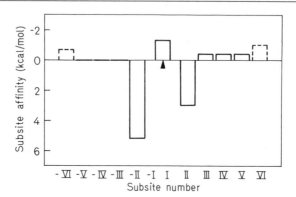

FIG. 5. Subsite interaction energies with xylosyl residues of linear 1,4-β-xylooligosaccharides in xylanase as evaluated by the method of Suganuma *et al.*[14] The hypothetic subsites around the catalytic groups (arrow) are numbered with Roman numerals. Only two subsites, −II and II, show a strong affinity to bind xylosyl residues, therefore the binding site appears to consist of four subsites. Reproduced with permission from Biely *et al.*[13]

lead to products larger than the starting substrates. For example, 40 m*M* xylotriose degradation affords xylotetraose which is hydrolyzed rapidly to xylobiose. Consequently, almost no xylose is found among initial products of xylotriose degradation. The degradation mechanisms of xylooligosaccharides at low and high substrate concentration are illustrated in Fig. 6.

Xylosyl transfer reactions are needed to initiate the degradation of aryl-β-D-xylopyranosides. Liberation of aglycons from phenyl or 4-ni-

TABLE IV
FORMATION OF [U-^{14}C]XYLOOLIGOSACCHARIDES AND
D-[U-^{14}C]XYLOSE[a,b]

Time (hr)	Radioactivity (%)			
	Xylose	Xylobiose	Xylotriose	Xylotetraose
0.5	2.7	8.9	5.7	2.2
1	2.7	14.5	10.1	3.3
2	4.4	28.9	15.2	4.1
3	5.6	38.9	13.8	3.9
4	7.8	38.4	12.5	3.4
5	9.4	47.2	8.5	2.1

[a] Reproduced with permission from Krátký *et al.*[16]
[b] Formed from 150 m*M* phenyl-β-D-[U-^{14}C]xylopyranoside during incubation with purified xylanase of *C. albidus* (3 U/ml).

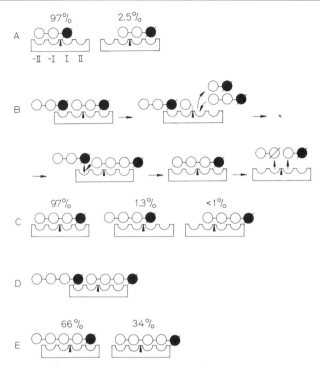

FIG. 6. Possible binding and degradation mechanisms of xylooligosaccharides by xylanase of *C. albidus*. (A) Productive enzyme–xylotriose complexes at low substrate concentration and probability of their formations; (B) shifted two–one binding of xylotriose at high substrate concentration (termolecular shifted complex) and subsequent xylosyl transfer reaction; (C) productive enzyme–xylotetraose complexes formed at low substrate concentration and probability of their formation; (D) termolecular shifted enzyme–xylotetraose complex occurring at high substrate concentration; (E) productive enzyme–xylopentaose complexes formed at low substrate concentration. Reproduced with permission from Biely *et al.*[15]

trophenyl-β-D-xylopyranosides is accompanied by the formation of xylooligosaccharides and only small amounts of xylose. It has been established that the reaction involves multiple xylosyl transfers leading first to aryl glycosides of xylooligosaccharides which are subsequently hydrolyzed to xylobiose and xylotriose.[12] These reactions are so significant that they can be used for preparing lower 1,4-β-xylooligosaccharides. Yields of products in the reaction with 150 mM phenyl-β-D-[U-^{14}C]xylopyranoside are shown in Table IV.[16]

[16] Z. Krátký, P. Biely, and M. Vršanská, *Carbohydr. Res.* **93**, 300 (1981).

In accordance with the ability of *C. albidus* xylanase to hydrolyze 1,3-β-xylosidic linkage, the transfer reactions lead also to the formation of a low proportion of products having 1,3-β-glycosidic linkage.[10] Of various saccharides examined cellobiose was found to be one of the best xylosyl acceptors, but was xylosylated at 6′-OH and not at 4′-OH as might be expected.[17]

[17] P. Biely and M. Vršanská, *Carbohydr. Res.* **123**, 97 (1983).

[81] Xylanases of *Streptomyces*

By TUNEO YASUI, MASAKI MARUI, ISAO KUSAKABE,
and KOTOYOSHI NAKANISHI

Streptomyces xylanase was first described by Sørensen.[1] Thereafter the xylanase system was extensively studied by many researchers.[2-5] The production of xylanase was studied by Nakanishi *et al.*[6] Most *Streptomyces* can produce inductively extracellular xylanase in the presence of xylan, but cannot secrete β-xylosidase into medium. When the xylanase acts on xylan, mainly xylose and xylobiose are accumulated in the hydrolyzate. Kusakabe *et al.*[7] utilized the properties of the enzyme for the preparation of xylobiose.

We found that *Streptomyces* can produce inductively the xylanase in response to nonmetabolizable β-xylosides. This chapter describes a convenient method for the preparation and purification and properties of the xylanase (1,4-β-D-xylan xylanohydrolase, EC 3.2.1.8, endo-1,4-β xylanase) from *Streptomyces*.

Materials and Methods

Organism. The strain of *Streptomyces* sp. No. 3137 was isolated from soil and maintained at the University of Tsukuba in Japan. The organism

[1] H. Sørensen, *Acta Agric. Scand. Suppl.* **1**, 5 (1957).
[2] H. Iizuka and T. Kawaminami, *Agric. Biol. Chem.* **29**, 520 (1965).
[3] I. Kusakabe, T. Yasui, and T. Kobayashi, *Nippon Nogei Kagaku Kaishi* **43**, 145 (1969).
[4] T. Nakajima, K. Tsukamoto, T. Watanabe, K. Kainuma, and K. Matsuda, *J. Ferment. Technol.* **62**, 269 (1984).
[5] M. Marui, K. Nakanishi, and T. Yasui, *Agric. Biol. Chem.* **49**, 3399 (1985).
[6] K. Nakanishi, T. Yasui, and T. Kobayashi, *J. Ferment. Technol.* **54**, 801 (1976).
[7] I. Kusakabe, T. Yasui, and T. Kobayashi, *Agric. Biol. Chem.* **39**, 1355 (1975).

can produce the extracellular xylanase effectively with some nonmetabolizable β-xyloside as well as with xylan. Stock cultures are maintained on Bennett's agar (0.1% glucose, 0.2% NZ-amine, 0.1 yeast extract, 0.1% beef extract, and 2.0% agar).

Materials. Xylan used for the determination of enzyme activity is prepared as follows. Xylan ("Sunpearl" Sanyo Kokusaku Pulp Co., Ltd.) is dissolved in 5% NaOH and precipitated with addition of an equal volume of ethanol. The precipitate is collected by centrifugation and suspended in water. The suspension is adjusted to pH 5 with acetic acid and washed with ethanol and acetone. After drying, the xylan is powdered by a mortar and pestle. Xylooligosaccharides are prepared from a xylan hydrolyzate by *Streptomyces* xylanase according to the method of Kusakabe *et al.*[7] Methyl-β-xyloside is synthesized from xylose and methanol using a cationic resin as catalyst by the method of Bollenback.[8]

Enzyme Assay. Xylanase activity is measured by incubating 0.2 ml of diluted enzyme solution with 1.0 ml of 2% xylan suspension and 0.8 ml of McIlvaine's buffer (pH 5.5) for 10 min at 55°, and determining the released reducing sugar by modification of Somogyi's method.[9] One unit of enzyme activity is defined as the amount causing production of 1 μmol of reducing sugar as a xylose equivalent per minute, under optimal conditions.

Purification[5]

Culture of Streptomyces. One liter of the first medium contains 30 g of glucose, 3 g of urea, 1 g of yeast extract, 10 g of KH_2PO_4, 0.5 g of $MgSO_4 \cdot 7H_2O$, and 5 g of corn steep liquor. One loopful of spores from a Bennett's agar slope is inoculated into a 500-ml shake flask containing 100 ml of the first medium, and the flask is incubated on a shaker at 120 strokes per minute at 36° for 40 hr. Mycelia obtained from the first culture are put into a 500-ml shake flask with 100 ml of the second medium containing 0.3 g of methyl-β-xyloside, 0.2 g of KH_2PO_4, and 0.01 g of $MgSO_4 \cdot 7H_2O$. The conditions of incubation are the same as that for the first culture except for the time of 24 hr.

Step 1. Ultrafiltration. The culture filtrate (2500 ml) is concentrated to one-tenth of the original volume (250 ml) by ultrafiltration with Diaflo UM10 (Amicon Far East Ltd., Tokyo, Japan).

Step 2. Chromatography on DEAE-Sephadex A-25. The concentrated filtrate (250 ml) is put on a DEAE-Sephadex A-25 column (2 × 60 cm)

---

[8] G. N. Bollenback, *Methods Carbohydr. Chem.* **2,** 326 (1962).
[9] M. Somogyi, *J. Biol. Chem.* **169,** 61 (1945).

previously buffered with McIlvaine's buffer (pH 5.5). The effluent (300 ml), decolorized by passing through this column, is concentrated to 30 ml by an evaporator under reduced pressure at 35°.

Step 3. Gel Filtration on BioGel P-100. The concentrated effluent (10 ml) is put on a BioGel P-100 column (2 × 90 cm) previously buffered with McIlvaine's buffer (pH 5.5). Filtration is performed with the same buffer at 4° at the flow rate of 7 ml/hr. Xylanase activities are recovered in two fractions (5 ml per tube), which are named X-I and X-II, respectively. The two fractions are concentrated to 1 mg of protein/ml by an evaporator under reduced pressure at 35°.

Step 4. Isoelectric Focusing. The concentrated X-I fraction (10 ml) is applied on the isoelectric focusing column (110-ml capacity, LKB Produkter AB, Bromma, Sweden) using 1% carrier ampholyte (pH 6–8, Servalyt 6–8, Serva Feinbiochemica GmbH and Co., Federal Republic of Germany) and run at 900 V for 48 hr at 4°. Xylanase activity of X-I is recovered in one peak, and only one protein peak is observed in the same position. For the isoelectric focusing of concentrated X-II fraction (10 ml), the 1% carrier ampholyte (pH 9–11, Servalyt 9–11) is used under the same conditions as for X-I. The isoelectric focusing pattern of X-II shows two fractions (X-II-A and X-II-B) of xylanase activity. The contaminating carrier ampholytes in the three fractions are removed by salting out with

TABLE I
PURIFICATION OF XYLANASES FROM *Streptomyces* sp. No. 3137

Fraction	Activity (units)	Protein[a] (mg)	Specific activity (units/mg)	Yield (%)
Culture filtrate	200,000	4900	41	100
Ultrafiltration	180,000	2700	67	90
DEAE-Sephadex A-25	168,000	2100	80	84
BioGel P-100				
X-I	46,000	340	135	23
X-II	66,000	460	143	33
Total				56
Isoelectric focusing				
X-I	42,000	260	161	21
X-II-A	14,000	132	106	7
X-II-B	45,000	290	155	23
Total				51

[a] Protein is determined by the method of Lowry *et al.* [O. H. Lowry, N. J. Rosebrough, A. L. Farr, and R. J. Randall, *J. Biol. Chem.* **193**, 265 (1951)].

ammonium sulfate (70% saturation). After that the enzyme solutions are dialyzed against distilled water with cellulose tubes.

The purity of the enzymes after isoelectric focusing is confirmed by performing slab polyacrylamide gel electrophoresis using 10% gel in β-alanine acetic acid buffer pH 4.3. Each xylanase shows a different protein band. The purification procedure is summarized in Table I.

Properties

The properties of xylanases from some *Streptomyces* sp. are compared in Table II.

Effects of pH and Temperature. The three xylanases are stable in the pH range 3.0–10.5. The optimal pH is 5.5–6.0.

The activity of the enzymes is maximal at 60–65°, but a sharp decline in activity is observed at higher temperatures, due to thermal denaturation of the enzymes. The enzymes are stable to 55° for 30 min at pH 5.5.

Molecular Weight and Isoelectric Point. X-I has an apparent molecular weight of 50,000 as derived from SDS–gel electrophoresis and gel filtration on BioGel P-100. The molecular weights of X-II-A and X-II-B are estimated to be approximately 25,000 by SDS–gel electrophoresis, and that of X-II-B calculated from the sedimentation analysis is 25,680. The results of molecular weight estimation demonstrate that the xylanases are monomeric enzymes.

The isoelectric points of X-I, X-II-A, and X-II-B are 7.10, 10.06, and 10.26, respectively.

TABLE II
PROPERTIES OF XYLANASES OF *Streptomyces* sp.

Property	Strain 3137			Strain E-86	Strain[4] KT-23
	X-I	X-II-A	X-II-B		
Molecular weight	50,000	25,000	25,000	40,500	43,000
Isoelectric point	7.10	10.06	10.26	7.3	6.9
Optimal pH	5.5–6.5	5.0–6.0	5.0–6.0	5.5–6.2	5.5
pH stability	3.0–10.5	1.5–11.5	1.5–11.5	4.5–10.5	4–10
Optimal temperature (°C)	60–65	60–65	60–65	55–60	55
Thermal stability (°C)	55	55	55	55	55
Inhibitors	Fe^{3+}, Hg^{2+} SDS, NBS	Fe^{3+}, Hg^{2+} SDS, NBS	Fe^{3+}, Hg^{2+} SDS, NBS	Cu^{2+}, Hg^{2+} Ag^{2+}, SDS $K_4Fe(CN)_6$	Hg^{2+}, Mn^{2+} SDS
Final products	Xylose Xylobiose	Xylose Xylobiose	Xylose Xylobiose	Xylose Xylobiose	Xylose Xylobiose

TABLE III

AMINO ACID COMPOSITION OF XYLANASES OF *Streptomyces* sp.

Amino acid	Strain 3137			Strain E-86	Strain KT-23
	X-I	X-II-A	X-II-B		
Aspartic acid	13.1	13.7	11.5	12.4	9.5
Threonine	7.7	14.0	11.6	7.8	5.4
Serine	7.9	13.9	10.4	8.2	17.5
Glutamic acid	8.8	4.5	5.7	8.1	14.9
Proline	2.5	2.1	3.3	4.2	2.8
Glycine	12.4	13.3	15.9	11.7	16.3
Alanine	9.4	3.3	3.8	9.4	9.0
Valine	6.0	5.2	6.1	5.4	4.8
Cysteine	2.7	0.9	1.3	1.9	0.6
Methionine	1.3	1.2	1.6	1.5	n.d.[b]
Isoleucine	3.5	2.5	2.2	4.0	2.8
Leucine	5.9	4.3	4.0	5.6	4.3
Tyrosine	3.4	8.6	8.2	3.9	1.2
Phenylalanine	2.8	3.1	3.3	4.1	2.0
Histidine	1.9	1.0	1.5	2.2	3.1
Lysine	3.1	2.3	2.7	3.8	3.1
Arginine	5.1	3.4	3.5	5.8	2.5
Tryptophan[a]	2.6	2.9	3.1	n.d.	n.d.

[a] Tryptophan is measured spectrophotometrically.
[b] n.d., not determined.

Amino Acid[10] and Carbohydrate Analysis. The N-terminal amino acid of X-I is an alanine residue and those of X-II-A and X-II-B are threonine residues. X-I has an aspartic acid residue as the C-terminal amino acid and X-II-B has an alanine residue. Amino acid compositions of the three xylanases are shown in Table III, in comparison with that of xylanases from other *Streptomyces* sp. The molar ratios of tyrosine in X-II-A and X-II-B are more than twice as much as that of X-I. Amino acid analysis indicates that X-I is a different protein from X-II-A and X-II-B in structure, and X-II-A and X-II-B are very similar proteins.

Carbohydrate analysis with the phenol–sulfuric acid method[11] and the orcinol–Fe^{3+} method[12] indicates that the enzymes are not glycoproteins.

[10] M. Marui, K. Nakanishi, and T. Yasui, *Agric. Biol. Chem.* **49**, 3409 (1985).
[11] M. Dubois, K. A. Gilles, J. K. Hamilton, P. A. Rebers, and F. Smith, *Anal. Chem.* **28**, 350 (1956).
[12] W. R. Fernell and H. K. King, *Analyst* (*London*) **78**, 80 (1953).

Inhibitors. Hg^{2+}, Fe^{3+}, and SDS are potent inhibitors for the enzymes, but *p*-chloromercuribenzoate is not. *N*-Bromosuccinimide inhibits the xylanase activity at high concentration (5 m*M*). In the concentration of 0.1 m*M*, it modifies the tryptophan residues of the enzymes, but the activity still remains.

Immunological Properties.[10] Antiserum to X-I shows no reaction with X-IIs and does not inhibit the activities of X-IIs. On the other hand, antiserum to X-II-B also shows no reaction with X-I and does not inhibit the activity of X-I. But antiserum to X-II-B cross-reacts to X-II-A. These results indicate that X-I is different from X-IIs and X-II-A and X-II-B are closely related in structure.

Specificity. Although the three xylanases are distinct enzymes, all of them show very similar action patterns on xylan and xylooligosaccharides. They liquefy rapidly hardwood xylan to produce xylose and xylobiose as main final products. In the course of hydrolysis, apparent transxylosidation reactions are not detected in paper chromatography. However, hydrolysis of xylooligosaccharides by xylanases reveals the presence of transxylosidation reaction.[13] The purified preparations of xylanases cannot hydrolyze xylobiose, *β*-xylosides, and other polysaccharides such as starch, cellulose, and maltose.

Additional Remarks

Streptomyces sp. No. 3137 and other strains produce inductively xylanases by various xylosides. Xylose and hydrolyzable *β*-xylosides such as phenyl-*β*-xylosides and xylobiose have no inducing ability, while nonmetabolizable *β*-xylosides such as aliphatic xylosides have higher inducing ability than do xylan as shown in Table IV.[14]

Xylanase from *Streptomyces* sp. E-86 has very low arabinofuranosidase activity. The following oligosaccharides are isolated from the enzymatic hydrolyzate of corncob xylan. These include *O*-*α*-L-arabinofuranosyl-(1 → 3)-D-xylopyranose, *O*-*α*-L-arabinofuranosyl-(1 → 3)-*O*-*β*-D-xylopuranosyl-(1 → 4)-D-xylopyranose, *O*-*β*-D-xylopyranosyl-(1 → 2)-*O*-*α*-L-arabinofuranosyl-(1 → 3)-*O*-*β*-D-xylopyranosyl-(1 → 4)-D-xylopyranose, *O*-*β*-D-xylopyranosyl-(1 → 4)-*O*-[*α*-L-arabinofuranosyl-(1 → 3)]-*O*-*β*-D-xylopyranosyl-(1 → 4)-D-xylopyranose, *O*-*β*-D-xylopyranosyl-(1 → 4)-*O*-[*β*-D-xylopyranosyl-(1 → 2)-*O*-*α*-L-arabinofuranosyl-(1 → 3)]-*O*-*β*-D-xylopyranosyl-(1 → 4)-D-xylopyranose, and *O*-*β*-D-xylopyranosyl-

[13] I. Kusakabe, T. Yasui, and T. Kobayashi, *Nippon Nogei Kagaku Kaishi* **51**, 439 (1977).
[14] T. Yasui, K. Nakanishi, and T. Kobayashi, *Hakkokogaku* **58**, 79 (1980).

TABLE IV
EFFECT OF VARIOUS INDUCERS ON XYLANASE PRODUCTION

Inducer	Inducer concentration (mg/ml)	Xylanase activity (units/ml)	Inducing activity	
			Activity/mg	Activity/mM
Xylan	3.00	91	30	4
Methyl-β-xyloside	1.00	50	50	8
Ethyl-β-xyloside	0.23	15	65	12
n-Propyl-β-xyloside	0.50	57	114	22
Isopropyl-β-xyloside	0.50	98	196	38
n-Butyl-β-xyloside	0.10	60	600	122
tert-Butyl-β-xyloside	0.10	12	120	25
Hexyl-β-xyloside	0.10	6	60	15
Allyl-β-xyloside	0.10	14	140	26
Glyceryl-β-xyloside	0.12	28	233	52
Benzyl-β-xyloside	0.10	20	200	47
Thiophenyl-β-xyloside	0.10	15	150	36
Cyclohexyl-β-xyloside	0.10	87	870	202
Epoxypropyl-β-xyloside	0.070	50	714	147
Ethylene cyanohydrin-β-xyloside	0.034	48	1410	282
Phenyl-β-xyloside	0.50	0.1	—	—
o-Nitrophenyl-β-xyloside	0.50	0.1	—	—

(1 → 2)-O-α-L-arabinofuranosyl-(1 → 3)-O-β-D-xylopyranosyl-(1 → 4)-O-β-D-xylopyranosyl-(1 → 4)-D-xylopyranose, which are arabinooligo-saccharides having a branch attached to the nonreducing end xylose residue.[15,16] On the other hand, *Aspergillus niger* xylanase I produces oligosaccharides having a branch attached to the reducing-end xylose residue in the hydrolysis of xylan from rice straw.[17]

In general, all of the action patterns of xylanases from different strains of *Streptomyces* are supposed to be the same.

[15] I. Kusakabe, T. Yasui, and T. Kobayashi, *Nippon Nogei Kagaku Kaishi* **51**, 669 (1977).
[16] I. Kusakabe, S. Ohgushi, T. Yasui, and T. Kobayashi, *Agric. Biol. Chem.* **47**, 2713 (1983).
[17] S. Takenishi and Y. Tsujisaka, *Agric. Biol. Chem.* **37**, 1385 (1973).

[82] Xylanases of Alkalophilic Thermophilic *Bacillus*

By Teruhiko Akiba and Koki Horikoshi

It was first reported in 1971 that an alkalophilic *Bacillus* produced alkaline enzymes under extreme conditions around pH 10.[1] Since then, various kinds of alkaline enzymes, including xylanases (1,4-β-xylan xylanohydrolase, EC 3.2.1.8, endo-1,4-β-xylanase), have been found in many alkalophilic microorganisms.[2] To date three alkalophilic *Bacillus* species have been shown to produce xylanase[3–5]; the purification of one from an alkalophilic thermophilic *Bacillus*[4,6] will be described here.

Assay

Principle. Xylanase is assayed by measuring the reducing sugars liberated from xylan as substrate. The reducing sugars are determined routinely by the well-known method using dinitrosalicylic acid (DNS), and expressed as micromoles of xylose per milliliter using xylose as a standard.

Reagents

McIlvaine's buffer, pH 6.0[7]

DNS reagent: there have been several modified compositions; among them, the method of Miller[8] is successfully used

Xylan: larchwood xylan purchased from either Sigma Chemical Co. (St. Louis, MO) or Fluka AG Chem. (Switzerland)

Procedure. The standard assay mixture contains 0.5 ml of 0.5% xylan suspended in McIlvaine's buffer and 0.05 ml of suitably diluted enzyme. Xylan is preliminarily suspended homogeneously in McIlvaine's buffer (pH 6.0) by ultrasonic treatment. The reaction mixture is incubated at 60° for 10 min. The reaction is stopped by adding 1.0 ml of DNS reagent and

[1] K. Horikoshi, *Agric. Biol. Chem.* **35**, 1407 (1971).
[2] K. Horikoshi and T. Akiba, *in* "Alkalophilic Microorganisms: A New Microbial World," p. 93. Springer-Verlag, Berlin, 1982.
[3] K. Horikoshi and Y. Atsukawa, *Agric. Biol. Chem.* **37**, 2097 (1973).
[4] W. Okazaki, T. Akiba, K. Horikoshi, and R. Akahoshi, *Appl. Microbiol. Biotechnol.* **19**, 335 (1984).
[5] H. Honda, T. Kudo, Y. Ikura, and K. Horikoshi, *Can. J. Microbiol.* **31**, 538 (1985).
[6] W. Okazaki, T. Akiba, K. Horikoshi, and R. Akahoshi, *Agric. Biol. Chem.* **49**, 2033 (1985).
[7] T. C. McIlvaine, *J. Biol. Chem.* **49**, 183 (1921).
[8] G. L. Miller, *Anal. Chem.* **31**, 426 (1959).

heating in a boiling water bath for 5 min, followed by cooling in water. Distilled water (4.45 ml) is then added to a final volume of 6.0 ml. The concentration of reducing sugars in the sample is measured spectrophotometrically at 500 nm.

Definition of Unit and Specific Activity. One unit of enzyme activity is defined as the amount that releases 1 μmol of reducing sugar as xylose equivalent per minute under the above conditions. Specific activity is expressed as units of enzyme per milligram of protein. Protein is determined by the method of Bradford[9] using Bio-Rad Protein Assay kit with bovine albumin as the standard protein.

Purification

Culture of Organisms. Bacillus sp. W1 (JCM 2888) and W2 (JCM 2889), from the Japan Collection of Microorganisms, The Institute of Physical and Chemical Research, Wako, Saitama 351-01, were grown at 45° for 20 hr on an alkaline medium (pH 10) consisting of (g/liter) xylan, 10.0; yeast extract, 5.0; polypeptone, 5.0; K_2HPO_4, 1.0; $MgSO_4 \cdot 7H_2O$, 0.2; and Na_2CO_3, 10.0 which was separately autoclaved and added to the medium. The prolonged culture period of over 24 hr may cause lysis of bacterial cells which will cause undesirable problems in purification. The cultures were centrifuged at 8000 rpm for 10 min to remove cells and the supernatant was collected.

All purification steps are carried out at 4°. Xylanases from both strains W1 and W2 are purified by the same procedures as follows.

Step 1. Ammonium Sulfate Fractionation. Prior to fractionation, the pH of the supernatant is adjusted to neutrality by adding concentrated HCl. To the supernatant, ammonium sulfate is added slowly to reach 90% saturation. After standing overnight, the resulting precipitate is collected by centrifugation at 9000 rpm for 15 min, dissolved in 50 mM phosphate buffer (pH 6.2), and dialyzed overnight against running water.

Step 2. Chromatography on DEAE-Toyopearl 650M. The dialyzed enzyme solution is added to a DEAE-Toyopearl 650M column (2.6 × 90 cm) previously equilibrated with 50 mM phosphate buffer (pH 6.2). Component I can be eluted with about 1 liter of the same buffer as a broad peak. The column is washed with about 1 liter of the buffer containing 0.2 M NaCl to elute protein impurities. Then component II can be eluted as a sharp peak with a linear gradient of from 0.2 to 0.6 M NaCl in the same buffer of about 900 ml. The two active fractions I and II are collected

[9] M. Bradford, *Anal. Biochem.* **72**, 248 (1976).

TABLE I
PURIFICATION OF XYLANASES I AND II FROM *Bacillus* sp. W1

Step	Volume (ml)		Total protein (mg)		Total activity (units)		Specific activity (units/mg)		Yield (%)	
Culture supernatant	3,950		32,200		162,300		5		100	
1. 90% ammonium sulfate fraction	770		7,390		120,400		15		74	
	(I)	(II)	(I)	(II)	(I)	(II)	(I)	(II)	(I)	(II)
2. First DEAE- Toyopearl 650M	705	305	140	220	48,000	9,500	340	43	30	6.0
3. Second DEAE- Toyopearl 650M	—	68	—	93	—	4,320	—	46	—	2.6
4. Toyopearl HW- 55S eluate	21	15	65	88	40,500	4,250	623	48	25	2.6

separately and concentrated about 100-fold with an Amicon PM10 (Amicon Co., Danvers, MA) membrane filter.

Step 3. Rechromatography. For further purification, component II is added to a column (2.2 × 55 cm) of the same exchanger as in step 2, previously equilibrated with the same buffer containing 0.2 *M* NaCl, and eluted with the NaCl linear gradient as described in step 2. The active fractions are collected and pooled.

Step 4. Gel Filtration on Toyopearl HW-55S. Component I obtained in step 2 and the component II obtained in step 3, both concentrated to about 4 ml or less by ultrafiltration on an Amicon PM10 membrane, are separately applied to a Toyopearl HW-55S column (1.5 × 55 cm) previously equilibrated with 50 m*M* phosphate buffer (pH 7.0) containing 0.2 *M* NaCl. Each of the active fractions is collected and dialyzed against the same buffer without NaCl. Gel filtration on dextrans (Sephadex series) and polyacrylamide (BioGel series) was not useful because of adsorption of the enzymes to those gels, which would cause delay of elution and loss of yield.

A typical purification procedure with xylanase from strain W1 is summarized in Table I.

Properties

Purity. The two components I and II purified as above are found to be homogeneous by disc and SDS–polyacrylamide gel electrophoresis.

Molecular Properties. The molecular weights of components I and II from strain W1 are 21,500 and 49,500, respectively. The molecular weights of components I and II from strain W2 are 22,500 and 50,000, respectively. These values were obtained from SDS–polyacrylamide gel electrophoresis. The molecular weights obtained from gel filtration on Toyopearl HW-55S are 48,500 for component II of strain W1 and 51,000 for component II of strain W2. The determination by gel filtration of the molecular weights of component I was unsuccessful with the two strains W1 and W2 due to the interaction between the enzymes and the gel. All four component are monomeric protein.

The isoelectric points, determined by electrofocusing on ampholine polyacrylamide gels, are 8.4 ± 0.1 for components I of the two strains, and 3.6 ± 0.1 for components II of the two strains.

pH Optima. The pH optima of components I of the two strains are both 6.0 in McIlvaine's buffer. On the other hand, the pH optima of components II of the two strains are broad ranging from 7.0 to 9.0.

Temperature Optima. The optimum temperatures of components I and II are 65 and 70°, respectively, in both strains W1 and W2.

Stability. Components I and II of both strains are stable between pH 4.5 and 10.0 against heat treatment at 45° for 1 hr. They are gradually inactivated by heating for 10 min over 60°.

Inhibitors. Components I and II of the two strains are completely inhibited by Hg^{2+} at 5 mM and partly inhibited by Ag^+, Cu^{2+}, and EDTA at a concentration of 5 mM. Other metals tested such as Cd^{2+}, Ca^{2+}, Ba^{2+}, Mg^{2+}, Ni^{2+}, and Zn^{2+} at 5 mM have no inhibitory effect on any of the components. *p*-Mercuribenzoate at 5 mM produces about 40% inhibition of components I in both strains.

Kinetic Properties. The K_m (mg of xylan/ml) for xylan is estimated to be 4.5 for the component I and 0.95 for component II of strain W1, and 4.0 for component I and 0.57 for component II of strain W2.

Hydrolysis Products from Xylan. The hydrolysis products from xylan were analyzed by high-performance liquid chromatography with a differential refractometer as detector, using a Shim-Pack SCR-101N column (Shimadzu Co., Kyoto) at a flow rate of 0.5 ml/min of degassed water. The hydrolysis products from xylan with components I from the two strains are xylobiose, xylotriose, xylotetraose, and xylopentaose; xylose is not formed. On the other hand, those from components II from the two strains are xylose, xylobiose, xylotriose, and xylopentaose; xylose but no xylotetraose is formed. Xylobiose is the predominant product and it is never hydrolyzed to xylose by any of the components; xylotriose is the minimum substrate to be cleaved.

Transferase Activity. The hydrolysis of xylotriose and xylotetraose by components I and II yields higher products of xylopentaose other than xylose and xylobiose. Xylotetraose is also formed from lower substrate of xylotriose. These facts suggest that all components may have transferase activity.

[83] Xylanase A of *Schizophyllum commune*

By L. JURASEK and M. G. PAICE

Xylanases (EC 3.2.1.8, endo-1,4-β-xylanase) are widespread in nature among microorganisms that decompose lignocellulosic materials. Wood-rotting fungi play a prominent role in this process and the fungus Common Split Gill (*Schizophyllum commune* Fr.) is a particularly strong producer of xylanases. Xylanases belong to a broader group of enzymes, hemicellulases, responsible for conversion of hemicelluloses to soluble sugars. Hemicellulases in general and xylanases in particular are also of some commercial interest. In the food industry they could be used as additives to fruit juices to facilitate their clarification.[1] In the pulp and paper industry, these enzymes could be used for treatment of pulp in order to reduce hemicellulose content. This could be of interest for dissolving pulp production[2] and for bleaching of chemical pulps.[3a]

Source of the Enzyme

Xylanase A originates from *S. commune,* strain #13 Delmar (ATCC 38548). The enzyme is probably identical with a xylanase previously obtained from another strain of *S. commune.*[3]

Enzyme Assay

The enzyme assay[4] is based on the hydrolysis of larch xylan. Due to the incomplete definition of the substrate and the fact that its properties change during the assay, the results are reproducible only under care-

[1] C. I. Beck and D. Scott, *Adv. Chem. Ser.* **136,** 1 (1974).

[2] M. G. Paice and L. Jurasek, *J. Wood Chem. Technol.* **4,** 187 (1984).

[3] J. Varadi, V. Necesany, and P. Kovacs, *Drev. Vysk.* **16,** 147 (1971).

[3a] L. Viikari, M. Ranua, A. Kantelinen, J. Sundquist, and M. Linko, *in* "Biotechnology in the Pulp and Paper Industry," p. 67. Swedish Forest Products Laboratory, Stockholm, Sweden, 1986.

[4] M. G. Paice, L. Jurasek, M. R. Carpenter, and L. B. Smillie, *Appl. Environ. Microbiol.* **36,** 802 (1978).

fully standardized conditions. Xylan from larchwood (molecular weight 20,000; Sigma Chemical Co.) was suspended in distilled water to make 1% consistency. After extensive agitation the suspension was centrifuged (30 min at 6000 g) and the clear supernatant freeze-dried. The soluble xylan was then redissolved in 0.1 M sodium acetate buffer (pH 5) to form a 1% solution which was then used for assays. One milliliter of the substrate solution mixed with 1 ml of an appropriately diluted enzyme was incubated for 10 min at 30°. The reaction was terminated by adding 2 ml of 2,4-dinitrosalicylic acid reagent[4a] and heating at 100° for 15 min. After diluting with 1% aqueous Rochelle salt to 20 ml, the optical density of the solution was read at 575 nm. The optical density was calibrated with a known concentration of xylose. For a standard assay, the enzyme solution was diluted so that 5 μmol of xylose equivalent was produced per tube under the assay conditions. The determination of the appropriate enzyme concentration requires a dilution series and the appropriate value is determined by interpolation. The international unit of xylanase is then defined as release of 1 μmol of xylose equivalent per minute.

Production of the Enzyme

Inoculum of the *S. commune* culture is prepared by adding a piece (approximately 7 mm in diameter) of an actively growing culture on malt agar (Difco Laboratories) to a 500-ml polypropylene Erlenmeyer flask containing 200 ml of mycological broth, low pH (Difco laboratories), and a 20-mm glass marble. The flask is then shaken at 250 rpm (New Brunswick Scientific Co. Gyratory Shaker A-33-500) at 30° for 4 days. The marble rolling inside the flask assures a finely suspended mycelial growth.

The production medium[5] has the following composition (g/liter): cellulose (Solka-Floc SW40, Brown Co.), 17; Bactopeptone (Difco Laboratories), 23; $Ca(NO_3)_2 \cdot 4H_2O$, 15.3; KH_2PO_4, 1.3; $MgSO_4 \cdot 7H_2O$, 0.5; and 1 ml of trace metal solution[6] consisting of 495 ml distilled water, 5 ml conc. HCl, 2.5 g $FeSO_4$, 0.98 g $MnSO_4 \cdot H_2O$, 0.83 g $ZnCl_2$, and 1.0 g $CoCl_2 \cdot 6H_2O$. After sterilization at 120° for 15 min, the production culture is started by adding 1% liquid inoculum (v/v).

The production culture is grown *either* in a series of 500-ml Erlenmeyer flasks, each containing 150 ml of medium, with shaking at 250 rpm,

[4a] G. L. Miller, *Anal. Chem.* **31**, 426 (1959).
[5] M. Desrochers, L. Jurasek, and M. G. Paice, *Dev. Ind. Microbiol.* **22**, 675 (1981).
[6] M. Mandels, D. Sternberg, and R. E. Andreotti, *in* "Symposium on Enzymatic Hydrolysis of Cellulose" (M. Bailey, T. H. Enari, and M. Linko, eds.), p. 81. Finnish National Fund for Research and Development, Helsinki, Finland, 1975.

or in a fermenter (Microferm; New Brunswick Scientific Co.) with 10 liters of medium agitated at 200 rpm and aerated (1.2 liters/min). In either case, the production culture is grown for 9 days at 30°. A culture supernatant is then obtained by centrifugation (30 min at 6000 g). Under the above fermentation conditions, xylanase A is the predominant protein produced and the usual yield is approximately 200 IU/ml.

Crude Enzyme Preparation

Crude xylanase is prepared by fractional precipitation.[2] Chilled culture supernatant is first precipitated by slow addition of 2 volumes of cold (−18°) ethanol. Most of the contaminating protein and viscous carbohydrate is removed by subsequent centrifugation (30 min at 6000 g). The clear supernatant is again precipitated by addition of another volume of ethanol which brings down crude xylanase. The precipitated crude preparation is centrifuged at 7000 g for 30 min.

Isolation of the Enzyme

An aliquot of the crude enzyme preparation corresponding to 2 liters of the original production culture is redissolved in 50 ml 300 mM acetate–50 mM N-ethylmorpholine ammonium buffer (pH 9).[4] The solution is clarified by centrifugation and applied to a DEAE-Sephadex A-50 column (5 × 85 cm) previously equilibrated with the same buffer. The column is then eluted with a linear gradient consisting of 4 liters of the starting buffer and 4 liters of 100 mM pyridine–50 mM N-ethylmorpholine acetate buffer (pH 5). The main xylanase peak, xylanase A, appears at approximately one-third of the elution profile. The fractions containing xylanase A are pooled and their total volume reduced to approximately 40 ml by partial freeze-drying. The concentrate is clarified by centrifugation and applied to a Sephadex G-50 column (5 × 85 cm) equilibrated with 200 mM pyridinium acetate buffer (pH 5). Fractions of the xylanase A peak appearing at approximately two-thirds of the column bed volume are pooled and freeze-dried. This should yield approximately 60 mg of pure enzyme. The enzyme is homogeneous on SDS–PAGE gel and gives a clean amino acid sequence upon sequencing on the automated sequencer.

Properties of Xylanase A

Although the pure enzyme preparation contains carbohydrate, the protein does not appear to be glycosylated, since the carbohydrate can be

TABLE I
PHYSICOCHEMICAL CONSTANTS FOR XYLANASE A[a]

Parameter	Units
Molecular weight	
SDS–PAGE	21,000
pH optimum	5
K_m (soluble xylan)	8.37 mg/ml
V_{max}	0.443 μmol/min
Specific activity	1.5×10^3 IU/mg
Temp. optimum (10 min assay)	50°
pH stability range	6–8
Isoelectric point	4.5

[a] Largely from Ref. 4. The molecular weight was previously reported as 33,000.

removed by treatment with 1% SDS.[4] The enzyme is a single chain polypeptide with serine at the amino terminus; the N-terminal sequence is homologous with xylanase from *Bacillus subtilis*.[7]

Xylanase A is assumed to be an endoxylanase although its specificity has not been studied in detail. The major hydrolysis products after 18-hr hydrolysis are xylobiose and xylose (approximate ratio 3 : 1). It appears to be strictly specific for xylan; no appreciable cellulase, β-glucosidase nor β-xylosidase activity has been found.

[7] M. G. Paice, R. Bourbonnais, M. Desrochers, L. Jurasek, and M. Yaguchi, *Arch. Microbiol.* **144**, 201 (1986).

[84] Xylanases and β-Xylosidase of *Trichoderma lignorum*

By MICHAEL JOHN and JÜRGEN SCHMIDT

$$1,4\text{-}\beta\text{-}D\text{-Xylan} + H_2O \xrightarrow{\text{xylanase}} 1,4\text{-}\beta\text{-}D\text{-xylooligosaccharides} + D\text{-xylose}$$

$$R\text{-}O\text{-}\beta\text{-}D\text{-Xylopyranoside} + H_2O \xrightarrow{\text{xylosidase}} R\text{-OH} + D\text{-xylose}$$

Trichoderma lignorum is a rich source of various extracellular poly- and oligosaccharide-degrading enzymes, e.g., xylanases,[1] cellulase,[2]

[1] M. John, J. Schmidt, H. Sahm, and C. Wandrey, *Biochem. Soc. Trans.* **9**, 166 (1981).
[2] D. Herr, G. Luck, and H. Dellweg, *J. Ferment. Technol.* **56**, 273 (1978).

METHODS IN ENZYMOLOGY, VOL. 160

amyloglucosidase,[3] β-xylosidase,[4,5] α-arabinosidase,[4,5] and β-glucosidase.[4] Described here is a simple procedure for the purification of two different xylanases, 1,4-β-D-xylan xylanohydrolase (EC 3.2.1.8, endo-1,4-β-xylanase) and a β-xylosidase, β-D-xyloside xylohydrolase (EC 3.2.1.37, xylan 1,4-β-xylosidase) from the culture supernatant of this fungus. Studies on the properties of these enzymes reveal that one xylanase has a remarkably high glycosyltransferase activity. The β-xylosidase is a glycoprotein and shows high thermal stability.

Assay Methods

Xylanase

Principle. Xylanase hydrolyzes arabinoxylan to yield a mixture of reducing sugars, which are determined with the alkaline dinitrosalicylic acid reagent.[6]

Reagents

Acetate buffer, 0.1 M, pH 4.5

Arabinoxylan from oat spelt, 1% (Roth, Karlsruhe, GFR) is purified from contaminating α-glucan (~16%) by successive treatment with α-amylase of *Bacillus subtilis* and amyloglucosidase of *Aspergillus niger* (Boehringer)

Arabinoxylan is removed from the digest by precipitation with ethanol (3 volumes), washed twice with ethanol, and lyophilized

Dinitrosalicylic acid reagent: dissolve 5.0 g of 2-hydroxy-3,5-dinitrobenzoic acid (Merck) in 200 ml of 1 M NaOH; add 150 ml of water and dissolve components at ~40°; finally dilute to 500 ml with water

Procedure. A homogeneous substrate suspension is prepared by sonication of the purified arabinoxylan (10 mg/ml) in 0.1 M acetate buffer (pH 4.5). Xylanase activity is determined by incubating 20 μl of enzyme solution with 2.0 ml of the buffered arabinoxylan suspension for 3–15 min at 37°. The reaction is stopped by adding 2.0 ml of the alkaline dinitrosalicylic acid reagent. After heating at 100° for 10 min, the solution is kept on ice for ~15 min and insoluble substrate is then removed by centrifugation (12,000 g for 5 min). The absorbance of the supernatant is measured at 546

[3] M. John and J. Schmidt, *Anal. Biochem.* **141**, 466 (1984).
[4] J. Schmidt, M. John, H. Sahm, and C. Wandrey, *Biochem. Soc. Trans.* **9**, 166 (1981).
[5] J. Schmidt, M. John, C. Wandrey, and H. Sahm, *DECHEMA–Monogr.* **95**, 223 (1984).
[6] G. L. Miller, *Anal. Chem.* **31**, 426 (1959).

nm using a 1-cm cuvette. The optical density is calibrated with known concentrations of xylose. One unit of xylanase activity is defined as the amount of enzyme catalyzing the release of 1 μmol of xylose/min under the specified conditions. Specific activity is defined as units/mg of protein. Protein is estimated by the method of Bradford[7] using the Bio-Rad protein assay with bovine γ-globulin as a standard.

β-Xylosidase

Principle. β-Xylosidase activity is determined by measuring the amount of *p*-nitrophenol released from the synthetic substrate *p*-nitrophenyl-β-D-xylopyranoside.

Reagents

Acetate buffer, 0.1 M, pH 4.5
p-Nitrophenyl-β-D-xylopyranoside, 25 mM
Na_2CO_3, 2 M

Procedure. The xylosidase assay mixtures contain the following components in a total volume of 1.22 ml: 1 ml of acetate buffer (pH 4.5), 200 μl of *p*-nitrophenyl-β-D-xylopyranoside, and 20 μl of enzyme solution. Assays are incubated for 5–10 min at 30° and the reaction is terminated by adding 2 ml of 2 M Na_2CO_3. The absorbance is measured at 405 nm using a 1-cm cuvette. A molar extinction coefficient of 17.5 cm^2/μmol is used to calculate the amount of *p*-nitrophenol in the solution. Units are defined as micromole of *p*-nitrophenol/minute. Specific activity is defined as units/ mg of protein.

Cell Growth

Trichoderma lignorum which was isolated from soil of the African equatorial forest (available from the authors) is grown in a medium containing 1.5% arabinoxylan, 0.68% KH_2PO_4, 0.25% $(NH_4)_2SO_4$, 0.1% $NaNO_3$, 0.03% $CaCl_2 \cdot 2H_2O$, 0.03% $MgSO_4 \cdot 7H_2O$, 0.15% tryptone, 0.05% yeast extract, and 3 ml/liter of a trace metal solution containing 0.15% $FeSO_4 \cdot 7H_2O$, 0.12% $MnCl_2 \cdot 4H_2O$, 0.1% $ZnSO_4 \cdot 7H_2O$, and 0.05% $CoCl_2 \cdot 6H_2O$. The initial pH of the medium is adjusted to 5.2. Large-scale growth (8 liters) in a fermentor is performed at 29° for 75 hr with moderate stirring and aeration.[8] The pH is kept constant with 1.5 M KOH and 1.5 M HCl, and polypropylene glycol 2000 (1.0 ml, Baker) is

---

[7] M. Bradford, *Anal. Biochem.* **72**, 248 (1976).
[8] M. John, B. Schmidt, and J. Schmidt, *Can. J. Biochem.* **57**, 125 (1979).

TABLE I

PURIFICATION OF XYLANASES AND β-XYLOSIDASE FROM *Trichoderma lignorum*

Step	Activity (units/ml)	Protein (mg/ml)	Specific activity (units/mg protein)	Purification (fold)
1. Culture supernatant				
Xylanase	123	0.6	205	1
Xylosidase	0.1	0.6	0.17	1
2. Precipitation with ammonium sulfate (50% saturation)				
Xylanase	374	1.53	244	1.2
Xylosidase	0.36	1.53	0.25	1.5
3. Chromatography on BioGel A0.5 m				
Xylanase A	165	0.16	1031	5
Xylanase B	110	0.08	1375	6.7
Xylosidase	0.14	0.32	0.43	2.5
4. Hydroxyapatite chromatography				
Xylanase A	590	0.4	1475	7.2
Xylanase B	633	0.4	1582	7.7
Xylosidase	3.1	1.3	2.38	14

used as an antifoaming agent. Since xylanases and β-xylosidases are inducible by soluble fragments of xylan, enzyme excretion starts during the first half of the logarithmic phase of growth, when the degradation of arabinoxylan is almost complete.[9]

Enzyme Purification

All steps are carried out at 4° unless otherwise indicated. Enzyme and protein assays are performed at each step in the purification procedure. Table I summarizes the steps of purification and the results of the assays.

Step 1. Culture Supernatant. The cells are removed by filtering the culture liquor through a nylon sieve (125 μm mesh size; Verseidag, Krefeld, GFR) and by centrifugation of the filtrate at 20,000 *g* for 20 min. The clear supernatant can be stored frozen at −20° for at least 1 year without considerable loss of activity.

Step 2. Ammonium Sulfate Precipitation. A 450-ml aliquot of the culture supernatant is concentrated to 60 ml by ultrafiltration (PM10 mem-

[9] J. Defaye, H. Driguez, M. John, J. Schmidt, and E. Ohleyer, *Carbohydr. Res.* **139**, 123 (1985).

brane, Amicon). Solid ammonium sulfate is slowly added to the stirred concentrate to achieve 50% saturation (2.05 M). After the mixture is stirred for 30 min, the precipitated protein is collected by centrifugation. The pellets can be stored over prolonged periods in a 3.2 M ammonium sulfate solution (pH 6.0). An aliquot of the drained pellet is dissolved in 0.1 M sodium phosphate (pH 5.5) and then centrifuged at 48,000 g for 10 min.

Step 3. Chromatography on BioGel A0.5 m. The supernatant enzyme solution (8 ml, 76 mg of protein) from step 2 is applied to a BioGel A0.5 m column (118 × 2.6 cm) equilibrated with 0.1 M sodium phosphate buffer (pH 5.5). The column is eluted at 22 ml/hr, with a fraction size of 3.2 ml. The fractions are assayed for protein, xylanase, and glycosidase activity. The elution profile in Fig. 1 shows two peaks with xylanase activity (xylanase A and B). Although the xylanases have nearly the same molecular weight (Table II), both enzymes can be completely separated on the agarose column. Xylanase B is retarded on the column since the enzyme interacts with the polysaccharide matrix. A similar effect is observed for some amylases on dextran gels.[10,11] A xylosidase and an arabinosidase peak are eluted prior to the xylanases. Peak fractions are pooled as indicated in Fig. 1, and are dialyzed against 5 mM sodium phosphate (pH 6.8) and then concentrated by ultrafiltration. At this point the two xylanase and the β-xylosidase preparations are only slightly contaminated by other proteins as judged by sodium dodecyl sulfate gel electrophoresis.[12]

Step 4. Hydroxyapatite Chromatography. High-resolution hydroxyapatite chromatography on fine-particle-sized DNA-grade BioGel HTP (Bio-Rad) is used as the final purification step. The procedure is similar to that previously reported.[3] A home-made glass-jacketed column (125 × 1.2 cm) equipped with two flow adaptors is packed with DNA-grade BioGel HTP (or BioGel HTP[3]) and equilibrated with 5 mM sodium phosphate (pH 6.8). The samples from the preceding step (~10 ml) are applied to the hydroxyapatite column, and the column is eluted with a linear gradient (250 ml in each chamber) of sodium phosphate (pH 6.8) from 5 to 200 mM. The flow rate is 16.6 ml/hr (~1.8 bar) and 2.5-ml fractions are collected. During this step each enzyme is eluted as a single symmetrical peak of activity and protein. The peak of xylanase A activity is eluted at about 15 mM phosphate concentration, whereas xylanase B activity emerged from the column at a concentration of ~65 mM sodium phosphate. β-Xylosidase is eluted at a phosphate concentration of ~75 mM. Chromatography of the

[10] H. Dellweg, M. John, and J. Schmidt, *Eur. J. Appl. Microbiol.* **1**, 191 (1975).
[11] J. Schmidt and M. John, *Biochim. Biophys. Acta* **566**, 88 (1979).
[12] U. K. Laemmli, *Nature (London)* **227**, 680 (1970).

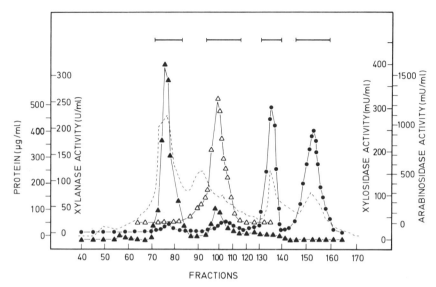

FIG. 1. Column chromatography of the 50% ammonium sulfate fraction from step 2 (see Table I) on BioGel A0.5 m. The column (118 × 2.6 cm) is eluted with 0.1 *M* sodium phosphate (pH 5.5) at a flow rate of 22 ml/hr. Protein concentration (--), xylanase activity (●), xylosidase activity (▲), and arabinosidase activity (△) are measured in the fractions. Fractions collected (——).

TABLE II

BIOCHEMICAL PROPERTIES OF XYLANASES AND β-XYLOSIDASE FROM
Trichoderma lignorum

Property	Xylanase A	Xylanases B	β-Xylosidase
Molecular weight	21,000[a]	20,000[a]	100,000[b]
Number of subunits	—	—	2
Isozymes	—	—	3
Glycoprotein (% carbohydrate moiety)	—	—	25
pH optimum	3.5	6.5	4.5
pH stability	2.5–6.0	3.5–8.0	3.5–7.5
Isoelectric point	5.1	8.7	5.9/6.1/7.4
K_m value (mM)	ND[c]	ND	5[d]
Specific activity (units/mg)	1,475	1,582	2.4

[a] Determined by gel electrophoresis in the presence of sodium dodecyl sulfate.

[b] Determined by column chromatography on Sephadex G-200.

[c] ND, not determined.

[d] Measured at 30° with *p*-nitrophenyl-β-D-xylopyranoside as substrate.

arabinosidase fraction from step 3 on hydroxyapatite gives a more complex elution pattern. By this procedure arabinosidase activity (eluted at ~60 mM phosphate) is completely separated from β-glucosidase and protease activity.[4] Peak fractions containing the bulk of enzyme activity are pooled and concentrated by ultrafiltration over a PM10 membrane (Amicon). The enzymes are stored in solution in 0.1 M sodium phosphate (pH 6.0) or after ammonium sulfate precipitation as a suspension in 3.2 M ammonium sulfate (pH 6.0).

Criteria of Purity. Both xylanases yield single bands after polyacrylamide gel electrophoresis in the presence of sodium dodecyl sulfate. Under the same conditions two protein bands can be detected for the β-xylosidase which indicates that this enzyme is composed of two different subunits. The specific activity of all three enzyme preparations is high and they emerge from the high-resolution hydroxyapatite column as a single symmetrical peak of activity and protein.

Properties

Some of the biochemical properties of the two xylanases and the β-xylosidase are listed in Table II.

Temperature Stability. Both xylanases show maximal activity at 45°. No considerable loss of activity after 1 hr at 40° occurs with xylanase A, while a 25% decrease of activity over the same time period is observed with the xylanase B preparation.

Maximal activity of the β-xylosidase occurs at 70°, and the enzyme shows high thermal stability with a residual activity of 70% after an incubation period of 1 hr at 70°. The xylosidase is rapidly inactivated at temperatures over 80°. The remarkable thermal stability may be due to the high carbohydrate content (25%) of this protein. Total acid hydrolysis and subsequent quantitative anion-exchange chromatography of the constituent monosaccharides show that the carbohydrate moiety of the xylosidase contains mannose, glucose, and xylose in the ratio of 9 : 3 : 1.

Effect of Metal Ions. The xylanases are strongly inhibited by 1 mM Cu^{2+} and 0.1 mM Hg^{2+}. Mg^{2+}, Ca^{2+}, Zn^{2+} (all at 1 mM), and EDTA (at 2.5 mM) do not affect xylanase activity.

The activity of β-xylosidase is strongly inhibited by 0.1 mM Hg^{2+}, whereas Cu^{2+}, Zn^{2+}, Ca^{2+}, Mg^{2+} (all at 1 mM), and EDTA (at 2.5 mM) do not stimulate or inhibit the enzyme reaction. Fe^{2+} and Mn^{2+} (both at 1 mM) stimulate xylosidase activity.

Inhibitors. Xylosidase activity can be inhibited by 10 mM xylose (60% inhibition), methyl-β-xylopyranoside (34%), and glucose (16%).

Specificity and Action Patterns of Xylanases. Using polyacrylamide gel chromatography it is possible to define the action patterns of polysac-

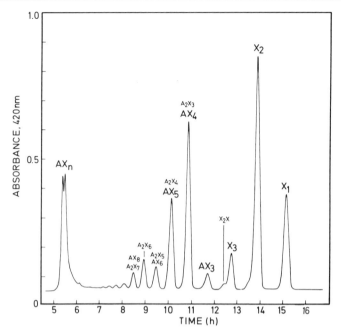

FIG. 2. Action of xylanase A on arabinoxylan and chromatographic analysis of the degradation products on a BioGel P-2 column (200 × 1.8 cm).[13] Carbohydrates in the column effluent are detected and quantitatively measured by the orcinol–sulfuric acid method using an autoanalyzer. Oligosaccharides obtained from the enzymatic degradation of arabinoxylan are acid hydrolyzed, and the constituent monosaccharides are identified; their concentrations are measured by anion-exchange chromatography on Animex A-27 resin (26 × 0.6 cm, 50°; 0.45 *M* borate buffer, pH 9.5).[8] Peaks identity: X_1, X_2, X_3 etc. denote xylose, xylobiose, xylotriose, etc.; AX_3, AX_4, etc. denote arabinoxylotriose, arabinoxylotetraose, etc.; A_2X_3 to A_2X_7, penta- to nonasaccharides with two arabinose units; AX_n, oligosaccharides DP > 25. X_2X denotes a branched xylose containing trisaccharide.

charide-degrading enzymes.[13,14] Figure 2 shows the action of the purified xylanase A from *T. lignorum* on arabinoxylan. The enzyme degrades arabinoxylan by an endo-mechanism producing mainly xylobiose (X_2) and arabinoxylotetraose (AX_4) with smaller amounts of xylose (X_1), arabinoxylopentaose (AX_5), xylotriose (X_3), and traces of branched arabinose–xylose oligosaccharides with a degree of polymerization from 4 to 9 (AX_3–A_2X_7). Digestion of arabinoxylan with xylanase B from *T. lignorum* and subsequent gel chromatography of the hydrolyzate shows a similar elution profile as obtained for xylanase A. Xylanase B produces less xylose but more xylotriose and xylotetraose than xylanase A. Both

[13] M. John, G. Trénel, and H. Dellweg, *J. Chromatogr.* **42**, 476 (1969).
[14] M. John, J. Schmidt, C. Wandrey, and H. Sahm, *J. Chromatogr.* **247**, 281 (1982).

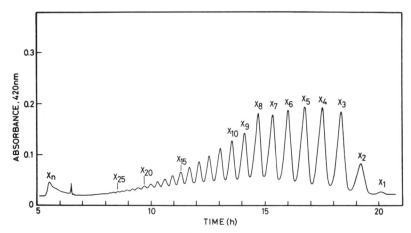

Fig. 3. Action of xylanase B on xylooctaose (X_8) and gel chromatography of the reaction products on a BioGel P-4 column (201 × 2.54 cm).[14] X_1, X_2, X_3, etc. denote xylose, xylobiose, xylotriose, etc.; X_n, oligosaccharides with DP > 50.

xylanases are unable to liberate arabinose from the arabinoxylan backbone or from the arabinoxylodextrins, so that the branched arabinose-containing xylooligosaccharides accumulate in the digest (Fig. 2). The ability of the xylanases A and B to act on linear xylooligosaccharides increases with increasing chain length of the substrate. Xylanase A hydrolyzes the linear xyloheptaose about three times faster than the branched arabinoxylan. Xylobiose, malto-, and cellooligosaccharides are not degraded by xylanase A and B.

Xylanase B shows high glycosyltransferase activity, and the action of this enzyme on xylooctaose (X_8) is shown in Fig. 3. The enzyme catalyzes the transfer of glycosyl residues from a xylooligosaccharide donor to an acceptor molecule producing a series of homologous 1,4-β-xylooligosaccharides. Successive transglycosylation leads to the formation of xylooligosaccharides with a degree of polymerization (DP) greater than 50 (X_n). Chain-lengthening transfer occurs with xylopentaose (X_5) to xylodecaose (X_{10}) as initial substrates. However, no essential chain-lengthening reaction is observed with xylooligosaccharides greater than DP 10. Evidence for the formation of 1,4-β-glycosidic linkages by xylanase B is obtained by treatment of the transglycosylation products with xylanase A. Subsequent chromatographic analyses of the hydrolyzates yield xylobiose and xylose as products. The transfer reaction of xylanase B resembles the transglycosylation reaction of bacterial amylomaltase.[15,16]

[15] T. N. Palmer, B. E. Ryman, and W. J. Whelan, *Eur. J. Biochem.* **69**, 105 (1976).
[16] J. Schmidt and M. John, *Biochim. Biophys. Acta* **566**, 100 (1979).

Substrate Specificity of β-Xylosidase. β-Xylosidase releases xylose residues from xylohexaose, xyloheptaose, and xylooctaose more rapidly than from xylooligosaccharides of DP2 to DP5. The following carbohydrates are not degraded by this enzyme: arabinoxylotetraose, methyl-β-xylopyranoside, and xylan.

[85] Xylanase of *Malbranchea pulchella* var. *sulfurea*

By MASARU MATSUO and TUNEO YASUI

Xylanase (1,4-β-xylan xylanohydrolase, EC 3.2.1.8, endo-1,4-β-xylanase) has been extensively studied and has been isolated from various kinds of microorganisms. More recently, the thermostable xylanase has been isolated and characterized from thermophilic fungi, such as *Malbranchea pulchella* var. *sulfurea*[1,2] and *Talaromyces byssochlamydoides*.[3,4] This chapter describes a purification procedure and some enzymatic properties of xylanase from the thermophilic microorganism *M. pulchella* var. *sulfurea*.

Assay Method

Principle. The method of determination of enzyme activity is based on the reduction of viscosity of xylan solution during the enzyme reaction.

Reagents

Soluble xylan (hardwood)[5,6] solution, 2%
Phosphate buffer solution, 10 mM
Enzyme, clear supernatant of culture broth or product of subsequent purification step is diluted with 10 mM phosphate buffer, pH 6.5, so that 1 ml of enzyme solution contains approximately 0.2 unit as defined below.

[1] M. Matsuo, T. Yasui, and T. Kobayashi, *Nippon Nogei Kagaku Kaishi* **49**, 263 (1975).
[2] M. Matsuo and T. Yasui, *Agric. Biol. Chem.* **49**, 839 (1985).
[3] H. Yoshioka, S. Chavanich, N. Nilubol, and S. Hayashida, *Agric. Biol. Chem.* **45**, 579 (1981).
[4] H. Yoshioka, N. Nagato, S. Chavanich, N. Nilubol, and S. Hayashida, *Agric. Biol. Chem.* **45**, 2425 (1981).
[5] M. Matsuo, T. Yasui, and T. Kobayashi, *Agric. Biol. Chem.* **41**, 1593 (1977).
[6] C. Doree, "The Methods of Cellulose Chemistry," p. 388. Chapman & Hall, London, 1953.

Procedure. All assays are at 45°. Two grams of soluble xylan suspended in 80 ml of water is autoclaved at 110° for 10 min to solubilize the xylan particle. Then the insoluble materials are removed by the centrifugation at 6000 g for 10 min. The supernatant of the above solution is diluted to 100 ml, and it is called "2% soluble xylan solution" in this chapter.

The reaction is initiated by the addition of 1 ml of enzyme solution to the substrate solution containing 5 ml of 2% soluble xylan solution and 4 ml of 10 mM phosphate buffer solution, pH 6.5. After 5 min incubation in an Ostwald-type viscometer, the flow time is measured.

Definition of Enzyme Unit. The value of viscosity reduction $\alpha_{(obs.)}$ is expressed as $\alpha_{(obs.)} = (t_r - t_w)/(t_0 - t_w)$, where t_0 is the flow time of the reaction mixture at zero time and t_w is the flow time of water. The activity is determined graphically by plotting the reciprocal values of the viscosity reduction against the amount of enzyme applied. The resulting slopes of the straight lines provide the values of the activity. A unit of enzyme activity is defined as $1 - \alpha_{(obs.)}$. Specific activity is defined as units of activity per milligram of protein measured by the method of Lowry *et al.*[7] with bovine serum albumin used as the reference standard.

Purification Procedure

Growth of Cultures. The strain of *Malbranchea pulchella* var. *sulfurea* No. 48 was isolated in our laboratory. Stocks are maintained in continuous vegetative culture on yeast–malt–glucose–agar: yeast extract 0.5%, malt extract 0.5%, glucose 0.5%, and agar 1.2%, pH 6.4. *Malbranchea pulchella* var. *sulfurea* No. 48 is cultivated for enzyme preparation in 10 liters of medium using a 20-liter jar-fermentor at 44° with aggitation of 470 rpm and aeration of 4.5 liters/min, as described.[1] The culture broth is harvested by vacuum filtration on a Büchner funnel lined with Toyoroshi (Toyoroshi Co. Ltd., Tokyo, Japan) No. 2 filter paper, and the culture filtrate is adjusted to pH 6.7 with 1 N HCl.

All subsequent steps are carried out at 4° unless otherwise stated.

Step 1. Ammonium Sulfate Fractionation. Solid ammonium sulfate (60.8 g/100 ml of solution) is added to the culture filtrate up to 0.8 saturation with continuous stirring, the pH of the solution being maintained at 6.7 with 1 N NaOH. After the filtrate is allowed to stand for 24 hr, the resulting precipitate is collected by centrifugation at 8000 g for 15 min,

[7] O. H. Lowry, N. J. Rosebrough, A. L. Farr, and R. J. Randall, *J. Biol. Chem.* **193**, 265 (1951).

TABLE I
PURIFICATION OF *Malbranchea* XYLANASE

Step and fraction	Volume (ml)	Total protein (mg)	Total activity (units)	Specific activity (unit/mg protein)	Recovery (%)
Culture filtrate	6,700	38,760	45,220	1.16	100
1. Ammonium sulfate	208	4,320	30,932	7.16	68.4
2. DEAE-Cellulose	660	552	22,204	40.2	49.1
3. CM-Sephadex	30	120	19,355	161	42.8

and redissolved in one-fiftieth the original volume of 10 mM phosphate buffer, pH 6.7, and dialyzed against several changes of the same buffer.

Step 2. DEAE-Cellulose Column Chromatography. After the insoluble material is removed by centrifugation at 3000 g for 10 min, the dialyzed solution is applied to a DEAE-cellulose (Serva) column (5 × 30 cm) equilibrated with 5 mM phosphate buffer, pH 6.7. The column is washed with 5 liters of the same buffer until no more xylanase activity is detected. The xylanase (viscometric activity) of this strain is not able to adsorb on DEAE-cellulose column whereas the β-xylosidase is adsorbed.[6]

Step 3. CM-Sephadex Column Chromatography. The eluted solution from DEAE-cellulose column is pooled and dialyzed by ultrafiltration with the XM100 (Amicon) membrane and followed by UM10 (Amicon) membrane. The dialyzed solution is applied to a CM-Sephadex (Pharmacia, Uppsala, Sweden) column (3.4 × 45 cm) equilibrated with 10 mM acetate buffer, pH 5.0. Fractions (10 ml) were collected at a flow rate of 16 ml/hr.

The method, summarized in Table I results in a 161-fold overall purification with about 43% recovery of xylanase activity.

Properties

Purity. The purified enzyme was shown to be homogeneous on alkaline polyacrylamide gel disc electrophoresis.

Isoelectric Point. Using an LKB electrofocusing column with pH 3–10 range ampholytes, the isoelectric point of the xylanase was pH 8.6.[2]

Stability. The purified xylanase is stable at 4° with no significant loss of activity over a 30-day period. The xylanase retained full activity by incubation at 50° for 30 min (pH 6.7) without substrate.

Optimum pH and Temperature. The purified xylanase had the opti-

TABLE II
HYDROLYSIS OF VARIOUS GLYCOSIDES BY
Malbranchea XYLANASE

Glycoside	Substrate concentration ($\times$ 10^{-2} M)	Relative activity (%)
Xylobiose	1	0
Phenyl-β-D-xyloside	1	0
Xylotriose	1	0.1
Xylotetraose	1	4
Xylopentaose	1	51
Xylan	1[a]	100

[a] Concentration (%).

mum pH in the range between 6.0 and 6.5, and then optimum temperature at 70°.

Substrate Specificity. Xylan was hydrolyzed approximately 51% by the purified xylanase and end products were found to be xylose, xylobiose,[8] and xylotriose.[7] The substrate specificity of purified xylanase toward various β-xylosides was tested, as shown in Table II. Xylobiose and phenyl-β-D-xyloside were not hydrolyzed by purified xylanase.[2] In the degradation of both xylotetraose[8] and xylopentaose[8] by purified xylanase, oligosaccharides that were of greater chain length than the substrate sugar were observed on the paper chromatogram. In degradation of xylotetraose, the first products to be formed were mainly larger products than the substrate sugar. Then their concentration continuously decreased. Xylobiose and xylotriose were formed after a short lag but not xylose. After prolonged incubation, the substrate sugar and larger products disappeared and xylose was formed. Xylose, xylobiose, and xylotriose were found as end products of xylotetraose degradation. This might indicate that there was a transglycosylation reaction in the degradation of both xylotetraose and xylopentaose that is similar to the reported degradation of both maltotetraose and maltopentaose with α-amylase.[9]

[8] I. Kusakabe, T. Yasui, and T. Kobayashi, *Agric. Biol. Chem.* **39**, 1355 (1975); see also I. Kusakabe, T. Yasui, and T. Kobayashi, *Nippon Nogei Kagaku Kaishi* **49**, 383 (1975).
[9] J. D. Allen and J. A. Thoma, *Biochemistry* **17**, 2338 (1978).

[86] Xylanase of *Talaromyces byssochlamydoides*

By Shinsaku Hayashida, Kazuyoshi Ohta, and Kaiguo Mo

The behavior of xylanase toward the various substrates shows that it contains at least two enzyme activities; one is an endoxylanase (β-1,4-xylan 4-xylanohydrolase; EC 3.2.1.8, endo-1,4-β-xylanase) which is responsible for the initial breakdown of the xylan chain and the formation of oligosaccharides, and the other is β-xylosidase (1,4-β-D-xylan xylohydrolase; EC 3.2.1.37, xylan 1,4-β-xylosidase) which causes a gradual fragmentation of the oligosaccharides leading finally to xylose as the end product. Many xylanases produced by various microorganisms have been investigated, and some have been obtained in homogeneous states on disc electrophoresis.[1-3] Thermostable xylanase of *Talaromyces byssochlamydoides* YH-50 is described here.

Assay Method

Principle. The assay is based on the release of xylose following the incubation of xylan with the enzyme.

Reagents

Sodium acetate buffer, 0.1 M, pH 5.5
Xylan, 10 mg/ml in butter
Bertrand reagents, solution A: $CuSO_4$, 0.16 M; solution B: potassium sodium tartrate, 0.7 M, NaOH, 3.8 M; solution C: $Fe_2(SO_4)_3$, 0.13 M; H_2SO_4, 2 M
$KMnO_4$, 3.2 mM

Procedure. Xylanase is assayed in the reaction mixture, containing 100 mg of xylan, 5 ml of buffer, and 5 ml of enzyme solution. The mixture is shaken in an L-shaped tube on a Monod shaker at 90 strokes per min at 60° for 1 hr. Liberated xylose is estimated as reducing sugar by Bertrand micromethod as described in an accompanying chapter.[4]

Definition of Unit and Specific Activity. One unit of xylanase activity is defined as the amount of enzyme releasing 1 μmol of xylose from the substrate per minute. Specific activity is expressed as units per milligram

[1] M. Inaoka and H. Soda, *Nature* (*London*) **178**, 202 (1956).
[2] M. Takahashi and Y. Hashimoto, *J. Ferment. Technol.* **41**, 181 (1963).
[3] S. Hashimoto, T. Muramatsu, and M. Funatsu, *Agric. Biol. Chem.* **35**, 501 (1971).
[4] S. Hayashida, K. Ohta, and K. Mo, this volume [34].

of protein. The protein content is determined by the method of Lowry et al.[5]

Organism and Culture Conditions

Talaromyces byssochlamydoides YH-50[6,7] was isolated from compost heaps in our laboratory (this microorganism can be obtained from Department of Agricultural Chemistry, Kyushu University, Fukuoka 812, Japan). The stock culture is maintained on slants of carrot juice agar medium (carrot 2000 g, chloramphenicol 50 mg, agar 20 g, tap water 1000 ml). The organism from a fresh slant culture is inoculated into the wheat bran medium (wheat bran 100 g, xylan 4 g, tap water 100 ml) in a 2-liter Erlenmeyer flask, and the 20 flasks are incubated at 50° for 4 days.

Purification Procedure

All purification steps are carried out at 0–4°. Centrifugation at each step is performed at 13,000 g for 20 min. Diluted enzyme solution from each chromatography is concentrated by lyophilization, followed by dissolving in a minimum volume of deionized water.

Step 1. Preparation of Crude Extract. To the fungal culture in each flask, 600 ml of tap water is added and mixed well. The mixture is left to stand for 3 hr and clarified by centrifugation. The supernatant is designated the crude extract.

Step 2. Ammonium Sulfate Fractionation. Solid ammonium sulfate is added to the crude extract (60 g/100 ml initial volume) with stirring. After being kept overnight, the precipitate is collected by filtration with a layer of Celite (food chemicals Codex grade).

Step 3. Sephadex G-50 Chromatography. The precipitate is dissolved in a minimum volume of deionized water and applied to a Sephadex G-50 column (4.5 × 103 cm). Filtration is carried out with deionized water at a rate of 24 ml/hr. Fractions of 10 ml are collected and those fractions containing xylanase activity are combined and concentrated.

Step 4. Batchwise Treatment with DEAE-Sephadex A-50. The concen-

[5] O. H. Lowry, N. J. Rosebrough, A. L. Farr, and R. J. Randall, *J. Biol. Chem.* **193**, 265 (1951).
[6] H. Yoshioka, S. Chavanich, N. Nilubol, and S. Hayashida, *Agric. Biol. Chem.* **45**, 579 (1981).
[7] H. Yoshioka, N. Nagato, S. Chavanich, N. Nilubol, and S. Hayashida, *Agric. Biol. Chem.* **45**, 2425 (1981).

trated enzyme solution (50 ml) is passed through a layer of DEAE-Sephadex A-50 (3 × 5 cm) equilibrated with 0.05 M phosphate buffer (pH 5.5). The DEAE-Sephadex A-50 is collected and washed three times with the same buffer. The xylanase adsorbed onto the DEAE-Sephadex A-50 is eluted batchwise with 1.0 M phosphate buffer (pH 5.5). The eluate is dialyzed against deionized water, combined, and concentrated.

Step 5. Sephadex G-100 Chromatography. The concentrated enzyme solution is applied to a Sephadex G-100 column (2.7 × 100 cm). Filtration is carried out with deionized water at a rate of 22 ml/hr. The xylanase activity is separated into two components, Xa and Xb. These fractions (6 ml) of Xa and Xb are independently combined and concentrated.

Step 6. DEAE-Sephadex A-50 Chromatography. The concentrate of each sample is applied to a DEAE-Sephadex A-50 column (2.3 × 57 cm) equilibrated with 0.05 M phosphate buffer (pH 5.5), and is eluted with a linear gradient of 0.05 to 1.0 M phosphate buffer (pH 5.5) in 500 ml. The flow rate is 22 ml/hr, and the eluate is collected in fractions (5 ml). Each fraction of Xa and Xb is combined and concentrated.

Step 7. Rechromatography on Sephadex G-100. The concentrated enzyme solution is rechromatographed on Sephadex G-100 column as described in Step 5. The xylanase activity of Xb is further separated into two components, XbI and XbII. Each fraction of Xa, XbI, and XbII is combined, concentrated, and lyophilized. The purification procedure is summarized in Table I.

TABLE I
PURIFICATION OF XYLANASE

Step	Volume (ml)			Protein (mg)			Activity (units)			Specific activity (units/mg)			Recovery (%)		
Crude extract	10,000			2,330			1,557			0.67			100		
Ammonium sulfate	1,400			1,730			1,557			0.90			100		
Sephadex G-50	310			492			1,232			2.50			79		
Batchwise treatment with DEAE-Sephadex	50			264			1,075			4.07			69		
First Sephadex G-100	Xa	Xb		Xa	Xb		Xa	Xb		Xa	Xb		Xa	Xb	
	90	80		129	109		440	492		3.41	4.51		28	31	
First DEAE-Sephadex A-50	35	35		30	32		302	346		10.2	10.7		19	22	
Second Sephadex G-100	Xa	Xb (I)	Xb (II)	Xa	Xb (I)	Xb (II)	Xa	Xb (I)	Xb (II)	Xa	Xb (I)	Xb (II)	Xa	Xb (I)	Xb (II)
	45	35	35	29	16	16	298	171	175	10.4	10.8	10.7	19	11	11

Properties

Homogeneity. Disc electrophoresis in 7.5% polyacrylamide gel at pH 8.3 in Tris-glycine buffer[8] indicates that these enzymes are electrophoretically homogeneous.

Molecular Properties. The molecular weights of Xa, XbI, and XbII are estimated as 76,000, 54,000, and 45,000, respectively. Amino acid compositions of Xa, XbI, and XbII show that aspartic and glutamic acids are characteristically abundant. The carbohydrate residues of Xa, XbI, and XbII are rich in glucose with lesser amounts of mannose and fucose. The total contents of carbohydrate in Xa, XbI, and XbII are 36.6, 31.5, and 14.2%, respectively.

Effect of Temperature. The enzyme activities are measured at various temperatures from 20 to 90°. Xa, XbI, and XbII exhibit the optimal temperature for activity at 75, 70, and 70°, respectively. The enzymes in 0.05 *M* phosphate buffer (pH 5.5) are incubated at the stated temperatures for 5 min, and the residual activities are assayed. Xa, XbI, and XbII are stable up to 70° and retain 65, 44, and 29% of the original activity, respectively, after heating at 95° for 5 min.

Effect of pH. The enzyme activity is measured in the pH range of 2.0–12.0. Xa, XbI, and XbII exhibit the optimal pH for activity at 5.5, 4.5, and 5.0, respectively. The enzyme solutions are incubated at various pH values at 4° for 24 hr, and the residual activities are assayed. Xa, XbI, and XbII are stable at pH 3.0–9.0, 3.5–8.5, and 3.0–9.0, respectively.

Hydrolysis of Xylan. Xa, XbI, and XbII hydrolyze xylan in 1% suspension to the respective extents of 38, 70, and 75% in 5 days. The enzyme mixture of Xa, XbI, and XbII hydrolyzes xylan in 1% suspension to 90% in 5 days incubation. The hydrolysis products by Xa are xylose, arabinose, glucose, xylobiose, xylotriose, and other oligosaccharides, whereas the hydrolysis products by XbII are xylose and xylobiose. Hydrolysis products from xylan by the enzyme mixture of Xa, XbI, and XbII are xylose, arabinose, glucose, and xylobiose.

Effect of Metal Ions. None of the tested metal ions markedly stimulates the activities of Xa, XbI, and XbII, whereas $HgCl_2$ and $KMnO_4$ significantly inhibit the activities of Xa, XbI, and XbII.

Action on Other Substrates. Xa, XbI, and XbII show weak activity toward CMC (carboxymethylcellulose), microcrystalline cellulose (Avicel), and starch, but no activity toward xylobiose, cellobiose, or maltose.

[8] B. J. Davis, *Ann. N.Y. Acad. Sci.* **121,** 404 (1964).

[87] 1,4-β-D-Xylan Xylohydrolase of *Sclerotium rolfsii*

By ANIL H. LACHKE

$$\beta\text{-}1,4\text{-Xyloside } (X_n) \rightarrow \beta\text{-}1,4\text{-xyloside } (X_{n-1}) + \text{D-xylose}$$
$$\text{D-Xylobiose} \rightarrow 2(\text{D-xylose})$$
$$p\text{-Nitrophenyl-}\beta\text{-D-xylopyranoside} \rightarrow p\text{-nitrophenol} + \text{D-xylose}$$

Endo-(1 → 4)-β-D-xylanase (1,4-β-D-xylan xylanohydrolase, EC 3.2.1.8) hydrolyzes in a random manner β-D-xylopyranosyl linkages of xylan to form xylooligosaccharides. β-D-Xylosidase (xylobiase; exo-1,4-β-D-xylosidase; 1,4-β-D-xylan xylohydrolase, EC 3.2.1.37) catalyzes from the nonreducing end the hydrolysis of xylobiose or xylooligosaccharides to form D-xylose as the ultimate end product.[1-4] Conclusive evidence of an exoxylanase which can produce mainly D-xylobiose is not available as yet.[2] Information on β-D-xylosidase is sparse for the basidiomycetous fungi which cause major rotting of wood. Partial characterization of an extracellular β-D-xylosidase is reported by Lachke *et al.*[5] from a basidiomycete fungus *Sclerotium rolfsii*. The methods used are summarized here.

Assay Method

Principle. β-D-Xylosidase releases *p*-nitrophenol or *o*-nitrophenol from *p*- or *o*-nitrophenyl-β-D-xylopyranoside (*p*-NPX, *o*-NPX) in the reaction mixture. At alkaline pH *o*- or *p*-nitrophenol exhibits an intense yellow color, which is proportional to the amount of enzyme, and can be determined spectrophotometrically at 410 nm.[6]

Reagents

Fresh solution of *p*-NPX (4 mM) prepared in citrate buffer (50 mM, pH 4.5)

Sodium carbonate, 2 M

[1] R. F. H. Dekker, "Biosynthesis and Biodegradation of Wood Components." Academic Press, New York, 1985.
[2] P. J. Reilly, "Trends in the Biology of Fermentation for Fuels and Chemicals." Plenum Press, New York, 1981.
[3] J. Comtat, *Carbohydr. Res.* **118,** 215 (1983).
[4] P. Biely, *Trends Biotechnol.* **3,** 286 (1985).
[5] A. H. Lachke, M. V. Deshpande, and M. C. Srinivasan, *Enzyme Microb. Technol.* **7,** 445 (1985).
[6] R. L. Nath and N. Rydon, *Biochem. J.* **57,** 1 (1954).

Procedure. A suitably diluted enzyme (0.1 ml) is introduced into a preincubated (50°, 1 min) solution of *p*-NPX (0.9 ml). The reaction is terminated by the addition of 1 ml of 2 *M* sodium carbonate after 30 min. For a blank reading the substrate solution (0.9 ml) is incubated for 30 min and diluted enzyme (0.1 ml) is introduced after the termination of the reaction.

Definition of Unit, Specific Activity, and Calculations. One unit of β-xylosidase activity is defined as the amount of enzyme that catalyzes the formation of 1 μmol of *p*-nitrophenol per minute under assay conditions. The specific activity is expressed as unit/mg protein.

The molar absorbancy index of *p*-nitrophenol is 18.5×10^3. The units can be calculated using the following formula:

$$\text{Unit/ml} = \frac{(\text{OD})_{\text{test}} - (\text{OD})_{\text{blank}} \times DV}{18.5t}$$

where V is volume of the reaction mixture (2.0 ml), D is dilution factor of the enzyme, and t is time in minutes.

Source of Enzyme

Growth of Microorganism and Preparation of Crude Culture Filtrate. *Sclerotium rolfsii* (NCIM 1084, NCL, India) maintained on potato–dextrose–agar (0.5% dextrose) slant is used for the production of extracellular β-D-xylosidase. The culture is grown in 500-ml Erlenmeyer flasks with 100 ml Mandels–Weber[7] medium in each flask at 28° for 12 days on a rotary shaker (180 rpm). The composition of the medium is (g/liter) KH_2PO_4, 2.0; $CaCl_2 \cdot 2H_2O$, 0.3; $MgSO_4 \cdot 7H_2O$, 0.3; urea, 0.3; $(NH_4)_2SO_4$, 1.4; proteose peptone (Difco), 0.25; yeast extract, 0.1; Tween 80, 0.1% (v/v). Trace element composition is (mg/liter) $FeSO_4 \cdot 7H_2O$, 5.0; $MnSO_4 \cdot H_2O$, 1.56; $ZnSO_4 \cdot 7H_2O$, 1.4; $CoCl_2 \cdot 6H_2O$, 2.0. Cellulose-123 (Schleicher and Schüll Co., FRG), 20 g/liter, is supplied as a sole carbon source.[8] The pH of the medium is 6.5 prior to autoclaving at 121° for 20 min. The crude broth is collected after 12 days of fermentation, filtered through a sintered funnel (G1), and clear culture filtrate is used for the purification of β-D-xylosidase.

Purification Procedure

All purification steps are described briefly and are carried out at 0–5°. In a typical β-D-xylosidase purification procedure, 3 liters of culture filtrate is used.

[7] M. Mandels and J. Weber, *Adv. Chem. Ser.* **95**, 391 (1969).
[8] J. C. Sadana, A. H. Lachke, and R. V. Patil, *Carbohydr. Res.* **133**, 297 (1984).

Step 1. Ammonium Sulfate Fractionation. Salting out of the enzyme from the culture filtrate is performed at 3.5 M ammonium sulfate (final concentration). The protein precipitate is collected by centrifugation (5000 g, 20 min) and suspended in citrate buffer (5.0 ml, 50 mM, pH 4.5).

Step 2. Gel Filtration. The concentrated suspension from Step 1 is subjected to gel filtration on a Sephadex G-200 (coarse) column (1.8 × 90 cm) which is preequilibrated with citrate buffer (50 mM, pH 4.5). Both desalting and fractionation of β-D-xylosidase is achieved by collecting 3.0-ml fractions with a flow rate of 10 ml/hr. The fractions with higher specific activity (0.6 U/mg) for β-D-xylosidase are pooled.

Step 3. Column Isoelectric Focusing (pH Range 3.5–10). The pooled fractions (9.0 ml) from Step 2 are dialyzed against 100 mM glycine solution (pH 7.2) for 16 hr at 5–7° using a collodion bag SM-13200 (Sartorius membrane filter, Göttingen). It must be mentioned that other dialyzing sacs often get damaged while using concentrated cellulolytic preparations. The dialyzed preparation is further purified by preparative isoelectric focusing (LKB 8100—1 column, 110 ml capacity) by mixing it directly into the dense gradient solution (50% sucrose, w/v) and light gradient solution (5% sucrose, w/v) containing 1% ampholyte generating a pH 3.5–10.0 gradient. The cathode solution (15 g sucrose in 0.24 M NaOH, 25 ml) is at the bottom of the column while the anode solution (0.15 M H$_3$PO$_4$, 10 ml) is at the top of the column. Electrofocusing is carried out at 5–7° for 65 hr employing 5 W power. The fractions, each of 1 ml, are collected at a flow rate of 40 ml/hr. The pH of each fraction is determined at 5°. The fractions with higher specific activity (0.92 U/mg) which show a pI of 6.8 are pooled.

Step 4. Column Isoelectric Focusing (pH Range 5–7). The collected fractions (5 ml) are again subjected to preparative column isoelectric focusing using a narrow range ampholyte carrier of pH 5–7 (1% final concentration). In contrast to Step 3, the cathode solution (0.25 M NaOH, 10 ml) is at the top of the column while the anode solution (15 g sucrose in 25 ml of 0.25 M H$_3$PO$_4$) is at the bottom of the column. Fractions with pI of 6.8 are pooled and made free of sucrose by dialyzing against citrate buffer (50 mM, pH 4.5) in a collodion bag for 16 hr. The specific activity of this preparation is 1.25 U/mg protein. Steps of the purification procedure for β-D-xylosidase are summarized in Table I.

Properties

Homogeneity. The 22-fold purified β-D-xylosidase preparation shows a single protein band in polyacrylamide gel electrophoresis (pH 8.9) as well as in analytical isoelectric focusing in polyacrylamide gel over a pH

TABLE I
PURIFICATION OF β-D-XYLOSIDASE FROM *Sclerotium rolfsii*

Steps	Total units (U)	Total protein (mg)	Specific activity (U/mg)	Yield (%)	Purification (fold)
Culture filtrate	165	2895	0.057	100	—
(NH₄)₂SO₄, 3.5 *M*	118	880	0.13	71.5	2.3
Sephadex G-200	41	68	0.60	24.8	10.5
Isoelectric focusing					
pH gradient 3.5–10.0	22	24	0.92	13.3	16.0
pH gradient 5.0–7.0	8.6	6.8	1.25	5.2	21.9

range of 3.5–10.0. Carboxyamidomethylation[9] of the reduced form of β-D-xylosidase on dodecyl sulfate polyacrylamide gel electrophoresis shows a single protein band indicating that it consists of one polypeptide chain. The enzyme exhibits traces of β-D-xylanase activity but carboxymethylcellulase (endo-1,4-β-D-glucanase, EC 3.2.1.4), β-D-glucosidase (EC 3.2.1.21), or cellobiase activities are absent.

Stability. Purified β-D-xylosidase can be stored at −15° at pH 4.5; no significant loss of activity is observed over a period of 6 months.

Effect of pH and Temperature. The optimum pH for enzyme activity is 4.5, and it is most stable in the pH range 4.0–5.0. The enzyme is not active below pH 2.5 (50 m*M* phthalate–HCl buffer) or above pH 6.5 (50 m*M* citrate–phosphate or phosphate buffer). The enzyme has optimum activity at 50° with *p*-NPX as a substrate. The E_a from Arrhenius plot is calculated to be 10.5 kcal/mol (44 kJ/mol). On incubation at 65° in citrate buffer (50 m*M*, pH 4.0), the enzyme loses 50, 65, and 88% of its original activity after 15, 30, and 60 min, respectively. The enzyme retains 92% of its activity on incubation at 50° for 90 min. The heat stability of *S. rolfsii* β-D-xylosidase is considered lower as compared to that from *Aspergillus niger*[10] which has a half-life of 75 hr at 65° (pH 4.0). The *S. rolfsii* enzyme is inactivated rapidly at 65° and had a half-life of only 15 min.

Kinetic Properties. The K_m value of 0.038 m*M* with *p*-NPX as substrate is very low as compared to other reported values for β-D-xylosidase preparations such as, for example, 0.362 m*M* for *A. niger*,[10] 0.121 m*M* for *Penicillium wortmanni*,[11] and 1.43 m*M* for *Bacillus pumilus*.[12]

[9] K. Weber and M. J. Osborn, *J. Biol. Chem.* **244**, 4407 (1969).
[10] G. B. Oguntimein and P. J. Reilly, *Biotechnol. Bioeng.* **22**, 1143 (1980).
[11] E. T. Reese, A. Maguire, and F. W. Parrish, *Can. J. Microbiol.* **19**, 1065 (1973).
[12] H. Kersters-Hilderson, F. G. Loontiens, M. Claeyssens, and C. K. De Bruyne, *Eur. J. Biochem.* **7**, 434 (1969).

Molecular Weight. The molecular weight of β-D-xylosidase by gel filtration on Sephadex G-200 and by SDS–polyacrylamide gel electrophoresis is found to be in the range of 170,000–180,000.

Effect of Metal Ions, Group-Specific Reagents, and D-Xylose on β-D-Xylosidase Activity. The effects of addition of inhibitors/activators of β-D-xylosidase activity are determined at 50° in citrate buffer (50 mM, pH 4.5). Addition of Hg^{2+}, Ag^+, or Cu^{2+} at 2 mM caused 100, 78, and 30% inhibition, respectively, of enzyme activity. At equivalent concentrations of Ca^{2+} or Fe^{3+}, the enzyme activity remained unchanged.

The group specific reagents such as *p*-hydroxymercuribenzoate (1–2 mM), *N*-bromosuccinimide (1–2 mM), and iodoacetamide are highly inhibitory to the enzyme activity. The data suggested the possible presence of tryptophan residues in the active center(s) of the enzyme.[5] D-Xylose inhibition in 1–3 mM concentration range is 8–15% and closely follows *A. niger* β-xylosidase.[10]

Distinct Nature of S. rolfsii β-D-Xylosidase. Four β-D-glucosidases (β-D-glucoside glucohydrolase, EC 3.2.1.21) purified from *S. rolfsii* could hydrolyze positional isomers of cellobiose [β-(1 → 4)], gentiobiose [β-(1 → 6)], laminaribiose [β-(1 → 3)], and sophorose [β-(1 → 2)]. These enzymes, however, did not hydrolyze D-xylosides.[13] In a brown rot fungus *Tyromyces palustris*, β-D-xylosidase and β-D-glucosidase activities are catalyzed by the same enzyme.[14] The purified *S. rolfsii* β-D-xylosidase is a distinct enzyme and has no activity toward carboxymethylcellulose, D-cellobiose, or *p*-NPG as substrates.

Functional Role of β-D-Xylosidase. The xylan component present in the biomass hemicellulose fraction can be used as feedstock for enzymatic production of D-xylose, and β-D-xylosidase can have an important role in this process. The enzyme cleaves short xylooligosaccharides and xylobiose to D-xylose. This enzyme could have practical applications in structural studies of glycoproteins or other carbohydrate-containing compounds as it splits off L-serine to D-xylose linkages.[11,15] The enzyme has substantial xylosyltransferase activity. In order to avoid this, low substrate concentration would be required. Moreover, β-xylosidase is not formed by microorganisms in high yields.[11] β-Xylosidase, in free form or in immobilized form, has roughly the same physicochemical properties, but their activities are rather lower than the native enzyme.[10,16] No data

[13] J. C. Sadana, J. G. Shewale, and R. V. Patil, *Carbohydr. Res.* **118**, 205 (1983).

[14] M. Ishihara and K. Shimizu, *Mokuzai Gakkaishi* **29**, 315 (1983).

[15] F. Deleyn, M. Claeyssens, J. Van Beeumen, and C. K. De Bruyne, *Can. J. Biochem.* **56**, 43 (1978).

[16] K. Shimizu and M. Ishihara, *Biotechnol. Bioeng.* **29**, 236 (1987).

are available on the relative levels of repression of xylan-degrading enzymes by xylobiose, unlike cellulose repression by cellobiose.

D-Xylose derived from xylans offers potential feedstock for generating food and fuel.[17] Therefore, higher β-xylosidase activities are necessary to obtain adequate levels of D-xylose in the hydrolyzates.[18]

[17] T. W. Jeffries, *Trends Biotechnol.* **3**, 208 (1985).
[18] A. H. Lachke, M. C. Srinivasan, S. S. Deshmukh, and M. V. Deshpande, *Biotechnol. Lett.* **9**, 147 (1987).

[88] β-Xylosidases of Several Fungi

By Masaru Matsuo and Tuneo Yasui

Recently, there has been increasing interest in the bioconversion of biomass to produce sugar. To date, enzymatic reactions have not been applied to the hydrolysis of xylan to xylose under high substrate concentration because microbial xylanase (β-xylosidase) has been known to possess a high degree of transglycosidase activity. On the other hand, few reports exist which deal with the substrate specificities of fungal β-xylosidase in its pure state. Hence, we describe here the β-xylosidase from different fungi, such as *Malbranchea pulchella* var. *sulfurea*, *Trichoderma viride*, and *Emericella nidulans*, to compare their properties and modes of action on various substrates.

β-Xylosidase of *Malbranchea pulchella* var. *sulfurea*

A thermophilic fungus, which could produce a xylan hydrolysis activity, was isolated from soil and identified as *Malbranchea pulchella* var. *sulfurea*.[1] In addition, the xylanase (culture filtrate) of this strain was most suitable for xylose production from xylan because the enzyme was thought to possess less transferase activity than other fungal xylanases tested. Therefore, we purified the xylanases to clarify the properties of the β-xylosidase (exo-1,4-β-D-xylosidase, EC 3.2.1. 37, xylan 1,4-β-xylosidase) of *Malbranchea pulchella* var. *sulfurea*.

[1] M. Matsuo, T. Yasui, and T. Kobayashi, *Nippon Nogei Kagaku Kaishi* **49**, 263 (1975).

Assay Methods

Principle. β-Xylosidase activity is assayed spectrophotometrically by measuring the amount of released phenol from phenyl-β-D-xyloside. Absorbance of phenol is measured at 660 nm by the addition of phenol reagent solution (Nakarai Chemicals, Tokyo, Japan) in alkaline solution.[2]

Reagents

Phenyl-β-D-xyloside, 20 mM in water
Phosphate buffer, 0.1 M, pH 6.7
Sodium carbonate, 50 mM in water

Procedure. The standard reaction mixture contained 3 ml of 20 mM phenyl-β-D-xyloside, 2 ml of 0.1 M phosphate buffer (pH 6.7), and 1 ml of enzyme solution. After incubation for 15 min at 45°, 1-ml aliquots are withdrawn and added to 5 ml of 50 mM sodium carbonate. By the addition of 1 ml of phenol reagent solution (1 N HCl) to the entire reaction mixture (6 ml), absorbance is measured at 660 nm against a blank in which water replaces phenyl-β-D-xyloside.

Definition of Enzyme Unit. One unit of activity is defined as the amount of enzyme required to release 1 μmol of phenol per minute under the assay conditions. Specific activity is defined as units of activity per milligram of protein as measured by the method of Lowry *et al.*[3]

Growth of Cultures. *Malbranchea pulchella* var. *sulfurea* No. 48 was used. Stock cultures are maintained in continuous vegetative culture on yeast–malt–glucose-agar: yeast extract 0.5%, malt extract 0.5%, glucose 0.5%, and agar 1.5% at pH 6.0. Large cultures of *Malbranchea pulchella* var. *sulfurea* No. 48 are grown on the medium, containing 1.5% xylan (hard wood), 1% polypeptone, 0.3% KH_2PO_4, 0.6% $(NH_4)_2HPO_4$, and 0.4% corn steep liquor. This stain is inoculated in a 20-liter jar-fermenter containing 10 liters of medium and the culture was grown at 44° for 29 hr with agitation of 470 rpm and aeration of 4.5 liters/min. The culture broth is harvested by vacuum filtration on a Büchner funnel lined with To-yoroshi (Toyoroshi, Tokyo, Japan) No. 2 filter paper, and the culture filtrate is adjusted to pH 6.0 with 1 N HCl. All purification steps are carried out at 4° unless otherwise stated.

Step 1. Ammonium Sulfate Factionation. Solid ammonium sulfate (22.8 g/100 ml of solution) is added to the culture filtrate to 0.3 saturation with continuous stirring, the pH of the solution being maintained at pH 6.7 with 1 N NaOH. The precipitate is removed by centrifugation at 8000 g for 15 min. The supernatant fraction is brought to 0.8 saturation

[2] O. Folin and V. Ciocalteu, *J. Biol. Chem.* **73**, 627 (1929).
[3] O. H. Lowry, N. J. Rosebrough, A. L. Farr, and R. J. Randall, *J. Biol.* **193**, 265 (1951).

with ammonium sulfate (33.9 g/ml of solution), collected by centrifugation at 8000 g for 10 min, dissolved in a small amount of water, and dialyzed against several changes of the 1 mM acetate buffer, pH 6.7.

Step 2. DEAE-Cellulose Column Chromatography. After the insoluble materials are removed by centrifugation at 3000 g for 10 min, the supernatant solution is applied to a DEAE-cellulose (Serva) column (5 × 30 cm) equilibrated with 5 mM phosphate buffer, pH 6.7. Xylanase (viscometric activity) of this strain does not adsorb on DEAE-cellulose column but β-xylosidase can be adsorbed. Therefore, DEAE-cellulose column chromatography is used to remove the xylanase activity from the culture filtrate. The column is washed with 5 liters of the same buffer until no more xylanase activity is detected. Elution is carried out by the application of a linear gradient of NaCl (0–0.3 M) with 2 liters of buffer. The volume of each fraction collected is 18 ml.

Step 3. First Column Electrophoresis. Fractions having β-xylosidase activity from step 2 are pooled, and are concentrated by ultrafiltration with an XM100 (Amicon) membrane. Concentrated solution (990 mg protein/15 ml) from step 2 is fractionated by electrophoresis using a column (Atto, Tokyo, Japan) (4.5 × 58 cm) of BioGel P-100 in 0.1 M phosphate buffer (pH 6.7) as the supporting medium. A voltage of 750 V (25 mA) is applied for 20 hr at 4°. After the electrophoresis run for 20 hr, the column is eluted with the same buffer. Fractions of 10 ml are collected at a flow rate of 15 ml/hr.

Step 4. Second Column Electrophoresis. The fractions showing β-xylosidase activity from step 3 are collected and concentrated. This fraction is subjected to the second electrophoresis under the same conditions. One-fourth of the concentrated fraction from step 3 is used in each electrophoretic run. The zone volume is 15 ml, containing 22 mg of protein. After the electrophoresis is run, the column is eluted with same buffer.

Step 5. Sephadex G-200 Column Chromatography. The fractions showing β-xylosidase activity from step 4 are collected and concentrated by ultrafiltration with a UM10 membrane. Three milliliters of a concentrated β-xylosidase-rich fraction is applied to a column (2.4 × 100 cm) of Sephadex G-200 (Pharmacia, Uppsala, Sweden) which was previously equilibrated with 10 mM phosphate buffer (pH 6.7) and eluted with the same buffer. The procedure summarized in Table I results in 99-fold overall purification with about 20.7% recovery on β-xylosidase activity.

Properties

Purity. The purified enzyme was shown to be homogeneous on polyacrylamide disc gel electrophoresis with pH 8.3 buffer system; a single

TABLE I
PURIFICATION OF *M. pulchella* var. *sulfurea* β-XYLOSIDASE

Purification step	Volume (ml)	Total protein (mg)	Total activity (U)	Specific activity (U/mg protein)	Yield (%)
Culture filtrate	4,000	23,080	2,440	0.106	100
(NH₄)₂SO₄-ppt	360	7,470	1,224	0.164	50.1
DEAE-Cellulose	324	1,315	1,105	0.840	43.5
Ultrafiltration	16	990	905	0.914	37.1
First electrophoresis	60	88	804	9.14	32.9
Second electrophoresis[a]	400	54	512	9.48	21.0
Ultrafiltration	16	51	506	9.92	20.7
Sephadex G-200[b]	50	9	94	10.5	

[a] The value indicated is the sum of the four parts of the second electrophoretic run.
[b] Three milliliters of the second ultrafiltration fraction was loaded onto a Sephadex G-200 column.

protein band was obtained. Homogeneity of the purified enzyme was also examined by ultracentrifugation using a synthetic boundary cell at 60,000 rpm. A discrete single sedimentation pattern was observed.

Molecular Properties. The molecular weight of the enzyme was estimated to be approximately 26,000[4] by gel filtration with Sephadex G-200. The sedimentation coefficient, $s_{20,w}$, of the enzyme was estimated to be 2.78 S by the method of Shachman.[5] The isoelectric point of the β-xylosidase was pH 4.8 using an LKB electrofocusing column with pH 3–10 range ampholytes. Carbohydrate content was determined by the phenol–sulfuric acid method and compared to a glucose standard. The purified enzyme contains either no or a few carbohydrate residues. The adsorption coefficient at 280 nm was $A_{280\,nm}^{1\%} = 13.2$ (g⁻¹ cm²).

Optimum pH and Temperature. The enzyme was most active in the pH range of 6.2–6.8. The optimal temperature for the enzyme reaction (pH 6.7) was about 50°, and 50% of maximal activity was found at 40 and 58°.

Stability. The purified enzyme was stable at a pH range of 6.3–6.7. The enzyme retained about 40% of its activity by incubation at 50° (pH 6.7) for 15 min without substrate.

Effect of Various Reagents on Enzyme Activity. Hg^{2+}, Zn^{2+}, Cu^{2+}, N-bromosuccinimide, p-chloromercuribenzoate, and sodium dodecyl sulfate strongly inhibited β-xylosidase activity. The enzyme retained 90% of

[4] M. Matsuo, T. Yasui, and T. Kobayashi, *Agric. Biol. Chem.* **41**, 1593 (1977).
[5] H. K. Schachman, this series, Vol. 4, p. 32.

TABLE II
HYDROLYSIS OF VARIOUS GLYCOSIDES BY *M. pulchella*
var. *sulfurea* β-XYLOSIDASE

Glycoside	Substrate concentration ($\times\ 10^{-2}\ M$)	Relative activity (%)
Xylobiose	1	100
Xylan[a]	1	30.5
Methyl-β-D-xyloside	1	0
Butyl-β-D-xyloside	1	0
Cyclohexyl-β-D-xyloside	1	0.009
Benzyl-β-D-xyloside	1	0
Thiophenyl-β-D-xyloside	1	0
Cellobiose	1	0

[a] Concentration (%).

its activity in the presence of 1 mM Ca^{2+} when it was heated at 50° for 1 hr but 60% of its activity was lost in the absence of Ca^{2+} under the same conditions.

Specificities. Table II shows the substrate specificities[6] of purified enzyme for various substrates. Enzyme activity was expressed in terms of the percentage of the activity with xylobiose. The activity on xylan at a concentration of 1% is expressed relative to that on xylobiose at a concentration of 10^{-2} M. *Malbranchea* β-xylosidase acted on xylan. Aryl-β-D-xylosidase was well hydrolyzed by the enzyme, whereas alkyl-β-D-xylosides remained intact. Cellobiose was not hydrolyzed and *o*-substituted phenyl-β-D-xylosides were hardly hydrolyzed as compared to *p*-substituted phenyl-β-D-xylosides. In hydrolysis reactions (e.g., *p*-NO$_2$-phenyl-, *p*-Cl-phenyl-, phenyl-, *p*-CH$_3$-phenyl-, and *p*-CH$_3$O-phenyl-β-D-xyloside) by *Malbranchea* β-xylosidase, the Michaelis constants, K_m, were almost independent of the aglycon, whereas maximum velocity, V_{max}, showed a marked dependence on the aglycon. In the later case, the molecular activities, k_0, were markedly increased by introducing both electron-withdrawing and electron-donating substituents in the para position of the phenyl ring. When electron-withdrawing substituents were introduced, the Hammett reaction constant, ρ, was $+1.27$, when electron-donating substituents were introduced, its value was -2.07. The K_m and k_0 (molecular activity) values for xylobiose were similar to those for xylotriose, xylotetraose, and xylopentaose. The apparent k_0 value for xylobiose was 15.0

[6] M. Matsuo, T. Yasui, and T. Kobayashi, *Agric. Biol. Chem.* **41**, 1601 (1977).

sec^{-1}; xylotriose, 20.6 sec^{-1}; xylotetraose, 19.7 sec^{-1}; and xylopentaose, 19.7 sec^{-1}. On enzymatic hydrolysis of *p*-nitrophenyl and phenyl-β-D-xyloside, the reaction products, in both cases, were found to be α-D-xylose with inversion of configuration.[7]

β-Xylosidase of *Trichoderma viride*

A strain of *Trichoderma viride* is well known as a fungus which produces outstanding xylan-degrading enzymes. The extracellular enzymes from this fungus were first described by Reese and Mandels.[8] Moreover, in the study of the production of xylose from xylan through the process of enzymatic hydrolysis, the culture filtrate of this fungus[9,10] was used as xylan-degrading enzymes. This chapter describes a purification procedure and some enzymatic properties for β-xylosidase (exo-1,4-β-D-xylosidase, EC 3.2.1.37) from *Trichoderma viride*.

Assay Methods

Principle. The principle, reagents, enzyme unit, and procedure are as described previously for *M. pulchella* var. *sulfurea* except that the reaction mixture contained, instead of phosphate buffer, 1 ml of 0.1 *M* acetate buffer (pH 4.5).

Purification Procedure

Growth of Cultures. A strain of *Trichoderma viride* was newly isolated from the rotting Grape vine in our laboratory. Stock cultures are maintained in continuous vegetative culture on yeast–malt–glucose–agar containing yeast extract 0.5%, malt extract 0.5%, glucose 0.5%, and agar 2%. The strain of *Trichoderma viride* is cultivated for enzyme preparation in a medium of 2% xylan (hard wood), 0.4% KH_2PO_4, 1% $(NH_4)_2HPO_4$, 1% peptone, and 0.3% corn steep liquor at 30° for 4 days on a reciprocal shaker (120 oscillation/min, 7 cm stroke). The culture broth is harvested by vacuum filtration on a Büchner funnel lined with Toyoroshi (Toyoroshi Co. Ltd., Tokyo, Japan) No. 2 filter paper, and the culture filtrate is adjusted to pH 5.0 with 1 *N* HCl. All subsequent steps are carried out at 4° unless otherwise stated.

[7] M. Matsuo and T. Yasui, *Agric. Biol. Chem.* **45,** 1603 (1981).
[8] E. T. Reese and M. Mandels, *J. Biol. Chem.* **7,** 378 (1959).
[9] K. Nomura, T. Yasui, S. Kiyooka, and T. Kobayashi, *J. Ferment. Technol.* **46,** 634 (1968).
[10] K. Nomura, T. Yasui, S. Kiyooka, and T. Kobayshi, *J. Ferment. Technol.* **47,** 313 (1969).

Step 1. Ammonium Sulfate Fractionation. Solid ammonium sulfate (22.8 g/100 ml of solution) is added to the culture filtrate (10 liters) to 0.3 saturation with continuous stirring, the pH of the solution being maintained at 5.0 with 1 N NaOH; the precipitate is removed by centrifugation at 8000 g for 15 min. The supernatant fraction is brought to 0.8 saturation with ammonium sulfate (33.9 g/100 ml of solution) and is allowed to stand at 24 hr with occasional stirring. The protein precipitate is collected by centrifugation at 8000 g for 10 min, dissolved in a small amount of water, and dialyzed against 5 m M acetate buffer (pH 4.5). The dialyzed solution was then centrifuged at 8000 g for 10 min to remove insoluble materials.

Step 2. DEAE-Sephadex Column Chromatography. The supernatant of the dialyzed solution (485 ml) is applied to a DEAE-Sephadex A-25 (Pharmacia, Uppsala, Sweden) column (3.5 × 45 cm) equilibrated with 0.05 M acetate buffer at pH 4.5. The column is washed with 1 liter of the same buffer until no more protein eluted. Elution is carried out by application of a linear gradient of NaCl (0–0.5 M) with 1 liter of buffer. Fractions of 5 ml are collected.

Step 3. Column Electrophoresis. The fractions showing β-xylosidase activity from step 2 are collected and concentrated by ultrafiltration with a PM10 membrane (Amicon, Danvers, MA). This sample (step 2) is applied to an electrophoresis column (4.5 × 80 cm) of BioGel P-100 in 0.1 M acetate buffer (pH 4.5) and run at 820 V for 29 hr and 4°. After electrophoresis, the column is eluted with the same buffer. Fractions of 10 ml are collected at a flow rate of 13 ml/hr.

Step 4. BioGel P-100 Column Chromatography. The fractions showing β-xylosidase activity from step 3 are collected and concentrated by ultrafiltration with a PM10 membrane. This sample is applied to a column of BioGel P-100 (3.5 × 45 cm) equilibrated with 0.05 M acetate buffer (pH 4.5), and eluted with the same buffer. The volume of each fraction collected is 5 ml.

Step 5. Electrofocusing. The fractions showing β-xylosidase activity from step 4 are pooled and concentrated by ultrafiltration (PM10 membrane). The amount of protein in the electrofocusing run is about 22 mg. Carrier ampholytes (pH 2.5–5) are used in a sucrose gradient and a voltage of 900 V is applied for 24 hr at 4°. The volume of each fraction collected is 2 ml. The methods, summarized in Table III, result in 24.7-fold overall purification with about 25.5% recovery on β-xylosidase activity.

Properties

Purity. The purified enzyme gave a single band on polyacrylamide disc gel electrophoresis and on sodium dodecyl sulfate–polyacrylamide gel

TABLE III
PURIFICATION OF *T. viride* β-XYLOSIDASE

Purification step	Volume (ml)	Total protein (mg)	Total activity (U)	Specific activity (U/mg protein)	Yield (%)
Culture filtrate	10,000	6,420	2,806	0.437	100
(NH$_4$)$_2$SO$_4$(ppt)	485	1,275	1,576	1.23	56.1
DEAE-Sephadex	350	573	1,360	2.37	48.4
Electrophoresis	302	124	816	6.58	28.1
BioGel P-100	40	81.6	750	9.19	26.7
Electrofocusing[a]	18	65.7	715	10.8	25.5

[a] Electrofocusing was performed three times, and the active fractions were pooled.

electrophoresis. This indicated that the purified sample was electrophoretically homogeneous both in the native state and under dissociating conditions.

Molecular Properties.[11] A molecular weight of about 101,000 was estimated by gel filtration with Sephadex G-200. The SDS–polyacrylamide gel electrophoresis gave a single band with a mobility corresponding to a molecular weight of about 102,000. The isoelectric point of the β-xylosidase was pH 4.86 using an LKB electrofocusing column with pH 2.5–5 range ampholytes. Carbohydrate content of purified β-xylosidase was about 4.5% by weight carbohydrate. Carbohydrate content was determined by the phenol–sulfuric acid method and compared to a glucose standard.

Optimum pH and Temperature. The optimal pH of *T. viride* β-xylosidase was about 3.5. The optimum temperature for the enzyme reaction was about 55° and 50% of maximal activity was found at 46 and 65°.

Stability. The purified enzyme was stable at a pH range of 3–4 at 50° for 30 min. The enzyme activity was stable up to 55° and was completely lost by incubation at 75° for 30 min without substrate.

Inhibitors. Hg^{2+} ion and *N*-bromosuccinimide strongly inhibited β-xylosidase activity.

Specificities. Table IV shows the substrate specificities[11] of *T. viride* β-xylosidase for various substrates. Enzyme activity was expressed in terms of the percentage of the activity with xylobiose. The activity on xylan at a substrate concentration of 1% is expressed relative to that of xylobiose at a concentration of 10^{-2} *M*. *T. viride* β-xylosidase scarcely acted on xylan and acted preferentially on a series of β-xylosides. The

[11] M. Matsuo and T. Yasui, *Agric. Biol. Chem.* **48,** 1845 (1984).

TABLE IV
HYDROLYSIS OF VARIOUS GLYCOSIDES BY
T. viride β-XYLOSIDASE

Glycoside	Substrate concentration ($\times$ 10^{-2} M)	Relative activity (%)
Xylobiose	1	100
Xylan	1[a]	0.5
Methyl-β-D-xyloside	1	2
Butyl-β-D-xyloside	1	21.6
Cyclohexyl-β-D-xyloside	1	23.7
Benzyl-β-D-xyloside	1	19.6
Thiophenyl-β-D-xyloside	1	0.03
p-Nitrophenyl-α-D-xyloside	1	0
Cellobiose	1	0

[a] Concentration (%).

apparent K_m value for xylobiose was similar to those for xylotriose, xylotetraose, and xylopentaose but the k_0 values (molecular activity) for each showed some differences. The apparent k_0 for xylobiose was 36.4 sec^{-1}; xylotriose, 34.5 sec^{-1}; xylotetraose, 17.4 sec^{-1}; and xylopentaose, 13.0 sec^{-1}. Both the p- and o-substituted phenyl-β-D-xylosides were well hydrolyzed at almost equal rates.

Anomeric Configuration of the Reaction Product D-Xylose. The anomeric forms of the products of the enzyme reaction were identified by GLC: the D-xylose moiety was released in its β-form[11] in the course of the enzymatic reaction using phenyl-β-D-xyloside as substrate.

β-Xylosidase of *Emericella nidulans*

In the study of the production of xylose from xylan through the process of enzymatic hydrolysis, Nomura *et al.*[9] investigated the influence of the substrate concentration on the enzymatic reaction by *Trichoderma viride*. In the earlier attempts to produce xylose from xylan by enzymatic hydrolysis, it is difficult to achieve the enzymatic reaction in a substrate concentration of 7% or above using any fungal β-xylosidases but the *Malbranchea* enzyme.[1] In order to obtain a β-xylosidase (exo-1,4-β-D-xylosidase, EC 3.2.1.37) other than the *Malbranchea* enzyme suitable for xylose production from xylan, a number of β-xylosidase-producing fungi were tested and finally the fungus *Emericella nidulans* was selected. This section describes the purification procedure and some properties of β-xylosidase from *Emericella nidulans*.

Assay Methods

The principle, reagents, procedure, and enzyme unit are as previously described.

Growth of Cultures. The strain of *Emericella nidulans* IFO 4340 was used. Stock cultures are maintained in continuous vegetative culture on yeast–malt–glucose–agar: yeast extract 0.5%, malt extract 0.5%, glucose 0.5% and agar 1.5% (pH 5.5). Large cultures of *Emericella nidulans* are grown on medium containing 2% xylan (hard wood), 0.4% KH_2PO_4, 1% $(NH_4)_2HPO_4$, 1% polypeptone, and 0.3% corn steep liquor in 500-ml flasks containing 100 ml of medium and incubated for 4 days at 30° on a reciprocal shaker (120 oscillation/min, 7 cm stroke). The culture broth is harvested by vacuum filtration on a Büchner funnel lined with Toyoroshi (Toyoroshi Co. Ltd., Tokyo, Japan) No. 2 filter paper and the culture filtrate is adjusted to pH 5.0 with 1 *N* HCl. All steps in the purification procedure are carried out at 4° unless otherwise stated.

Step 1. Ammonium Sulfate Fractionation. Solid ammonium sulfate (22.8 g/100 ml solution) is added to the culture filtrate to 0.3 saturation with continuous stirring, the pH of the solution being maintained at 5.0 with 1 *N* NaOH, and the precipitated protein is removed by centrifugation at 8000 *g* for 15 min. The supernatant fraction is brought to 0.8 saturation with ammonium sulfate (33.9 g/100 ml of solution) and is allowed to stand for 24 hr with occasional stirring. The protein precipitate is collected by centrifugation at 8000 *g* for 15 min, dissolved in a small amount of water, and dialyzed against several changes of the 5 m*M* acetate buffer (pH 4.5). The dialyzed solution is then centrifuged at 8000 *g* for 10 min to remove insoluble materials.

Step 2. BioGel P-100 Column Chromatography. The dialyzed solution from step 1 is concentrated by ultrafiltration with a PM10 membrane (Amicon, Danvers, MA). The samples are applied to a column (4.5 × 80 cm) of BioGel P-100 (Bio-Rad Laboratories, Richmond, CA) equilibrated with 50 m*M* acetate buffer (pH 4.5) and eluted with same buffer. Fractions of 5 ml are collected at a flow rate of 10 ml/hr.

Step 3. DEAE-Sephadex Column Chromatography. The fractions showing β-xylosidase activity from step 2 are collected and concentrated by ultrafiltration with a PM10 membrane, applied on a DEAE-Sephadex column (2.64 × 45 cm) equilibrated with 50 m*M* acetate buffer at pH 4.5. The column is washed with same buffer until no more protein eluted. Elution of 1 liter is carried out with a linear gradient of NaCl (0–0.6 *M*). Fractions of 5 ml are collected at a flow rate of 10 ml/hr.

Step 4. First Electrofocusing. The β-xylosidase-rich fractions from step 3 are dialyzed and concentrated by ultrafiltration with a PM10 membrane. One-third of the concentrated solution (about 9 mg is electrofo-

TABLE V
PURIFICATION OF *E. nidulans* β-XYLOSIDASE

Purification step	Volume (ml)	Total protein (mg)	Total activity (U)	Specific activity (U/mg protein)	Yield (%)
Culture filtrate	5,700	22,087	2,109	0.095	100
(NH₄)₂SO₄(ppt)	510	4,207.5	1,443	0.343	68.4
BioGel P-100	306	191.6	1,202	6.27	57.0
DEAE-Sephadex	108	25.3	995.7	39.4	47.2
First electrofocusing[a]	24	11.9	714.8	60.1	33.9
Second electrofocusing	4	7.2	453.3	62.9	21.5

[a] Electrofocusing was performed three times, and the active fractions obtained were pooled.

cused in each run). The experiments are performed three times in a 110 ml column (LKB-Produkter) at 4°. Carrier ampholytes (pH 2.5–4) is used in a sucrose gradient and a voltage of 900 V is applied for 42 hr. The volume of each fraction collected is 2 ml.

Step 5. Second Electrofocusing. The β-xylosidase-rich fractions from step 4 in each run are pooled, dialyzed, and concentrated by ultrafiltration (PM-10 membrane). A second electrofocusing is carried out by the same method as described for the first electrofocusing step. The volume of each fraction collected is 2 ml. The procedure, summarized in Table V, resulted in 662-fold overall purification with about 21.5% recovery on β-xylosidase activity from the culture filtrate.

Properties

Purity. The purified enzyme gave a single band on sodium dodecyl sulfate-polyacrylamide gel electrophoresis.

Molecular Properties.[12] From the preparative electrofocusing experiment, the isoelectric point of *E. nidulans* β-xylosidase was estimated to be 3.25. The molecular weight of the purified enzyme was estimated to be about 240,000 by chromatography on a calibrate column of Sephadex G-200 (Pharmacia). The SDS–polyacrylamide gel electrophoresis gave a single band with a mobility corresponding to a molecular weight of about 116,000. Thus the native enzyme is estimated to be as a dimer composed of two subunits, having the same molecular weight. The carbohydrate content of purified enzyme was about 4% by weight. Carbohydrate con-

---

[12] M. Matsuo and T. Yasui, *Agric. Biol. Chem.* **48**, 1853 (1984).

TABLE VI
HYDROLYSIS OF VARIOUS GLYCOSIDES BY
E. nidulans β-XYLOSIDASE

Glycoside	Substrate concentration ($\times 10^{-2}$ M)	Relative activity (%)
Xylobiose	1	100
Xylan	1[a]	0.3
Methyl-β-D-xyloside	1	0.5
Butyl-β-D-xyloside	1	13.4
Cyclohexyl-β-D-xyloside	1	4.2
Benzyl-β-D-xyloside	1	4.8
Thiophenyl-β-D-xyloside	1	0
p-Nitrophenyl-α-D-xyloside	1	0
Cellobiose	1	0

[a] Concentration (%).

tent was determined by the phenol–sulfuric acid method and compared to glucose standard.

Optimum pH and Temperature. The optimum pH of the enzyme was found to be in the range of 4.5–5.0. The optimum temperature for the enzyme reaction was about 55°.

Stability. The purified enzyme was stable at 50° for 30 min in the pH range of 4.5–5.0. The activity was stable up to 50° and was completely lost by incubation at 65° for 30 min without substrate.

Effect of Various Reagents on Enzyme Activity. Hg^{2+}, SDS, and N-bromosuccinimide inhibited the enzyme activity completely and p-mercuribenzoate also showed strong inhibition at a concentration of 10^{-4} M.

Specificities. Table VI shows the substrate specificities[12] of *E. nidulans* β-xylosidase. Enzyme activity was expressed in terms of percentage against that of xylobiose. The action on xylan at a concentration of 1% is expressed relative to that of xylobiose at a substrate concentration of 10^{-2} M. The K_m values for xylooligosaccharides were almost equal but the k_0 values appeared to be affected by the chain length of the β-1,4-linked D-xylose residues. The apparent k_0 for xylobiose was 992.4 sec^{-1}; xylotriose, 1321.9 sec^{-1}; xylotetraose, 789.7 sec^{-1}; and xylopentaose, 508.0 sec^{-1}. Both the p- and o-substituted phenyl-β-D-xylosides were well hydrolyzed at almost equal rates.

Anomeric Configuration of the Reaction Product D-Xylose. The anomeric forms of the products of the enzyme reaction were identified by GLC and the D-xylose moiety was released in its β-form[12] in the course of the enzymatic reaction using β-D-xyloside as substrate.

[89] β-Xylosidase/β-Glucosidase of *Chaetomium trilaterale*

By Tuneo Yasui and Masaru Matsuo

β-Xylosidase of *Chaetomium trilaterale* has been isolated and studied from the strain (*Chaetomium trilaterale* strain B) by Kawaminami and Iizuka.[1] The purification procedure[1] involved initial steps on DEAE-Sephadex A-25 and Sephadex G-75 column chromatography. This chapter describes a purification procedure and some enzymatic properties for β-xylosidase from *Chaetomium trilaterale* strain B. Although the *C. trilaterale* enzyme may be a β-glucosidase with β-xylosidase activity. it is called the enzyme "xylosidase" in this chapter.

Assay Methods

Principle. β-Xylosidase activity is assayed spectrophotometrically by measuring the release of phenol from phenyl-β-D-xyloside. Absorbance of phenol is measured at 660 nm by the addition of phenol reagent solution (Nakarai Chemicals, Tokyo, Japan) in alkaline solution.[2]

Reagents

Phenyl-β-D-xyloside, 20 mM in water
Acetate buffer, 0.1 M, pH 4.5
Sodium carbonate, 50 mM in water

Procedure. All assays are at 45°. The reaction is initiated by the addition of 0.5 ml of 20 mM phenyl-β-D-xyloside and 0.4 ml of 10 mM acetate buffer, pH 4.5, to 0.1 ml of enzyme solution. At the end of 15 min, the reaction is terminated by the addition of 5 ml of 50 mM sodium carbonate. By the addition of 1 ml of phenol reagent solution (1 N HCl) to the entire reaction mixture (6 ml), absorbance is measured at 660 nm against a blank in which water replaces phenyl-β-D-xyloside. When a β-glucosidase activity is assayed using phenyl-β-D-glucoside as substrate instead of phenyl-β-D-xyloside, released phenol from the substrate is measured by the same procedure as described in the β-xylosidase assay method.

Definition of Enzyme Unit. One unit of activity is defined as the amount of enzyme which is required to release 1 μmol of phenol per minute under the assay conditions. Specific activity is defined as units of

[1] T. Kawaminami and H. Iizuka, *J. Ferment. Technol.* **48**, 169 (1970).
[2] O. Folin and V. Ciocalteu, *J. Biol. Chem.* **73**, 627 (1927).

activity per milligram of protein as measured by the method of Lowry *et al.*[3]

Purification Procedure

Growth of Cultures. The strain of *Chaetomium trilaterale* strain B was obtained from Kawaminami and Iizuka (Kyowa Miles Co. Ltd. and Tokyo Science University, Japan). Stock cultures are maintained in continuous vegetative culture on yeast–malt–glucose–agar: yeast extract 0.5%, malt extract 0.5%, glucose 0.5%, and agar 2%, pH 5.5. Large cultures of *C. trilaterale* strain B are grown on the medium, as described earlier,[4] containing 0.05% yeast extract, 0.1% corn steep liquor, and 7% wheat bran in 500-ml shaking flasks. *C. trilaterale* strain B is inoculated into 500-ml flasks containing 100 ml of medium and incubated for 7 days at 30° on a reciprocal shaker (120 oscillations/min, 7 cm stroke). The culture broth is harvested by vacuum filtration on a Büchner funnel lined with Toyoroshi No. 2 filter paper (Toyoroshi Co. Ltd., Tokyo, Japan), and the culture filtrate is adjusted to pH 5.0 with 1 *N* HCl.

All subsequent steps are carried out at 4° unless otherwise stated.

Step 1. Ammonium Sulfate Fractionation. Solid ammonium sulfate (22.8 g/100 ml of solution) is added to the culture filtrate to 0.3 saturation with continuous stirring, the pH of the solution being maintained at 5.0 with 1 *N* NaOH, and the precipitate protein is removed by centrifugation at 8000 *g* for 15 min. The supernatant fraction is brought to 0.8 saturation with ammonium sulfate (33.9 g/100 ml of solution) and is allowed to stand at 24 hr with occasional stirring. The protein precipitate is collected by centrifugation at 8000 *g* for 10 min, dissolved in a small amount of water, and dialyzed against several changes of the 1 m*M* acetate buffer, pH 6.0.

Step 2. DEAE-Sephadex Column Chromatography. The dialyzed solution from step 1 is applied to a DEAE-Sephadex A-25 (Pharmacia, Uppsala, Sweden) column (2.5 × 45 cm) equilibrated with 5 m*M* acetate buffer, pH 6.0. The column is washed with 1 liter of the same buffer until no more protein is eluted. Elution is carried out with a linear gradient of the same buffer containing 0–1 *M* NaCl (total volume 1 liter). Fractions of 5 ml are collected at a flow rate of 16 ml/hr.

Step 3. Sephadex G-200 Column Chromatography. The fractions showing β-xylosidase activity from step 2 are collected and concentrated by ultrafiltration with a UM10 membrane (Amicon, Danvers, MA), and

[3] O. H. Lowry, N. J. Rosebrough, A. L. Farr, and R. J. Randall, *J. Biol. Chem.* **193,** 265 (1951).

[4] M. Uziie, M. Matsuo, and T. Yasui, *Agric. Biol. Chem.* **49,** 1159 (1985).

TABLE I

PURIFICATION OF β-XYLOSIDASE FROM *Chaetomium trilaterale*

		Xylosidase			Glucosidase		
Step and fraction	Total protein (mg)	Total activity (units)	Specific activity (unit/mg)	Recovery (%)	Total activity (units)	Specific activity (unit/mg)	Recovery (%)
Culture filtrate	917.8	1652	1.8	100.0	8924	9.7	100.0
1. Ammonium sulfate	152.0	661	4.3	40.0	4622	30.4	51.8
2. DEAE-Sephadex	24.5	340	13.9	20.6	2891	118.0	32.4
3. Sephadex G-200	8.7	152	17.5	9.2	1642	188.7	18.4
4. Isoelectric focusing	1.5	76	50.7	4.6	1178	785.3	13.2

applied on a Sephadex G-200 column (2 × 100 cm) equilibrated with 10 mM acetate buffer, pH 6.0, containing 0.1 M KCl. Fractions of 5 ml are collected at a flow rate of 8 ml/hr. Fractions having β-xylosidase activity are pooled and concentrated by ultrafiltration with a Minicon B15 (Amicon).

Step 4. Isoelectric Focusing. Fractions having β-xylosidase activity from step 3 are further fractionated by isoelectric focusing in a 110 ml column at 4° according to the LKB manual (LKB, Stockholm, Sweden). Carrier ampholite (pH 3.5–10) is used in a sucrose density gradient and a voltage of 900 V is applied for 48 hr. The volume of each fraction collected is 2 ml.

The methods, summarized in Table I, result in 50.7-fold overall purification with about 4.6% recovery on β-xylosidase activity and 785.3-fold overall purification with about 13.2% recovery on β-glucosidase activity.

Properties

Purity. The purified enzyme was shown to be homogeneous on polyacrylamide disc gel electrophoresis with pH 4.3 and 8.3 buffer system and a single protein band was obtained.

Molecular Properties. A molecular weight of about 205,000[5] was calculated from equilibrium sedimentation experiments. Gel filtration on calibrate Sephadex G-200 columns gave a molecular weight estimate of 240,000.[4] The SDS–polyacrylamide gel electrophoresis gave a single band with a mobility corresponding to a molecular weight of 118,000.[4] Thus the native enzyme exists as a dimer composed of subunit, having the same

[5] M. Uziie, Ph.D. thesis. University of Tsukuba, Tsukuba, Japan (1984).

TABLE II
INHIBITION TYPES AND K_i VALUES OF VARIOUS INHIBITORS FOR *C. trilaterale* β-XYLOSIDASE AND
β-GLUCOSIDASE ACTIVITIES

Inhibitor	Xylosidase activity		Glucosidase activity	
	K_i (M)	Type of inhibition	K_i (M)	Type of inhibition
Glucono-1,5-lactone	3.88×10^{-4}	Noncompetitive	5.34×10^{-4}	Competitive
Nojirimycin	3.27×10^{-6}	Noncompetitive	1.95×10^{-5}	Competitive
Methyl-β-D-xyloside	2.37×10^{-2}	Competitive	3.86×10^{-2}	Noncompetitive
Methyl-β-D-glucoside	1.44×10^{-2}	Noncompetitive	1.43×10^{-2}	Competitive
1-Thiophenyl-β-D-xyloside	1.25×10^{-2}	Competitive	3.03×10^{-2}	Competitive
1-Thiophenyl-β-D-glucoside	0.83×10^{-2}	Competitive	1.00×10^{-2}	Competitive

molecular weight. The isoelectric point of the β-xylosidase was pH 4.86 using an LKB electrofocusing column with pH 3.5–10 range ampholytes.[4] The carbohydrate content of purified β-xylosidase was about 20.7% by weight carbohydrate.[4] Carbohydrate content was determined by the phenol–sulfuric acid method and compared to a glucose standard. The absorption coefficient at 280 nm was $A_{280\ nm}^{1\%} = 16.3$ (g^{-1} cm^2).[4]

Optimum pH. The pH dependence of the β-xylosidase and that of the β-glucosidase activities is not similar. The optimum pH of the β-glucosidase activity was 4.2, while that of the β-xylosidase activity was in a relatively wide pH range from 3.8 to 5.5.[4]

Stability. The β-xylosidase and β-glucosidase activities retain full activity for several months when stored at 4° in acetate buffer, pH 6.0. After preincubation for 3 hr at 30°, the β-xylosidase retained full activity in the pH range from 4.0 to 11.0, while β-glucosidase was stable only from 4.5 to 8.0.[4] The β-xylosidase activity was stable in alkaline pH range but the β-glucosidase activity was unstable under the same conditions. After heating for 30 min at 60° and pH 4.5, the β-xylosidase activity retained about 80% of its original activity but the β-glucosidase activity was about 58%.

Inhibitors. Hg^{2+} ion and *N*-bromosuccinimide strongly inhibited both of the β-xylosidase and β-glucosidase activities. Both of the β-xylosidase and β-glucosidase activities were inhibited by glucono-1,5-lactone[6] and nojirimycin (5-amino-5-deoxy-D-glucopyranose).[6] β-Xylosidase activity was inhibited noncompetitively by the above two inhibitors, but β-glucosidase activity was competitive. Methyl-β-D-xylopyranoside[6] inhibited the β-xylosidase activity competitively but the β-glucosidase activity was

[6] M. Uziie, M. Matsuo, and T. Yasui, *Agric. Biol. Chem.* **49**, 1167 (1985).

TABLE III

K_m and V_{max} VALUES FOR SOME SUBSTRATES OF
β-XYLOSIDASE FROM *Chaetomium trilaterale*

Substrate	K_m (M)	V_{max} (mol/mg protein/min)
β-1,4-Xylobiose	8.0×10^{-2}	0.93
Phenyl-β-D-xyloside	2.8×10^{-3}	2.6
Cellobiose	1.9×10^{-3}	12.2
Phenyl-β-D-glucoside	5.1×10^{-3}	73.3

noncompetitive. Whereas methyl-β-D-glucopyranoside[6] inhibited the β-xylosidase activity noncompetitively but the β-glucosidase activity was competitive. 1-Thiophenyl-β-D-xylopyranoside[6] and 1-thiophenyl-β-D-glucopyranoside[6] behaved as competitive inhibitors. Table II summarizes kinetic data obtained for some inhibitors, suggesting two nonidentical binding sites.[6]

Specificity. The purified enzyme showed both the β-xylosidase and β-glucosidase activities, and could catalyze the hydrolysis of cellobiose, gentiobiose, and laminaribiose, under the substrate concentration of 5 mM, at almost the same reaction rates.[5] K_m and V_{max} values on various substrates are shown in Table III. Xylan (hard wood) was not degraded. From the kinetic study on the β-xylosidase and β-glucosidase activities, it could indicate that the β-xylosidase/β-glucosidase of *C. trilaterale* has a single site and two nonidentical binding sites.[6]

[90] Acetylxylan Esterase of *Schizophyllum commune*

By P. BIELY, C. R. MACKENZIE, and H. SCHNEIDER*

$$O\text{-Acetyl-4-}O\text{-methyl-D-glucurono-D-xylan} \xrightarrow{\text{H}_2\text{O}}$$
$$4\text{-}O\text{-methyl-D-glucurono-D-xylan} + \text{acetic acid}$$

$$O\text{-Acetyl-1,4-}\beta\text{-xylooligosaccharides} \xrightarrow{\text{H}_2\text{O}}$$
$$1,4\text{-}\beta\text{-xylooligosaccharides} + \text{acetic acid}$$

Acetylxylan esterases produced by fungi catalyze the removal of acetyl groups from xylan.[1] Acetyl groups occur commonly on xylans,[2] with

---

* All three authors are affiliated with the National Research Council of Canada.
[1] P. Biely, J. Puls, and H. Schneider, *FEBS Lett.* **186,** 80 (1985).
[2] T. E. Timell, *Wood Sci. Technol.* **1,** 45 (1967).

the content being generally higher in material from hardwood. The esterases act cooperatively with xylanases in hydrolyzing glycosidic bonds of acetylxylans in that the extent of liberation of reducing sugar is limited unless the acetyl groups are first removed.[3]

Appreciable levels of acetylxylan esterase are produced by the fungus *Schizophyllum commune*.[1] The enzyme seems to be under common regulatory control with cellulases and xylanases.[4] The appearance of high levels in the medium is associated with growth on cellulose, conditions that also result in the secretion of appreciable amounts of cellulase and xylanase. Coproduction of the esterase with enzymes involved in saccharification is consistent with the view that esterase function is associated with the degradation of lignocellulosics.

The present chapter describes some methods used in the study of acetylxylan esterases.

Assay Methods

Reagents

O-Acetyl-4-*O*-methyl-D-glucurono-D-xylan from beechwood is obtained by extraction of the holocellulose with dimethyl sulfoxide.[5,6] The polymer can also be prepared from birchwood by solubilizing the hemicellulose by steam treatment under controlled conditions, followed by isolation of the higher molecular fraction by dialysis against water.[1] The acetyl content of these materials is about 10% w/w.

Potassium phosphate buffer, 0.4 and 0.2 M, pH 6.5

Acetic acid, 0.1–1.0% w/v

Glycerol, 1% w/v

4-Nitrophenyl acetate, saturated solution in 0.2 M phosphate buffer, pH 6.5 (freshly prepared)

4-Nitrophenol, 10 mM in 0.2 M phosphate buffer, pH 6.5, a calibration solution

Procedure Using Acetylxylan as Substrate. A 20% w/v solution of either beechwood or birchwood acetylxylan in 0.4 M phosphate buffer is mixed with an equal volume of appropriately diluted enzyme solution, a few drops of toluene are added as an antimicrobial agent, and the mixture

[3] P. Biely, C. R. MacKenzie, J. Puls, and H. Schneider, *Bio/Technology* **4,** 731 (1986).
[4] P. Biely, C. R. MacKenzie, and H. Schneider, *Can. J. Microbiol.,* in press.
[5] E. Hägglund, B. Lindberg, and J. McPherson, *Acta Chem. Scand.* **10,** 1160 (1956).
[6] H. O. Bouveng, P. J. Garegg, and B. Lindberg, *Acta Chem. Scand.* **14,** 742 (1960).

is incubated in a closed vessel for 1–14 hr. Immediately prior to the time chosen for analysis, insoluble material is removed by a short centrifugation. The acetate content of the supernatant is then determined by HPLC using a Bio-Rad Aminex HPX-87H column at 60° with 0.01 N sulfuric acid as eluant. A calibration curve is constructed using solutions of acetic acid in 0.2 M phosphate buffer. Alternatively, glycerol added to a final concentration of 0.2% w/v is used as an internal standard.

Units. One unit of acetylxylan esterase is defined as the amount of enzyme liberating 1 μmol of acetic acid per minute at 25° from 10 mg of substrate contained in 0.1 ml of reaction mixture.

Notes. Reactions can be terminated by freezing and samples stored in the frozen state for analysis at a later time. Acetate can also be determined by more sensitive enzymatic methods, which allow for the use of lower concentrations of substrate. Gas chromatography is unsuitable for following deacetylation, because acetylxylan decomposes thermally to yield acetic acid.

Procedure Using 4-Nitrophenyl Acetate as Substrate. Assays can also be carried out using 4-nitrophenyl acetate, a chromogenic substrate used for the determination of acetylesterase activity (EC 3.1.1.6).[7] A volume of 10–50 μl of appropriately diluted enzyme solution is pipetted into a test tube, mixed with a saturated solution of the substrate, and incubated at 25°. Absorbance at 410 nm in a 1-cm cuvette is measured after an appropriate time interval, using as a blank a control that contains all reagents but the enzyme. Alternately, absorbance against the substrate blank can be followed continuously. The amount of substrate hydrolyzed is determined photometrically on the basis of a standard curve prepared using 4-nitrophenol in the 0.05–0.3 μmol/ml range in 0.2 M phosphate buffer. The reaction cannot be terminated chemically or thermally because of the labile nature of the substrate.

Units. One unit of acetylxylan esterase or acetylesterase (EC 3.1.1.6) hydrolyzes 1 μmol of 4-nitrophenyl acetate in 1 min at 25° under the conditions specified above.

Methods for Detecting Acetylxylan Esterase Activity in Electrophoresis Gels

Materials

Acetylxylan from beechwood or birchwood
Potassium phosphate buffer, 0.1 and 0.3 M, pH 6.5

---

[7] C. Huggins and J. Lapides, *J. Biol. Chem.* **170**, 467 (1947).

4-Methylumbelliferyl acetate
Sodium azide
Low melting point agarose (Bethesda Research Laboratories, MD)
Glass plates, 15 × 30 cm
Plastic film (e.g., polyester), 15 × 30 cm
Space bars, plastic, 1 × 30 × 0.075 cm

Detection Using Acetylxylan as Substrate

Principle. Deacetylation of acetylxylan in agar gels leads to the formation of a visible precipitate, because of differences in solubility. The position of the precipitate in detection gels that contain acetylxylan is used to localize the position of acetylxylan esterase in separation gels.

Procedure. The solution for the detection gel is prepared by adding, at 40–50°, 0.6 g of birchwood or beechwood acetylxylan in 10 ml of 0.1 *M* phosphate buffer to 20 ml of 2.5% w/v of low melting agarose in the same buffer. The gel is cast by pouring this solution at 40–50° between two sheets of plastic film mounted with water on the glass plates, which are separated by the plastic space bars. If the gels are not to be used immediately, they are prepared with 0.01% azide and stored at 4°. Visualization of the position of the esterase is carried out by superimposing the detection and separation gels, placing them in a plastic bag, and then incubating at 30° until a cloudy area appears. A positive result may require incubation for several hours or even overnight. The precipitate forms in both the separation and detection gels, because of cross-diffusion of enzyme and substrate. The method fails when the esterase is close to xylanase, because of hydrolysis and resultant solubilization of deacetylated xylan. Gels with precipitates are recorded photographically using diffuse light, as with immunoprecipitation bands (Fig. 1).

Detection Using 4-Methylumbelliferyl Acetate as Substrate

Principle. Acetylxylan esterase of *S. commune* hydrolyzes 4-methylumbelliferyl acetate, a fluorogenic substrate for acetylesterases.[8] Hydrolysis of this substrate in detection gels is indicated by the intense fluorescence at 360 nm of the liberated 4-methylumbelliferone.

Procedure. The detection gel is prepared as above using 30 mg of substrate suspended in 10 ml of water which is added at 40–50° to 20 ml of 2.5% w/v low melting point agarose in 0.3 *M* phosphate buffer. For visualization purposes, the detection gel is placed in contact with the separation gel and incubated at room temperature for 5–30 min. A positive reaction

[8] T. J. Jacks and H. W. Kircher, *Anal. Biochem.* **21**, 279 (1967).

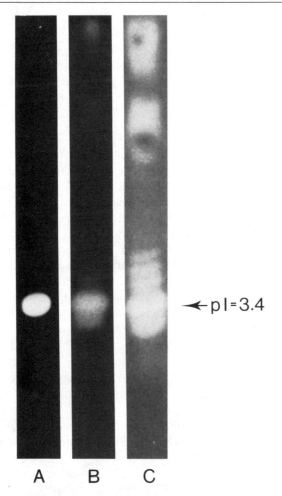

$\leftarrow$pI=3.4

A   B   C

FIG. 1. Visualization of acetylxylan esterase activity in *S. commune* extracellular proteins after electrophoresis. (A) Detection gel made with birchwood acetylxylan. (B) Detection gel made with 4-methylumbelliferyl acetate. (C) Separation gel that was in contact with the 4-methylumbelliferyl acetate detection gel. The greater sensitivity for esterase detection in the separation gel when using 4-methylumbelliferyl acetate is attributed to the more rapid diffusion of this compound than enzyme, and the higher concentration of enzyme present. The proteins were obtained after growth on 1% cellulose and were separated on a thin polyacrylamide gel by isoelectric focusing in the pH range 2.5–5.0. The pI of the acetylxylan esterase is indicated as 3.4.

is indicated by a bright fluorescent zone under ultraviolet light on both the separation and detection gels (Fig. 1). Since esterases of differing specificity can hydrolyze 4-methylumbelliferyl acetate, bands that fail to exhibit activity against acetylxylan may also be detected.

Isolation of Acetylxylan Esterase

Growth of the Fungus. *Schizophyllum commune* ATCC 35848 is cultured at 30° for 11 days in 100 ml volumes in 500-ml Erlenmeyer flasks kept on a gyratory shaker operating at 150 cycles/min. The medium, optimized for cellulase production,[9] contains 23 g/liter Bactopeptone (Difco), 15.3 g/liter $Ca(NO_3) \cdot 4H_2O$, 0.5 g/liter $MgSO_4 \cdot 7H_2O$, 1.3 g/liter KH_2PO_4, 10 g/liter cellulose (Solka-Floc, Brown Co., Berlin, NH), 20 mg/liter $CoCl_2 \cdot 6H_2O$, 5.0 mg/liter $FeSO_4 \cdot 7H_2O$, 2.0 mg/liter $MnSO_4 \cdot H_2O$, and 1.6 mg/liter $ZnCl_2$. The inoculum is grown in 1.5% w/v malt extract broth.[9]

Concentration of Culture Fluid. After centrifugation, the culture fluid is desalted and concentrated by a factor of five using an Amicon ultrafiltration system and a PM10 membrane. During the filtration, a portion of the extracellular polysaccharide produced by the fungus accumulates on the membrane and is discarded. The retentate is stored in a lyophilized form. About 230 mg of material is obtained from 100 ml of culture fluid. One milligram contains 0.08 units of acetylxylan esterase and 0.4 units of acetylesterase, the latter activity being measured with 4-nitrophenyl acetate.

Ion-Exchange High-Performance Liquid Chromatography. Ten milligrams of freeze-dried material is suspended in 0.01 *M* phosphate buffer, pH 6, centrifuged, and 2 ml of the supernatant then applied to a 7.5 × 150 mm TSK DEAE-3SW HPLC column (LKB, Sweden). Elution is performed using initially 10 ml of 0.01 *M* phosphate buffer followed by 30 ml of a 0–1.0 *M* linear NaCl gradient in the same buffer. Sample application and elution is carried out at a flow rate of 1 ml/min. The eluant is monitored continuously for protein at 280 nm and collected as 1-ml fractions, each of which is assayed for activity of endo-1,4-β-glucanase,[10] β-xylosidase,[3] endo-1,4-β-xylanase,[10] acetylxylan esterase, and acetylesterase (Fig. 2). Acetylxylan esterase elutes at a concentration of 0.5 *M* NaCl in a fraction free of the above mentioned glycanase activities. Acetylesterase and acetylxylan esterase activities coelute. One chromatographic run affords a fraction containing 0.25 units of acetylxylan esterase activity and

[9] M. Desrochers, L. Jurasek, and M. G. Paice, *Dev. Ind. Microbiol.* **22,** 675 (1981).
[10] P. Biely, D. Mislovičová, and R. Toman, *Anal. Biochem.* **144,** 142 (1985).

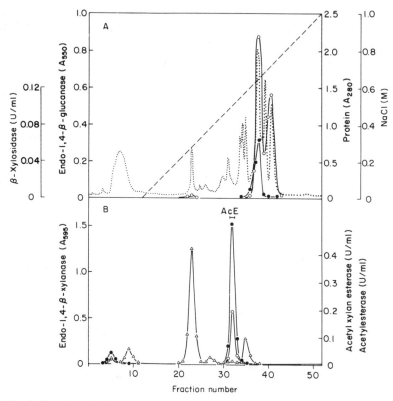

FIG. 2. Fractionation by ion-exchange HPLC of *S. commune* proteins produced during growth on 1.0% cellulose. (A) Dotted line, protein; dashed line, NaCl gradient, O, endo-1,4-β-glucanase activity; ●, β-xylosidase activity. (B) △, endo-1,4-β-xylanase activity; ●, acetylesterase activity; O, acetylxylan esterase activity. The bar labeled AcE refers to the acetylxylan esterase fraction used in studies of properties of this enzyme.

0.75 units of acetylesterase activity. The fractionation results in a 30-fold purification of acetylxylan esterase activity.

Properties

The enzyme is more active against acetylxylan than two other esterases tested. The ratios of acetic acid release are 34 : 1 : 0.01 for the *S. commune* enzyme, orange peel acetylesterase, and porcine liver esterase, respectively, when compared on the basis of equal activities against 4-nitrophenyl acetate.[3] In contrast to the other two esterases, the fungal enzyme also increases appreciably the extent of glycosidic bond hydrolysis of acetylxylan catalyzed by an esterase-free xylanase fraction.[3]

In addition to increasing the extent to which glycosidic bonds are hydrolyzed by xylanase, the presence of esterase changes the nature of the products formed. Acetylated xylooligosaccharides are produced by an esterase-free xylanase fraction acting on acetylxylan, but xylose, xylobiose, or xylotriose is not detected. However, these sugars are produced when the fungal esterase is present.[3]

The acetylxylan esterase acts more rapidly on shorter acetylxylan chains, as indicated by the enhancement in acetic acid release when xylanase is added to esterase.[3]

In addition to being active on acetylxylan, the fungal enzyme also releases acetic acid from tri-*O*-acetylglycerol and penta-*O*-acetyl-D-glucose.

[91] α-L-Arabinofuranosidase from *Aspergillus niger*

By KIYOSHI TAGAWA and AKIRA KAJI

α-L-Arabinofuranosidase (α-L-arabinofuranoside arabinofuranohydro-lase, EC 3.2.1.55) catalyzes the hydrolysis of nonreducing terminal α-L-arabinofuranosidic linkages in L-arabinan, arabinoxylan, and other L-arabinose-containing polysaccharides.[1-10] In *Aspergillus niger*, this enzyme is produced inducibly in a medium containing L-arabinose or L-arabinan.[11]

Assay Method

Principle. Phenyl- or *p*-nitrophenyl-α-L-arabinofuranoside is hydro-lyzed by this enzyme action, thus, the amount of reducing group or ni-

[1] A. Kaji, K. Tagawa, and K. Matsubara, *Agric. Biol. Chem.* **31,** 1023 (1967).
[2] A. Kaji, K. Tagawa, and T. Ichimi, *Biochim. Biophys. Acta* **171,** 186 (1969).
[3] K. Tagawa and A. Kaji, *Carbohydr. Res.* **11,** 293 (1969).
[4] A. Kaji and K. Tagawa, *Biochim. Biophys. Acta* **207,** 456 (1970).
[5] K. Tagawa, *J. Ferment. Technol.* **48,** 730 (1970).
[6] A. Kaji and O. Yoshihara, *Biochim. Biophys. Acta* **250,** 367 (1971).
[7] A. Kaji, *Adv. Carbohydr. Chem. Biochem.* **42,** 383 (1984).
[8] H. Neukom, L. Providoli, H. Gremi, and P. A. Hui, *Cereal Chem.* **44,** 238 (1967).
[9] M. Tanaka and T. Uchida, *Biochim. Biophys. Acta* **522,** 531 (1978).
[10] J.-M. Brillouet, J.-C. Moulin, and E. Agosin, *Carbohydr. Res.* **144,** 113 (1985).
[11] K. Tagawa and G. Terui, *J. Ferment. Technol.* **46,** 693 (1968).

trophenolate ion liberated is a measure of the enzyme activity. In this chapter, the colorimetoric assay method using p-nitrophenyl-α-L-arabinofuranoside is described.

Reagents

25 mM p-nitrophenyl-α-L-arabinofuranoside (Sigma Chemical Co.) solution, pH 3.8
0.1 M citrate-phosphate buffer, pH 3.8
Suitably diluted enzyme solution
0.2 M Na_2CO_3

Procedure. The reaction mixture, containing 0.5 ml each of p-nitrophenyl-α-L-arabinofuranoside, buffer, and enzyme solution, is incubated for 30 min at 30°. After incubation, 2.5 ml of Na_2CO_3 solution is added to stop the further reaction and to develop the color of nitrophenolate ion. The absorbancy at 420 nm is measured spectrophotometrically. A linear relationship between enzyme concentration and nitrophenolate ion liberated is obtained up to 0.65 μg of crystalline enzyme per ml of reaction mixture during a 30-min incubation.

Definition of Unit. One unit of α-L-arabinofuranosidase activity is defined as the amount of enzyme which will produce 1 μmol of nitrophenolate ion (or L-arabinose) per minute at 30°.

Preparation and Purification Procedure

Cultivation of Organism. *Aspergillus niger* is cultured in a medium composed of 5 g of peptone, 2 g of $NaNO_3$, 1 g of K_2HPO_4, 0.5 g of $MgSO_4 \cdot 7H_2O$, and 0.4 ml of 2% $FeCl_3$ in 1 liter of sugar beet pulp extract which is made by extracting 25 g of beet pulp with 0.1 N NaOH solution for 1 hr at 100°. The initial pH of the culture medium is adjusted to 5.4 with 1 N HCl. Cultivation is carried out for 72 hr at 28° under aerobic conditions in a jar-fermentor. The mycelia are removed by filtration through cloth and the filtrate is centrifuged at about 10,000 g for 20 min. The clear solution, thus obtained, is purified as follows.

Purification. All steps are carried out below 5°, unless otherwise indicated.

Step 1. Salting Out with Ammonium Sulfate. To 10 liters of culture filtrate, 5.61 kg of $(NH_4)_2SO_4$ (0.8 saturation) is added with constant stirring. After the solution has stood overnight, the resulting precipitate is collected by centrifugation at 15,000 g for 20 min and dissolved in water (800 ml). The solution is dialyzed for 48 hr against water, and the precipitate removed by centrifugation.

Step 2. Chromatography on DEAE-Cellulose. The dialyzed solution (960 ml) is applied to a column (3.8 × 60 cm) of DEAE-cellulose which is prepared with 0.02 *M* sodium phosphate buffer, pH 7.0. The column is eluted by a linear gradient of 0.05–0.4 *M* NaCl in a 0.02 *M* sodium phosphate buffer, pH 7.0. The fraction rich in α-L-arabinofuranosidase is collected and dialyzed against water for 24 hr. The solution is concentrated by the adsorption-elution method on a small column (2 × 40 cm) of DEAE-cellulose under conditions described above. The concentrated solution is dialyzed against water for 24 hr.

Step 3. Heat Treatment. The pH of the dialyzed concentrate (50 ml) is adjusted to 7.0 with 1 *N* HCl and then the solution is heated for 10 min at 75°. It is immediately cooled in ice water and solid NaCl is added to a concentration of 0.1 *M*.

Step 4. First Gel Filtration on Sephadex. A column (2.2 × 70 cm) of Sephadex G-100 is equilibrated with 0.02 *M* sodium phosphate buffer containing 0.1 *M* NaCl, pH 7.0. The heat-treated enzyme solution is applied to the column and eluted with the same buffer. Fraction rich in the enzyme is collected and dialyzed against 0.02 *M* sodium phosphate buffer, pH 6.0.

Step 5. Chromatography on DEAE-Sephadex. The dialyzed enzyme solution (63 ml) is added to a column (2 × 50 cm) of DEAE-Sephadex A-50 equilibrated previously with 0.02 *M* sodium phosphate buffer, pH 6.0. The enzyme is eluted stepwise with an equal volume (50 ml) of the same buffer containing 0.1, 0.2, 0.3, 0.4, and 0.5 *M* NaCl. The fraction rich in the enzyme is collected and dialyzed against water for 24 hr, and then lyophilized.

Step 6. Chromatography on SE-Sephadex. The lyophilized enzyme is dissolved in 0.05 *M* sodium acetate–HCl buffer, pH 2.5 (11 ml) and added to a column (2 × 20 cm) of SE-Sephadex C-50 which was previously equilibrated with 0.05 *M* sodium acetate–HCl buffer, pH 2.5. The enzyme is eluted successively with equal volumes (20 ml) of the following buffers: 0.05 *M*, pH 2.5; 0.05 *M*, pH 4.0; and 0.1 *M*, pH 4.0 of sodium acetate–HCl. The fraction rich in the enzyme is collected, dialyzed against water, and lyophilized as above.

Step 7. Second Gel Filtration on Sephadex. The enzyme is dissolved in 5 ml of 0.02 *M* sodium phosphate buffer, pH 6.0, and applied to a column (2 × 25 cm) of Sephadex G-100 equilibrated with the same sodium phosphate buffer. Elution is performed under gravity with the buffer. A fraction with α-L-arabinofuranosidase activity is coincidentally eluted with that of protein, showing that the enzyme is highly purified.

Step 8. Crystallization. The purified enzyme solution is concentrated

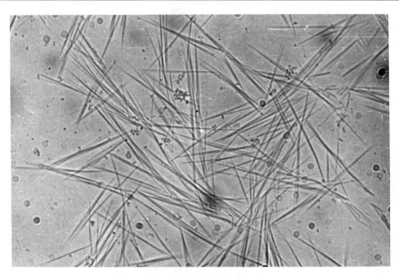

FIG. 1. Photomicrograph of crystalline α-L-arabinofuranosidease from *Aspergillus niger* (×600).

by lyophilization to 1 ml (about 1% protein) and pulverized $(NH_4)_2SO_4$ is added gradually until the solution becomes cloudy, followed by addition of one or two drops of the saturated ammonium sulfate solution once a day for 4 days. Fine needle-like crystals (Fig. 1) are formed at 0.45 saturation of ammonium sulfate and collected by centrifugation (about 8 mg as the enzyme protein). The crystalline enzyme has a specific activity of 390 units/mg of protein and no elevation of specific activity is observed in this step.

A summary of the above purification procedure is presented in Table I.

Comment on Purification Procedure. Sephadex gel is greatly affected by the ionic strength of suspending solution, that is, it swells in a lower ionic strength solution and conversely it shrinks in a higher solution. Thus, the difficulty is often encountered during operation of an ion-exchange Sephadex gel column, because the flow rate of eluate is greatly reduced when the ionic strength of solution is varied. Because of this we recommend substitution of affinity chromatography on arabinan cross-linked Sepharose 6B prepared by reacting epichlorohydrin or epoxy-activated Sepharose 6B with arabinan as described by Waibel *et al.*[12] for the step following first gel filtration on Sephadex G-100.

[12] R. Waibel, R. Amado, and H. Neukom, *J. Chromatogr.* **197**, 86 (1980).

TABLE I
PURIFICATION OF α-L-ARABINOFURANOSIDASE FROM *A. niger*

Step	Volume (ml)	Activity (units/ml)	Protein (mg/ml)	Specific activity (units/mg protein)	Yield (%)
Culture filtrate	10,000	2.5	2.3	1.1	100.0
Salting out with ammonium sulfate	960	17.8	4.8	3.7	68.4
DEAE-cellulose chromatography	133	98.0	5.6	17.5	52.1
Heat treatment	50	209.6	10.4	20.2	41.9
First Sephadex G-100 gel filtration	63	149.8	1.2	124.8	37.7
DEAE-Sephadex A-50 chromatography	13	498.5	2.0	249.3	25.9
SE-Sephadex C-50 chromatography	11	430.0	1.4	307.1	18.9
Second Sephadex G-100 gel filtration	5	816.0	2.1	388.6	16.3

Properties

Physicochemical Properties. The crystalline enzyme preparation is homogeneous in the following tests: sedimentation, electrophoresis in various buffers, and chromatography on Sephadex gel. The molecular weight is estimated by the molecular sieve method to be 53,000.[4] The isoelectric point of enzyme protein is pH 3.6.[4]

Optimum pH. The pH optimum for the enzyme is 3.8 and at pH 2.0 and at pH 5.5, 50% of the maximum activity is attained.

Stability. The enzyme is extremely stable at pH 7.0; for instance, after 10 min exposure at 70°, more than 90% of the original activity is retained, and at pH between 2.5 and 8.5, it has considerable stability. Over a 3-month period, if asceptically stored at 5°, no significant loss of activity is noted. The enzyme activity is completely inhibited by 10^{-3} M mercurous acetate and about 60% of the activity is inhibited by 10^{-3} M AgNO$_3$,[13] but these inhibitions are reduced by the presence of SH compounds. The enzyme is completely denatured in 6 M urea solution.[14]

Substrate Specificity. The enzyme is highly specific for nonreducing terminal α-L-arabinofuranosidic linkages and will not attack internal α-L-

[13] A. Kaji and K. Tagawa, *J. Agric. Chem. Soc. Jpn.* **38,** 580 (1964).
[14] K. Tagawa, *J. Ferment. Technol.* **48,** 740 (1970).

arabinofuranosyl linkages. L-Arabinan hydrolysis proceeds in two kinetic processes; the rapid process with up to 30% hydrolysis is attributable to the cleavage of one unit L-arabinofuranose side chains which are attached along a main chain; the slow process is responsible for the cleavage of the main chain composed of α-L-1,5-linked arabinofuranose units, which is designated as 1,5-L-arabinan.[3] K_m value for L-arabinan is 0.26 g/liter and for 1,5-L-arabinan 20.4 g/liter, respectively.[5] Phenyl- and p-nitrophenyl-α-L-arabinofuranosides are easily hydrolyzed and K_m values for these glycosides are about 5×10^{-3} M. L-Arabinose is a competitive inhibitor of enzyme action and K_i value is 2×10^{-2} M. The enzyme hydrolyzes arabinoxylan and polysaccharides containing L-arabinofuranose residues, such as gum arabic, and liberates L-arabinose from these polysaccharides.[3,8,15]

[15] H. Gremli and H. Neukom, *Carbohydr. Res.* **8**, 110 (1968).

[92] α-L-Arabinofuranosidase from *Scopolia japonica*

By Tsuneko Uchida and Michio Tanaka[1]

Assay[2]

This assay procedure measures the absorbance at 420 nm of the liberated p-nitrophenol from the p-nitrophenylglycosides. For following the enzymatic activity during the purification procedure, p-nitrophenyl-α-L-arabinofuranoside[3] was used as substrate. The amount of enzyme which can liberate 1 μmol of p-nitrophenol per 1 min is defined as 1 unit of enzyme. The protein concentration is determined by the method of Lowry *et al.*,[5] using crystalline bovine serum albumin as a standard.

Reagents

Reaction mixture, total volume 300 μl, contains
50 mM sodium citrate/phosphate buffer, pH 4.8

[1] Dr. Michio Tanaka sadly passed away on October 29, 1983.
[2] M. Tanaka and T. Uchida, *Biochim. Biophys. Acata* **522**, 531 (1978).
[3] p-Nitrophenyl-α-L-arabinofuranoside and p-nitrophenyl-α-L-arabinopyranoside were synthesized according Fielding *et al.*[4]
[4] A. H. Fielding and L. Hough, *Carbohydr. Res.* **1**, 327 (1965).
[5] O. H. Lowry, N. J. Rosebrough, A. L. Farr, and R. J. Randall, *J. Biol. Chem.* **193**, 265 (1951).

1 mM substrate (p-nitrophenylglycosides)
Enzyme (properly diluted), 100 μl
1 M Na$_2$CO$_3$ solution, 2 ml
Procedure. Reaction mixture was incubated at 37°. After a known time 2 ml of 1 M Na$_2$CO$_3$ was added and the absorbance of the liberated p-nitrophenol was measured at 420 nm.

Purification Procedure[2]

Culture of Plant Calluses. The calluses of *Scopolia japonica* were cultured in suspension in Murashige's and Skoog's medium[6] containing 5% sucrose/0.5% casamino acid/1 ppm 2,4-dichlorophenoxyacetic acid at 27° in the dark, using a rotary shaker at 120 rpm. The growth of the calluses began to increase after the lag phase of 5 days and reached a plateau after 30 days. The enzyme was mainly localized in the culture medium (about 65% of the total activity). The enzymatic activity in the medium increased with the growing of callus cells after 10 days and began to decrease after 40 days in the stationary phase of the growth.

Ultrafiltration. The purification procedure below was carried out at 4° unless otherwise specified.

To minimize the effect of degrading enzymes, the day 21 culture medium (12 liters) in the logarithmic phase of the growth (70% growth) was harvested by centrifugation at 16,000 g for 20 min, concentrated, and dialyzed against 50 mM sodium citrate/phosphate buffer (pH 7.0) with a Diaflo PM10 membrane (Amicon Corp., Danvers, MA). This culture medium contained various glycosidase activities, relatively high activities of β-galactosidase, α-L-arabinofuranosidase, and α-mannosidase as shown in Table I.

Heat Treatment. The enzyme solution (protein, 3.7 mg/ml) was heated at 55° for 20 min, and then was centrifuged at 46,000 g for 1.5 hr. The supernatant was lyophilized.

In this procedure, 97% of β-galactosidase activity was lost, but the activities of α-L-arabinofuranosidase and α-mannosidase were almost completely retained (Table I). The heat treatment was essential for removal of most of the β-galactosidase activities because no effective separation of them from α-L-arabinofuranosidase was available by various chromatographies.

Ammonium Sulfate Precipitation. The lyophilized sample (protein, 464 mg) was solubilized in 50 ml of 10 mM sodium citrate/phosphate buffer, pH 6.5. The resulting viscous solution was brought to 40% satura-

[6] T. Murashige and F. Skoog, *Physiol. Plant.* **15**, 473 (1962).

TABLE I
GLYCOSIDASES AND THEIR HEAT STABILITY IN CULTURE MEDIUM
OF *Scopolia japonica* CALLUSES[a]

Glycosidases	Total activity[b] (units/liter of medium)	Relative activity	Relative activity after heat treatment[c]
α-L-Arabinofuranosidase	4.42	100	92
α-L-Arabinopyranosidase	0.285	6.45	5.02
α-Galactosidase	0.274	6.20	1.43
β-Galactosidase	5.94	134	4.42
α-Glucosidase	0.063	1.4	1.11
β-Glucosidase	0.037	0.84	0.42
α-Mannosidase	0.775	17.5	15.4
β-Mannosidase	0.011	0.25	0.14
α-L-Fucosidase	0.026	0.59	0.38

[a] Day 21 of culture.
[b] Assay was done with the corresponding p-nitrophenylglycoside at 37° and pH 4.5 for adequate periods.
[c] Heat treatment was done at 55° and pH 7.0 for 20 min.

tion at 0° with 12 g of solid $(NH_4)_2SO_4$ per 50 ml of solution centrifuged at 5700 g for 20 min. $(NH_4)_2SO_4$ (200 g/liter) was added to the supernatant at 0° bringing it to 70% saturation and the supernatant was again centrifuged at 5700 g for 20 min. The precipitate was dissolved in 10 mM sodium citrate/phosphate buffer, pH 6.5, containing 0.2 M NaCl. The resulting solution was dialyzed against the same buffer and concentrated with a Diaflo PM10 membrane. The viscous components in the solution were removed in this procedure, which was of great advantage in the next step.

Gel Filtration on Sephadex G-150. A 5.0 ml portion (protein, 28 mg) of the enzyme solution was applied to a column (1.8 × 125 cm) of Sephadex G-150 equilibrated with 10 mM sodium citrate/phosphate buffer, pH 6.5, containing 0.2 M NaCl. The column was eluted with the buffer at a flow rate of 14.3 ml/hr and the eluate was collected in 5.0-ml fractions. The fractions containing α-L-arabinofuranosidase were combined and concentrated with a Diaflo PM10 membrane (Fig. 1).

Chromatography on DEAE-Sephadex A-50. A portion (protein, 15 mg) of the enzyme solution, previously dialyzed against 10 mM sodium citrate/phosphate buffer, pH 7.5, was loaded on a column (1.2 × 20 cm) of DEAE-Sephadex A-50 equilibrated with the same buffer. The material passing through the column was collected, dialyzed against distilled water, and then concentrated.

Isoelectric Focusing Fractionation. The enzyme solution (protein, 3.0

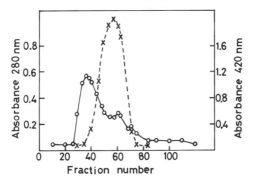

FIG. 1. Gel filtration of α-L-arabinofuranosidase on Sephadex G-150. The dotted line shows enzymatic activity and the solid line the absorbance at 280 nm of protein.

mg) after DEAE-Sephadex chromatography was electrophoresed at 900 V for 92 hr in a column (LKB 8101, 110 ml) of 1.5% carrier ampholyte with a pH gradient of 6–9, prepared by mixing ampholytes of pH range 6–8 and 7–9 (LKB Produkter AB, Bromma, Sweden) with a linear gradient 0–47% of sucrose. After electrophoresis, the content of the column was fractionated in 1.94-ml fractions. Then, the pH, protein concentration (A_{280}), and enzyme activity were measured. As shown in Fig. 2, three active fractions were obtained; the main fraction was located at p*I* 8.0 and two minor fractions were at p*I* 6.0 and 5.7, respectively. The main fraction was stored as the most purified α-L-arabinofuranosidase preparation.

Comments. A typical purification is summarized in Table II. α-L-Arabinofuranosidase thus obtained is purified about 163-fold in a good yield

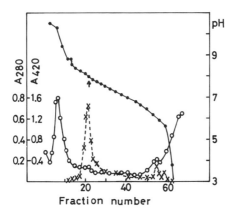

FIG. 2. Isoelectric focusing fractionation of α-L-arabinofuranosidase. The activity of α-L-arabinofuranosidase (×), pH (●), and the absorbance at 280 nm of protein (○) was measured.

TABLE II
PURIFICATION OF α-L-ARABINOFURANOSIDASE

Purification step	Total protein (mg)	Total activity (units)	Specific activity (units/mg protein × 10³)	Yield (%)
Culture medium	14,144	53.0	3.75	100
Ultrafiltration	3,031	51.2	16.9	96.6
Heat treatment	1,282	52.4	40.9	98.9
Ammonium sulfate (40–70% sat.) fraction	469	41.4	88.3	78.1
Sephadex G-150	277	40.2	145	75.8
DEAE-Sephadex A-50	213	39.8	187	75.1
Isoelectric focusing, main fraction	37.0	22.6	611	42.6

and still contains small amounts of α-mannosidase, α-galactosidase, and β-glucosidase, and a definite amount of α-L-arabinopyranosidase (Table III), which seemed to be included as an intrinsic activity of α-L-arabinofuranosidase, taking into consideration the heat stability (Table I) and the inhibition by a specific inhibitor of α-L-arabinofuranosidase as described below.

TABLE III
GLYCOSIDASE ACTIVITIES IN ISOELECTRIC FOCUSING
FRACTIONS

	Relative activity	
Glycosidases	Main fraction (pI 8.0)	Minor fraction (pI 6.0)
α-L-Arabinofuranosidase	100	100
α-L-Arabinopyranosidase	6.9	7.0
α-Galactosidase	1.0	4.8
β-Galactosidase	0	8.2
α-Glucosidase	0	0
β-Glucosidase	1.4	7.2
α-Mannosidase	2.1	358
β-Mannosidase	0	0
α-L-Fucosidase	0	0
β-L-Arabinopyranosidase[a]	0	—
β-D-Arabinopyranosidase[a]	0	—

[a] The enzyme sample was incubated with phenyl-β-L- or D-arabinopyranoside, followed by measuring the liberated arabinose.

As shown in Fig. 2, α-L-arabinofuranosidase is a basic protein. However, no cation-exchange chromatography such as CM-Sephadex C-50 and SP-Sephadex C-50 was useful for the purification of this enzyme, because the enzyme was too strongly adsorbed on the cation exchanger to be eluted even with a high concentration of salt.

Properties of Enzyme[2]

Stability. α-L-Arabinofuranosidase is a stable enzyme. After storage at various pH of pH 2.5–11 for 20 hr at 4 or 37°, the activity was completely retained at both temperatures above pH 3.5, and was rapidly lost below pH 3.5. When the enzyme is stored at pH 5.5 and 4° for a month, the loss of activity was less than 5%. As shown in Table I, α-L-arabinofuranosidase was more stable than β-D-galactosidase which was the major glycosidase in the culture medium of *Scopolia japonica*. After heat treatment at pH 7.0 for 3 min, α-L-arabinofuranosidase lost 50% of its activity at 63°, while β-D-galactosidase was lost completely at 46°.

Molecular Weight. The molecular weight was estimated to be 62,000 by gel filtration on a Sephadex G-150 column in the presence of 0.1 M NaCl.

pH Optimum. α-L-Arabinofuranosidase is optimally active around pH 4.8, but 50% of the activity can be detected at pH 3.5 and pH 6.3.

Reaction Requirements[2]

Kinetics. Michaelis constant (K_m) and maximum velocity (V_{max}) for the synthetic substrate and arabinodisaccharides are summarized in Table IV. This enzyme requires no metal ions.

Inhibitors. Addition of up to 0.2 M NaCl had no effect on the enzymatic activity. This enzyme was inhibited about 45 and 90% by 10 mM Zn^{2+} and Cu^{2+}, respectively. EDTA and also SH reagents were unable to

TABLE IV
KINETIC CONSTANTS OF α-L-ARABINOFURANOSIDASE

Substrate	K_m (mM)	V_{max} (units/mg protein)
p-Nitrophenyl-α-L-arabinofuranoside	6.7	5.26
α-Arabinofuranosyl-(1 → 3)-arabinose	9.22	0.022
α-Arabinofuranosyl-(1 → 5)-arabinose	8.06	0.013

reduce the enzymatic activity, suggesting that the enzyme had no SH groups at the active site.

Effect of Plant Growth Hormones and Antibiotics. Plant growth hormones, 1 ppm 2,4-dichlorophenoxyacetic acid, indole-3-acetic acid, and kinetin, and antibiotics, 50 ppm kanamycin and chloramphenicol, showed no direct effect on the enzymatic activity.

Substrate Specificity[2,7]

This enzyme was concluded to be highly specific for the configuration of L-arabinofuranose from the results of the inhibition test by some analogs of α-L-arabinofuranoside.[2] Addition of 10 mM L-arabinose and D-arabinose showed some degree of inhibition, but D-galactose, L-arabitol, and D-galactiol (each 10 mM) showed only a few percent of inhibition. Further, phenyl-β-L-arabinopyranoside and phenyl-β-D-arabinopyranoside showed no effect under the same conditions. The most structurally related compound, L-arabono-1,4-lactone (1 mM) strongly inhibited about 90% of the enzymatic activity, while the structurally different lactones, such as D-arabono-1,4-lactone and L and D-galactono-1,4-lactone had no effect. Incidentally, the activity of α-L-arabinopyranosidase included in the most purified enzyme preparation was also inhibited by L-arabono-1,4-lactone, a specific inhibitor of α-L-arabinofuranosidase.

The specificity of α-L-arabinofuranosidase for arabinoside linkages was examined using three kinds of arabinodisaccharides,[7] α-arabinofuranosyl-(1 → 3)-arabinose, α-arabinofuranosyl-(1 → 5)-arabinose,[8] and α-arabinopyranosyl-(1 → 5)-arabinose.[8] The relative activity at 16 hr digestion was 0.592 for α-arabinofuranosyl-(1 → 5)-arabinose and 0.014 for α-arabinopyranosyl-(1 → 5)-arabinose, assuming that the activity for α-arabinofuranosyl-(1 → 3)-arabinose was 1.00. Practically this enzyme can hydrolyze arabinofuranosylarabinose having either an α-(1 → 3) or an α-(1 → 5) linkage. In addition, the enzyme shows a slight preference for α-(1 → 3) linkage, although the values of K_m for both linkages are very similar as shown in Table IV.

Enzyme action of α-L-arabinofuranosidase on natural substrates was examined using beet araban prepared from beet pulp chips,[9] and arabi-

---

[7] M. Tanaka, A. Abe, and T. Uchida, *Biochim. Biophys. Acta* **658**, 377 (1981).

[8] α-Arabinofuranosyl-(1 → 3)-arabinose and α-arabinofuranosyl-(1 → 5)-arabinose were prepared from the hydrolyzate of acetylated beet araban and beet araban, respectively.[7] α-Arabinosylpyranoside was chemically synthesized.[7] The arabinodisaccharides thus obtained were identified by paper chromatography, optical rotation, and methylation analysis.

[9] J. K. N. Jones and Y. Tanaka, *Methods Carbohyd. Chem.* **5**, 74 (1965).

nooligosaccharide and glycopeptide from the cell walls of *Scopolia japonica*.[7] Beet araban was presumed to have a similar structure to that of araban from *Pinus pinaster*,[10] which was composed of long linear chains of arabinose in (1 → 5) linkages with numerous arabinose monomers attached through (1 → 3) linkages.[11] However, beet araban was degraded exoenzymatically but incompletely by excess amounts of α-L-arabinofuranosidase even after long periods of incubation, though this enzyme ought to hydrolyze both linkages in a slightly different rate (Table IV). The reasons for this incomplete degradation were explained by the novel finding of (1 → 2) linkages and arabinopyranosides, and by the inclusion of a trace amount of galactose into the carbohydrate chain of beet araban.[7] Accordingly, this enzyme is regarded as an exoenzyme having a preference for arabinofuranosylarabinose with α-(1 → 3) or α-(1 → 5) linkage.

Two arabinooligosaccharides (degree of polymerization[12]; 17 and 10) prepared from plant cell walls were relatively good substrates which released about 40% of arabinose by a α-L-arabinofuranosidase, while a trace amount of arabinose was released from the glycopeptides containing an average of 0.5 hydroxylarabinose linkages per single molecule. Practically, no hydroxyprolylarabinose linkage in glycopeptides of plant cell walls appears to be cleaved by this enzyme.

[10] R. F. H. Dekker and G. N. Richards, *Adv. Carbohydr. Chem. Biochem.* **32**, 277 (1976).
[11] G. O. Aspinall, *in* "The Carbohydrates, Chemistry and Biochemistry" (W. Pigman and D. Horton, eds.), 2nd Ed. p. 515. Academic Press, New York, 1970.
[12] M. Tanaka, *Carbohydr. Res.* **88**, 1 (1981).

[93] Arabinogalactanase of *Bacillus subtilis* var. *amylosacchariticus*

By TAKEHIKO YAMAMOTO and SHIGENORI EMI

Many strains of *Bacillus subtilis* and *Bacillus amyloliquefaciens*[1,2] are good producers of several enzymes and some of them or their mutant strains have been industrially utilized for production of α-amylase and proteinase. These bacteria are also a good source of the enzymes[3–5] which

[1] J. Fukumoto, *Nippon Nogei Kagaku Kaishi* **19**, 487 (1943).
[2] N. E. Welker and L. L. Campbell, *J. Bacteriol.* **94**, 1124 and 1131 (1967).
[3] S. Emi, J. Fukumoto, and T. Yamamoto, *Agric. Biol. Chem.* **35**, 1891 (1971).
[4] S. Emi, J. Fukumoto, and T. Yamamoto, *Agric. Biol. Chem.* **36**, 991 (1972).
[5] S. Emi and T. Yamamoto, *Agric. Biol. Chem.* **36**, 1945 (1972).

hydrolyze "hemicelluloses" of plant cell wall such as arabinogalactan, galactomannan, and arabinoxylan.

This chapter deals with the method of purification and crystallization of arabinogalactanase starting from the cultured liquor of a strain of saccharifying α-amylase producing *Bacillus subtilis* var. *amylosacchariticus*,[6,7] showing the separation stages of α-amylase and proteinase which are simultaneously produced in the cultured liquor. The properties of the enzyme are also described indicating that it is a metal ion requiring enzyme for the stability and that it hydrolyzes endwise the arabinogalactan of soybean seed to produce galactose and several galactooligosaccharides with or without arabinose residues. The properties of arabinogalactanase obtained from another bacterial strain are briefly commented.

Assay Method

Principle. The arabinogalactanase of *Bacillus subtilis* var. *amylosacchariticus* catalyzes the hydrolysis of soybean seed arabinogalactan according to the following equation: arabinogalactan + n $H_2O \rightarrow a$ galactose + b galactobiose + c galactotriose + d intermediary products + e arabinogalactooligosaccharides (where $n = \Sigma a + b + c + d + e$ and $c > b > a > e$ after a long incubation of the enzyme reaction mixture).

Determination of the Arabinogalactan Hydrolyzing Activity. The arabinogalactan as substrate prepared as described below is added to deionized water and the mixture is gently heated to dissolve the polysaccharide. After cooling, the solution is adjusted to 0.5% substrate and 0.05 M acetate buffer by adding deionized water and 1.0 M acetate buffer, pH 6.0. To 4.0 ml of this solution is added 1.0 ml of the enzyme solution at 40° and, after 5 min incubation, the reducing sugar formed is determined as galactose by the method of Shaffer–Somogyi.[8] One unit of arabinogalactanase activity is defined as the amount of enzyme that produces 5.0 μmol of galactose under the conditions (1.0 μmol of reducing sugar as galactose per minute).

Preparation of Arabinogalactan as Substrate

The arabinogalactan is prepared by a slight modification of the method of Morita.[9,10] The soybean cake meal obtained by defatting with hexane is suspended in 10 times the weight of 0.1 N sodium hydroxide solution with

[6] A. Nishida, J. Fukumoto, and T. Yamamoto, *Agric. Biol. Chem.* **31**, 682 (1967).

[7] K. Umeki and T. Yamamoto, *J. Biochem.* (*Tokyo*) **72**, 101 and 1219 (1972).

[8] P. A. Shaffer and M. Somogyi, *J. Biol. Chem.* **100**, 695 (1933).

[9] M. Morita, *Agric. Biol. Chem.* **29**, 564 (1965).

[10] M. Morita, *Agric. Biol. Chem.* **29**, 626 (1965).

gentle stirring at 25°. One hour later, the suspension is centrifuged and the precipitate is treated three times with fresh 0.1 N sodium hydroxide solution under the same conditions. The precipitate finally obtained is washed with water and then boiled with 5-fold the weight of water on a dry basis for 1 hr. The extraction of the precipitate with boiling water is repeated three times. The extracts are combined and concentrated under reduced pressure to be slightly cloudy; the concentrate is allowed to stand at 4° for 3 weeks. This preserved solution is then centrifuged and to the supernatant is added 2 volumes of cold acetone. The resulting precipitate is washed with acetone and subsequently with ether to bring to dryness *in vacuo*. The hydrolysis of the arabinogalactan (moisture content, 11.5%) with 1.0 N hydrochloric acid at 100° for 2 hr produces 80% of reducing sugar as galactose. Paper chromatographic analysis reveals that the arabinogalactan consists of one part of arabinose and two parts of galactose on a molar basis.

Production of Arabinogalactanase by Culture

Many strains of *Bacillus subtilis* and *Bacillus amyloliquefaciens* produce arabinogalactanase. One of the strains of *Bacillus subtilis* named as var. *amylosacchariticus* because of producing α-amylase of saccharifying type[2] is cultured in a submerged manner in a medium of 5.0% soybean cake extract (the cake is gently boiled with 20-fold the weight of 0.1 N sodium hydroxide solution for 1 hr and the susupernatant is neutralized with hydrochloric acid to pH 7.0), 3.0% dextrin, 1.2% $(NH_4)_2HPO_4$, 0.02% each of $MgSO_4 \cdot 7H_2O$ and KCl, and 0.05% $CaCl_2$, pH 7.2. The culture is grown at 37° for 2 days with aeration (half volume per volume of medium per minute) and with continuous stirring at 400 rpm. The arabinogalactanase is an inducible enzyme and the enzyme is produced only in the presence of arabinogalactan.

Purification of Arabinogalactanase

After growth, the culture medium is centrifuged and ammonium sulfate is added to the supernatant to 0.75 saturation at 4°. After standing overnight, the resulting precipitate is collected by centrifugation and dissolved in water to one-tenth the original volume of culture medium. This solution is centrifuged and to the supernatant is added calcium chloride in an amount equimolar to the ammonium sulfate present. The resulting precipitate is removed by centrifugation and 2 volumes of cold 2-propanol is added to the supernatant to precipitate the enzyme. The enzyme precipitate is collected by centrifugation, dissolved in 0.002 M calcium acetate, and dialyzed against the same solution.

After adding acetate buffer, pH 5.8, to be 0.05 M, the dialyzed enzyme is applied to a column of Duolite A-2, an anion exchanger equilibrated with 0.05 M acetate buffer, pH 5.8, to remove α-amylase and coloring substance. The arabinogalactanase in the effluent is precipitated by adding ammonium sulfate to 0.75 saturation. The resulting precipitate is collected, dissolved in a minimal amount of 0.002 M calcium acetate, and dialyzed against the same solution.

The decolorized and dialyzed enzyme is buffered with 0.01 M Tris–maleate buffer containing 0.002 M calcium acetate, pH 6.4, and filtered through a column of CM-cellulose equilibrated with the same buffer to remove proteinase. The arabinogalactanase in the effluent is precipitated with 0.75 saturation of ammonium sulfate, the precipitate is dissolved in 0.002 M calcium acetate, and dialyzed against the same solution.

The dialyzed enzyme obtained above is subjected to chromatography using a column of SP-Sephadex C-50 equilibrated with 0.01 M acetate buffer containing 0.002 M calcium acetate, pH 5.0. The enzyme is completely adsorbed and desorbed from the column by a linear gradient of sodium chloride up to 0.35 M. However, the enzyme activity in the eluate separates into three fractions. The protein estimated by the absorption at 280 nm also appears to separate into three fractions almost in proportion to the enzyme activity. The enzyme in each fraction (numbered I, II, and III in the order of elution with the sodium chloride solution) is precipitated with ammonium sulfate of 0.75 saturation and after being dialyzed against 0.002 M calcium acetate, subjected to gel chromatography using a column of Sephadex G-75 equilibrated with 0.002 M calcium acetate. The enzyme specific activity of each fraction is substantially increased by this chromatography. The arabinogalactanase preparation even after purification to this stage, however, generally still contains a slight amount of proteinase as a contaminant.

The enzyme in each fraction is precipitated by adding 2 volumes of

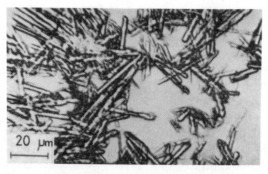

Fɪɢ. 1. Crystalline arabinogalactanase I of *Bacillus subtilis* var. *amylosacchariticus*.

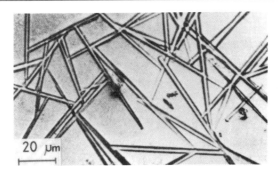

FIG. 2. Crystalline arabinogalactanase II of *Bacillus subtilis* var. *amylosacchariticus*.

cold acetone and after being dialyzed against 0.002 M calcium acetate, purified respectively by the method of isoelectric focusing using sucrose and carrier ampholytes as the density and pH gradients. The pH of the main activity is 6.9, 8.0, and 8.9 for arabinogalactanase I, II, and III, respectively. Each arabinogalactanase purified by this method becomes free from proteinase activity.

The enzymes I and II purified above usually have better yields and are obtained in a crystalline state. The respective fractions of the enzyme I and II are separately combined, and the solutions are dialyzed against 0.002 M calcium acetate to remove sucrose and carrier ampholytes. The enzymes in the dialyzed solution are precipitated by adding 2 volumes of cold acetone and the precipitated enzymes are dissolved in a minimal amount of deionized water. To these solutions are added dropwise the saturated solution of calcium sulfate to be slightly cloudy and the mixtures are allowed to stand for several days in the cold whereupon the enzymes crystallize out, as shown in Figs. 1 and 2.

Properties

Molecular Weight and Activity. The molecular weights estimated by the gel chromatographic method on a column of Sephadex G-75 are 35,000 and 36,000 for arabinogalactanase I and II, respectively. The molecular weight of arabinogalactanase III is about 32,000. $E_{1\,cm}^{280\,nm}$ (1%), pH 6.0, is 18.2 and 17.2 with the activity of 3000 and 3400 for arabinogalactanase I and II, respectively.

The optimum pH is 7.0, 6.0, and 6.5 for arabinogalactanase I, II, and III, respectively. Also, the enzyme activities increase linearly with temperature up to 55° in the presence of Ca^{2+} (0.002 M).

K_m values estimated on soybean seed arabinogalactan are 0.043, 0.11, and 0.86% for arabinogalactanase I, II, and III, respectively.

Stability. The arabinogalactanases all are stable at a pH range from 5.5 to 10.5 (24 hr treatment at 30°). They are also stable up to 50° in the presence of 0.002 M Ca^{2+} (15 min treatment at pH 8.0). Their thermal stabilities are greatly depressed in the presence of 0.002 M EDTA and they lose their activities at temperatures higher than 37°.

Specificity and Action Pattern

Arabinogalactanase I, II, and III hydrolyze soybean seed arabinogalactan to the extent of 30.0, 33.6, and 28.2%, respectively, after a long incubation. They show negligible activity toward coffee bean arabinogalactan. Also, they show no activity on arabinoxylan and galactomannan isolated from soybean seed coat.

The hydrolysis products of soybean seed arabinogalactan by the enzymes are galactose (Gal_1), galactobiose (Gal_2), and galactotriose (Gal_3), and their amounts are $Gal_3 > Gal_2 > Gal_1$, the degree somewhat differing depending on arabinogalactanase I, II, or III. Galactotetraose appears as an intermediary product but it is hydrolyzed into Gal_1 and Gal_3 upon long incubation. Trace amounts of two kinds of oligosaccharides which consist of arabinose and galactose appear from the early stage of the enzyme reaction, but arabinose is not liberated even after a long incubation.

The structures of arabinogalactan of soybean seed and coffee bean have been reported to be as follows:

two residues (average)

$$\text{Ara}f\ 1 \rightarrow 5\ \text{Ara}f$$
$$1$$
$$\downarrow$$
$$3$$
$$\left[\overset{\beta}{\rightarrow} 4\ \text{Gal}p\ 1 \overset{\beta}{\rightarrow} 4\ \text{Gal}p\ 1 \overset{\beta}{\rightarrow} 4\ \text{Gal}p\ 1 \overset{\beta}{\rightarrow} 4\ \text{Gal}p\ 1 \overset{\beta}{\rightarrow}\right]_n$$

four or five residues (average)

Soybean seed arabinogalactan[10]

$$\text{Ara}f\ 1 \overset{\alpha}{\rightarrow} 3\ \text{Gal}p$$
$$1$$
$$\downarrow$$
$$6$$
$$\text{Ara}p\quad \text{Gal}p$$
$$\downarrow\qquad \downarrow$$
$$\rightarrow \text{Gal}p \rightarrow \text{Gal}p \rightarrow \left[\overset{\beta}{\rightarrow} 3\ \text{Gal}p\ 1 \overset{\beta}{\rightarrow} 3\ \text{Gal}p\ 1 \overset{\beta}{\rightarrow}\right]_n \rightarrow 3\ \text{Gal}p \rightarrow$$

Coffee bean arabinogalactan[11,12]

---

[11] M. L. Wolfrom and D. L. Patin, *J. Org. Chem.* **30**, 4060 (1965).
[12] Y. Hashimoto, *Nippon Nogei Kagaku Kaishi* **45**, 147 (1971).

(where Gal$p$ = D-galactopyranosyl residue, Ara$f$ = L-arabinofuranosyl residue, and Ara$p$ = L-arabinopyranosyl residue).

The arabinogalactanase of *Bacillus subtilis* is thus concluded to hydrolyze endwise β-1,4-galactosidic linkages of the main chain of soybean seed arabinogalactan to produce Gal$_4$, Gal$_3$, Gal$_2$, and arabinogalactooligosaccharides and degrade the Gal$_4$ intermediately produced into Gal$_3$ and Gal$_1$. It is also clear that the arabinogalactanase has no ability to degrade β-1,3-galactosidic linkages of the main chain of coffee beam arabinogalactan.

Bacillus amyloliquefaciens is also a potent producer of arabinogalactanase. The bacterial culture method described is available for production of arabinogalactanase of this bacterium. Also, the arabinogalactanase produced can be purified and crystallized out by the method described above.

The arabinogalactanase isolated from the culture liquor of a strain of *Bacillus amyloliquefaciens* is higher in specific activity than that from *Bacillus subtilis* var. *amylosacchariticus*. Also, the former enzyme is characteristic in that the main product of the hydrolysis of soybean seed arabinogalactan is galactobiose.[3]

Author Index

Numbers in parentheses are footnote reference numbers and indicate that an author's work is referred to although the name is not cited in the text.

Subject Index

Z